MP 1143979 3

AF598625

RODD'S CHEMISTRY OF CARBON COMPOUNDS

ELSEVIER SCIENTIFIC PUBLISHING COMPANY
335 JAN VAN GALENSTRAAT
P.O. BOX 211, AMSTERDAM, THE NETHERLANDS

DISTRIBUTORS FOR THE U.S.A. AND CANADA:

ELSEVIER NORTH-HOLLAND INC.
52 VANDERBILT AVENUE, NEW YORK, N.Y. 10017

WITH 56 TABLES

ISBN: 0-444-40664-6 (SERIES)
ISBN: 0-444-41644-7 (VOL. IVG)

LIBRARY OF CONGRESS CARD CATALOG NUMBER 64-4605

PRINTED IN THE NETHERLANDS

RODD'S CHEMISTRY OF CARBON COMPOUNDS

ADVISORS

RODD'S CHEMISTRY OF CARBON COMPOUNDS

VOLUME I

GENERAL INTRODUCTION

ALIPHATIC COMPOUNDS

*

VOLUME II

ALICYCLIC COMPOUNDS

*

VOLUME III

AROMATIC COMPOUNDS

*

VOLUME IV

HETEROCYCLIC COMPOUNDS

*

VOLUME V

MISCELLANEOUS

GENERAL INDEX

*

RODD'S CHEMISTRY OF CARBON COMPOUNDS

A modern comprehensive treatise

SECOND EDITION

Edited by

S. COFFEY

M.Sc. (London), D.Sc. (Leyden), C.Chem., F.R.I.C.
formerly of
I.C.I. Dyestuffs Division, Blackley, Manchester

VOLUME IV PART G

HETEROCYCLIC COMPOUNDS

Six-membered heterocyclic compounds with a single nitrogen atom in the ring to which are fused two or more carbocyclic ring systems, and six-membered ring compounds where the hetero-atom is phosphorus, arsenic, antimony or bismuth. Alkaloids containing a six-membered heterocyclic ring system

ELSEVIER SCIENTIFIC PUBLISHING COMPANY
AMSTERDAM OXFORD NEW YORK
1978

CONTRIBUTORS TO THIS VOLUME

R. E. ATKINSON, B.SC., PH.D.
Proctor & Gamble Ltd., Newcastle upon Tyne NE12 9TS

K. W. BENTLEY, M.A., D.SC., D.PHIL., F.R.S.E.
Department of Chemistry, Loughborough University of Technology, Loughborough

N. CAMPBELL, O.B.E., PH.D., D.SC., F.R.S.E.
Department of Chemistry, The University, Edinburgh 9

J. D. HUNT, PH.D.
I.C.I. Plant Protection Division, Bracknell, Berkshire RG12 6EY

A. MCKILLOP, PH.D., D.SC.
School of Chemical Sciences, University of East Anglia, Norwich NR4 7TJ

A. R. PINDER, D.SC., PH.D., D.PHIL.
Department of Chemistry, Clemson University, Clemson, S. Carolina, U.S.A.

M. SAINSBURY, A.C.T.(BRIST.), PH.D., F.R.I.C.
School of Chemistry, University of Bath, Bath BA2 7AY

B. P. SWANN, PH.D.
Lilly Research Centre Ltd., Windlesham, Surrey GU20 6PH

R. E. FAIRBAIRN, B.SC., PH.D., C.CHEM., F.R.I.C.
formerly of Research Department, Dyestuffs Division, I.C.I. Ltd., Manchester 9 *(Index)*

PREFACE TO VOLUME IVG

The first four sub-volumes of Volume IV of this edition of *Rodd's Chemistry of Carbon Compounds* are concerned with the chemistry of those organic compounds in which an isolated three-, four- or five-membered ring containing a single hetero-atom is present. Six-membered ring compounds containing a single atom of oxygen, sulphur, selenium, tellurium or the less frequently encountered hetero-atoms silicon, germanium, tin, lead and iodine are the subject of Volume IVE, while Volume IVF deals with the chemistry of the basic mono- and bi-cyclic six-membered ring compounds in which the single hetero-atom is nitrogen, namely pyridine, quinoline and isoquinoline.

The present volume, IV G, is a continuation of Volume IV F. Thus Chapter 28, by Professor NEIL CAMPBELL, deals with the chemistry of compounds containing a pyridine ring carrying two or more fused carbocyclic rings which includes acridine and phenanthridine, benzo[*f*,*g*]quinolines, benzoisoquinolines, naphtho-quinolines and -isoquinolines, benzacridines and benzophenanthridines and their derivatives, while Chapter 29, by Dr. R.E. ATKINSON, describes six-membered heterocyclic compounds in which the hetero-atom is phosphorus, arsenic, antimony or bismuth.

As mentioned in the Preface to Volume IV F, the introduction of a nitrogen atom into the fundamental aromatic carbocyclic structures results in the production in plants and mammalian organs of a most interesting and diverse series of bases grouped together under the general classification as alkaloids. These interesting and, in many cases, highly physiologically active compounds are considered in the remaining chapters of Volume IVG as follows: Chapter 30, by Dr. J.D. HUNT and Dr. A. MCKILLOP, deals with pyridine and piperidine alkaloids, Chapter 31, by Dr. MALCOLM SAINSBURY, with quinoline alkaloids, Chapter 32, by Dr. B.P. SWANN and Dr. MCKILLOP, with acridine alkaloids, Chapter 33, by Professor K.W. BENTLEY, with alkaloids of the morphine group and, finally, Chapters 34 and 35, both by Professor A.R. PINDER, with diterpenoid and steroidal alkaloids, respectively. An account of the isoquinoline alkaloids will follow as Chapter 36 in Volume IV H.

The study of these alkaloids has provided one of the most interesting and fascinating themes of modern chemistry and is still being pursued in all its many aspects with sustained vigour. Many of the compounds described and their special properties are so well known that it seems superfluous to emphasise their interest, not only for organic chemists but also for biologists, pharmacologists, biochemists and kindred scientists.

The editor is glad to place on record his appreciation of the help and cooperation he has received from all the contributors to this volume and to Mr. E.B. ROBINSON for his help in preparing the manuscripts for the press.

May 1977 S. COFFEY

CONTENTS

VOLUME IV G

Heterocyclic Compounds: Six-membered heterocyclic compounds with a single nitrogen atom in the ring to which are fused two or more carbocyclic ring systems, and six-membered ring compounds where the hetero-atom is phosphorus, arsenic, antimony or bismuth. Alkaloids containing a six-membered heterocyclic ring system.

PREFACE VII
OFFICIAL PUBLICATIONS; SCIENTIFIC JOURNALS AND PERIODICALS XIV
LIST OF COMMON ABBREVIATIONS AND SYMBOLS USED XVI

Chapter 28. Polycyclic Compounds Comprising a Pyridine and Two or More Carbocyclic Rings
by N. CAMPBELL

1. Acridine and its derivatives 1
   a. General introduction to the chemistry of acridine 1
      (*i*) Methods of preparation, 4 – (*ii*) Physical properties, 7 – (*iii*) Chemical properties and reactions, 8 –
   b. Acridine and substituted acridines 15
      (*i*) Acridine and homologues, 15 – (*ii*) Halogenoacridines, 18 – (*iii*) Nitroacridines, 19 – (*iv*) Aminoacridines, 19 – (*v*) Hydroxyacridines, 23 –
   c. 9,9′-Biacridines and related compounds 25
   d. Hydrogenated acridines and their derivatives 28
      (*i*) Hydroacridines, 28 – (*ii*) 9-Acridone (acridanone), 31 – (*iii*) 9-Acridonequinones, 35 – (*iv*) Thioacridone, 35 –
   e. Acridine dyes 36
      (*i*) Basic dyestuffs, 36 – (*ii*) Vat dyestuffs, 37 – (*iii*) Pigments. Quinacridines, quinacridones, 38 –
2. Phenanthridine and its derivatives 40
   a. Methods of synthesis 41
   b. Phenanthridines 43
      (*i*) General physical properties, 43 – (*ii*) Chemical properties, 44 –
   c. Substituted phenanthridines 49
      (*i*) Alkylphenanthridines, 49 – (*ii*) Halogenophenanthridines, 51 – (*iii*) Nitro- and amino-phenanthridines, 52 – (*iv*) Hydroxyphenanthridines or phenanthridinols, 53 – (*v*) Phenanthridones, 53 –
   d. Hydrogenated phenanthridines 57
3. Benzo[*f*], [*g*] and [*h*]quinolines 60
   a. Structure and synthesis 60
   b. Properties 62
4. Benzisoquinolines 66
   a. Benz[*f*]isoquinolines 66
   b. Benz[*g*]isoquinoline 68
5. Naphthoquinolines and naphthoisoquinolines 71
6. Benzacridines and benzophenanthridines 75
   a. Benzacridines 75
   b. Benzophenanthridines 78

Chapter 29. Six-membered Heterocycles Containing Phosphorus, Arsenic, Antimony and Bismuth as a Single Heteroatom
by R. E. Atkinson

1. Phosphorus compounds 83
   a. Phosphorinane and its derivatives 83
      (*i*) Phosphorinanes, 83 – (*ii*) Phosphorinanones, 86 – (*iii*) Phosphorinanols, 88 –
   b. Tetrahydrophosphorins and related compounds 89
      (*i*) Tetrahydrosphosphorins, 89 – (*ii*) Tetrahydrophosphinolines, 91 –
   c. Dihydrophosphorins 93
   d. Phosphorins, phosphabenzenes 97
   e. Miscellaneous phosphorus heterocycles 104
      (*i*) Heterocycles containing bridgehead phosphorus, 104 – (*ii*) 5-Phosphoniaspiro[4.5]decane derivatives, 105 –
2. Arsenic compounds 105
   Arsenanes and related compounds 105
      (*i*) Arsenanes, 105 – (*ii*) 4-Arsenones, 106 – (*iii*) Derivatives of tetrahydroarsenin, tetrahydroarsinoline and tetrahydroisoarsinoline, 108 – (*iv*) 5,10-Dihydroacridarsines and 5,6-dihydroarsanthridines, 110 – (*v*) 1*H*-Naphth[1,8-*c*,*d*]arsenins, 112 – (*vi*) Arsenins, 112 –
3. Antimony compounds 113
4. Bismuth compounds 114

Chapter 30. Pyridine and Piperidine Alkaloids
by J. D. Hunt and A. McKillop

a. Alkaloids of areca nut 119
b. Alkaloids of hemlock 120
c. Alkaloids of pomegranate root bark 126
d. Lobelia and sedum alkaloids 128
e. Alkaloids of tobacco 137
f. Miscellaneous alkaloids 144

Chapter 31. The Quinoline Alkaloids
by M. Sainsbury

1. Quinoline alkaloids of the Rutacea family 171
   a. Simple quinolines 193
   b. Simple quinolones 195
   c. Furoquinolines and furoquinolones 201
   d. Pyranoquinolines 209
   e. Acridone alkaloids 210
   f. Biosynthesis of the rutaceous quinoline alkaloids 214
2. Miscellaneous alkaloids of probable biogenetic derivation from anthranilic acid 216
   a. Echinorine and echinine 216
   b. The alkaloids of *Macrorungia longistrobus* 217
   c. Cryptolepine 220

3. Miscellaneous alkaloids containing the quinoline ring system 220
a. Quinolines from fungi and micro-organisms 221

(*i*) Pseudanes, 221 – (*ii*) Viridicatine, viridicatol, cyclopenine and cyclopenol, 222 –

4. Quinoline alkaloids derived from tryptophan 225
a. The cinchona alkaloids 225

(*i*) Occurrence, 225 – (*ii*) Individual quinoline alkaloids and related compounds, 226 – (*iii*) Cinchona alkaloids derived from indole, 232 – (*iv*) Stereochemistry of the cinchona bases, 234 – (*v*) Physical properties of cinchona bases, 239 – (*vi*) Synthetical studies in the cinchona series, 239 – (*vii*) Biosynthesis of cinchona alkaloids, 244 –

b. Camptothecin 246
c. Calycanthine, calycanthidine, folicanthine, chimonanthine and meso-chimonanthine 249

(*i*) Calycanthine, 249 – (*ii*) Folicanthine and chimonanthine, 250 – (*iii*) calycanthidine, 250 – (*iv*) Biosynthesis of the calycanthaceous alkaloids, 251 –

d. 2-Quinolone alkaloids of *Melodinus scandens* 252

(*i*) Meloscine and epimeloscine, 252 – (*ii*) Scandine and meloscandonine, 252 – (*iii*) Perloline and perlolidine, 253 – (*iv*) Perhydroquinolines, 254 –

Chapter 32. The Acridine Alkaloids
by B. P. SWANN AND A. MCKILLOP

Spectral characteristics 258
Reactions of acridone alkaloids 259
The acronycine group of alkaloids 263

Chapter 33. The Alkaloids of the Morphine Group
by K. W. BENTLEY

1. The morphine–thebaine group 268
a. Structure 270
b. Total synthesis of morphine 275
c. Stereochemistry 278

(*i*) Morphine alkaloids, 278 – (*ii*) B/C-*trans*-morphine and its derivatives, 281 –

d. Reactions and derivatives of the morphine–thebaine group of alkaloids 283

(*i*) The halogeno-morphides and -codides, 283 – (*ii*) Exhaustive methylation of codeine, 284 – (*iii*) The deoxy compounds, 285 – (*iv*) The reduction of thebaine, 286 – (*v*) The thebainones, 288 – (*vi*) The thiocodides, 290 – (*vii*) 14-Bromo- and 14-hydroxy-codeinone, 290 – (*viii*) Migration of nitrogen from C-9 to C-14, 291 – (*ix*) Diels–Alder reactions of thebaine, 292 – (*x*) Molecular rearrangements involving the C-13 side-chain, 294 – (*xi*) Molecular rearrangements not involving the C-13 side-chain, 297 – (*xii*) Flavothebaone, 299 – (*xiii*) Reaction of thebaine with metal carbonyls, 300 – (*xiv*) ψ-Morphine, 301 –

e. The relationship between structure and analgesic activity 301
2. The sinomenine–salutaridine group 302
3. The hasubanonine group 309
4. The homomorphine alkaloids 319

Chapter 34. Diterpenoid Alkaloids
by A. R. PINDER

1. C_{20}-diterpene alkaloids 324
   a. The *Garrya* alkaloids 324
   b. The atisine alkaloids 334
   c. Alkaloids with a modified atisane skeleton 341
2. C_{19}-diterpene alkaloids 350
   a. Lycoctonine-type alkaloids 350
   b. Aconitine-type alkaloids 358
   c. Lactonic alkaloids 370
   d. *Daphniphyllum* alkaloids 374
      (*i*) Daphniphylline subgroup, 374 – (*ii*) Secodaphniphylline subgroup, 376 – (*iii*) Yuzurimine subgroup, 376 –
3. Bis-diterpene alkaloids 378

Chapter 35. Steroidal Alkaloids
by A. R. PINDER

1. Normal steroidal alkaloids 381
   a. Bases related to 3-amino-5α-pregnane 382
   b. Bases related to 20-amino-5α-pregnane 389
   c. Bases related to 3,20-diamino-5α-pregnane 391
   d. Bases related to 3-aminoconanine 399
   e. Bases related to 20-piperidyl-5α-pregnane 411
      (*i*) Simple bases, 411 – (*ii*) Solanidanes, 415 – (*iii*) Spirosolane alkaloids, 422 –
   f. Biosynthesis of normal steroidal alkaloids 426
   g. Biological properties of normal steroidal alkaloids 427
2. Alkaloids with a modified steroidal structure 427
   a. Salamander alkaloids 430
      (*i*) Alkaloids containing an oxazolidine ring, 431 – (*ii*) Alkaloids not containing an oxazolidine ring, 434 – (*iii*) Biosynthesis of salamander bases, 435 – (*iv*) Physiological action, 435 –
   b. C-Nor-D-homosteroidal alkaloids 436
      (*i*) Alkamines, 437 – (*ii*) Ester-alkaloids, 444 – (*iii*) Fritillaria alkaloids, 453 –
   c. Buxus alkaloids 454
      (*i*) Monoacidic Buxus alkaloids, 455 – (*ii*) Diacidic Buxus alkaloids, 459 –

INDEX 467

Titles of other parts of Volume IV

HETEROCYCLIC COMPOUNDS

Vol. IV A: Three-, four- and five-membered heterocyclic compounds with a single hetero-atom in the ring

Vol. IV B: Five-membered heterocyclic compounds with a single hetero-atom in the ring: alkaloids, dyes and pigments

Vol. IV C: Five-membered heterocyclic compounds with two hetero-atoms in the ring from Groups V and/or VI of the Periodic Table

Vol. IV D: Five-membered heterocyclic compounds with more than two hetero-atoms in the ring

Vol. IV E: Six-membered monoheterocyclic compounds containing oxygen, sulphur, selenium, tellurium, silicon, germanium, tin, lead or iodine as the heteroatom

Vol. IV F: Six-membered heterocyclic compounds with a single nitrogen atom in the ring: pyridine, polymethylenepyridine, quinoline, isoquinoline and their derivatives

OFFICIAL PUBLICATIONS

B.P.	British (United Kingdom) Patent
F.P.	French Patent
G.P.	German Patent
Ger. Offen.	German Patent Application, open for inspection
Sw.P.	Swiss Patent
U.S.P.	United States Patent
U.S.S.R.P.	Russian Patent
B.I.O.S.	British Intelligence Objectives Sub-Committee Reports, H.M. Stationery Office, London.
C.I.O.S.	Combined Intelligence Objectives Sub-Committee Reports
F.I.A.T.	Field Information Agency, Technical Reports of U.S. Group Control Council for Germany
B.S.	British Standards Specification
A.S.T.M.	American Society for Testing and Materials
A.P.I.	American Petroleum Institute Projects
C.I.	Colour Index Number of Dyestuffs and Pigments

SCIENTIFIC JOURNALS AND PERIODICALS

With few obvious and self-explanatory modifications the abbreviations used in references to journals and periodicals comprising the extensive literature on organic chemistry, are those used in the World List of Scientific Periodicals.

LIST OF ABBREVIATED NAMES OF CHEMICAL FIRMS MENTIONED IN PATENT REFERENCES

A.G.F.A., Agfa A.G.	Aktiengesellschaft für Anilinfabrikation (Berlin)
B.A.S.F.	Badische Anilin- und Soda-Fabrik (Ludwigshafen)
Bayer	Farbenfabriken vorm. Friedrich Bayer und Co. (Leverkusen)
Cassella	Leopold Cassella und Co. (Frankfurt am Main)
C.F.M.	Compagnie française des Matières Colorantes (Paris)
CIBA	Gesellschaft für chemische Industrie (Basel)
Du Pont	E.I. Du Pont de Nemours and Co. (U.S.A.)
G.A.F.	General Anilin and Film Corporation (U.S.A.)
Geigy A.G.	J. R. Geigy S.A. (Basel)
Hoechst	Hoechst A.G. (see M.L.B.)
I.C.I.	Imperial Chemical Industries, Ltd. (London)
I.G.	(= Interessen Gemeinschaft Farbenindustrie) of the principal dyestuffs manufacturers in Germany
Kalle	Kalle und Co., A.G. (Biebrich am Rhein)
M.L.B.	Farbwerke vormals Meister, Lucius und Brüning (Hoechst)
Sandoz	Sandoz A.G. Chemische Fabrik (Basel)

LIST OF COMMON ABBREVIATIONS AND SYMBOLS USED

A	acid
Å	Ångström units
Ac	acetyl
a	axial
as, *asymm.*	asymmetrical
at.	atmosphere
B	base
Bu	butyl
b.p.	boiling point
C, mC and μC	curie, millicurie and microcurie
c, *C*	concentration
c.d.	circular dichroism
conc.	concentrated
crit.	critical
D	Debye unit, 1×10^{-18} e.s.u.
D	dissociation energy
D	dextro-rotatory; dextro configuration
DL	optically inactive (externally compensated)
d	density
dec. or decomp.	with decomposition
deriv.	derivative
E	energy; extinction; electromeric effect
E1, E2	uni- and bi-molecular elimination mechanisms
E1cB	unimolecular elimination in conjugate base
e.s.r.	electron spin resonance
Et	ethyl
e	nuclear charge; equatorial
f	oscillator strength
f.p.	freezing point
G	free energy
g.l.c.	gas liquid chromatography
g	spectroscopic splitting factor, 2.0023
H	applied magnetic field; heat content
h	Planck's constant
Hz	hertz
I	spin quantum number; intensity; inductive effect
i.r.	infrared
J	coupling constant in n.m.r. spectra
K	dissociation constant
k	Boltzmann constant; velocity constant
kcal.	kilocalories
L	laevorotatory; laevo configuration
M	molecular weight; molar; mesomeric effect
Me	methyl

m	mass; mole; molecule; *meta-*
ml	millilitre
m.p.	melting point
Ms	mesyl (methanesulphonyl)
[M]	molecular rotation
N	Avogadro number; normal
n.m.r.	nuclear magnetic resonance
N.O.E.	Nuclear Overhauser Effect
n	normal; refractive index; principal quantum number
o	*ortho-*
o.r.d.	optical rotatory dispersion
P	polarisation; probability; orbital state
Pr	propyl
Ph	phenyl
p	*para-*; orbital
p.m.r.	proton magnetic resonance
R	clockwise configuration
S	counterclockwise config.; entropy; net spin of incompleted electronic shells; orbital state
S_N1, S_N2	uni- and bi-molecular nucleophilic substitution mechanisms
S_Ni	internal nucleophilic substitution mechanisms
s	symmetrical; orbital
sec	secondary
soln.	solution
symm.	symmetrical
T	absolute temperature
Tosyl	*p*-toluenesulphonyl
Trityl	triphenylmethyl
t	time
temp.	temperature (in degrees centigrade)
tert	tertiary
U	potential energy
u.v.	ultraviolet
v	velocity
α	optical rotation (in water unless otherwise stated)
$[\alpha]$	specific optical rotation
α_A	atomic susceptibility
α_E	electronic susceptibility
ε	dielectric constant; extinction coefficient
μ	microns (10^{-4} cm); dipole moment; magnetic moment
μ_B	Bohr magneton
μg	microgram (10^{-6} g)
λ	wavelength
υ	frequency; wave number
χ, χ_d, χ_μ	magnetic, diamagnetic and paramagnetic susceptibilities

~	about
(+)	dextrorotatory
(−)	laevorotatory
⊖	negative charge
⊕	positive charge

Chapter 28

Polycyclic Compounds Comprising a Pyridine and Two or More Carbocyclic Rings

N. CAMPBELL

1. Acridine and its derivatives

Few compounds and their derivatives have been so extensively reviewed as the acridines, the subject of two outstanding monographs each with its distinctive merits. The first* written by Professor Adrien Albert, who has made the acridine field very much his own, gives a conspectus of the subject, while the second**, carefully edited by Dr. R. M. Acheson, contains eighteen contributions by experts on various facets of acridine chemistry. A pleasing item in the first book is a short biography of Kurt Lehmstedt, whose "vital contributions" to acridine chemistry have not been fully recognised.

(a) General introduction to the chemistry of acridine

Acridine, $C_{13}H_9N$, is a linear dibenzopyridine, resembling anthracene in structure with one of the *meso*-methine groups replaced by a nitrogen atom. Two systems of numbering the formula have been most commonly used:

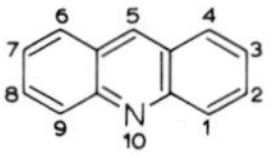

6 5 4 7 3 8 2 9 N 10 1

Richter

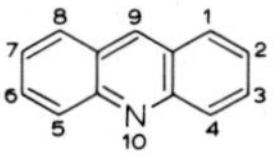

8 9 1 7 2 6 3 5 N 10 4

Graebe

* *A. Albert*, "The Acridines", 2nd Edn., Arnold, London, 1966.
** *R. M. Acheson* (ed. and contributor), "Acridines", 2nd Edn., Wiley, New York, 1973.

This has given rise to some confusion, but the Richter system has almost entirely given way to that proposed by Graebe. This system is the one adopted by the I.U.P.A.C. in its Rules on Nomenclature.

C. Graebe and *H. Caro* (Ber., 1870, **3,** 746; Ann., 1871, **158,** 265) discovered acridine in the crude anthracene fraction of coal-tar and gave it the name on account of its irritating action which causes sneezing and smarting of the skin and eyes. These authors assigned to acridine the empirical formula $C_{12}H_9N$, but later workers advanced the correct formula, $C_{13}H_9N$, which was confirmed by the careful analysis of acridine and its chloroplatinate (*C. Riedel, ibid.*, 1883, **16,** 1609; *A. Bernthsen* and *F. Bender*, Ann., 1884, **224,** 1). The formulation of acridine in terms of Kekulé structures, modified in the light of the resonance theory (see below), is the one generally used.

Constitution. The formulation of acridine as a tricyclic compound I follows from its synthesis and chemical properties. Especially informative is the oxidation of acridine by potassium permanganate to quinoline-2,3-dicarboxylic acid and by chromic acid to 9-acridone (II), the structure of which is established by the ring-closure of diphenylamine-2-carboxylic acid:

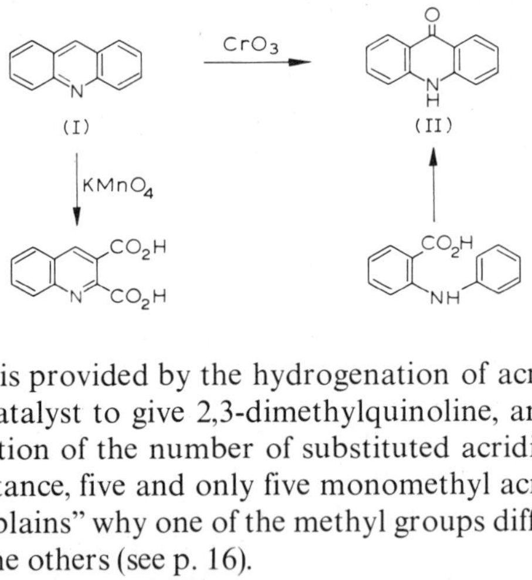

CrO_3

(I) (II)

$KMnO_4$

Further evidence is provided by the hydrogenation of acridine in the presence of a nickel catalyst to give 2,3-dimethylquinoline, and by the correctness of the prediction of the number of substituted acridines which can be prepared. For instance, five and only five monomethyl acridines are known and formula I "explains" why one of the methyl groups differs in its chemical behaviour from the others (see p. 16).

By analogy it is to be expected that acridine will resemble anthracene to a considerable extent and this proves to be so. Both have essentially planar molecules and their ultraviolet spectra are very similar. In anthracene carbon atoms $C_{(1)}$ and $C_{(4)}$ are identical as are $C_{(2)}$ and $C_{(3)}$. This is reflected in the acridine series by the similarity of the ultraviolet spectra of 2- and 3-hydroxyacridine and of 1- and 4-hydroxyacridine.

Anthracene has a resonance energy of 105 kcal/mole and acridine 106 kcal/mole (*A. Albert* and *J. B. Willis*, Nature, 1946, **157**, 341). It may be noted here that the resonance energy for acridine, calculated from the heat of hydrogenation of acridine by lithium tetrahydridoaluminate is 84 kcal/mole (*L. M. Jackman* and *D. I. Packman*, Proc. chem. Soc., 1957, 349). Acridine accordingly is a resonance hybrid with contributing structures III–VI along with other minor structures containing an excited pyridine ring:

(III) (IV) (V) (VI)

The most important of these conjugated structures are III and IV, each containing two aromatic rings in contrast to V and VI, which contain two quinonoid nuclei. This rather simplistic picture of the acridine molecule is supported by the bond lengths as measured by X-ray methods (*D. C. Phillips*, Acta Cryst., 1956, **9**, 237). In Table 1 the double-bond order, derived on the assumption that the four contributing structures III–VI make equal contributions to the hybrid, and the bonds lengths are shown. The correlation between bond length and double-bond character is as expected with the exception of the $C_{(4a)}-N$ linkage which is unexpectedly short.

TABLE 1

BOND CHARACTERISTICS IN THE ACRIDINE MOLECULE

Bond	*Length (Å)*	*Double bond character*
$C_{(1)}-C_{(2)}$	1.37	3/4
$C_{(2)}-C_{(3)}$	1.42	1/4
$C_{(3)}-C_{(4)}$	1.37	3/4
$C_{(4)}-C_{(4a)}$	1.43	1/4
$C_{(4a)}-N$	1.34	1/2
$C_{(9)}-C_{(9a)}$	1.39	1/2
$C_{(9a)}-C_{(4a)}$	1.43	1/4

Clearly acridine is a conjugated resonating compound which will possess pronounced aromatic character. Formula III gives little insight into the position of attack by electrophilic reagents. This has been accomplished with some success by the more sophisticated approach based on electrophilic localisation energy calculations (p. 9).

Chemically, anthracene and acridine show some similarity. Both add hydrogen at the *meso*-positions and both on oxidation yield carbonyl products. On the other hand acridine does not yield the transannular adducts with maleic anhydride which are so characteristic of anthracene, while anthracene, lacking a nitrogen atom, fails to undergo many of the reactions encountered in the acridine series.

(i) Methods of preparation

(*1*) *Bernthsen reaction.* In the first synthesis of acridine and its derivatives diphenylamine and a carboxylic acid were heated with zinc chloride (*Bernthsen* and *Bender, loc. cit.*). With formic acid acridine is obtained in very poor yield (7.5%), but with many other acids fair yields result. Thus diphenylamine and benzoic acid give a 48% yield of 9-phenylacridine. The use of polyphosphoric acid instead of zinc chloride has certain advantages although the yields are unsatisfactory (*F. D. Popp*, J. org. Chem., 1962, **27**, 2658).

(*2*) *From acridones.* Acridines (VIII) can be prepared by reduction of the readily accessible acridones (VII) (p. 31), but the reaction is accompanied by further reduction to the 9,10-dihydroacridine (IX).

Reduction can be effected with sodium and amyl alcohol or by sodium or aluminium amalgam, while in the subsequent oxidation potassium dichromate and dilute sulphuric acid or ferric chloride are used. The use of a reducing reagent may be avoided by converting the acridone (VII) into the 9-chloroacridine by phosphoryl chloride. The chloroacridine is then converted by hydrazine into 9-hydrazinoacridine, which with oxygen in the presence of platinum and a little alkali yields the acridine (*Albert*, J. chem. Soc., 1965, 4653):

(VII) (VIII) (IX)

[H] [O] $POCl_3$ $H_2N{\cdot}NH_2$ $NH{\cdot}NH_2$

In another method the chloro-compound is treated with tosylhydrazine and the tosylhydrazide so formed when heated in glycol with sodium carbonate yields acridine and *p*-toluenesulphinic acid (*Albert* and *R. Royer*, *ibid.*, 1949, 1148).

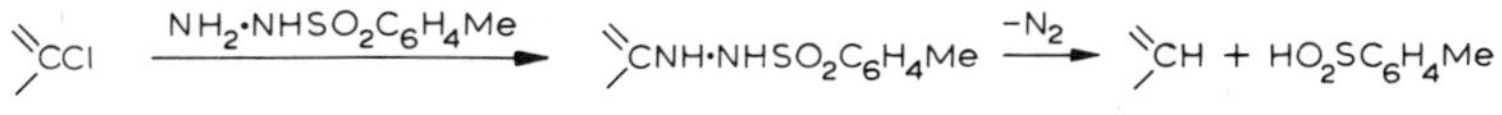

$$\gt CCl \xrightarrow{NH_2{\cdot}NHSO_2C_6H_4Me} \gt CNH{\cdot}NHSO_2C_6H_4Me \xrightarrow{-N_2} \gt CH + HO_2SC_6H_4Me$$

The conversion of acridones to 9-alkylacridines is outlined on p. 33.

(*3*) *Pfitzinger synthesis.* This synthesis is illustrated by 1-*o*-tolyl-7-methylisatin (X), which is conveniently prepared by heating di-*o*-tolylamine with oxalyl chloride and aluminium chloride (*M. S. Newman* and *W. H. Powell*, J. org. Chem., 1961, **26,** 812). The isatin with alkali yields the corresponding isatinic acid, which undergoes ring-closure to give 4,5-dimethylacridine-9-carboxylic acid (XI). The isatin can also be oxidised with alkaline hydrogen peroxide to give di(2,2′-methylphenyl)amine-2-carboxylic acid, which with phosphoryl chloride gives 9-chloro-4,5-dimethylacridine (XII, R = Cl). The chloro compound with hydrogen and Raney nickel yields 4,5-*dimethylacridine* (XII, R = H), m.p. 77.5–77.9°, also obtained by the thermal decarboxylation of the 9-carboxylic acid (XI). Yields are excellent and the method is capable of wide application:

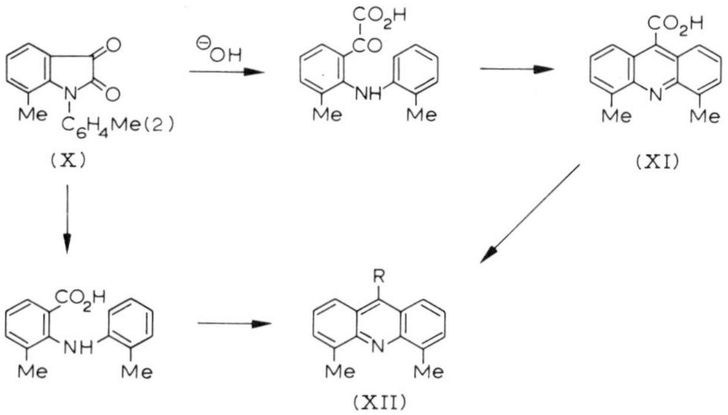

(X) (XI) (XII)

(*4*) *From anthranils.* Anthranil and substituted anthranils can be converted into acridine and its derivatives, and although the methods are not always of preparative value they are interesting and instructive.

Anthranil with cyclohexanone and mercuric sulphate yields 1,2,3,4-tetrahydroacridine by a series of reactions summarised in the sequence:

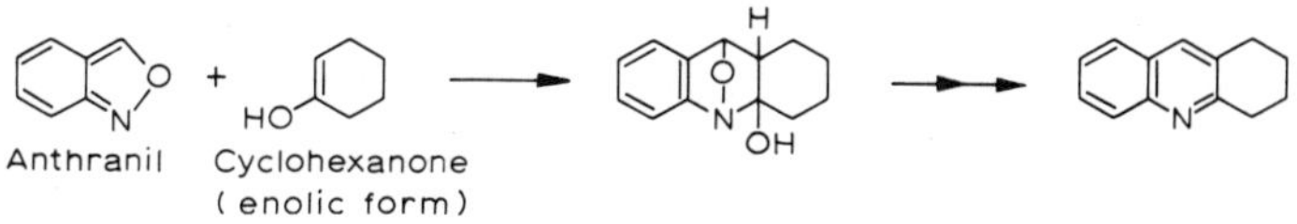

Anthranil Cyclohexanone (enolic form)

With cyclohexene the product is acridine (*M. Wilk, H. Schwab* and *R. Rochlitz*, Ann., 1966, **698,** 149).

Anthranil and benzyne (generated by oxidisng 1-aminobenzotriazole with lead tetra-acetate) yield a betaine, XIII, from which a 5% yield of acridine is obtained (*C. D. Campbell* and *C. W. Rees*, J. chem. Soc., C, 1969, 748):

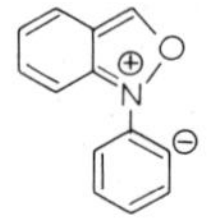

(XIII)

3-Phenylanthranils isomerise, often quantitatively, to acridones when heated or when treated with nitrous acid at room temperature, a reaction described by Bamberger in 1909, but usually known as the *Lehmstedt–Tanasescu* reaction. Bamberger pictured the reaction as involving a nitrosoamine and this accounts for the conversion of 5-chloro-3-*p*-chlorophenylanthranil into 3,7-dichloroacridone (*R. B. Davis* and *L. C. Pizzini*, J. org. Chem., 1960, **25,** 1884):

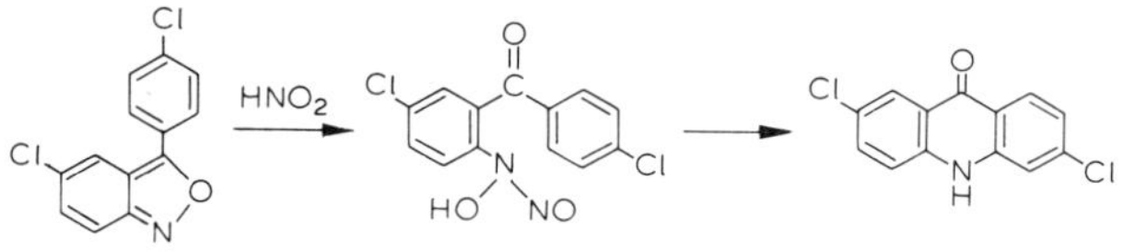

$\xrightarrow{HNO_2}$ $\longrightarrow$

On the other hand the pyrolysis of 3-(2,4-dimethoxyphenyl)anthranil gives 2,4-dimethoxyacridone and not the 1,3-dimethoxyisomer required by this route (*R. Kwok* and *P. Pranc*, *ibid.*, 1968, **33,** 2880). Another course *via* an intermediate nitrene has been postulated. It is noteworthy that methoxyphenylanthranils with nitrous acid yield azoxybenzoic acids. The isomerisation and chemical properties of the substituted 3-phenylanthranils call for further investigation.

An example of nitrene insertion is observed in the formation of acridine (2.8%) and 9,10-dihydroacridine (10%) among the products obtained by the decomposition of 2-azido-2′-methoxydiphenylmethane (*G. R. Cliff* and *G. Jones*, Chem. Comm., 1970, 1705).

(5) *Miscellaneous methods.* The thermolysis or photolysis of 3-phenylbenzotriazin-4-one results in the formation of acridone and nitrogen (*D. H. Hey, C. W. Rees* and *A. R. Todd*, Chem. and Ind., 1962, 1332; *G. Ege*, Ber., 1968, **101,** 3079):

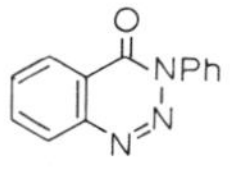

3-Phenylbenzotriazin-4-one

9-Methylacridine is obtained when *N*-acetyldibenzo[*b,f*]azepine (XIV, R = COMe) is heated with hydrobromic acid (*H. Schindler* and *H. Blattner*, Helv., 1961, **44,** 753) or when the parent compound (XIV, R = H) is heated with hydrochloric acid (*P. Rumpf* and *R. Reynaud*, Bull. Soc. chim. Fr., 1962, 2241):

(XIV)

(ii) Physical properties

One of the most striking properties of the acridines is their fluorescence. Acridine salts in solution exhibit a green fluorescence, while further dilution results in hydrolysis and the fluorescence changes to the violet colour of free acridine. Some acridines are used as fluorescent indicators.

Some acridines, but not the simple ones, exhibit chemiluminescence (p. 27).

The ultraviolet spectra of acridine and its derivatives are characteristic and can be used to differentiate between the acridine, acridinium salt and 9,10-dihydroacridine (acridan*) species. Acridine has a sharp band at 252 nm (logε 5.24) and a broader band with a maximum at 347 nm (logε 3.90), very similar to those of anthracene (*G. M. Badger et al.*, J. chem. Soc., 1951, 3199; *D. P. Craig* and *L. N. Short*, *ibid.*, 1945, 419), whereas 9,10-dihydroacridine has a broad band at 290 nm (logε ~4.3) (*V. Zanker* and *B. Schneider*, Z. phys. Chem. neue Folge, 1969, **68,** 19). The acridinium ion spectrum is displaced towards the red with a maximum at about 410 cm^{-1} and the short-wave band is intensified.

Electron bombardment of acridine results in a considerable number of fragments as the result of loss of hydrogen, hydrogen cyanide, etc. with molecular ion as the base ion.

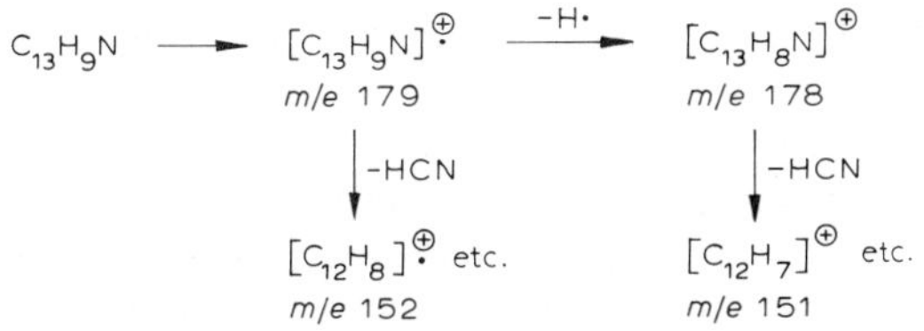

$$C_{13}H_9N \longrightarrow [C_{13}H_9N]^{\oplus\cdot} \ (m/e\ 179) \xrightarrow{-H\cdot} [C_{13}H_8N]^{\oplus} \ (m/e\ 178)$$

$$[C_{13}H_9N]^{\oplus\cdot} \xrightarrow{-HCN} [C_{12}H_8]^{\oplus\cdot} \text{ etc.} \ (m/e\ 152)$$

$$[C_{13}H_8N]^{\oplus} \xrightarrow{-HCN} [C_{12}H_7]^{\oplus} \text{ etc.} \ (m/e\ 151)$$

* 9,10-Dihydroacridine is frequently described and sometimes indexed under the trivial name "acridan", which is a convenient abbreviation; the latter name is not acceptable, however, in the I.U.P.A.C. Rules. Similarly, acridone, which is the accepted term for 9-oxo-9,10-dihydroacridine, is sometimes referred to as "acridanone".

The p.m.r. spectrum shows, as expected, an ABCD system with the chemical shift of the $C_{(9)}$-proton considerably lower than the shifts of the other protons (*J. P. Kokko* and *J. H. Goldstein*, Spectrochem. Acta, 1963, **19,** 1119).

(iii) Chemical properties and reactions

The acridines form salts such as the hydrochlorides, nitrates and picrates and their tertiary amino-structure is shown by the formation of *N*-alkylacridinium salts by reaction with alkyl halides and of *N*-oxides when oxidised by perbenzoic acid. The acridines are chemically and thermally stable. Acridine itself, for example, boils at a high temperature without decomposition and is not attacked when fused with potassium hydroxide.

(*1*) *Oxidation.* Acridine retains its ring-structure when oxidised by many reagents, but with potassium permanganate quinoline-2,3-dicarboxylic acid is obtained. This acid is the main product (~75%) of ozonolysis in methanol (*E. J. Moriconi* and *F. A. Spano*, J. Amer. chem. Soc., 1964, **86,** 38). Acridine with potassium dichromate in acetic acid gives some of the dicarboxylic acid in poor yield and 10,10′-*biacridonyl* (XV), a bright yellow compound, m.p. 251°. The nitrogen–nitrogen linkage in this compound is shown by the smooth cleavage by sodium amalgam in ethanol to yield 9,10-dihydroacridine. In contrast to the parent compound the nitroacridines are oxidised by chromic anhydride and acetic acid to give good yields of the corresponding nitroacridones (p. 32).

As already noted, acridine with perbenzoic acid yields the *N*-oxide, but in addition a small quantity of another substance is obtained which was originally regarded as the trans-annular peroxide (*K. Lehmstedt* and *H. Klee*, Ber., 1936, **69,** 1514).

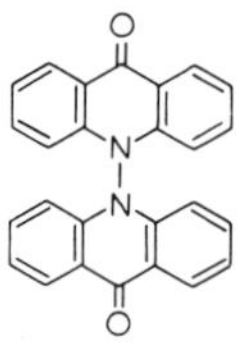

(XV) 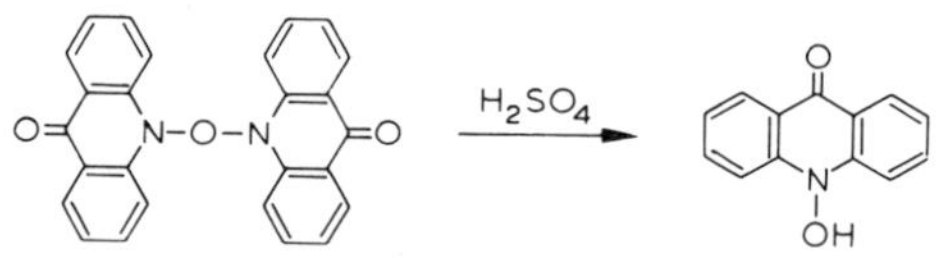 (XVI) $\xrightarrow{H_2SO_4}$ (XVII)

When heated in pyridine it yields acridone (9-oxo-9,10-dihydroacridine) and with sulphuric acid gives N-*hydroxyacridone* (XVII), m.p. 256° (decomp.), and its physical properties, notably mol. wt. (mass spectrographic) of 404 and typical carbonyl absorption in the infrared region shows it to be 10,10′-*bisacridone ether* (XVI), m.p. 175° (*R.M. Acheson* and *B. Adcock*, J. chem. Soc., C, 1968, 1045).

(*2*) *Reduction.* Reduction of acridine with some reagents gives mainly

9,10-dihydroacridine (acridan), but more highly hydrogenated products can also be obtained (p. 26).

(*3*) *Comparison with anthracene.* Some of the properties of acridine, as already pointed out (p. 2), suggest a similarity between the chemical behaviour of acridine and that of anthracene and this is reinforced by the addition of sodium to acridine to form the 9,10-disodio derivative which with ethanol yields 9,10-dihydroacridine, thereby recalling the reaction of sodium with anthracene (*W. Schlenk* and *E. Bergmann*, Ann., 1928, **463,** 281). It is easy, however, to push this comparison too far and there are notable differences between the behaviour of the two compounds. Two characteristic properties of anthracene which are not observed in acridine are the ready formation of a transannular peroxide and a photo-dimer through the 9,10,9′,10′-positions. Again a striking feature of anthracene is the ease with which it forms transannular adducts with maleic anhydride or dimethyl acetylenedicarboxylate to form Diels–Alder adducts across the 9,10-positions. Acridine on the other hand does not react with maleic anhydride but with dimethyl acetylenedicarboxylate in ether gives *trans*-10-(1′,2′-bis-methoxycarbonyl)vinyl-9-acridone (XIX) as well as the *cis*-form, possibly through the ylid XVIII (*Acheson* and *M. L. Burstall*, J. chem. Soc., 1954, 3240).

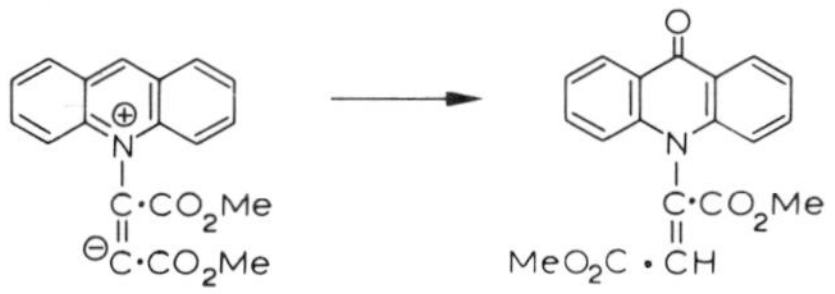

(XVIII) (XIX)

In methanol the product is largely the bright red 10-(*trans*-1,2-dimethoxy-carbonylvinyl)acridinium methoxide which is in equilibrium with the corresponding 9-methoxy-9,10-dihydroacridine derivative:

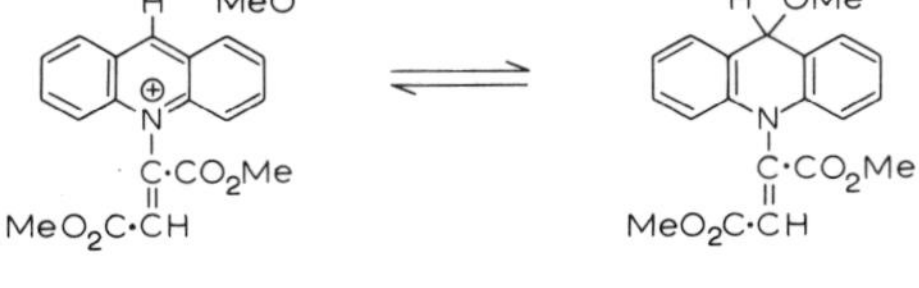

The acridinium methoxide in neutral solution is oxidised by air or hydrogen peroxide to the orange acridone ester XIX, and in this way behaves as a typical 9-alkoxy-10-alkyl-9,10-dihydroacridine (*H. Decker*, J. pr. Chem., 1892, [ii], **45,** 161).

A transannular acridine derivative has been obtained, however, by using dimethylketene (*S. A. Proctor* and *G. A. Taylor*, J. chem. Soc., 1967, 1937;

Chem. and Ind., 1969, 1019). With this reagent acridine in ether gives the dihydroacridine derivative XXa, but 9-methylacridine gives not only a product of this type XXb, but also the bridged compound XXI:

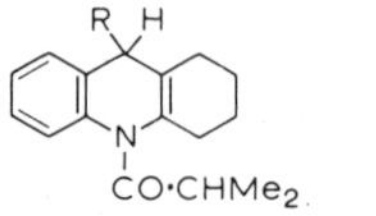

(XX, a, R = H
b, R = Me)

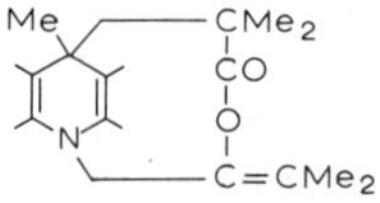

(XXI)

A characteristic of acridine is the formation of bi(9,10-dihydroacridine) when reduced by various reagents (p. 25) including *p*-toluenethiol (*H. Gilman et al.*, J. Amer. chem. Soc., 1954, **76**, 2920) or when photolysed (p. 27) (*Zanker* and *P. Schmid*, Z. physik. Chem., 1958, **17**, 11).

(*4*) *Acridinium salts*; pseudo-bases. N*-Alkylacridinium salts are the analogues of the pyridinium and quinolinium compounds, but are not so easily prepared. Acridine does not react with methyl iodide at 60°, but does so when heated at 100° (*Albert* and *B. Ritchie*, J. chem. Soc., 1943, 458) or in dimethylformamide at room temperature (*F. Kröhnke* and *H. L. Honig*, Ber., 1957, **90,** 2215).

Alkylation is best effected by heating acridine in nitrobenzene with methyl *p*-toluenesulphonate or dimethyl sulphate and saturating the mixture with sodium chloride to give *N*-methylacridinium chloride (*A. Kaufmann* and *A. Albertini*, *ibid.*, 1909, **42**, 1999).

The acridinium salts dissolve in various solvents to give yellow solutions with a green fluorescence. They are frequently coloured compounds, the *methiodide*, m.p. 220–222°, for example, having a red colour, and the colours of these and other heterocyclic salts have been studied (*Kröhnke*, *ibid.*, 1957, **90**, 2237).

The acridinium salts, *e.g.*, *N*-methylacridinium halides (XXII), contain resonating cations and are consequently very susceptible to nucleophilic attack at $C_{(9)}$. This is shown by the reaction of *N*-alkyl- (or *N*-aryl-)acridinium iodides with alkylmagnesium halides to give 9,10-dialkyl (or 9-alkyl-10-aryl)-9,10-dihydroacridines. With potassium cyanide they give *N*-alkyl-9-cyano-9,10-dihydroacridines (*Kaufmann* and *Albertini*, *ibid.*, 1911, **44,** 2052) and with alkali yield pseudo-bases (see below).

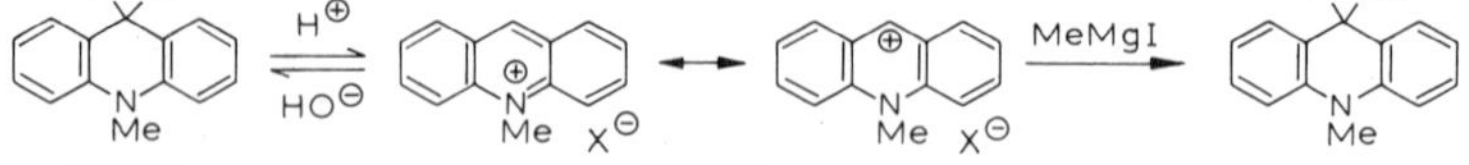

(XXV) (XXII)

* *I. A. Selby*, in "Acridines" (ed. *R. M. Acheson*), Wiley, New York, 1973, p. 433.

Reactive methylene compounds in the presence of alkali react with acridinium salts to give 9,10-disubstituted 9,10-dihydroacridines, *N*-methylacridinium iodide and acetophenone, for instance, yielding 10-methyl-9-phenacyl-9,10-dihydroacridine (XXIII) (*Kröhnke* and *Honig*, *loc. cit.*). An example of nucleophilic attack accompanied by rearrangement is given by the interaction of *N*-methylacridinium iodide and hydroxylamine-*O*-sulphonic acid, NH_2OSO_3H, in methanol to give 5-*methyldibenzo*[b,e][1,4]*diazepine* (XXIV), m.p. 96° (*M. Hirobe* and *T. Ozawa*, Tetrahedron Letters, 1971, 4493). Attack by the cation $NH_4SO_3\overset{\ominus}{N}OH$ is postulated.

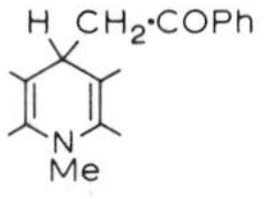

H CH₂·COPh

N
Me

(XXIII)

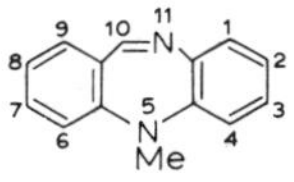

9 10 11 N 1 8 2 7 5 N 3 6 4 Me

(XXIV)

In view of the sensitivity of the 9-position to nucleophilic agents it is not surprising that the acridinium hydroxides (*e.g.* XXII, X=OH), formed by the addition of alkali to acridinium salts, isomerise in a few minutes to the *pseudo-base* (XXV) (*A. Hantzsch* and *M. Kalb*, Ber., 1899, **32**, 3109). The problem of the constitution of the pseudo-bases is discussed elsewhere (C.C.C., Vol. IVF, pp. 272 *et seq.*) and is simplified in the acridine series by the fact that possible open-chain structures (2-acyldiphenylamine derivatives) do not call for consideration. Evidence suggests that in solution an equilibrium between the ammonium base and the tertiary alcohol base may obtain or the latter form may predominate. This is illustrated by *N*-methyl-9-phenylacridinium iodide which with sodium hydroxide yields the ammonium base, which isomerises to the tertiary alcohol base, a change which can be followed visually:

Ph

⊕N
Me OH⊖

Acridinium base

Ph OH

N
Me

tertiary alcohol pseudo-base

A dilute aqueous solution of the methiodide with an equimolecular quantity of sodium hydroxide gives a clear solution of ammonium base, but on warming the tertiary alcohol or pseudo-base separates. It is soluble in ether and chloroform and the ultraviolet spectra of such solutions resemble that of 9-phenyl-9,10-dihydroacridine and differ from that of *N*-methyl-9-phenylacridinium iodide. The tertiary alcohol form of the pseudo-base is thus established (*J. J. Dobbie* and *C. K. Tinkler*, J. chem. Soc., 1905, **87**, 269). Further evidence for this structure is supplied by the preparation of the pseudo-base by the action of phenylmagnesium bromide on *N*-methylacridone.

The matter, however, is not always so clear-cut. In aqueous methanol, for example, the ultraviolet spectrum is similar to that of a solution containing 75% of 9-phenyl-9,10-dihydroacridine and 25% of *N*-methyl-9-phenylacridinium iodide, thereby clearly indicating an equilibrium mixture of the ammonium salt and the tertiary alcoholic base.

The pseudo-bases are readily oxidised by air to *N*-alkylacridones and in absence of air dismutation on steam-distillation occurs (*A. Pictet* and *E. Patry*, Ber., 1902, **35,** 2534). Thus the pseudo-base from *N*-methylacridinium iodide yields 10-methyl-9,10-dihydroacridine and 10-methylacridone:

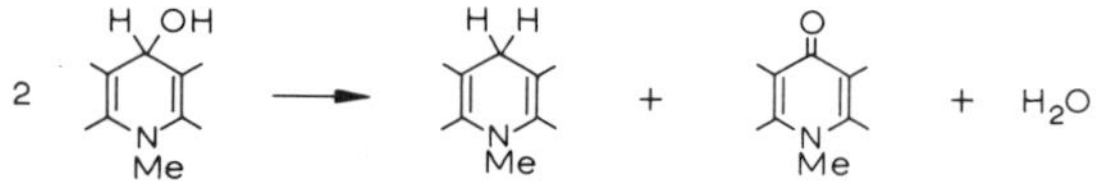

Such reactions confirm the constitution of the pseudo-bases.

The pseudo-bases with mineral acids immediately yield the corresponding acridinium salts. Like many tertiary alcohols they are prone to form ethers such as 9-*ethoxy*-10-*methyl*-9-*phenyl*-9,10-*dihydroacridine*, m.p. 108°, which is formed when the pseudo-base is recrystallised from ethanol or even when treated with cold ethanol (*H. Decker* and *P. Becker*, *ibid.*, 1913, **46,** 969).

Pseudo-bases with a methyl or methylene group in the 9-position give **methides** when treated with a base, 9,10-dimethylacridinium iodide and alkali yielding 9-*methylene*-10-*methyl*-9,10-*dihydroacridine* (XXVI), pale yellow crystals, m.p. 93° (*W. König*, *ibid.*, 1923, **56,** 1543):

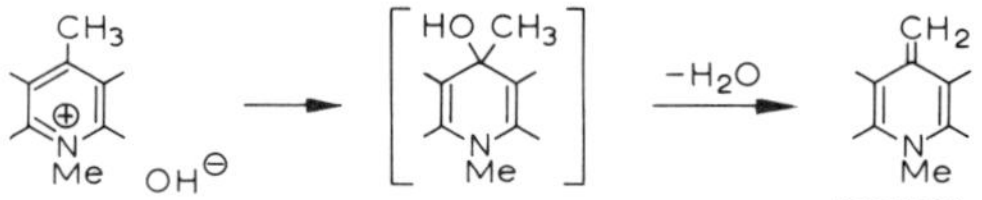

(XXVI)

9-Benzylidene-10-methylacridinium salts likewise yield 9-benzylidene-10-methyl-9,10-dihydroacridine (XXVI, CH_2 replaced by CHPh), which is oxidised by moist air to benzaldehyde and *N*-methylacridone.

(5) *Acridine* N-*oxide*. The *N*-oxide is obtained in 19% yield by the oxidation of acridine in chloroform with perbenzoic acid (*Acheson et al.*, J. chem. Soc., C, 1968, 1045). It forms hygroscopic, yellow needles, m.p. 169°, and unlike the free base it undergoes electrophilic attack at $C_{(9)}$. Bromine in acetic acid yields 9-*bromoacridine* N-*oxide*, m.p. 174°, identical with a sample obtained by treating 9-bromoacridine with perbenzoic acid. With nitric acid and sulphuric acid it gives 9-*nitroacridine* N-*oxide*, orange needles, m.p. 223°, converted by phosphorus trichloride into 9-chloroacridine.

The *N*-oxide is reduced with sodium amalgam to 9,10-dihydroacridine and with phosphorus pentachloride yields 9-chloroacridine (*A. Kliegl* and *A. Fehrle*, Ber., 1914, **47,** 1629). With acetic anhydride it is converted into 9-acridone, possibly through an intermediate bridged compound:

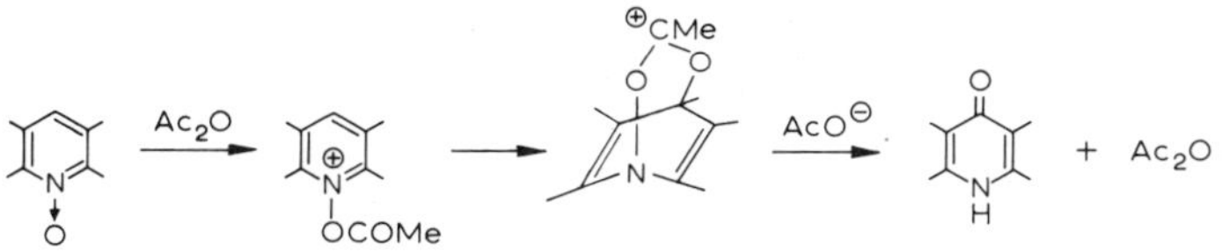

Irradiation in benzene or dichloromethane of acridine *N*-oxide with ultraviolet light yields *cyclohept* [b]*indol*-10(5H)-*one* (XXVII), m.p. 285–286° (70–80% yield) *M. Ishikawa, C. Kaneko* and *S. Yamada,* Tetrahedron Letters, 1968, 4519). The product can be reduced to the known tetrahydro compound and its constitution is thereby established. Irradiation in methanol yields 5,11-*dihydrodibenz*[b,e]-1,4-*oxazepine* (XXVIII), m.p. 111°, ν_{max} 3348 cm^{-1} (NH) and 1078 cm^{-1} (C-O-C), readily hydrolysed by acid to 2-*hydroxy-2'-formyldiphenylamine*, m.p. 122° (*A. Mantsh et al.*, Ann., 1969, **723**, 95):

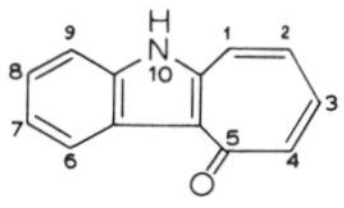

(XXVII) 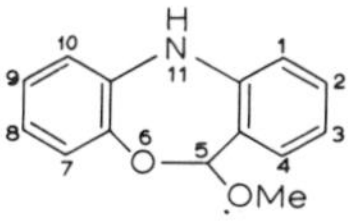 (XXVIII)

(*6*) *Substitution.* Electrophilic substitution in acridine occurs mainly at carbon atoms 2 and 4 (p. 19), a result not unexpected in view of the *ortho* and *para* positions of these carbon atoms with respect to the nitrogen atom.

Attempts to account for the substitution pattern in quantum mechanical terms have been made (*M. J. S. Dewar* and *P. M. Maitlis*, J. chem. Soc., 1957, 2521; *D. A. Brown* and *Dewar, ibid.*, 1953, 2410; *K. Fukui et al.*, J. chem. Phys., 1954, **22,** 1433).

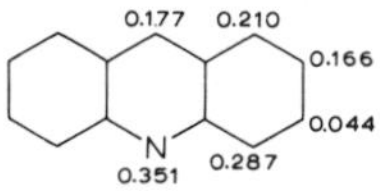

The frontier electron density pattern for the electrophilic substitution of acridine has been calculated and as the diagram shows $C_{(4)}$ is the predicted point of electrophilic attack followed by $C_{(1)}$. In fact, however, nitration occurs mainly at $C_{(2)}$ and to a lesser extent at $C_{(4)}$, while bromination occurs almost entirely at $C_{(2)}$ (see p. 18). Fukui's claim that agreement between theory and experiment is "almost satisfactory" may apply to other heterocyclic compounds but clearly not to acridine.

The presence of the pyridine nucleus in acridine leads to the prediction that $C_{(9)}$ will be the site of nucleophilic substitution and there is abundant evidence that this is indeed the case. Amination by ammonium carbonate in phenol or sodamide in dimethylaniline gives 9-aminoacridine (p. 19) and acridine with potassium cyanide gives 9-cyano-9,10-dihydroacridine, which is readily oxidised by air to 9-cyanoacridine (*K. Lehmstedt* and *F. Dostal*, Ber., 1939, **72,** 804):

H CN, CN⊖, [O], CN, −PhCHO, H CN, N, N H, N, N COPh

9-Cyanoacridine is also obtained by the action of potassium cyanide and benzoyl chloride on acridine, probably by a Reissert reaction in which the unstable intermediate, 10-benzoyl-9-cyano-9,10-dihydroacridine spontaneously breaks down into the cyano compound and benzaldehyde (*K. Bauer*, *ibid.*, 1950, **83,** 10).

Acridine with Grignard reagents or preferably organolithium compounds yields 9-substituted 9,10-dihydroacridines (p. 24). Thus phenyllithium adds to the conjugated system to give a lithium product which on hydrolysis yields 9-phenyl-9,10-dihydroacridine:

LiPh, H Ph, H_2O, H Ph, N, N Li, N H

Acridine 9-Phenyl-9,10-dihydroacridine

Acridine with methyl phenyl sulphone in hexamethylphosphoric triamide (HMPA) gives 9-methylacridine in 60% yield (*H. Nozaki*, *Y. Yamamoto* and *T. Nisimura*, Tetrahedron Letters, 1968, 4625), probably by nucleophilic attack:

$PhSO_2CH_2^{\ominus}$, CH_2SO_2Ph, $-PhSO_2^{\ominus}$, CH_3, N, N H, N

Compounds with reactive methylene groups in the presence of alkali react with acridine at the 9-position, nitromethane, for instance, readily yielding 9-(nitromethyl)-9,10-dihydroacridine (*F. Kröhnke* and *H. K. Honig*, Ann., 1959, **624,** 97).

Acridine resembles anthracene in undergoing free radical attack at the 9- and 10-positions. The benzyl radical, generated from toluene by *tert*-butyl butoxide, thus gives 9-*benzylacridine*, m.p. 173° (17% yield) and 9,10-*dibenzyl-9,10-dihydroacridine*, m.p. 143° (18%) (*W. A. Waters* and *D. H. Watson*, J. chem. Soc., 1957, 253). Interestingly 9-phenylacridine yields not only 9,10-dibenzyl-9-phenyl-9,10-dihydroacridine (2.5%), but also 4-*benzyl-9-phenylacridine*, m.p. 144° (10%).

Acyl radicals likewise attack at $C_{(9)}$, acridine and benzaldehyde, for example, in the presence of *tert*-butyl hydroperoxide and ferrous sulphate giving 9-*benzoylacridine*, m.p. 225°, in 55% yield (*T. Caronna et al.*, *ibid.*, Perkin II, 1972, 2035):

$$\mathrm{Bu^tO\cdot} + \mathrm{PhCHO} \rightarrow \mathrm{PhCO\cdot} + \mathrm{Bu^tOH}$$

9-*Acetylacridine*, m.p. 111°, is likewise obtained with acetaldehyde.

Decomposition of diacyl peroxides, $[\mathrm{MeO_2C(CH_2)_nCO\cdot O\cdot}]_2$, results in radical formation and attack as instanced by the formation of *methyl β-acridine-9-propionate*, m.p. 95° (*S. Goldschmidt* and *L. Beer*, Ann., 1961, **641,** 40).

$$[\mathrm{MeO_2C(CH_2)_2CO\cdot O\cdot}]_2 \longrightarrow 2\ (\mathrm{MeO_2C\cdot CH_2\cdot CH_2\cdot})\cdot + \mathrm{CO_2}$$

$$\text{(pyridine ring)} + \mathrm{MeO_2C\cdot CH_2\cdot CH_2\cdot} \longrightarrow \text{(ring bearing } \mathrm{CH_2\cdot CH_2\cdot CO_2Me}\text{)}$$

(b) Acridine and substituted acridines

(i) Acridine and homologues

Acridine is prepared in 85% yield by the reduction of acridone with sodium and amyl alcohol, followed by oxidation of the resulting 9,10-dihydroacridine by potassium dichromate and sulphuric acid (*A. Albert* and *J. B. Willis*, J. Soc. chem. Ind., 1946, **65,** 26). In another useful method 9-chloroacridine, readily obtained from acridone, is reduced with hydrogen and Raney nickel to give 9,10-dihydroacridine, which is then dehydrogenated as above.

Acridine crystallises from benzene or ethanol in almost colourless needles, m.p. 110–110.5°; sublimes at about 350°, and boils at 345–346°. It exists in five polymorphic forms (*A. Korfler*, Ber., 1943, **76**, 871), is slightly volatile in steam and in dilute solution has a blue fluorescence. It is a weak base, pK_a 5.60 in water at 20° (*Albert et al.*, J. chem. Soc., 1948, 2240); compare pyridine pK_a 4.86.

Acridine forms a number of salts and molecular compounds including the *hydrobromide*, m.p. 267°, very soluble in ethanol, *nitrate*, m.p. 186°, and *picrate*, m.p. 261–262° (decomp.) (*R. M. Acheson* and *M. L. Burstall*, *ibid.*, 1954, 3240), but m.p. 208° is reported by other workers (*e.g. M. Wilk, H. Schwab* and *R. Rochlitz*, Ann., 1966, **698,** 149). It may be noted that 9-phenylacridine also forms two picrates. Acridine forms double compounds with other aromatic compounds, *e.g.* with diphenylamine, m.p. 84–86°.

By the action of sunlight acridine undergoes dimerisation to a pale yellow substance, subliming at 276° and giving acridone when heated (*W. R. Orndorff* and *F. K. Cameron*, Amer. chem. J., 1895, **17,** 658). Later investigations show that in solvents such as isopropyl alcohol or ethanol acridine in ultraviolet light undergoes photoreduction to yield 9,9′,10,10′-tetrahydro-9,9′-biacridine, "biacridan" (p. 26) (*V. Zanker* and *H. Schnith*, Ber., 1959, **92,** 2210; *A. Kellmann*, J. Chim. phys., 1957, **54,** 468; *R. Noyori et al.*, Tetrahedron, 1959, **25,** 1125; *Zanker* and *F. Mader*, Ber., 1964, **97,** 2418).

9-**Phenylacridine,** m.p. 186–187°, polymorphic forms m.p. 177° and 175°, is formed in good yield by heating diphenylamine and benzoic acid with zinc chloride (*B. Staskun*, J. org. Chem., 1964, **29,** 2856; *F. D. Popp, ibid.*, 1962, **27,** 2658) or in 92% yield by the action of phenyllithium on acridone in benzene (*K. Lehmstedt* and *F. Dorstal*, Ber., 1939, **72,** 804):

O ... N H —2 PhLi→ LiO Ph ... N Li —H_2O→ Ph ... N

It forms two *picrates*, m.p. 227.7° (1:1 mol. phenylacridine and picric acid) and 185–186° (1:2 mol. phenylacridine and picric acid) (*H. Bassett* and *T. A. Simmons*, J. chem. Soc., 1921, **119,** 416). In organic solvents it gives solutions with a blue fluorescence. It has a dipole moment of 2.40 D, and interference with the hydrogen atoms at $C_{(1)}$ and $C_{(8)}$ causes the benzene ring to be tilted at an angle greater than 40° to the plane of the acridine rings (*C. W. N. Cumper, R. F. A. Ginman* and *A. I. Vogel*, J. chem. Soc., 1962, 4518).

9-**Methylacridine,** m.p. 117–118°, is obtained in 50–60% yield by heating diphenylamine with zinc chloride and acetic acid for 14 hours at 220° (*O. Blum*, Ber., 1929, **62,** 881). A double compound, 9-*methylacridine-diphenylamine*, m.p. 99–100°, is also formed. Acridines and hydrogenated acridines frequently form similar products, *e.g.* the double compound, 9-*methylacridine*-9-*methyl*-9,10-*dihydroacridine*, m.p. 99–100°.

As expected 9-methylacridine exhibits the reactivity observed in 4-picoline and 4-methylquinoline. Aromatic aldehydes yield α-aryl-β-(9-acridyl)ethanols, XXIX, or styrene derivatives, XXX or both according to the aldehyde used and the conditions of condensation (*W. Sharp, M. M. J. Sutherland* and *F. J. Wilson*, J. chem. Soc., 1943, 5):

$CH_2 \cdot CHOHPh$ (XXIX)   CH=CHPh (XXX) $CH{=}\overset{O}{\overset{\uparrow}{N}}C_6H_4NMe_2$ (XXXI) $CH{=}NC_6H_4NMe_2$ (XXXII)

9-Methylacridine condenses with *C*-nitroso compounds to yield anils or nitrones, 4-nitrosodimethylaniline, for example, giving a mixture of *nitrone* XXXI, dark red crystals, m.p. 247–248°, and 9-(*dimethylphenyliminomethyl*)*acridine* (XXXII), garnet coloured rhombs, m.p. 255–257° (*T. D. Perrine* and *L. J. Sargent*, J. org. Chem., 1949, **14,** 583; *L. Chardonnes* and *P. Heinrich*, Helv., 1949, **32,** 656). XXXI is reduced by sulphur dioxide to XXXII, identical with the product obtained by condensing 9-formylacridine with 4-dimethylaminoaniline.

With phenyllithium 9-methylacridine gives 9-lithiomethyleneacridine, which with cyanomethylaniline yields 9-*acridinylacetonitrile*, m.p. 227° (*H. Lettre, P. Jungmann* and *J.-C. Salfeld*, Ber., 1952, **85**, 397):

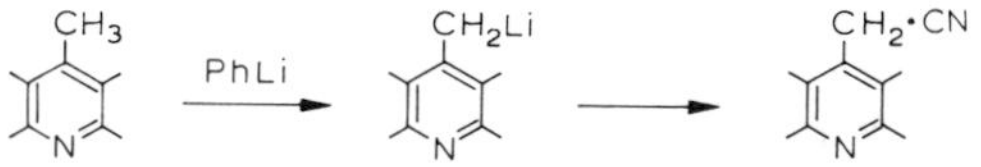

CH_3 —PhLi→ CH_2Li → $CH_2{\cdot}CN$

Diazonium salts couple with 9-methylacridine to give phenylhydrazones, idential with the products obtained from 9-formylacridine and the appropriate phenylhydrazine (*A. E. Porai-Koshits* and *A. A. Kharkharov*, Bull. Acad. Sci. U.R.S.S., Cl. Sci. Chim., 1944, 90; C.A., 1944, **39**, 1631):

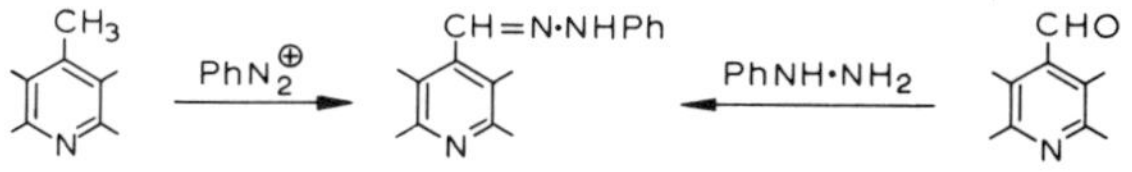

CH_3 —$PhN_2^{\oplus}$→ $CH{=}N{\cdot}NHPh$ ←$PhNH{\cdot}NH_2$— CHO

The melting points of some alkyl- and phenyl-acridines are listed in Table 2.

TABLE 2

ALKYL- AND ARYL-ACRIDINES

	m.p. (°C)	*Picrate m.p. (°C)*	*Ref.*
1-Methyl	98–99		1
2-Methyl	134	226	1,2
3-Methyl	129–130	225–226	1,8
4-Methyl	90		3
9-Methyl	117–119	218–220	4,7
1-Ethyl	89–90		5
2-Ethyl	77–78.5		5
3-Ethyl	90–91.5		5
4-Ethyl	37–38.5		5
9-Ethyl	110–111	203	6
2-Phenyl	127–128	244–245	8
4-Phenyl	122–122.5		9
9-Phenyl	184–185	185–186* 222.7	10

* See text p. 16.

References
1 *R. A. Reed*, J. chem. Soc., 1944, 679.
2 *H. Jensen* and *F. Rethwisch*, J. Amer. chem. Soc., 1928, **50,** 1144.
3 *Jensen* and *M. Friedrich*, *ibid.*, 1927, **49,** 1049.
4 *S. M. H. van der Krogt* and *B. M. Webster*, Rec. Trav. chim., 1955, **74,** 161.
5 *L. J. Sargent*, J. org. Chem., 1954, **19,** 599.
6 *N. Fisher et al.*, J. chem. Soc., 1958, 1411.
7 *H. Schindler* and *H. Blattner*, Helv., 1961, **44,** 753.
8 *V. A. Petrov*, J. chem. Soc., 1942, 693.
9 *G. Wittig* and *K. Niethammer*, Ber., 1960, **93,** 944.
10 *F. D. Popp*, J. org. Chem., 1962, **27,** 2658.

(ii) Halogenoacridines

Acridine in carbon tetrachloride with bromine yields a product with an ultraviolet spectrum similar to that of acridine hydrobromide and is probably N-*bromoacridinium bromide*, orange needles, m.p. 108° (*Acheson, T. G. Hoult* and *K. A. Barnard*, J. chem. Soc., 1954, 4142). In acetic acid, however, bromine acts in a different manner to yield 2-**bromoacridine**, m.p. 175°, and 2,7-*dibromoacridine*, m.p. 249°.

1-*Bromoacridine*, m.p. 110–111°, 3-*bromoacridine*, m.p. 137–138° and 4-*bromoacridine*, m.p. 107–108°, are prepared by synthesis (*G. S. Chandler, R. A. Jones* and *W. H. F. Sasse*, Austral. J. Chem., 1965, **18,** 108), while 9-*bromoacridine*, m.p. 116°, is obtained by the action of bromine and phosphorus tribromide on acridone (*Acheson et al.*, J. chem. Soc., 1960, 3367).

The bromo compounds cannot be prepared from the amines by the Sandmeyer reaction, but are obtained in 15% yield by the thermal decomposition of the mercuric bromide complexes of the diazotised amines (*Chandler, Jones* and *Sasse, loc. cit.*).

2,4,5,7-*Tetrabromoacridine*, m.p. 286–287°, is obtained from the corresponding 9-acridone, the product of bromination of 9-acridone in acetic acid (*Acheson, Hoult* and *Barnard, loc. cit.*).

1-**Chloroacridine,** m.p. 85° (*I. Tanasescu* and *V. Farcasan*, C.A., 1952, **46,** 8118). 2-*Chloroacridine*, m.p. 170° (*Tanasescu* and *E. Ramontianu*, Bull. Soc. chim. Fr., 1934, [v], **1,** 547). 3-*Chloroacridine*, is prepared by the selective dechlorination of 3,9-dichloroacridine (see below) and occurs in two interconvertible forms, m.p. 129 and 134° (*Albert* and *R. Royer*, J. chem. Soc., 1949, 1148). 4-*Chloroacridine* likewise occurs in two forms, one separating in needles, m.p. 79°, when a light petroleum solution is quickly cooled, but changing on standing into the stable form, prisms, m.p. 90° (*G. R. Clemo et al.*, *ibid.*, 1924, **125,** 1751).

9-*Chloroacridine*, m.p. 120°, is prepared in excellent yield by heating *N*-phenylanthranilic acid with phosphoryl chloride (*Albert* and *B. Ritchie*, Org. Synth., 1942, **22**, 5). It can be obtained also from the *N*-oxide (p. 13) or from 9-bromoacridine in chloroform by the action of phosphorus trichloride. The chlorine in 9-chloroacridine is readily replaced by other atoms or radicals. It may be replaced by hydrogen by heating with *p*-toluenesulphonylhydrazide and treating the product with dilute alkali (*Albert* and *Royer, loc. cit.*). This method of converting 9-halogenoacridines into the parent acridines can be applied to 9-halogenoacridines which also contain easily reducible groups such as nitro or cyano. Chlorine may also be removed by heating 9-chloroacridine with hydrazine to give 9-*acridinylhydrazine*, m.p. 171–172°, and treating the product with oxygen in the presence of platinum and a little alkali (*Albert*, J. chem. Soc., 1965, 4653). The reactivity of the 9-position is emphasised by the hydrogenation of 3,9-dichloroacridine over Raney nickel to give 3-chloroacridine, although this is attended by partial removal of both chlorine atoms (*Albert* and *Royer*, *loc. cit.*).

9-Chloroacridine is converted by hydrolysis into acridone, a change also effected by heating with benzoic acid when benzoyl chloride is the other product (*H. Inoue et al.*, Chem. Lett., 1972, 207; C.A., 1972, **76**, 140, 465z. With ammonium carbonate in phenol it yields 9-aminoacridine and when heated at 160–170° with a mixture of potassium and cuprous cyanide it gives 9-*cyanoacridine*, m.p. 181° (*Lehmstedt*, Ber., 1931, **64**, 1232).

(*iii*) *Nitroacridines*

Acridine with fuming nitric acid undergoes nitration at all four positions in the benzene rings, but $C_{(9)}$ appears to be completely inert to electrophilic substitution. 2-Nitroacridine is the main product with the relative quantities of 1-, 2-, 3- and 4-nitroacridines being 5:130; 1; and 25 (*idem*, *ibid*., 1938, **71**, 808). This does not provide a preparative method since separation of the isomers is not easy, particularly since 2- and 4-nitroacridine form mixed crystals, as do 1- and 3-nitroacridine. The ease with which 2- and 4-nitroacridine can be oxidised to the acridones by chromic acid in contrast to the comparative resistance to oxidation of the other two isomers is sometimes used in separating the nitroacridines.

The nitroacridines are prepared by a variety of methods as, for example, from nitroacridones by conversion into 9-chloronitroacridines and subsequent dehalogenation. 1-**Nitroacridine**, m.p. 154°; 2-*nitroacridine*, m.p. 215°; 3-*nitroacridine*, m.p. 183°; 4-*nitroacridine*, m.p. 167°. 2,4-*Dinitroacridine*, orange needles, m.p. 284° (corr.) (*A. Albert* and *W. H. Linnell*, J. chem. Soc., 1938, 22) and 1,3-*dinitroacridine*, m.p. 218° (*S. Secareanu*, Ber., 1931, **64**, 837).

Nitroacridines are reduced with stannous chloride to amino-9,10-dihydroacridines.

(*iv*) *Aminoacridines*

The aminoacridines are of considerable importance: they are themselves bactericidal and compounds derived from them have valuable antiseptic and chemotherapeutic properties (p. 22). Other derivatives are commercial dyestuffs.

The monoamines are obtained in a variety of ways. 2-Aminoacridine is prepared by the alkaline reduction of 2-nitroacridone to give 2-amino-9,10-dihydroacridine, which is then oxidised with ferric chloride to the amine (*Albert* and *Ritchie*, J. Soc. chem. Ind., 1941, **60**, 120). 3-Aminoacridine is best prepared by the interaction of 3-aminodiphenylamine and formic acid (*Albert*, "The Acridines", 2nd Edn., Arnold, London, 1966, p. 101), while 4-aminoacridine is prepared in the same manner as the 2-isomer. 9-Aminoacridine, according to *Albert* (*ibid*., p. 292), is best obtained by the action of ammonium carbonate on 9-chloroacridine dissolved in phenol. It is also obtained in 72% yield by the interaction of sodamide and acridine in dimethylaniline at 150° (*K. Bauer*, Ber., 1950, **83**, 10).

The monoaminoacridines are yellow or orange coloured compounds; some m.p.s. and pK_a values are given in Table 3.

TABLE 3

AMINOACRIDINES

	Colour	*m.p. (°C)*	*pK_a in H_2O* at 20°*	*Colour of hydrochloride*
1-Aminoacridine	Orange	180	6.04	Violet
2-Aminoacridine	Orange	213–214	5.88	Red
3-Aminoacridine	Yellow	218	8.04	Scarlet
4-Aminoacridine	Orange	108	4.40	Red
9-Aminoacridine	Pale yellow	234	9.99	Pale yellow
Acridine	Colourless	110–110.5	5.60	Pale yellow

**A. Albert*, J. chem. Soc., 1965, 4653.

The amines, with the exception of 4-aminoacridine (XXXIII), are obviously stronger bases than acridine and this has been explained by intramolecular hydrogen bonding which makes protonation of the ring nitrogen more difficult (*L. N. Short*, J. chem. Soc., 1952, 4585). The basicity of 9-aminoacridine is striking and is comparable to that of aliphatic primary amines. A reason advanced for this is the additional cationic resonance stability which results from salt formation XXXIV and XXXV:

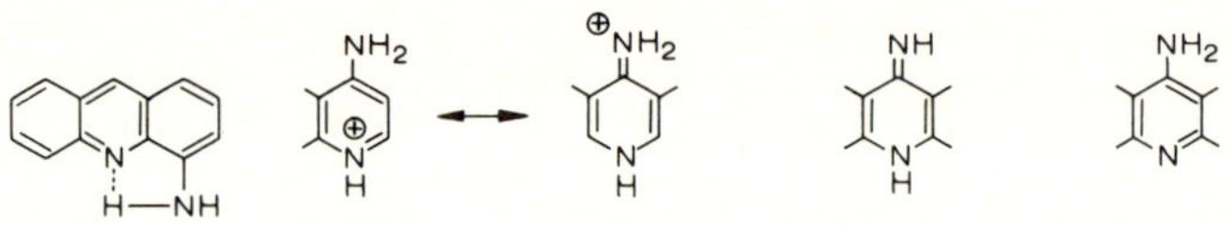

(XXXIII) (XXXIV) (XXXV) (XXXVI) (XXXVII)

The salts of the aminoacridines are deeply coloured (see Table). Many of the aminoacridines fluoresce, the 2-, 3- and 9-isomers, for example, having a yellowish-green fluorescence in aqueous or ethanolic solution when exposed to ultraviolet light. 1-Aminoacridine has a bright orange fluorescence. 4-Aminoacridine does not fluoresce, but gives a bright yellowish-green fluorescence in concentrated sulphuric acid.

The spectra of the aminoacridines have been measured and applied particularly to the tautomerism of 9-aminoacridine. The similarity of the ultraviolet spectra of 9-aminoacridine and 9-imino-10-methylacridine in methanolic or ethyl acetate solution and the difference between those of the amino compound and 9-dimethylaminoacridine support the imino form XXXVI (*Acheson et al., ibid.*, 1954, 3742). On the other hand infrared spectra have been interpreted as supporting the amino form XXXVII (*Short, loc. cit.; S. F. Mason*, J. chem. Soc., 1959, 1281; *cf.* however, *A. V. Karyakin* and *A. V. Shablya*, Doklady Akad. Nauk, S.S.S.R., 1957, **116,** 969; C.A., 1958, **52,** 4324. See also *S. J.* and *C. L. Angyal*, J. chem. Soc., 1952, 1461). Neat decisive evidence has been provided by the infrared spectra of partially deuterated 9-aminoacridine (*N.*

Bacon et al., *ibid.*, 1965, 5230). Primary amines give rise to two NH stretching frequencies in the 3500 cm^{-1} region as the result of coupling. If one of the protons is replaced by deuterium, coupling is weak and the NH stretching occurs at a frequency lying between the symmetrical and asymmetrical NH_2 stretching modes. 9-Aminoacridine in chloroform or carbon tetrachloride shows bands at 3527 cm^{-1} (NH asymm.) and 3442 cm^{-1} (NH symm.), while the deuterated compound band occurs at 3488 cm^{-1} (NHD) in dioxane; conclusive proof that the compound in dioxane exists in the amino form XXXVII. The other aminoacridines in chloroform exhibit the two i.r. bands at $\sim$3500 cm^{-1} characteristic of primary amines *(Short, loc. cit.)*.

The p.m.r. spectrum of 9-aminoacridine shows a broad peak in the NH_2 region and there is no evidence of coupling between a ring NH proton and ring protons as in 9,10-dihydroacridine (*J. P. Kokko* and *J. H. Goldstein*, Spectrochimica Acta, 1963, **19,** 1119). These results confirm the amine structure XXXVII.

The aminoacridines with the exception of the 9-isomer behave in many ways as typical aromatic amines. 1-, 2- and 3-Aminoacridine undergo diazotisation to give diazonium salts which couple with phenols, but 4-aminoacridine with sodium nitrite and hydrochloric acid forms a buff precipitate which does not couple (*Albert* and *Ritchie*, J. chem. Soc., 1943, 458). The diazonium salts do not undergo the Sandmeyer reaction (p. 18).

The aminoacridines with acetic anhydride yield *N*-acetyl compounds. 9-Aminoacridine readily yields a diacetamino compound in which both hydrogen atoms of the primary amino group are replaced by acetyl groups (*J. H. Wilkinson* and *I. L. Finar*, *ibid.*, 1946, 115; *A. I. Gurevich* and *Y. N. Sheinker*, Zhur. fiz. Khim., 1962, **30,** 734; C.A., 1962, **57,** 7231a).

1- and 2-Aminoacridines react with benzaldehyde to give quantitative yields of the phenylimino derivatives in contrast to the 3- and 4-isomers which give poor yields and 9-aminoacridine which fails to react, possibly as the result of steric hindrance.

The aminoacridines are stable to alkali except 9-aminoacridine which when boiled with 5 *N* potassium hydroxide gives ammonia and acridone. The amines are converted into the hydroxyacridines by heating with concentrated hydrochloric acid at 180° *(Albert* and *Ritchie, loc. cit.)* (*cf.*, however, p. 23).

Visible and ultraviolet spectral evidence shows that except for 9-aminoacridine aminoacridines are protonated at the ring nitrogen and not at the amino group (*D. P. Craig* and *Short*, J. chem. Soc., 1945, 419; *N. H. Turnbull*, *ibid.*, 1945, 441). It is known that protonation of aromatic amino compounds such as aniline results in a spectrum very similar to that of the parent hydrocarbon since protonation removes the possibility of interaction between the nitrogen atom and the ring. It is also known that protonation of acridine gives rise to a band at about 410 nm as the result of a shift to the longer wavelength. If therefore aminoacridines are protonated first at the ring-nitrogen the resulting cation should exhibit a shift of the long wave band towards the red. Further protonation should give a dication whose spectrum is similar to that of the acridinium ion. Both these effects are indeed observed. 1-Aminoacridine with a broad band at 410 nm gives a cation with this band shifted to 510 nm and a dication with a spectrum similar to that exhibited by acridine hydrochloride with a broad band at $\sim$410 nm.

With methyl iodide the aminoacridines (with the exception of the 9-isomer) undergo methylation at the primary amino groups. 9-Aminoacridine, on the other hand, yields 9-*amino*-10-*methylacridinium iodide* (XXXVIII), m.p. 305° (*Albert* and *Ritchie*, *loc. cit.*), which with sodium hydroxide yields 9-*imino*-10-*methylacridan* (XXXIX), orange, m.p. 134–136° (sealed tube):

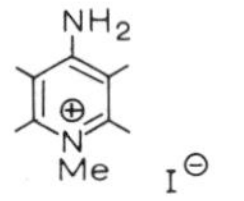

(XXXVIII) 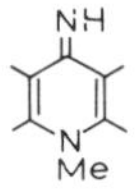 (XXXIX)

Aminoacridines of therapeutic importance. 9-Aminoacridine has antibacterial properties which were recognised only during the Second World War. The hydrochloride under the name Aminacrine Hydrochloride is listed in the British Pharmacopoeia. Some of its derivatives which have found wide use are discussed below.

At the beginning of the century Ehrlich and Brenda demonstrated the trypanocidal activity of certain 3,6-diaminoacridine derivatives. One of these, 3,6-diamino-10-methylacridinium chloride, is the drug **acriflavine** or **trypaflavine**, prepared by acetylating 3,6-diaminoacridine, forming the methyl *p*-toluenesulphonate and finally removing the acetyl groups by hydrolysis with hydrochloric acid. The commercial product contains a fair proportion of 3,6-diaminoacridine. Browning and Gilmore in 1913 found that 3,6-diaminoacridine, **Proflavine**, is an antiseptic. Acriflavine serves the same purpose, but is inferior as an antibactericide to 3,6-diaminoacridine, which is very much less toxic and has superseded acriflavine as an antiseptic. 3,6-*Diaminoacridine*, m.p. 277° (sealed tube) or 288° (corr.), is obtained in 70% yield by heating *m*-phenylenediamine with oxalic or formic acid and is manufactured in this way (*A. Albert*, J. chem. Soc., 1941, 121). It dissolves in water to give a yellow solution with a bright green fluorescence. Diazotisation followed by replacement of the diazonium groups yields acridine.

Rivanol, ethacridine, 3,9-*diamino*-7-*ethoxyacridine lactate*, light yellow crystals, m.p. 235°, is used to cure amoebic dysentery. A good method of preparation is given by *Albert* and *W. Gledhill* (J. Soc. chem. Ind., 1942, **61,** 159).

It has long been established that the antibacterial properties of the aminoacridines require both the acridine nucleus and the attached amino groups. A closer insight into the relationship between the chemical constitution of these compounds and their physiological activity has been obtained by a number of workers, particularly, Professor Adrien Albert, and for the conclusions reached the reader is referred to the appropriate chapters by that worker and Dr. A.C.R. Dean in the books cited on p. 1.

As already noted 9-aminoacridine is a strong antibacterial which is used in many preparations. **Tacrine,** 9-*amino*-1,2,3,4-*tetrahydroacridine* (XL), colourless crystals, m.p. 179–180°, is a strong base, pK_a 10.0, which is used in "petit mal" epilepsy and for other purposes:

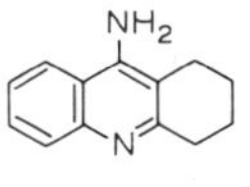

(XL) 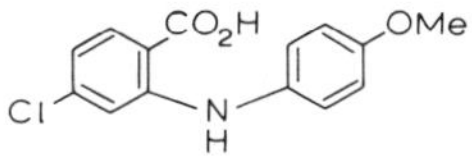 (XLI)

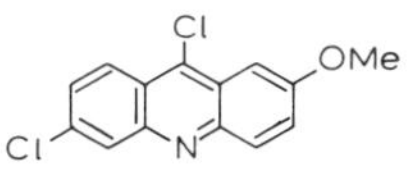

(XLII) 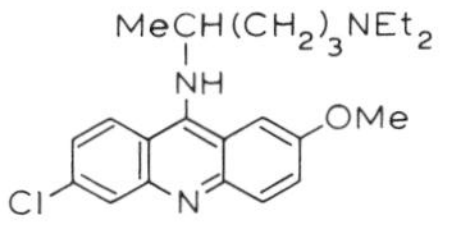 (XLIII)

A 9-aminoacridine derivative which is a valuable antimalarial is **mepacrine, atebrin** or **quinacrine,** 6-*chloro*-9(4′-*diethylamino*-1′-*methylbutyl*)*amino*-2-*methoxyacridine dihydrochloride* (XLIII), yellow crystals, m.p. 248–250° (decomp.). The free *base* crystallises as the *monohydrate*, m.p. 81–86°, depending on the rate of heating. Mepacrine was widely used in the prevention and treatment of malaria during the Second World War although it is not a complete causal prophylactic. It has the disadvantage that prolonged use colours the skin and frequently causes gastric irritation.

Mepacrine and its congeners are prepared by condensing the appropriate 9-chloroacridine and amine in the presence of phenol (*O. J. Magidson* and *A. M. Grigorowski*, Ber., 1936, **69,** 396). 2,4-Dichlorobenzoic acid and *p*-anisidine give 5-chloro-4′-methoxydiphenylamine-2-carboxylic acid (XLI), which with phosphoryl chloride yields 6,8-dichloro-2-methoxyacridine (XLII). This acridine with 1-diethylamino-4-aminopentane in phenol and subsequent basification gives the free base of mepacrin.

The structural relationship of mepacrin to the quinoline antimalarials pamaquin and chloroquine is noteworthy. Mepacrin contains an asymmetric carbon atom and has been resolved into its optically active forms.

Mepacrine forms a di-*N*-oxide and is characterised by the ease with which the side-chain at $C_{(9)}$ can be removed. This is accomplished merely by heating the hydrochloride in boiling water to give 2-methoxy-6-chloroacridone (XLIV) and 4-amino-1-diethylaminopentane:

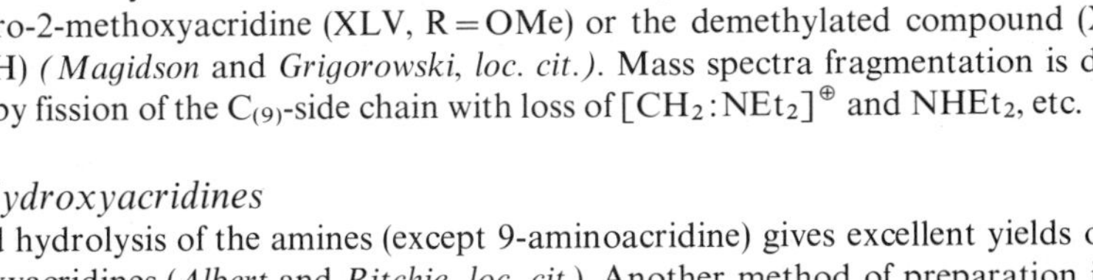

(XLV) + $CH_2{:}CH(CH_2)_3NEt_2$ ← XLIII → (XLIV) + $MeCH(NH_2)(CH_2)_3NEt_2$

Concentrated hydrochloric acid also effects bond rupture, but the product is 9-amino-6-chloro-2-methoxyacridine (XLV, R = OMe) or the demethylated compound (XLV, R = OH) (*Magidson* and *Grigorowski, loc. cit.*). Mass spectra fragmentation is dominated by fission of the $C_{(9)}$-side chain with loss of $[CH_2{:}NEt_2]^{\oplus}$ and $NHEt_2$, etc.

(*v*) *Hydroxyacridines*

Acid hydrolysis of the amines (except 9-aminoacridine) gives excellent yields of the hydroxyacridines (*Albert* and *Ritchie, loc. cit.*). Another method of preparation is demethylation of the methoxyacridines by hydrobromic acid or hydriodic acid. These methoxyacridines are obtained by synthesis. 2-Aminobenzaldehyde and the methyl ethers of halogenophenols condense in the presence of copper to give diphenylamine derivatives which with sulphuric acid undergo ring-closure. 2-Aminobenzaldehyde and 2-bromoanisole, for example, yield 2-(2′-methoxyanilino)benzaldehyde, which after ring-closure can be demethylated to give 4-hydroxyacridine (*H. Jensen* and *F. Rethwisch*, J. Amer. chem. Soc., 1928, **50,** 1144):

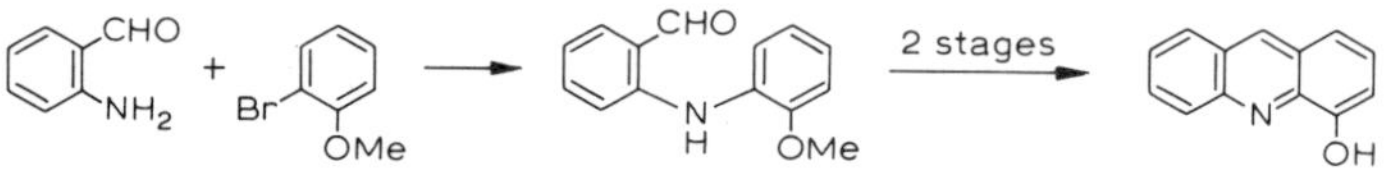

CHO NH2 + Br OMe → CHO N H OMe —2 stages→ N OH

4-Hydroxyacridine

1-, 2-, 3- and 4-Hydroxyacridines are as basic as acridine and have pronounced phenolic properties. They thus differ from 9-hydroxyacridine (acridone) which will be considered separately.

The tautomerism of the hydroxyacridines has been investigated by ultraviolet and visible spectroscopy (*Albert* and *L. N. Short*, J. chem. Soc., 1945, 760; *S. F. Mason*, *ibid.*, 1957, 5010), and infrared spectroscopy (*idem*, *ibid.*, 1957, 4874). In dioxane the four compounds exhibit ultraviolet spectra closely resembling those of the corresponding methoxyacridines. In this solvent they therefore exist solely in the hydroxylic form (*e.g.* XLVII). In absolute and aqueous ethanol the spectra of 2- and 4-hydroxyacridine are unaltered and in these solvents both compounds exist in the hydroxylic form. On the other hand, the ultraviolet spectrum of 1-hydroxyacridine (XLVII) as observed in ethanol undergoes a change as water is added and a broad band at ~570 nm appears, a band also found in the spectrum of *N*-methylacrid-1-one (XLVI). It thus appears that in aqueous ethanol the compound exists predominately in the oxo form XLVIII. This conclusion is confirmed by the observation that 1-hydroxyacridine in the solid state is yellow, but gives a blue colour in 20% ethanol reminiscent of the blue colour of *N*-methylacrid-1-one (XLVI) (*S. Nitzsche*, Ber., 1943, **76,** 1187). 3-Hydroxyacridine

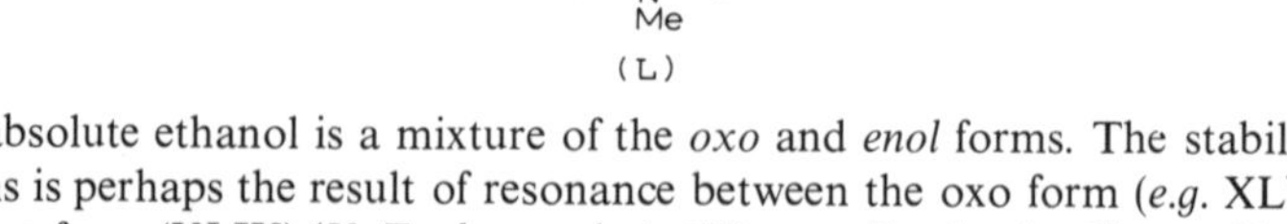

(XLVI) (XLVII) (XLVIII) ⟷ (XLIX)

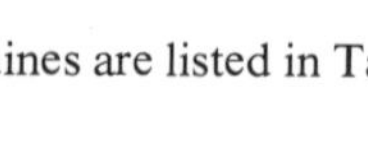

(L)

even in absolute ethanol is a mixture of the *oxo* and *enol* forms. The stability of the oxo forms is perhaps the result of resonance between the oxo form (*e.g.* XLVIII) and the betaine form (XLIX) (*V. Zanker* and *A. Wittwer*, Z. physik. Chem., Neue Folge, 1960, **24**, 183).

Some hydroxy- and methoxy-acridines are listed in Table 4.

TABLE 4

HYDROXYACRIDINES. METHOXYACRIDINES

	m.p. (°C)	*m.p. (°C) of methoxyacridine*	*Ref.*
1-Hydroxy	250 (decomp.)	122	1,6
2-Hydroxy	284–285 (decomp.)	105	1,2
3-Hydroxy	292 (decomp.)	91	1
4-Hydroxy	116–117	134	1,3
9-Hydroxy	354	103 (hydrate) 65 (anhydrous)	4,5

References
1 *S. Nitzsche*, Ber., 1943, **76,** 1187.
2 *K. Lehmstedt et al., ibid.*, 1936, **69,** 2399.
3 *H. Jensen* and *F. Rethwisch*, J. Amer. chem. Soc., 1928, **50,** 1144.
4 *Lehmstedt*, Ber., 1935, **68,** 1455.
5 *H. Irving, E. J. Butler* and *M. F. Ring*, J. chem. Soc., 1949, 1489.
6 *A. Albert* and *B. Ritchie, ibid.*, 1943, 458.

A striking feature of Table 4 is the low melting-point of 4-hydroxyacridine as the result of internal hydrogen bonding. This is in harmony with a broad band in the infrared spectrum at 3398 cm^{-1} and with its solubility in the common organic solvents in contrast to the insolubility of the other isomers.

The *N*-methyl derivatives of the hydroxyacridines have been prepared and exhibit interesting properties *(Nitzsche, loc. cit.)*. They are highly coloured substances in the solid state or in aqueous or ethanolic solutions. For the 2- and 4-compounds betaine structures (*e.g.* L) have to be assigned and Nitzsche in fact assigned dipolar structure to all the isomers (he did not consider acridone).

N-**Methyl-1-acridone,** m.p. 280° (decomp.), blue; N-*methyl-2-acridone*, m.p. 188° (decomp.), black, giving a red aqueous solution; N-*methyl-3-acridone*, m.p. 157°, red; N-*methyl-4-acridone*, black, unstable, violet aqueous solution.

*(c) 9,9′-Biacridines and related compounds**

In discussing these compounds the nomenclature adopted in I.U.P.A.C. Rule C 71 and in Chemical Abstracts will be used. However, they are referred to in the literature also by a variety of trivial names which are indicated in appropriate places in parentheses.

The parent compound, 9,9′-**biacridine,** 9,9′-**biacridinyl** (LI), m.p. 385°, can be prepared conveniently from 9-chloroacridine in a number of ways including treatment with chromous chloride in nitrogen (*R. Royer*, J. chem. Soc.,

* *F. McCapra*, "Acridines", Wiley, New York, 1973, p. 519.

1949, 1663), phenylmagnesium bromide (quantitative yield claimed) (*K. Gleu* and *A. Schubert*, Ber., 1940, **73,** 805) or copper bronze at 140° in 92% yield (*K. Lehmstedt* and *H. Hundertmark*, *ibid.*, 1929, **62,** 1065). It is also obtained in 60% yield by heating thioacridone, (9-thioxo-9,10-dihydroacridine) with copper at 280° (*idem, ibid.*, 1930, **63,** 1229):

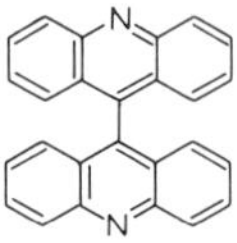

(LI) 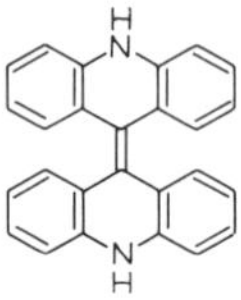 (LII) 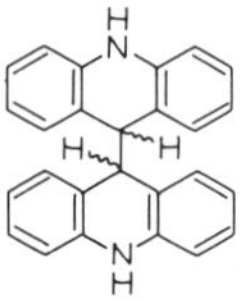 (LIII)

9,9′-Biacridine is a weak base, insoluble in most solvents, which forms in dilute hydrochloric acid with potassium iodide a characteristic red diiodide by which it may be detected. In boiling toluene with dimethyl sulphate it yields a monomethosulphate, but in boiling dimethyl sulphate both nitrogen atoms are alkylated (*H. Decker* and *W. Petsch*, J. pr. Chem., 1935, [ii], **143,** 211). The dimethonitrate has interesting properties (see below).

Reduction of 9,9′-biacridine with zinc and hydrochloric acid gives 9,9′-**bi**(9,10-**dihydroacridinylidene**), (*9,9′-biacridene*, $\Delta^{9,9'}$*-biacridan*) (LII), yellow plates, m.p. 392°, and 9,9′,10,10′-**tetrahydro**-9,9′-**biacridine,** (*9,9′-biacridan, 9,9′-bisacridanyl*) (LIII), m.p. 214° (*Albert* and *G. Catterall*, J. chem. Soc., 1965, 4657). The latter compound is formed also by reduction of 9,9′-biacridine with zinc and acetic acid (*Lehmstedt* and *Hundertmark*, *loc. cit.*, 1930) or by heating 9,10-dihydro-9-cyanoacridine in ethanol (*Lehmstedt* and *E. Wirth*, Ber., 1928, **61,** 2044).

That the substance, LIII, is a derivative of 9,10-dihydroacridine is shown by its ultraviolet spectrum, λ_{max} 250 and 287 nm (*cf.* 9,10-dihydroacridine 250 and 290 nm). Another *form*, m.p. 247°, results when acridine is reduced with sodium in moist ether (*E. Bergmann* and *O. Blum-Bergmann*, Ber., 1930, **63,** 757), with a specially prepared Raney nickel (along with 9,10-dihydroacridine) (*G. M. Badger* and *W. H. F. Sasse*, J. chem. Soc., 1956, 616), or with zinc dust and acetic acid (*I. W. Elliott* and *R. B. McGriff*, J. org. Chem., 1957, **22,** 515). Some workers reported a higher m.p., 265°, but *Lehmstedt* and *Hundertmark* showed that the m.p. depends on the rate of heating. They recommended preheating the bath to 230° and by raising the temperature at a certain rate were able to obtain reproducible results. Both forms give identical X-ray diagrams and polymorphism is therefore extremely unlikely. Possibly they are geometrical isomers resulting from the *cis*- and *trans*-positioning of the hydrogen atoms at $C_{(9)}$ and $C_{(9')}$.

When boiled in nitrobenzene, 9,9′,10,10′-tetrahydro-9,9′-biacridine undergoes carbon–carbon fission to yield acridine.

9,9′,10,10′-Tetrahydro-9,9′-biacridine is also obtained as the main product by the photoreduction of acridine in solvents such as methanol or ethanol (*H. Goth, P. Cerutti* and *H. Schmid*, Helv., 1965, **48,** 1395; *A. Kellmann* and *J. T. Dubois*, J. chem. Phys., 1965, **42,** 2518). Side-products are 9,10-dihydroacridine (2.6%) and 9-*hydroxymethyl*-9,10-*dihydroacridine* (LIV), (8.5%), m.p. 135–136°, ν_{max} 3546 cm^{-1} (OH) and 3425 cm^{-1} (NH) and an u.v. spectrum similar to that of 9,10-dihydroacridine; O-*acetyl* deriv., m.p. 157–158°:

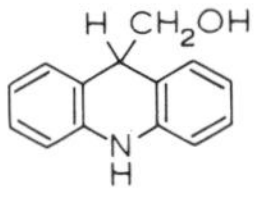

(LIV) 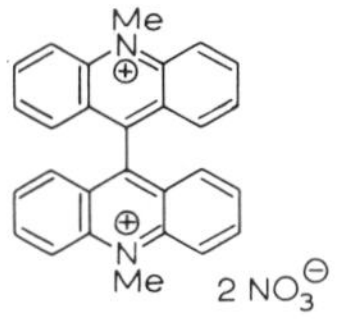 (LV) 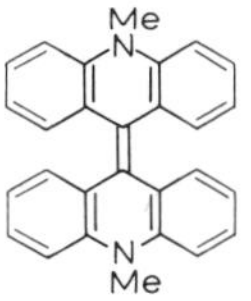 (LVI)

LIV is also prepared by the reduction of ethyl acridine-9-carboxylate with lithium tetrahydroaluminate (*A. Campbell* and *E. Morgan*, J. chem. Soc., 1958, 1712). It undergoes rearrangement to 5H-*dibenzo*[b,f]*azepine*, m.p. 196.5–198°, when heated with phosphorus pentoxide (*P. N. Craig et al.*, J. org. Chem., 1961, **26,** 135), a reaction comparable to the conversion of 9-hydroxymethyl-9,10-dihydroanthracene to dibenzocycloheptatriene:

(LIV) Dibenzo[*b,f*]azepine

Lucigenin, 10,10′-**dimethyl-9,9′-bisacridinium dinitrate** (LV), is prepared by reducing 9-chloro-10-methylacridinium salts with phenylmagnesium bromide or zinc dust in acetone followed by oxidation with dilute nitric acid (*K. Gleu* and *R. Schaarschmidt*, Ber., 1940, **73,** 909) or by reduction of 10-methyl-9-acridonephosphoryl chloride complex with zinc and acetic acid to give 10,10′-dimethyl-9,9′-bi(9,10-dihydroacridinylidene) (LVI) and treatment with dilute nitric acid (*Gleu* and *S. Nitsche*, J. pr. Chem., 1939, **153,** 233).

Lucigenin gives a yellow solution with a green fluorescence in water, but its most striking property is the intense chemiluminescence which results when the compound is treated with a base and hydrogen peroxide. There is evidence that luminescence is the result of oxidation by singlet oxygen to give a 1,2-dioxetane, which decomposes to yield *N*-methylacridone (*F. McCapra* and *R. A. Hann*, Chem. Comm., 1969, 442):

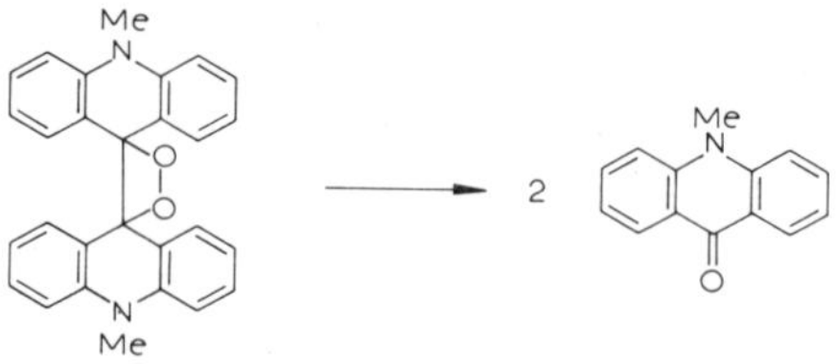

It is claimed that lucigenin is the most highly chemiluminescent synthetic substance known with the glow visible even at a dilution of one part in 20 million (*H. Decker* and *W. Petsch, loc. cit.*). It is reduced by zinc and hydrochloric acid to 10,10′-dimethyl-9,9′-dihydrobiacridine.

The chemiluminescence of the 10,10′-dialkyl-9,9′-biacridinium dinitrates was first noted by *Gleu* and *Petsch* (Z. angew. Chem., 1935, **48,** 57).

(*d*) *Hydrogenated acridines and their derivatives*

(*i*) *Hydroacridines*

Acridine is readily reduced and is described as "one of the most reactive compounds towards hydrogen ever encountered in our laboratory" (University of Wisconsin) (*H. Adkins* and *H. L. Coonradt*, J. Amer. chem. Soc., 1941, **63,** 1563). The products depend on the reagents used. Only one dihydro product can be obtained in this way, namely, 9,10-**dihydroacridine** (**acridan***), m.p. 172°, *acetyl* deriv., m.p. 151.5–153°; *benzoyl* deriv., m.p. 179°. It is prepared by reducing acridine with hydrogen and Raney nickel (*K. Bauer*, Ber., 1950, **83,** 10); in 87% yield with lithium tetrahydridoaluminate (*E. A. Braude et al.*, J. chem. Soc., 1960, 3249); or lithium and liquid ammonia. It is also obtained from acridone by reduction with sodium and boiling amyl alcohol.

9,10-Dihydroacridine (LVII) is colourless and non-fluorescent and is a very weak base, pK_a −0.93, *cf.* diphenylamine 0.86. Its ultraviolet and infrared spectra resemble those of diphenylamine: ν_{max} 3390 cm^{-1} and λ_{max} 290 nm compared to 3428 cm^{-1} and 285 nm for diphenylamine. The structural resemblance between the two substances is thus emphasized:

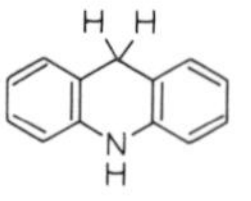

(LVII) 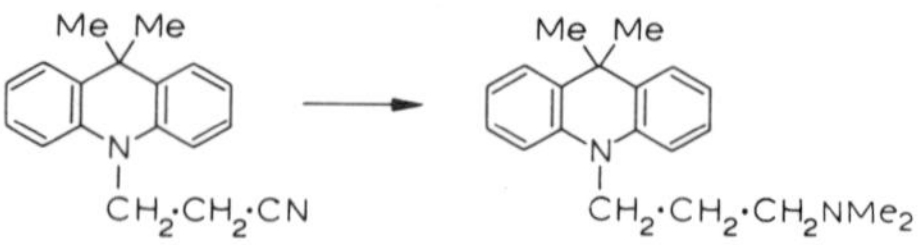 (LVIII) Dimethacrin

* See footnote p. 7.

9,10-Dihydroacridine is readily oxidised to acridine by chromic acid or milder reagents such as ferric chloride, chloranil *(Braude et al., loc. cit.)*, or potassium ferricyanide (*E. Hayashi et al.*, Yakugaku Zasshi, 1959, **79,** 967; C.A., 1959, **53,** 21946). The dehydrogenation is also effected by exposure to sunlight (*F. Kröhnke* and *H. L. Honig*, Ann., 1959, **624,** 97).

Reduced acridines tend to form double compounds with the parent compounds (*E. Bamberger*, Ber., 1891, **24,** 2473, 2648) and this is illustrated by the bright yellow 9-*methylacridine*-9,9′-*dimethyl*-9,10-*dihydroacridine* complex, m.p. 124°.

9-Substituted 9,10-dihydroacridines have already been encountered in the discussion on the pseudo-bases (p. 10). Alkyl derivatives are prepared by the action of Grignard reagents on acridines, acridones and acridinium salts (*W. L. Semon* and *D. Craig*, J. Amer. chem. Soc., 1936, **58**, 1278). Acridine and methylmagnesium iodide, for instance, yield 9-methylacridine and 9,9-*dimethyl*-9,10-*dihydroacridine*, m.p. 125–126° or 92–93° (metastable), while 9,10-dimethylacridinium iodide and the same reagent give 9,10-*dihydro*-9,10-*dimethylacridine*, m.p. 138–140°, and 9,10-*dihydro*-9,9,10-*trimethylacridine*, m.p. 100–102°.

9,10-Dihydro-9,9-dimethylacridine condenses with vinyl cyanide in the presence of alkali to give the 10-β-cyanoethyl derivative LVIII, which when reduced and the product methylated gives the antidepressant drug **Dimethacrin**.

Reduction of acridine or 9,10-dihydroacridine with lithium in liquid ammonia and ethanol yields 1,4,5,8-**tetrahydroacridine** (LIX), m.p. 130.5°, which on hydrogenation with tris(triphenylphosphine)rhodium chloride gives 1,2,3,4,5,6,7,8-octahydroacridine (LXIX), m.p. 70°, (*A. J. Birch* and *H. H. Mantsch*, Austral. J. Chem., 1969, **22,** 1103).

1,2,3,4-**Tetrahydroacridine** (LXI), m.p. 55–56°, *picrate*, m.p. 222° (decomp.), is obtained by the Pfitzinger reaction in which isatin and cyclohexanone are condensed in the presence of alkali to give 1,2,3,4-tetrahydroacridine-9-carboxylic acid (LX), which loses carbon dioxide when heated above its melting point to yield LXI. For another preparation from anthranil see p. 5.

1,2,3,4-Tetrahydroacridine condenses with benzaldehyde in the presence of alkali to give the 4-*benzylidene* deriv., m.p. 103–104°. Hydrogenation of 1,2,3,4-tetrahydroacridine gives a mixture of octahydroacridines (see below):

CO_2H heat

(LIX) (LX) (LXI)

(LXVIII) (LXIV) (LXV)

(LXVII) (LXVI)

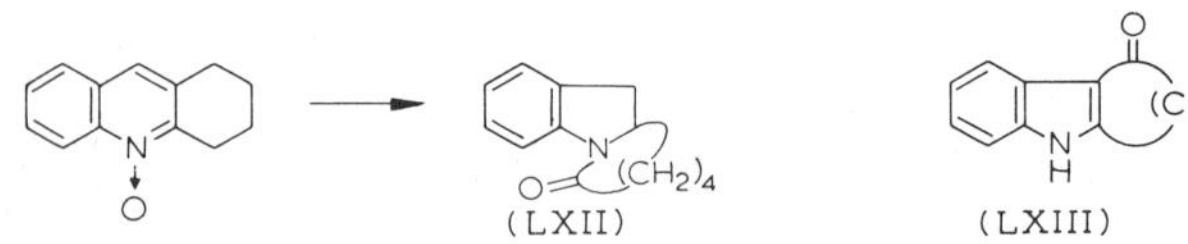

(LXII) (LXIII)

1,2,3,4-Tetrahydroacridine *N*-oxide when irradiated in aproptic solvents undergoes rearrangement to yield the *indoline* derivative LXII, m.p. 150°, (70% yield) and 5,6,7,8,9,10-*hexahydrocyclohept*[b]*indol*-10[5H]-*one* (LXIII), m.p. 221°, (10% yield) (*C. Kaneko et al.*, Tetrahedron Letters, 1967, 1873; C.A., 1969, **71**, 81119m).

Reduction of 1,2,3,4-tetrahydroacridine with tin and hydrochloric acid yields a mixture of *cis*-1,2,3,4,4a,9,9a,10-**octahydroacridine** (LXIV), m.p. 72°, *acetyl* deriv., m.p. 137–138°, and the trans-*isomer* (LXV), m.p. 82°, *acetyl* deriv., m.p. 185–187°, with the *cis*-compound predominating (*D. A. Archer et al.*, J. chem. Soc., 1963, 330). The *cis*-compound LXIV is also obtained from 1,2,3,4-tetrahydroacridine by hydrogenation in acetic acid with a platinum catalyst (*T. Masamune* and *S. Wakamatsu*, C.A., 1958, **52,** 11850) or a platinum–palladium–carbon catalyst (*E. Hayashi* and *T. Nagao*, Yakugaku Zasshi, 1964, **84,** 198; C.A., 1964, **61,** 3071). The *trans*-isomer is obtained by heating acridine with iodine and phosphorus (*C. Graebe*, Ber., 1883, **16,** 2828). The configurations of the two compounds have been established by p.m.r. spectroscopy by relating the coupling of the two benzylic protons at $C_{(9)}$ to the proton at $C_{(9a)}$, and the proton at $C_{(4a)}$ to three protons, two at $C_{(4)}$ and one at $C_{(9a)}$, to the *cis*- and *trans*-configurations of the isomers (*H. Booth*, Tetrahedron, 1963, **19,** 91).

1,2,3,4,5,6,7,8-**Octahydroacridine** (LXIX), m.p. 69–71°, *picrate*, m.p. 198–200°, has already been mentioned (p. 29) and can be prepared in a number of ways including the action of hydroxylamine on bis(2-oxocyclohexyl)methane (*N. S. Gill et al.*, J. Amer. chem. Soc., 1952, **74,** 4923):

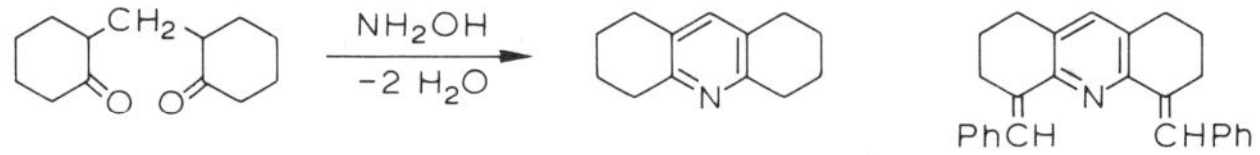

$$\xrightarrow[-2\,H_2O]{NH_2OH}$$

(LXIX) (LXX)

This octahydroacridine is prepared in 100% yield also by the hydrogenation of acridine in trifluoroacetic acid using a platinic oxide catalyst (*F. W. Viehapper* and *E. L. Eliel*, J. org. Chem., 1975, **40,** 2729) and with sodium in ethanol yields α-perhydroacridine, m.p. 89° (*idem, ibid.*, p. 2734). The selective hydrogenation of the aromatic rings is noteworthy.

The 4- and 5-positions are reactive and yield with benzaldehyde and acetic anhydride 4,5-*dibenzylidene*-1,2,3,4,5,6,7,8-*octahydroacridine* (LXX), m.p. 227° (*V. I. Vysotskii* and *M. N. Tilichenko*, C.A., 1969, **70,** 77756d).

The *trans*-octahydroacridine (LXV) when hydrogenated with a platinum catalyst yields α-**perhydroacridine** (LXVI) in 60% yield, m.p. 90.5–91.5°, *picrate*, m.p. 192–194°, and β-**perhydroacridine** (LXVII), m.p. 51° (15% yield), *benzoyl* deriv., m.p. 87–89° (*Masamune* and *S. Wakamatsu*, C.A., 1958, **52,** 11850). Similar hydrogenation of the *cis*-octahydroacridine (LXIV) yields *β*-perhydroacridine (LXVII), 66% yield, and *γ*-**perhydroacridine** (LXVIII), m.p. 72.5–74.5°, *picrate*, m.p. 211–213° (*Masamune et al.*, Bull. chem. Soc. Japan, 1968, **51,** 2458). *α*-Perhydroacridine may also be obtained by hydrogenation of acridine or 1,2,3,4-tetrahydroacridine with a Raney catalyst at 240° (*Masamune* and *Wakamatsu*, *loc. cit.*; *Adkins* and *Coonradt*, *loc. cit.*).

The stereochemistry of the perhydroacridines has been studied by *Masamune* and his co-workers (*loc. cit.*, 1958, 1968) who concluded from spectroscopic and other evidence that the *α*-, *β*- and *γ*-compounds have the configurations LXVI, LXVII and LXVIII, respectively (dots indicate hydrogen atoms lying above the plane of the molecule). In other words the three compounds have respectively the *trans,syn,trans*- LXVI *trans, anti,cis*- LXVII, and *trans,syn,cis*- LXVIII -configurations.

(ii) 9-Acridone (acridanone*)

The most general method for the preparation of 9-acridone and substituted derivatives is the ring-closure of diphenylamine-2-carboxylic acids by means of sulphuric acid (*C. F. H. Allen* and *G. H. W. McKee*, Org. Synth., 1939, **19,** 6), polyphosphoric acid (*H. Brockmann* and *H. Muxfeldt*, Ber., 1956, **89,** 1379), or phosphoryl chloride followed by the facile acid hydrolysis of the resulting 9-chloracridine. For a critical review of these and other methods see *J. M. F. Gagan*, "Acridines", Interscience, New York, 1973, p. 141. The acid-catalysed ring-closure probably occurs by intramolecular carbonium ion attack:

9-Acridone

Ring closure of *m*-substituted diphenylamine-2-carboxylic acids gives a mixture of 1- and 3-substituted acridones, 3′-nitrodiphenylamine-2-carboxylic acid, for instance, giving a mixture of 1- and 3-nitroacridone (*G. S. Chandler*, *W. H. F. Sasse* and *R. A. Jones*, Austral. J. Chem., 1965, **18,** 108).

In a useful modification of this method (*M. M. Jamieson* and *E. E. Turner*, J. chem. Soc., 1937, 1954) use is made of *A. W. Chapman*'s diphenylamine synthesis (*ibid.*, 1929, 569). An imidoyl chloride, *e.g.* from a benzanilide, reacts quantitatively with the sodium salt of methyl salicylate to give a methoxycarbonylphenyl ester LXXI, which when heated above 200° rearranges to an *N*-benzoyldiphenylamine derivative.

This substance with sulphuric acid at 160° or when heated at 300° yields 10-benzoyl-2-chloro-9-acridone:

---

* The trivial names acridan (for 9,10-dihydroacridine) and acridanone (for 9-acridone) are not approved for general use (see I.U.P.A.C. Rule 315.3).

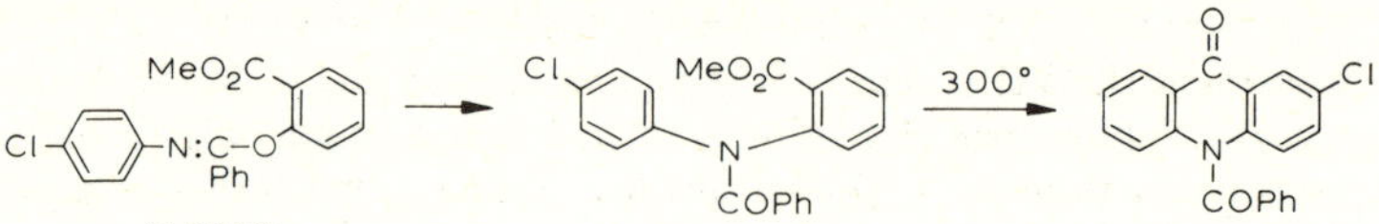

(LXXI)

The oxidation of acridines to acridones is often unsatisfactory, but nitroacridines are oxidised by chromic acid in acetic acid to the corresponding nitroacridones in good yield (*K. Lehmstedt*, Ber., 1938, **71,** 808). Oxidation may be effectively used by converting acridines into 9-thioacridones (p. 35) and subsequent oxidation with potassium ferricyanide or sodium hypochlorite.

Acridone, m.p. 354°, sublimes and distils without decomposition. It is best crystallised from acetic acid and then amyl acetate, and forms yellow needles which are only slightly soluble in the common organic solvents. It exhibits in ethanol an intense blue fluorescence.

Certain alkaloids such as melicopidine and melicopine are substituted *N*-methylacridones.

Acridone is a very weak base, pK_a −0.32, for although it dissolves in warm hydrochloric acid it separates unchanged when the solution cools. The formula for acridone can be written either in the oxo-form LXXII or the hydroxy-form LXXIII:

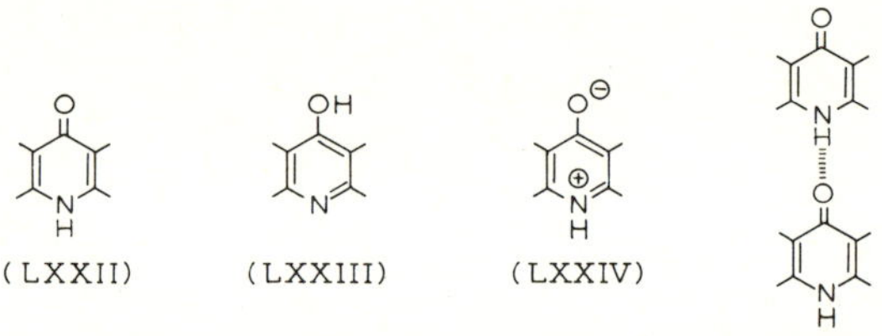

(LXXII) (LXXIII) (LXXIV) (LXXV)

The solubility of acridone in cold ethanolic potassium hydroxide in contrast to its insolubility in cold ethanol and the interaction of acridone and phosphorus pentachloride to yield 9-chloroacridine have been used as evidence in favour of the hydroxy-form LXXIII, but such evidence is not compelling. On the other hand *O*-acetyl derivatives cannot be obtained directly from acridones and spectroscopic evidence shows unmistakably that in solution the oxo-form LXXII predominates. The ultraviolet spectrum of acridone in methanol is very similar to that of 10-methyl-9-acridone (LXXII with NH replaced by NMe) and appreciably different from that of 9-methoxyacridine (LXXIII, with OH replaced by OMe) (*Acheson et al.*, J. chem. Soc., 1954, 3742). In the p.m.r. spectrum coupling occurs between the N-proton and the $C_{(1)}$-proton, thus suggesting localisation of the N-proton (*J. P. Kokko* and *J. H. Goldstein*, Spectrochim. Acta, 1963, **19,** 1119). As expected, the protons at $C_{(1)}$ and $C_{(8)}$ are deshielded by the carbonyl group and give signals at τ 1.73 in DMSO. The oxo-form may resonate with the polar form LXXIV (*Albert* and *R. Goldacre*, J. chem. Soc., 1943, 454). That this polar form makes a considerable contribution to the hybrid is suggested by the high melting-point and slight solubility of acridone in organic solvents.

9-Acridone and *N*-methylacridone absorb in the infrared at 1632 cm^{-1}. This means that if, as is often postulated, acridone is strongly hydrogen-bonded (LXXV) the carbonyl frequency is not affected by such bonding.

Reduction of 9-acridone with sodium and amyl alcohol or sodium amalgam (*Albert*, "The Acridines", 1st Edn., Arnold, London, 1966, p. 18) yields 9,10-dihydroacridine (acridan). The initial reduction product is probably the dihydro compound LXXVI, which loses water to form acridine:

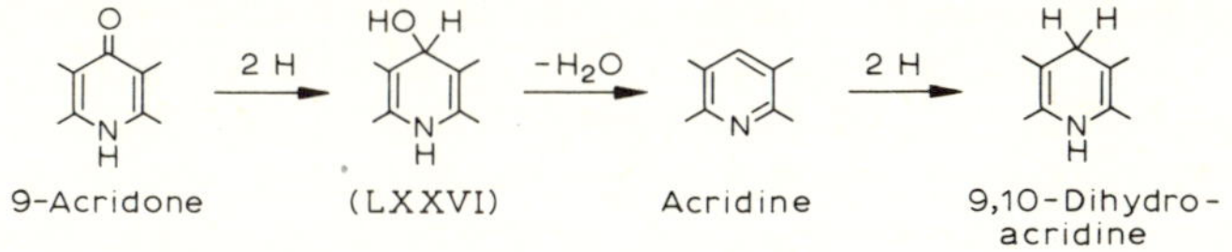

9-Acridone —2 H→ (LXXVI) —$-H_2O$→ Acridine —2 H→ 9,10-Dihydro-acridine

This in turn is reduced to 9,10-dihydroacridine.

Unlike its isomers 9-acridone with methyl iodide or dimethyl sulphate and alkali yields 10-*methyl*-9-*acridone*, m.p. 203°, which is also obtained from 9-*methoxyacridine*, m.p. 65° (anhyd.), by heating at 200° or better by warming with methyl iodide (*cf.* the conversion of 4-methoxyquinoline to *N*-methylquinoline) (*K. Gleu* and *S. Nitzsche*, J. pr. Chem., 1939, **153,** 200). 9-Methoxyacridine is obtained by treating 9-chloroacridine with sodium methoxide (*Lehmstedt*, Ber., 1935, **68,** 1455). 10-Methyl-acridone is also prepared by the action of alkali on 9-chloro-10-methylacridinium methosulphate when a smooth conversion occurs (*Gleu* and *Nitzsche*, J. pr. Chem., 1939, **153,** 233):

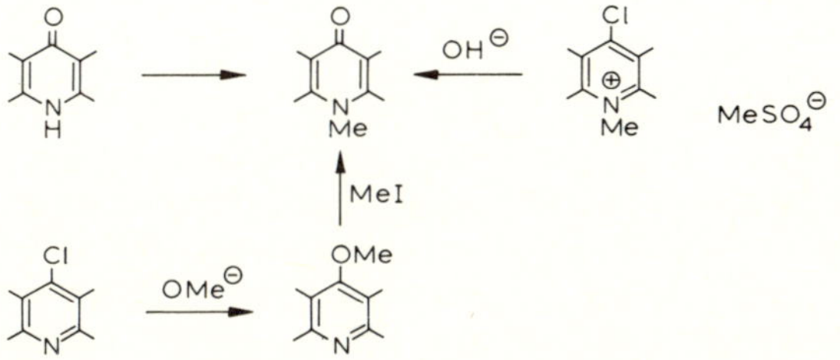

$OH^{\ominus}$ MeI $OMe^{\ominus}$ $MeSO_4^{\ominus}$

9-Acridones react with Grignard reagents or better with alkyl- or aryl-lithium compounds to give good yields of alkyl- or aryl-acridines (*e.g.* 9-phenylacridine, p. 16), or 9,10-dihydroacridines (p. 28).

Protonation of 9-acridone occurs at the oxygen atom rather than at the nitrogen atom as might be anticipated from the contribution the polar form LXXIV makes to the resonance hybrid. Otherwise electrophilic reagents attack the benzene rings, and nitration, bromination and sulphonation all follow the same pattern by substituting mainly at $C_{(2)}$ and to a lesser extent at $C_{(4)}$. Nitration gives 2-*nitro*- (85%) and 4-*nitro*-

9-*acridone* (15%), m.ps. >370° (decomp.) and 262°, respectively (*Lehmstedt*, Ber., 1931, **64**, 2381), and 2,7-dinitro-9-acridone has also been obtained as one of the products (*S.M. Scherlin et al.*, Ann., 1935, **516**, 218).

1-**Nitro**-9-**acridone**, m.p. 364°, and 3-*nitro*-9-*acridone*, m.p. >400°, have been synthesised (*Goldberg* and *Kelly*, *loc. cit.*). Further nitration of the mono-nitroacridones is summarised in the following sequences:

1-Nitro-9-acridone → 1,7-dinitro-9-acridone
2-Nitro-9-acridone → 2,7-dinitro-9-acridone (main product) and 2,5-dinitro-9-acridone
3-Nitro-9-acridone → 3,7-dinitro-9-acridone
4-Nitro-9-acridone → 2,5-dinitro-9-acridone

The nitroacridones are reduced by stannous chloride in hydrochloric acid or by photoreduction in oxygen-free protonated solvents such as alcohols (*V. Zanker* and *E. Cmiel*, Ann., 1975, 1576). 1-*Amino*-, m.p. 290–292°; 2-*amino*-, m.p. 300–302°; 3-*amino*-, m.p. 300–302°; 4-*amino-acridone*, m.p. 342°.

Bromine and 9-acridone in acetic acid give a number of products, but under suitable conditions an excellent yield of 2,7-*dibromo*-9-*acridone*, m.p. >400° may be obtained (*Acheson* and *M. J. T. Robinson*, J. chem. Soc., 1953, 232; *S. Kruger*, *ibid.*, 1952, 3648; *A. Kliegl* and *L. Schaible*, Ber., 1957, **90,** 60). Further bromination gives 2,4,7-*tribromo*-9-*acridone*, m.p. 340° and 2,4,5,7-*tetrabromo*-9-*acridone*, m.p. 308°.

Chlorine in carbon tetrachloride with acridone gives 2,7-*dichloro*-9-*acridone*, m.p. >350° and in acetic acid the product is 2,4,5,7-*tetrachloro*-9-*acridone*, m.p. 253° (*M. Ionescu* and *J. Goia*, C.A., 1961, **55,** 533).

Sulphonation occurs readily to give mainly 9-**acridone**-2-**sulphonic acid** and a little of the 4-isomer (*K. Matsumura*, J. Amer. chem. Soc., 1935, **57,** 1533). With oluem the product is acridone-2,7-disulphonic acid.

The hydroxyacridones are prepared synthetically and some of their derivatives occur naturally as alkaloids. Of possible biosynthetic importance is the conversion of the naturally occurring 2-methylaminobenzophenone derivative tecleanone into 1,3-dimethoxy-10-methylacridone, a constituent of *Acronychia baueri* and *Vepris ampody*, by treatment with sodium hydride in dimethyl sulphoxide at room temperature (*M. S. Khan*, *J. R. Lewis* and *R. A. Watt*, Chem. and Ind., 1975, 744).

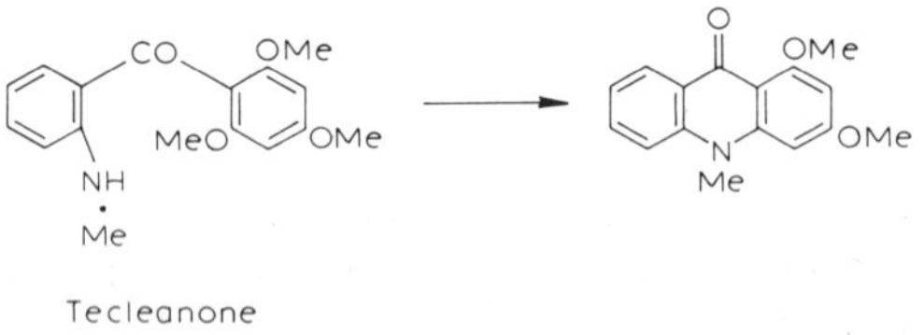

Tecleanone

The m.ps. of most acridones are too high to use for identification purposes. The difficulty may be overcome by converting the acridones into the comparatively low melting 9-(4′-dimethylaminophenylacridines by treatment of the acridones with phosphoryl chloride and 4-dimethylaminoaniline (*Acheson* and *Robinson*, J. chem. Soc.,

1956, 484). The method is useful only if the acridone is reasonably pure, since mixtures of the dimethylaminophenyl derivatives cannot be purified by crystallisation on account of mixed crystal or lattice compound formation, while chromatographic separations are tedious.

9-Acridones hydrogenated in the benzene rings are prepared synthetically, anthranilic acid and cyclohexanone, for instance, yielding when heated at 220°, 1,2,3,4-*tetrahydro*-9-*acridone*, m.p. 370°, *picrate*, m.p. 198° (*R. A. Reed, ibid.*, 1944, 425). It is markedly more basic than 9-acridone and forms crystalline salts with mineral acids. Dehydrogenation with copper powder at 360° gives 9-acridone.

(iii) 9-Acridonequinones

9-Acridonequinones are obtained by the oxidation of dihydroxy-9-acridones, 1,4-dihydroxy-9-acridone and nitric acid, for example, yielding 9-**acridone**-1,4-**quinone,** orange-red needles, m.p. >250° (decomp.). A preferable method is to oxidise monohydroxyacridones with potassium nitrosodisulphonate (*H. Brockmann et al.*, Ber., 1957, **90**, 44); in this way 2-hydroxyacridone gives 9-**acridone**-1,2-**quinone**, red rhombs.

The most important of these quinones is **Actinomycinol,** red needles decomposing at 300°, *diacetate*, m.p. 190°, *triacetate*, m.p. 180° (*Brockmann* and *H. Muxfeldt*, Ber., 1956, **89,** 1379; *S. J. Angyal et al.*, Chem. and Ind., 1955, 1295):

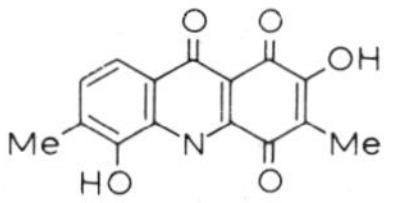

Actinomycinol

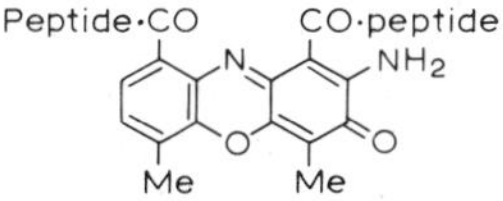

(LXXVII)

It is obtained from the group of antibiotics known as actinomycins by hydrolysis and rearrangement with barium hydroxide. These bright red compounds contain a phenoxazinone chromophore LXXVII, which under the influence of alkali undergoes rearrangement to the quinone.

(iv) Thioacridone

9,10-**Dihydroacridine**-9-**thione,** *thioacridone*, *acridine*-9-*thiol*, red needles, m.p. 275°, S-*benzoyl* deriv., m.p. 209°, S-*methyl* deriv., m.p. 117–118°, is prepared by melting acridine with sulphur at 190° (*A. Edinger* and *W. Arnold*, J. pr. Chem., 1901, [ii], **64,** 182) or by heating 9-chloroacridine with sodium sulphide or thiourea (*E. Ochiai* and *I. Iwai*, J. pharm. Soc., Japan, 1949, **69,** 413; C.A., 1950, **44,** 1988). Other methods for preparing the thiones include treatment of 9-aminoacridines in ammonia with hydrogen sulphide (*F. Wild* and *J. M. Young*, J. chem. Soc., 1965, 7261; *R. S. Asquith, D. Ll. Hammick* and *P. L. Williams*, *ibid.*, 1948, 1181).

9,10-Dihydro-9-acridinethione is more acidic than 9-acridone, but does not dissolve in aqueous sodium carbonate. In methanol the ultraviolet spectrum of 9,10-dihydroacridine-9-thione closely resembles that of 9,10-dihydro-10-methylacridine-9-thione (LXXVIII, with NMe replacing NH) and differs from that of the methylthio derivative LXXIX (SMe replacing SH), thus showing that in neutral solution the thione form LXXVIII is predominant (*Acheson et al., ibid.*, 1954, 3742):

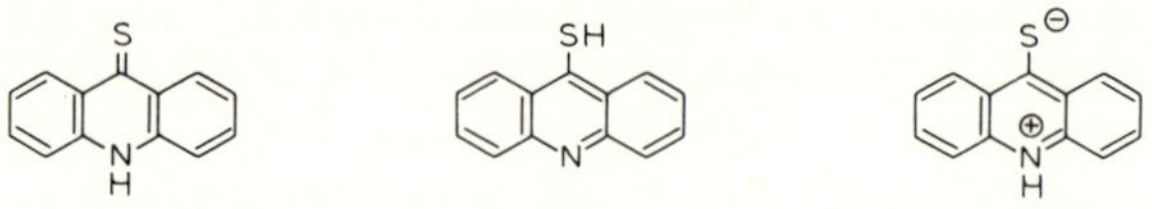

(LXXVIII) (LXXIX) (LXXX)

Its structure is thus comparable to that of 9-acridone. The large dipole moment, 5.2 D, indicates that the polar form LXXX contributes to the resonance hybrid.

Some of the properties of 9,10-dihydroacridine-9-thione suggest the thiol form LXXIX. It dissolves in warm alkali or ammonia, is converted by alkyl or acyl halides into *S*-alkyl or *S*-acyl derivatives (*Edinger* and *Arnold, loc. cit.* and also p. 471), and yields 9-chloroacridine by reaction with phosphorus pentachloride.

9,10-Dihydroacridine-9-thione (acridanthione) is oxidised by a variety of reagents including sulphuric acid to 9-acridone.

$$>\!\!=S + 3\,H_2SO_4 \longrightarrow\ >\!\!=O + 3\,H_2O + 4\,SO_2$$

Since acridine is easily converted into the thione this oxidation provides a method for converting acridines into acridones when direct oxidation gives poor results (p. 32) (*A.P. Wolf* and *R.C. Anderson*, J. Amer. chem. Soc., 1955, **77**, 1608).

10-*Phenyl*-9,10-*dihydroacridine*-9-*thione*, red needles, m.p. 227–228°, is prepared by heating 10-phenylacridone with phosphorus pentasulphide (*A. Schönberg et al.*, Ber., 1928, **61**, 1375).

*(e) Acridine dyes**

There are three types of dye in use derived from acridine or containing the acridine nucleus: (*a*) basic dyestuffs derived from 3-amino- and 3,6-diamino-acridine, (*b*) vat dyestuffs which contain the acridone nucleus generally fused with anthraquinone, and (*c*) pigments derived from quinacridine.

(i) Basic dyestuffs

These are salts of the coloured aminoacridines, yellow, orange or brown, which dye protein fibres such as silk and leather and tannin-mordanted cotton. The intensity of colour can be associated with the resonance of the acridinium ion (see, *e.g.* formula of Acridine Orange L, below), the positive charge being distributed between the ring nitrogen atom and the amino nitrogens, and possibly to some extent the bridge

* *B. D. Tilak* and *R. N. Ayyangar*, "Acridines", Wiley, New York, 1973, p. 579.

carbon. Amino groups in the 3- and 6-positions are conjugated both with the ring nitrogen and the bridge carbon. As has already been mentioned protonation of the aminoacridines occurs at the ring nitrogen. This proton can be replaced by an alkyl group as in acriflavine and the stability of the dyestuff is thereby increased.

Both proflavine and acriflavine are dyestuffs, but more important as a dyestuff is **Acridine Orange L,** (C.I. Basic Orange 14), 3,6-tetramethyldiaminoacridine hydrochloride, which is formed by heating 2,2′-diamino-4,4′-dimethylaminodiphenylmethane to give the leuco-base, oxidation of which yields the dyestuff.

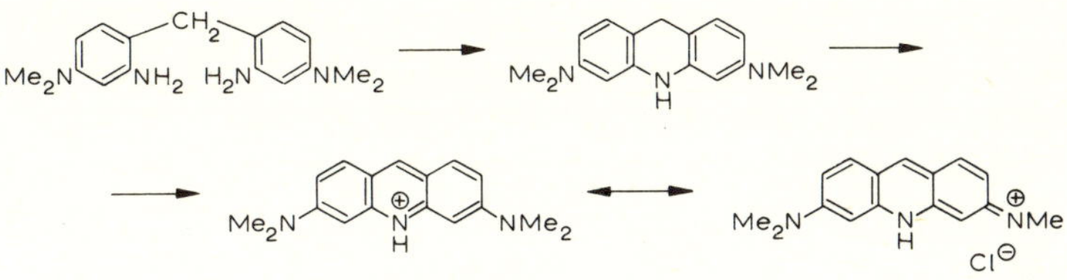

Acridine Orange L

(ii) Vat dyestuffs

The 3,4-phthaloylacridone vat dyes are very fast to light and provide directly red, violet and blue vat shades. They possess other desirable qualities and can be effectively applied to cotton, rayon and silk. 3,4-Phthaloylacridone (LXXXI) is obtained by reacting 1-chloroanthraquinone with anthranilic acid followed by cyclisation. It is not used as a dyestuff, but chlorination yields dyestuffs including the 6,10,12-trichloro compound, **Indanthrene Red Violet RRK,** (C.I. Vat Violet 14) (LXXXII). The best known of the simpler dyestuffs is **Caledon Red BN** or **Indanthrene Red RK,** (C.I. Vat Red 35) which is prepared from 1-chloroanthraquinone-2-carboxylic acid and 2-naphthylamine and cyclising the product (*F. Ullmann* and *H. Bincer*, Ber., 1916, **49**, 732):

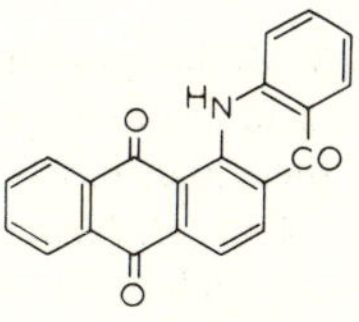

(LXXXI)

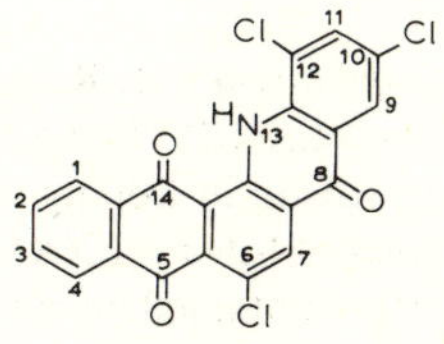

(LXXXII)

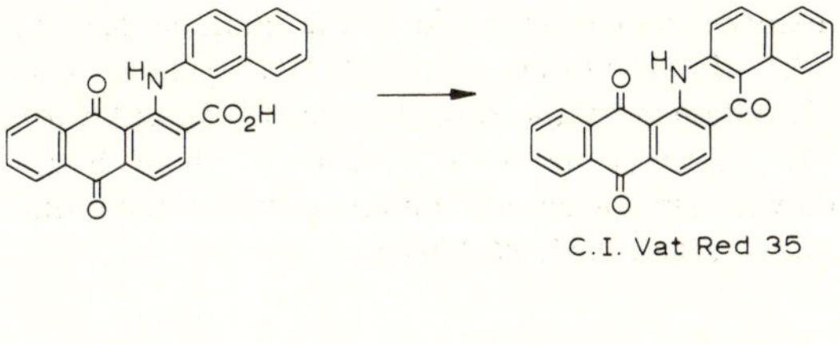

C.I. Vat Red 35

It has a distinct reddish hue lacking in the anthraquinone dyestuffs.

More complex vat dyestuffs which contain both the acridone and mesobenzanthrone systems are known. The accessible 3-bromomesobenzanthrone (LXXXIII), for example, condenses with 1-aminoanthraquinone in the presence of sodium carbonate and copper to give the intermediate LXXXIV, which is cyclised by alkali to **Indanthrene Olive Green B,** (C.I. Vat Green 3), a fast cotton dyestuff. By a similar process 3,9-dibromo-mesobenzanthrone yields **Indanthrene Olive T,** (C.I. Vat Black 25):

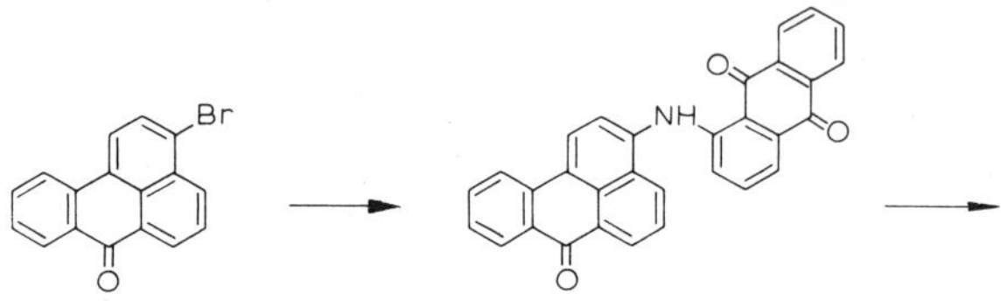

(LXXXIII) (LXXXIV)

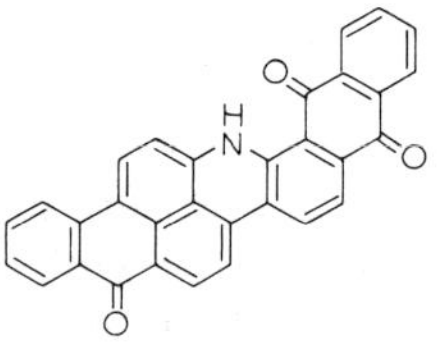

C.I. Vat Green 3

(iii) Pigments. Quinacridines, quinacridones

Substances which contain two acridine nuclei are known as quinacridines. They are of minor importance, but some of the corresponding quinacridones containing acridone rings have valuable tinctorial properties. Both types are described in the literature under a variety of names, the simplest of which are quinacridines and quinacridones, and these are used in the text. Each type may be divided into *linear* and *angular* types and examples of both are given in the sequel, although the angular compounds are not dyestuffs.

(*1*) *Linear* trans-*quinacridones.* Ring-closure of *N,N'*-diphenyl-*p*-phenylenediamine-2,5-dicarboxylic acid (LXXXV) by heating with boric acid or other cyclising agent such as phosphorus pentoxide in boiling cymene yields the violet-red *trans*-**quinacridone,** *quino*[2,3-b]*acridine*-7,14-*dione* (LXXXVI), decomp. >400° (*H. Liebermann et al.*, Ann., 1935, **518,** 245). It is obtained in polymorphic forms each with its own tinctorial properties and its derivatives are valuable — *Monastral*-type — pigments comparable to the phthalocyanines in properties but with a wider colour range. The dione LXXXVI can be oxidised by chromic acid to the quinone, 5,6,7,12,13,14-*hexahydro*-6,7,13,14-*tetraoxo*-5,12-*diazapentacene* (LXXXVIII), yellow crystals, m.p. 400° (decomp.), which is obtained by the ring-closure of 2,5-bis(2'-carboxyanilino)benzo-quinone (LXXXVII) (*Acheson* and *B. F. Sansom*, J. chem. Soc., 1955, 4440). The quinone is reduced with zinc or aluminium and alkali to the quinacridone LXXXVI (*W. Braun* and *R. Mecke*, Ber., 1966, **99,** 1991):

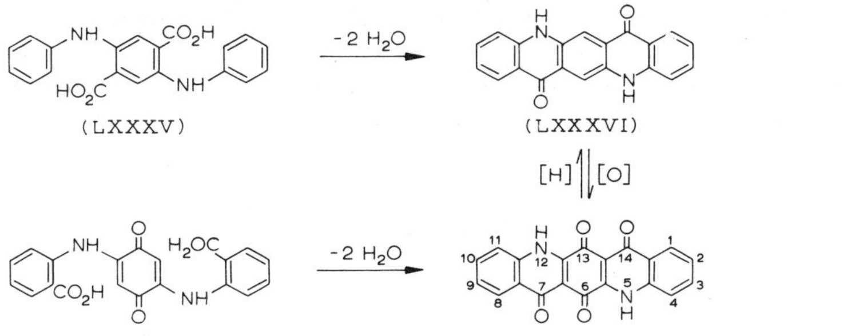

(LXXXV) $\xrightarrow{-2\,H_2O}$ (LXXXVI)

[H] ⇅ [O]

(LXXXVII) $\xrightarrow{-2\,H_2O}$ (LXXXVIII)

The linear *cis*-compound, *quino*[3,2-b]*acridine*-12,14-*dione* (LXXXIX), is prepared by ring-closure of the 4,6-dianilinoisophthalic acid by phosphorus pentachloride in boiling xylene (*A. Eckert* and *F. Seidel*, J. pr. Chem., 1921, **102**, 338). It is obtained in a number of different crystalline forms and is insoluble in most solvents:

(LXXXIX)

(*2*) *Angular quinacridines.* Ring-closure of *N,N'*-di(2'-carboxyphenyl)-*p*-phenylenediamine (XC, R = H) may occur in either of two ways and the earlier workers assumed that a *linear* quinacridone resulted. *G. M. Badger* and *R. Pettit* (J. chem. Soc., 1952, 1874) showed, however, that the product is in fact the angular *quin*[3,2-b]*acridine*-13,14-*dione* (XCI), which can be converted into the parent *dibenzo*[b,j] [4,7]*phenanthroline* (XCII), pale yellow, m.p. 235°, with an ultraviolet spectrum similar to that of pentaphene and quite different from that of pentacene:

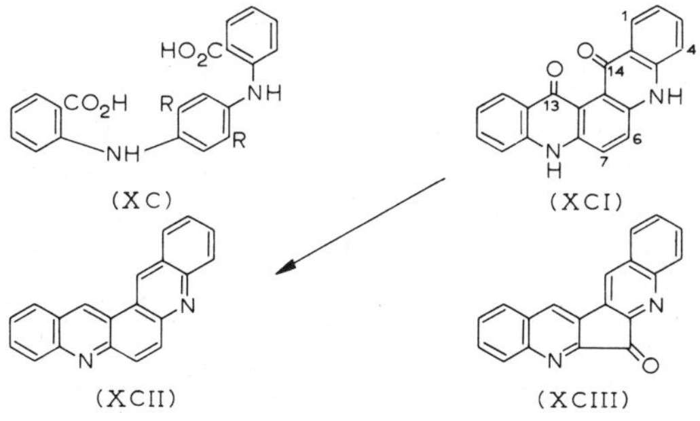

(XC) (XCI)

(XCII) (XCIII)

Oxidation of XCII yields not the expected 6,7-quinone, but the orange 1,8-*diaza-dibenzo*[b,h]*fluorenone* (XCIII), m.p. 374–376°, presumably by oxidation to the 6,7-quinone followed by a benzilic acid type of rearrangement. The *p*-phenylenediamine derivative (XC, R = Me), yields the *linear* quinacridone.

Dibenzo[b,j] [1,10]**phenanthroline** (XCIV), m.p. 310°, orange crystals, is prepared from a 2,2′-biquinoline derivative (*E. Uhlemann* and *P. Kurze*, J. pr. Chem., 1970, **312,** 1105), while *quino*[2,3-a]*acridine, dibenzo*[b,j] [1,7]*phenanthroline* (XCV), colourless, m.p. 221°, is prepared by distilling the corresponding quinone with zinc dust (*S. von Niementowski*, Ber., 1919, **52,** 461):

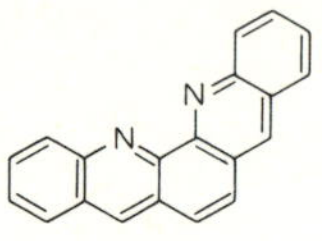

(XCIV) 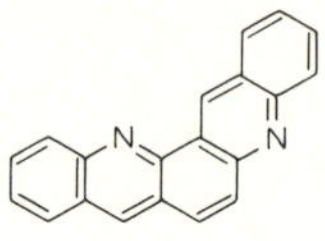 (XCV)

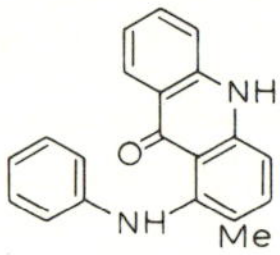

(XCVI) 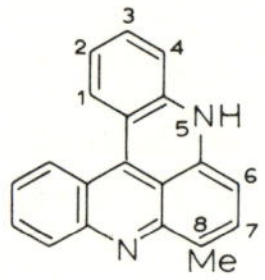 (XCVII)

A pentacyclic structure in which acridine and quinoline nuclei are annealed together in a different manner from that encountered above is illustrated by the deep blue 8-*methyl*-5H-*quino*[2,3,4-kl]*acridine* (XCVII), m.p. >335°, which is obtained by heating 1-phenylamino-2-methylacridone (XCVI) with phosphoryl chloride (*R. Weiss* and *J. L. Katz*, Monatsh., 1928, **50,** 225).

2. Phenanthridine and its derivatives*

Phenanthridine is related to phenanthrene in the same way as acridine is to anthracene, a nitrogen atom taking the place of the 9- (or 10-)methine group in phenanthrene. From this relationship to phenanthrene and to pyridine its name is derived. It may be designated dibenzo[*b,d*]pyridine, 3,4-benzo[*c*]quinoline or -isoquinoline, or 9-azaphenanthrene. Phenanthridine is found in the higher coal-tar distillation fraction and the phenanthridine nucleus occurs in certain alkaloids such as lycorin.

The pioneer work on the chemistry of phenanthridine was carried out by *Pictet* and his co-workers (1889–1905), but interest in the compound and its derivatives flagged until *G. T. Morgan* and *L. P. Walls* in 1931 at the Chemical Research Laboratory, Teddington, showed that 6-phenylphenanthridinium compounds have useful therapeutic properties (p. 46).

The *structure* of phenanthridine (I or II) follows from its synthesis from

** L. P. Walls*, in Elderfield's Heterocyclic Compounds (Wiley, New York, 1952), Vol. IV, p. 564; *R. S. Theobald* and *K. Schofield*, Chem. Reviews, 1950, **46,** 171; *J. Eisch* and *H. Gilman*, *ibid.*, 1957, **57,** 525; *B. R. T. Keene* and *P. Tissington*, in Advances in Heterocyclic Chemistry (Academic Press, New York and London), Vol. 13, p. 315.

biphenyl derivatives, quinoline or benzylideneaniline; from degradations such as oxonolysis to quinoline-3,4-dicarboxylic acid or Emde breakdown of *N*-methyl-5,6-dihydrophenanthridine to 2-dimethylamino-2′-methylbiphenyl; and from physical evidence such as nuclear magnetic resonance spectroscopy. The ring-numbering I, officially approved in the I.U.P.A.C. Rules for nomenclature and used in Chemical Abstracts, is used in the text, but many important papers adopt the numbering shown in II.

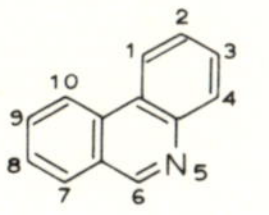

(I) 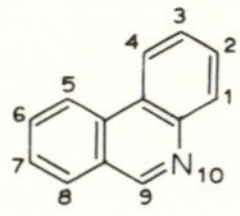 (II)

This has given rise to some confusion and even Chemical Abstracts is not always consistent.

(a) Methods of synthesis

(*1*) Many phenanthridines are prepared from substituted biphenyls and one very convenient method is very similar to the Bischler–Napieralski synthesis of isoquinoline derivatives (C.C.C., Vol. IV F, p. 359). 2-Acylaminobiphenyls undergo ring-closure when heated with phosphoryl chloride in high boiling solvents, preferably nitrobenzene, to give phenanthridines in good yield (*Morgan* and *Walls*, J. chem. Soc., 1932, 2225; *cf. Walls*, *ibid.*, 1947, 67). The cyclisation is reasonably regarded as an intramolecular Friedel–Crafts reaction in which the catalyst gives a reactive carbonium ion (*e.g.* III). Ring-closure then occurs by electrophilic attack. *H.J. Barber et al.* (J. Soc. chem. Ind., 1950, **69**, 82) isolated the imonio-chloride III and proved that an inorganic catalyst such as ferric chloride, phosphoryl chloride or aluminium chloride is required for cyclisation. Yields of 80% are obtained. The favourable influence of nitrobenzene is due not only to its high b.p. (a reaction temperature of at least 205° is necessary) but also to its ionising properties.

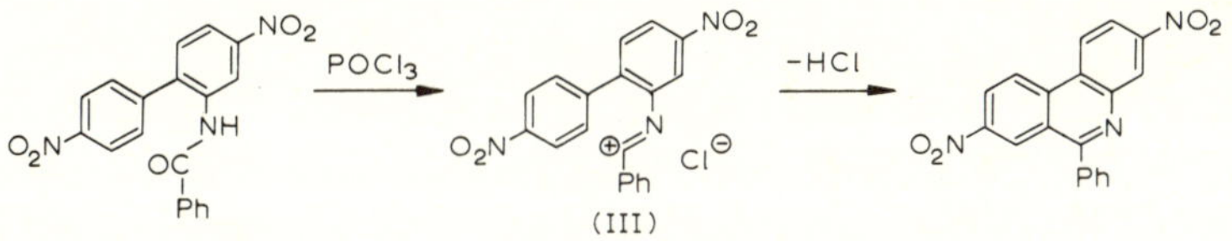

$\xrightarrow{POCl_3}$ (III) $\xrightarrow{-HCl}$

(*2*) Fluorene derivatives can be converted into phenanthridines by a number of methods involving ring-enlargement of the cyclopentadiene ring. In the Stieglitz rearrangement, for example, 9-chloroamino derivatives of fluorene react with alkali to give phenanthridines (*L. A. Pinck* and *G. E. Hilbert*, J. Amer. chem. Soc., 1937, **59,** 8). 6-Phenylphenanthridine is prepared in this way:

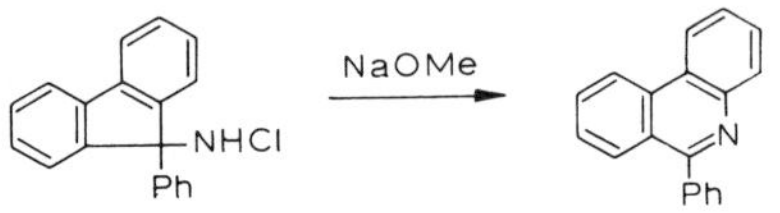

NaOMe

In a related reaction 9-phenyl-9-fluorenylamine with potassamide in liquid ammonia gives 6-aminophenanthridine (61% yield) (*H. C. White* and *F. W. Bergstrom*, J. org. Chem., 1942, **7,** 499), possibly by the following mechanism:

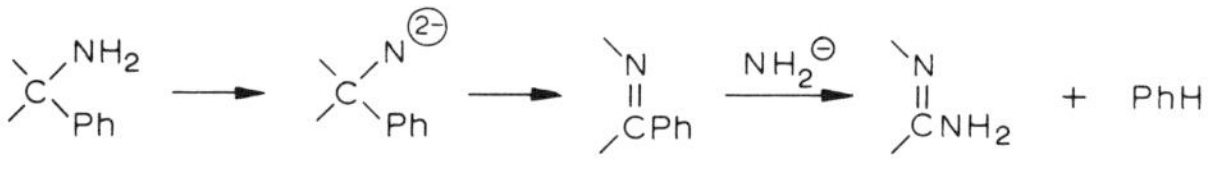

$NH_2^{\ominus}$ + PhH

Noteworthy is the expulsion of the phenyl radical as benzene.

(*3*) By the reduction of phenanthridones (p. 56).

(*4*) Phenanthridine can be prepared from dibenzothiazepines by the extrusion of sulphur (*A. D. Jarrett* and *J. D. Loudon*, J. chem. Soc., 1957, 3818). 2-Nitrodibenzo[*b,f*]1,4-thiazepine (IV) when heated with copper in diethyl phthalate thus yields 8-nitrophenanthridine:

(IV)

(*5*) Irradiation of Schiff's bases by ultraviolet light gives dihydro-intermediates (*e.g.* V) which are readily dehydrogenated by iodine to phenanthridines (*F. B. Mallory* and *C. S. Wood*, Tetrahedron Letters, 1965, 2643). Thus phenyliminobenzophenone yields 6-phenylphenanthridine in good yield.

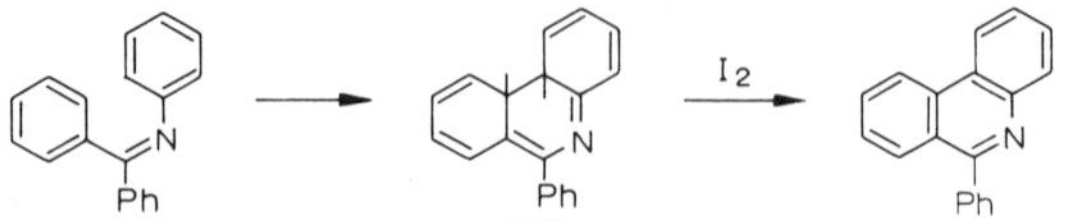

I_2

(V)

The yield from benzylideneaniline is very small, probably because the required *cis*-compound is unstable to ultraviolet light.

(6) Benzyne participation is used to synthesise phenanthridines. *o*-Chlorobenzylideneaniline, for example, with potassamide in liquid ammonia gives an excellent yield of phenanthridine *via* the benzyne intermediate VI (*S. V. Kessar* and *M. Singh*, *ibid.*, 1969, 1155):

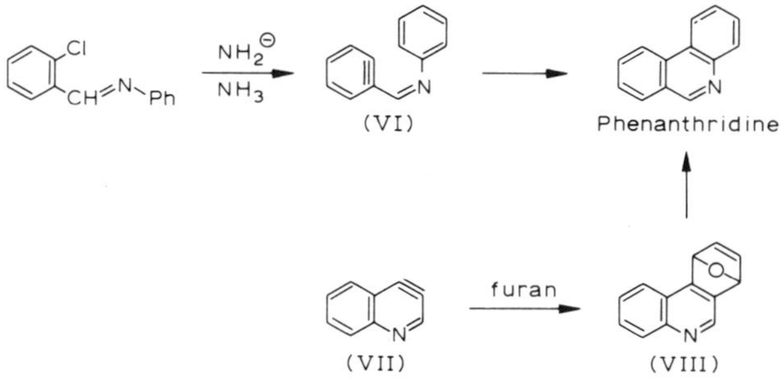

(VI) Phenanthridine

furan

(VII) (VIII)

3-Bromo-4-chloroquinoline reacts in characteristic fashion with lithium amalgam and furan to give phenanthridine *via* the quinolyne intermediate VII and adduct VIII (*T. Kauffmann, F. P. Boettcher* and *J. Hansen*, Ann., 1962, **659,** 102). In another type of Diels–Alder reaction, benzyne, generated by the thermal decomposition of diazotised anthranilic acid in benzene reacts with phenyl isocyanate to give phenanthridone and 9-phenoxyphenanthridine (*J. C. Sheehan* and *G. D. Daves*, J. org. Chem., 1965, **30,** 3247).

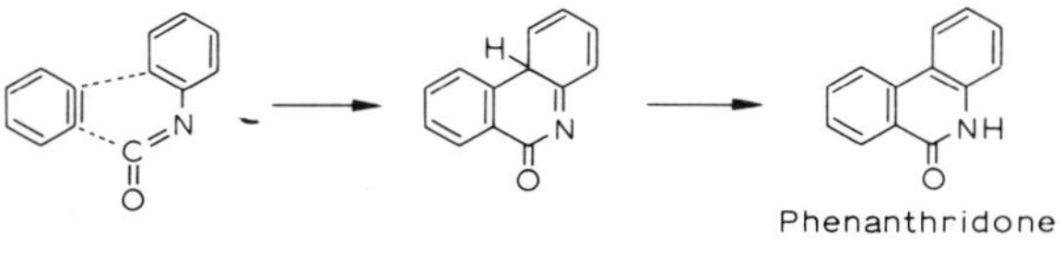

Phenanthridone

(b) Phenanthridines

(i) General physical properties

Phenanthridine is a weak base, pK_a 3.30 in 50% aqueous ethanolic solution and 4.52 in water and is thus weaker than quinoline and acridine (4.94 and 5.60, respectively: *A. R. Osborn* and *K. Schofield*, J. chem. Soc., 1956, 4191; *A. Albert* and *J. N. Philips*, *ibid.*, 1956, 1294). It has a dipole moment of 2.39 D (*C. W. N. Cumper* and *R. F. A. Ginman*, J. chem. Soc., 1962, 4524). Its ultraviolet spectrum resembles those of phenanthrene and *o*-phenanthroline with three main absorption bands at ~250, 295 and 345 nm (*G. M. Badger, R. S. Pearce* and *R. Pettit*, *ibid.*, 1951, 3199). Useful information about the constitution of substituted phenanthridines is obtained from their infrared spectra, 10-chlorophenanthridine, for example, showing bands at 760 cm^{-1} (4 adjacent CH groups) and 790 cm^{-1} (3 adjacent CH groups) (*H.-L. Pan* and *T. L. Fletcher*, J. heterocyclic Chem., 1970, **7,** 597; *G. S. Chandler et al.*, Austral. J. Chem., 1967, **20,** 2037; *R. C. Petterson et al.*, J. org. Chem., 1974, **39,** 1841).

P.m.r. shows the characteristic deshielding (τ 0.88) at $C_{(6)}$ adjacent to the ring nitrogen atom, at $C_{(4)}$ and at $C_{(1)}$ and $C_{(10)}$ (τ 1.59), the result of the proximity of the carbocyclic rings (*R. H. Martin et al.*, Tetrahedron Supplement, 1966, **8,** 181; *B. R. T. Keene* and *P. Tissington*, J. chem. Soc., 1965, 3032).

The mass spectrum of phenanthridine contains the base peak at *m/e* 179 with other peaks as shown underneath (relative abundance in brackets) (*P. M. Brown et al., ibid.*, 1968, 842):

Phenanthridine *m/e* 179 (100)
- $\xrightarrow{-C_2H_2}$ *m/e* 153 (6)
- $\xrightarrow{-HCN}$ *m/e* 152 (12)
- $\xrightarrow{-H}$ *m/e* 178 (16) $\xrightarrow{-HCN}$ 151 (14)

(ii) Chemical properties

(1) Phenanthridines. The chemical properties of phenanthridine and its derivatives can be predicted with a fair degree of certainty. As a pyridine derivative it reacts with alkyl halides to give phenanthridinium salts and with hydrogen peroxide or peracids yields the *N*-oxide. Electrophilic substitution occurs in the carbocyclic rings and substituents in these rings behave as they do in phenanthrene. The amines, for example, can be diazotised and the diazonium salts undergo the Sandmeyer reaction, etc. while the hydroxyl compounds have phenolic properties, form methyl ethers with diazomethane, and have a band at about 3600 cm^{-1} in the infrared spectrum. On the other hand, nucleophilic substitution occurs at $C_{(6)}$, the carbon atom contiguous to the nitrogen atom. Phenanthridine thus reacts with the hydroxyl ion to give phenanthridone, the lactam of 6-hydroxyphenanthridine. Substituents in the 6-position differ from those in other parts of the molecule, 6-aminophenanthridine, for example, with nitrous acid yielding phenanthridone, while the 6-hydroxy compound exists as already indicated predominately in the lactam form. Other examples are the reactivity of 6-chlorophenanthridine towards amines, the oxidation of 6-methylphenanthridine with selenium dioxide to 6-formylphenanthridine and the ready decarboxylation of phenanthridine-6-carboxylic acid merely by heating.

The azomethine linkage is the site of a number of reactions recalling those of quinoline. Phenanthridine is reduced by hydrido compounds to 5,6-dihydrophenanthridine, reacts with lithium alkyls to give 6-alkyl-5,6-

dihydrophenanthridines, forms Reissert compounds and reacts characteristically with dimethyl acetylenedicarboxylate. These reactions are discussed later.

The phenanthridines are particularly sensitive to attack at $C_{(6)}$ by alkali and pseudo-bases are formed. 5-Methylphenanthridinium iodide with alkali forms an insoluble pseudo-base whose structure follows from its direct oxidation with alkaline potassium ferricyanide to *N*-methylphenanthridone, identical with the compound obtained by the methylation of phenanthridone (*A. Pictet et al.*, Ber., 1902, **35,** 2534):

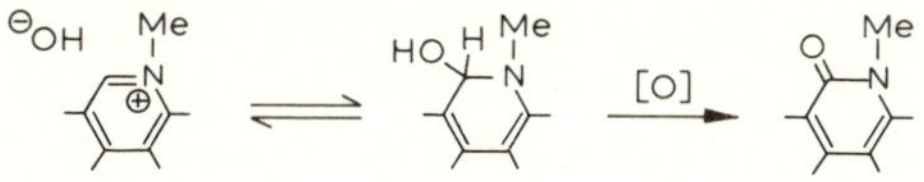

It is interesting that the alkaline oxidation of 6-phenyl-*N*-methylphenanthridinium methosulphate by potassium ferricyanide yields 5-methylphenanthridone, another example of elimination of the phenyl radical from $C_{(6)}$ (see p. 42) (*V. Petrov* and *W. R. Wragg*, J. chem. Soc., 1947, 1410).

Like quinoline, phenanthridine reacts with two molecules of dimethyl acetylenedicarboxylate in benzene to form a yellow "labile" *tetramethyl* [13bH]-*dibenzo*[a,c]*quinolizinetetracarboxylate* (IX), m.p. 284°, λ_{max} 410 nm, which when heated in quinoline is converted into the red 4H-*dibenzoquinolizine* derivative X, m.p. 245° (decomp.), λ_{max} 466 nm (*cf.* the corresponding quinolizine derivatives) (*Acheson* and *A. O. Plunkett*, *ibid.*, 1962, 3758). In methanol or aqueous methanol the reaction takes a different course.

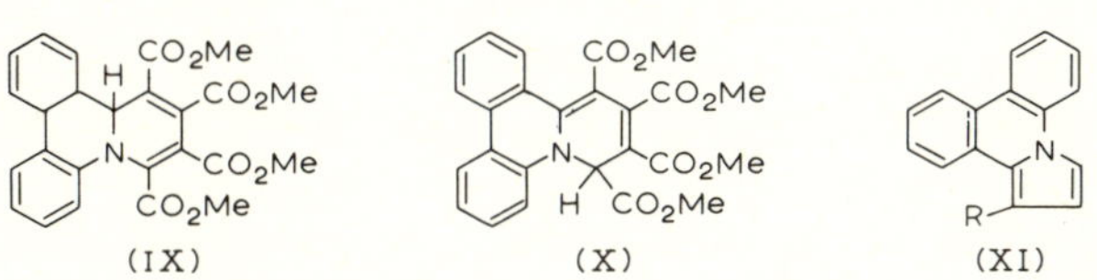

(IX) (X) (XI) (a) R = CO_2Me
(b) R = CO_2H
(c) R = H

Phenanthridine with methyl propiolate yields several products including *methyl pyrrolo*[1,2-f]*phenanthridine*-1-*carboxylate* (XIa), m.p. 137–138°, which when hydrolysed by sodium hydroxide gives the *carboxylic acid* XIb, m.p. 195–196°. The acid when heated with soda-lime is decarboxylated to give *pyrrolo*[1,2-f]*phenanthridine* (XIc), m.p. 151° (*Acheson* and *M. S. Verlander*, J. chem. Soc., C, 1969, 2311).

Phenanthridine is stable to many of the common oxidising agents such as chromic acid, alkaline potassium permanganate and dilute nitric acid. With acidified potassium dichromate or potassium permanganate it yields phenanthridone. Substituted phenanthridines behave similarly; but under certain conditions ring-fission occurs. Nitrophenanthridines, for example, are oxidised by potassium permanganate and sulphuric acid to the nitrophenanthridones (*C. L. Arcus* and *M. M. Coombs*, J. chem. Soc., 1954, 4319), but the resulting nitrophenanthridones with the same reagent may be oxidised further. Thus 8-nitrophenanthridone yields 4-nitrophthalic acid, while 3-nitrophenanthridone gives phthalic acid *(Jarrett* and *Loudon, loc. cit.)*. It appears that in these oxidations ring A is attacked while ring B remains intact. 6-Methylphenanthridines are oxidised to phenanthridones, 3-chloro-6-methylphenanthridine thus yielding 3-chlorophenanthridone (75% yield) (*B. L. Hollingsworth* and *Petrov*, J. chem. Soc., 1961, 3771).

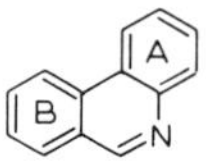

Ozonolysis of phenanthridine yields phenanthridone (23% yield) and quinoline-3,4-dicarboxylic acid (2%) and much unchanged phenanthridine (*E. J. Moriconi* and *F. A. Spano*, J. Amer. chem. Soc., 1964, **86,** 38). Oxidation with chloride of lime (calcium hypochlorite) in the presence of a cobalt salt is reported to give phenanthridone, but later efforts to repeat the method were unsuccessful (*J. Sielisch* and *R. Sandke*, Ber., 1933, **66,** 433).

The reduction of phenanthridine is discussed later.

Dimethylketene adds across the azomethine linkage to give the adduct XII, the structure of which is supported by four singlets in the p.m.r. spectrum attributable to the methyl groups and by rupture of the lactone ring to give the amido acid XIII (*R. N. Pratt, G. A. Taylor* and *S. A. Proctor*, J. chem. Soc., C, 1967, 1569):

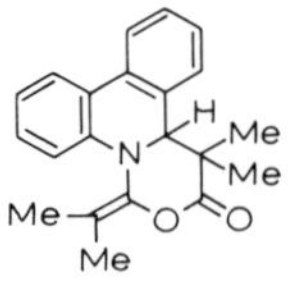

(XII)

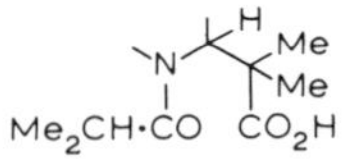

(XIII)

(2) *Phenanthridinium compounds.* The phenanthridinium salts, obtained by the interaction of phenanthridine and alkyl halides, were first prepared by *Amé Pictet* (Ber., 1896, **29,** 1182). Many phenanthridines readily form phenanthridinium salts with methyl iodide, but if a large group such as

phenyl occupies the 6-position it is necessary first to add dimethyl sulphate or methyl *p*-toluenesulphonate to a hot solution of the phenanthridine in nitrobenzene (*Morgan* and *Walls*, J. chem. Soc., 1931, 2447; 1945, 294). The corresponding chloride, bromide or iodide may be obtained by further interaction with hydrochloric acid, potassium bromide or potassium iodide. With alkali the phenanthridinium salts form pseudo-bases, those from nitrophenanthridinium salts being well defined crystalline substances (*C. K. Tinkler*, J. chem. Soc., 1906, **89,** 856; *Walls*, *ibid.*, 1945, 294). 6-*Hydroxy*-3,8-*dinitro*-6-*phenyl*-5,6-*dihydrophenanthridine* (XIV), for instance, forms red crystals, m.p. 186–188° (decomp.):

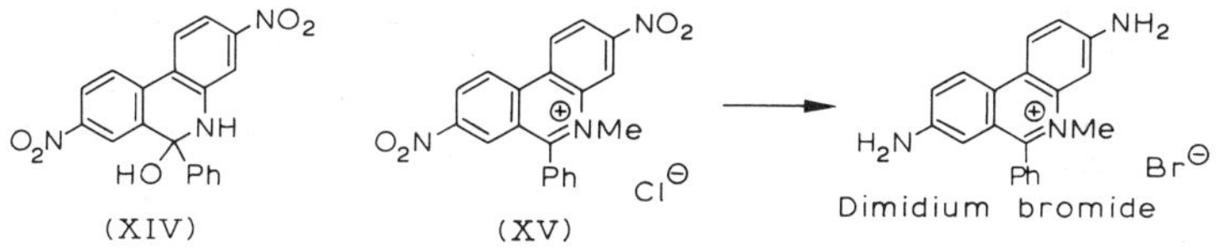

(XIV) (XV) Dimidium bromide

Some of the amino-9-arylphenanthridinium salts are trypanocides and to a lesser extent antiseptics.

The best known is **Dimidium bromide, phenanthridinium 1553,** 3,8-*diamino*-5-*methyl*-6-*phenylphenanthridinium bromide*, purple-black prisms, m.p. 241°. It is prepared by reducing 3,8-dinitro-5-methyl-6-phenylphenanthridinium chloride (XV) with iron filings and water followed by the addition of potassium bromide (*Walls et al., loc. cit.*; *Barber et al., loc. cit.*). Dimidium bromide is used in treating African cattle suffering from infections due to *Trypanosoma congolense*.

Other effective compounds are **Phenidium chloride (S.897),** 8-amino-6-*p*-aminophenyl-5-methylphenanthridinium chloride, which is also effective against *T. congolense*; 6-*p*-dimethylaminophenyl-5-methylphenanthridinium chloride which inhibits *E. coli* at a concentration of 1:200,000 and 3,8-diamino-6-*p*-aminophenyl-5-methylphenanthridinium chloride, a very potent drug against *T. congolense* infections (*Walls*, Chem. and Ind., 1951, 606).

(*3*) *Phenanthridine* N-*oxide*, m.p. 220°, is prepared by the action of perphthalic acid on phenanthridine (*P. Mamalis* and *Petrov*, J. chem. Soc., 1950, 703). It yields phenanthridine with phosphorus tribromide or hydrogen and Raney nickel, and with lithium tetrahydridoaluminate in ether gives 5,6-dihydrophenanthridine (*W. Traber*, *M. Hubmann* and *P. Karrer*, Helv., 1960, **43,** 265). It is converted by phosphoryl chloride or sulphuryl chloride to 9-chlorophenanthridine (*E. Hayashi*, J. pharm. Soc. Japan, 1961, **81,** 1030). The oxide when irradiated with ultraviolet light gives an 86% yield of phenanthridone, probably *via* an oxaziridine intermediate (*E. C. Taylor*, *B. Furth* and *M. Pfau*, J. Amer. chem. Soc., 1965, **87,** 1400):

$$-CH{=}\underset{\downarrow}{N}- \quad \longrightarrow \quad -CH{-}N- \text{ (bridged by O)} \quad \longrightarrow \quad -\underset{\|}{C}{-}NH-$$

(4) *Electrophilic substitution.* Phenanthridine with bromine, silver sulphate and sulphuric acid gives much unchanged phenanthridine, some dibromo-products, and monobromophenanthridines in the order 10-Br > 4-Br > 2-Br in the ratio of 9:6:1 (*G. S. Chandler*, Austral. J. Chem., 1969, **22**, 1105). With *N*-bromosuccinimide only the 2-isomer is obtained (*J. J. Eisch* and *H. Gilman*, J. Amer. chem. Soc., 1955, **77**, 6379).

Nitration with mixed acids gives mononitrophenanthridines in the order of 1 > 10 > 8 > 3 > 2 > 4 with the respective percentage yields of 26, 21, 11, 6, 3 and 1 (*A. G. Caldwell* and *L. P. Walls*, J. chem. Soc., 1952, 2156). The predicted order of electrophilic attack based on localisation energy calculations is 10 > 1 > 8 > 3 > 4 > 2 > 7 > 9 (*M. J. S. Dewar*, and *P. M. Maitlis*, *ibid.*, 1957, 2521). Clearly this agrees moderately well with the nitration, but not with the bromination results.

The ease of nitration increases in passing from pyridine to phenanthridine: pyridine < quinoline < acridine < phenanthridine. Pyridine under harsh conditions yields 22% of 3-nitropyridine, while phenanthridine nitrate in sulphuric acid gives an excellent yield of six mononitrophenanthridines.

(5) *Nucleophilic substitution.* Phenanthridine undergoes nucleophilic substitution at position 6. When heated at 225° with potassium and barium hydroxides it gives a 67% yield of phenanthridone (6-hydroxyphenanthridine) (*Gilman* and *Eisch*, J. Amer. chem. Soc., 1957, **79**, 5479). Tschitschibabin amination is effected with sodamide in liquid ammonia in the presence of nitrate (*H. C. White* and *F. W. Bergstrom*, J. org. Chem., 1942, **7**, 497) or when heated in dimethylaniline or xylene under nitrogen (*B. R. T. Keene* and *P. Tissington*, *ibid.*, 1965, 3032). The product, 6-aminophenanthridine, is also obtained from 6-phenylphenanthridine with potassamide in liquid ammonia, benzene being liberated in the process; yet another example of ejection of the phenyl radical from position 6 (p. 42) (*White* and *Bergstrom*, *loc. cit.*). 6-Chlorophenanthridine gives 6-aminophenanthridine when heated with ammonia and methanol at 180° (*C. B. Reese*, J. chem. Soc., 1958, 895) and it is therefore somewhat curious that it fails to give phenanthridone when heated with potassium hydroxide.

Organometallic compounds attack at the 6-position to give 6-alkyl- or 6-aryl-5,6-dihydrophenanthridines, which can be dehydrogenated with nitrobenzene to yield 6-alkyl- or 6-aryl-phenanthridines. Propylmagnesium bromide thus gives an 84% yield of 6-propylphenanthridine (*Gilman* and *Eisch*, J. Amer. chem. Soc., 1957, **79**, 4423) and phenyllithium in ether under nitrogen gives an 85% yield of 6-phenylphenanthridine (*Gilman* and *R.D.*

Nelson, ibid., 1948, **70**, 3316). The dihydrointermediates are so easily oxidised that they are only occasionally isolated.

Attack by the methylsulphinyl carbanion gives an excellent yield of 6-methylphenanthridine (*G. A. Russell* and *S. A. Weiner*, J. org. Chem., 1966, **31,** 248):

$$RH + CH_3SOCH_2^{\ominus} \rightarrow [RH\cdot CH_3SOCH_2]^{\ominus} \rightarrow RCH_3 + CH_3SO^{\ominus}$$

Phenanthridine undergoes the Reissert reaction by reacting with benzoyl chloride and hydrogen cyanide to give a 94% yield of 5-*benzoyl*-6-*cyano*-5,6-*dihydrophenanthridine* (XVI), m.p. 141°, which with sulphuric acid yields benzaldehyde and *phenanthridine*-6-*carboxylic acid*, m.p. 155° (*G. Wittig, M. A. Jesaitis* and *M. Glos*, Ann., 1952, **577**, 1):

NCOPh
H CN
(XVI)
N
CO_2H
+ PhCHO

The 6-position in the phenanthridine molecule is also the site of attack by free radicals, the benzoyl radical, for example, giving 6-*benzoylphenanthridine*, m.p. 135° (*T. Caronna et al.*, J. chem. Soc., Perkin II, 1972, 2035).

Phenanthridine, m.p. 108°, has a blue fluorescence in aqueous solution. It is identified by its *picrate*, m.p. 235°, and forms a *mercuric chloride adduct*, m.p. 199–200°, which is used to obtain pure phenanthridine into which it is converted by the use of sodium hydroxide. As a tertiary amine it forms an *N-oxide*, m.p. 220° (see p. 47) and phenanthridinium salts with alkyl halides. The *methiodide*, orange rhombs, m.p. 202–203°, is best obtained by the interaction of phenanthridine and dimethyl sulphate in boiling nitromethane, followed by conversion of the methosulphate to the iodide by the addition of potassium iodide (*Acheson* and *G. J. F. Bond*, J. chem. Soc., 1956, 246).

Phenanthridine is thermally stable and is unchanged when distilled at about 350°.

(c) Substituted phenanthridines

(i) Alkylphenanthridines

The alkylphenanthridines are prepared by synthesis, or in the case of 6-alkyl compounds by the action of alkyl-lithium compounds on phenanthridine. Other methods are sometimes used, 6-ethylphenanthridine, for example, resulting from irradiation of phenanthridine in ethanol (*F. R. Stermitz, R. P. Seiber* and *D. E. Nicodem*, J. org. Chem., 1968, **33,** 1136).

Only the 6-alkylphenanthridines call for special comment and especially 6-methylphenanthridine with its reactive methyl group recalling that in 2-picoline or quinaldine. It is oxidised by sodium dichromate and acetic acid to phenanthridone (*Walls*, J. chem. Soc., 1935, 1405) and by selenium dioxide to 6-*formylphenanthridine*, m.p. 139°, *phenylhydrazone*, m.p. 166°.

$$\underset{\text{CHO}}{-\text{C}=\text{N}-} \xleftarrow{SeO_2} \underset{\text{Me}}{-\text{C}=\text{N}-} \xrightarrow{Na_2Cr_2O_7} -\text{CO}-\text{NH}-$$

With the first reagent 3,6-dimethylphenanthridine is oxidised to 3-*methylphenanthridone*, m.p. 251°, thus contrasting the reactivity of the 6-methyl

TABLE 5

ALKYL- AND ARYL-PHENANTHRIDINES

Substituent	*m.p. (°C)*	*Picrate m.p. (°C)*	*Ref.*
1-Methyl	64–65	232	3
2-Methyl	89	271–272	1
3-Methyl	81	251	2,3
4-Methyl	95	234	2,4
6-Methyl	85–86	249	1,5,9
8-Methyl	88	237	2
10-Methyl	109–111	205–207	3
1-Ethyl	57	233	6
2-Ethyl	62–63	230	6
3-Ethyl		216	6
4-Ethyl		231	6
6-Ethyl	55.5	210	8,11
4-Phenyl	102–103	227–228	12
6-Phenyl	106	251–252	13,14

References
1 *C. L. Arcus* and *M. M. Coombs*, J. chem. Soc., 1954, 4319.
2 *J. J. Eisch* and *H. Gilman*, J. Amer. chem. Soc., 1955, **77**, 6379.
3 *B. A. T. Keene* and *P. Tissington*, J. chem. Soc., 1965, 3032.
4 *A. Risaliti* and *A. Stener*, Ann. Chim. (Rome), 1967, **57**, 3.
5 *G. R. Proctor* and *W. C. Peaston*, J. chem. Soc., C, 1969, 2151.
6 *H. Kondo* and *S. Uyeo*, Ber., 1937, **70**, 1094.
7 *B. R. T. Keene* and *G. L. Turner*, Tetrahedron, 1971, **27**, 3405.
8 *A. Pictet* and *A. Hubert*, Ber., 1896, **29**, 1182.
9 *G. A. Russell* and *S. A. Weiner*, J. org. Chem., 1966, **31**, 248.
10 *G. T. Morgan* and *L. P. Walls*, J. chem. Soc., 1931, 2447.
11 *E. Hayashi* and *Y. Hoshi*, Yakugaku Zasshi, 1967, **87**, 651; C.A., 1968, **68**, 21818.
12 *A. Risaliti* and *S. Bozzini*, Ann. Chim. (Rome), 1964, **54**, 685.
13 *F. B. Mallory* and *C. S. Wood*, Tetrahedron Letters, 1965, 2643.
14 *B. Staskun*, J. org. Chem., 1964, **29**, 2856.

group with methyl groups in other positions (*E. Ritchie*, J. Proc. roy. Soc. N.S.W., 1945, **78,** 169).

6-Methylphenanthridine condenses with 4-dimethylaminobenzaldehyde; undergoes deuterium exchange when dissolved in C_2H_5OD; and reacts with phenyllithium to give the 6-lithiomethylene compound ($-CH_2Li$), no addition to the azomethine linkage occurring (*A. M. Jones, C. A. Russell* and *S. Skidmore*, J. chem. Soc., C, 1969, 2245).

In 1,10-dimethylphenanthridine the methyl groups must overlap causing twisting of the 10a–10b bond and non-planarity of the molecule. Evidence of non-planarity is afforded by the partial resolution of 6-amino-1,10-dimethylphenanthridine by (+)-camphorsulphonic acid and by the anomalous spectra of the parent compound. In the p.m.r. spectrum of 1,10-dimethylphenanthridine the methyl resonance (τ 7.54) is much greater than in 1-methyl- (τ 6.99) and 10-methylphenanthridine (τ 7.01). The ultraviolet spectrum is not typical and shows only two main bands (*B. R. T. Keene* and *P. Tissington*, J. chem. Soc., 1965, 3032, 4426).

The melting points of some alkyl- and phenyl-phenanthridines and their picrates are listed in Table 5.

(ii) Halogenophenanthridines

The halogenophenanthridines (see Table 6) are prepared by synthetic methods, but 6-*chlorophenanthridine*, m.p. 116°, is obtained by the action of phosphoryl chloride on phenanthridone. Phosphorus tribromide similarly yields 6-bromophenanthridine. 6-Chlorophenanthridine shows marked reactivity towards nucleophilic reagents and although inactive towards sodium hydroxide it reacts with sodium methoxide to give 6-*methoxyphenanthridine*, m.p. 52°, and with ammonia to give 6-aminophenanthridine.

The bromophenanthridines are recorded in Table 6.

TABLE 6
BROMOPHENANTHRIDINES

	m.p. (°C)	*Ref.*
1-Bromo-	97–99	1
2-Bromo-	162	2,3,4
3-Bromo-	122–123	4
4-Bromo-	129–130	3,5
6-Bromo-	125–126	1
7-Bromo-	160–161	6
8-Bromo-	86–87	4
9-Bromo-	118–119	6
10-Bromo-	144–145	1

References
1 *G. S. Chandler et al.*, Austral. J. Chem., 1967, **20,** 2037.
2 *J. J. Eisch* and *H. Gilman*, J. Amer. chem. Soc., 1955, **77,** 6379.
3 *Chandler*, Austral. J. Chem., 1969, **22,** 1105.
4 *G. M. Badger* and *W. F. H. Sasse*, J. chem. Soc., 1957, 4.
5 *J. L. Huppatz* and *Sasse*, Austral. J. Chem., 1965, **18,** 206.
6 *Huppatz* and *Sasse*, *ibid*, 1964, **17,** 1406.

(iii) Nitro- and amino-phenanthridines

Like the halogeno compounds the nitrophenanthridines are obtained by synthesis, for example, by the Schmidt reaction on fluorenols. 2-Nitrofluorenol (XVII) with hydrazoic acid thus yields a mixture (92%) of 3-nitro-(XVIII) (3%) and 8-nitro-phenanthridine (XIX) (97%) (*C. L. Arcus* and *M. M. Coombs*, J. chem. Soc., 1954, 4319):

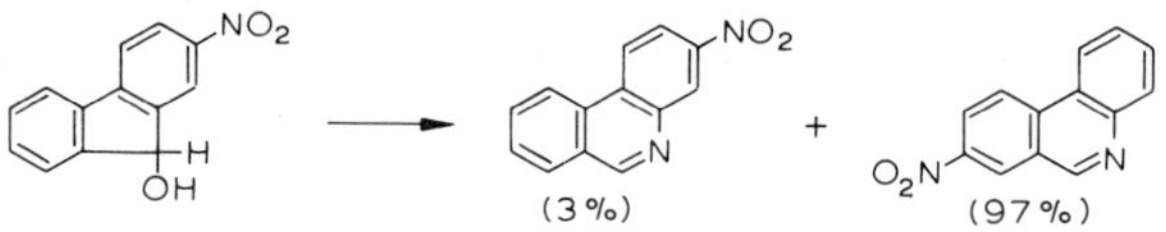

NO_2 H OH (XVII) → NO_2 N (3%) (XVIII) + O_2N N (97%) (XIX)

The nitro compounds are reduced to the aminophenanthridines by iron powder in hot aqueous suspension (*A. G. Caldwell* and *L. P. Walls*, J. chem. Soc., 1952, 2156). 6-Aminophenanthridine is prepared by the interaction of 6-chloro- or 6-bromo-phenanthridine and ammonia (*Reese, loc. cit.*) or by nucleophilic attack on phenanthridine by the amide ion. The properties of the aminophenanthridines have already been mentioned (p. 44).

Two tautomeric forms (XXa and XXb) for 6-aminophenanthridine can be envisaged:

$$\underset{NH_2}{\underset{|}{-C}}=N- \qquad \underset{NH}{\underset{\|}{-C}}-NH-$$

(XXa) (XXb)

Spectroscopic evidence shows that the compound exists predominately in the amino form (XXa) (*S. F. Mason*, J. chem. Soc., 1959, 1281). The i.r. spectrum exhibits in carbon tetrachloride the characteristic NH_2 bands at 3409 cm^{-1} (NH symm.) and 3513 cm^{-1} (NH antisymm.). 6-Amino- and 6-*N*-methylamino-phenanthridines (pK_a in ethanol 6.30 and 6.70, respectively) are stronger bases than phenanthridine, but the pK_a value of 6-*N*,*N*-dimethylaminophenanthridine rather surprisingly is 4.91. The weakness of the base thus indicated perhaps results from steric interference between the *N*-methyl groups and the $C_{(7)}$-hydrogen atom.

The melting points of mono-nitro- and mono-amino-phenanthridines are recorded in Table 7.

TABLE 7

NITRO- AND AMINO-PHENANTHRIDINES

	m.p. (°C)		*m.p. (°C)*	*Ref.*
1-Nitro	161	1-Amino	116–117	1,2
2-Nitro	268	2-Amino	152–153	1,2
3-Nitro	196–197	3-Amino	144–145	1,2
4-Nitro	191–192			2
		6-Amino	190	3
		7-Amino	208–210	2
8-Nitro	180	8-Amino	203–204	1,2
9-Nitro	192–194	9-Amino	192–194	1,2
10-Nitro	166		144	2

References
1 *C. L. Arcus* and *M. M. Coombs*, J. chem. Soc., 1954, 4319.
2 *A. G. Caldwell* and *L. P. Walls*, *ibid.*, 1952, 2156.
3 *H. C. White* and *F. W. Bergstrom*, J. org. Chem., 1942, **7**, 499.

(iv) Hydroxyphenanthridines or phenanthridinols

The hydroxyphenanthridines (with the exception of the 6-isomer which is discussed separately under phenanthridone) are phenolic compounds which can be methylated by diazomethane and in carbon tetrachloride exhibit an OH stretch at ~3620 cm^{-1} (*Mason, ibid.*, 1957, 4874). The methoxyphenanthridines can be prepared from the methoxy-9-fluorenols by the Schmidt reaction, 2-methoxyfluorenol thus giving a mixture (32%) containing 3-methoxyphenanthridine (32%) and 8-methoxyphenanthridine (68%) (*Arcus* and *Coombs, loc. cit.*).

3-**Phenanthridinol,** m.p. 245°, *methyl ether*, m.p. 58°; 8-*phenanthridinol*, m.p. 282°, *methyl ether*, m.p. 90°; 9-*phenanthridinol*, m.p. 271–272° (*Arcus* and *Coombs, loc. cit.*); 4-*phenanthridinol*, m.p. 186–188°, *methyl ether*, m.p. 140–142° (*A. Ya. Berlin* and *T. P. Sycheva*, Zhur. obshcheĭ Khim., 1952, **22**, 2003; C.A., 1953, **47**, 933).

(v) Phenanthridones

6-Phenanthridinol is generally known as phenanthridone since it exists mainly in the lactam form (see below), thereby resembling 2-pyridone and 2-quinolone.

The phenanthridones can be prepared by the oxidation of suitably substituted phenanthridines (p. 46). They are frequently prepared from fluorenones by the Schmidt reaction, *i.e.* by the action of hydrazoic acid generated from sodium azide by the action of sulphuric acid, trichloracetic acid (*P. A. S. Smith*, J. Amer. chem. Soc., 1948, **70**, 320) or polyphosphoric acid (*R. T.

Conley, J. org. Chem., 1958, **23,** 1330). By the last method a 92% yield of phenanthridone is obtained from fluorenone. Symmetrically substitued fluorenones such as 4,5-dimethylfluorenone yield single products, in this instance 1,10-dimethylphenanthridone, but unsymmetrically substituted fluorenones give mixtures of isomeric phenanthridones, particularly when the substituent is electron-donating such as the methoxyl group (*H.-L. Pan* and *T. L. Fletcher*, J. heterocyclic Chem., 1970, **7,** 313, 597). When the substituent is electron-attracting, NO_2, CO_2H or $NH_3^{\oplus}$, one product results, namely the isomer in which the substituted benzene ring is attached to the lactam nitrogen atom. 2-Nitrofluorenone (XXI), for example, yields 3-nitrophenanthridone (XXIII) and not 8-nitrophenanthridone (XXII).

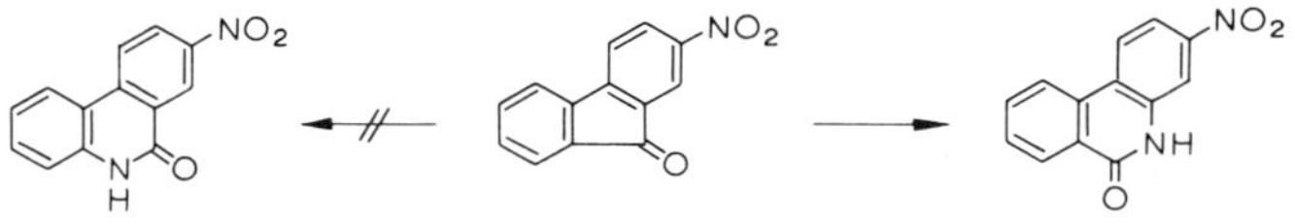

(XXII) (XXI) (XXIII)

Phenanthraquinone is converted by hydrazoic acid into phenanthridone and diphenamic acid (*E. F. M. Stephenson*, J. chem. Soc., 1949, 2620; *Caldwell* and *Walls*, *loc. cit.*). The last named authors suggest the reaction follows the course outlined below:

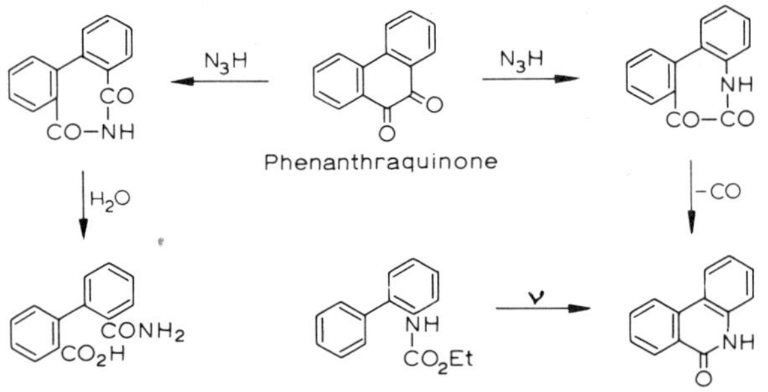

Phenanthraquinone

Diphenamic acid (XXIV) Phenanthridone

Fluorenones can be converted into phenanthridones *via* the oximes by means of the Beckmann transformation in the presence of phosphorus pentachloride or polyphosphoric acid (*E. C. Horning*, *V. L. Stromberg* and *H. A. Lloyd*, J. Amer. chem. Soc., 1952, **74,** 5153). Here again unsymmetrical fluorenones may yield mixtures of isomeric phenanthridones, but single products are sometimes obtained; thus 2-nitrofluorenone oxime yields both 3- and 8-nitrophenanthridone (*A. J. Nunn*, *K. Schofield* and *R. S. Theobald*,

J. chem. Soc., 1952, 2797), whereas 3-nitrofluorenone oxime gives only 2-nitrophenanthridone and 7-chloro-2-nitrofluorenone oxime gives an 87% yield of 8-chloro-3-nitrophenanthridone (*Pan* and *Fletcher*, J. heterocyclic Chem., 1970, **7,** 313).

Photolysis of benzanilide in benzene in the presence of iodine gives a 20% yield of phenanthridone (*B. S. Thyagarajan et al.*, Chem. Comm., 1967, 614; *Y. Kanaoka* and *K. Itoh*, *ibid.*, 1973, 647):

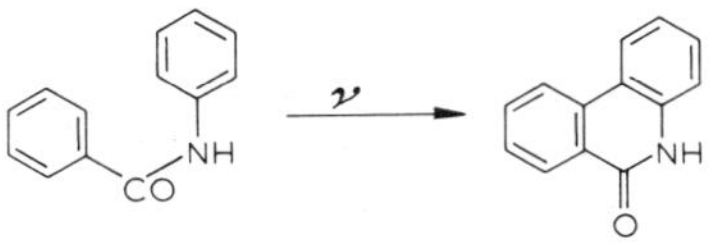

NH CO $\xrightarrow{\nu}$ NH O

Phenanthridone itself is most conveniently prepared by the cyclisation of biphenyl 2-isocyanate by aluminium chloride or preferably polyphosphoric acid (*E. C. Taylor Jr.* and *N. W. Kalenda*, J. Amer. chem. Soc., 1954, **76,** 1699). It is obtained in 85% yield when ethyl 2-ethoxycarbonylamino-biphenyl (XXIV) is irradiated with ultraviolet light in methanol (*N.C. Yang*, *A. Shani* and *G.R. Lenz*, J. Amer. chem. Soc., 1966, **88**, 5369).

N-Substituted phenanthridones are obtained by an extension of the Pschorr method (*R. A. Heacock* and *D. H. Hey*, J. chem. Soc., 1953, 3) by the decomposition of diazonium sulphates or tetrafluoroborates obtained from 2-aminobenzanilides in aqueous solution or in acetone with copper powder. N-*Methylphenanthridone* (XXVII, R = H), m.p. 109°, is thus obtained by the inner cyclisation of diazotised 2-amino-*N*-methylbenzanilide (XXV, R = H). Other products such as the substituted benzanilides XXVI, may be obtained from 2-amino-*N*-methylbenzanilides with *ortho*-substituents (with reference to the NMe group), which undergo both deamination and demethylation (*Hey* and *D. G. Turpin*, *ibid.*, 1954, 2471).

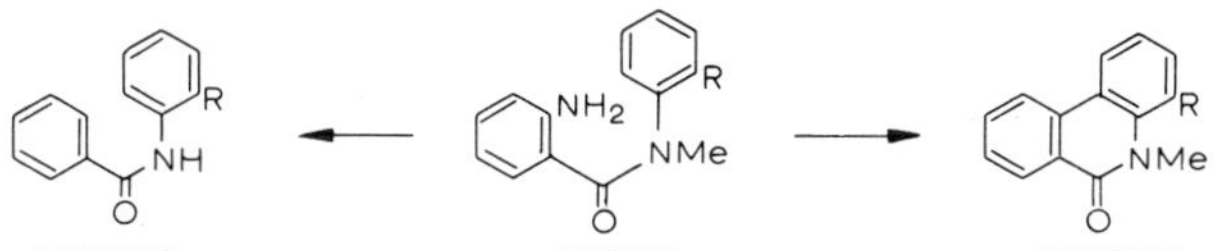

(XXVI) R = Me, Cl, NO_2, CO_2H ← (XXV) → (XXVII) R = H

Phenanthridone, 6(5H)-*phenanthridone*, m.p. 293°, occurs in coal tar (*O. Kruber*, Ber., 1939, **72,** 771). In solution it exists mainly in the lactam form XXVIII since the ultraviolet spectrum is the same as that of *N*-methylphenanthridone and is different from that of 5-methoxyphenanthridine (*C. B. Reese*, J. chem. Soc., 1958, 895). This is confirmed by infrared absorption measurements in $CHCl_3$ which show bands at 1669 cm^{-1} (C=O) and 3402 cm^{-1} (NH) (*Mason*, *ibid.*, 1957, 4874). Phenanthridones, however, react tautomerically, for although with alkali and methyl iodide or dimethyl sulphate phenanthridone forms only N-*methylphenanthridone*, m.p. 108.5°, 2-nitrophenanthridone with diethyl sulphate yields 27% of 6-ethoxy-2-nitrophenanthridine (XXX) as well as the *N*-ethyl derivative XXIX (*Hey et al.*, *ibid.*, 1961, 232);

$$\underset{\text{(XXIX)}}{-\text{NEt}-\text{CO}-} \longleftarrow \underset{\text{(XXVIII)}}{-\text{NH}-\text{CO}-} \longrightarrow \underset{\text{(XXX)}}{-\text{N}=\text{C(OEt)}-}$$

Phenanthridone is insoluble in most organic solvents, but is sufficiently acidic to dissolve in ethanolic potassium hydroxide. It is unaffected by potassium hydroxide at 300° and by most oxidising agents, but with alkaline potassium permanganate yields phthalic acid. It is reduced by lithium tetrahydridoaluminate to phenanthridone and with phosphorus pentasulphide yields **phenanthridinethione,** yellow needles, m.p. 281–283°, which is converted by Raney nickel (*Taylor* and *A. E. Martin*, J. Amer. chem. Soc., 1952, **74**, 6295) or Raney cobalt (*G.M. Badger*, *N. Kowanko* and *W.H.F. Sasse*, J. chem. Soc., 1959, 440) into phenanthridine.

The sodium salt of phenanthridone, obtained by the action of sodium hydride, with benzoyl chloride at −20° yields the *O*-benzoate, but at higher temperatures (*e.g.* that of boiling toluene) gives the *N*-benzoyl derivative, which is also obtained by heating the *O*-benzoate in methylene dichloride (*D. Y. Curtin* and *J. H. Engelmann*, Tetrahedron Letters, 1968, 3911):

−CO·NH− → −C(OCOPh)=N− ; −CO·NH− → −CO·N(COPh)− ; −C(OCOPh)=N− → −CO·N(COPh)−

Inspection of the formula of phenanthridone and comparison with that of benzanilide leads to the expectation that electrophilic substitution will occur mainly at $C_{(2)}$ and $C_{(4)}$, the benzene ring attacked being activated by the −NHCOR grouping whereas the other benzene nucleus is deactivated by the −CO·NHR′ fragment:

Phenanthridone

Benzanilide

Experiment shows that $C_{(2)}$ is indeed the most reactive site followed by $C_{(4)}$ (*H. Gilman* and *J. J. Eisch*, J. Amer. chem. Soc., 1957, **79**, 5479). Phenanthridone with concentrated nitric acid yields 2-*nitrophenanthridone*, m.p. 382–383° (83%) and 4-*nitrophenanthridone*, m.p. 264–265° (14%) (*Nunn*, *Schofield* and *Theobald*, J. chem. Soc.,

1952, 2797). Bromine in acetic acid gives 2-*bromophenanthridone*, m.p. 329° (91%) (*W. L. Mosby*, J. Amer. chem. Soc., 1954, **76**, 936), and chlorine in acetic acid with iron as a catalyst gives 2-*chlorophenanthridone*, m.p. 327–328° (71%). Potassium iodide and iodate in acetic acid yield 2-*iodophenanthridone*, m.p. 323–325° (88%) and concentrated sulphuric acid gives phenanthridone-2-sulphonic acid (98%).

N-Bromosuccinimide or *N*-chlorosuccinimide with phenanthridone in dimethylformamide yield 2-halogenophenanthridones, which on further treatment give 2,4-dihalogenophenanthridones (*H.-L. Pan* and *T. L. Fletcher*, J. heterocyclic Chem., 1970, **7**, 597). 2-*Chloro*-, m.p. 322–323°, 2,4-*dichloro*-, m.p. 270–272° and 2,4-*dibromophenanthridone*, m.p. 266–267°.

Nitrophenanthridones are obtained by the nitration of phenanthridone (see above) and by the oxidation of nitrophenanthridines by potassium permanganate and sulphuric acid (*C. L. Arcus* and *M. M. Coombs*, J. chem. Soc., 1954, 4319) or by synthesis (*R. A. Labriola* and *A. Felitte*, J. org. Chem., 1943, **8**, 536). 1-*Nitro*-, m.p. 325–327°, 3-*nitro*-, m.p. 357–358°, 8-*nitro*-, m.p. 329°, 9-*nitro*-, m.p. 368° and 10-*nitro-phenanthridone*, m.p. 316–318°. Oxidation of the nitrophenanthridones has already been mentioned (p. 46).

Aminophenanthridones can be obtained by synthesis, application of the Schmidt reaction to phenanthridonecarboxylic acids (*G. S. Chandler et al.*, Austral. J. Chem., 1967, **20**, 2037), or reduction of the nitrophenanthridones by hydrogen and an Adams' catalyst (*D. H. Hey*, *J. A. Leonard* and *C. W. Rees*, J. chem. Soc., 1963, 5251).

1-*Amino*-, m.p. 319–323°, 2-*amino*-, m.p. 280°, 3-*amino*-, m.p. 312°, 8-*amino*-, m.p. 319–320° and 10-*amino-phenanthridone*, m.p. 330–334°.

(d) Hydrogenated phenanthridines

Reduction of phenanthridine with lithium tetrahydridoaluminate gives 5,6-**dihydrophenanthridine,** m.p. 125°, *acetyl* deriv., m.p. 128°; *benzoyl* deriv., m.p. 90.5–92.5° (*W. C. Wooten* and *R. L. McKee*, J. Amer. chem. Soc., 1949, **71**, 2946). It is also obtained by the reduction of phenanthridone by the same reagent (*P. de Mayo* and *W. Rigby*, Nature, 1950, **166**, 1075). It is basic, exhibits a strong blue fluorescence in the solid state or in solution, and is readily dehydrogenated to phenanthridine by chloranil or picric acid. The *picrate*, m.p. 224–225°, orange-red crystals, reverts to phenanthridine when recrystallised and dehydrogenation of the dihydro compound is also effected with triphenylmethyl perchlorate to yield *phenanthridinium perchlorate* (91%), m.p. 223–225° (decomp.) (*W. Bonthrone* and *D. H. Reid*, J. chem. Soc., 1959, 2772).

With lithium in tetrahydrofuran phenanthridine gives 5,6-dihydrophenanthridine and bimolecular reduction products (*Eisch* and *R. M. Thomson*, J. org. Chem., 1962, **27**, 4171).

Phenanthridinium salts are reduced by lithium tetrahydridoaluminate or hydrogen and a platinum catalyst to 5,6-dihydro products, *N*-methylphenanthridinium bromide, for example, giving N-*methyl*-5,6-*dihydrophenanthridine* (XXXI), m.p. 48°. With dimethyl sulphate this is converted into the methosulphate (XXXII), which with Raney nickel undergoes a modified Emde ring-opening to give 2-dimethylamino-2′-methylbiphenyl (XXXIII) (*S. Sugasawa* and *H. Matsus*, Chem. and pharm. Bull., Tokyo, 1958, **6,** 601; C.A., 1960, **54,** 14162):

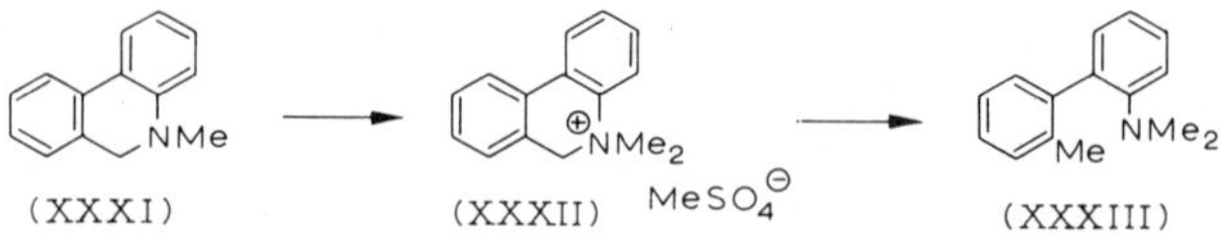

(XXXI) (XXXII) (XXXIII)

More highly hydrogenated phenanthridines are prepared by synthesis as well as by hydrogenation of phenanthridine or of partially hydrogenated phenanthridines. Hydrogenation of phenanthridine at high temperatures and pressures in the presence of alkali metals yields a mixture of octahydrophenanthridines (*S. Friedman, M. L. Kaufman* and *I. Wender*, *ibid*., 1971, **36,** 694). The synthetic approach is illustrated by the cyclisation of *trans*-formylamino-2-phenylcyclohexane with phosphorus pentoxide or phsophoric acid to give trans-1,2,3,4,4a,10b-*hexahydrophenanthridine* (XXXIV), m.p. 52–53°, *picrate*, m.p. 192–193°, which loses hydrogen to give 1,2,3,4-*tetrahydrophenanthridine* (XXXV), an oil, *picrate*, m.p. 192–193°, λ_{max} 260 nm and with hydrogen and a platinum catalyst yields trans-1,2,3,4,4a,5,6,10b-*octahydrophenanthridine* (XLII), m.p. 89.5–90.5°, *picrate*, m.p. 178–179° (*T. Masamune et al.*, *ibid*., 1964, **29,** 1419; *G. van Binst* and *D. Tourwe*, J. heterocyclic Chem., 1972, **9,** 895):

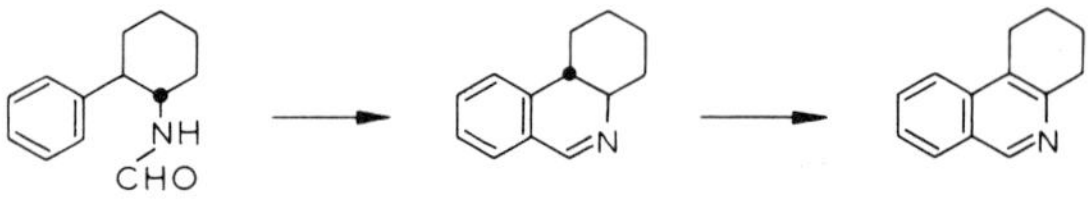

Formylamino-2-phenylcyclohexane (XXXIV) (XXXV)

cis-1,2,3,4,4a,10b-*Hexahydrophenanthridine*, m.p. 52.5–54°, *hydrochloride*, m.p. 210–212°, and cis-1,2,3,4,4a,5,6,10b-*octahydrophenanthridine* (XLIII), b.p. 141–145°/17 mm, *picrate*, m.p. 165–166°, have been prepared similarly.

7,8,9,10-**Tetrahydrophenanthridine** (XXXIX), m.p. 63°, *picrate*, m.p. 212–214° (decomp.), is prepared by condensing 2-hydroxymethylcyclohexanone with aniline and aniline hydrochloride in ethanol (*B. L. Hollingsworth* and *V. Petrov*, J. chem. Soc., 1948, 1537) or by reduction of 7,8,9,10-tetrahydrophenanthridone (*T. Masamune et al.*, J. org. Chem., 1964, **29,** 681). In contrast to 1,2,3,4-tetrahydrophenanthridine it exhibits an absorption minimum at 260 nm in the ultraviolet spectrum and may be regarded as a 3,4-dialkylquinoline. Consequently it is not surprising that it undergoes nitration to yield 1-*nitro*-, m.p. 160–161°, and 4-*nitro*-7,8,9,10-*tetrahydrophenanthridine*, m.p. 142–143° (*E. Hayashi* and *R. Goto*, C.A., 1965, **63,** 9913).

5,6,6a,7,8,9,10,10a-**Octahydrophenanthridines** (*e.g.* XL) have been prepared by reduction of the appropriate hexahydrophenanthridones (*Masamune et al., loc. cit.*): *cis*- XL, m.p. 41–42°, *picrate*, m.p. 173–174°; *trans*- XL, m.p. 69–70°, *picrate*, m.p. 171–172°, which is also obtained by the reduction of phenanthridone with sodium and ethanol (*O. Kruber*, Ber., 1939, **72,** 771) or by synthesis (*E. A. Braude* and *J. S. Fawcett*, J. chem. Soc., 1951, 3113). 5,6,6a,7,8,10a-*Hexahydrophenanthridine* (XXXVI), m.p. 104°, has been prepared synthetically *(idem, loc. cit.)*:

(XXXVI)

The conformations of the octahydrophenanthridines have been established (*Masamune, S. Ohuchi* and *S. Shimokawa*, Tetrahedron, 1966, **22,** 773).

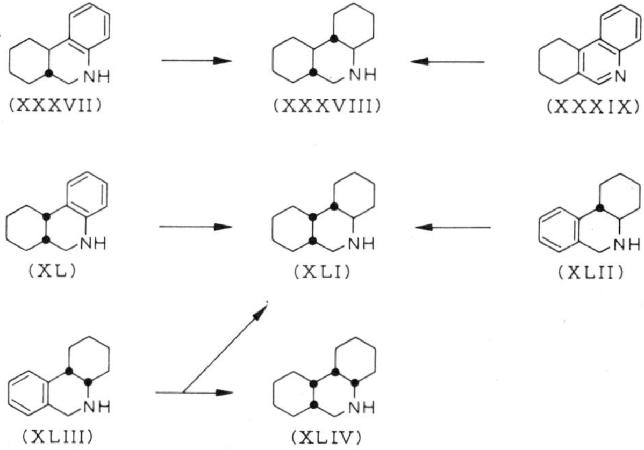

Formation of perhydrophenanthridines

Hydrogenation of the octahydrophenanthridines gives the perhydrophenanthridines (*Masamune et al.*, Bull. chem. Soc., Japan, 1968, **41,** 2458). *trans*-5,6,6a,7,8,9,10,10a-Octahydrophenanthridine (XXXVII) in acetic acid with hydrogen and a platinum catalyst yields α-**perhydrophenanthridine** (XXXVIII), m.p. 103–104°, *picrate*, m.p. 231–233°, with the *trans, anti-trans*-configuration, which is also obtained by hydrogenation of 7,8,9,10-tetrahydrophenanthridine (XXXIX) in the presence of nickel.

β-**Perhydrophenanthridine** (XLI), m.p. 122–123°, *picrate*, m.p. 223–225°, with the *cis-syn,trans*-configuration, is prepared by the hydrogenation of *trans*-1,2,3,4,4a,5,6,10b-octahydrophenanthridine (XLII) or *cis*-5,6,6a,7,8,9,10,10a-octahydrophenanthridine (XL) with a platinum catalyst.

γ-**Perhydrophenanthridine** (XLIV), m.p. 37–38°, *picrate*, m.p. 196°, *cis-syn,cis*-configuration, is obtained along with the β-compound (XLI) by the hydrogenation of *cis*-1,2,3,4,4a,5,6,10b-octahydrophenanthridine (XLIII) in acetic acid and hydrochloric acid with a platinum catalyst. The configurational change necessary to convert XLIII into the β-compound is probably effected by the hydrochloric acid present.

3. Benzo[*f*], [*g*] and [*h*]quinolines

Of the five benzoquinolines, two, acridine and phenanthridine with the additional benzene ring annealed to the pyridine nucleus have already been discussed. The other isomers in which the "extra" benzene ring is annealed to the carbocyclic ring of the quinoline molecule are 5,6-, 6,7- and 7,8-benzoquinoline, or to use the alternative "*a*" nomenclature, used in Chemical Abstracts, benzo[*f*]-, benzo[*g*]- and benzo[*h*]-quinoline, respectively. The latter formulation will be used in this text.

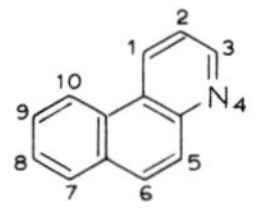

Benzo[*f*]quinoline [5,6] 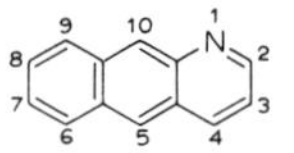 Benzo[*g*]quinoline [6,7] 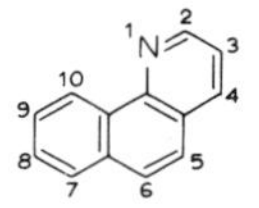 Benzo[*h*]quinoline [7,8]

(a) Structure and synthesis

The structure of the benzoquinolines often follows unequivocally from their synthesis by methods such as those of Skraup or von Miller. 1-Naphthylamine, for example, gives benzo[*h*]quinoline. Care must be exercised, however, when 2-naphthylamine or a substituted 2-naphthylamine is used since ring-formation may yield either a *linear* or an *angular* product. The Skraup reaction usually results in ring-closure at the 1-position to give an *angular* product, benzo[*f*]quinoline (I) being thus obtained from 2-naphthylamine (*C. A. Knueppel*, Ber., 1896, **29,** 703). Even when the α-position is occupied by substituents such as the bromo- or nitro-group the same angular product is obtained with simultaneous expulsion of the substituent (*E. Lellmann* and *O. Schmidt*, *ibid*., 1887, **20,** 3154). On the other hand, 1-methyl-2-naphthylamine yields the linear 10-methylbenzo[*g*]quinoline and 1-chloro-2-naphthylamine gives not only the angular benzo[*f*]quinoline (I) by elimination of chlorine but also, in equal quantity, the linear 10-chlorobenzo[*g*]quinoline [II] (*G.R. Clemo* and *G.W. Driver*, J. chem. Soc., 1945, 829; *F. H. Gerhardt* and *C. S. Hamilton*, J. Amer. chem. Soc., 1944, **66,** 479):

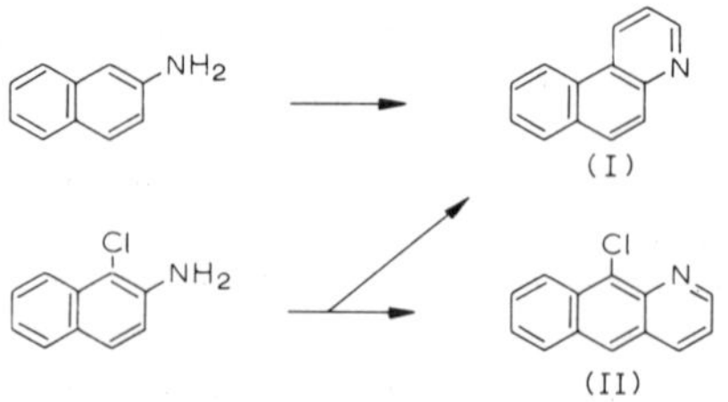

(I)

(II)

The Döbner–Miller synthesis likewise yields angular products (*W. Dilthey* and *A. Kaiser*, Ann., 1949, **563**, 11). This applies to the Knorr and Conrad–Limpach syntheses, but if the 1-position in the naphthylamine is blocked linear products result without difficulty. 2-Acetoacetnaphthalide, for instance, with concentrated hydrochloric acid yields the angular compound III, while 1-methyl-2-acetoacetnaphthalide with concentrated sulphuric acid gives the linear compound IV:

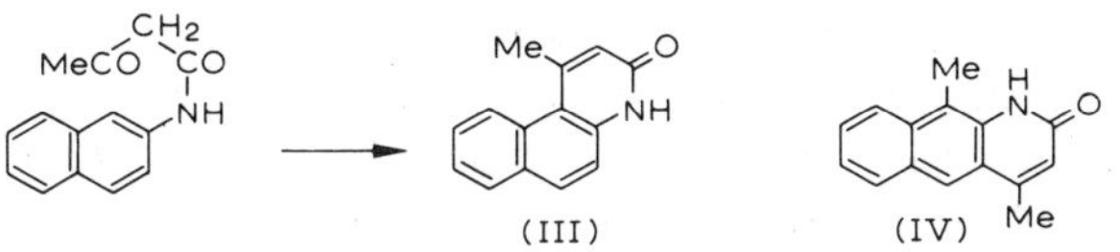

(III) (IV)

Both linear and angular quinolines have been obtained by the Combes synthesis. 2-Naphthylamine and acetylacetone condense to form the naphthylimino ketone V, which with hydrofluoric acid gives a 96% yield of linear 2,4-dimethylbenzo[*g*]quinoline (VI) (*W. S. Johnson, E. Woroch* and *F. J. Mathews*, J. Amer. chem. Soc., 1947, **69,** 566; *Clemo* and *N. Legg*, J. chem. Soc., 1947, 545):

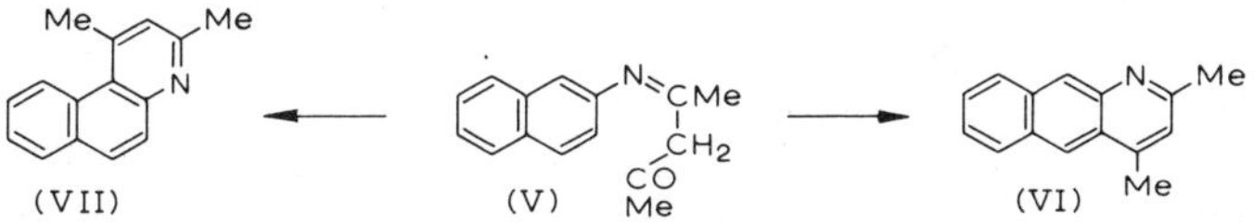

(VII) (V) (VI)

The naphthylimino compound V, however, when heated with 2-naphthylamine hydrochloride in ethanol gives angular 1,3-dimethylbenzo[*f*]quinoline (VII). Dibenzoylmethane and 2-naphthylamine in sulphuric acid give a mixture of linear (VIII) and angular (IX), diphenylbenzoquinolines, in the ratio 10:1 (*R. Huisgen*, Ann., 1949, **564**, 16):

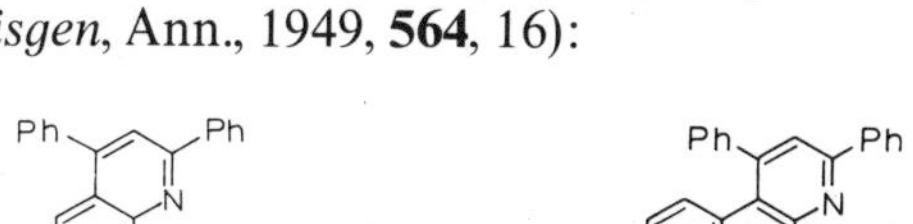

(VIII) (IX)

Irradiation of the styrylpyridines yields benzoquinolines and benzoisoquinolines (*C. E. Loader, M. V. Sargent* and *C. J. Timmons*, J. chem. Soc., C, 1966, 1078; *P. Bortolus, G. Cauzzo* and *G. G. Galiazzo*, Tetrahedron Letters, 1966, 3717; *P. L. Kummler* and *R. A. Dybas*, J. org. Chem., 1970, **35,** 125). Thus *cis*-3-styrylpyridine (X) yields benzo[*f*]isoquinoline (XI) and benzo-[*h*]quinoline (XII) in the ratio of 4:1 (p. 68):

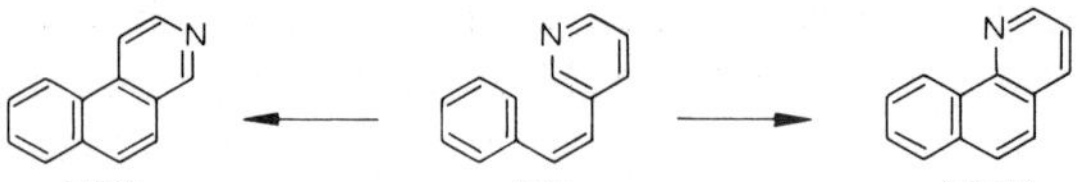

(XI) (X) (XII)

Benzo[*f*]quinoline is found among the products of pyrolysis of 2-phenylethylpyridine (*J. W. Sweeting* and *J. F. K. Wilshire*, Austral. J. Chem., 1962, **15**, 800).

Ultraviolet spectra differentiate between the *angular* and *linear* benzoquinolines, whose spectra strikingly resemble those of anthracene and phenanthrene, respectively (*G. M. Badger*, *R. S. Pearce* and *R. Pettit*, J. chem. Soc., 1951, 3199). Benzo[*h*]quinoline and benzo[*h*]isoquinoline have the same general absorption curve, but with differences sufficient to distinguish between the two compounds (*Bortolus*, *Cauzzo* and *Galiazzo*, *loc. cit.*).

P.m.r. spectra are of service in structural studies (*R. H. Martin et al.*, Tetrahedron, 1964, **20**, 1495; Tetrahedron Supplement, 1966, **8**, 181; *W. W. Paudler* and *T. J. Krees*, J. org. Chem., 1967, **32**, 2616). Protons *ortho* to the nitrogen atom are strongly deshielded; protons in the *para* or *peri* positions to a lesser extent. In the angular compounds the protons at $C_{(10)}$ are also strongly deshielded. In benzo[*h*]quinoline, for example, the order of deshielding is $H_{(10)} > H_{(2)} > H_{(4)} > H_{(3)}$.

Oxidation is frequently used to establish the structures of the benzoquinolines. Benzo[*f*]quinoline, for instance, is oxidised by potassium permanganate to the dibasic acid XIII, which on decarboxylation forms 3-phenylpyridine (XIV). Further oxidation gives nicotinic acid (XV):

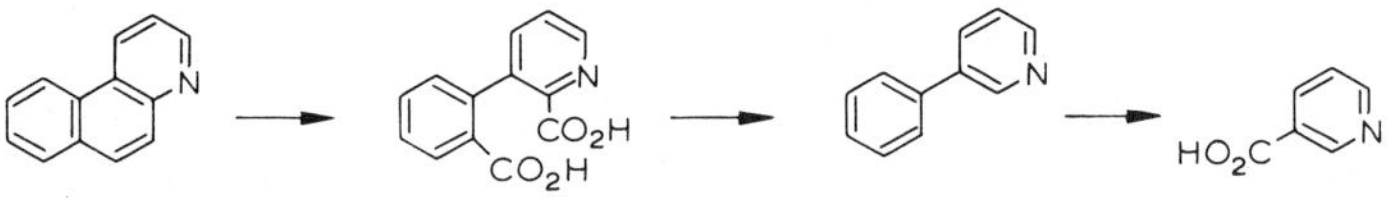

CO_2H CO_2H HO_2C

(I) (XIII) (XIV) (XV)

(b) Properties

The benzoquinolines behave very much as their formulae would lead us to expect. Benzo[*g*]quinoline and its derivatives, for example, are oxidised to the corresponding aza-anthraquinones (*e.g.* 1-aza-anthraquinone, XVI), the constitution of which follows from its synthesis from benzene and quinolinic acid (*J. von Braun* and *H. Gruber*, Ber., 1922, **55**, 1710). Reduction results in hydrogenation of the pyridine ring, and benzo[*g*]quinoline is reduced, for example, by tin and hydrochloric acid to 1,2,3,4-tetrahydrobenzo[*g*]quinoline (XVII) (*idem*, *loc. cit.*) (see, however, p. 65). The same product is obtained by the hydrogenation of 2,4-dichlorobenzo[*g*]quinoline with a palladium–carbon catalyst (*K. Schofield* and *D. E. Wright*, J. chem. Soc., 1965, 6074). The other benzoquinolines with tin and hydrochloric acid also give 1,2,3,4-tetrahydro compounds:

(XVII) ← Benzo[*g*]quinoline → (XVI)

Reduction is sometimes accompanied by double compound formation and consequent confusion. Benzo[*f*]quinoline when reduced with tin and hydrochloric acid yields a *compound*, m.p. 93–93.5°, which was regarded as 1,2,3,4-tetrahydrobenzo[*f*]quinoline (*E. Bamberger*, Ber., 1891, **24,** 1648), but is in fact a double compound of the benzoquinoline and 1,2,3,4-tetrahydrobenzo[*f*]quinoline (*T. Masamune* and *M. Koshi*, Bull. chem. Soc. Japan, 1957, **30,** 307). 5,6-Dihydrobenzo[*f*]quinoline and 1,2,3,4-tetrahydrobenzo[*f*]quinoline likewise form a molecular *compound*, m.p. 88–90°. Such double compound formation is commonly encountered in the heterocyclic series (see, *e.g.* p. 76).

The benzoquinolines form "quaternary" salts with alkyl halides, yield *N*-oxides, and salts which may be used to distinguish between the *linear* and *angular* compounds. The linear benzoquinolines form bright yellow hydrochlorides in contrast to the colourless hydrochlorides of the angular compounds. Fluorescence is also useful. 1,3-Diphenylbenzo[*f*]quinoline (IX), for instance, is colourless, forms colourless salts, and has a moderately strong blue-violet fluorescence, while the yellow 2,4-diphenylbenzo[*g*]quinoline (VIII) forms orange-red salts and exhibits a very intense light blue fluorescence *(Huisgen, loc. cit.)*. Finally it may be noted that the linear benzoquinolines are readily sulphonated, yield quinones when oxidised with chromic acid, and form Diels–Alder adducts. In contrast the angular compounds are sulphonated only under drastic conditions, do not yield quinones on oxidation and do not form Diels–Alder adducts with maleic anhydride, etc.

Benzo[*f*]quinoline (I), 5,6-*benzoquinoline*, m.p. 94°, b.p. 350°/721 mm, *methiodide*, m.p. 200–205°, *picrate*, m.p. 259°, is prepared from 2-naphthylamine by the Skraup reaction (*P. M. G. Bavin*, Canad. J. Chem., 1960, **38,** 1227). It occurs in the high-boiling bases of coal-tar (*O. Kruber* and *R. Oberkobusch*, Ber., 1953, **86,** 309).

In this series oxidative ring-contraction is observed, 3-phenylbenzo[*f*]quinoline-1-carboxylic acid (XVIII), for example, with potassium permanganate giving 2-phenyl-1-

azafluorenone-4-carboxylic acid (XIX) (*R. Ciusa* and *D. Buono*, Gazz., 1950, **80,** 864). Presumably a 5,6-quinone is first formed and undergoes a benzilic acid type of rearrangement:

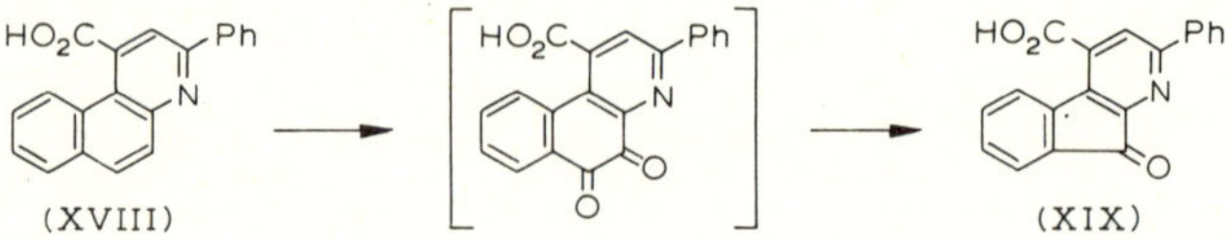

HO_2C Ph N (XVIII) → [HO_2C Ph N O O] → HO_2C Ph N O (XIX)

Reduction of benzo[*f*]quinoline generally yields mixtures of products, all of which contain a reduced pyridine ring with the exception of 5,6-dihydrobenzo[*f*]quinoline, *picrate*, m.p. 205–207° (*Masamune* and *Koshi*, *loc. cit.*; *Masamune et al.*, J. org. Chem., 1964, **29,** 1419). Hydrogenation with a Raney nickel catalyst at 100° gives 1,2,3,4-*tetrahydrobenzo*[f]*quinoline*, b.p. 200–202°/14 mm, m.p. 93.5°, *picrate*, m.p. 167–168°; N-*nitroso* deriv., m.p. 104.5°. At 150° the main product is 1,2,3,4,7,8,9,10-*octahydrobenzo*[f]*quinoline* (XX), m.p. 64–65°, *picrate*, m.p. 150–152°, and *hydrochloride*, m.p. 220–222°; *benzoyl* deriv., m.p. 144–146°, along with some cis-1,2,3,4,4a,5,6,10b-*octahydrobenzo*[f]*quinoline* (XXIa), b.p. 160–165°/5 mm, *picrate*, m.p. 197–198.5°, *benzoate*, m.p. 96–97°, *hydrochloride*, m.p. 258–259°. trans-1,2,3,4,4a,5,6,10b-*Octahydrobenzo*[f]-*quinoline* (XXIb), m.p. 88–89°, *picrate*, m.p. 189–190°, *benzoate*, m.p. 100–102°, *hydrochloride*, m.p. 251–252°, is obtained as a minor product when the benzoquinoline is reduced by sodium and amyl alcohol. Both octahydro compounds yield carbonates when exposed to air and may thus be separated since the *cis*-isomer forms the carbonate more readily.

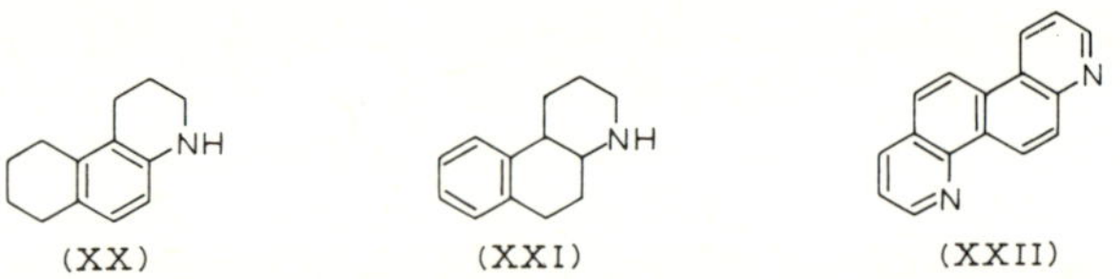

NH (XX) NH (XXI) N N (XXII)

a *cis*, b *trans*

Nitration gives 7-*nitrobenzo*[f]*quinoline*, m.p. 174°, the constitution of which is established by reduction to the amine and subsequent conversion into 1,10-diazachrysene (XXII) (*W. J. Clem* and *C. S. Hamilton*, J. Amer. chem. Soc., 1940, **62,** 2349). Further nitration yields 7,9-*dinitrobenzo*[f]*quinoline*, m.p. 250°.

Benzo[*g*]quinoline, 6,7-*benzoquinoline*, 1-*aza-anthracene*, b.p. 200–203°/14 mm, m.p. 117–118°, *picrate*, m.p. 247–248° (258° is also reported), *methiodide*, m.p. 225–226°, is best prepared from 9-*chlorobenzo*[g]*quinoline* (see above), m.p. 140–141°, by oxidation to 1-*aza-anthraquinone* (XVI), yellow needles, m.p. 282–284°, ν_{max} 1693 and 1675 cm^{-1} (CO at $C_{(5)}$ and $C_{(10)}$, respectively), and then converting this into the benzoquinoline by Clar's zinc dust fusion method (*A. Etienne*, Ann. Chim., 1946, [xii], **1,** 1). The N-*oxide*, m.p. 128–130°, is obtained by oxidation of the benzoquinoline with *tert*-amyl hydroperoxide and molybdenem pentachloride (*G. A. Tolstikov et al.*, Tetrahedron Letters, 1971, 2807).

Other properties of the benzoquinoline have already been noted. 2,4-*Dimethylbenzo*[g]*quinoline* (VI), m.p. 92°, containd reactive methyl groups (*A. B. Lal* and *N. Singh*, Ber., 1965, **98,** 2427). It is perhaps surprising that it condenses with benzaldehyde to yield 4-styryl-2-methylbenzo[*g*]quinoline (XXIV), since in 2,4-dimethylquinoline it is the 2-methyl group which is attacked (*C. E. Kaslow* and *R. D. Slayner*, J. Amer. chem. Soc., 1945, **67,** 1716):

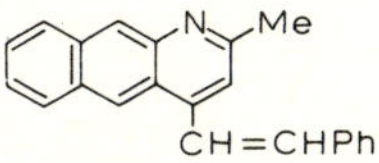

N Me
CH=CHPh
(XXIV)

1,2,3,4-*Tetrahydrobenzo*[g]*quinoline*, m.p. 149°, N-*nitroso* deriv., m.p. 129°.

A product obtained by the dehydrogenation of the alkaloid selagine with selenium is thought to be 6,8-dimethylbenzo[*g*]quinoline, *picrate*, m.p. 195–198° (*Z. Valenta et al.*, Tetrahedron Letters, 1960, **10,** 26).

Phomazarin, an orange pigment, m.p. 196°, obtained from *Phoma terrestris* Hansen, the fungus responsible for the "pinkroot disease" of onions has the formula XXV (*A. J. Birch, D. N. Butler* and *R. W. Rickards, ibid.*, 1964, 1853):

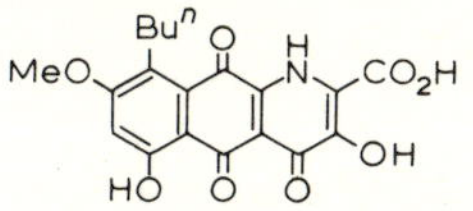

Bu^n O H
MeO N CO_2H
OH
HO O O
(XXV)

Chemical and spectroscopic evidence points to a quinonoid structure and its stability to alkali suggests a *para*-quinone formula.

Benzo[*h*]**quinoline**, m.p. 52°, *picrate*, m.p. 191–192°, *hydrochloride*, m.p. 213°, occurs in anthracene oil and is prepared from 1-naphthylamine by the Skraup reaction (*A. Pinenas* and *J. Zdanavicius*, C.A., 1964, **61,** 6988). It fails to react with methyl iodide even after being heated at 100° in nitrobenzene for 48 hours. The proton at $C_{(10)}$ is strongly deshielded (*cf.* the proton at $C_{(1)}$ in benzo[*c*]acridine, p. 77).

Hydrogenation of benzo[*h*]quinoline in trifluoroacetic acid with a platinum oxide catalyst gives 5,6,6a,7,8,9,10,10a-octahydrobenzo[*h*]quinoline (XXVI) (62.5%) and 1,2,3,4,7,8,9,10-isomer (XXVII) (18%), *picrate*, m.p. 161–162° (*F. W. Vierhapper* and *E. E. Eliel*, J. org. Chem., 1975, **40,** 2734). It is noteworthy that hydrogenation occurs mainly on the aromatic rings thus leaving the pyridine ring intact. Reduction of XXVI with sodium and ethanol yields trans,anti,trans- (30%), m.p. 41–42°, trans,syn,cis- (52%), m.p. 76°, and trans,anti,cis-*perhydrobenzo*[h]*quinoline* (18%) (*idem, ibid.*, 2734):

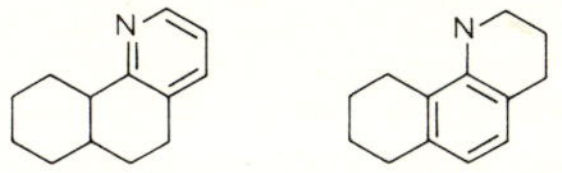

N N
(XXVI) (XXVII)

4. Benzisoquinolines*

The three isomers are:

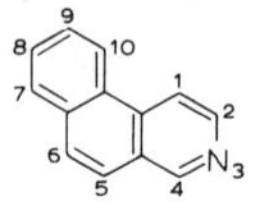

Benz[*f*]isoquinoline	Benz[*g*]isoquinoline	Benz[*h*]isoquinoline
or	or	or
5,6-Benzoisoquinoline	6,7-Benzoisoquinoline	7,8-Benzoisoquinoline
or	or	or
2-Azaphenanthrene	2-Azaanthracene	3-Azaphenanthrene
m.p. 97-98°	m.p. 168-170°	m.p. 53-54°

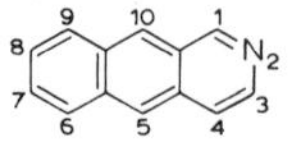

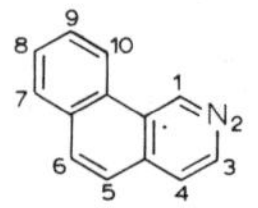

Comparatively little is known about these ring systems, although recently with the discovery of some biological activity in benz[*g*]isoquinolines, some more interest has been awakened. The subject was reviewed in 1969 (*N. S. Prostakov*, Russian chem. Reviews, 1969, **38**, 774).

(*a*) *Benz*[f]*isoquinoline*

Early attempts to utilise the usual isoquinoline ring syntheses, such as the Bischler–Napieralski and the Pomeranz–Fritsch synthesis using α- and β-naphthaldehydes failed (*G. Coppens*, Bull. Soc. chim. Belg., 1960, **69**, 413), although *W. R. Schleigh* found that Bischler–Napieralski cyclisations could be achieved (J. heterocyclic Chem., 1970, **7**, 1157) using phosphorus pentoxide in tetralin:

R^1	R^2	%
H	H	21
H	Me	56
Me	H	44
Me	Me	30

The activated 2-(6′-methoxy-1-naphthyl)ethylamide (I) has been cyclised (*N. A. Nelson* and *R. S. P. Hsi*, J. org. Chem., 1961, **26**, 3086) to the 8-methoxybenz[*f*]isoquinoline derivative II and *S. V. Kessor et al.* (J. chem. Soc., C, 1971, 266) have found that polyphosphoric acid is the reagent of choice III → IV.

* This section has been contributed by Dr. S. F. Dyke.

(I) → (II)

R = Me, CO_2Et

(III) —PPA→ (IV)

R^1	R^2	%
H	Me	90
H	H	94
OMe	H	42
H	Ph	86

An attempt to employ the Pictet–Gams reaction with (V, R = NHCOMe) failed (*C. F. Koelsch* and *R. M. Lindquist*, J. org. Chem., 1956, **21,** 657). These workers were also unable to repeat the previously reported cyclisation of V (R = H) (*A. Pictet* and *B. Manevitch*, Arch. Sci. phys. nat., 1913, **35,** 46).

A useful recent synthesis of benz[*f*]isoquinoline, which has not been exploited, involves thermal cyclisation of the azide (VI) to the amide (VII) (*F. Eloy* and *A. Deryckere*, J. heterocyclic Chem., 1970, **7,** 1191; Chim. ther., 1971, **6,** 48; C.A., 1972, **75,** 5664):

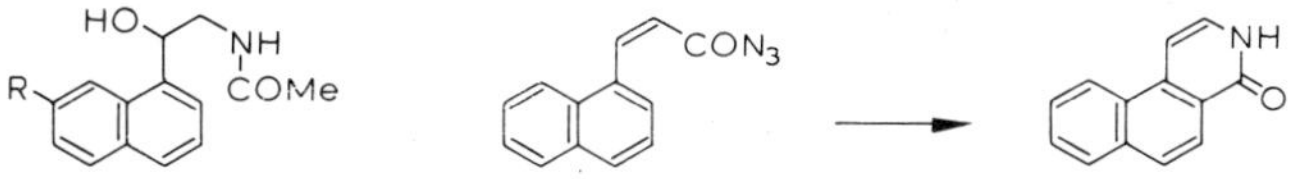

(V) (VI) → (VII)

Pyrylium salts (*A. T. Balaban, W. Schroth* and *G. Fischer*, Adv. in heterocyclic Chem., 1969, **10,** 241) have also been used (*G. I. Zhungietu* and *E. M. Perepelitsa*, Zhur. obshcheĭ Khim., 1966, **36,** 1858; C.A., 1967, **66,** 5563) in the preparation of benz[*f*]isoquinoline derivatives VIII → XII:

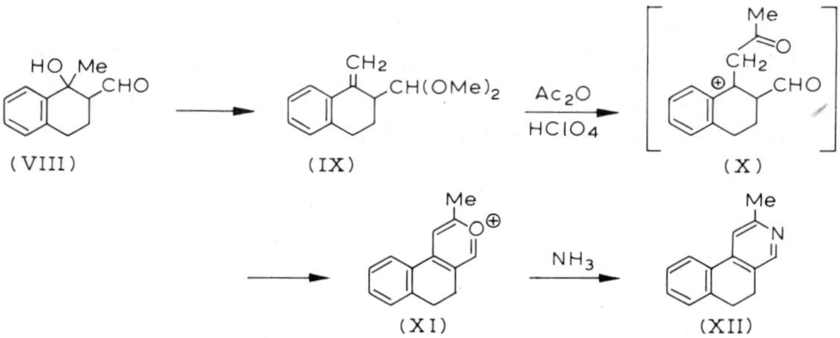

(VIII) → (IX) —Ac_2O, $HClO_4$→ [(X)] → (XI) —NH_3→ (XII)

An alternative approach to the use of naphthalene derivatives is illustrated by the application of the Pschorr cyclisation when XIV was obtained in 50% yield from XIII (*W. Herz* and *D. R. K. Murty*, J. org. Chem., 1961, **26,** 418):

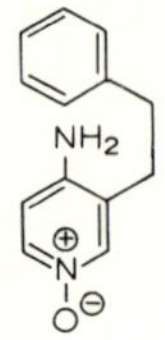

(XIII) 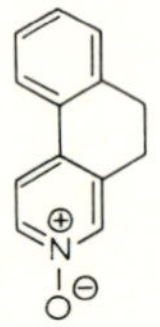 (XIV)

This product was easily reduced, then dehydrogenated to benz[*f*]isoquinoline. Photolysis of 3-styrylpyridines XV proceeds in good yield to the benz[*f*]isoquinolines XVI (*C. F. Loader* and *C. T. Timmons*, J. chem. Soc., C, 1966, 1078). The isomeric compound XVII was not reported by these workers, although later workers (*G. Galiazzo, P. Bortolus* and *G. Cauzzo*, Tetrahedron Letters, 1966, 3171) found a ratio of XVI:XVII of 7:1:

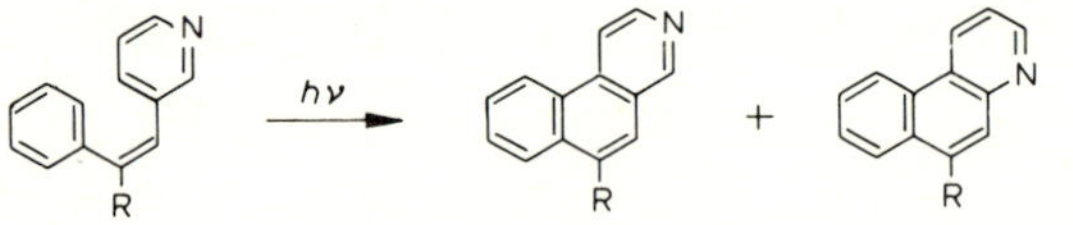

(XV) (XVI) (XVII)

R	%
H	66
CN	41
Me	57

(*b*) *Benz*[g]*isoquinoline*

One of the first methods of preparation involved the pyrolysis of a mixture of 2-methylbenzyl bromide and pyridine (*J. von Braun* and *J. Nelles*, Ber., 1937, **70,** 1760; *A. Etienne*, Ann. Chim., 1946, **1,** 1). Presumably the 4-benzylpyridine XVIII is formed, which then cyclises to benz[*g*]isoquinoline. A more useful preparation from pyridine derivatives provides a route to the 2-aza analogue of anthraquinone (*A. Philips*, Ber., 1895, **28,** 1658). A Friedel–Crafts reaction between cinchomeronic acid and benzene gives 4-benzoylpyridine-3-carboxylic acid (XIX), which can be cyclised with sulphuric acid to benz[*g*]isoquinoline (XX):

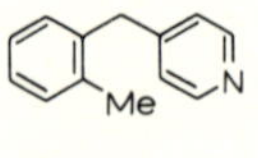

(XVIII) 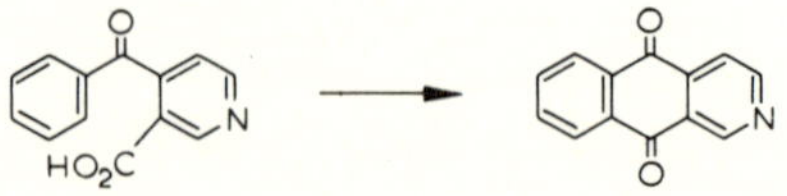 (XIX) (XX)

In another pyrolytic method (*N. S. Prostakov* and *V. V. Dorogov*, Khim. geterotsiki Soedin., 1971, **7**, 373; C.A., 1973, **76**, 25076) the tetrahydropyridine derivative XXI is heated at 460° when de-*N*-methylation and dehydrogenation occurs to give XXII as the major product (48–80%) with small amounts (2–5%) of the benz[*g*]isoquinoline XXIII produced as a by-product:

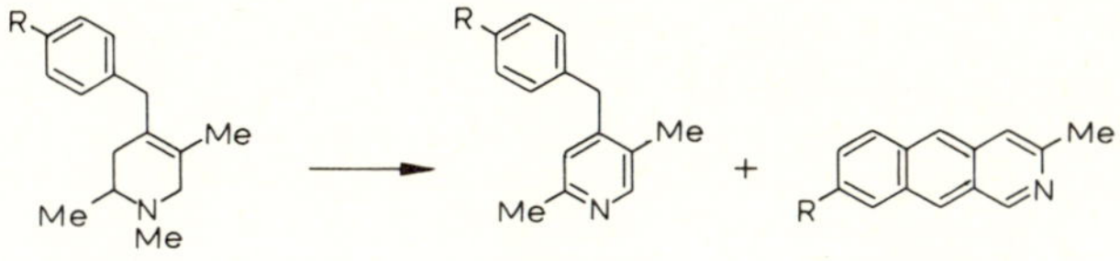

(XXI) (XXII) (XXIII)

R = Me, Et, Pr^i

The Bischler–Napieralski reaction has been successfully used for the preparation of benz[*g*]isoquinoline derivatives (*K. Kindler*, *W. Peschke* and *G. Pluddemann*, Arch. Pharm., 1939, **227**, 25; *J. Robert*, Ann. Chim. Fr., 1952, **7**, 722). The Pictet–Gams reaction was used to provide the compound XXV from the 1-naphthylethylamine derivative XXIV (*T. R. Govindachari* and *B. R. Pai*, J. org. Chem., 1953, **18**, 1253):

(XXIV) (XXV)

Benz[*g*]isoquinolines are high melting solids which fluoresce in some solvents; the u.v. spectra of derivatives are very similar to those of anthracene compounds. Benz[*g*]isoquinoline itself, pale yellow needles, m.p. 168–170°, forms an adduct with maleimide (*Robert*, *loc. cit.*) and also like anthracene, it undergoes some photochemical reactions. Thus, when a solution of the base in carbon disulphide is exposed to sunlight, an unstable photo-oxide is formed, which is easily oxidised by air to 2-*aza-anthraquinone*, m.p. 265–266°:

(*c*) *Benz*[h]*isoquinoline*

The parent compound (*picrate*, m.p. 229–231°) was synthesised in poor yield by the method outlined in Scheme 1 (*Koelsch* and *Lindquist*, *loc. cit.*). An alternative, equally unattractive method (*Coppens*, *loc. cit.*), is summarised in Scheme 2:

Scheme 1

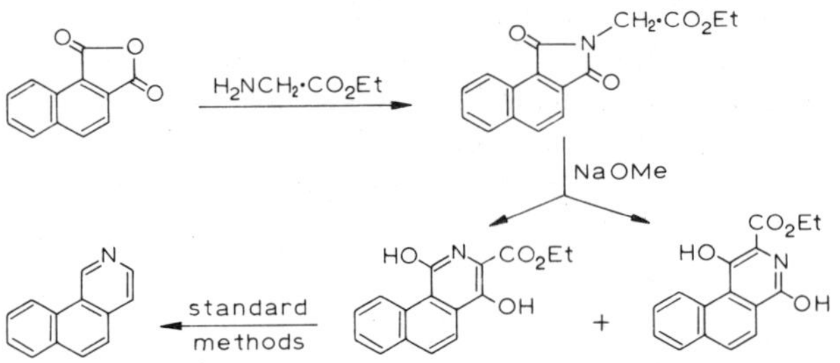
CH₂·CO₂Et
H₂NCH₂·CO₂Et
NaOMe
HO
CO₂Et
OH
CO₂Et
HO
OH
standard
methods
+

Scheme 2

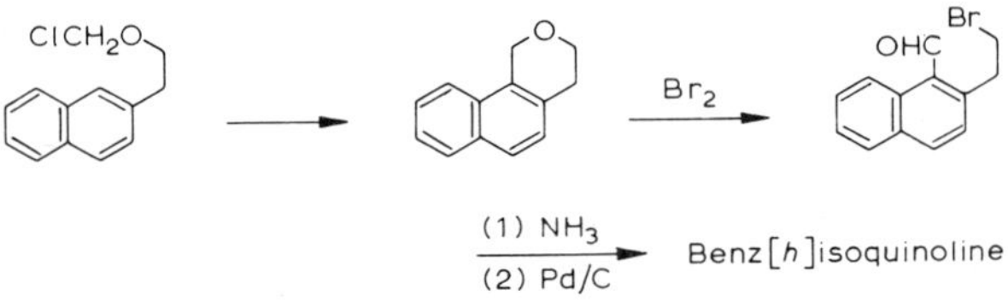
ClCH₂O
Br₂
Br
OHC
(1) NH₃
(2) Pd/C
Benz[h]isoquinoline

A third synthesis of benz[*h*]isoquinoline (*J. N. Chatterjea* and *K. Prasad*, J. Indian chem. Soc., 1960, **37,** 357) (Scheme 3) also starts from an inaccessible naphthalene derivative:

Scheme 3

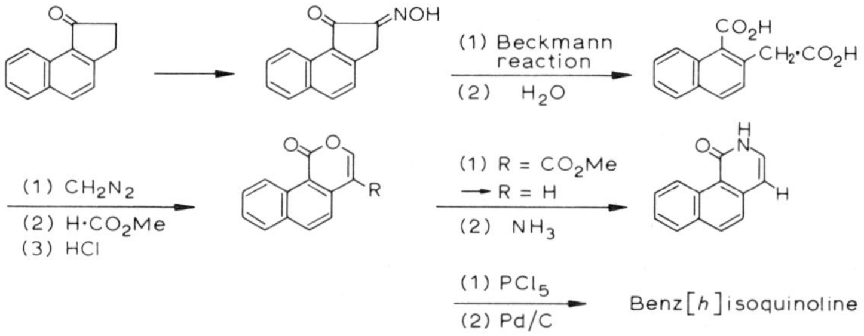
NOH
(1) Beckmann
reaction
(2) H₂O
CO₂H
CH₂·CO₂H
(1) CH₂N₂
(2) H·CO₂Me
(3) HCl
R
(1) R = CO₂Me
→ R = H
(2) NH₃
H
(1) PCl₅
(2) Pd/C
Benz[h]isoquinoline

The Pictet–Spengler method has been used (*F. Mayer* and *O. Schnecko*, Ber., 1923, **56,** 1408) to prepare 1,2,3,4-tetrahydrobenz[*h*]isoquinoline itself and (*K. Y. Zee-Cheng, W. N. Wynberg* and *C. C. Cheng*, J. heterocyclic Chem., 1972, **9,** 805) the corresponding 8,9-dimethoxy derivative. However, yields are not good, and the starting materials are not readily available. A more promising method (*Eloy* and *Deryckere*, Bull. Chim. ther., 1970, **5,** 121; C.A., 1970, **73,** 77019) involves thermolysis of the azide XXVI when the amide XXVII can be obtained in 60% yield:

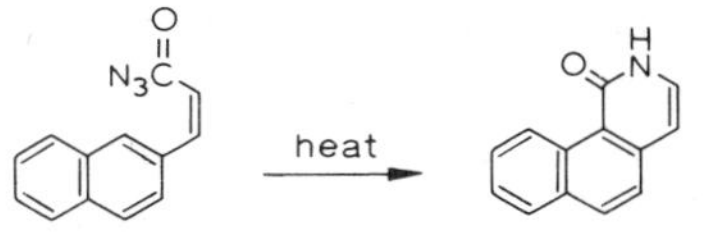

The Pschorr reaction has been applied *(Herz* and *Murty, loc. cit.)*, to 3-amino-4-phenethylpyridine (XXVIII), when a 40% yield of 5,6-*dihydrobenz*[h]*isoquinoline*, b.p. 140°/1 mm, *picrate*, m.p. 195–196°, was obtained, which is easily dehydrogenated to benz[*h*]isoquinoline.

A promising route to benz[*h*]isoquinolines involves photochemical cyclisation of the appropriate azastilbene *(Loader* and *Timmons, loc. cit.)*. Thus irradiation of XXIX gave the parent heterocycle in 21% yield.

5. Naphthoquinolines and naphthoisoquinolines

In the naphthoquinolines and naphthoisoquinolines the naphthalene nucleus may be fused either to a benzene ring or to a pyridine ring. The benzoacridines and benzophenanthridines belong to the second type and are discussed separately.

The compounds described in this chapter have often been regarded and designated as *aza* compounds, *i.e.* as derivatives of tetracyclic aromatic hydrocarbons in which a ring methine group is replaced by a nitrogen atom. Thus compound I is 1-azanaphthacene. This system of nomenclature has the advantage of simplicity, but it has its limitations and in the sequel the formulae and official I.U.P.A.C. and Chemical Abstracts nomenclature will be used.

The naphthoquinolines have not been exhaustively investigated, but the methods of preparation and the chemical properties contain few surprises. They are often prepared by the Skraup or Döbner–von Miller methods from suitable amino compounds or preferably from the corresponding acetamido derivatives (*F. Geerts-Evrard* and *R. H. Martin*, Tetrahedron Supplement, 1966, **7**, 287).

It is known that replacement of methine groups in aromatic hydrocarbons by nitrogen atoms is not accompanied by profound changes in ultraviolet spectra and in this regard the naphthoquinolines are no exception. They exhibit ultraviolet spectra very similar to those of their polycyclic aromatic analogues (*G. M. Badger et al.*, J. chem. Soc., 1951, 3199), the spectrum of naphtho[1,2-*g*]quinoline (IV), for instance, resembling that of 1,2-benzanthracene (IV, with N replaced by CH).

The p.m.r. spectra have been carefully studied (*E. Vander-Donckt, Martin* and *Geerts-Evrard*, Tetrahedron, 1964, **20,** 1495) and conform to expectation. Points of interest are noted in the sequel.

The naphthoquinolines can be oxidised to quinones, and where the product may be an *ortho*- or a *para*-quinone, the latter seems to be preferred.

Naphtho[2,3-*g*]**quinoline,** 1-*azanaphthacene* (I), m.p. 245.5°, is found in heavy gas oil. The 6,11-*quinone*, yellow needles, m.p. 322°, is obtained with other products by the Skraup reaction on 2-aminoanthraquinone (*O. Bally* and *R. Scholl*, Ber., 1911, **44,** 1656) or from 12-chloro-1-azanaphthacene-6,11-dione (obtained by the Skraup reaction on 1-chloro-2-aminoanthraquinone) by the action of sodium dithionite. Reduction of the quinone followed by dehydrogenation (Pd/C) of the resulting 1,2,3,4,7,8,9,10-*octahydro*-1-*azanaphthacene*, m.p. 209°, gives a number of products including 1-azanaphthacene (I) (*N. B. Desai, V. Ramanathan* and *K. Venkataraman*, J. Sci. Indian Research, India, 1956, **15B,** 279):

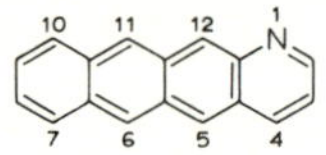

(I) 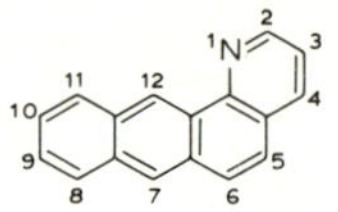 (II) 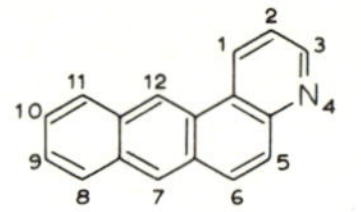 (III)

Naphtho[2,3-*h*]**quinoline** (II), yellow crystals, m.p. 126.5–128°, is obtained by the distillation with zinc dust of the 7,12-*quinone*, m.p. 169°, which results from the Skraup reaction on 1-aminoanthraquinone. It is a weak base and shows a blue fluorescence in ethanol. 2,4-*Dimethylnaphtho*[2,3-h]*quinoline*, m.p. 176°, is prepared in 30% yield by heating the anil from 1-aminoanthracene and acetylacetone with polyphosphoric acid (Combes synthesis) (*G. Saint-Ruf, A. De* and *J. C. Perche*, Bull. Soc. chim. Fr., 1973, 2514).

Naphtho[2,3-*f*]**quinoline** (III), m.p. 170°, *methiodide*, m.p. 221°, is obtained by fusing Alizarin Blue (Vol. IVF, p. 344) with zinc dust or by the Skraup reaction on 2-amino- or 2-acetamino-anthracene (*Geerts-Evrard* and *Martin, loc. cit.*). It is also formed by the Elbs pyrolysis of 5-*o*-toluoylquinoline (*L. F. Fieser* and *E. B. Hershberg*, J. Amer. chem. Soc., 1940, **62,** 1640). It is oxidised to the 7,12-*quinone*, m.p. 185°, thus behaving as an anthracene rather than as a phenanthrene derivative.

Naphtho[1,2-*g*]quinoline (IV), m.p. 159–160° (corr.), *hydrochloride*, m.p. 280–295°, is prepared by the Skraup reaction on 2-amino-9,10-dihydrophenanthrene (*E. Mosettig* and *J. W. Krueger*, J. org. Chem., 1939, **3**, 317), and subsequent dehydrogenation of the dihydro product. The ring-closure at position 3 is radically different from that observed in the Skraup reaction on 2-aminophenanthrene, when ring-closure occurs at $C_{(1)}$ to give naphtho[2,1-*f*]quinoline (X). The structure of IV follows from the Emde degradation of the methiodide of the tetrahydro-*N*-methyl derivative to 3-(3′-dimethylamino-*n*-propyl)phenanthrene (V) and the similarity of the ultraviolet spectrum to that of 1,2-benzanthracene:

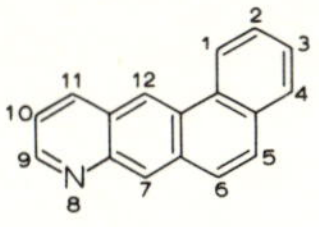

(IV) 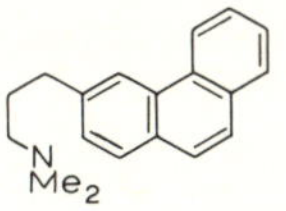 (V) 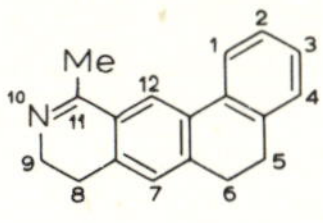 (VI)

Naphtho[2,1-*g*]**isoquinoline** has not yet been prepared, but the cyclisation of the acetyl derivative of 2-(2′-aminoethyl)-9,10-dihydrophenanthrene by means of phosphoryl chloride is believed to be 11-*methyl*-5,6,8,9-*tetrahydronaphtho*[2,1-g]*isoquinoline* (VI) (*A. H. Stuart* and *E. Mossettig*, J. Amer. chem. Soc., 1940, **62**, 1110). It is an oil, *methiodide*, m.p. 267–278, *hydrochloride*, m.p. 230–232°.

Naphtho[2,1-*g*]**quinoline,** 11-*azabenz*[a]*anthracene*, likewise is not known, but 8,10-*dimethylnaphtho*[2,1-g]*quinoline* (VII), yellow needles, m.p. 135–136°, *picrate*, m.p. 305–307° (evac. tube), is prepared by ring-closure of the "anil", formed from 3-aminophenanthrene and acetylacetone, with hydrofluoric acid (*W. S. Johnson et al.*, J. Amer. chem. Soc., 1947, **69**, 566). Its spectrum is similar to that of 1,2-benzanthracene. It has a blue fluorescence in dilute ethanolic solution and an intense yellow fluorescence in dilute hydrochloric acid:

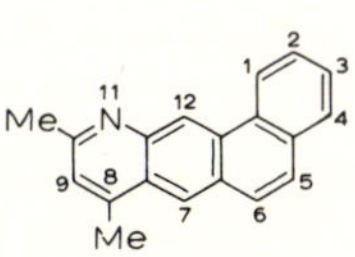

(VII) 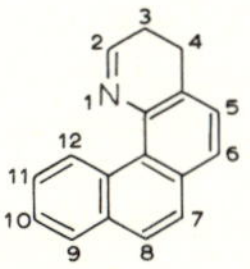 (VIII) 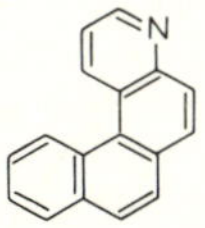 (IX)

Naphtho[2,1-*h*]**quinoline** (VIII), colourless crystals, m.p. 95–96°, *picrate*, m.p. 200–201°, is the product of the Skraup reaction on 4-amino- or 4-acetamino-phenanthrene. The proton at $C_{(12)}$ is exceptionally strongly deshielded.

Naphtho[1,2-*f*]**quinoline** (IX), m.p. 97–98°, is obtained in 64% yield from 3-acetaminophenanthrene by the Skraup reaction (*Geerts-Evard* and *Martin*, *loc. cit.*). It can be reduced and degraded to 4-(3-dimethylamino-*n*-propyl)phenanthrene and its constitution is thus established.

Naphtho[2,1-*f*]**quinoline,** 1-*azachrysene* (X), m.p. 230°, *picrate*, m.p. 278–281°, *hydrochloride*, m.p. 296–300° (evac. tube), is prepared in 80% yield from 2-acetamidophenanthrene by the Skraup reaction. It is oxidised to the 5,6-*quinone*, crimson needles, m.p. 285–286°, which with alkali yields *indeno*[2,1-f]*quinolin*-11-*one* (XI), m.p. 178–179° (*N. Campbell* and *A. Temple*, J. chem. Soc., 1957, 207):

(X) (XI) (XII)

Naphtho[1,2-*h*]**quinoline,** 4-*azachrysene* (XII), colourless crystals, m.p. 137–138°, *picrate*, m.p. 256–258°, is obtained from 1-aminophenanthrene by the Skraup reaction (*J. W. Cook* and *W. H. S. Thomson*, *ibid.*, 1945, 395) or by heating 5-chloro-6-azachrysene with tetralin and palladium black.

Thebenidine (XIII), 4-*azapyrene*, m.p. 157–159°, *picrate*, m.p. 250–253°, was first isolated by the zinc dust distillation of the alkaloid thebenine (*E. Vongerichten*, Ber., 1901, **34,** 767). It is also obtained from coal-tar (*R. Oberkobusch*, *ibid.*, 1953, **86,** 975). The ketone, 4*H*-cyclopenta[*def*]phenanthrene-4-one, with hydrazoic acid yields the lactam, 4,5-dihydro-5-oxo-4-azapyrene, which on distillation with zinc dust gives the parent compound (XIII):

Cyclopenta-phenanthrene-4-one $\xrightarrow{N_3H}$ Lactam $\longrightarrow$ (XIII)

Naphtho[2,1,8-*def*]**isoquinoline,** 2-*azapyrene* (XVI), pale yellow, m.p. 162–165°, *picrate*, m.p. 280°, is prepared by the interaction of phenalene and the salt XIV, followed by heating the resulting intermediate XV (*R. Kirchlechner* and *Ch. Jutz*, Angew. Chem., internat. Edn., 1968, **7,** 376):

Phenalene $\xrightarrow[(XIV)]{Me_2\overset{\oplus}{N}:CHN:CHNMe_2}$ (XV) $\xrightarrow{-Me_2NH}$ (XVI)

The compound has an intense blue fluorescence in aqueous solution and a blue-green fluorescence in aqueous mineral acid. The ultraviolet spectrum resembles that of pyrene and a singlet at τ 0.87 in the p.m.r. spectrum is attributed to the deschielded protons at $C_{(1)}$ and $C_{(3)}$.

6. Benzacridines and benzophenanthridines

(a) Benzacridines

There are three benzoacridines whose formulae and numbering are given below:

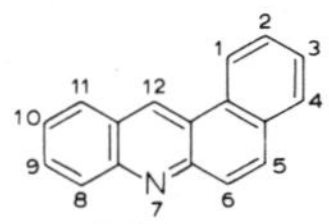

Benz[a]acridine [1,2] Benz[b]acridine [2,3] Benz[c]acridine [3,4]

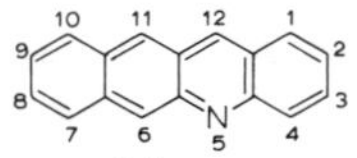

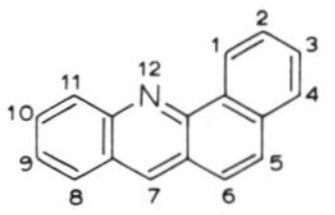

They can be prepared in a number of ways including the distillation of the corresponding benzacridones with zinc dust and by other methods based on those used in the acridine series. 2-Chlorobenzoic acid and 2-naphthylamine condense to yield *N*-2-naphthylanthranilic acid (I), ring-closure of which gives benz[*a*]acridone (II) and hence benz[*a*]acridine (III). The same product is obtained by heating the benzisatin (IV) with alkali and thermally decarboxylating the resulting benz[*a*]acridine-12-carboxylic acid (V) (*J. Martinet* and *A. Dansette*, Bull. Soc. chim. Fr., 1929, **45,** 101; *A. Etienne* and *A. Staehelin*, *ibid.*, 1954, 748):

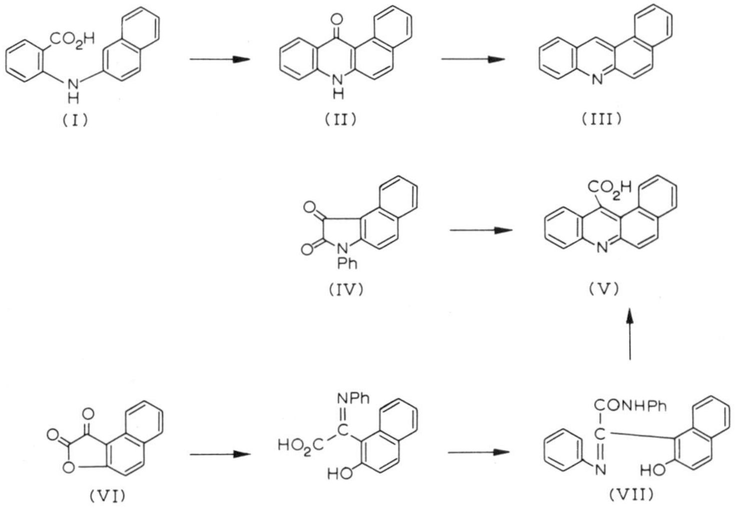

(I) (II) (III)

(IV) (V)

(VI) (VII)

In another, less usual method 4,5-benzocoumarandione (VI) is heated with aniline to give the phenylimino derivative of 2-hydroxy-1-naphthoylformanilide (VII), which loses aniline and yields the carboxylic acid V (*K. Saftien*, Ber., 1925, **58,** 1958).

Substituted benzacridines can be prepared by the Bernthsen method (p. 4), but it is interesting that when trialkylacetic acids such as trimethylacetic acid (pivalic acid) are used, unsubstituted benzacridines are obtained (*N. P. Buu-Hoi et al.*, J. chem. Soc., C, 1966, 1793); thus *N*-phenyl-1-naphthylamine, pivalic acid and zinc chloride yield benzo[*c*]acridine.

Other methods are mentioned in the sequel.

The ultraviolet spectra of the benzacridines resemble those of the corresponding carbocyclic analogues. The p.m.r. spectra present some interesting features such as the greater deshielding (τ 0.53) of the proton at $C_{(1)}$ in benz[*c*]acridine facing the ring nitrogen atom, compared to that of the corresponding proton (τ 1.25) in benz[*a*]anthracene (*R. H. Martin et al.*, Tetrahedron, 1964, **20,** 1495).

The benzacridines are reduced by lithium tetrahydridoaluminate to meso-dihydrobenzacridines, but oxidation with chromic acid is less predictable. Benz[*b*]acridine gives the *para*-6,11-quinone; benzo[*c*]acridine is oxidised to the *ortho*-5,6-quinone; while benz[*a*]acridine curiously enough fails to yield a quinone with this reagent.

The *meso* positions are susceptible to nucleophilic attack, chlorine in that position, for example, being readily replaced by the amino group (*A. Albert, D. J. Brown* and *H. Duewell*, J. chem. Soc., 1948, 1284). Methyl groups in the *meso* positions are reactive as shown by their condensation with benzaldehydes to yield styryl derivatives. These methyl groups are also oxidised much more readily than methyl groups in other parts of the molecule and this is illustrated by the oxidation of 7,9,10-trimethylbenz[*c*]-acridine to 7-formyl-9,10-dimethylbenz[*c*]acridine by means of selenium dioxide (*Buu-Hoi, M. Dufour* and *P. Jacquignon, ibid.*, 1964, 5622). The *meso* positions are the sites of attack by free radicals and in this way benz[*c*]acridine with the benzyl radical, generated from toluene by *tert*-butyl peroxide, yields 7-*benzylbenz*[c]*acridine*, m.p. 144° (64% yield) and a dibenzyl-7,12-dihydrobenz[*c*]acridine (1%) (*W. A. Waters* and *D. H. Watson, ibid.*, 1959, 2082). 7,12-*Dibenzyl*-7,12-*dihydrobenz*[a]*acridine*, m.p. 157–159° (7.5% yield) is produced similarly.

Benz[*a*]acridine, 1,2-*benzacridine*, m.p. 132°, *picrate*, m.p. 260° (decomp.), has an ultraviolet spectrum very similar to that of 1,2-benzanthracene (*G. M. Badger et al., ibid.*, 1951, 3199). Reduction of benz[*a*]acridone with lithium tetrahydridoaluminate gives an orange molecular *complex*, m.p. 140°, composed of benz[*a*]acridine and 7,12-dihydrobenz[*a*]acridine (*Badger, J. H. Seidler* and *B. Thomson, ibid.*, 1951, 3207).

Benz[*b*]acridine, 2,3-*benzacridine* (X), orange crystals, m.p. 225–226°, forms violet salts. It is prepared from benz[*b*]acridone by reduction with sodium amalgam and subsequent dehydrogenation of the resulting 5,12-dihydrobenzacridine (*Albert*, *Brown* and *Duewell*, *loc. cit.*) or by conversion into the 12-chlorobenz[*b*]acridine, followed by hydrogenation and finally oxidation with potassium dichromate and sulphuric acid (*Waters* and *Watson*, *loc. cit.*). It is also formed from benzisatin (VIII) and cyclohexanone to give the tetrahydrocarboxylic acid IX. Decarboxylation and dehydrogenation then yield the benzacridine X (*Etienne* and *Staehelin*, *loc. cit.*).

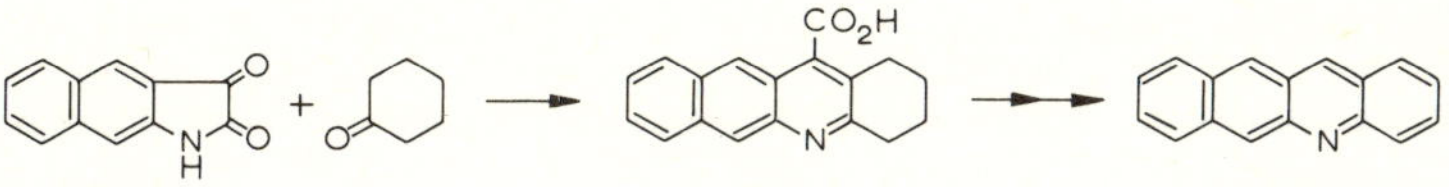

(VIII) (IX) (X)

The ultraviolet spectrum of benz[*b*]acridine resembles that of tetracene, but lacks the fine structure of the latter compound at longer wavelengths (*V. Zanker* and *P. Schmid*, Ber., 1959, **92**, 2253).

Benz[*b*]acridine is oxidised by chromic acid to the 6,11-quinone XI, which is also formed when the benzacridine is exposed in carbon disulphide or benzene to sunlight, the unstable transannular peroxide first formed decomposing to yield the quinone (*Zanker* and *F. Mader*, *ibid.*, 1960, **93**, 850). Exposure to sunlight in ether gives a *photodimer*, m.p. 369–370°:

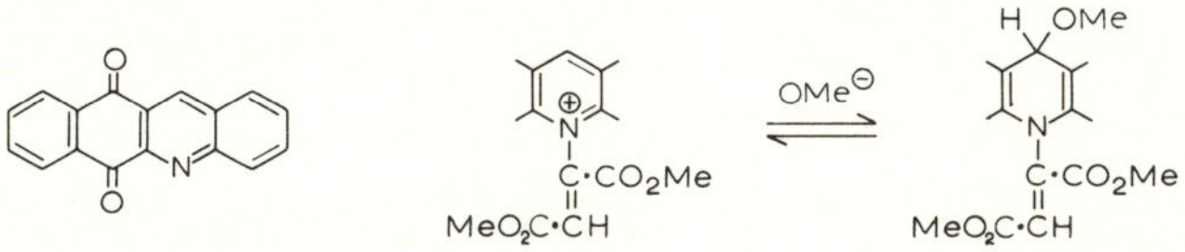

(XI) (XII) (XIII)

Benz[*b*]acridine with maleic anhydride and dimethyl azadicarboxylate forms transannular products, but with dimethyl acetylenedicarboxylate in methanol the product is a benzacridinium salt (XII) which is in equilibrium with a dihydrobenzacridine form (XIII) (*R. M. Acheson* and *C. W. Jefford*, J. chem. Soc., 1956, 2676).

Benz[*c*]acridine, 3,4-*benzacridine*, lemon-yellow crystals, m.p. 108°, *picrate*, m.p. 240°, is prepared like benz[*b*]acridine from the 7-chloro compound and also by the Pfitzinger method from isatin and 1-tetralone and subsequent decarboxylation and dehydrogenation (see benz[*b*]acridine) (*J. von Braun* and *P. Wolff*, Ber., 1922, **55**, 3675). It has a blue fluorescence in solution The p.m.r. spectrum is characterised by the strongly deshielded proton (τ 0.5 in $CDCl_3$) at $C_{(1)}$. The p.m.r. of the chromium tricarbonyl complex has been measured (*E. O. Fischer*, Organometal. Chem., 1968, **14**, 359). 5,6-Dihydrobenz[*c*]acridine, unlike the parent compound, is oxidised by chromic acid to the 5,6-*quinone*, orange powder, m.p. 242° (*von Braun* and *Wolff*, *loc. cit.*).

12-Aminobenz[*b*]acridine in chloroform or carbon tetrachloride unlike 9-amino-acridine exists mainly in the imino form (XIV) since it exhibits an amino-NH band at 3448 cm^{-1} (9,10-dihydroacridine, 3434 cm^{-1}) and an imino-NH band at 3298 cm^{-1} (9-imino-10-methyl-9,10-dihydroacridine, 3294 cm^{-1}) (*S. F. Mason*, J. chem. Soc., 1959, 1281):

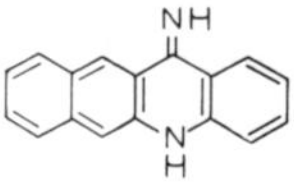

(XIV) 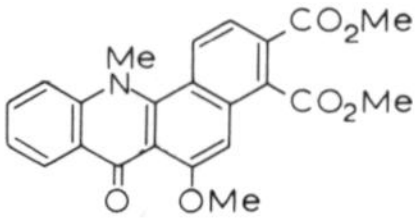 (XV)

A benz[*c*]acridone derivative XV is obtained by the interaction of acronycine and dimethyl acetylenedicarboxylate.

7*H*-**Benz[*kl*]acridine** (XVII), yellowish green needles, m.p. 125° (decomp.), 2,4,7-*tri-nitrofluorenone adduct*, m.p. 183–185°, has a blue-green fluorescence in solution. Nitrosation of *N*-phenyl-1,8-diaminonaphthalene gives the triazine XVI, which loses nitrogen when heated to yield the benzacridine XVII (*H. Waldmann* and *S. Back*, Ann., 1940, **545,** 52; *H. Sieper*, Ber., 1967, **100,** 1646):

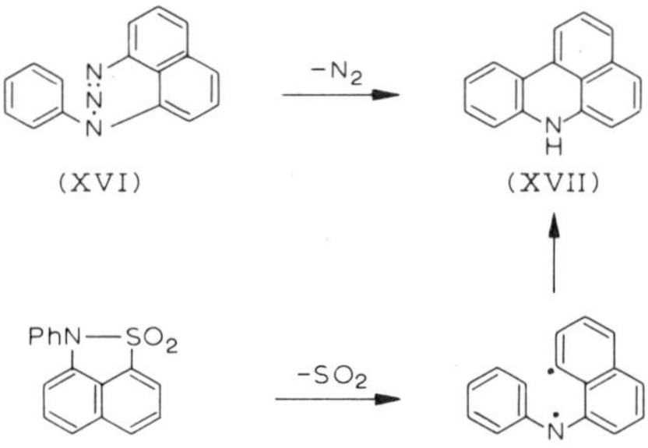

(XVI) → $-N_2$ → (XVII)

(XVIII) → $-SO_2$ → (XIX)

Benzo[*x*]phenanthridines

It is also formed in 43% yield by pyrolysis at 680° of 2-phenyl-2*H*-naphth[1,8-*cd*]-isothiazole 1,1-dioxide (XVIII), plausibly through the diradical XIX (*D. C. de Jongh* and *G. N. Evensen*, J. org. Chem., 1972, **37,** 2152).

(b) Benzophenanthridines

The benzophenanthridines have been formulated in a variety of ways, the position of the "benzo" ring being indicated sometimes by numbers, sometimes by the now officially approved "a"-method; the compound formulated frequently in the literature as 1,2-benzophenanthridine or benzo[1,2]phen-anthridine, for example, being listed in Chemical Abstracts according to the I.U.P.A.C. rules as benzo[*a*]phenanthridine. The latter system, used in the text, is outlined in the following formulae, with the corresponding numerical nomenclature in parentheses:

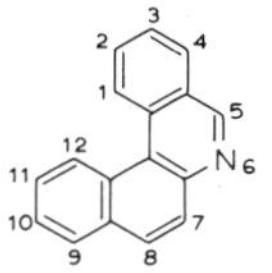

[a]
[1,2]

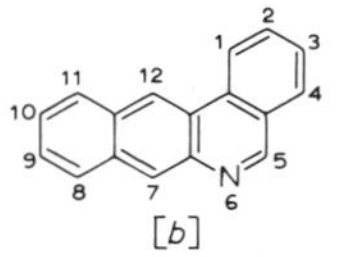

[b]
[2,3]

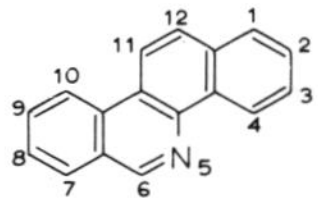

[c]
[3,4]

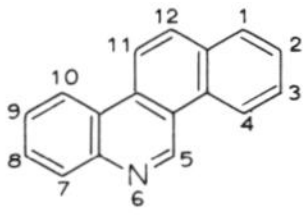

[i]
[7,8]

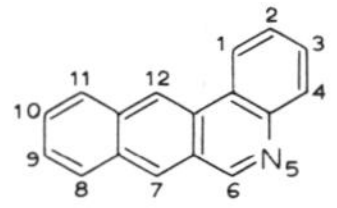

[j]
[8,9]

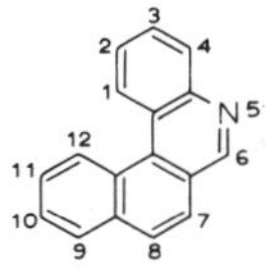

[k]
[3,4]

The methods of preparation for the most part recall those used in the quinoline and isoquinoline series and some representative examples are given in the sequel.

trans-α-(4-Isoquinolyl)-*o*-aminocinnamic acid (XX) when diazotised and treated with copper undergoes the Pschorr reaction by ring-closure on the pyridine nucleus to give benzo[*c*]phenanthridine-11-carboxylic acid (XXI) in poor yield (*R. A. Abramovitch* and *G. Tertzakian*, Canad. J. Chem., 1963, **41,** 2265). The acid is decarboxylated when heated in quinoline with copper powder to give benzo[*c*]phenanthridine (XXII):

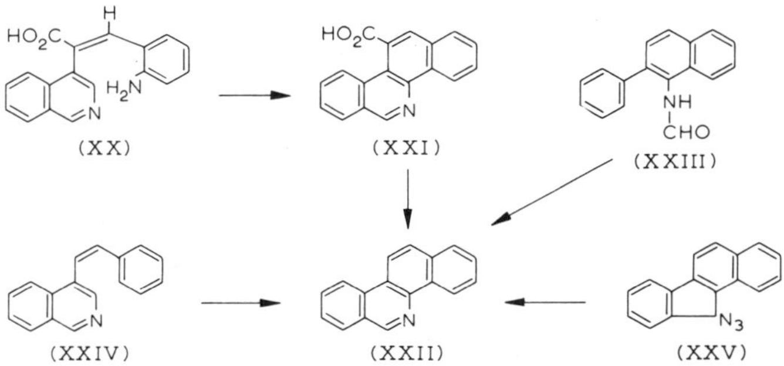

(XX) (XXI) (XXIII)

(XXIV) (XXII) (XXV)

The benzophenanthridine is also obtained by the ring-closure of 1-formamido-2-phenylnaphthalene (XXIII) by phosphoryl chloride and stannic chloride in nitrobenzene (*W. M. Whaley* and *M. Meadow*, J. org. Chem., 1954, **19,** 661) or in 43% yield by irradiation of 4-*trans*-styrylisoquinoline (XXIV, *cis*-form shown) (*C. E. Loader* and *C. J. Timmons*, J. chem. Soc., C, 1968, 330). Finally may be mentioned the decomposition of the azide of

1,2-benzofluorene (XXV) which gives benzo[*c*]phenanthridine and benzo-[*i*]phenanthridine in a ratio of 2.9:1 (*C.L. Arcus, R.E. Marks* and *M.M. Coombs, ibid.*, 1957, 4064).

The chemical and physical properties of the benzophenanthridines are predictable. The ultraviolet spectra of benzo[*c*]- and benzo[*i*]-phenanthridines resemble that of chrysene, while the spectra of benzo[*a*]- and benzo-[*k*]-phenanthridine are similar to that of benzo[*c*]phenanthrene.

The benzophenanthridines have a characteristic blue fluorescence in solution and they form picrates, the melting-points of which are useful for identification purposes. P.m.r. spectra and particularly the deshielding of some of the protons provide useful evidence of structure (*R. H. Martin et al.*, Tetrahedron Supplement, 1966, **8,** 181).

The chemical properties of the benzophenanthridines have not been thoroughly investigated, but it is known that nucleophilic attack occurs, as would be expected, at the carbon atom continguous to the nitrogen atom in the ring.

Benzo[*a*]phenanthridine, m.p. 108–110°, *picrate*, 229–231°, but 250° also reported (*Loader* and *Timmons, loc. cit.)*, is obtained by the ring-closure of 2-formamido-1-phenylnaphthalene by polyphosphoric acid (*D. N. Brown, D. H. Hey* and *C. W. Rees*, J. chem. Soc., 1961, 3873) or phosphoryl chloride and stannic chloride (*B. Mills* and *K. Schofield, ibid.*, 1956, 4213), and in 34% yield by ultraviolet irradiation of 3-styryl-isoquinoline *(Loader* and *Timmons, loc. cit.)*. With methyl iodide and subsequent oxidation with alkaline potassium ferricyanide it yields N-*methylbenzo*[a]*phenanthridone* (XXVI), m.p. 119–120°, identical with a sample prepared by the decomposition of the diazonium tetrafluoroborate of the *β*-naphthalide XXVII:

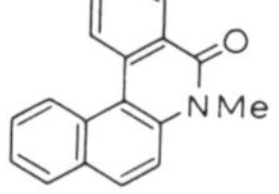

O NMe

(XXVI)

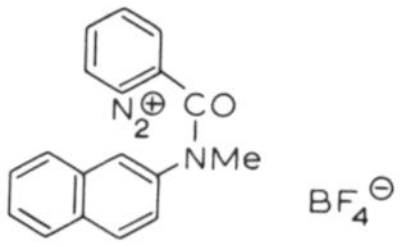

$N_2^{\oplus}$ CO NMe $BF_4^{\ominus}$

(XXVII)

Benzo[*b*]phenanthridine, m.p. 202°, is prepared from *benzo*[b]*phenanthrid-6-one* (XXVIII), yellow needles, m.p. 304.5–306°, which is obtained along with *benzo*[j]-*phenanthrid-6-one* (XXIX, R = H), red needles, m.p. 354–356°, from benzo[*b*]fluorenone either by the Schmidt reaction or by the Beckmann rearrangement of its oxime (*L. H. Klemm* and *A. Weisert*, J. heterocyclic Chem., 1965, **2**, 15). Reduction of benzo[*b*]phenanthria-6-one with lithium tetrahydridoaluminate and subsequent dehydrogenation with palladium–charcoal yields the benzo[*b*]phenanthridine which can be prepared by ring closure of 2-formamido-3-phenylnaphthalene using polyphosphoric acid (Brown, Hey and Rees, *loc. cit.*).

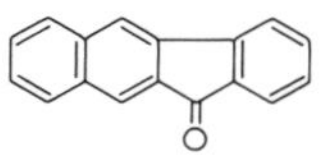

2,3-Benzofluorenone 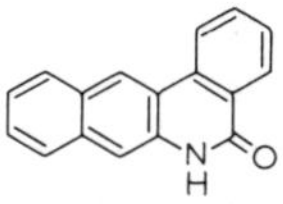 (XXVIII) 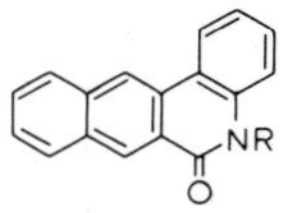 (XXIX)

Benzo[c]phenanthridine (XXXI, R = H), m.p. 135°, *picrate*, m.p. 256°, can be prepared as already outlined. In another preparation chrysenequinone with excess hydrazoic acid gives *benzo*[c]*phenanthrid-6-one* (XXX), prisms, m.p. 332° (*G. M. Badger* and *J. H. Seidler*, J. chem. Soc., 1954, 2329), which is converted by phosphoryl chloride into 6-*chlorobenzo*[c]*phenanthridine* (XXXI, R = Cl), m.p. 15.65°. The chloro compound is reduced by hydrogen and Raney nickel or by lithium tetrahydridoaluminate to products which when dehydrogenated yield benzo[c]phenanthridine (XXXI, R = H).

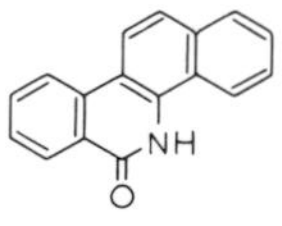

(XXX) 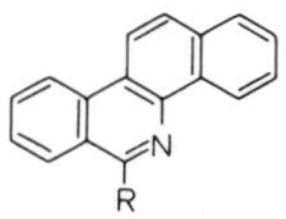 (XXXI)

The phenanthridone is also obtained by the Beckmann rearrangement of chrysenequinone monoxime, the other product being *benzo*[i]*phenanthrid-5-one*, m.p. 163°.

A group of alkaloids including chelidonine and sanguinarine are benzo[c]phenanthridine derivatives. These alkaloids have features in common, namely, terminal aromatic nuclei and a methylated nitrogen atom (*R. F. H. Manske*, "The Alkaloids", Academic Press, New York, Vol. IV, p. 253).

Benzo[i]phenanthridine (XXXIII), m.p. 182°, *picrate*, m.p. 277–278°, is obtained along with benzo[c]phenanthridine from 9-azido-1,2-benzofluorene (XXXII) by the action of trichloroacetic acid *(Arcus et al., loc. cit.)* or in 25% yield by the ultraviolet irradiation of 4-styrylquinoline (XXXIV) (*Loader* and *Timmons*, J. chem. Soc., C, 1967, 1457):

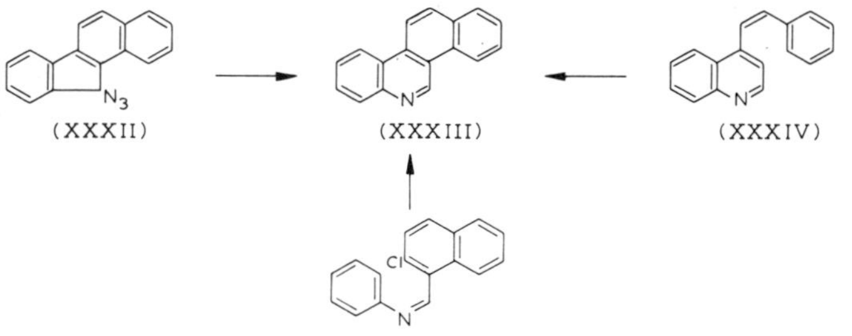

(XXXII) (XXXIII) (XXXIV)

(XXXV)

Benzo[*i*]phenanthridine is also synthesised through a benzyne intermediate by treating the chloro-compound XXXV with potassamide in liquid ammonia (*S. V. Kessar*, *B. S. Dhillon* and *G. S. Joshi*, Indian J. Chem., 1973, **11,** 624). For other syntheses of this benzophenanthridine see *N. Arumugam et al.*, *ibid.*, 1974, **12,** 664; *Klemm* and *Weisert*, *loc. cit.*).

Benzo[*k*]phenanthrid-6-one (XXXVI, R = H), m.p. 275–276°, N-*methyl* compound (XXXVI, R = Me), m.p. 141–142°, is one of the products of the action of hydrazoic acid on benzo[*c*]fluorenone (XXXV) (*B. R. T. Keene* and *K. Schofield*, J. chem. Soc., 1958, 2609). Reduction of benzo[*k*]phenanthridone with lithium tetrahydridoaluminate followed by dehydrogenation with palladium gives *benzo*[k]*phenanthridine*, m.p. 109–110°, *picrate* m.p. 252–254°. The benzophenanthridone (XXXVI, R = H) is also obtained as the sole isolable product by treating the oxime of benzo[*c*]fluorenone with polyphosphoric acid (*Keene* and *P. Tissington*, *ibid.*, 1965, 3032) and the *N*-methylbenzo[*k*]-phenanthrid-6-one (XXXVI, R = Me) is obtained in poor yield by the decomposition of the diazonium tetrafluoroborate obtained from the methylanilide (XXXVII) (*D. M. Collington et al.*, *ibid.*, C, 1968, 1021). With sodamide in boiling xylene benzo[*k*]phenanthridine forms the 6-*amino* compound, m.p. 190–192°. 1-Methylbenzo[*k*]phenanthridine gives an ultraviolet spectrum whose characteristics indicate non-planarity of the molecule as the result of "molecular overcrowding". Further evidence for this comes from the partial resolution of 1,12-dimethylbenzo[*k*]phenanthridine.

Benzo[*j*]phenanthridine, m.p. 143°, yellow needles, is obtained by the reduction of *benzo*[j]*phenanthridone* (XXIX, R = H), m.p. 355°, followed by dehydrogenation (*Klemm* and *Weisert*, *loc. cit.*).

NR O NMe H_2N O

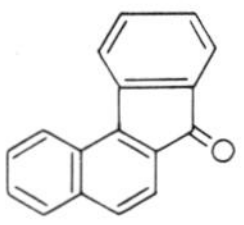

3,4-Benzofluorenone 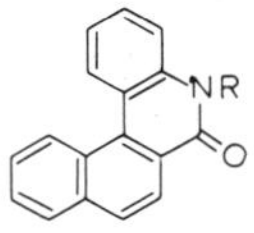 (XXXVI) 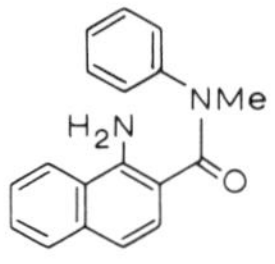 (XXXVII)

Chapter 29

Six-membered Heterocycles Containing Phosphorus, Arsenic, Antimony and Bismuth as a Single Heteroatom

R. E. ATKINSON

Although the majority of published reports on heterocyclic derivatives of Group V elements are concerned with those derived from nitrogen, in many instances analogous ring systems containing the other Group V elements are known. In particular the rapid progress made in organophosphorus chemistry since the discovery of the Wittig reaction, has led to the development of successful syntheses for a wide range of phosphorus-containing heterocycles. A large number of organo-arsenic, -antimony and -bismuth compounds containing a single heteroatom in a six-membered ring system, have also been reported.

1. Phosphorus compounds*

(a) Phosphorinane and its derivatives

(i) Phosphorinanes

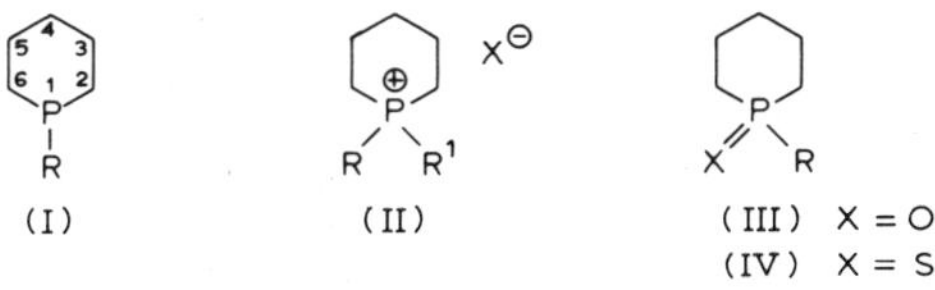

* *Nomenclature*. The phosphorus analogue of pyridine, phosphabenzene, has the alternative trivial name of *phosphorin*; the hexahydro derivative, corresponding with piperidine, is named *phosphorinane*, as well as *phosphacyclohexane*.

Phosphacyclohexanes, pentamethylenephosphines, phosphorinanes, the phosphorus analogues of piperidine contain a ring system of type I and in general are malodorous liquids which react slowly with air and with elemental sulphur to give the corresponding phosphine oxides III (X=O) and phosphine sulphides IV (X=S), respectively. They are basic and form addition salts with acids (II, R^1=H) and are readily quaternised to the phosphonium salts of type II. The ring substituted derivatives of I, II, III and IV provide geometric isomers similar to those observed in the cyclohexane and piperidine series.

1-Substituted phosphorinanes are prepared by the reaction of 1,5-dibromopentane with a secondary phosphine (method 1) or alkali metal phosphine (method 2), or by the reaction of a dichlorophosphine, and the di-Grignard reagent from 1,5-dibromopentane (method 3) and are listed in Table 1.

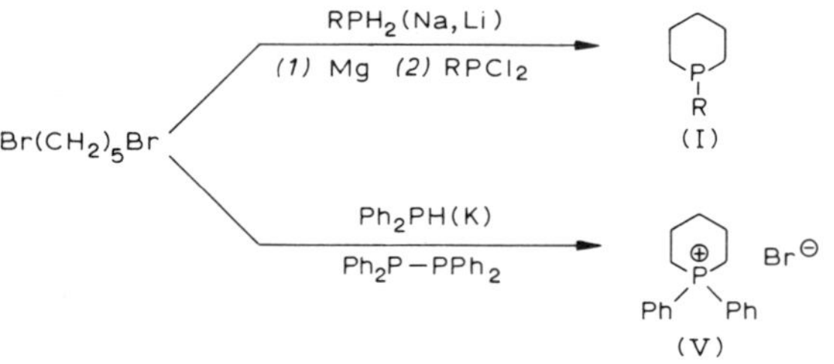

TABLE 1

1-SUBSTITUTED PHOSPHORINANES (I)

R	*Method (Yield)*	*B.p. (°C/mm Hg)*	*Derivatives*	*M.p. (°C)*	*Ref.*
Et	2 (20%)	170	methiodide	293–296	1
			sulphide	67	
Bu^n	2				2
n-C_5H_{11}	2				2
C_6H_{11}	1 (57%)	115/3 Torr	methiodide	230–232	3,1
	2 (20%)				
Ph	3	154–155/24	ethiodide	188	
			mercurichloride	172	4
	2 (31%)	119/3 Torr	methiodide	176	
			sulphide	86	1
4-MeC_6H_4	3	167–168/24	ethiodide	163–164	4
			mercurichloride	157	
4-EtC_6H_4	2				2
4-BrC_6H_4	2				2
$PhCH_2$	3	153–156/30	methiodide	123	5
			oxide	149–150	

References
1 *K. Issleib* and *S. Hausler*, Ber., 1961, **94,** 113.
2 *R. I. Wagner*, U.S.P. 3,086,053/1957 and 3,086,056/1960; C.A., 1963, **59,** 10124; 1964, **60,** 559.
3 *Issleib, K. Krech* and *K. Gruber*, Ber., 1963, **96,** 2186.
4 *G. Gruttner* and *M. Wiernik*, *ibid.*, 1915, **48,** 1473.
5 *P. J. Slota*, Dissertation, Temple University, Philadelphia (U.S.A.), Jan. 1954, "Studies in the Synthesis of Organophosphorus Heterocycles".

Similar ring-closure reactions using 1,5-dibromopropane and either diphenylphosphine (*K. Issleib, K. Krech* and *K. Gruber*, Ber., 1963, **96,** 2186), tetraphenylbiphosphine (*G. Markl*, Angew. Chem. intern. Edn., 1963, **2,** 620) or potassium diphenylphosphine (*idem, ibid.*, 1963, **2,** 479) gave 1,1-*diphenylphosphorinanium bromide* (V), m.p. 262–263°, which gave the corresponding *phosphinane oxide* III (R = Ph), m.p. 128°, with aqueous silver oxide. Similar reactions were used to prepare the corresponding *iodide*, m.p. 278–279° (*S. O. Grim* and *R. Schaaff, ibid.*, 1963, **2,** 486) and 4-*methoxy*-1,1-*diphenylphosphorinanium bromide*, m.p. 220–221°, which was converted to 4-*hydroxy*-1,1-*diphenylphosphorinanium bromide*, m.p. 148–149° (*Markl, ibid.*, 1963, **2,** 479).

Hydrolysis of phosphorinanium salts, II, leads to formation of phosphorinane oxides. Hydrolysis of 1-methyl-1-phenylphosphorinanium iodide (II, R = Ph, R^1 = Me, X = I) leads to hydrolytic fission of the exocyclic group giving only 1-methylphosphorinane-1-oxide (III, R = Me) (*D. W. Allen* and *I. T. Millar*, J. chem. Soc., B, 1969, 263).

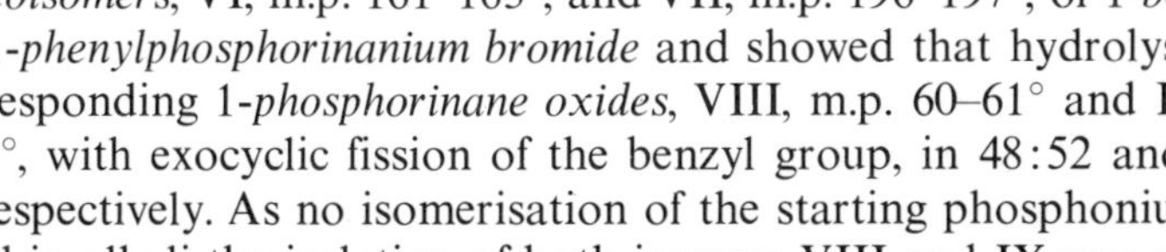

(VI) (VII) $\xrightarrow{OH^{\ominus}/H_2O}$ (VIII) (IX)

K. L. Marsi and *R. T. Clark* (J. Amer. chem. Soc., 1970, **92,** 3791) isolated the *stereoisomers*, VI, m.p. 161–163°, and VII, m.p. 196–197°, of 1-*benzyl*-4-*methyl*-1-*phenylphosphorinanium bromide* and showed that hydrolysis gave the corresponding 1-*phosphorinane oxides*, VIII, m.p. 60–61° and IX, m.p. 148–149°, with exocyclic fission of the benzyl group, in 48:52 and 22:78 ratios, respectively. As no isomerisation of the starting phosphonium salts occurred in alkali the isolation of both isomers VIII and IX was ascribed to rearrangement at the hydrolysis stage.

The corresponding phosphine sulphides, IV, are prepared from the di-Grignard reagent from 1,5-dibromopentane. Reaction with methylthiophosphonic dibromide gave 1,1′-biphosphorinane-1,1′-disulphide (X), which gave 1,1,1-*trifluorophosphorinane* (XII), b.p. 64–65°/40 mm with antimony trifluoride and on oxidation gave the phosphinic acid, 1-*hydroxyphosphorinane*-1-*oxide* (XI), m.p. 128–129° (*R. Schmutzler*, Inorg. Chem., 1964, **3**, 421). The latter was prepared also from 1,5-pentane di-magnesium bromide and dichloro-*N*,*N*-diethylphosphoramidate (*G. M. Kosolapoff*, J. Amer. chem. Soc., 1955, **77**, 6658):

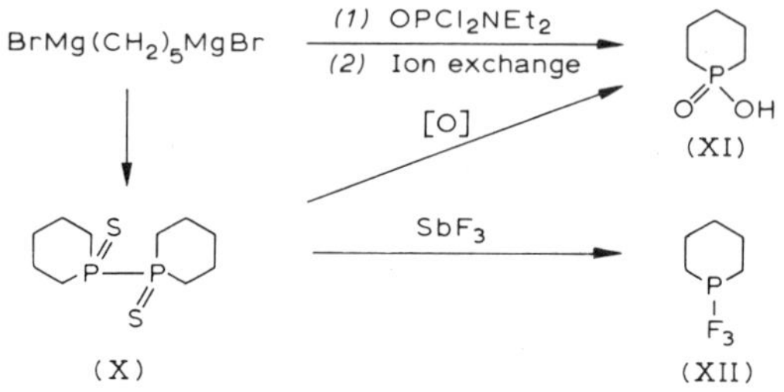

$BrMg(CH_2)_5MgBr$ —(1) $OPCl_2NEt_2$ (2) Ion exchange→ (XI)

[O]

SbF_3

(X) (XI) (XII)

M. Braid (Dissertation Abstracts, 1962, **23**, 434) converted the phosphinic acid XI to the acid chloride 1-*chlorophosphorinane*-1-*oxide* (XIII), b.p. 115–116°/2 mm, m.p. 80°. Reduction of XIII or the corresponding *n*-butyl ester XIV gave the parent compound **phosphorinane** (XV) as a malodorous low melting solid, b.p. 121°, m.p. 19° [*hydriodide*, m.p. 239–240° (dec.), *methiodide*, m.p. 232° (dec.)]. It is reported to give 1,1′-biphosphorinane-1,1′-dioxide (XVI), described as a liquid, on careful oxidation with oxygen under anhydrous conditions and a hydrated phosphorinane-1-oxide, XVII, under aqueous conditions, whereas oxidation with hydrogen peroxide gives solely the phosphinic acid XI.

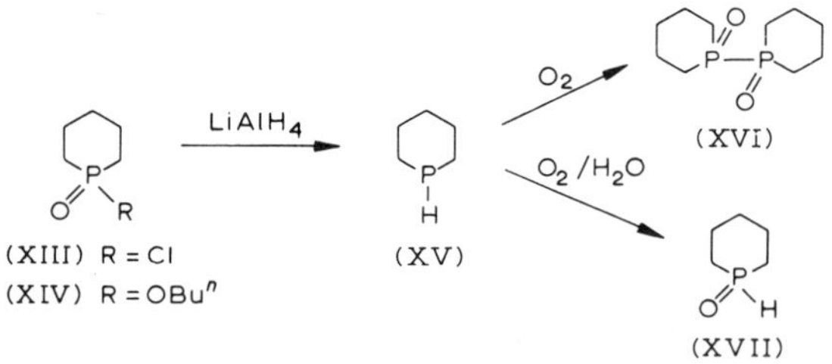

$LiAlH_4$ O_2 O_2/H_2O

(XIII) R = Cl
(XIV) R = OBu^n
(XV) (XVI) (XVII)

Phosphorinane is also reported to arise from the reaction of monosodium phosphine and 1,5-dibromopentane (*R. I. Wagner*, *loc. cit.*; U.S.P. 3,086,053/1957; 3,086,056/1960; C.A., 1963, **59**, 10124; 1964, **60**, 559).

(ii) Phosphorinanones

4-Phosphorinanones (XIX) are a versatile class of phosphorinane compounds which have been used to synthesise a variety of phosphorus heterocycles. They were first synthesised by the hydrolysis of the 4-amino-3-cyanophosphorinanes (XVIII) [prepared from di(cyanoethyl)phosphines]:

$$R{-}P(CH_2CH_2CN)_2 \xrightarrow{KOBu^t} \text{(XVIII)} \xrightarrow{\Delta/HCl} \text{(XIX)}$$

(XVIII) (XIX)

Prepared by this route were 4-*amino*-3-*cyano*-1-*ethyl*-1,2,5,6-*tetrahydrophosphorin* (XVIII, R = Et), m.p. 75°, and the *phenyl* analogue (XVIII, R = Ph), m.p. 140°, which were converted into 1-*ethyl*-4-*phosphorinanone* (XIX, R = Et), b.p. 92°/7 mm (*methiodide*, m.p. 213–214°, *semicarbazone*, m.p. 167–169°), and the *phenyl* analogue (XIX, R = Ph), b.p. 185–190°/1 mm, m.p. 43–44° (*methiodide*, m.p. 155–156°, *semicarbazone*, m.p. 156°) (*R. P. Welcher, G. A. Johnson* and *V. P. Wystrach*, J. Amer. chem. Soc., 1960, **82,** 4437). The *methyl* analogue has been prepared similarly (XIX, R = CH_3), b.p. 55–57°/1.2 mm (*H. E. Shook* and *L. D. Quin, ibid.*, 1967, **89,** 1841).

Alternatively 2,6-disubstituted-4-phosphorinanones, XX, result from the reaction of a primary phosphine with a dienone and are listed in Table 2 (*Welcher* and *N. E. Day*, J. org. Chem., 1962, **27,** 1824):

$$RPH_2 + (R^1R^2C{=}CH_2)_2CO \longrightarrow \text{(XX)}$$

(XX)

TABLE 2

4-PHOSPHORINANONES (XX)

R	*R¹*	*R²*	*B.p. (°C/mm Hg)*	*M.p. (°C)*	*Derivative*	*M.p. (°C)*
$(CH_2)_2CN$	H	Ph	Subl. 180/0.6	126–128		
$(CH_2)_2CN$	Me	Me	140–150/1.0	48–50	methiodide	290–295
sec-Bu	H	Ph	Subl. 150/1.0	118–119		
sec-Bu	Me	Me	90–97/0.3	—	methiodide	263–265
Ph	H	Ph	Subl. 200/0.5	177	methiodide	137
					semicarbazone	270
Ph	Me	Me	130–140/0.5	91–92	methiodide	229–230
					semicarbazone	198–199.5
C_6H_{11}	H	Ph	190/1.0	120–121		
C_6H_{11}	Me	Me	135–140/1.0	63–65	methiodide	275–278
n-C_8H_{17}	H	Ph	—	80–81		
n-C_8H_{17}	Me	Me	146–148/0.8	—	methiodide	150–152

The ketones XIX and XX give normal reactions of carbonyl compounds and form phenylhydrazones, semicarbazones and thiosemicarbazones. With hydroxylamine, however, oxime formation and oxidation occur simultaneously to give the 4-*hydroxyiminiophosphorinane oxides* (XXI, R = Me, m.p. 175–183°; R = Et, m.p. 190–193°) (*M. D. Martz* and *Quin*, *ibid*., 1969, **34**, 3195):

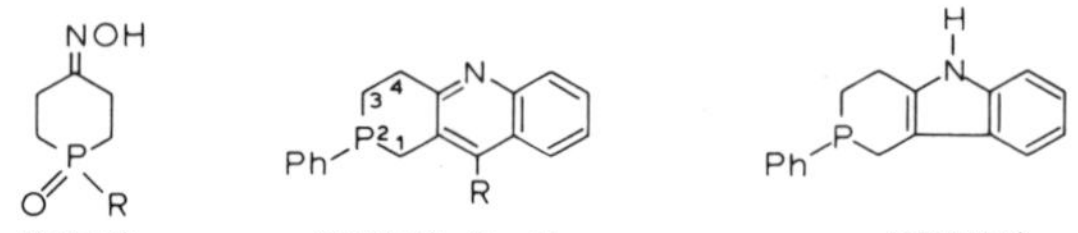

(XXI) (XXII) R = H (XXIII) R = CO_2H (XXIV)

1-Phenyl-4-phosphorinanone has also been reacted with 2-aminobenzaldehyde to give 2-*phenyl*-1,2,3,4-*tetrahydrophosphorino*[4,3-b]*quinoline* (XXII), m.p. 67° (*methiodide*, m.p. 212–213°) and with isatin to the corresponding *carboxylic acid* (XXIII), m.p. 248–249° (decomp.) (*benzylthiouronium salt*, m.p. 212.5–213.5°), while the phenylhydrazone gave 2-*phenyl*-1,2,3,4-*tetrahydro*-1H-*phosphorino*[4,3-b]*indole* (XXIV), m.p. 113–114° (*methiodide*, m.p. 205–206°) (*M. J. Gallagher* and *F. G. Mann*, J. chem. Soc., 1962, 5110).

(iii) Phosphorinanols

1-Alkyl-4-phosphorinanols (XXV) have been prepared by reduction of the corresponding phosphorinanones, which give mixtures of *cis*- and *trans*-isomers:

(XXV)

Reduction with lithium tetrahydridoaluminate gives XXV (R^1 = H), while reaction with either a Grignard reagent or in the Reformatsky reaction gave the corresponding 4-substituted-4-phosphorinanols (XXV, R^1 = alkyl, aryl), which in some instances could be resolved by careful fractionation into their geometric isomers. These are listed in Table 3.

Although the parent 2,6-dihydroxyphosphorinanes are at present unknown, an elegant synthesis of the 2,6-dihydroxyphosphorinanium analogues XXVI has been developed involving the acid catalysed nucleophilic addition of glutaraldehyde and a secondary phosphine:

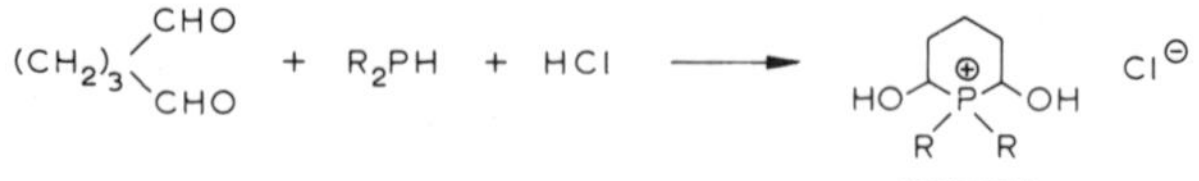

$$(CH_2)_3(CHO)_2 + R_2PH + HCl \longrightarrow \text{(XXVI)}\ Cl^{\ominus}$$

(XXVI)

TABLE 3

4-PHOSPHORINANOLS (XXV)

	R^1	*B.p. (°C/mm Hg)*	*Derivatives*	*M.p. (°C)*	*Ref.*
1e	H	65/0.5	benzyl bromide	222–225	1
1e	*cis*-Et	62/0.55	benzyl perchlorate	199–200	1
1e	*trans*-Et	68–69/0.6	benzyl perchlorate	166–168	1
1e	$-CH_2 \cdot CO_2Et$	75–90/0.15	benzyl bromide	146	1
1e	$-CH_2 \cdot CH_2OH$	107–120/0.15	benzyl tetraphenylborate	170–171	1
1e	Ph		methiodide	228–230	2
1e	*cis*-Ph	102–110/0.2	benzyl perchlorate	228–229	1
1e	*trans*-Ph	114–116/0.2	benzyl perchlorate	215–217	1
1e	$4\text{-}FC_6H_4\text{-}$		methiodide	227.5–229	2
1e	$4\text{-}BrC_6H_4\text{-}$		methiodide	233–235.5	2
1e	$4\text{-}CH_3C_6H_4\text{-}$		methiodide	182–183	2
1e	$4\text{-}MeOC_6H_4\text{-}$		methiodide	212–214	2
1e	$3\text{-}F_3C \cdot C_6H_4\text{-}$		methiodide	142–148	2
t	Et	47–63/0.1	benzyl perchlorate	139	1

References

1 *H. E. Shook* and *L. D. Quin*, J. Amer. chem. Soc., 1967, **89**, 1841; Tetrahedron Letters, 1965, 2193.
2 *D. Lednicer*, J. org. Chem., 1970, **35**, 2307.

By this route were prepared 1,1-*di*-n-*butyl*-, (XXVI, $R = Bu^n$), and 1,1-*di-isobutyl*-2,6-*dihydroxyphosphorinanium chloride*, m.p. 115–117° and 150–152°, respectively; 1,1-*di*-n-*octyl*, m.p. 100–105°, and *dicyclohexyl*, m.p. 160–164°, analogues. Glutaraldehyde and 1-(propylbutyl)phosphine oxide react atypically to give 2,6-*dihydroxy*-1-(1-*propylbutyl*)*phosphorinane*-1-*oxide* (XXVIII), m.p. 226–227° (*S. A. Buckler* and *M. Epstein*, Tetrahedron, 1962, **18**, 1221):

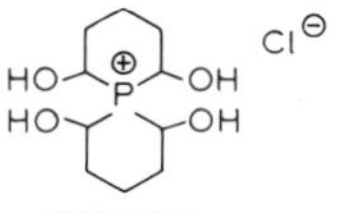

(XXVII) 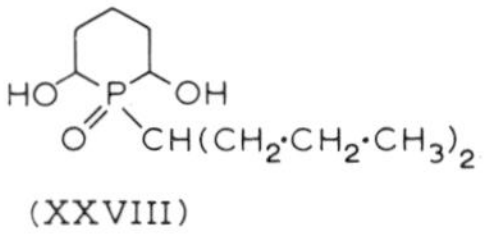 (XXVIII)

The use of phosphine (PH_3) in this reaction gave 6-*phosphoniaspiro*[5.5]*undecane*-1,5,7,11-*tetraol chloride* (XXVII), m.p. 167–168° (*Buckler* and *Wystrach*, J. Amer. chem. Soc., 1958, **80**, 6454; 1961, **83**, 168).

(b) Tetrahydrophosphorins and related compounds

(i) Tetrahydrophosphorins

Dehydration of 4-aryl-1-methylphosphorinan-4-ols has been used to give **4-aryl**-1,2,5,6-**tetrahydro**-1-**methylphosphorins** (XXIX) (Table 4). This reaction did not however work for 1-methyl-4(3′-trifluoromethylphenyl)phosphorinan-4-ol.

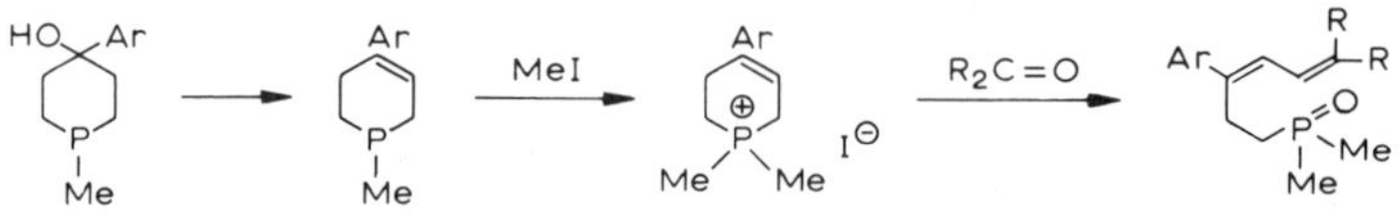

(XXIX) (XXX)

They are readily quaternised to the methiodides which were employed in the Wittig reaction and are unusual in that retention of phosphorus occurs to give the acyclic phosphine oxides XXX (Table 4, Ref. 2). Similar reactions were employed to prepare 1,2,5,6-*tetrahydro*-1,1-*diphenylphosphorinanium perchlorate* (XXXI), m.p. 177–179°, which was converted *via* the dibromo *adduct* XXXII, m.p. 206–208° to 1,1-*diphenyl*-4-*bromo*-1,3,5,6-*tetrahydrophosphorinium chlorate* (XXXIII), m.p. 175–176° (*G. Markl*, Angew. Chem., intern. Edn., 1963, **2,** 478):

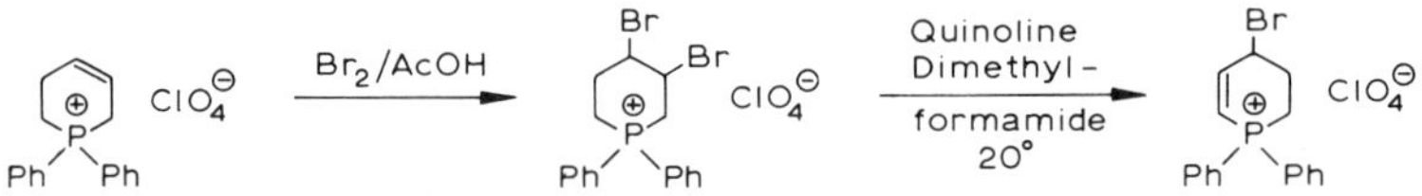

(XXXI) (XXXII) (XXXIII)

TABLE 4

TETRAHYDROPHOSPHORIN DERIVATIVES

Ar	*Derivative*	*M.p. (°C)*	*Ref.*
Ph	benzyl bromide	222–223	1,2
	methiodide	186–189	
4-FC_6H_4-	methiodide	222–224	2
	4′-tosylate	161–162	
4-MeC_6H_4-	methiodide	184–185	2
4-$MeOC_6H_4$-	methiodide	217–219	2
4-BrC_6H_4-	tosylate	210–215	2

References
1 *H. E. Shook* and *L. D. Quin*, J. Amer. chem. Soc., 1967, **89,** 1841.
2 *D. Lednicer*, J. org. Chem., 1970, **35,** 2307.

One unusual tetrahydrophosphorin recently reported is *compound* XXXIV, m.p. 43–45° (*oxide*, m.p. 101–102°), prepared photochemically from dichlorophenylphosphine and cyclo-octatetraene dianion (*T. J. Katz et al.*, J. Amer. chem. Soc., 1970, **92,** 734):

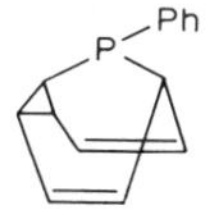

(XXXIV)

(ii) Tetrahydrophosphinolines

Closely related to the tetrahydrophosphorins are the tetrahydrophosphinolines (XXXV) and tetrahydroisophosphinolines (XXXVI) which are phosphorus analogues of tetrahydroquinoline and tetrahydroisoquinoline, respectively:

(XXXV) (XXXVI)

2-**Phenyl**-1,2,3,4-**tetrahydroisophosphinoline** (XXXVI, R = Ph), b.p. 130–160°/0.2 mm (*methiodide*, m.p. 116–118°), was originally obtained in negligible yield by the Wurtz–Fittig reaction (*F. G. Holliman* and *Mann*, J. chem. Soc., 1943, 547):

$$C_6H_4(CH_2)_2Br(CH_2Br) + PhPCl_2 + 4\,Na \longrightarrow \text{(XXXVI, R = Ph)} + 2\,NaBr + 2\,NaCl$$

A more satisfactory method is to form a quaternary salt by the intramolecular *P*-alkylation process shown below:

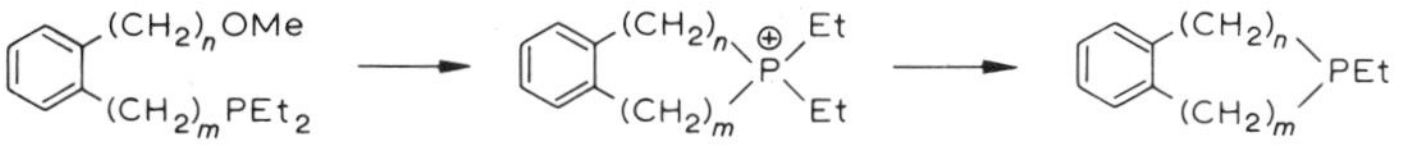

$$C_6H_4(CH_2)_nOMe\,(CH_2)_mPEt_2 \longrightarrow C_6H_4(CH_2)_n(CH_2)_m\overset{\oplus}{P}Et_2 \longrightarrow C_6H_4(CH_2)_n(CH_2)_mPEt$$

(XXXV; R = Et, n = 3, m = 0)
(XXXVI; R = Et, n = 2, m = 1)

The resultant compound decomposes smoothly to 1-**ethyl**-1,2,3,4-**tetrahydrophosphinoline** (XXXV, R = Et), b.p. 141–143/18 mm (*methiodide*, m.p. 184–185°, *methopicrate*, m.p. 121°). 2-*Ethyl*-1,2,3,4-*tetrahydroisophosphinoline* (XXXVI, R = Et), b.p. 129–132°/15 mm (*ethiodide*, m.p. 93–94°) (*M. H. Beeby* and *Mann*, *ibid.*, 1951, 411), was obtained by a similar process.

2-(4′-*Bromophenyl*)-2-*phenyl*-1,2,3,4-*tetrahydroisophosphinolinium bromide* (XXXVII), m.p. 218–220°, and the corresponding 4′-*hydroxy* compound

(XXXVIII), m.p. 287°, were similarly obtained but do not give isophosphinolines on pyrolysis. Both compounds contain an asymmetric P atom and fractional crystallisation of the (+)camphorsulphonate allowed the isolation of (+)XXXVIII, m.p. 268–270°, $[M]_D$ +32.9° (EtOH–H_2O):

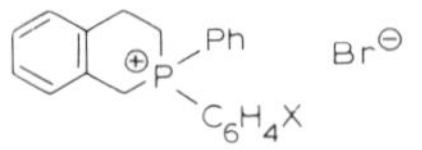

(XXXVII, X = Br)
(XXXVIII, X = OH)

The separation of a partial racemate prevented isolation of the (−)-form. The 4′-bromo analogue XXXVII could not be resolved.

Quaternisation also provided a route to the disymmetric (+)*P*-**spirobis-1,2,3,4-tetrahydrophospholinium bromide** (XXXIX), which was resolved as the (−)-menthyloxyacetates. The enantiomorphic *iodides*, m.p. 246–248°, have $[M]_D$ ±66° ($CHCl_3$) (*F. A. Hart* and *Mann*, *ibid.*, 1955, 4107):

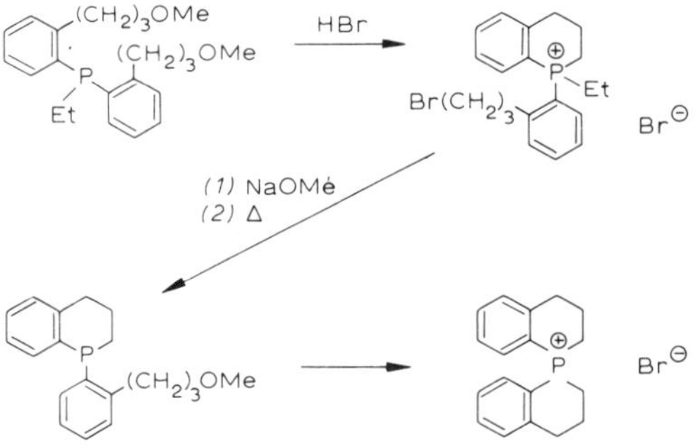

(XXXIX)

Using similar reactions, *Markl* prepared 1,2,3,4-*tetrahydro*-1,2-*diphenylphosphinolinium bromide* (XL, R = H) isolated as the *tetrafluoroborate*, m.p. 193–195°, and similarly 1,2,3,4-*tetrahydro*-1,1,2-*triphenylphosphinolinium tetrafluoroborate* (XL, R = Ph), m.p. 248–250° (Angew. Chem., intern. Edn., 1963, **2,** 153). *C. H. Chem* and *K. D. Berlin* (J. org. Chem., 1971, **36,** 2791) have extended this synthesis to give 1-*ethyl*-1,2,3,4-*tetrahydro*-1-*phenylbenzo*-[h]-*phosphinolinium bromide* (XLI), m.p. 227.5–228.5°. Resolution as the hydrogen dibenzoyltartrate was used to give the (+)*bromide* (XLI), m.p. 259° (decomp.), $[M]_D$ +28° ($CHCl_3$):

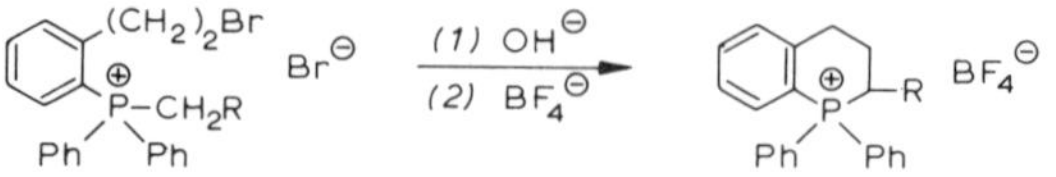

(XL)

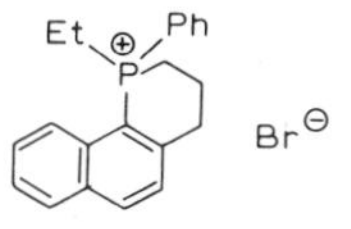

(XLI)

After numerous attempts at preparation by intramolecular acylation of the benzene nucleus (*Mann* and *I. T. Millar*, J. chem., Soc., 1952, 4453; *R. C. Hinton*, *Mann* and *D. Todd*, *ibid*., 1961, 5454), 1,2,3,4-*tetrahydro*-1-*phenyl*-4-*phosphinolone* (XLIII), b.p. 143–145°/0.05 mm, m.p. 46–47° (*phosphine oxide*, m.p. 124–126°, *phenylsemicarbazone*, m.p. 225–226°), has been prepared using a similar route to that used for the phosphorin-4-ones *via* the *amino-nitrile* XLII, b.p. 205–206°/0.2 mm, m.p. 180–181°. An intramolecular acylation process gave a very low yield of the *dimethyl* analogue XLIV characterised as the *methopicrate*, m.p. 170–171° (*M. J. Gallagher*, *E. C. Kirby* and *Mann*, *ibid*., 1963, 4846).

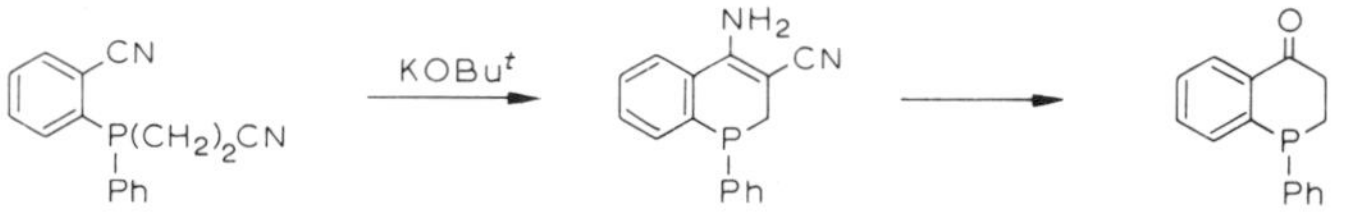

(XLII) (XLIII)

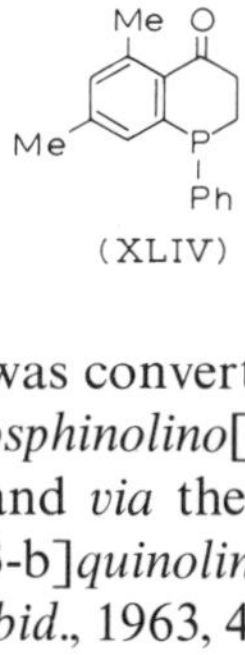

(XLIV) 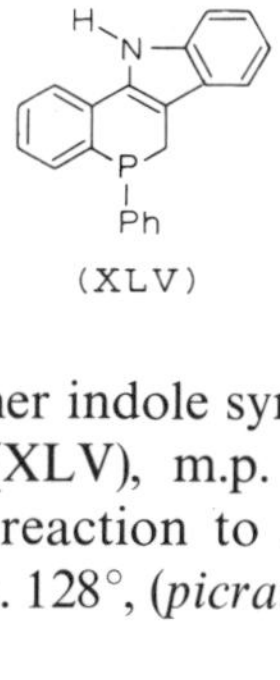 (XLV) 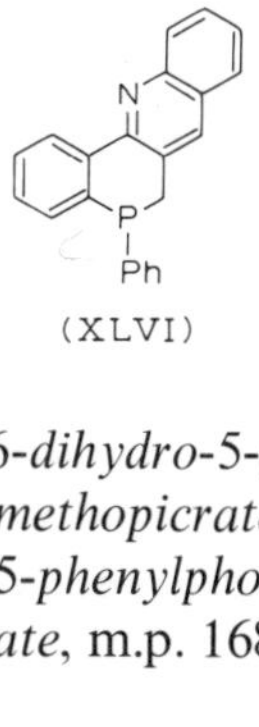 (XLVI)

XLIII was converted *via* a Fischer indole synthesis to 5,6-*dihydro*-5-*phenyl*-5H-*phosphinolino*[4,3-b]*indole* (XLV), m.p. 191–192° (*methopicrate*, m.p. 214°), and *via* the Friedländer reaction to 5,6-*dihydro*-5-*phenylphosphinolino*[4,3-b]*quinoline* (XLVI), m.p. 128°, (*picrate monohydrate*, m.p. 168–169°) (*idem*, *ibid*., 1963, 4855).

(c) Dihydrophosphorins

1,2-**Dihydro**-1,1-**diphenylphosphorinium perchlorate** (XLVII), m.p. 117–119°, was prepared by the dehydrobromination outlined below (*Markl*, Angew. Chem., intern. Edn., 1963, **2**, 479). The 1,2-*dihydrophosphinolinium* analogues XLVIII (R = H, X = $BPh_4^{\ominus}$), m.p. 220–221°, XLVIII (R = Ph, X = $BF_4^{\ominus}$), m.p. 178–180°, were prepared similarly by a dehydrobromination process (*Markl*, *ibid*., 1963, **2**, 153):

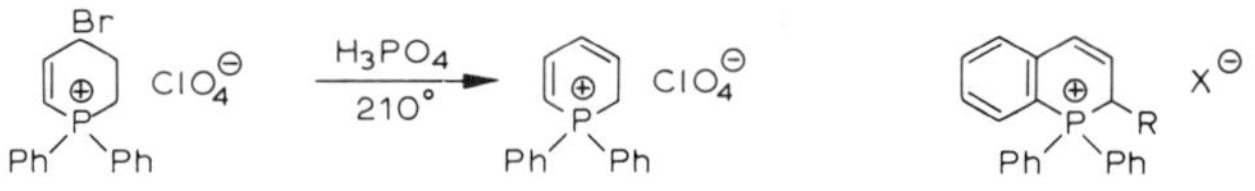

(XLVII) (XLVIII)

Alternatively, 2,4,6-triphenylphosphorin has been reduced to 1,2-*dihydro*-1,2,4,6-*tetraphenylphosphorin* (XLIX), m.p. 144–145° (*methiodide*, m.p. 160–162°), using phenyl-lithium. Oxidation gave 1,2-*dihydrotetraphenylphosphorin oxide* (L), m.p. 144–145°, which was shown by ultraviolet absorption studies to exist in equilibrium with 1-hydroxy-1,2,4,6-tetraphenylphosphorin (LI). 1-Butyl-1,2-dihydro-2,4,6-triphenylphosphorin was similarly prepared from butyllithium as a yellow oil (*Markl*, *ibid*., 1967, **6,** 87):

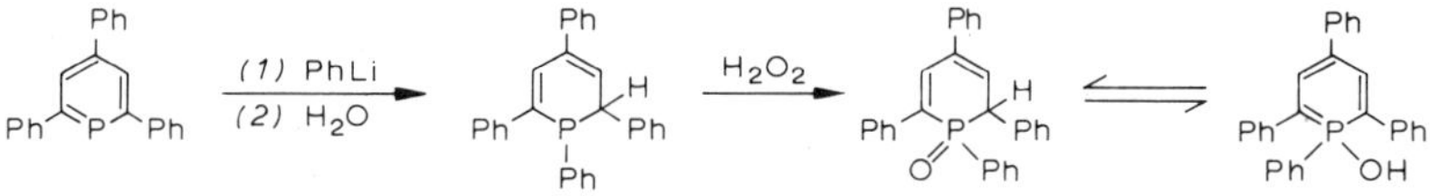

(XLIX) (L) (LI)

Oxidation of 4-phosphinanones with selenium dioxide has been used to prepare 1,4-*dihydro*-4-*oxo*-1-*phenylphosphorin*-1-*oxide* (LII, R = H), m.p. 130°–131°, and 1,4-*dihydro*-4-*oxo*-1,2,6-*triphenylphosphorin*-1-*oxide* (LII, R = Ph), m.p. 162°. The *quaternary salts* LV (R = Et), m.p. 173–175°, LV (R = $CH_2 \cdot Ph$), m.p. 218–220°, were prepared similarly.

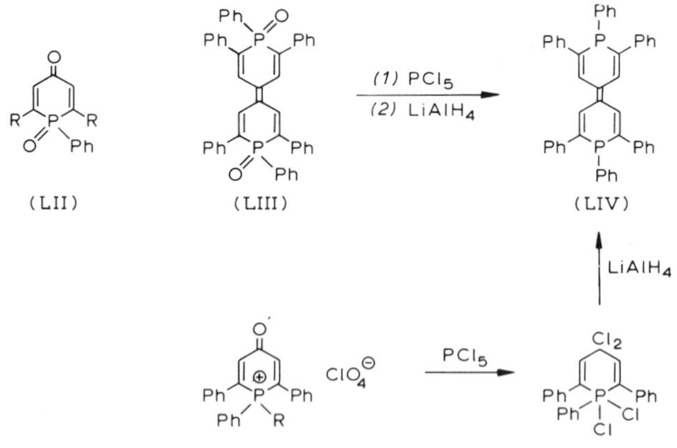

(LII) (LIII) (LIV)

(LV) (LVI)

Reaction of LII (R = Ph) with triethyl phosphite gave 1,1′,4,4′-*tetrahydro*-1,1′,2,2′,6,6′-*hexaphenyl*-4,4′-*biphosphorinylidene*-1,1′-*dioxide* (LIII), m.p. 425–428°, which on chlorination and reduction gave the parent **biphos-**

phorinylidene (LIV), m.p. 335–337° (1,1'-*disulphide*, m.p. 403–405°), which could also be prepared by reduction of the *tetrachlorodihydrophosphorin* (LVI), m.p. 128–131° (*Markl* and *H. Olbrich*, *ibid.*, 1966, **5**, 588). Reaction of LII (R = Ph) with diphenylketene gave 4-*(diphenylmethylene)*-1,4-*dihydro*-1,2,6-*triphenylphosphorin*-1-*oxide* (LVII, R = R^1 = Ph), m.p. 319–321°, which was converted to the parent *dihydrophosphorin* (LVIII, R = R^1 = Ph), m.p. 238–240° (*methiodide*, m.p. 257–258°):

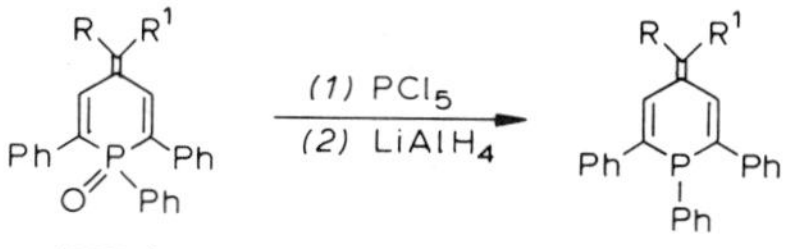
R
R1
(1) PCl5
(2) LiAlH4
Ph
P
O
Ph
Ph
(LVII)
(LVIII)

Similarly condensation of LII (R = Ph) with ethyl cyanoacetate or malononitrile gave the cyano derivatives LVII (R = CN, R^1 = CO_2Et), m.p. 205–206°, and LVII (R = R^1 = CN), m.p. 259–260°. Alternatively reaction of the tetrachloro compound LVI with a cyclopentadiene was used to prepare the analogous phosphine *oxides* LIX, m.p. 261–263°, from tetraphenylindone and LX, m.p. 264–266°, from fluorene which were converted to the parent *dihydrophosphorins*, *e.g.* LXI, m.p. 223–225° (*Markl* and *H. Olbrich*, Angew. Chem., intern. Edn., 1966, **5**, 589).

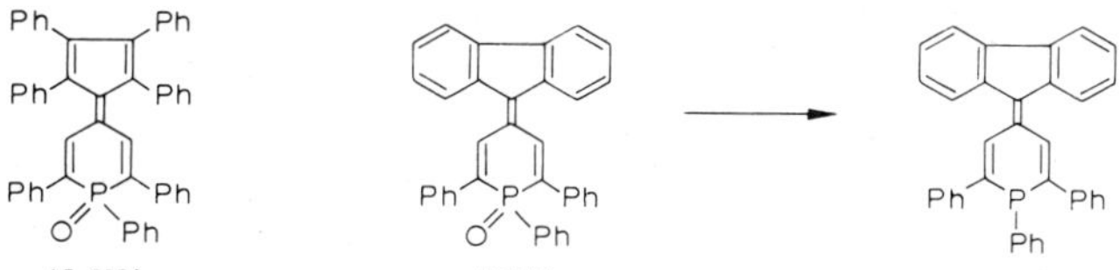
Ph
P
O
(LIX)
(LX)
(LXI)

5,6-**Dihydro**-5,5-**diphenyldibenzo**[*b,d*]**phosphorinium bromide** (LXII), m.p. 335–337°, was prepared by an intramolecular ring closure reaction (*E. A. Cookson* and *P. C. Crofts*, J. chem. Soc., C, 1966, 2003):

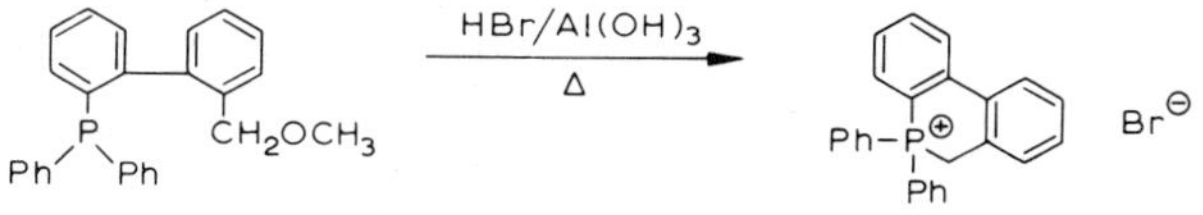
HBr/Al(OH)3
Δ
P
Ph
CH2OCH3
Ph-P⊕
Br⊖
(LXII)

(LXIII) (LXIV) (LXV)

The phosphinic acid analogue of 5,6-*dihydro*-5-*phosphaphenanthrene* (LXIII), m.p. 236–238° (*anhydride*, m.p. 272–276°, *methyl ester*, m.p. 171.5–172.5°, *phenyl ester*, m.p. 75–77°), was prepared by an intramolecular condensation process. This was converted to the *acid chloride* (LXIV, R = Cl, R^1 = H), m.p. 125–126°, which could be reduced to the *phosphine oxide* (LXIV, R = R^1 = H), m.p. 99–100° (*E. R. Lynch*, J. chem. Soc., 1962, 3729) and converted to 5,6-*dihydro*-5-*methyldibenzo*[b,d]*phosphorin*-5-*oxide* (LXIV, R = Me, R^1 = H), as an unrecrystallisable oil and the *phenyl* analogue, m.p. 131–132° (*E. M. Richards* and *J. C. Tebby*, *ibid.*, C, 1971, 1064).

Ring expansion reactions of 9-phosphafluorene have also been used to prepare 5,6-dihydro-5-methyldibenzo[*b*,*d*]phosphorin-5-oxide (LXIV, R = Me, R^1 = H) (uncrystallisable gum) and the 5-*phenyl* analogue (LXIV, R = Ph, R^1 = H), m.p. 127–130°, *via* the phosphonium iodide as outlined above (*D. W. Allen* and *I. T. Millar*, *ibid.*, C, 1969, 252), whilst reaction with methyl propiolate gave LXIV (R = Me, $R^1 = CH_2 \cdot CO_2Me$), m.p. 145–148°, hydrolysed to the *carboxylic acid*, m.p. 231–235°. Similarly prepared were the 5-phenyl- (LXIV, R = Ph, $R^1 = CH_2 \cdot CO_2Me$), as an unrecrystallisable glass, and the benzyl- (LXIV, R = CH_2Ph, $R^1 = CH_2 \cdot CO_2Me$), m.p. 142–146°, analogues (*Richards* and *Tebby*, Chem. Comm., 1967, 957).

An intramolecular condensation of 2-dichlorophosphinodiphenylmethane with zinc chloride gave 5-*chloro*-5,10-*dihydrodibenzo*[b,e]*phosphorin* (LXV), m.p. 78–86°, which could not be dehydrohalogenated and was converted to the *phosphinic acid* with alkaline hydrogen peroxide (decomp. > 225°) (*G. O. Doak*, *L. D. Freedman* and *J. B. Levy*, J. org. Chem., 1964, **29**, 2382; *P. De Koe* and *F. Bickelhaupt*, Angew. Chem., intern. Edn., 1967, **6**, 567).

(d) Phosphorins, phosphabenzenes

Phosphorin (LXVI, $R^1 - R^5 = H$) the phosphorus analogue of pyridine, has been prepared by the reaction of 1,4-dihydro-1,1-dibutylstannabenzene with phosphorus tribromide. It was isolated by preparative gas–liquid chromatography at 110° as a volatile liquid which reacts with air but is stable in an inert atmosphere. The structure was confirmed using spectroscopic techniques (*A. J. Ashe*, J. Amer. chem. Soc., 1971, **93**, 3293).

(LXVI)

Substituted phosphorins are prepared by the addition of the elements of phosphine to a pyrylium salt. This has been accomplished by the reaction of pyrylium fluoroborates with tris(hydroxymethylene)phosphine in pyridine (Method 1). Higher yields are obtained from pyrylium iodides and tris(trimethylsilyl)phosphine (Method 2), but not from perchlorates and tetrafluoroborates due to the low nucleophilicity of the anions. These methods do not work for 2,6-diaryl-4-methylpyrylium salts where attack of the phosphine derivatives occurs at the C-4 carbon due to greater steric hindrance at the C-2 carbon. A general synthesis involves the decomposition of phosphonium iodide in the presence of a pyrylium salt (Method 3). The smaller steric requirements of phosphine leads to phosphorin formation without side reactions. Phosphorins prepared by these methods are listed in Table 5.

The corresponding phosphaphenanthrene derivatives, LXVII, were similarly prepared from the pyrylium tetrafluoroborates; LXVII (R = H), m.p. 99–101°; LXVII (R = Cl), m.p. 150–152°; LXVII (R = Me), m.p. 115–117°; LXVII (R = OMe), m.p. 160–163° (*W. Fischer et al.*, Tetrahedron Letters, 1968, 6227). Attempts to prepare a 1,1-diarylphosphorin from 2,4,6-triphenylpyrylium tetrafluoroborate and phenylphosphine were unsuccessful as the phosphorium intermediate LXVIII was not formed and gave instead two products formulated as 1,2,5,6-*tetrahydro-2-hydroxy*-1,2,4,6-*tetraphenylphosphorin* 1-*oxide* (LXIX), m.p. 256–257° and 1-*hydroxy*-1,2,4,6-*tetraphenylphosphorin hydrate* (LXX), m.p. 70–80° (*C. C. Price, T. Parasaran* and *T. V. Lakshimarayan*, J. Amer. chem. Soc., 1966, **88**, 1034):

(LXVII) (LXVIII) (LXIX) (LXX)

TABLE 5

PHOSPHORIN DERIVATIVES (LXVI)

R^1	R^2	R^3	R^4	R^5	*M.p. (°C)*	*Method*	*Yield (%)*	*Re*
Me	H	Ph	H	Ph	79–81	3	61	1
Bu^t	H	Bu^t	H	Bu^t	88	1		2
Bu^t	H	Ph	H	Bu^t	104–105	1		2
Bu^t	H	4-MeC_6H_4	H	Bu^t	116–116.5	1		2
Ph	H	Ph	H	Ph	172–173	1	30	3
						2	45	4
						3	61	1
C_6D_5	H	C_6D_5	H	C_6D_5	168–171	1		2
C_6D_5	H	Ph	H	C_6D_5	167	1		2
Ph	H	4-$MeOC_6H_4$	H	Ph	106	2	35	4
						3	63	1
					110.5–112	1		2
Ph	H	Bu^t	H	Ph	87.5–88.5	1		2
Ph	H	Me	H	Ph	118–120	3	63	1
Ph	Ph	Ph	H	Ph	209–210	2	41	4
					188.5–189.5	1		2
Ph	Ph	Ph	Ph	Ph	253–254	2	51	4
						3	67	1
					216–217	1		2
4-MeC_6H_4	H	Ph	H	4-MeC_6H_4	133–134	2	62	4
						3	81	1
4-MeC_6H_4	H	Ph	H	Ph	155–156.5	1		2
4-$MeOC_6H_4$	H	Ph	H	4-$MeOC_6H_4$	136–137	2	45	4
						1		2
4-$MeOC_6H_4$	H	4-$MeOC_6H_4$	H	4-$MeOC_6H_4$	105–106	2	36.5	4
4-$MeOC_6H_4$	H	Ph	H	Ph	161.5–163	1		2
4-$MeOC_6H_4$	H	4-$MeOC_6H_4$	H	Ph	134–136	1		2
4-$MeOC_6H_4$	H	4-PhC_6H_4	H	Ph	148–150.5	1		2
α-$C_{10}H_7$	H	Ph	H	Ph	163–164	1		2

References
1 *G. Markl, F. Lieb* and *A. Merz*, Angew. Chem., intern. Edn., 1967, **6**, 944.
2 *K. Dimroth, N. Grief, W. Stade* and *F. W. Steuber*, *ibid.*, p. 711.
3 *Markl*, *ibid.*, 1966, **5**, 846.
4 *Markl*, *Lieb* and *Merz*, *ibid.*, 1967, **6**, 458.

^{31}P n.m.r. studies (*G. Markl*, Angew. Chem., intern. Edn., 1966, **5**, 846) and X-ray measurements (*J. C. Bart* and *J. C. Daby*, *ibid.*, 1968, **7**, 811; *Fischer et al.*, *loc. cit.*) support a planar aromatic structure for the phosphorins LXVI. This is further supported by the general unreactivity of these materials which do not react with methyl iodide or autoxidise in air. In ethanol, however, 2,4,6-tri-*tert*-butylphosphorin is oxidised by hydrogen peroxide to *tert*-butyl 3,5-di-*tert*-but-2-ylfuryl ketone (LXXI) (*K. Dimroth* and *W. Mach*, Angew. Chem., intern. Edn., 1968, **7**, 460):

(LXXI)

Reaction of phosphorins of type LXVI with alkoxy and phenoxy radicals (from mercuric acetate and alcohols or phenols) gave the 1,1-dialkoxyphosphorins LXXII (Table 6), which were shown by X-ray studies to contain a planar heterocyclic ring whose bond lengths (P–C = 1.72 Å) and bond angles are consistent with a delocalised aromatic system (*U. Thewalt, ibid.*, 1969, **8**, 769). These compounds can also be prepared by the photochemical addition of halogen to give 1,1-*dihalogeno*-2,4,6-*triphenylphosphorins* (1,1-*dichloride*, m.p. 100–103°; 1,1-dibromide, uncharacterised; 1,1-*difluoride*, m.p. 129–131°) (Table 6, ref. 3), followed by reaction with an appropriate alcohol (Scheme 1):

Scheme 1

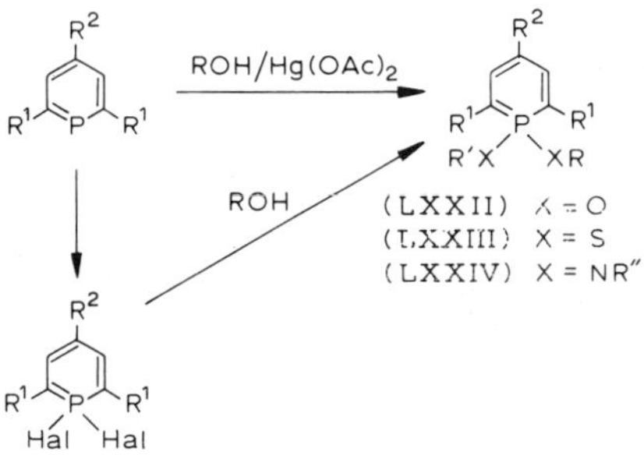

Scheme 2

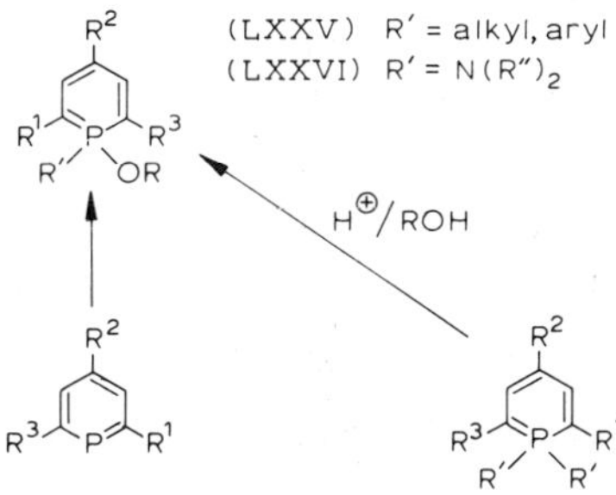

(LXXII) X = O
(LXXIII) X = S
(LXXIV) X = NR″

(LXXV) R′ = alkyl, aryl
(LXXVI) R′ = N(R″)$_2$

Reaction of phosphorins with diazonium tetrafluoroborates and alcohols gives rise to the related 1-alkoxy-1-alkylphosphorins LXXV. These materials can also be prepared from 1,1-dialkylphosphorins (Table 6) (Scheme 2).

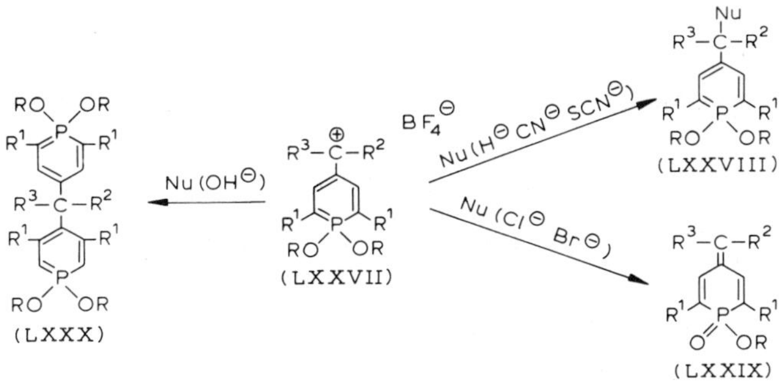

(LXXX) (LXXVII) (LXXVIII) (LXXIX)

Reaction of 1,1-dialkoxyphosphorins LXXII with trityl tetrafluoroborate gives the 1,1-dialkoxyphosphorinium salts LXXVII (Table 7). Reaction of these salts with nucleophilic reagents gives rise to the various 1-alkoxyphosphorins LXXVIII–LXXX outlined above. The 1-dialkylamino analogues LXXIV–LXXVI have been prepared by a similar series of reactions. These are listed in Tables 6–8.

TABLE 6

1,1-DISUBSTITUTED PHOSPHORINS, (LXXII)

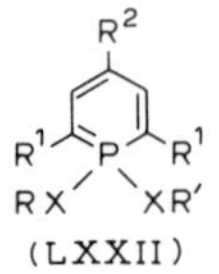

XR′	*XR*	*R¹*	*R²*	*M.p. (°C)*	*Ref.*
OMe	OMe	Ph	Ph	112	1,8
OMe	OMe	C_6D_5	C_6D_5	112	1
OEt	OEt	Ph	Ph	107	1
OEt	OEt	4-$MeOC_6H_4$	Ph	127	1
OPh	OPh	Ph	Ph	152–154	1
4-$NO_2C_6H_4O$	4-$NO_2C_6H_4O$	4-$MeOC_6H_4$	Ph	192	1
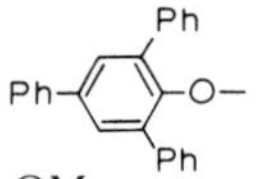 2,4,6-triphenylphenyl-O–	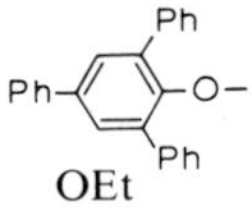 2,4,6-triphenylphenyl-O–	Ph	Ph	249–251	2
OMe	OEt	Ph	Ph	109–110	3
OMe	$OCH_2\cdot CH_2OH$	Ph	Ph	—	3
OMe	OMe	Ph	$CH_2\cdot CN$	102–103	4
OMe	OMe	Ph	Me	141–142	4
OMe	OMe	Ph	CH_2SCN / CH_2NCS	140	4
OMe	OMe	Ph	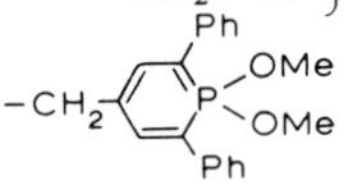 –CH_2–(2,6-diphenyl-1,1-dimethoxyphosphorin-4-yl)	141–142	5
OEt	OEt	Ph	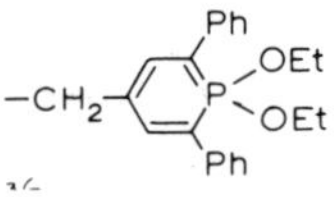 –CH_2–(2,6-diphenyl-1,1-diethoxyphosphorin-4-yl)	138	5
OMe	OMe	Ph	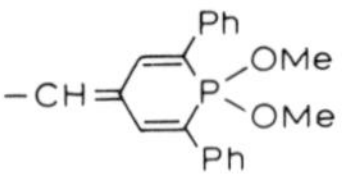 –CH=(2,6-diphenyl-1,1-dimethoxyphosphorin-4-ylidene)	190	5
OMe	OMe	Ph	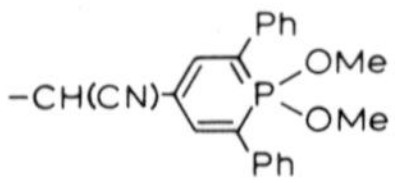 –CH(CN)–(2,6-diphenyl-1,1-dimethoxyphosphorin-4-yl)	185	5

(continued)

Table 6 *(continued)*

XR′	*XR*	*R¹*	*R²*	*M.p. (°C)*	*Ref.*
OMe	OMe	Ph	−CH(OMe)− (ring: Ph, P(OMe)(OMe), Ph)	176–177	5
OMe	OMe	Ph	−CH(OEt)− (ring: Ph, P(OMe)(OMe), Ph)	114–115	5
OMe	OPr^{β}	Ph	Ph	149–150	3
OMe	OMe	Ph	−CH= (ring: Ph, P(OEt)(OEt), Ph)	110–111	5
OMe	OAc	Ph	Ph	127–129	7
OEt	OAc	Ph	Ph	120–123	7
OPr^{β}	OAc	Ph	Ph	121–122	7
OBu^{t}	OAc	Ph	Ph	123–125	7
OMe	−O(MeO)P (ring: Ph, −Ph, Ph)	Ph	Ph	152–154	7
OEt	−O(EtO)P (ring: Ph, −Ph, Ph)	Ph	Ph	120–122	7
OPr^{β}	−O($Pr^{\beta}O$)P (ring: Ph, −Ph, Ph)	Ph	Ph	161–163	7
OBu^{t}	−O($Bu^{t}O$)P (ring: Ph, −Ph, Ph)	Ph	Ph	129–130	7
OPr^{β}	−O($Bu^{t}O$)P (ring: Ph, −Ph, Ph)	Ph	Ph	144–146	7
SMe	SMe	Ph	Ph	146–147	4
SEt	SEt	Ph	Ph	104–105	4
NMe_2	NMe_2	Ph	Ph	121–122	7
NEt_2	NEt_2	Ph	Ph	126–127	7
$NHPr^{\beta}$	$NHPr^{\beta}$	Ph	Ph	181–182	7
NPh_2	NPh_2	Ph	Ph	179 (dec.)	2,8
$MeNCH_2 \cdot CH_2NMe$		Ph	Ph	106–107	7
$HNCH_2 \cdot CH_2NH$		Ph	Ph	194–196	7

TABLE 7

1,1-DISUBSTITUTED PHOSPHORINIUM TETRAFLUOROBORATES

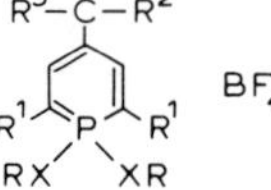

R	*R[1]*	*R[2]*	*R[3]*	*M.p. (°C)*	*Ref.*
Me	Ph	H	H	160	4
Et	Ph	H	H	120–122	4
Me	Ph	H	Me	134–135	4
Me	Ph	H	Ph	148–149	4
Me	Ph	Me	Me	157–158	4
NMe_2	Ph	H	H	133–135	4

TABLE 8

1,1-DISUBSTITUTED PHOSPHORINS OF THE TYPE:

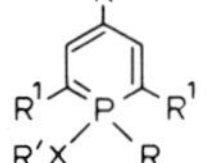

R	*XR'*	*R[1]*	*R[2]*	*M.p. (°C)*	*Ref.*
Ph	OMe	Ph	Ph	135–136	6
Ph	OPr	Ph	Ph	123–124	6
Ph	OMe	Bu^t	Ph	137–139	6
$4\text{-}MeC_6H_4$	OMe	Ph	Ph	121–123	6
$4\text{-}MeC_6H_4$	OMe	Ph	Me	134–136	6
$4\text{-}MeOC_6H_4$	OMe	Ph	Ph	147–149	6
$4\text{-}MeOC_6H_4$	OC_6H_{11}	Ph	Ph	171–172	6
$4\text{-}ClC_6H_4$	OEt	Ph	Ph	149–151	6
Ph	OPr^β	Ph	Ph	158–160	6
*	OMe	Ph	Ph	203–205	6
*	OEt	Ph	Ph	198–200	6
*	OEt	Bu^t	Ph	217–219	6
*	OMe	Ph	**	249–251	6
*	OEt	Bu^t	**	237–239	6
Me	NEt_2	Ph	Ph	83–86	3
Me	NMe_2	Ph	Ph	140	3
Me	$NHPr^\beta$	Ph	Ph	155–156	3
Et	NEt_2	Ph	Ph	98–100	3
Me	$N(C_6H_4Me\text{-}[4])_2$	Ph	Ph	195	3
$MeNCH_2 \cdot CH_2O$ (bridging R and XR')		Ph	Ph	149–150	7

* 2-Methyl-4-nitrophenyl-
** 2′-Methyl-4′-nitro-4-biphenyl-

References

1 *K. Dimroth* and *W. Stade*, Angew. Chem., intern. Edn., 1968, **7,** 881.
2 *Dimroth, A. Hettche, Stade* and *F. W. Steuber, ibid.*, 1969, **8,** 770.
3 *Dimroth, Hettche* and *Stade*, Tetrahedron Letters, 1972, 835.
4 *Dimroth, W. Schafer* and *W. Pohl, ibid.*, 1972, 839.
5 *Schafer* and *Dimroth, ibid.*, 1972, 843.
6 *O. Schaffer* and *Dimroth, ibid.*, 1972, **11,** 1091.
7 *Hettche* and *Dimroth*, Tetrahedron Letters, 1972, 829.
8 *H. Kauter* and *Dimroth*, Angew, Chem., intern. Edn., 1972, **11,** 1090.

Unlike pyridine, with 2,4,6-triphenylphosphorin, lithium alkyls and aryls add directly on to the phosphorus atom to give the stable radical anion LXXXI, which reacts with alkyl iodides to give the 1,1-disubstituted 2,4,6-triphenyl phosphorins LXXXII (Table 9):

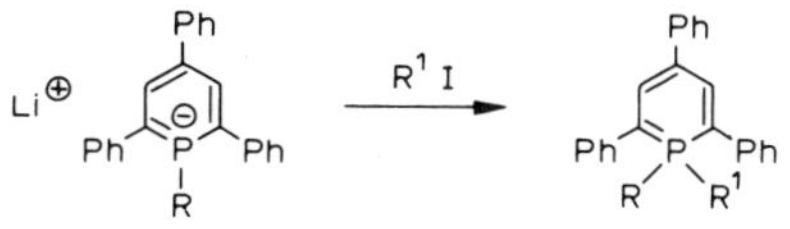

(LXXXI) (LXXXII)

TABLE 9

1,1-DISUBSTITUTED PHOSPHORINS OF THE TYPE: (LXXXII)

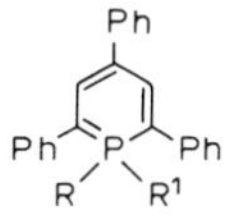

R	*R¹*	*M.p. (°C)*	*Yield (%)*	*Ref.*
Ph	Me	169–170	76	1,2
Ph	Et	151–152	72	2
Ph	$-CH_2Ph$	201–203	42	2

References
1 *G. Markl, F. Lieb* and *A. Merz*, Angew. Chem., intern. Edn., 1967, **6,** 87.
2 *Markl* and *Merz*, Tetrahedron Letters, 1968, 3611.

1,1-**Diphenylphosphorin** (LXXXIII) ($R = R^1 = Ph$), amorphous, non-crystalline, has been prepared by the action of bases on a dihydrophosphorinium salt (*Markl*, Angew. Chem., intern. Edn., 1963, **2,** 478). Prepared similarly were LXXXIII ($R = Ph$, $R^1 = Me$) and LXXXIII ($R = Bu^n$, $R^1 = Me$) (*Markl, Lieb* and *Merz, ibid.*, 1967, **6,** 287).

(LXXXIII)

1,1-**Diphenylphosphinoline** (LXXXIV, R = H) and the 2-phenyl analogue LXXXIV (R = Ph) (amorphous non-crystalline solids), which exist as tautomers with the ylides were prepared by a similar method. They gave the salts LXXXV (R = H, $X^{\ominus} = BPh_4^{\ominus}$), m.p. 220–221° and LXXXV (R = Ph, $X^{\ominus} = BF_4^{\ominus}$), m.p. 178–180° (*Markl*, *ibid.*, 1963, **2,** 153):

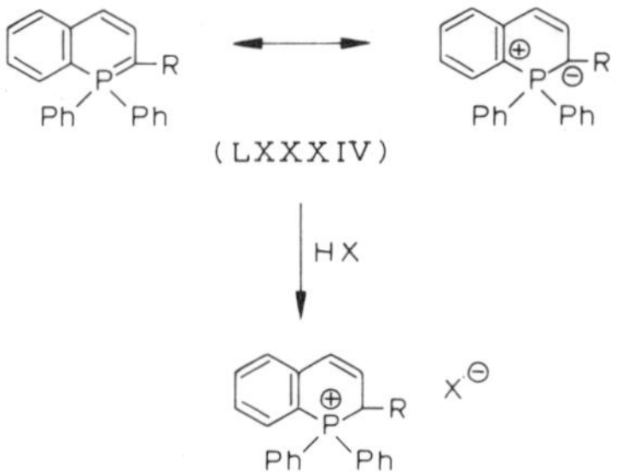

(LXXXIV)

(LXXXV)

(e) Miscellaneous phosphorus heterocycles

(i) Heterocycles containing bridgehead phosphorus

4-**Hydroxy**-1-**phosphabicyclo**[2.2.2]**octane ethobromide** (LXXXVI), m.p. 220° (decomp.), was prepared by the intramolecular alkylation reaction outlined below (*L.D. Quin* and *D.A. Methewes*, Chem. and Ind., 1963, 210).

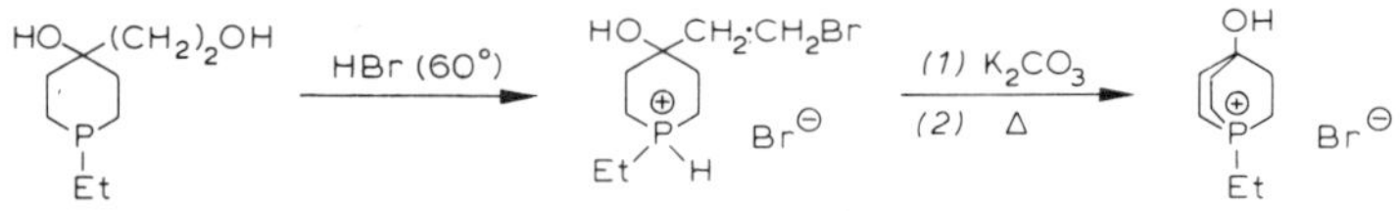

(LXXXVI)

Phosphabicyclo[2.2.2]**octa**-2,5,7-**trienes (phosphabarrelenes)** LXXXVII were prepared from the 1,4-cycloaddition of hexafluoro-2-butyne and phosphorins: LXXXVII ($R^1 = R^2 = Ph$), m.p. 189°; LXXXVII ($R^1 = Bu^t$, $R^2 = Me$), m.p. 92–93°; LXXXVII ($R^1 = Me$, $R^2 = Ph$), m.p. 105–106° (*Markl* and *Lieb*, Angew. Chem., intern. Edn., 1968, **7**, 733):

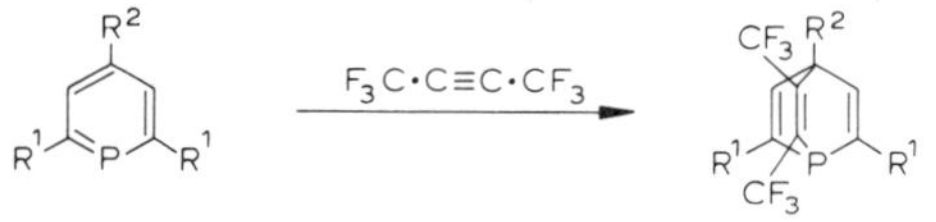

(LXXXVII)

(LXXXVIII)

The cycloaddition of dimethyl acetylenedicarboxylate and 1,2,5-triphenylphosphole is reported to give 1,6,9-**triphenyl**-1-**phosphatricyclo**[4.3.0]**nona**-

1,3,8-**triene**-2,3,4,5-**tetracarboxylate** (LXXXVIII), m.p. 180–182°, *perchlorate*, m.p. 188–190° (*N. E. Waite* and *J. C. Tebby*, J. chem. Soc., C, 1970, 386).

(ii) 5-Phosphoniaspiro[4.5]*decane derivatives*

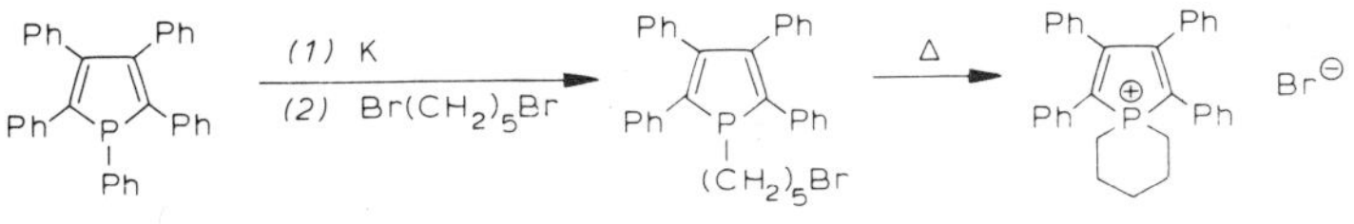

(LXXXIX)

1,2,3,4 - **Tetraphenyl** - 5 - **phosphoniaspiro** [4.5] **deca** - 1,3 - **diene bromide** (LXXXIX), m.p. 357°, prepared as outlined above, is the only reported example of this unusual ring system (*E. H. Braye* reported in "The Heterocyclic Derivatives of Phosphorus, Arsenic, Antimony and Bismuth", ed. *F. G. Mann*, Wiley–Interscience, Chichester, 1970, p. 120).

2. Arsenic compounds

Arsenanes and related compounds

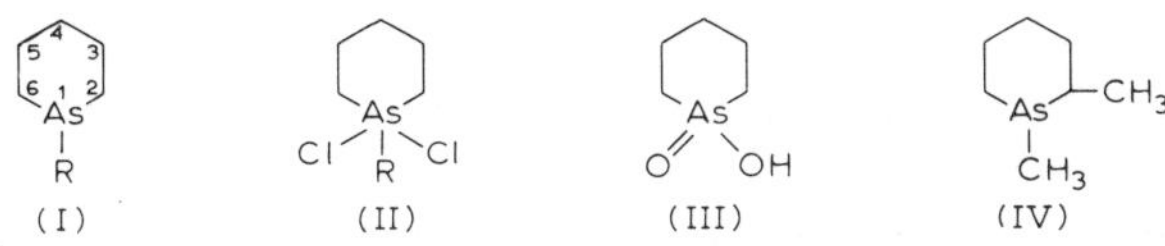

A large number of arsenane derivatives are known which are prepared in general by similar techniques and bear a close chemical resemblance to the phosphorinanes.

(i) Arsenanes

1-Alkyl- and 1-aryl-arsenanes (I), described also as (cyclo-)pentamethylenearsines, arsenidines, have been prepared from pentane-1,5-dimagnesium bromide and a dichloroarsine. They are readily quaternised with alkyl iodides, and form addition compounds (II) with chlorine. The properties of these compounds are listed in Table 10. A similar method was used to prepare 1,2-**dimethylarsenane** (IV), b.p. 169° (*methopicrate*, m.p. 231°) (*E. V. Zappi* and *L. M. Simonin*, Ciencia Mex., 1942, **3**, 160; C.A., 1943, **37**, 2009).

TABLE 10

1-ALKYL- AND 1-ARYL-ARSENANES (I)

R	*B.p. (°C/mm Hg)*	*Derivatives*	*M.p. (°C)*	*Ref.*
Me	156/760	methiodide	290	1
		dichloride	133	1
		methopicrate	258	1
Et	62–64/12.5	methiodide	276	2
Ph	153–154/20	methiodide	179.5	3
		ethiodide	185	3
		dichloride	139	3
4-MeC_6H_4	162–163/20	mercurichloride	175	3
		dichloride	134	3
CH=CHCl	89–91/5	dichloride	121	4

References
1 *E. V. Zappi*, Bull. Soc. chim. Fr., 1916, [4], **19,** 151, 290.
2 *W. Steinkopf, H. Donat* and *P. Jaeger*, Ber., 1922, **55,** 2579.
3 *G. Gruttner* and *M. Wiernik, ibid.*, 1915, **48,** 1473.
4 *I. Gorski, W. Schpanski* and *L. Muljav, ibid.*, 1934, **67,** 730.

The dichloro adduct of 1-methylarsenane eliminates methyl chloride when heated to 166° giving 1-*chloroarsenane* (I, R=Cl), b.p. 84–86°/13 mm, which gives the 1-thiocyanato analogue (I, R=SCN) with sodium thiocyanate (*W. Steinkopf, I. Schubart* and *J. Roch*, Ber., 1932, **65,** 409), and is converted to 1-*hydroxyarsenane*-1-*oxide* (III), m.p. 200.5–202°, on treatment with sodium methoxide followed by hydrogen peroxide (*I. Gorski, W. Schpanski* and *L. Muljav, ibid.*, 1934, **67,** 730).

1-Chloroarsenane (I, R=Cl) can be reduced with lithium tetrahydridoaluminate to form **arsenane** (I, R=H), m.p. 11–13°. This is unstable in air and is converted with iodine to 1-*iodoarsenane* (I, R=I), m.p. 27° (*E. Wiberg* and *K. Modritzer*, Z. Naturforsch., 1957, **12b,** 135).

(ii) 4-Arsenones

M. H. Gallagher and *F. G. Mann* (J. chem. Soc., 1962, 5110) have compared the properties of 1-phenyl-4-arsenone (VI) with the nitrogen and phosphorus analogues. 1-Alkylarsenones of type VI are at present unknown. They prepared 3-**ethoxycarbonyl**-1-**phenyl**-4-**arsenone** (V), b.p. 153–154°/1.25 mm, n_D^{20} 1.580 (*dinitrophenylhydrazone*, m.p. 179–180°), by cyclisation of bis(2-carboxyethyl)phenylarsine, which was hydrolysed to 1-*phenyl*-4-*arsenone* (VI), m.p. 9–10°, b.p. 114–115°/0.3 mm (*phenylsemicarbazone*, m.p. 175–176°):

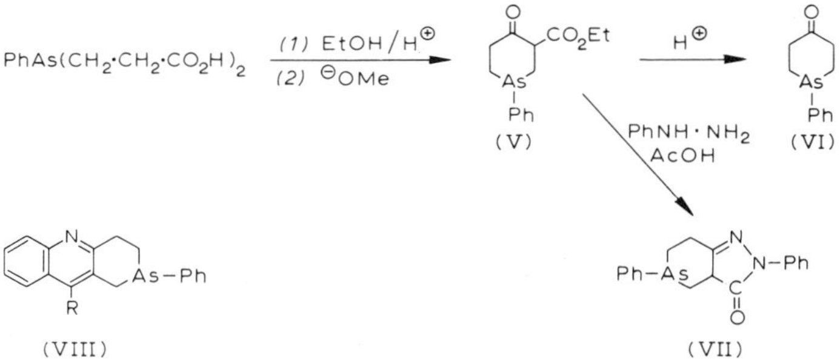

The 4-arsenones have been used to prepare a variety of heterocyclic systems. Reaction of the oxo-ester V with phenylhydrazine gave the unusual heterocycle 4,5,6,7-*tetrahydro*-2,5-*diphenylarsenino*[4,3-c]*pyrazol*-3(2H)-*one* (VII), m.p. 173–174°. Unlike the 4-phosphorinanone analogue, 1-phenyl-4-arsenone (VI) did not give an indole derivative with phenylhydrazine. Normal reaction was shown, however, with isatin (Pfitzinger reaction) and 2-aminobenzaldehyde to give 1,2,3,4-*tetrahydro*-2-*phenylarsenino*[4,5-b]*quinoline* (VIII, R = H), m.p. 76–77° (*hydrochloride hemihydrate*, m.p. 156–157°) and the *carboxylic acid* (VIII, R = CO_2H), m.p. 266–268° (*benzylthiouronium salt*, m.p. 150–151°), respectively.

The behaviour with methyl iodide distinguishes 1-phenyl-4-arsenone (VI) from the corresponding phosphorus compound which forms a simple methiodide only. 1-*Methyl*-4-*oxo*-1-*phenylarsenanium iodide* (IX), m.p. 145–146°, 4,4-*dimethoxy*-1-*methyl*-1-*phenylarsenanium iodide* (X), m.p. 146–147°, and 4,4-*dihydroxy*-1-*methyl*-1-*phenylarsenanium iodide* (XI), m.p. 120–121°, may be formed under the different reaction conditions outlined below:

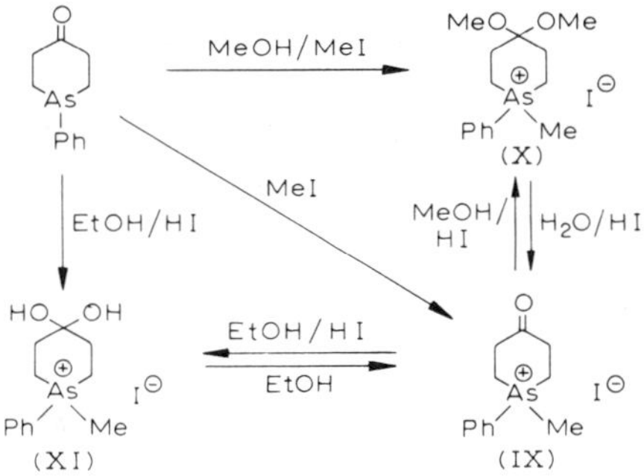

(iii) Derivatives of tetrahydroarsenin, tetrahydroarsinoline and tetrahydroiso-arsinoline

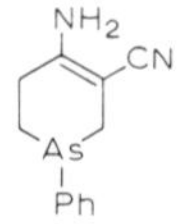

(XII) 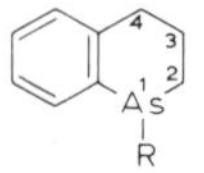 (XIII) 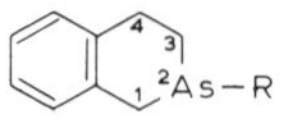 (XIV)

4-*Amino*-1,2,5,6-*tetrahydro*-1-*phenylarsenin*-3-*carbonitrile* (XII), m.p. 65–67°, prepared by the cyclisation of bis(2-cyanoethyl)phenylarsine, is the only reported example of a simple tetrahydroarsenin (*R. P. Welcher*, *G. A. Johnson* and *V. P. Wystrach*, J. Amer. chem. Soc., 1960, **82,** 4437). The tetrahydro-arsinoline (XIII) and tetrahydro-isoarsinoline (XIV) ring systems are well known, however.

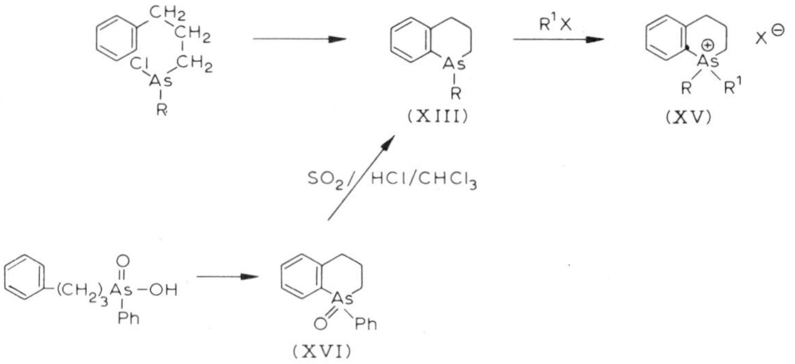

(XIII) (XV)

SO_2 // $HCl/CHCl_3$

(XVI)

1,2,3,4-**Tetrahydro**-1-**methylarsinoline** (XIII, R = Me), b.p. 140°/14 mm (*methiodide*, m.p. 235°), was prepared by the internal Friedel–Crafts reaction shown above (*G. J. Burrows* and *E. E. Turner*, J. chem. Soc., 1921, **119,** 426). 1,2,3,4-Tetrahydro-1-methylarsinoline also forms a dichloro-adduct which on heating under reduced pressure eliminated methyl chloride to give 1-*chloro*-1,2,3,4-*tetrahydroarsinoline* (XIII, R = Cl), m.p. 22°, b.p. 155°/16 mm. This material could not be dehydrohalogenated to a dihydroarsinoline (*E. Roberts*, *Turner* and *W. Bury*, *ibid*., 1926, 1443).

An alternative synthesis of the tetrahydroarsinoline ring was developed by *F. G. Holliman* and *D. A. Thornton* (reported in "The Heterocyclic Derivatives of Phosphorus, Arsenic, Antimony and Bismuth", ed. *F. G. Mann*, Wiley–Interscience, Chichester, 1970, p. 402). Cyclisation of phenyl-(3-phenylpropyl)arsinic acid gave the arsine oxide XVI, which be reduced to the oily 1,2,3,4-*tetrahydro*-1-*phenylarsinoline* (XIII, R = Ph), (*methiodide*, m.p. 165–166°). This material was quaternised to the 4-*chlorophenacyl bromide*, m.p. 210–211°. Unlike the corresponding isoarsinolinium salt, attempts to resolve this salt into optically active forms were unsuccessful due to the formation of a partial racemate.

In the case of the spirobisarsinolinium derivatives XVII, however, successful resolution was achieved. 3,3′,4,4′-**tetrahydro**-1,1′-2*H*,2′*H*-**spirobi(arsinolinium) bromide** (XVII, X = Br), m.p. 253–254°, *iodide* (XVII, X = I), m.p. 277–278°, was prepared by the intramolecular quaternisation outlined below and resolved as the (−)-menthyloxyacetate. The enantiomeric *iodides*, m.p. 223°, had $[M]_D$ ±132° ($CHCl_3$) (*Holliman, Mann* and *Thornton*, J. chem. Soc., 1960, 9).

Br−$(CH_2)_3$ → (XVII) $X^\ominus$

(1) o-$C_6H_4(CH_2Br)_2$
(2) Δ, MeBr
As–Me → (XVIII) $X^\ominus$

Similarly, 1′,3,3′,4-tetrahydro-1,2′(2*H*)-spiro(arsinoline-1,2′-isoarsindolium) bromide (XVIII, X = Br) was prepared from the reaction of *o*-xylylene dibromide and 1-*methyltetrahydroarsinoline* (XIII, R = Me) and characterised as the *iodide* (XVIII, X = I), m.p. 241–243° and *picrate*, m.p. 163–164°.

Isoarsinoline derivatives, XIV, have been prepared from 2-(bromomethyl)-phenethyl bromide *via* the Wurtz–Fittig reaction, which gave 2-**methyl**-1,2,3,4-**tetrahydroisoarsinoline** (XIV, R = Me), b.p. 131°/18 mm (*methiodide*, m.p. 179–181°) and the 2-*phenyl* analogue (XIV, R = Ph), b.p. 110–112°/0.01 mm (*methiodide*, m.p. 136–137°) (*Holliman* and *Mann*, *ibid*., 1943, 547). Alternatively, the phenyl derivative XIV (R = Ph), has been made in higher yield (70%) from the dibromide and phenylarsinebismagnesium bromide (*Mann et al., ibid*., 1950, 1917). Quaternisation gave the *bromide* (XIX), m.p. 190–191°, which was resolved as the (+)-bromocamphorsulphonates to give the optically pure *picrates*, $[M_D]$ ∓450° ($CHCl_3$) (*idem, ibid*., 1943, 550). 3,3′,4,4′-**Tetrahydro**-2,2′-1*H*,1′*H*-**spirobi(iso-arsinolinium) bromide** (XX), m.p. 280–290°, was prepared similarly to the arsinolinium analogue XVII. The enantiomeric *iodides*, m.p. 226–228°, have $[M]_D$ ∓342° ($CHCl_3$) (*idem, ibid*., 1945, 45).

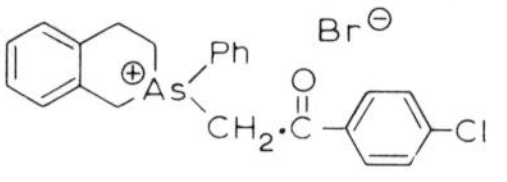

(XIX) 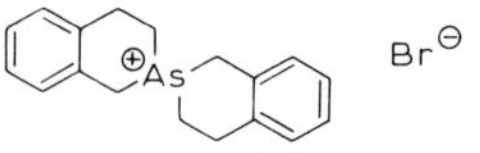 (XX)

In addition to the arsinoline derivatives outlined above, 1,2,3,4-**tetrahydro**-7-**methoxy**-1-**methyl**-4-**arsinolone** (XXI), m.p. 52° (*methiodide*, m.p. 239°, *phenylhydrazone*, m.p. 121–122°) and 2,3,5,6-*tetrahydro*-8-*methoxy*-1,7-*dioxo*-1H,7H-*benz*[i,j]*arsinolizine* (XXII), m.p. 150° (*bisphenylhydrazone*, m.p. 169–171°), were prepared by ring closure of an appropriate carboxylic acid (*Mann* and *A. J. Wilkinson*, *ibid.*, 1959, 3336).

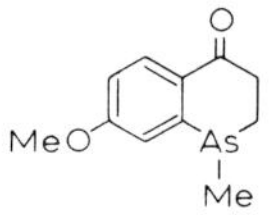

(XXI) 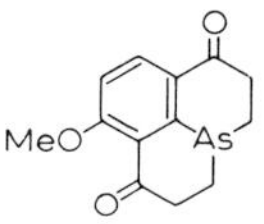 (XXII)

Indole and quinoline derivatives have been prepared from XXI and XXII using standard reactions (*idem*, *ibid.*, p. 3346).

(iv) 5,10-*Dihydroacridarsines and* 5,6-*dihydroarsanthridines*

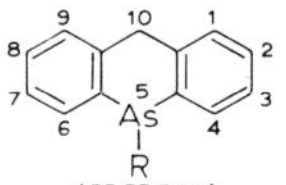

(XXIII) 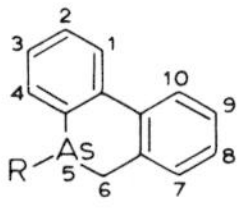 (XXIV)

5,10-**Dihydroacridarsines** XXIII are prepared using similar routes to the cyclisation reaction outlined below and are listed in Table 11.

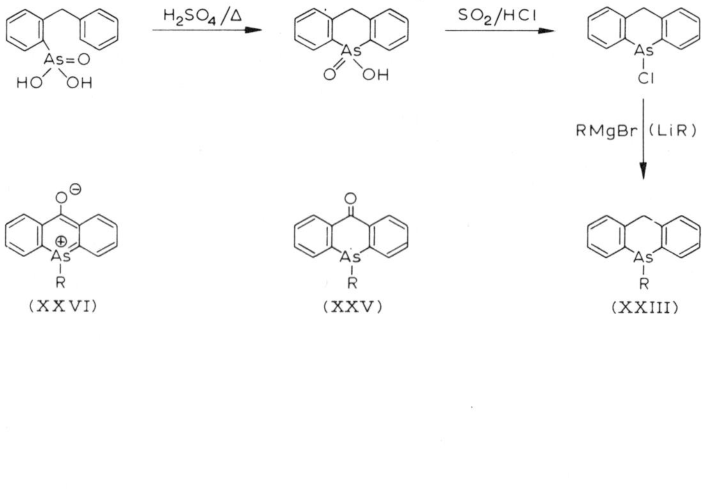

H_2SO_4/Δ SO_2/HCl RMgBr (LiR)

(XXVI) (XXV) (XXIII)

TABLE 11
5,10-DIHYDROACRIDARSINE DERIVATIVES

	M.p. (°C)	*Ref.*
5,10-Dihydro-5-phenylacridarsine	74–75 (methopicrate 260 dec.)	1
5-Chloro-5,10-dihydroacridarsine	114–115	2,3
5-Cyano-5,10-dihydroacridarsine	114–115	3
5-Chloro-5,10-dihydro-2-methylacridarsine	87	4
5-Chloro-5,10-dihydro-3-methylacridarsine	65.5–66.5	3
2,5-Dichloro-5,10-dihydroacridarsine	116	4
2-Chloro-5-cyano-5,10-dihydroacridarsine	113–114	4
5(4′-Carboxyphenyl)-5,10-dihydro-3-methylacridarsine	167–168 ($[\alpha]_D \mp 84$)* (arsine oxide, m.p. 187–190)	5
Acridarsinic acid	235–236	2,3
2-Chloroacridarsinic acid	210–212 dec.	4
2-Methylacridarsinic acid	184 dec.	4

*Dissymetry attributed to the conformational stability of the tertiary arsenic atom.

References

1 *E. R. H. Jones* and *F. G. Mann*, J. chem. Soc., 1958, 294
2 *W. Gump* and *H. Stoltzenberg*, J. Amer. chem. Soc., 1931, **53**, 1428.
3 *A. R. Todd et al.*, J. chem. Soc., 1948, 292.
4 *Idem, ibid.*, p. 295.
5 *K. Mislow*, *A. Zimmerman* and *J. T. Mellilo*, J. Amer. chem. Soc., 1963, **85**, 594.

Attempts to prepare the **acridarsone** system XXV, using similar cyclisation reactions have been unsuccessful (*J. A. Aeschlimann* and *N. P. McClelland*, J. chem. Soc., 1924, **125**, 2025; *E. Sakellarios*, Ber., 1926, **59**, 2552). 5-*Phenyl-10-acridarsone-5-oxide*, m.p. 238°, from the oxidation of the parent acridarsine XXIII, (R = Ph), is smoothly reduced in hydrochloric acid to the *acridarsone* XXV (R = Ph), m.p. 138–139°, with sulphur dioxide. On the basis of chemical and spectral evidence this compound was formulated as the semi-aromatic zwitterion XXVI (*Jones* and *Mann*, J. chem. Soc., 1958, 294).

5,6-**Dihydroarsanthridines** are prepared by the intramolecular cyclisation route outlined below. In this way were prepared 5,6-*dihydro-5-methylarsanthridine* (XXIV, R = Me), b.p. 145–200°/0.1 mm (*methiodide*, m.p. 212–215°) and the 5-*phenyl* analogue (XIV, R = Ph), b.p. 150–200°/0.1 mm (*methiodide*, m.p. 195°) (*G. H. Cookson* and *Mann*, *ibid.*, 1949, 2888):

CH_2AsRCl → CH_2—As—R

(XXIV)

(v) 1H-Naphth[1,8-c,d]arsenins

This system is known as the dihydro derivatives (XXVII). 2,3-**Dihydro**-2-**phenyl**-1*H*-**naphth**[1,8-*c,d*]**arsenin** (XXVII, R = Ph) is an oil which could not be obtained crystalline. It was characterised as the orange *bis(arsine) dichloropalladium* derivative, $(C_{18}H_{15}As)_2PdCl_2$, m.p. >240°, and as the *methiodide*, m.p. 196–199°.

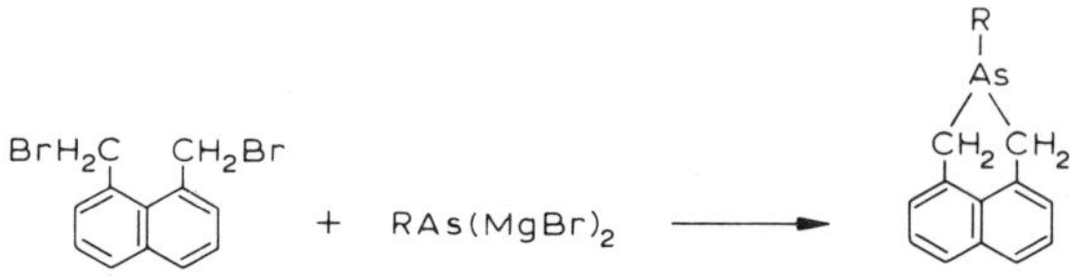

BrH$_2$C CH$_2$Br + RAs(MgBr)$_2$ ⟶ (XXVII)

Boiling hydriodic acid converts the 2-phenyl derivative to the 2-*iodo* derivative (XXVII, R = I), m.p. 117–119° (*M. H. Beeby, Cookson* and *Mann, ibid.*, 1950, 1917).

(vi) Arsenins

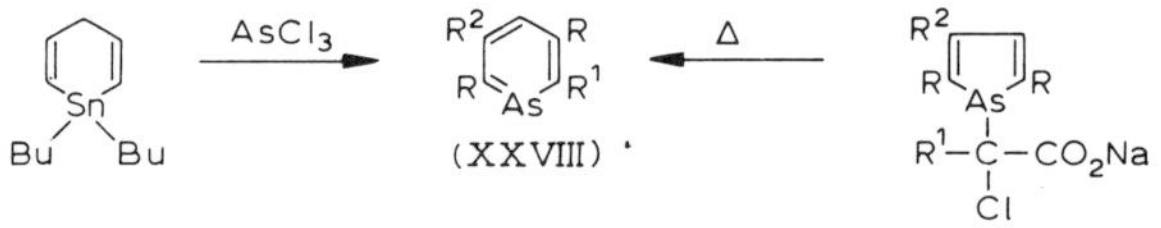

AsCl$_3$ ⟶ (XXVIII) ⟵ Δ

The parent compound, **arsenin** (XXVIII, $R = R^1 = R^2 = H$), has been prepared from 1,4-dihydro-1,1-dibutylstannabenzene and arsenic trichloride. It is a colourless liquid, isolated by preparative gas–liquid chromatography and degrades rapidly in air. The structure was established by spectroscopic methods (*A.J. Ashe*, J. Amer. chem. Soc., 1971, **93**, 3293). Aryl substituted arsenins are more stable and may be prepared by insertion of carbene into an arsole ring (Table 12).

TABLE 12

ARYLARSENINS

R	*R^1*	*R^2*	*M.p. (°C)*
Ph	Ph	H	146
Ph	$C_{10}H_7$	H	170–172
Ph	Ph	Ph	223–225

Reference
G. Markl, H. Hauptmann and *J. Advena*, Angew. Chem., intern. Edn., 1972, **11**, 441.

(XXIX)

The related acridarsine system XXIX is also known. The parent **acridarsine** (XXIX, R = H) is unstable and has only been detected spectroscopically (*P. Jutzi* and *K. Deuchert*, Angew. Chem., intern. Edn., 1969, **8**, 991; *H. Vermeer* and *F. Bickelhaupt*, *ibid*., p. 992). 6-*Phenylacridarsine* (XXIX, R = Ph) has however been isolated as a crystalline solid, m.p. 172–176° (*idem*, Tetrahedron Letters, 1970, 3255).

3. Antimony compounds

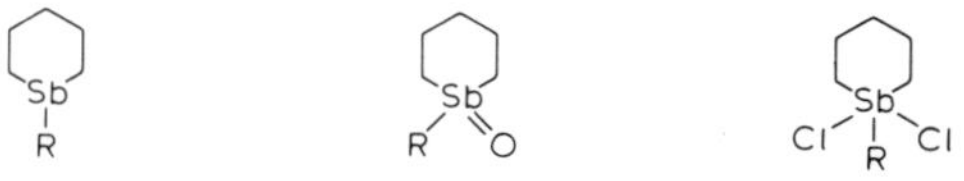

(I) (II) (III)

Antimonanes I are prepared by a similar route to the phosphorus and arsenic analogues from the reaction of a pentane-1,5-dimagnesium halide and a dichlorostibene.

1-**Phenylantimonane** (I, R = Ph), b.p. 169–171°/18 mm, is readily oxidised to the 1-*oxide* (II, R = Ph), m.p. > 280°, and forms a 1,1-*dichloro adduct* (III, R = Ph), m.p. 141–142°, with chlorine in carbon tetrachloride. It does not however form an ethiodide derivative (*G. Gruttner* and *W. Wiernik*, Ber., 1915, **48**, 1473).

1-*Methylantimonane* (I, R = Me) prepared similarly, b.p. 73–73.5°/17 mm, gave a dichloro adduct (III, R = Me) which decomposes on heating at 180° to 1-*chloroantimonane* (I, R = Cl), b.p. 110–111°/13 mm (*W. Steinkopf*, *I. Schubart* and *J. Roch*, *ibid*., 1932, **65**, 409).

The only known compounds containing antimony in a condensed ring system are the 5-substituted-5,10-dihydrodibenz[*b*,*e*]antimonins (V):

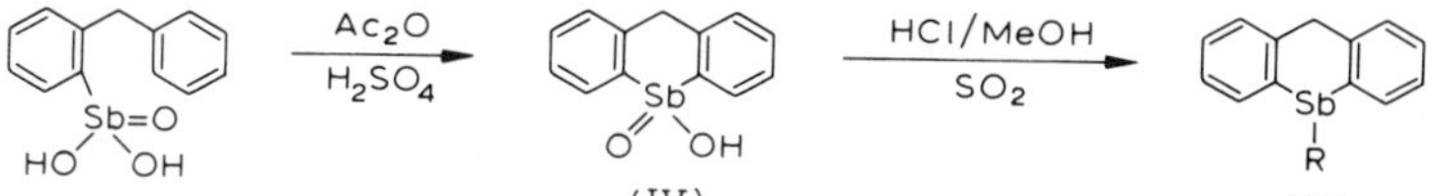

$\xrightarrow[H_2SO_4]{Ac_2O}$ (IV) $\xrightarrow[SO_2]{HCl/MeOH}$ (V)

Cyclisation of 2-benzylphenylstibonic acid gives the cyclic acid IV as a brown powder, which was converted to 5-*chloro*-5,10-*dihydrodibenz*[b,e]-*antimonin* (V, R=Cl), m.p. 105°, characterised by conversion to the 5-*iodo*- (V, R=I), m.p. 160–162°, and 5-*bromo*- (V, R=Br), m.p. 112°, analogues. Treatment of the 5-iodo compound with methylmagnesium bromide gave 5,10-*dihydro*-5-*methyldibenz*[b,e]*antimonin* (V, R=Me), m.p. 101° (5,5-*dichloro adduct*, m.p. 177–178° decomp.):

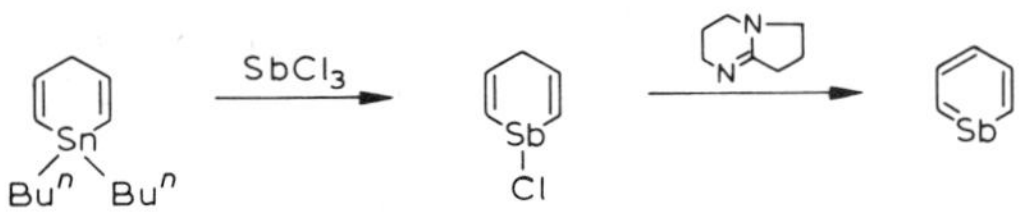

(VI) (VII)

Similar attempts to prepare the 5-*phenyl* compound gave an impure product which was isolated as the 5,5-*dichloro adduct*, m.p. 221–223° (*G. T. Morgan* and *G. R. Davies*, Proc. roy. Soc., 1933, **143A,** 38). Reaction of 1,1-dibutyl-1,4-dihydrostannabenzene with antimony trichloride gives the crystalline 1-**chloro**-1,4-**dihydroantimonin** (VI), m.p. 115–117° decomp. Treatment in "diglyme" with 1,5-diazabicyclo[4.3.0]non-5-ene gives the unstable **antimonin** (VII), which readily polymerises at −80°. The structure was confirmed spectroscopically.

4. Bismuth compounds

The only example of a six membered heterocycle containing bismuth is the pyrophoric 1-**ethylbisminane** (I), b.p. 108–112°/18 mm, prepared by *Gruttner* and *Wiernik* (Ber., 1915, **48,** 1473) by an analogous route to that used for the antimonins.

(I)

Chapter 30

Pyridine and Piperidine Alkaloids

J. D. HUNT AND A. McKILLOP

The parent bases, pyridine and piperidine, and some of their simple derivatives are found in nature. Pyridine has been isolated from *Aplopappus hartwegi* (*T. F. Buehrer et al.*, Amer. J. Pharm., 1939, **111,** 105) and 3-methoxypyridine from *Equisetum arvense* L. (*R. H. F. Manske* and *L. Marion*, Canad. J. Research, 1942, B, **20,** 87) and from *Thermopsis rhombifolia* (*Manske, ibid.*, p. 265). Piperidine is found in pepper (*Piper nigrum* L.) (*E. Späth* and *G. Englaender*, Ber., 1935, **68,** 2218), tobacco (*Späth* and *C. E. Zajic, ibid.*, 1936, **69,** 2448) and several other plants. *N*-Methylpiperidine is a constituent of *Girgensohnia* diptera (*N. K. Yurashevskii* and *S. I. Stepanov*, J. gen. Chem. U.S.S.R., 1939, **9,** 2203).

Piperine, $C_{17}H_{19}NO_3$, occurs in *Piper nigrum* L. (*V. Regnault*, Ann., 1911, **383,** 152). It is a very weak base which is hydrolysed by acid or alkali to piperidine and piperic acid (*R. Fittig* and *W. H. Mielch*, Ann., 1874, **172,** 134). **Piperic acid,** $C_{12}H_{10}O_4$, m.p. 210°, is produced when *trans*-3,4-methylenedioxycinnamylidenemalonic acid (I) is heated with acetic anhydride (*L. Higginbotham* and *A. Lapworth*, J. chem. Soc., 1922, **121**, 2829). Piperine is formed by the action of piperidine on piperyl chloride (*L. Rugheimer*, Ber., 1882, **15**, 1390; *C. Feugas*, Bull. Soc. chim. Fr., 1964, 1892) and is, therefore, *N*-piperylpiperidine:

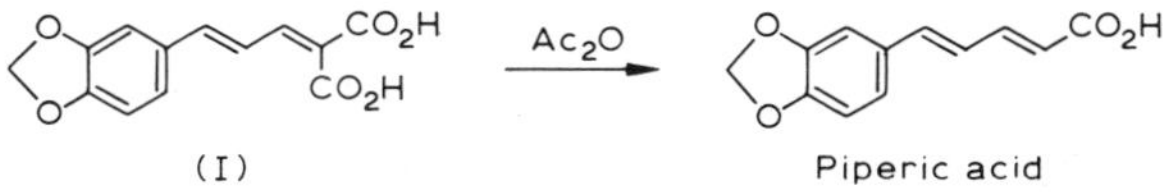

CO_2H CO_2H (I) $\xrightarrow{Ac_2O}$ CO_2H Piperic acid

Piperine   Chavicine

Piperine has m.p. 128–129.5°; *hydrobromide* m.p. 170°; 1,3,5-*trinitrobenzoate* complex, red needles, m.p. 130°.

Pepper also contains the amorphous alkaloid **chavicine,** which is isomeric with piperine (*E. Ott et al.*, Ber., 1922, **55,** 2653; 1924, **57,** 214). Hydrolysis yields piperidine and chavicic acid, which is a geometrical isomer of piperic acid (*H. Lohaus* and *H. Gall*, Ann., 1935, **517,** 282). Chavicine is therefore, the piperide of chavicic acid.

Piperlongumine, $C_{17}H_{19}NO_5$, m.p. 124°, is found in the root of *Piper longum* (*A. Chatterjee* and *C. P. Dutta*, Tetrahedron, 1967, **23,** 1769). It is a monoacidic, weak base and forms no crystalline salts either with the hydrogen halides or with methyl iodide. The u.v. spectrum in ethanolic alkali is comparable with that of 3,4,5-trimethoxycinnamic acid, while the i.r. spectrum indicates the presence of an amide carbonyl group and olefinic unsaturation. This latter point is verified by the n.m.r. spectrum, which shows the presence of four vinyl protons. Catalytic hydrogenation gives *tetrahydropiperlongumine*, m.p. 84°, which shows two separate amide carbonyl bands in the i.r. spectrum but no olefinic unsaturation.

Hydrolysis of piperlongumine gives 3,4,5-trimethoxycinnamic acid and *piperlongumic acid*, $C_{17}H_{21}NO_6$, m.p. 174°, the u.v. spectrum of which is still similar to that of 3,4,5-trimethoxycinnamic acid. The i.r. spectrum is basically similar to that of the parent compound except for the bands attributable to the acid function. Catalytic reduction of piperlongumic acid gives the *tetrahydro derivative*, $C_{17}H_{25}NO_6$, m.p. 114°, the i.r. spectrum of which shows the presence of -NH, $-CO_2H$ and -N-C=O groups. The amide band moves to lower wavelength on reduction, thus indicating that it was originally conjugated with one of the double bonds. Hydrolysis of piperlongumic acid leads to 3,4,5-trimethoxyphenylpropionic acid and α-piperidone. From this information it follows that piperlongumine must be represented as either II or III.

Ozonolysis of piperlongumic acid gives 3,4,5-trimethoxybenzaldehyde and succindialdehyde; hence piperlongumine is III and piperlongumic acid is IV:

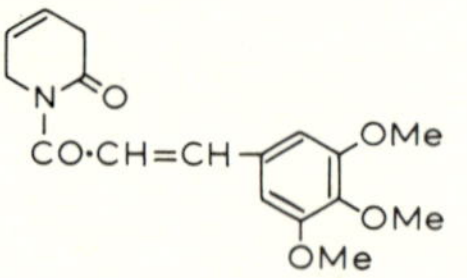

(II)

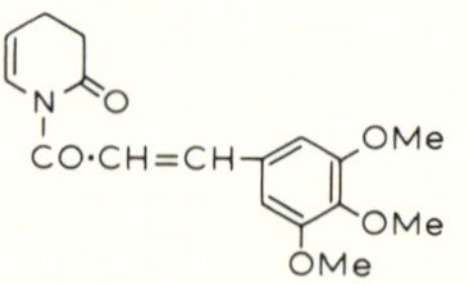

(III)
Piperlongumine

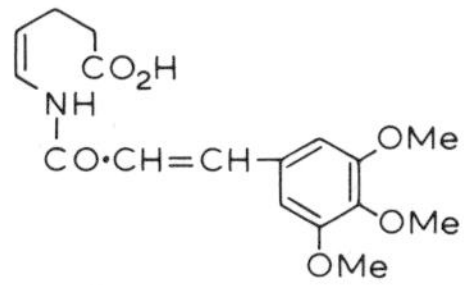

(IV)
Piperlongumic acid

Piplartine, isolated from *Piper longum,* was originally thought to be the piperide of 3,4,5-trimethoxycinnamic acid (*C. K. Atal* and *S. S. Banga,* Current Sci. India, 1963, **32,** 354), but more recently it has been shown to be identical with piperlongumine (*Chatterjee* and *Dutta, loc. cit.*).

Trigonelline, $C_7H_7NO_2$, occurs in a wide variety of plants including coffee (*K. Gorter,* Ann., 1910, **372,** 237), and potato (*K. Yoshimura,* Biochem. Z., 1934, **274,** 408). When heated with alkali it affords methylamine, and with hydrochloric acid it gives nicotinic acid and methyl chloride. Its formulation as 1-methylpyridinium-3-carboxylate, the methyl betaine of nicotinic acid:

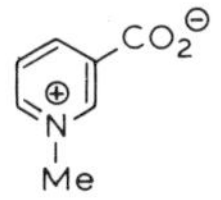

has been confirmed by its synthesis by the action of moist silver oxide on nicotinic acid methiodide (*A. Hantzsch,* Ber., 1886, **19,** 31).

Trigonelline monohydrate melts at 130°, while anhydrous *trigonelline* has m.p. 218° (decomp.); *hydrochloride,* m.p. 260°; *picrate,* m.p. 198–200°.

Ricinine, $C_8H_8N_2O_2$, is an alkaloid which is found in the seeds of the castor oil plant, *Ricinus communis* L. (*R. V. Tuson,* J. chem. Soc., 1864, **17,** 195). It is a neutral, optically inactive compound which yields pyridine on zinc dust distillation. Alkaline hydrolysis affords methanol and ricininic "acid", $C_7H_6N_2O_2$, which, with hydrochloric acid at 150°, is converted into carbon dioxide, ammonia and a base $C_6H_7NO_2$, which was proved by synthesis (*Späth* and *E. Tachelnitz,* Monatsh., 1921, **42,** 251) to be 4-hydroxy-1-methyl-2-pyridone. Treatment of ricininic "acid" with phosphoryl chloride yields a chloro compound $C_7H_5ClN_2O$, reducible to ricinidine, $C_7H_6N_2O$, proved to be 3-cyano-1-methyl-2-pyridone by synthesis (*Späth* and *G. Koller,* Ber., 1923, **56,** 880):

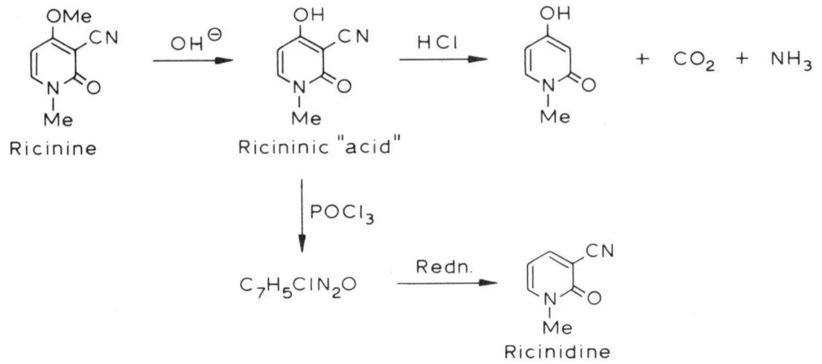

Ricinine $\xrightarrow{OH^{\ominus}}$ Ricininic "acid" $\xrightarrow{HCl}$ (1-methyl-4-hydroxy-2-pyridone) + CO_2 + NH_3

Ricininic "acid" $\xrightarrow{POCl_3}$ $C_7H_5ClN_2O$ $\xrightarrow{\text{Redn.}}$ Ricinidine

The structure of ricinine has been proved by several syntheses: from 4-chloroquinoline (*idem, ibid.*, p. 2454); from 2,4-dihydroxy-6-methylpyridine-3-carboxylic acid (*idem, ibid.*, 1925, **58,** 2124); from *β*-picoline (*E. C. Taylor* and *A. J. Crovetti*, J. Amer. chem. Soc., 1956, **78,** 214); and from cyanoacetyl chloride, which polymerises spontaneously to 6-chloro-3-cyano-2,4-dihydroxypyridine; methylation gives 6-chlororicinine which is reduced to the alkaloid with zinc and acid (*G. Schroeter et al.*, Ber., 1932, **65,** 432; 1938, **71,** 671).

Ricinine has m.p. 201.5° and sublimes at 170°/20 mm. It is so weakly basic that it forms no salts. *Ricininic "acid"* has m.p. 320° (decomp.).

Leucaenine, (Leucinol), $C_8H_{10}N_2O_4$, occurs in the seeds of *Leucaena glauca* Benth. (*M. Mascre*, Compt. rend., 1937, **204,** 890). It is an α-amino acid and contains a phenolic hydroxyl group. *J. P. Wibaut* and co-workers (Rec. Trav. chim., 1946, **65,** 65; 1947, **66,** 24; Helv., 1946, **29,** 1669) have shown that it yields pyridine on distillation with zinc dust, and on treatment with dimethyl sulphate it is converted into a methyl ether, $C_7H_{11}NO_3$, which is a monohydrate, $C_7H_9NO_2 \cdot H_2O$. The latter compound, when heated with hydrochloric acid affords 3-hydroxy-1-methyl-4-pyridone, (V), identified by synthesis. Pyrolysis of leucaenine yields 3,4-dihydroxypyridine. The alkaloid is, therefore, a 3-hydroxy-4-pyridone with an α-aminopropionic acid side chain attached either to oxygen or to nitrogen. Recent investigations suggest that the point of attachment is the nitrogen atom, so that leucaenine is represented by VI (*R. Adams et al.*, J. Amer. chem. Soc., 1947, **69,** 1803, 1806, 1810):

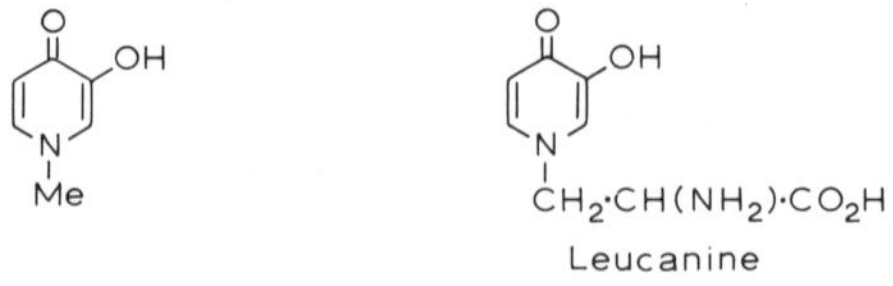

(V)

$CH_2 \cdot CH(NH_2) \cdot CO_2H$

Leucanine

(VI)

This structure has been confirmed by a simple synthesis of leucaenine by the addition of 3-methoxy-4-pyridone to α-acetamidoacrylic acid, followed by hydrolysis of the product with hydrochloric acid (*Adams* and *J. L. Johnson*, *ibid.*, 1949, **71,** 705).

Although it was originally reported as being optically inactive, more recent research has shown that leucaenine is laevorotatory with $[\alpha]_D$ $-22°$ (water) (*Wibaut* and *J. P. Schumacker*, Rec. Trav. chim., 1952, **71**, 1017). Although l-leucaenine can be racemised by prolonged boiling in water, *Mascré*'s original sample was, fortunately, still available to the later workers and shown to be optically active. Furthermore, a sample of **mimosine** (C.C.C., 1st Edn., 1960, Vol. IV C, p. 1805) isolated from *Mimosa pudica* L. (*J. Renz*, Z. Physiol. Chem., 1936, **244**, 153) was also shown to be identical with l-leucaenine.

Leucaenine has m.p. 236°; *hydrochloride*, m.p. 174–175°; *hydrobromide*, m.p. 179–180° (decomp.); *hydriodide*, m.p. 183°.

(a) Alkaloids of areca nut

The areca or betel nut (the fruit of *Areca catechu* L., a palm tree of the Far East) contains several alkaloids. The most important is **arecoline**, $C_8H_{13}NO_2$, which is optically inactive and is hydrolysed to methanol and **arecaidine**, (arecaine), $C_7H_{11}NO_2$, another natural base. Arecaidine is amphoteric and contains an *N*-methyl group and a carboxylic acid, the methyl ester of which is identical with arecoline. Both arecoline and arecaidine, which are reduced catalytically to dihydro derivatives, are tetrahydropyridines, as catalytic reduction of methyl nicotinate methiodide (VIII) yields a hexahydro derivative identical with dihydroarecaidine. As the

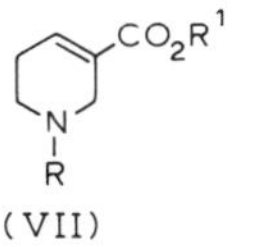

(VII) 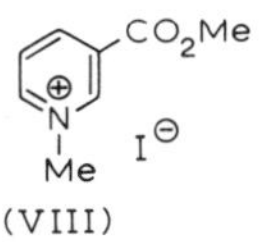 (VIII)

alkaloids are optically inactive, the double bond must be in the 2,3- or the 3,4-position.

The u.v. spectrum of arecoline is typical for an α,β-unsaturated ester, but different to that of a tertiary β-aminocrotonic ester, suggesting that arecoline is methyl 1,2,5,6-tetrahydro-1-methylnicotinate, (VII, $R = R^1 = Me$) and that arecaidine is the free acid, VII ($R = Me$, $R^1 = H$).

This has been confirmed by several syntheses (*A. Wohl* and *A. Johnson*, Ber., 1907, **40,** 4712; *E. Merck* and *H. Maeder*, G.P. 485,139/1926). An interesting synthesis of arecaidine aldehyde consists in heating methylamine hydrochloride, formaldehyde, acetaldehyde and water at pH 3 and 70° for some hours (*C. Mannich*, Ber., 1942, **75,** 1480). The aldehyde can be converted, *via* the oxime and nitrile, to arecaidine.

Guvacine, $C_6H_9NO_2$, a third natural base is also amphoteric and contains an imino group. It has been proved by degradation and synthesis (*E. Jahns*, Arch. Pharm., 1892, **229,** 669; *E. Winterstein* and *A. Weinhagen*, Z. physiol. Chem., 1918, **104,** 48; *Wohl* and *M. S. Lasanitsch*, Ber., 1907, **40,** 4698; *S. M. McElvain* and *G. Stork*, J. Amer. chem. Soc., 1946, **68,** 1049) to be 1,2,5,6-tetrahydronicotinic acid, (VII, $R = R^1 = H$). **Guvacoline,** $C_7H_{11}NO_2$, is the methyl ester (VII, $R = H$, $R^1 = Me$) of guvacine (*K. Hess*, Ber., 1918, **51,** 1004). The physical properties of the bases and some of their derivatives are recorded in Table 1.

TABLE 1

ARECA NUT ALKALOIDS

Base	*m.p. (°C)*	*Hydrochloride m.p. (°C)*	*Hydrobromide m.p. (°C)*
Arecoline	b.p. 209	157–158	170–171
Arecaidine	223–224	257–258	248–249
Guvacine	271–272	316	280
Guvacoline	27 b.p. 114/13 mm	121–122	144–145

(b) Alkaloids of hemlock

The toxic properties of the common hemlock, *Conium maculatum* L., are attributable to the hemlock alkaloids, the most important of which is coniine. Other bases are *N*-methylconiine, conhydrine, pseudoconhydrine and γ-coniceine. The alkaloids occur in the plant as salts of malic and caffeic acids; various methods for isolation and separation have been described (*J. von Braun*, Ber., 1905, **38,** 3108; *F. Chemnitius*, J. pr. Chem., 1928, **118,** 25).

Coniine, $C_8H_{17}N$, is found as (+)- and (−)-forms. It is a secondary base which, when dehydrogenated with zinc dust or silver acetate is converted into **conyrine,** $C_8H_{11}N$, oxidation of which affords picolinic acid while it is reconverted into coniine by reduction with hydriodic acid. Conyrine is, therefore, an α-substituted pyridine, and is, in fact, 2-*n*-propylpyridine (*A. Ladenburg*, Ber., 1885, **18,** 1587). Hence coniine is 2-*n*-propylpiperidine. This structure has been confirmed by several syntheses, notable amongst which is that of *Ladenburg* (*ibid.*, 1886, **19,** 439, 2579) which is the first recorded synthesis of an alkaloid. Condensation of α-picoline with paraldehyde at 250° gives 2-propenylpyridine (IX) which, on reduction with sodium and ethanol, gives (±)-coniine. Resolution with (+)-tartaric acid afforded (+)-coniine. The synthesis has been improved by later workers (*Hess* and *W. Wellzien*, *ibid.*, 1920, **53**, 139; *Koller*, Monatsh., 1926, **47**, 393):

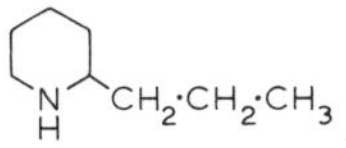

Coniine 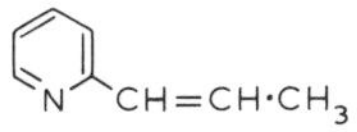 (IX)

(±)-Coniine has also been synthesised *via* a diene addition reaction between pyridine and dimethyl acetylenedicarboxylate (*O. Diels* and *K. Alder*, Ann., 1932, **408,** 1; *E. Ochiai* and *K. Tauda*, Ber., 1934, **67,** 1011), and from 2-(1′-hydroxypropyl)pyridine (X), which is obtained by treatment of pyridine 2-carbaldehyde with ethylmagnesium bromide (*C. Engler* and *F. M. Bauer*, *ibid.*, 1891, **24,** 2530; *L. Lautinschläger* and *A. G. T. Onsager*, *ibid.*, 1918, **51,** 602). More recently, coniine has been prepared by catalytic and Wolff–Kishner reduction of 2-propionylpyridine, formed by the action of ethylmagnesium bromide on 2-cyanopyridine (*K. B. Prasad* and *S. C. Shaw*, *ibid.*, 1965, **98,** 2822):

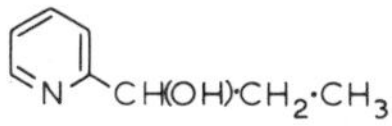

(X) 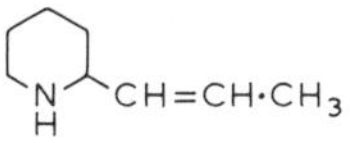 β-Coniceine

Labelled coniine[^{15}N] has been prepared by the reductive amination of 5-oxo-octanal with ammonium[^{15}N] bromide and sodium cyanotrihydridoborate (*E. Leete et al.*, J. Label. Compounds, 1971, **7**, 313).

(+)-Coniine is racemised when heated with alcoholic baryta (*Hess* and *A. Eichel*, Ber., 1917, **50,** 1192).

(−)-Coniine is also found in hemlock. It has been synthesised by reduction of (−)-β-coniceine (*K. Loffler* and *F. Friedrich*, *ibid.*, 1909, **42,** 107).

(+)- and (−)-*N*-Methylconiines, which occur in hemlock in small quantities, are obtained by methylation of the coniines by the action of potassium methyl sulphate (*M. Passon*, *ibid.*, 1891, **24,** 1678) or formaldehyde and formic acid (*Hess* and *Eichel*, Ber., 1917, **50**, 1386).

Conhydrine, $C_8H_{17}NO$, is a secondary amine containing a secondary alcoholic group. On mild oxidation it yields the corresponding ketone **conhydrinone,** $C_8H_{15}NO$; more vigorous oxidation affords pipecolinic acid (piperidine-2-carboxylic acid). Treatment of conhydrine with hydriodic acid converts it into an iodoconiine, which on reduction gives coniine. Coniine is also obtained by dehydration of conhydrine with phosphorus pentoxide to β-coniceine, followed by reduction. Conhydrine is, therefore, a hydroxyconiine (*Späth* and *Adler*, Monatsh., 1933, **63,** 127; *A. W. Hofmann*, Ber., 1885, **18,** 5). The position of the secondary alcoholic group is established by the dehydrogenation of conhydrine to 2-(1-hydroxypropyl)pyridine (X) (*Hess* and *R. Grau*, Ann., 1925, **441,** 101), and by the *N*-methylation and oxidation of the alkaloid to the ketone 1-methyl-2-propionylpiperidine (XI), isomeric with methyl isopelletierine (*Hess et al.*, Ber., 1917, **50,** 351, 1386; 1919, **52,** 964; Ann., 1925, **441,** 101; *Späth* and *Adler*, *loc. cit.*):

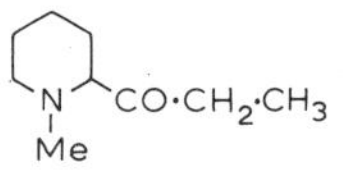

$CO{\cdot}CH_2{\cdot}CH_3$ (N–Me) (XI) — $CHOH{\cdot}CH_2{\cdot}CH_3$ (N–H) Conhydrine (XII)

Conhydrine, therefore, has the structure shown, which has been confirmed by synthesis (*F. Galinovsky* and *H. Mully*, Monatsh., 1948, **79,** 426). Ethyl picolinate and ethyl propionate condense to give the β-keto-ester XIII which, on hydrolysis and decarboxylation, affords 2-propionylpyridine:

$CO{\cdot}CH(CH_3){\cdot}CO_2Et$ (XIII) ⟶ $CO{\cdot}CH_2{\cdot}CH_3$

Catalytic hydrogenation of the latter yields 2-(1′-hydroxypropyl)piperidine (XII) as a mixture of two racemates which are separable by fractional crystallisation. One of these on resolution with 6,6′-dinitro-2,2′-diphenic acid gives (+)- and (−)-conhydrine, the former of which is identical with the natural alkaloid. The absolute configuration of conhydrine was determined by degradation to octane-3,4-diol (Scheme 1).

Scheme 1

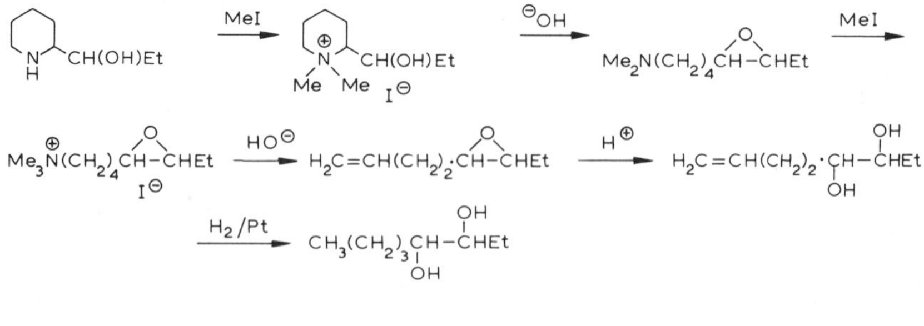

$\text{CH(OH)Et (NH)} \xrightarrow{MeI} \overset{\oplus}{N}Me_2\ \text{CH(OH)Et}\ I^{\ominus} \xrightarrow{^{\ominus}OH} Me_2N(CH_2)_4CH\text{–}CHEt\ (\text{epoxide}) \xrightarrow{MeI}$

$Me_3\overset{\oplus}{N}(CH_2)_4CH\text{–}CHEt\ (\text{epoxide})\ I^{\ominus} \xrightarrow{HO^{\ominus}} H_2C{=}CH(CH_2)_2CH\text{–}CHEt\ (\text{epoxide}) \xrightarrow{H^{\oplus}} H_2C{=}CH(CH_2)_2{\cdot}CH(OH)\text{–}CH(OH)Et$

$\xrightarrow{H_2/Pt} CH_3(CH_2)_3CH(OH)\text{–}CH(OH)Et$

Comparison of the diol obtained from the alkaloid with synthetic samples showed that the diol had the *erythro* configuration, from which it was deduced that the absolute configuration of conhydrine is as indicated in XIV (*R. K. Hill*, J. Amer. chem. Soc., 1958, **80,** 1609; *J. Sicher* and *M. Tichy*, Chem. Ind., 1958, 16; Chem. Listy, 1958, **52,** 2316):

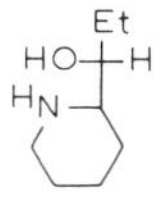

(XIV)

(+)-Conhydrine has been synthesised by hydrogenation of (+)-ethyl-(2-pyridyl)methanol (X) (*G. Fodor* and *E. Bauerschmidt*, J. heterocycl. Chem., 1968, **5,** 205).

Pseudoconhydrine is isomeric with conhydrine, and also contains a secondary alcoholic and a secondary amino group. It is converted into coniine by reactions similar to those described for conhydrine (*Löffler*, Ber., 1909, **42,** 116; *Späth et al., ibid.*, 1933, **66,** 591). The hydroxyl group is not, however, in the side chain since oxidation of the alkaloid does not yield pipecolinic acid. The methine base, $C_{10}H_{21}NO$, obtained by Hofmann degradation, can be reduced to a dihydro derivative, $C_{10}H_{23}NO$, which affords *n*-heptanoic acid on oxidation. Thus the alkaloid contains an unbranched chain of eight carbon atoms and the dihydromethine base is represented by either XV or XVI. The latter structure is inconsistent with the properties observed for the methine base, which has, therefore, structure XV, and pseudoconhydrine is as shown by the confirmatory synthesis

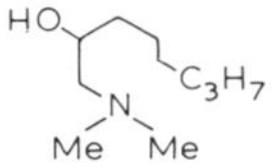

(XV) 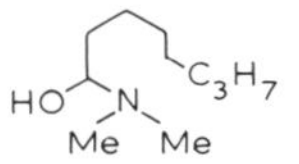 (XVI)

outlined in Scheme 2 (*L. Marion* and *W. F. Cockburn*, J. Amer. chem. Soc., 1949, **71,** 3402):

Scheme 2

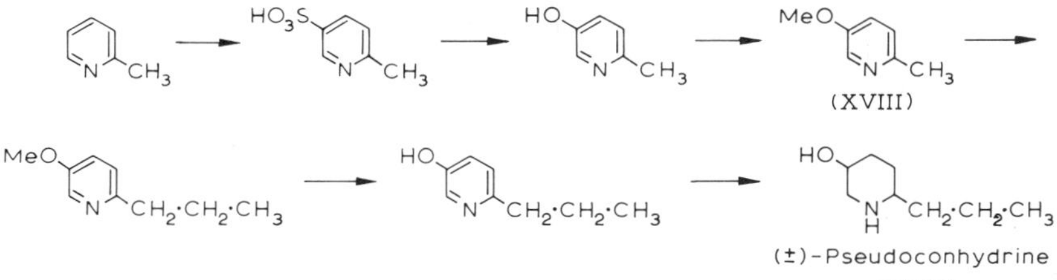

(XVIII)

(±)-Pseudoconhydrine

(XVII)

α-Picoline on sulphonation yields the 5-sulphonic acid, which, on fusion with potassium hydroxide followed by treatment with diazomethane, is converted into 5-methoxy-α-picoline, (XVIII). On ethylation with potassium amide and ethyl chloride the latter affords 5-methoxy-2-*n*-propylpyridine, (XIX); demethylation with hydrobromic acid gives the corresponding alcohol which, on catalytic hydrogenation, yields 5-hydroxy-2-*n*-propylpiperidine (XVII) as a mixture of two racemates separable by vacuum distillation. One racemate, like the natural alkaloid, forms a hydrate very readily; this form was resolved with 6,6′-dinitro-2,2′-diphenic acid to give a (+)-form identical with natural (+)-pseudoconhydrine. Another synthesis (Scheme 3) starting with 2-chloro-5-nitropyridine has been described (*W. Gruber* and *K. Schlögl*, Monatsh., 1949, **80,** 499):

Scheme 3

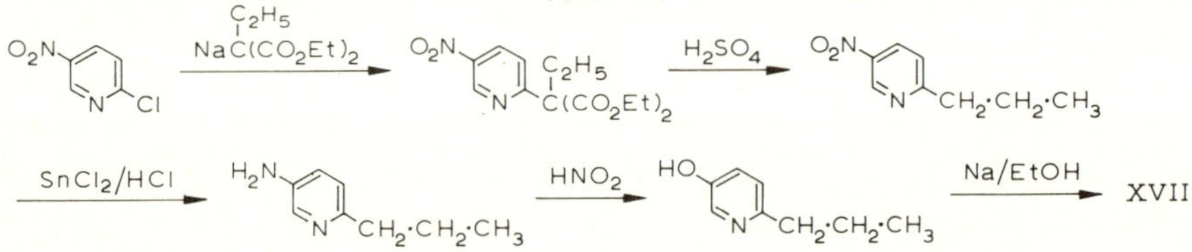

The absolute stereochemistry of pseudoconhydrine has been deduced by *Hill* (J. Amer. chem. Soc., 1958, **80,** 1611) to be:

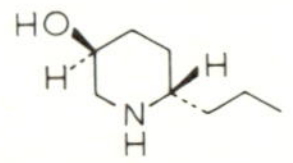

This has subsequently been confirmed by an X-ray diffraction study of the hydrobromide (*H.S. Yanai* and *W. N. Lipscomb*, Tetrahedron, 1959, **6**, 103).

γ-**Coniceine,** $C_8H_{15}N$, is the major alkaloid of *Conium maculatum* (*B. F. Cromwell*, Biochem. J., 1956, **64,** 259) and it occurs in crude mixtures with coniine from which it can be separated by fractional crystallisation of the hydrochlorides. It is an optically inactive base which is easily reduced to (±)-coniine and which yields conyrine on dehydrogenation. γ-Coniceine is, therefore, a dehydroconiine and the double bond is evidently endocyclic, as no small aldehyde, ketone or acid fragments are obtained on oxidation. Structure XX is consistent with the lack of optical activity, and for many years it was thought to represent the true structure of γ-coniceine. The i.r. spectrum of γ-coniceine shows no N–H absorption, however, but does show a moderately intense band attributable to a carbon–nitrogen double

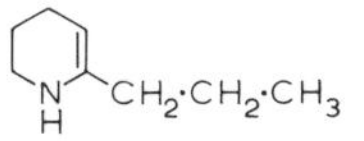

(XX)

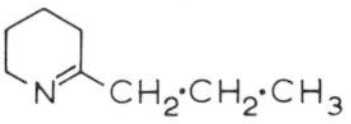

γ-Coniceine (XXI)

bond. The n.m.r. spectrum shows no evidence of vinyl protons and hence structure XXI, which is also compatible with the lack of optical activity, would appear to be the correct structure of γ-coniceine (*H. C. Beyerman et al.*, Rec. Trav. chim., 1961, **80**, 513).

γ-Coniceine has been synthesised by the acid hydrolysis and cyclisation of δ-phthalimidobutyl *n*-propyl ketone (XXII) (*S. Gabriel*, Ber., 1909, **42,** 4059). Alternatively, base-catalysed condensation of *N*-butyryl-2-piperidone

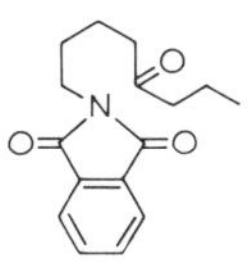

(XXII)

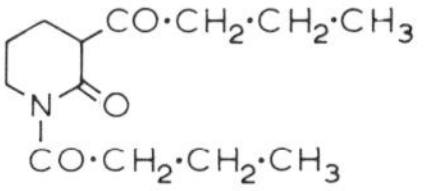

(XXIII)

with ethyl butyrate yields XXIII, which undergoes acid-catalysed migration and decarboxylation to give γ-coniceine, (XXI) (*K. H. Büchel* and *F. Korte*, *ibid.*, 1962, **95,** 2460). The synthesis, however, that confirms structure XXI beyond doubt is due to *B. E. Reynolds* (J. chem. Soc., 1964, 2445). *N*-Chloro-2-*n*-propylpiperidine, prepared by the action of *N*-chlorosuccinimide on the 2-alkylpiperidine, is dehydrochlorinated to γ-coniceine with ethanolic potassium hydroxide.

The three bases α-, β- and δ-**coniceine,** $C_8H_{15}N$, do not occur naturally, but are formed by the action of acidic dehydrating agents on conhydrine. β-Coniceine is 2-propenylpiperidine (*cf.* p. 120) (*Löffler et al.*, Ber., 1905, **38**, 3326; 1909, **42,** 107, 929). δ-Coniceine is obtained by the acid dehydrobromination of *N*-bromoconiine (formed by the action of sodium hypobromite on coniine). It is a saturated tertiary base and the structure XXV has been confirmed by synthesis (*Löffler* and *H. Kaim*, *ibid.*, 1909, **42,** 94). α-Picoline and chloral react to give the aldol XXIV, which, on successive hydrolysis, dehydration and reduction, yields β-(2-piperidyl)propionic acid. This acid lactamises readily to XXV, reduction of which with sodium and ethanol gives δ-coniceine XXV. Other syntheses have been described more

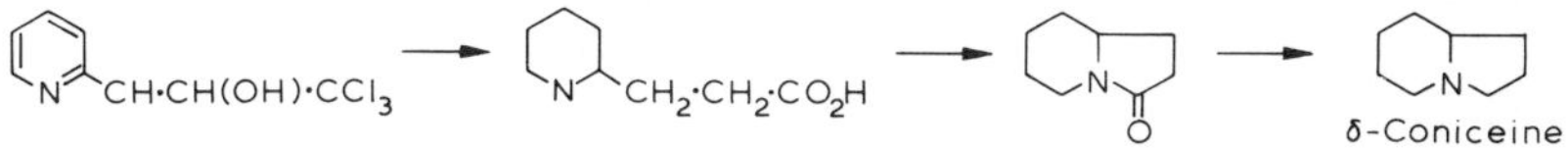

(XXIV) (XXV) δ-Coniceine (XXVI)

recently (*K. Arh-Lipovac* and *R. Seiwerth*, Monatsh., 1953, **84,** 992; *W. J. Linn et al.*, J. Amer. chem. Soc., 1953, **75,** 3243; *K. Winterfeld* and

E. Müller, Arch. Pharm., 1951, **284,** 269; *R. Lukeš* and *Z. Veseiý*, Coll. Czech. chem. Comm., 1959, **24,** 944; *Lukeš* and *M. Janda, ibid.*, 1959, **24,** 2717).

α-Coniceine is of somewhat uncertain structure, but appears to be a 2-(2-propenyl)piperidine.

ε-**Coniceine,** obtained by the action of bases on *N*-iodoconiine, is a mixture of racemates represented by formula XXVII (*Löffler* and *M. Kirschner*, Ber., 1905, **38,** 3329). These bases, termed 2-methylconidine and iso-2-methylconidine, have been separated by fractional crystallisation of their tartrates (*Löffler, ibid.*, 1909, **42,** 948). A stereospecific synthesis of *erythro*-ε-coniceine has been achieved by reduction of optically active 1-(2′-pyridyl)-2-propanol (XXVIII) (*Fodor* and *G. A. Cooke*, Tetrahedron, 1966, Suppl. 8, Part 1, 113):

N CH_3
ε-Coniceine
(XXVII)

N $CH_2{\cdot}CH(OH)CH_3$
(XXVIII)

Pseudoconiceine is formed by the dehydration of pseudoconhydrine and is represented by formula XXIX or XXX (*Späth et al.*, Ber., 1933, **66,** 591):

N H $CH_2{\cdot}CH_2{\cdot}CH_3$
(XXIX)

N H $CH_2{\cdot}CH_2{\cdot}CH_3$
(XXX)

TABLE 2

HEMLOCK ALKALOIDS AND RELATED COMPOUNDS

Base	*m.p. (°C)*	*b.p. (°C)*	$[\alpha]_D$	*B.HCl m.p. (°C)*	*B.HBr m.p. (°C)*	*Picrate*
Connine[a]	−2	166–167	±15.7°	220	211	75
N-Methylconiine		174–175	±81.5°	188–190	189–190	121–122
Conhydrine[b]	121	226	+10°			
Pseudoconhydrine[c]	105–106	236	+11°	212–213		
γ-Coniceine		171–172 /746 mm	Inactive	143	139	62
α-Coniceine		158	+18.4°			224
β-Coniceine	41	168	−53°	206–207		113–114.5
δ-Coniceine		158	−7.8°			226
2-Methylconidine		152–154	+67.4°	167–168[d]		
Iso-2-methylconidine		143–145	−87.3°	198–199[d]		
Pseudoconiceine		171–172	+122.6	205–206		
Conyrine		166–168	Inactive			70

[a] 2,4-*Dinitrobenzoyl* derivative, m.p. 139°; 3,5-*dinitrobenzoyl* derivative, m.p. 108–109°.
[b] (±)-Base, m.p. 69–70°; *benzoyl* derivative, m.p. 132°.
[c] *Hydrate*, m.p. 80°.
[d] $B.HCl.AuCl_3$

The physical properties of the hemlock alkaloids and related compounds are listed in Table 2. All are very poisonous.

(c) Alkaloids of pomegranate root bark

The bark of the pomegranate tree, *Punica granatum* L., contains several alkaloids, the most important of which is pseudopelletierine, which is closely related to the tropane alkaloids and is considered with them (see C.C.C., Vol. IVB, pp. 233 *et seq.*). The minor alkaloids have been the subject of dispute for a great many years and elucidation of their structures has only recently been possible with the aid of modern spectroscopic methods. Early controversy about the existence of four minor alkaloids, **pelletierine, methylpelletierine, isopelletierine** and **methyl-isopelletierine** (*C. Tanret*, Compt. rend., 1878, **86**, 1270; 1880, **90**, 695; 1920, **170**, 1118; *K. Hess* and *A. Eichel*, Ber., 1917, **50**, 368, 1192, 1386; 1918, **51**, 741; *A. Piccinini*, Gazz., 1899, **2**, 29, 311; *J. P. Wibaut et al.*, Rec. Trav. chim., 1954, **73**, 102; *K. V. Giri*, J. Indian Inst. Sci., 1955, **37**, 1) was finally settled by *R. E. Gilman* and *L. Marion* (Bull. Soc. chim. Fr., 1961, 1993). These workers even managed to obtain some of *Tanret*'s original samples, and showed by i.r. and n.m.r. spectroscopy that pelletrierine and isopelletierine were identical, as were the corresponding methyl compounds. The structure that they confirmed, XXXI,

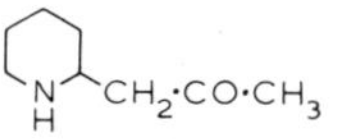

$CH_2{\cdot}CO{\cdot}CH_3$ (piperidin-2-yl, N–H)

Pelletierine
(XXXI)

was that originally assigned to isopelletierine (*J. Meisenheimer* and *E. Mahler*, Ann., 1928, **462**, 301; *M. G. J. Beets*, Rec. Trav. chim., 1944, **63**, 120), but for the sake of simplicity the prefix iso- is not generally now used.

A variety of syntheses of pelletierine and *N*-methylpelletierene have been reported. 1-(2′-Pyridyl)propan-2-ol (XXVIII) was converted into the methosulphate and reduced catalytically to the corresponding piperidine derivative, which gave *N*-methylpelletierine on oxidation with chromic acid *(Meisenheimer* and *Mahler, loc. cit.)*. Alternatively, 2-pyridylacetone may be prepared by condensation of α-picolyllithium with ethyl acetate, and catalytic reduction affords pelletierine *(Beets, loc. cit.)*. Syntheses have also been reported which start from δ-aminovaleraldehyde (XXXII) (*E. Anet et al.*, Nature, 1949, **164,** 501), cadaverine (XXXIII) (*H. Tuppy* and *M. S.*

Faltaous, Monatsh., 1960, **91,** 167) and from lysine (XXXIV) (*R. N. Gupta* and *I. D. Spenser*, Canad. J. Chem., 1969, **47,** 445):

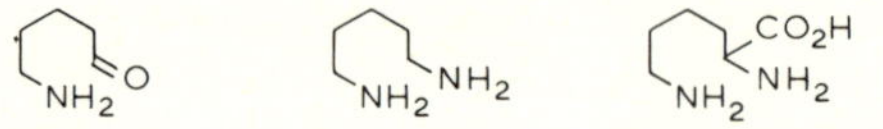

(XXXII) (XXXIII) (XXXIV)

This latter synthesis from lysine closely parallels the biogenetic route (*M. F. Keogh* and *D. G. O'Donovan*, J. chem. Soc. C, 1970, 1792).

(−)-Pelletierine possesses the (*R*) configuration (*H. C. Beyerman* and *L. Matt*, Rec. Trav. chim., 1963, **82,** 1033; 1965, **84,** 385):

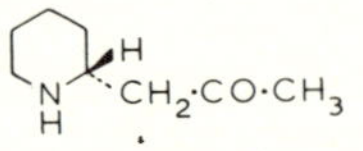

(−)-Pelletierine

Pelletierine has b.p. 102–107°/11 mm; *hydrochloride*, m.p. 143°; *hydrobromide*, m.p. 149°; *picrate*, m.p. 127–129°.

N-*Methylpelletierine* has b.p. 114–117°/26 mm., *hydrochloride*, m.p. 156°; *hydrobromide*, m.p. 151–152°; *methiodide*, m.p. 156°; *picrate*, m.p. 158°.

The pomegranate bark alkaloids have anthelmintic properties.

(d) Lobelia and sedum alkaloids

Species of lobelia and sedum have been the source of a large number of closely related alkaloids and *Lobelia inflata* L. alone has provided at least fourteen different bases. In an attempt to instil some order into the classification of these alkaloids a systematic nomenclature has been introduced (*C. Schöpf et al.*, Ann., 1957, **608,** 88) and this will be used here; where a trivial name had previously gained acceptance both this and the systematic name will be indicated.

The alkaloids are conveniently divided into two basic structural types, XXXV and XXXVI, with name stems "lobeli" and "lobel", respectively:

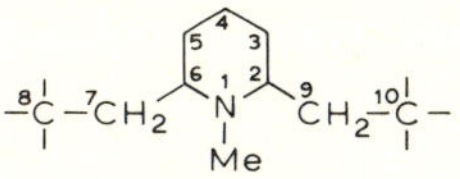

lobeli (XXXV)

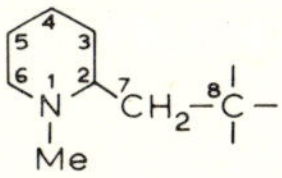

lobel (XXXVI)

In cases of "lobeli" compounds where there are different carbon substituents at C-8 and C-10 the smaller group is placed at C-8; alcohol or ketone functions at C-8 and C-10 are indicated by the suffixes "-ol" and "-one". The relative stereochemistry at C-2 and C-6 in "lobeli" compounds is denoted by the terms *cis* or *trans*. As an example, pelletierine, (XXXI), becomes 8-methylnorlobelone in this system.

(−)-**Lobeline,** $C_{22}H_{27}NO_2$, is the major *Lobelia* alkaloid. It is tertiary base containing an *N*-methyl group, and forms an *O*-benzoyl derivative; it therefore contains a hydroxy group. On reduction with sodium amalgam lobeline yields lobelanidine, and on oxidation with chromic acid it affords lobelanine.

Lobelanine, $C_{22}H_{25}NO_2$, is optically inactive and yields a crystalline dioxime. It is a diketone which on reduction gives lobeline and then lobelanidine.

Lobelanidine, $C_{22}H_{29}NO_2$, is also optically inactive, and gives a di-*O*-benzoyl derivative and a di-*O*-acetyl derivative; it is therefore a dihydric alcohol. On oxidation with chromic acid it affords lobeline and lobelanine.

From these transformations it is evident that lobeline and lobelanidine are respectively the monosecondary and disecondary alcoholic reduction products of the diketone lobelanine. All three bases give acetophenone on dry distillation, and lobelanine yields more than one molar equivalent. Lobelanine is oxidised by chromic acid to 1-methylpiperidine-2,6-dicarboxylic acid (scopolinic acid) and benzoic acid, and its dioxime undergoes the Beckmann rearrangement with the formation of the dianilide of 1-methylpiperidine-2,6-diacetic acid (lobelinic acid) (*H. Wieland et al.*, Ann., 1925, **444,** 40). When lobelanine is subjected to a Hofmann degradation it yields eventually an unsaturated diketone which is reduced catalytically to 1,7-dibenzoyl-*n*-heptane (*Wieland* and *O. Dragendorff, ibid.*, 1929, **473,** 83). On the basis of these and other observations lobelanine is formulated as *cis*-8,10-diphenyllobelidione, lobeline as *cis*-8,10-diphenyllobelionol and lobelanidine as *cis*-8,10-diphenyllobelidiol. These structures have been confirmed by a variety of syntheses, one of the earliest of which is shown in Scheme 4 (*G. Scheuing* and *L. Winterhalder, ibid.*, 1929, **473,** 126).

Scheme 4

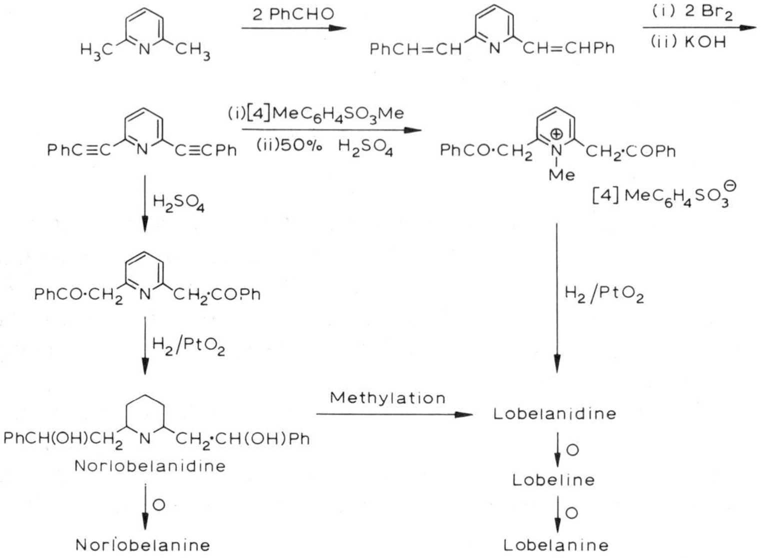

H_3C–(pyridine)–CH_3 → (2 PhCHO) → PhCH=CH–(pyridine)–CH=CHPh → ((i) 2 Br_2; (ii) KOH) →

PhC≡C–(pyridine)–C≡CPh → ((i)[4]$MeC_6H_4SO_3Me$; (ii)50% H_2SO_4) → PhCO·CH_2–(N-Me pyridinium)–CH_2·COPh [4] $MeC_6H_4SO_3^{\ominus}$

PhC≡C–(pyridine)–C≡CPh → (H_2SO_4) → PhCO·CH_2–(pyridine)–CH_2·COPh → (H_2/PtO_2) → PhCH(OH)CH_2–(piperidine)–CH_2·CH(OH)Ph (Norlobelanidine) → (O) → Norlobelanine

Norlobelanidine → (Methylation) → Lobelanidine

Pyridinium salt → (H_2/PtO_2) → Lobelanidine → (O) → Lobeline → (O) → Lobelanine

Another early synthesis, starting from diethyl glutarate and acetophenone is shown in Scheme 5 (*Wieland* and *I. Drishaus, ibid.*, 1929, **473**, 102):

Scheme 5

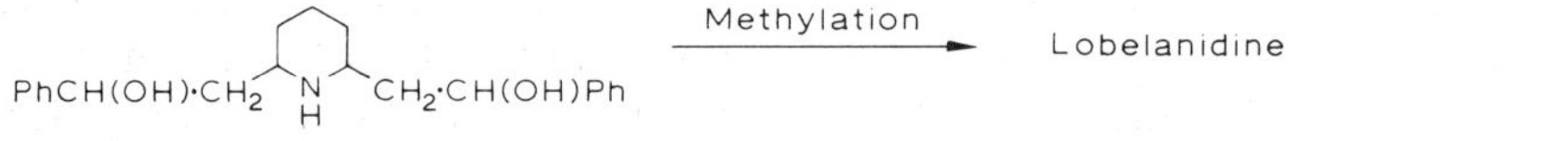

EtO_2C–$(CH_2)_3$–CO_2Et → (PhCOMe/$NaNH_2$) → PhCO·CH_2·CO–$(CH_2)_3$–CO·CH_2·COPh → (NH_3) →

PhCO·CH=(piperidine, NH)=CH·COPh → (H_2/PtO_2) → PhCH(OH)·CH=(piperidine, NH)=CH·CH(OH)Ph → (Al/Hg) →

PhCH(OH)·CH_2–(piperidine, NH)–CH_2·CH(OH)Ph → (Methylation) → Lobelanidine

More recently the absolute configuration of natural (−)-*cis*-8,10-diphenyllobelionol has been established as (2*S*,6*R*,8*S*) by synthesis from *O*-acetyl-(−)-phenyllobelol *via* condensation of the immonium salt with benzoylacetic acid followed by hydrolysis (*Schöpf, E. Müller* and *E. Schenkenberger, ibid.*, 1965, **687**, 241).

(±)-**Lelobanidine,** $C_{18}H_{29}NO_2$, is a minor tertiary base which forms a dibenzoate and which, on oxidation, yields benzoic acid and methylgranatic acid (2-carboxy-6-carboxymethyl-1-methylpiperidine) (XXXVII). Milder oxidation affords the diketone **lelobanine,** $C_{18}H_{25}NO_2$; other oxidation products are acetic and propionic acids. Lelobanine on Hofmann degradation affords an unsaturated diketone which can be reduced to 1-benzoyl-7-propionyl-*n*-heptane (XXXVIII) (*Wieland et al., ibid.*, 1939, **540,** 103). Lelobanidine therefore is 8-ethyl-10-phenyllobelidiol:

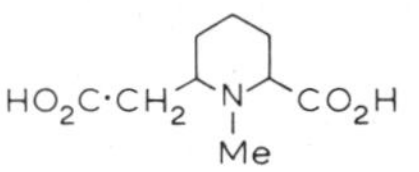

$HO_2C{\cdot}CH_2$ N CO_2H Me

(XXXVII)

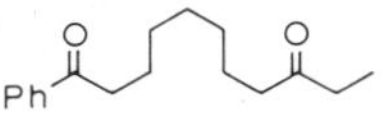

Ph

(XXXVIII)

(−)-**Lelobanidine-I** and (−)-**Lelobanidine-II** both $C_{18}H_{29}NO_2$, are two optically active bases closely related to (±)-lelobanidine. The former is an optically active form of the racemic base and the latter is almost certainly an epimer. Both bases yield (−)-lelobanine on oxidation, so that the configurations at C-2 and C-6 must be the same in each, and the stereochemical difference must be located at C-8 or C-10, or both.

An optically inactive minor alkaloid, $C_{13}H_{27}NO_2$, isolated from *Lobelia inflata* was assigned as 10-ethyl-8-methyllobelidiol on the basis of its oxidation by chromic acid to 10-ethyl-8-methyllobelidione, identical with that prepared by condensation of glutaraldehyde, methylamine, propionylacetic acid and acetoacetic acid (*Schöpf et al., loc. cit.*).

Lobinine, $C_{18}H_{25}NO_2$, is an oily *N*-methyl tertiary base which forms an *O*-benzoyl derivative and an oxime. It is a keto-alcohol which on mild oxidation affords the corresponding diketone, **lobinanine,** $C_{18}H_{23}NO_2$. The diketone forms a methiodide which, when subjected to Hofmann degradation, yields an unsaturated diketone, $C_{17}H_{18}O_2$; on hydrogenation this is converted into 1-benzoyl-7-propionyl-*n*-heptane (XXXIX). The unsaturated diketone is deep yellow in colour, and gives a violet colour with aqueous alkali. It contains three double bonds and is probably represented by XL:

$PhCO(CH_2)_7CO{\cdot}CH_2{\cdot}CH_3$

(XXXIV)

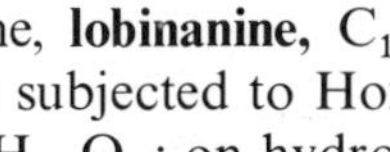

Ph

(XL)

More vigorous oxidation of the alkaloid gives benzoic acid and an unidentified acid not derived from piperidine. Lobinine contains one ethylenic bond, since, on hydrogenation, it yields a tetrahydrobase $C_{18}H_{29}NO_2$ which is isomeric with lelobanidine. The structure best representing the properties of lobinine is *trans*-3,4-dehydro-8-ethyl-10-phenyllobelionol (*Wieland et al., loc. cit.*).

Lobinanidine, $C_{18}H_{27}NO_2$, is converted by mild oxidation into lobinanine, and is therefore a disecondary alcohol. On hydrogenation it absorbs one molar equivalent of hydrogen and gives another lelobanidine isomer. Degradation of the alkaloid yields the same products as lobinine; lobinanidine is therefore *trans*-3,4-dehydro-8-ethyl-10-phenyllobelidiol.

(−)-*trans*-**3,4-Dehydro-8,10-diethyllobelidiol,** $C_{14}H_{27}NO_2$, occurs in *Lobelia syphilitica* (*R. Tschesche et al.*, Ber., 1961, **94,** 3327). The position of the double bond is established by oxidation to the unsaturated lobelidione, which is oxidised to a dicarboxylic acid by means of permanganate–periodate. Hofmann degradation of this dicarboxylic acid, followed by hydrogenation, yields β-propionylpropionic acid. Catalytic hydrogenation of *trans*-3,4-dehydro-8,10-diethyllobelidiol followed by chromic acid oxidation gives *cis*-8,10-diethyllobelidione, identical with that synthesised from glutaraldehyde, methylamine and propionylacetic acid (*C. Schöpf et al.*, Ann., 1957, **608,** 88).

Various isomers of **sedamine,** $C_{14}H_{21}NO$, are found in species of *Sedum.* It is a secondary alcoholic, *N*-methyl tertiary base which on oxidation yields benzoic and *N*-methylpipecolinic acids. The structure 8-phenyllobelol proposed for the base has been proved by the synthesis of (±)-sedamine shown in Scheme 6 (*L. Marion et al.*, Canad. J. Chem., 1951, **29,** 347).

Scheme 6

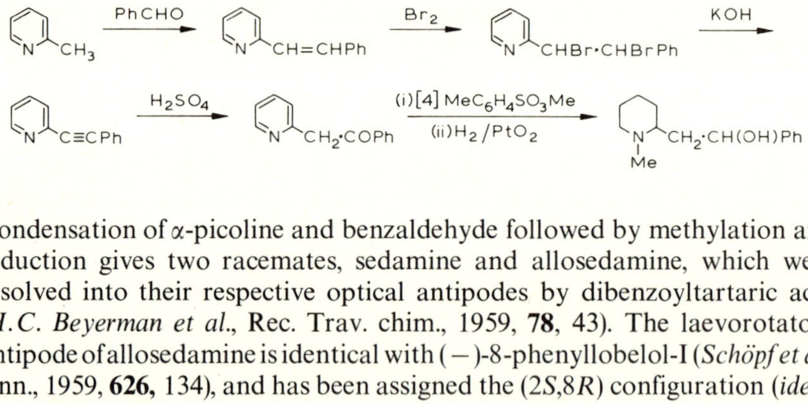

PhCHO; CH=CHPh; Br_2; CHBr·CHBrPh; KOH

C≡CPh; H_2SO_4; CH_2·COPh; (i)[4] $MeC_6H_4SO_3Me$ (ii) H_2/PtO_2; CH_2·CH(OH)Ph; N–Me

Condensation of α-picoline and benzaldehyde followed by methylation and reduction gives two racemates, sedamine and allosedamine, which were resolved into their respective optical antipodes by dibenzoyltartaric acid (*H.C. Beyerman et al.*, Rec. Trav. chim., 1959, **78**, 43). The laevorotatory antipode of allosedamine is identical with (−)-8-phenyllobelol-I (*Schöpf et al.*, Ann., 1959, **626,** 134), and has been assigned the (2*S*,8*R*) configuration (*idem, ibid.*, 1959, **628,** 101). (−)-Sedamine ((−)-8-phenyllobelol) has been assigned the (2*S*,8*S*) configuration.

Reduction of the condensation product from Δ′-piperideine and benzoylacetic acid gives two racemates, norsedamine and norallosedamine. Norallosedamine, resolved by means of 6,6′-dinitro-2,2′-diphenic acid, gives a dextrorotatory enantiomer which is identical with (+)-8-phenylnorlobelol isolated from *Lobelia inflata*, and to which the (2*S*,8*R*) configuration has been assigned *(idem, loc. cit.)*.

Sedridine, $C_8H_{17}NO$, isolated from *Sedum acre* L., is identical with synthetic (+)-8-methylnorlobelol. Racemic 8-methylnorlobelol is prepared by reduction of (2-piperidyl)-2-propanone with lithium tetrahydridoaluminate and resolution achieved by the use of *N*-acetyl-L-leucine *(Beyerman et al., loc. cit.)*. Since (−)-8-methylnorlobelol affords (*R*)-(+)-pipecolic acid on oxidation and the absolute configuration of this latter compound is known, the configuration of naturally occurring (+)-8-methylnorlobelol is (2*S*,8*S*) (*idem*, Rec. Trav. chim., 1965, **84,** 1367).

The properties of these and other lobelia and sedum alkaloids are summarised in Tables 3, 4 and 5.

One lobelia alkaloid of different structural form to the others is **lobinaline,** the major alkaloid of *Lobelia cardinalis* L., which was first isolated by *R. H. F. Manske* (Canad. J. Res., 1938, **16B,** 445) and assigned the empirical formula $C_{28}H_{38}N_2O$. It was later recognised that this formula represented a monohydrate of the alkaloid, and the empirical formula was modified to $C_{28}H_{36}N_2$ (*F. Kaczmarch* and *E. Steinhegger*, Pharm. Acta Helv., 1958, **33,** 852). More recently, on the basis of analytical and mass spectral data, *M. M. Robinson* and co-workers amended this latter formula to $C_{27}H_{34}N_2$ (J. org. Chem., 1966, **31,** 3206).

The u.v. spectrum shows only the presence of benzene absorption, while the i.r. spectrum indicates the presence of a monosubstituted benzene ring and the absence of NH groups. There is also a strong absorption at 1665 cm^{-1}. The alkaloid is dibasic, with pK_a values of 5.7 and 8.2. The n.m.r. spectrum shows the presence of two monosubstituted benzene rings and one *N*-methyl group, but no vinyl protons. These spectroscopic data, together with evidence available from degradation studies, led to the suggestion of the following structure for lobinaline:

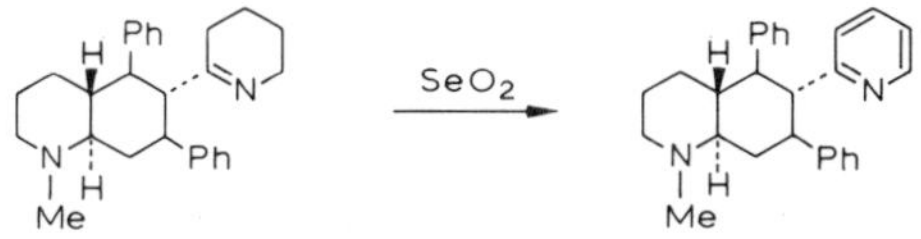

Lobinaline → (SeO_2) → (XLI)

This was confirmed by synthesis of the dehydro degradation product, XLI, as outlined in Scheme 7 (*Robinson et al., ibid.*, 1966, **31,** 3213, 3220).

A detailed mass spectrometric analysis has provided further confirmation of the above

TABL

LOBELIA AND

R^2

Base	R	R^1	R^2
(−)-*cis*-8,10-Diethyllobelionol	Me	EtCHOH	EtCO
cis-8,10-Diethyllobelidiol	Me	EtCHOH	EtCHOH
Lobelanine (*cis*-8,10-diphenyllobelidione)	Me	PhCO	PhCO
Lobeline { ((−)-*cis*-8,10-diphenyllobelionol) ((+)-*cis*-8,10-diphenyllobelionol)	Me	PhCO	PhCHOH
Lobelanidine (*cis*-8,10-diphenyllobelidiol)	Me	PhCHOH	PhCHOH
8-Methyl-10-ethyllobelidiol	Me	MeCHOH	EtCHOH
8-Methyl-10-phenyllobelionol (sedinone)	Me	MeCHOH	PhCO
(+)-*cis*-8-Methyl-10-phenyllobelidiol	Me	MeCHOH	PhCHOH
Lelobanidine (±)-*cis*-8-ethyl-10-phenyllobelidiol (−)-*cis*-8-ethyl-10-phenyllobelidiol-I (−)-*cis*-8-ethyl-10-phenyllobelidiol-II	Me	EtCHOH	PhCHOH
cis-8,10-Diethylnorlobelidione	H	EtCO	EtCO
cis-8,10-Diethylnorlobelionol	H	EtCHOH	EtCO
Norlobelanine (*cis*-8,10-diphenyllobelidione)	H	PhCO	PhCO
Norlobelanidine (*cis*-8,10-diphenylnorlobelidiol)	H	PhCHOH	PhCHOH

[a] B.HCl in ethanol.

TABL

LOBELIA AND

R^2

Base	R	R^1
trans-3,4-Dehydro-8,10-diethyllobelidiol	Me	EtCHOH
trans-3,4- or 4,5-Dehydro-8-methyl-10-ethyllobelidiol	Me	MeCHOH or EtCHOH
Lobinanine [(−)-*trans*-3,4-dehydro-8-ethyl-10-phenyllobelidione]	Me	EtCO
Isolobinanine [(−)-*cis*-3,4-dehydro-8-ethyl-10-phenyllobelidione]		
Lobinine [(−)-*trans*-3,4-dehydro-8-ethyl-10-phenyllobelionol]	Me	EtCHOH
Isolobinine [(−)-*cis*-3,4-dehydro-8-ethyl-10-phenyllobelionol]		
Lobinanidine [(−)-*trans*-3,4-dehydro-8-ethyl-10-phenyllobelidiol]	Me	EtCHOH
Isolobinanidine [(−)-*cis*-3,4-dehydro-8-ethyl-10-phenyllobelidiol]		
trans-4,5-Dehydro-8-methyl-10-phenylllobelidiol (sedinine)	Me	PhCHOH

[a] B.HCl in ethanol.

OIDS

1

(°C)	$[\alpha]_D$	B.HCl	B.HBr m.p. (°C)	B.HI m.p. (°C)	Other derivatives
		120–121			
	Inactive				Picrate, m.p. 138°
	Inactive	188	188	169–172	Dioxime, m.p. 209°
-131	−42.85°	182			
	Inactive	170			
	Inactive	135–138	188–190		Dibenzoate, m.p. 109–110°
	Inactive				Picrate, m.p. 133–134°
		175			
	+3	91–92			Chloroaurate, m.p. 170°
	Inactive	79		159	
	−41.5°[a]	86 (hydrate)		171	Perchlorate, m.p. 176°
	−41.7°[a]	102–105 (hydrate)		165	Perchlorate, m.p. 158°
		183–184			
		183–184			
-121	Inactive	201–202			*N*-benzyl, m.p. 125–126°
	Inactive	244		211	

OIDS

	m.p. (°C)	$[\alpha]_D$	B.HCl m.p. (°C)	B.HI m.p. (°C)	Other derivatives
HOH		−114°	128		Perchlorate, m.p. 123°
HOH		−110°[a]	120		
CHOH					
CO		−65°[a]	94		Perchlorate, m.p. 133°
		−11°[a]	150		
CO		−106°[a]	144	130	Oxime.HCl, m.p. 182°
	78	−76°[a]	154		Oxime.HCl, m.p. 186°
HOH	95	−120°	169	200	
		−28.3°[a]	110 (dihydrate)	164	Perchlorate, m.p. 169°
CHOH	121	−105°			

TABL

LOBELIA AND S

Base	R	R^1	m.p. (°C
(−)-Sedamine [(−)-8-phenyllobelol]	Me	PhCHOH	89
(−)-Allosedamine [(−)-8-phenyllobelol-I]			81
(+)-8-Methylnorlobelol (sedridine)	H	MeCHOH	83–84
(±)-8-Methylnorlobelol	H	MeCHOH	74–75
(+)-8-Ethylnorlobelol-I	H	EtCHOH	87
(+)-8-Phenylnorlobelol-I [(+)-norallosedamine]	H	PhCHOH	102–10

structure for lobinaline (*D. M. Clugston, D. B. MacLean* and *Manske*, Canad. J. Chem., 1967, **45,** 39). The base melts at 108–110° after recrystallisation from hexane and has $[\alpha]_D^{26}$ +38° (chloroform).

Syphilobine-A, $C_{28}H_{26}N_2O_2$, an optically inactive base isolated from *Lobelia syphilitica* appears to be related to lobinaline. The following structures have been proposed on the basis of spectroscopic evidence for syphilobine-A and a similarly related compound **syphilobine-F**, $C_{28}H_{26}N_2O_3$.

Scheme 7

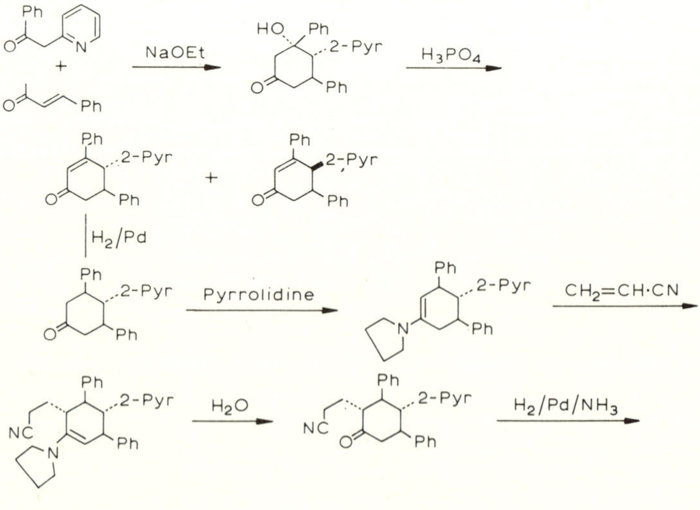

NaOEt

H_3PO_4

H_2/Pd

Pyrrolidine

CH_2=CH·CN

H_2O

H_2/Pd/NH_3

$[\alpha]_D$	*B.HCl m.p. (°C)*	*B.HBr m.p. (°C)*	*Other derivatives*
−56.75°	205		
−18.6°	118–119		Chloroaurate, m.p. 181°
+29.3°			
+22.3°	135	115–116	Picrate, m.p. 116–117°
+49.3°	133		Chloroaurate, m.p. 142–143°

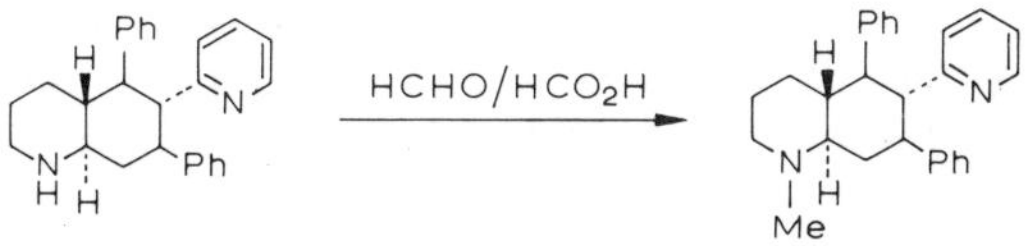

HCHO/HCO_2H

...(XLI)

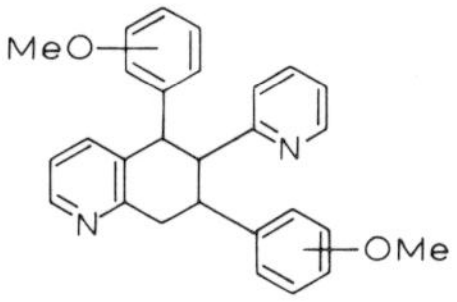

Syphilobine-A

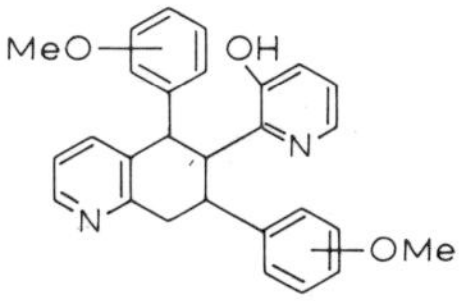

Syphilobine-F

Syphilobine-A has m.p. 192–194° and *syphilobine-F*, m.p. 222–223°, $[\alpha]_D$ −1.64° (pyridine) (*Tschesche, D. Klöden* and *H. W. Fehlhaber*, Tetrahedron, 1964, **20,** 2885).

(e) Alkaloids of tobacco

In addition to piperidine, pyrrolidine and other simple bases, *Nicotiana* species contain a variety of more complex alkaloids. As a result of the commercial interest in tobacco and growing concern over the physiological effects of smoking, the tobacco alkaloids have been the subject of considerable recent interest. ("Tobacco Alkaloids and Related Compounds", ed. *U. S. von Euler*, MacMillan, New York, 1965). The alkaloids are isolated from the leaves, roots, stems and flowers of the tobacco plant, mainly *Nicotiana tabacum* L. and *Nicotiana rustica* L., by the usual extraction technique (*R. J. Hyatt*, U.S.P. 2,117,558/1938).

Nicotine, $C_{10}H_{14}N_2$, is the major tobacco alkaloid. It is a laevorotatory liquid, ditertiary base which is completely miscible with water below 60° and above 210°. Oxidation with permanganate or dichromate gives nicotinic acid, identical with that prepared by the oxidation of β-picoline, the formation of which established the presence of a pyridine nucleus substituted in the 3 position. A further product of the oxidation is methylamine, which indicates that the second nitrogen atom is present as a methylimino group and hence nicotine may be written as $C_5H_4N \cdot C_4H_7NCH_3$. Nicotine forms a crystalline zincichloride, $B \cdot ZnCl_2 \cdot 2HCl \cdot H_2O$, which when distilled with lime yields pyridine, pyrrole and methylamine, suggesting that the residue may be *N*-methylpyrrolidine. Conclusive proof for this hypothesis is provided by the reaction of nicotine with methyl iodide; two methiodides are formed from which trigonelline (p. 117) and hygrinic acid (see C.C.C., Vol. IVB, p. 1) may be obtained on oxidation (*P. Karrer* and *R. Widmer*, Helv., 1925, **8**, 364). Bromination of nicotine in acetic acid solution gives dibromocotinine (XLII), while in hydrobromic acid solution dibromoticonine (XLIII) is obtained (*A. Pinner*, Ber., 1892, **25**, 2807; 1893, **26**, 292, 765):

Nicotine → (Br_2, AcOH) → (XLII)

Nicotine → (Br_2, HBr) → (XLIII)

A variety of syntheses of nicotine have been reported. In one of the earliest (*A. Pictet* and *P. Crépieux*, *ibid.*, 1895, **28,** 1904) *N*-(3-pyridyl)pyrrole is isomerised at red heat to 2-(3-pyridyl)pyrrole (nornicotyrine), which on methylation yields **nicotyrine,** itself a tobacco alkaloid convertible to (±)-nicotine by reduction. The racemic base can be resolved with tartaric acid (*Pictet* and *A. Rotschy*, *ibid.*, 1904, **37,** 1225):

→ Nornicotyrine → Nicotyrine

In an alternative synthesis, nicotinonitrile is condensed with γ-ethoxypropylmagnesium bromide to give the ketone (XLIV) (*L. C. Craig*, J. Amer. chem. Soc., 1933, **55,** 2854). Reduction of the corresponding oxime affords the amine XLV which, on heating with hydrobromic acid at 150°, cyclises to (±)-nornicotine, the racemic form of another tobacco alkaloid. This racemic base has been resolved (*E. Späth* and *F. Kesztler*, Ber., 1936, **69,** 2725); methylation of the (−)-base with formaldehyde and formic acid gives (−)-nicotine, identical with the natural alkaloid:

(XLIV) → (XLV) → (±)-Nornicotine

In a third synthesis, 3-bromopyridine was treated with 1-methyl-2-pyrrolidine in the presence of butyllithium to give the base XLVI, which was readily reduced to (±)-nicotine by the action of formic acid and fused potassium formate (*R. Lukeš* and *O. Cervinka*, Coll. Czech. chem. Comm., 1961, **26,** 1893):

Bu^nLi

(XLVI)

Four other synthetic routes to nicotine have been reported which start from ethyl nicotinate (XLVII). Thus, condensation of ethyl nicotinate with *N*-methyl-2-pyrrolidone (XLVIII) in the presence of potassium in ether gave *N*-methyl-3-nicotinyl-2-pyrrolidone, (XLIX), from which (±)-nicotine was obtained on hydrolysis and hydrogenation (*F. Korte* and *H. J. Schulze-Steinen*, Ber., 1962, **95,** 244). This was an improvement on a similar, earlier route (*Späth* and *H. Bretschneider*, *ibid.*, 1928, **61,** 327):

(XLVII) + (XLVIII) —K→ (XLIX)

Alternatively, base-catalysed condensation of ethyl nicotinate with ethyl acetate gave ethyl nicotinylacetate which reacted with ethyl bromoacetate to give diethyl nicotinylsuccinate, (L). Stepwise hydrolysis, decarboxylation and hydroxyimination gave LI, which was hydrogenated to LII and finally cyclised to (±)-norcotinone on heating. Reduction and methylation then yielded (±)-nicotine (*F. Zymalkowski* and *B. Trenkfog*, Arch. Pharm., 1959, **292,** 9):

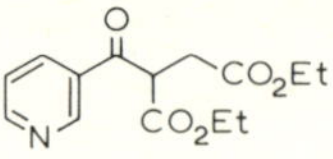

(L) 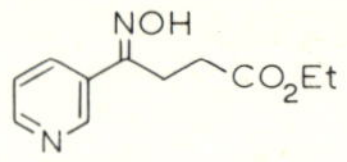 (LI)

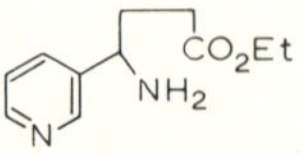

(LII) 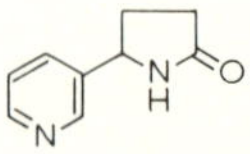 (±)-Norcotinone

The pyridine derivatives LIII (*T. Mizoguchi*, Chem. Pharm. Bull., Tokyo, 1961, **9**, 818; C.A., 1962, **57**, 166673), and LIV (*H. Hellmann* and *D. Dieterich*, Ann., 1964, **672**, 97) have both been prepared by condensation reactions on ethyl nicotinate and both can be cyclised to give nicotine precursors:

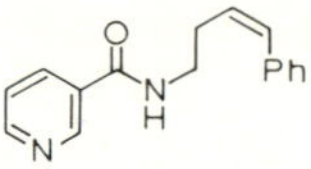

(LIII) 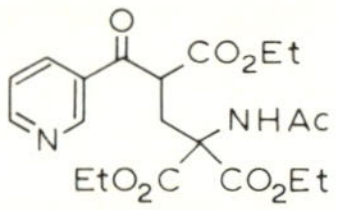 (LIV)

Nornicotine, $C_9H_{12}N_2$, occurs in tobacco as its (−)- and (±)-forms, and as its (+)-form in *Duboisia hopwoodii.* Its constitution, 2-(3-pyridyl)pyrrolidine follows since it is a secondary base which on methylation yields nicotine, from which it is formed by demethylation. On dehydrogenation it yields **nornicotyrine,** $C_9H_8N_2$, 2-(3-pyridyl)pyrrole. A synthesis of nornicotyrine has been achieved by means of a Gomberg reaction between *N*-nitroso-*N*-(3-pyridyl)isobutyramide and 1-ethoxycarbonylpyrrole. This leads to *N*-ethoxycarbonylnornicotyrine from which nornicotyrine can be obtained by hydrolysis (*H. Rapoport* and *M. Look*, J. Amer. chem. Soc., 1953, **75,** 4605).

Myosmine, $C_9H_{10}N_2$, which has been isolated from tobacco smoke, is optically inactive (not racemic). On dehydrogenation it gives 2-(3-pyridyl)-pyrrole, so that the alkaloid has a nornicotine framework. It is a diacidic base, both nitrogen atoms being tertiary; though the alkaloid yields an *N*-benzoyl derivative, $C_{16}H_{16}N_2O_2$, formation of this compound involves ring scission and hydration. By comparison of molecular formulae myosmine is a dehydronornicotine, and since it is one of the products formed when nicotine is dehydrogenated it must be formulated as LV or LVI (*Späth et al.*, Ber., 1936, **69,** 393):

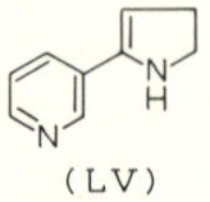

(LV) 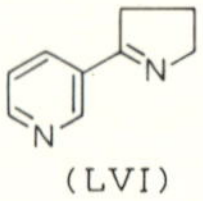 (LVI)

The three alternative positions for the double bond in the five-membered ring are ruled out because the alkaloid has no asymmetric centre. A synthesis of myosmine has been described by *Späth* and *L. Mamoli* (*ibid.*, 1936, **69,** 757), who condensed ethyl nicotinate and *N*-benzoyl-2-pyrrolidone to give the ketone LVII, which when heated at 130° with hydrochloric acid was hydrolysed, decarboxylated and cyclised to a product identical with myosmine, which was considered to have structure LV. This synthesis, however, does not establish the position of the double bond in the five-membered ring, and i.r. studies (*B. Witkop*, J. Amer. chem. Soc., 1954, **76,** 5597) show that the spectrum of myosmine contains a strong band characteristic of a C=N group conjugated with an aromatic ring, and no band in the N–H region. Myosmine is therefore correctly formulated as LVI, which structure is more in line with the observation that myosmine is readily hydrolysed to γ-aminopropyl 3-pyridyl ketone (LVIII) (*C. F. Woodward et al., ibid.*, 1944, **66,** 911), from which it may also be synthesised (*M. L. Stein* and *A. Burger*, *ibid.*, 1957, **79,** 154):

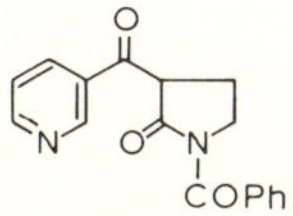

(LVII) 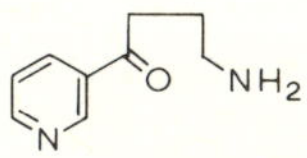 (LVIII)

A further elegant synthesis has been reported (*R. V. Stevens et al., ibid.*, 1968, **90,** 5576), in which condensation of 3-pyridyllithium and cyclopropanecarbonitrile gave cyclopropyl 3-pyridyl ketimine:

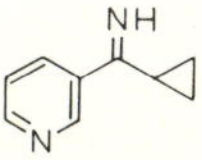

acid-catalysed thermal rearrangement of which gave myosmine in 70% yield.

Two convenient new syntheses of myosmine have recently been reported from nicotinyl chloride and 2-pyrrolidone (*B. P. Mundy et al.*, J. org. Chem., 1972, **37,** 1635), and from pyridine-3-carbaldehyde (*R. Leete, M. R. Chedekel* and *G. B. Boden, ibid.*, 1972, **37,** 4465).

Anabasine, $C_{10}H_{14}N_2$, is an optically active, diacidic base found in *Anabasis aphylla* and in tobacco. It is both a secondary and a tertiary base, and forms a monobenzoyl derivative. On oxidation with nitric acid or potassium permanganate anabasine yields nicotinic acid (*A. Orechoff* and *G. Menschikoff*, Ber., 1932, **65,** 232), and the alkaloid can be degraded in a stepwise manner to pipecolinic acid (*Orechoff et al., ibid.*, 1934, **67,** 1398). By analogy with nicotine the formula LXII, 2-(3-pyridyl)piperidine, was suggested for anabasine; the structure is supported by the mild oxidation of the alkaloid with silver acetate to 3′,2-dipyridyl, and its reduction to a mixture of bases containing 3′,2-dipiperidyl.

Späth and *Mamoli* (*ibid*., 1936, **69**, 1082) have synthesised anabasine by a route resembling that used in an earlier synthesis of nicotine (p. 139). Ethyl nicotinate is condensed with *N*-benzoyl-2-piperidone (LIX) to give the ketone LX, which when heated with hydrochloric acid at 130° is converted into anabaseine, 2-(3-pyridyl)-1,4,5,6-tetrahydropyridine (LXI). Catalytic hydrogenation of the latter yields (±)-anabasine (LXII), resolved with 6,6′-dinitro-2,2′-diphenic acid (*Späth* and *Kesztler*, *ibid*., 1937, **70,** 70):

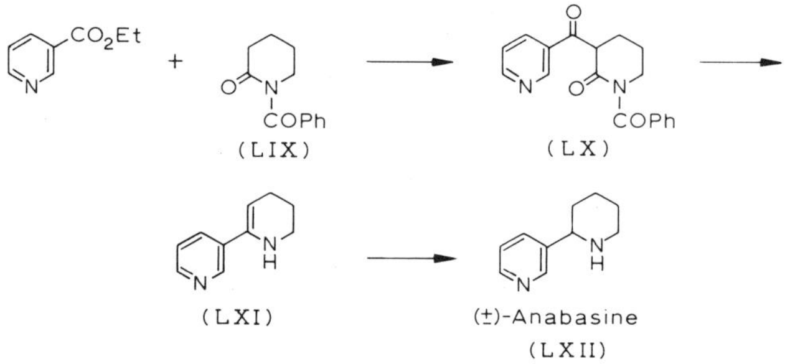

CO_2Et + COPh (LIX) → COPh (LX) →

(LXI) → (±)-Anabasine (LXII)

The absolute configuration of natural (−)-anabasine is (2*S*)-2-(3-pyridyl)-piperidine, (LXIII) (*Lukeš et al*., Coll. Czech. chem. Comm., 1962, **27**, 751):

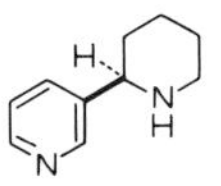

(LXIII)

N-Methylanabasine is readily obtained by methylation of anabasine with formaldehyde and formic acid.

Anatabine, $C_{10}H_{12}N_2$, is found as the (−)- and racemic forms in the low boiling fractions of crude nicotine. On dehydration it affords 3′,2-dipyridyl, and on hydrogenation anabasine, with the uptake of one mole of hydrogen. Anatabine is, therefore, a dehydroanabasine. Oxidation of the *N*-benzoyl derivative of anatabine yields nicotinic acid and hippuric acid (*Späth* and *Kesztler*, Ber., 1937, **70**, 704) which rules out the Δ^5 structure, and the Δ^2 structure is untenable as the alkaloid is optically active. Anatabine must, therefore, be represented by LXIV or LXV, and the latter has been proved to be correct by synthesis (*P. M. Quan*, *T. K. B. Karns* and *L. D. Quin*, J. org. Chem., 1965, **30,** 2769):

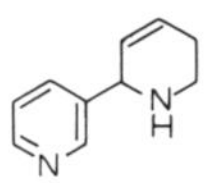

(LXIV)

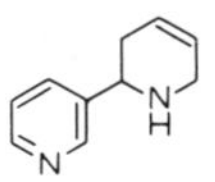

Anatabine
(LXV)

3-(Bisethoxycarbonylamino)methylpyridine (LXVI) reacts with butadiene in the presence of boron trifluoride and acetic acid, presumably *via* the intermediate LXVII, to give LXVIII, the *N*-ethoxycarbonyl derivative of anatabine. This latter compound gives (±)-*N*-methylanatabine on reduction with lithium tetrahydridoaluminate. (−)-Anatabine has the 2*S* configuration (*Lukeš et al., loc. cit.*):

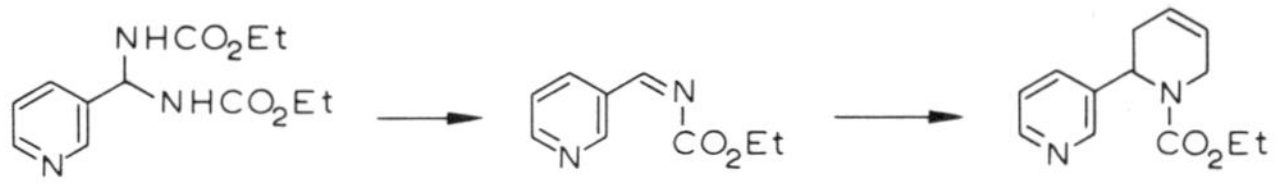

(LXVI) (LXVII) (LXVIII)

N-**Methylanatabine,** $C_{11}H_{14}N_2$, is also obtained from crude nicotine, and is formed by the methylation of anatabine.

Nicotelline, $C_{15}H_{11}N_3$, is a minor tobacco alkaloid which gives nicotinic acid and pyridine-2,4-dicarboxylic acid on oxidation and was assigned structure LXIX (*F. Kuffner* and *N. Faderl*, Monatsh., 1956, **87,** 71). This structure has been confirmed by synthesis (*J. Thesing* and *A. Müller*, Ber., 1957, **90,** 711).

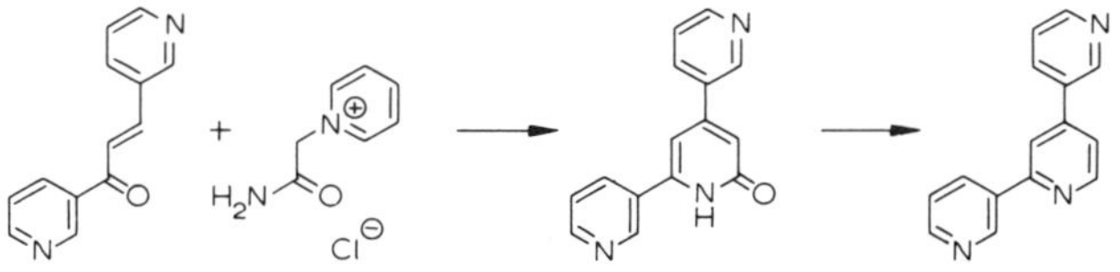

(LXIX)

Anatalline, $C_{15}H_{17}N_3$, has been isolated from *Nicotiana tabacum* and is readily dehydrogenated to nicotelline. This, coupled with i.r., u.v., n.m.r. and mass spectral data led to the assignment of the following structure (*T. Kisaki, S. Mizusaki* and *E. Tamaki*, Phytochem., 1968, **7,** 323):

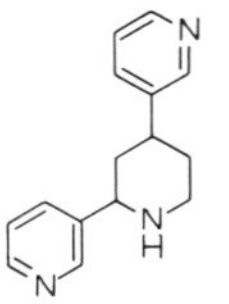

Anatalline

Isonicoteine, $C_{10}H_8N_2$, 3′,2-dipyridyl, is a dehydrogenation product of anabasine and anatabine and occurs in the lower boiling fraction of crude nicotine (*Späth* and *S. Biniecki*, Ber., 1939, **72,** 1809).

Three new alkaloids have been isolated recently from *Nicotiana tabacum*; spectroscopic studies have shown them to be 5-methyl-2,3′-bipyridine, *N*-formylnornicotine, and *N*-acetylnornicotine; these structures have been confirmed by synthesis (*A. Warfield et al.*, Phytochem., 1972, **11,** 3371).

Nicotine and related alkaloids have little pharmacological value. The bases are very poisonous, especially to insects, and crude nicotine preparations find extensive use as insecticides. The properties of the tobacco alkaloids are summarised in Table 6.

TABLE 6

TOBACCO ALKALOIDS

Alkaloid	*b.p. (°C)*	$[\alpha]_D$	*Picrate m.p. (°C)*	*Picrolonate m.p. (°C)*
(−)-Nicotine	246	−169.3°	226	228 Salts are dextrorotatory
(±)-Nicotine	246	Inactive	229	239
(+)-Nornicotine		+88.8°	194	240
(−)-Nornicotine	267	−88.8°	192	253
(±)-Nornicotine	267	Inactive		
Nicotyrine	280	Inactive	171	Methiodide, m.p. 211–213°
Myosmine	45[a]	Inactive	185	213
(−)-Anabasine	276	−82.4°	207	237
(±)-Anabasine	282	Inactive	214	259
(−)-*N*-Methylanabasine	268	−143.8°	238	236
(−)-Anatabine	146/10 mm	−177.8°	193	235
(±)-Anatabine		Inactive	202	235 *p*-Nitrobenzoyl, m.p. 95–
3′,2-Dipyridyl	294	Inactive	168	
Nicotelline	147[a]	Inactive	219	
Anatalline	225/3 mm		258	Perchlorate, m.p. 244–252°

[a] m.p.

(*f*) *Miscellaneous alkaloids*

Ammodendrine, $C_{12}H_{20}N_2O$, is an alkaloid closely related to the tobacco alkaloids and is found in the leaves of the tree *Ammondendron Conollyi* Bunge. It is a crystalline, monoacidic, secondary base containing an *N*-acetyl group and is optically inactive. Catalytic hydrogenation gives the corresponding dihydrobase, $C_{12}H_{22}N_2O$, hydrolysis and dehydration of which yields 3′,2-dipyridyl. Dihydroammodendrine is, therefore, LXX or LXXI. *N*-Methylammondendrine, on hydrogenation and hydrolysis yields an *N*-methyl-3′,2-dipyridyl which is identical with the product obtained by the hydrogenation of *N*-methylanabasine. Dihydroammodendrine must there-

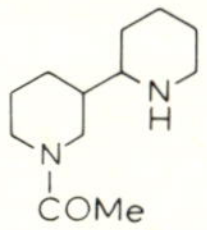

(LXX) 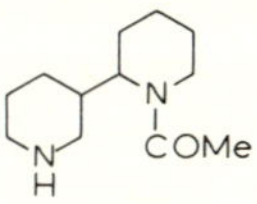 (LXXI)

fore by LXX, and ammodendrine has the same structure with an ethylenic bond (*A. Orechoff* and *N. Proskurnia*, Ber., 1935, **68,** 1807; Bull. Soc. chim. Fr., 1938, [v], **5,** 29). The position of this double bond has been established by a synthesis from *N*-acetyltripiperideine, and the structure of ammodendrine is as shown (*C. Schöpf* and *F. Braun*, Naturwissenschaften, 1949, **36,** 377):

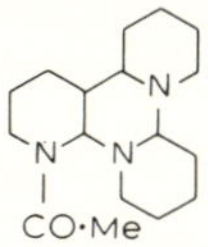

N-Acetyltripiperideine 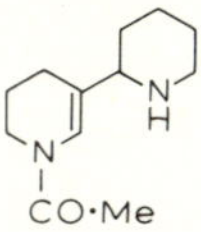 Ammodendrine

Ammodendrine monohydrate has m.p. 73–74°; *perchlorate*, m.p. 199–200°; *hydriodide*, m.p. 218–220°.

Hystrine, $C_{10}H_{16}N_2$, was first isolated from *Genista hystrine* in 1967 (*E. Steinhegger* and *C. Moser*, Pharm. Acta Helv., 1967, **42,** 177). It forms a crystalline, water-soluble, optically inactive *hydrochloride*, m.p. 209°, which exhibits a maximum in the u.v. spectrum at λ 323 nm. From its relatively simple spectroscopic properties it was assigned structure LXXII (*Steinhegger*, *Moser* and *P. Weber*, Phytochem., 1968, **7,** 849), confirmed by synthesis of hystrine by introduction of a double bond into the piperidine nucleus of ammodendrine, followed by deacetylation and purification *via* the *N*-nitroso derivative (*Steinhegger* and *Weber*, Helv., 1968, **51,** 206):

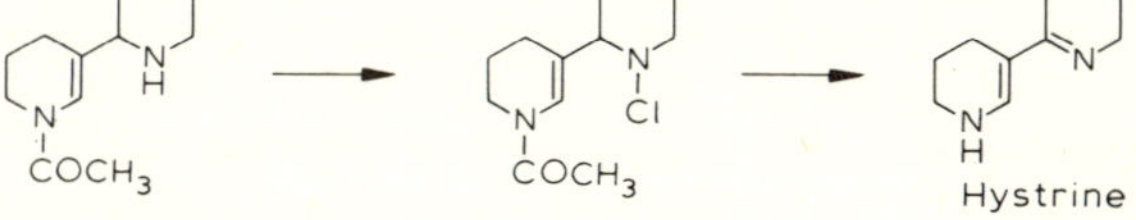

Hystrine (LXXII)

Gentianine, $C_{10}H_9NO_2$, was long thought to be a naturally occurring alkaloid isolated from species of *Gentiana* (C.C.C., 1st Edn., 1960 Vol. IV C, p. 1825). Ammonium hydroxide had always been used in the isolation, however, and using labelled ammonium hydroxide it has recently been shown that gentianine is in fact formed as an artefact during the isolation procedure from such compounds as swertiamarin and gentiapicrin (*H. G. Floss, U. Mothes* and *A. Rettig*, Z. Naturforsch., 1964, **19B,** 1106).

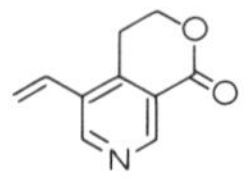

Gentianine 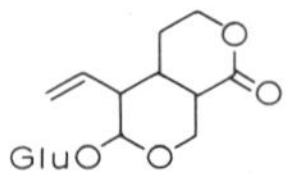 Swertiamarin 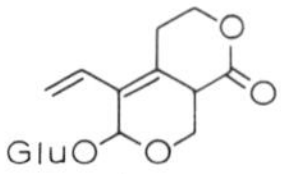 Gentiapicrin

Carpaine, $C_{14}H_{25}NO_2$, is an alkaloid which occurs in the leaves of *Carica papaya* L. It is a lactonic secondary base which yields the corresponding hydroxy-acid, carpamic acid, $C_{14}H_{27}NO_3$ on hydrolysis. Carpamic acid cannot be reconverted into the alkaloid, and hence the lactone cannot be γ- or δ- in type. Both carpaine and carpamic acid contain one *C*-methyl group (Kuhn–Roth estimation) and oxidation of each with nitric acid yields azelaic acid. On the basis of these and other observations, earlier workers (*G. Barger et al.*, J. chem. Soc., 1937, 711) favoured a pyrrolidine-based structure for carpaine, but more recent investigations have proved that the alkaloid is in fact derived from piperidine. When carpaine is subjected to a two-stage Hofmann degradation, myristic acid, $CH_3(CH_2)_{12}CO_2H$, is formed indicating that there is in the alkaloid an unbroken chain of fourteen carbon atoms. When carpaine is dehydrogenated, two moles of hydrogen are lost and deoxycarpyrinic acid (LXXIII) is produced which, on oxidation with permanganate, affords pyridine-2,6-dicarboxylic acid (*H. Rapoport* and *H. D. Baldridge*, J. Amer. chem. Soc., 1951, **73,** 343; 1952, **74,** 5365). Dehydrogenation of ethyl carpamate results in loss of three moles of hydrogen and formation of ethyl carpyrinate (LXXIV), which has the properties of a 3-hydroxypyridine (*T. R. Govindachari* and *N. S. Narasimhan*, J. chem. Soc., 1953, 2635) and which has been synthesised (*Govindachari et al.*, Chem.

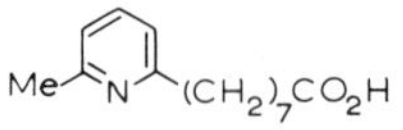

(LXXIII) 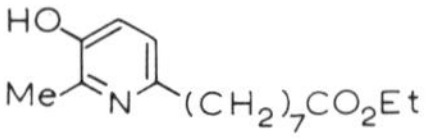 (LXXIV) $H_5C_2 \cdot CO(CH_2)_{10}CO_2H$ (LXXV)

Ind., 1956, 53). Finally, methyl *N*-methylcarpamate methiodide, when subjected to exhaustive Hofmann methylation, with hydrogenation at each stage, yields a nitrogen-free product which on oxidation affords 12-oxotetradecanoic acid (LXXV). Carpaine was, therefore, formulated as LXXVI (*Rapoport et al.*, J. Amer. chem. Soc., 1953, **75,** 5290); more recently, however, the molecular weight has been shown to be 478 by mass spectroscopy and hence the true structure is that of a dimer (*M. Spiteller-Friedmann* and *G. Spiteller*, Monatsh., 1964, **95,** 1234) and the absolute stereochemistry is as shown in structure LXXVII ($x=y=7$) (*J. L. Coke* and *W. Y. Rice Jr.*, J. org. Chem., 1965, **30,** 3420):

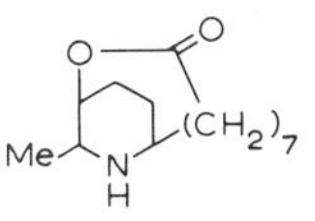

(LXXVI)

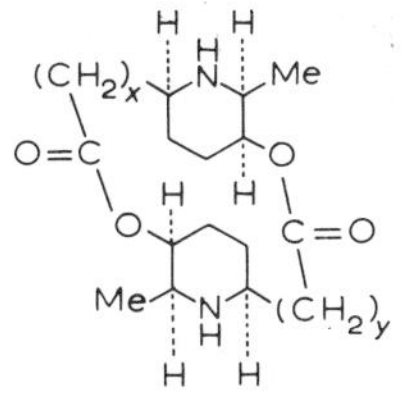

Carpaine
(LXXVII)

Carpaine has m.p. 121°, b.p. 215–235°, and $[\alpha]_D$ +21.9° (ethanol); N-*nitroso* deriv., m.p. 145°; N-*acetyl* deriv., m.p. 114°; N-*benzoyl* deriv., m.p. 100°

ψ-Carpaine accompanies carpaine in the leaves of *Carica papaya.* It is diastereoisomeric with carpaine, probably differing only in the stereochemistry of one of the methyl groups (*Govindachari, K. Nagarajan* and *N. Viswanathan,* Tetrahedron Letters, 1965, 1907).

Two new piperidine alkaloids isolated from the leaves of *Azima tetracantha* have been shown by mass spectroscopic and n.m.r. comparison studies to be of similar structure to carpaine. **Azimine,** $C_{24}H_{42}N_2O_4$, m.p. 112–113°, $[\alpha]_D^{20}$ 0.8° (ethanol), has structure LXXVII, $x=y=5$; *hydrochloride,* m.p. 284–287°; *di*-N-*methyl* deriv., m.p. 92–93°.

Azocarpine, $C_{26}H_{46}N_2O_4$, has the related structure where $x=5$ and $y=7$ (*G. J. H. Rall et al., ibid.,* 1967, 3465).

Cassine, $C_{18}H_{35}NO_2$, is an optically active base isolated from *Cassia excelsa* Shrad. The i.r. spectrum shows ketonic carbonyl absorption at 1720 cm^{-1} and a peak at 3530 cm^{-1} in the region of alcohol or amine absorption. Acetylation gives an O,N-*diacetyl* derivative, $C_{23}H_{41}NO_4$, as a neutral oil the i.r. spectrum of which shows peaks at 1725 and 1630 cm^{-1}, but no NH or OH peaks. Reduction of cassine with sodium tetrahydridoborate gives a *dihydro* derivative, m.p. 53–57°

The n.m.r. spectrum of cassine shows no signals characteristic of olefinic or aldehydic protons; a multiplet at δ 3.45 is attributable to the hydroxyl proton of a secondary alcohol. A singlet at δ 2.05 was assigned to a methyl ketone, and this is verified by a positive iodoform test. A doublet centred at δ 1.02 (3H) corresponds to a methyl group with one proton on the α-carbon atom.

Dehydrogenation of cassine gives an optically inactive derivative, $C_{18}H_{29}NO_2$, in which the acetyl group is retained and which shows the characteristic u.v. spectrum of a 3-hydroxypyridine derivative. The n.m.r. spectrum of this product shows a new peak at δ 2.50 corresponding to an aromatic methyl group, while the doublet at δ 1.02 in the spectrum of cassine is now absent. The presence of two aromatic protons mutually *ortho* is also clear from the n.m.r. spectrum and hence the dehydrogenation product is LXXVIII or LXXIX (*R. J. Highet*, J. org. Chem., 1964, **29,** 471):

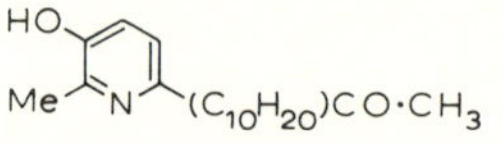

(LXXVIII) 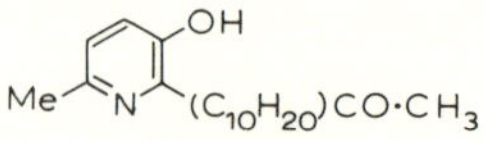 (LXXIX)

Examination of the mass spectrum of cassine indicates a skeletal form of the type LXXVIII (*Highet* and *P. F. Highet*, *ibid.*, 1966, **31,** 1272). This has been verified independently and the stereochemistry of cassine assigned on the basis of a synthesis of the *N*-methyl derivative *via* a homologation sequence from carpaine, the stereochemistry of which is known. This synthesis gives the mirror image from of the *N*-methyl derivative compared to that derived from the natural product (*Rice Jr.*, and *Coke*, *ibid.*, 1966, **31,** 1010):

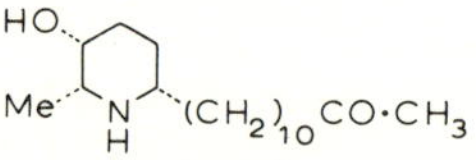

Cassine 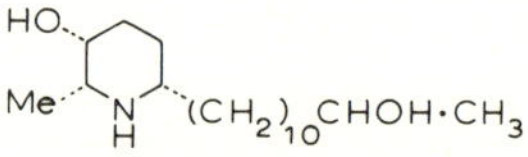 Carnavoline

Cassine has m.p. 57–58.5° and $[\alpha]_D^{25}$ −0.6°; *hydrochloride*, m.p. 173–175°; *hydronitrate*, m.p. 116–117°. Reduction of cassine with sodium tetrahydroborate yields the diol **carnavoline,** $C_{18}H_{37}NO_2$, which also occurs naturally in *Cassia carnavol* Spreg. (*D. Lythgoe* and *M. J. Vernengo*, Tetrahedron Letters, 1967, 1133).

Prosopine, $C_{18}H_{37}NO_3$, is an optically active alkaloid isolated from *Prosopis africana*. Acetylation reveals the presence of three alcohol and one secondary amine functions. Oppenhauer oxidation of prosopine gives a ketone, **prosopinone,** $C_{18}H_{35}NO_3$, which is isomeric with another alkaloid, **prosopinine,** extracted from the same species. The presence of a piperidine ring in prosopinine is indicated by dehydrogenation experiments, which give a 3-hydroxypyridine derivative, readily identifiable by its characteristic u.v. spectrum. The mass spectra of the two alkaloids and prosopinone all show peaks attributable to the ions LXXX and LXXXI.

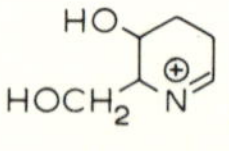

(LXXX) 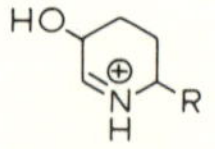 (LXXXI)

The position and stereochemistry of the hydroxy and hydroxymethyl groups on the piperidine nucleus were confirmed by reduction of the side chain to give desoxyprosopine, followed by the action of benzaldehyde. Detailed examination of the 220 MHz n.m.r. spectrum of this derivative shows it to have the double chair configuration LXXXII:

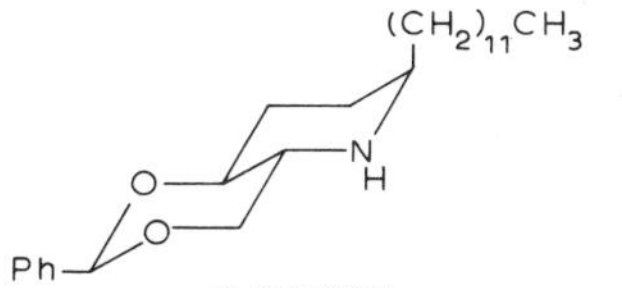

(LXXXII) 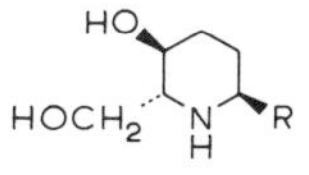 (LXXXIII)

This, and other evidence, allows the assignment of the absolute configuration (2*R*,3*S*,6*R*) to structure LXXXIII (R = $-(CH_2)_{10}CHOH \cdot CH_3$) for prosopine and LXXXIII (R = $-(CH_2)_{10}CO \cdot CH_3$ for prosopinine (*G. Ratle et al.*, Bull. Soc. chim. Fr., 1966, 2945; Bull. Soc. chim. Belges, 1972, **81,** 425).

Isoprosopinine A and **Isoprosopinine B** are isomers of prosopinine and both give desoxyprosopine on Wolff–Kishner reduction. The position of the ketone function in the side chain was determined by Baeyer–Villiger oxidation; isoprosopinine A gives *n*-valeric acid and is assigned structure LXXXIII [R = $-(CH_2)_6CO(CH_2)_4CH_3$], while isoprosopinine B gives caproic acid and thus has structure LXXXIII [R = $-(CH_2)_7CO(CH_2)_3CH_3$].

A further isomeric alkaloid from the same source is the optically inactive ethyl ketone (±)-**prosophylline,** $C_{18}H_{35}NO_3$. Wolff–Kishner reduction of prosophylline, however, does not give desoxyprosopine, but the geometrical isomer with absolute configuration (2*R*,3*S*,6*S*). Hence prosophylline has structure LXXXIV:

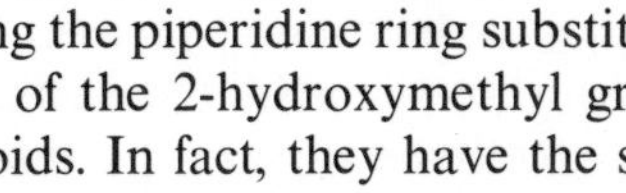

(LXXXIV) 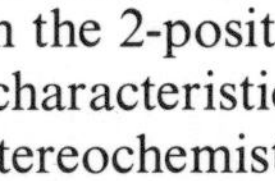 (LXXXV)

Prosafrine, $C_{18}H_{37}NO_2$, and **prosafrinine,** $C_{18}H_{35}NO_2$, are related to cassine in having the piperidine ring substituted in the 2-position by methyl groups instead of the 2-hydroxymethyl groups characteristic of the other prosopis alkaloids. In fact, they have the same stereochemistry as cassine, (2*R*,3*R*,6*S*); prosafrinine is the ethyl ketone LXXXV [R = $-(CH_2)_9CO \cdot CH_2 \cdot CH_3$] and prosafrine the alcohol LXXXV [R = $-(CH_2)_9CHOH \cdot CH_2 \cdot CH_3$] (*idem, ibid.*, 1972, **81,** 443). The properties of the prosopis alkaloids are listed in Table 7.

Two related, minor alkaloids have recently been isolated from *Cassia carnavol.* **Prosopinone,** $C_{18}H_{35}NO_3$, has been assigned structure

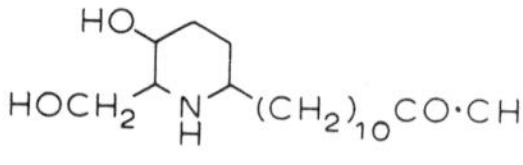

HO, $HOCH_2$, N, H, $(CH_2)_{10}CO{\cdot}CH_3$

and **Alkaloid D** the provisional structure

HO, Me, N, H, $(CH_2)_3CHOH(CH_2)_{10}CHOH{\cdot}CH_3$

following spectroscopic investigation (*D. Lythgoe et al.*, An. Asoc. quim. Argentina, 1972, **60**, 317; C.A., 1972, **77**, 164901k).

TABLE 7

THE PROSOPIS ALKALOIDS

	m.p. (°C)	$[\alpha]_D$
Prosopine	126–127	+25°
Prosopinine	94.5	+12°
Isoprosopinine A } Isoprosopinine B }	95	+16°
(±)-Prosophylline	79	Inactive
Prosafrine	68	−8.5°
Prosafrinine	56	−8°

Pinidine, $C_9H_{17}N$, is isolated from *Pinus sabiniana* along with (+)-α-pipecoline. It contains a secondary amine group, and the i.r. spectrum indicates that it is a substituted piperidine. Ozonolysis yields acetaldehyde, suggesting the presence of the $CH_3 \cdot CH{=}C\lt$ group, but the i.r. spectrum shows that this group is not conjugated with the amine function (*W. H. Tallent, V. L. Stromberg* and *E. C. Horning*, J. Amer. chem. Soc., 1955, **77**, 6361). Dehydrogenation gives a 2,6-disubstituted pyridine, $C_9H_{13}N$, which can be oxidised to pyridine-2,6-dicarboxylic acid. Hence the dehydrogenation product was thought to be 2-methyl-6-*n*-propylpyridine and this was confirmed by synthesis. Catalytic hydrogenation of this synthetic product gave (±)-*cis*-2-methyl-6-*n*-propylpiperidine which has an i.r. spectrum identical to that of optically active dihydropinidine obtained from the natural product by mild reduction. Pinidine, therefore, is a 2-methyl-6-(2-propenyl)piperidine (*Tallent* and *Horning, ibid.*, 1956, **78**, 4467). This conclusion has been confirmed by later workers who also showed that the complete structure is 2-(*R*)-methyl-6-(*R*)-(2-*trans*-propenyl)piperidine. The relative configuration was determined by stepwise degradation of pinidine to *cis*-2,6-dimethyl-*N*-benzylpiperidine, identical with a synthetic sample.

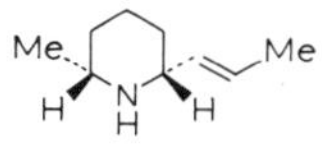

Pinidine 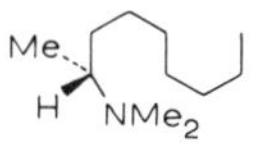 (LXXXVI)

The geometry of the double bond was established spectroscopically, the peak at 966 cm^{-1} in the i.r. spectrum and the vinylic splitting pattern in the n.m.r. spectrum being typical of *trans*-1,2-disubstituted olefins. The absolute configuration was elucidated by removal of one of the asymmetric centres of pinidine and relation of the other to a substance of known configuration, (+)-2-dimethylaminononane, LXXXVI, (*R. K. Hill, T. H. Chan* and *J. A. Joule*, Tetrahedron, 1965, **21**, 147).

Pinidine is a colourless oil, b.p. 176–177°/751 mm, $[\alpha]_{435}^{25}$ −10.5° (absolute ethanol); *hydrochloride*, m.p. 244–246°; N-*methyl hydriodide*, m.p. 214–218°.

Adenocarpine, $C_{19}H_{24}N_2O$, is an optically active base which occurs in the (+)-form in *Adenocarpus complicatus* and in the (−)-form in *Adenocarpus commutatis*. The racemic form is known as **orensine** and also occurs in *Adenocarpus commutatis*; it has been resolved into the (+)- and (−)-enantiomers with the camphorsulphonic acids. Acid hydrolysis cleaves adenocarpine to give cinnamic acid and a basic moiety $C_{10}H_{18}N_2$, which yields 2,3′-bipiperidyl on catalytic hydrogenation. Adenocarpine is therefore an *N*-cinnamoyl derivative of a mono-unsaturated 2,3′-bipiperidyl. Oxidation of adenocarpine with chromic acid followed by hydrolysis gives oxalic acid and the diamine LXXXVII, hence adenocarpine is as shown:

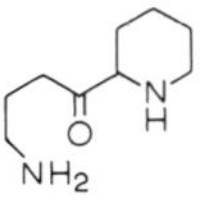

(LXXXVII) 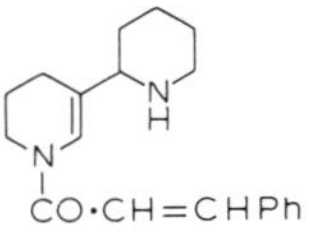 Adenocarpine

This has been verified by synthesis from tripiperideine as outlined in Scheme 8 (*C. Schöpf* and *K. Kreibich*, Naturwissensch., 1954, **41**, 335).

Scheme 8

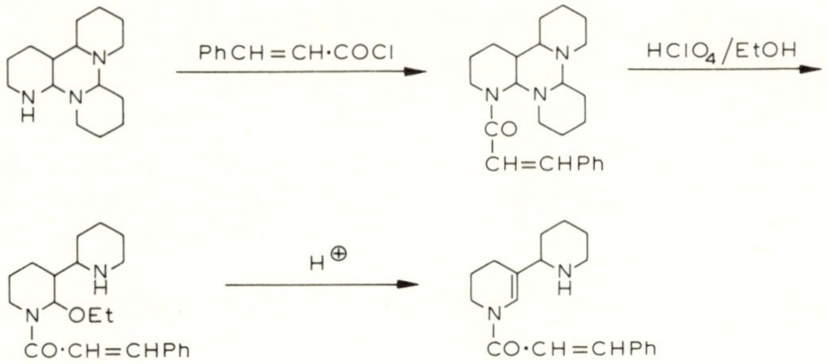

PhCH=CH·COCl
HClO$_4$/EtOH
CH=CHPh
N OEt
CO·CH=CHPh
H$^{\oplus}$
CO·CH=CHPh

This synthesis showed that the cinnamic acid part of adenocarpine has *trans*-stereochemistry about the double bond; the *cis* isomer proved to be identical with **iso-orensine**, a minor alkaloid from *Adenocarpus complicatus* (*Schöpf* and *W. Markel, ibid.*, 1966, **53,** 274).

Santiaguine, is a further alkaloid with the same empirical formula, $C_{19}H_{24}N_2O$, which occurs in a species of *Adenocarpus*, but it exists, in fact, as a dimer $C_{38}H_{48}N_4O_2$. Hydrolysis of santiaguine yields α-truxilic acid and the same base, $C_{10}H_{18}N_2$, as obtained from adenocarpine (*L. Castro* and *I. Ribas*, Anales Real Soc. Espan. Fis. y Quim., 1953, **B49,** 777). Synthesis by a method similar to that shown in Scheme 8 showed santiaguine to have the following structure (*J. Dominquez, M. R. Mendez* and *I. Ribas, ibid.*, 1956, **B52,** 133):

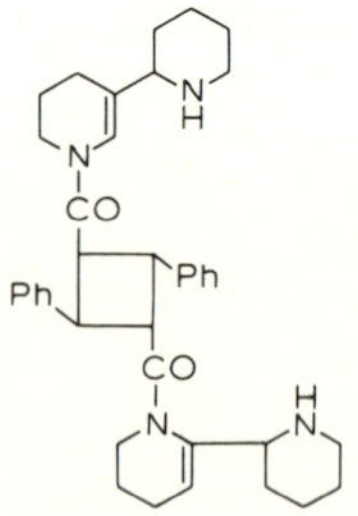

Santiaguine

The properties of the *Adenocarpus* alkaloids are listed in Table 8.

TABLE 8

ADENOCARPUS ALKALOIDS

	m.p. (°C)	*[α]D*	*B.HCl m.p. (°C)*	*B.HBr m.p. (°C)*	*Picrate m.p. (°C)*
(+)-Adenocarpine	65–66	+30.9°	133–147	191–192	213
(−)-Adenocarpine	64–65	−30.1°			
Orensine	82–83	Inactive	208–210		210–211
Iso-orensine			209–210	207–209	204–205
Santiaguine	235–236	+3.3°	241[a]	244–253[a]	285[a]

[a] Di-salts.

The alkaloids isolable from *Trypterygium wilfordii* have not yet been fully characterised, but it has been shown that a number of them are esters of **wilfordic acid,** $C_{11}H_{13}NO_4$, m.p. 195–196°, or of **hydroxywilfordic acid,** $C_{11}H_{13}NO_5$, m.p. 178–179° (*M. Beroza*, J. Amer. chem. Soc., 1952, **74,** 1585). An n.m.r. and degradative study of these acids showed them to have the following structures (*idem*, J. org. Chem., 1963, **28**, 3562):

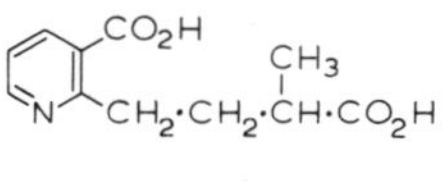

Wilfordic acid

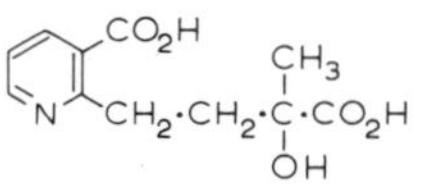

Hydroxywilfordic acid

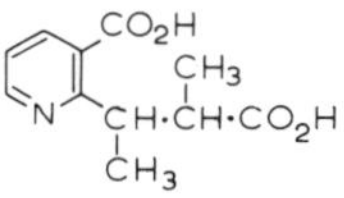

Evoninic acid

An isomer of wilfordic acid, **evoninic acid,** m.p. 127–133°, has similarly been shown to be the acidic component of a variety of alkaloids isolated from *Euonymus europaeus.* It has the structure shown (*M. Pailer* and *R. Libiseller*, Monatsh., 1962, **93**, 403).

An elegant and detailed chemical and spectroscopic study has resulted in the complete structural elucidation of a series of esters of evoninic acid and wilfordic acid isolated from the *Celastraceae* family, *Euonymus sieblodiana,* and *E. alatus* (*K. Yamada et al.*, Tetrahedron Letters, 1971, 2655, 2659, 2733, 3131; 1973, 113). The structures and properties of these esters are summarised in Table 9.

TABLE 9

CELASTRACEAE ALKALOIDS

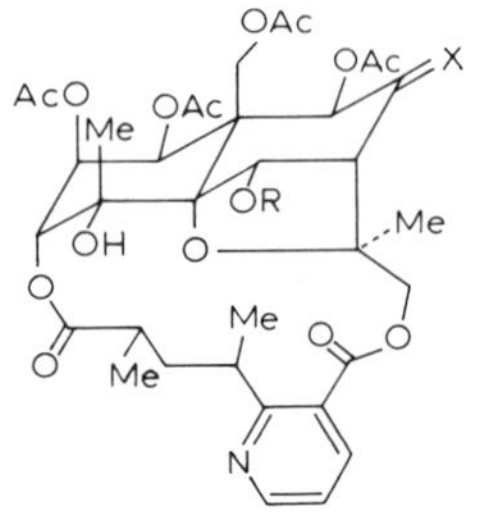

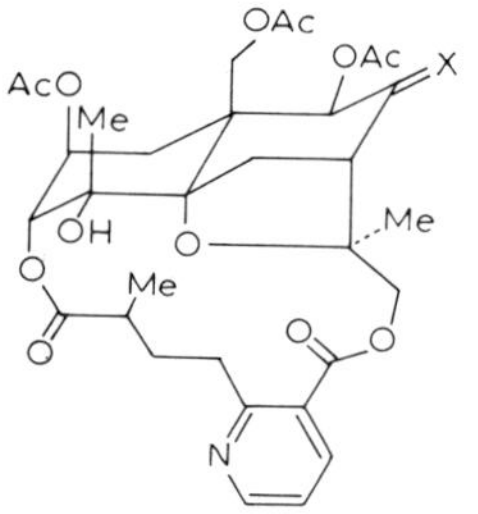

	R	*X*	*m.p. (°C)*	$[\alpha]_D$	*Picrate m.p. (°C)*
Evonine	Ac	O	184–190	+8.4°	
Neo-evonine	H	O	264–265	+24.9°	
Euonymine	Ac	OAc/H		−20°	140–146
Neo-euonymine	H	OAc/H	259–262	−11°	
Evonimine	Ac	O		+21°	
Euonine	Ac	OAc/H	149–153	−2.5°	
Wilfordine		OAc/H	174–183	+15.4°	
Alatamine		O	185–193	+44°	

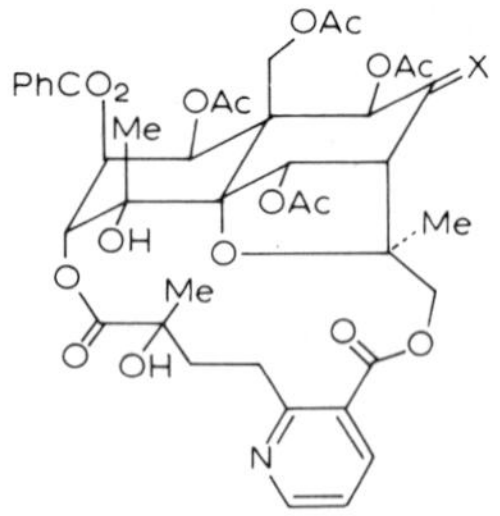

Similar esters of nicotinic acid have been isolated from *Maytenus ovatus* (*S. M. Kupchan, R. M. Smith* and *R. F. Bryan*, J. Amer. chem. Soc., 1970, **92,** 6667). **Maytoline** and **maytine** have the following structures:

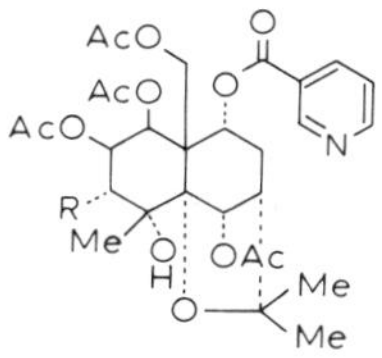

Maytoline R = OH
Maytine R = H

Anaferine, $C_{13}H_{24}N_2O$, was first isolated as a dihydrochloride from an alcoholic extract of the root of *Withania somnifera* by ion-exchange chromatography and countercurrent distribution. The i.r. spectrum shows a broad peak at 3270 cm^{-1} which is absent in the non-basic *N,N'*-diacetyl derivative and, as the mass spectrum clearly shows the presence of piperidine fragments, it was assumed that anaferine contains two piperidine moieties. The i.r. spectrum also shows a carbonyl absorption at 1710 cm^{-1} and the n.m.r. spectrum has a doublet corresponding to four hydrogen atoms on carbon atoms α to a carbonyl group at δ 2.35. There is also a six proton multiplet at δ 2.92 and a twelve proton multiplet at δ 1.45. On the basis of these data anaferine was assigned the structure:

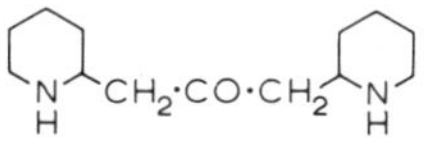

$CH_2 \cdot CO \cdot CH_2$

and this was verified by a synthesis involving condensation of α-picolyllithium with ethyl chloroformate, followed by hydrogenation and oxidation (*A. Rothes, J. M. Bobbitt* and *A. E. Schwarting*, Chem. Ind., 1962, 654). Anaferine as isolated is optically inactive and, although the conditions of isolation could cause racemisation of the (+)- and (−)-forms, it has been shown that anaferine occurs naturally as the *meso* form (*C. Schöpf et al.*, Ann., 1970, **737,** 1).

Anaferine is an oily liquid, b.p. 55°/0.01 mm; it forms a *dihydrochloride*, m.p. 223° and a *dipicrate*, m.p. 184–185° (decomp.).

Anahygrine, $C_{13}H_{24}N_2O$, was isolated from *Withania somnifera* in the same manner as its isomer anaferine. It is also isomeric with cuscohygrine (XII; C.C.C. Vol. IVB, Chap. 7), but it has one *N*-methyl group, whereas cuscohygrine has two and anaferine none. The major peak in the i.r. spectrum at 1725 cm^{-1} shows that anahygrine also contains a ketone group. The mass spectrum shows the presence of a piperidine and an *N*-methylpyrrolidine ring and hence the formula below was suggested for anahygrine. This was confirmed by the synthesis shown (*J. D. Leary et al.*, Chem. Ind., 1964, 283):

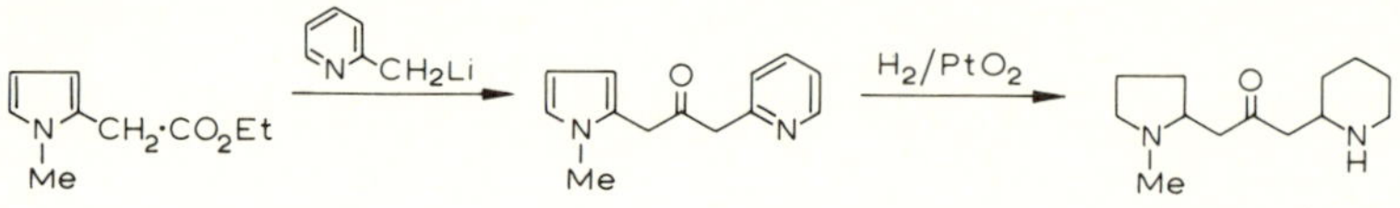

Anahygrine

Anahygrine is an oily liquid, b.p. 106°/0.2 mm; it forms a *dihydrochloride*, m.p. 216–217°; and a *dipicrate*, m.p. 173–174.5°.

Astrocasine, $C_{20}H_{26}N_2O$, was isolated from *Astrocasia phyllantaides* and assigned the following structure on the basis of chemical and spectroscopic evidence:

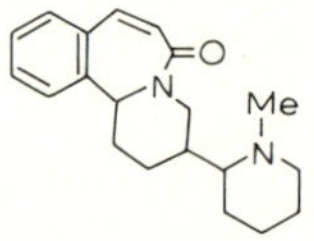

Astrocasine

Thus, the i.r. (1645, 1610, 1597 and 1570 cm^{-1}), u.v. (λ_{max} 263 nm, log ε 4.09) and n.m.r. (four aromatic protons, δ 7.31–7.49; two olefinic protons as two doublets centred at δ 7.08 and δ 6.43, $J = 12$ cps) spectra best fit a nearly planar *cis*-cinnamoyl-*N*-dialkylated lactam structure. Reduction of the double bond gives dihydroastrocasine which shows the spectroscopic properties of a simple aromatic system and an *N*-dialkylated lactam. The properties of astrocasine are consistent with those of an *ortho*-disubstituted aromatic ring, and this is confirmed by the isolated of phthalic acid from the oxidation of the alkaloid with potassium permanganate. Phthalonic, oxalic, malonic, succinic and glutaric acids were also isolated from this oxidation. Hofmann degradation of dihydroastrocasine yields a compound whose properties are compatible with structure LXXXVIII and which gives the carbaldehyde LXXXIX on mild oxidation:

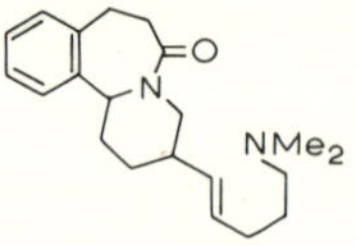

(LXXXVIII)

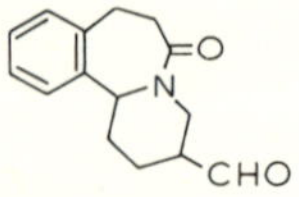

(LXXXIX)

This evidence, coupled with biosynthetic considerations and the fact that the mass spectra of astrocasine and its dihydro derivative show the presence of an *N*-methylpiperidine fragment, confirmed the structure shown for astrocasine.

Astrocasine has m.p. 171–172° and $[\alpha]_D$ −270° (ethanol); *hydroperchlorate*, m.p. 149–151°; *methiodide*, m.p. 227–228° (*H. A. Lloyd*, Tetrahedron Letters, 1965, 1761).

Astrocasia phyllanthoides yields another alkaloid, **astrophylline,** $C_{19}H_{26}N_2O$. The n.m.r. spectrum (five aromatic protons; two olefinic protons as two doublets centred at δ 6.64 and δ 6.01, *J* 12 cps) and the u.v. spectrum suggest a *cis*-cinnamoyl disubstituted amide system and, in fact, cinnamic acid is obtained on hydrolysis. The other product of hydrolysis is a base, $C_{10}H_{20}N_2$, which was identified as (+)-2,3-bipiperidyl. Astrophylline is readily isomerised to its *trans*-cinnamoyl isomer and the *N*-methyl derivative of this latter compound is identical with a sample prepared from the base XC and cinnamoyl chloride. As the absolute configuration of (+)-2,3-bipiperidyl is known, it follows that astrophylline has the structure and configuration shown:

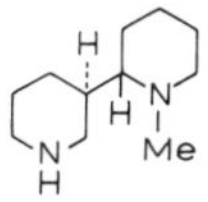

(XC)

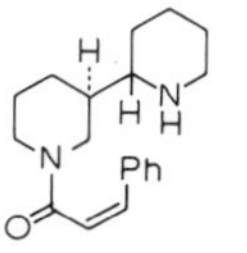

Astrophylline

Astrophylline is a viscous oil, b.p. 115°/0.001 mm, $[\alpha]_D$ +23°, which forms a *perchlorate*, m.p. 172–174° (*H. A. Lloyd*, Tetrahedron Letters, 1965, 4537).

Nudifluorine, $C_7H_6N_2O$, isolated from the leaves of *Trewia nudifloria*, is a neutral alkaloid. Its u.v. spectrum is typical of that of a 2-pyridine derivative, (λ_{max} 206, logε 4.27; 254, logε 4.32 and 306, logε 3.89, unchanging in acid or alkali) while the i.r. spectrum shows the presence of a nitrile group (2200 cm^{-1}), a conjugated amide (1670 cm^{-1}) and aromatic C=C bonds (1610 cm^{-1}). The presence of an *N*-methyl group is confirmed by the n.m.r. spectrum (δ 3.57, three-proton singlet), which also shows the presence of three aromatic protons; the splitting pattern of these protons suggests the following structures, which is consistent with the other observed properties:

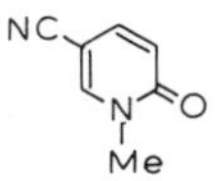

This structure was confirmed by a synthesis of nudifloric acid, the hydrolysis product of nudiflorine, by *N*-methylation of 6-oxonicotinic acid (XCI):

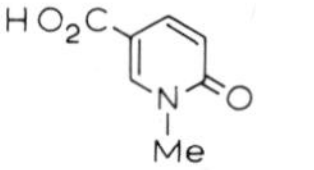

Nudifloric acid

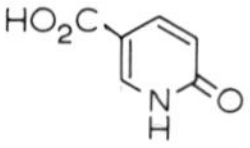

(XCI)

Nudiflorine has m.p. 161°, and *nudifloric acid* m.p. 238° (*R. Mukherjee* and *A. Chatterjee*, Tetrahedron, 1966, **22**, 1461).

Actinidine, $C_{10}H_{13}N$, isolated from *Actinidia polygama*, shows i.r. and u.v. spectra typical of a pyridine derivative and, on oxidation with permanganate gives 5-methylpyridine-3,4-dicarboxylic acid. Milder oxidation gives a base $C_{10}H_{13}NO$, which was assigned structure XCII, indicating structure XCIII for actinidine:

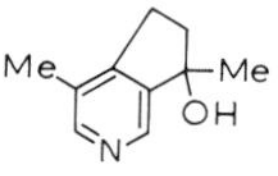

(XCII)

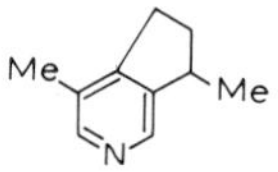

Actinidine
(XCIII)

This structure has been verified by several syntheses, two of which are shown in Scheme 9 (*T. Sakan et al.*, Bull. chem. Soc., Japan, 1959, **32**, 315, 1155).

The racemic actinidine obtained by these routes was resolved by means of dibenzoyl-L-tartaric acid. (+)-Actinidine has also been prepared indepen-

Scheme 9

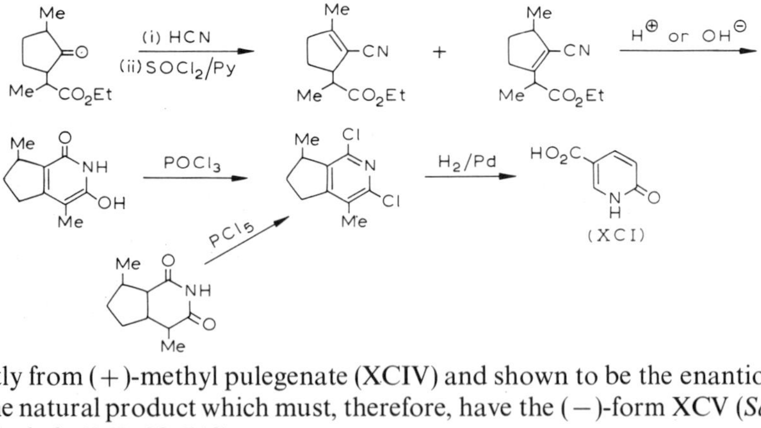

(i) HCN
(ii) $SOCl_2$/Py

$H^{\oplus}$ or $OH^{\ominus}$

$POCl_3$

H_2/Pd

PCl_5

(XCI)

dently from (+)-methyl pulegenate (XCIV) and shown to be the enantiomer of the natural product which must, therefore, have the (−)-form XCV (*Sakan et al., ibid.*, 1960, **33**, 712):

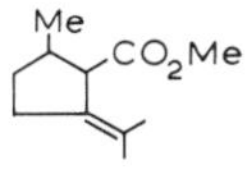

(XCIV) 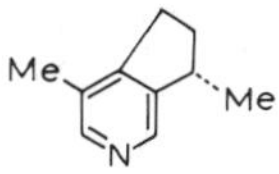 (XCV)

An alkaloid, $C_{18}H_{22}NOCl$, isolated from *Valeriana officinalis* proved to be *N*-β-(4-hydroxyphenyl)-ethylactinidium chloride, (XCVI) (*K. Torssell* and *K. Wahlberg*, Tetrahedron Letters, 1966, 445; Acta Chem. Scand., 1967, **21,** 53):

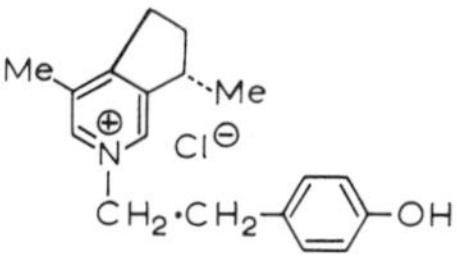

(XCVI)

Tecostidine, $C_{10}H_{13}NO$, isolated from the leaves of *Tecoma stans*, has i.r. (3200, 3400, 1600, 1375, 1210, 1043, 910, 845 and 820 cm^{-1}) and u.v. (λ_{max} 262, 270 nm logε 3.27 and 3.21 in neutral ethanol; λ_{max} 231, 262 nm, logε 3.49 and 3.61 in acidic solution) spectra similar to those of actinidine except for the presence of bands attributable to a hydroxyl group. The n.m.r. spectrum contains two peaks at δ 8.22 and δ 8.27 corresponding to two *ortho* pyridinic protons. The absence of other peaks in this region confirms the trisubstituted pyridine skeleton of actinidine. The rest of the spectrum is identical with that of actinidine except that the pyridine methyl group is missing and there are, instead, peaks at δ 4.65 and 4.62 attributable to the three protons of a primary alcohol. These data led to the suggestion of the structure shown for tecostidine (*Y. Hammouda* and *J. LeMen*, Bull. Soc. chim. Fr., 1963, 2901):

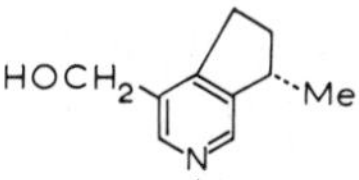

Tecostidine 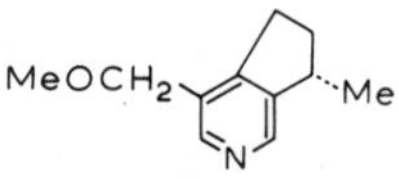 Valarianine

This structure has been confirmed by a synthesis which also showed the configuration of tecostidine to be the same as that of natural (L)-(−)-actinidine (*G. W. K. Cavill* and *A. Zeitlin*, Aust. J. Chem., 1967, **20,** 349).

Two closely related alkaloids have been obtained from *Valeriana officinalis*, *N*-β-(4-**hydroxyphenyl)-ethyltecostidinium chloride**, $C_{19}H_{22}ClNO_2$, (*Torssell* and *Wahlberg. loc. cit.*) and **valerianine,** $C_{11}H_{15}NO$, which proved to be the methyl ether of tecostidine (*B. Franck, U. Peterson* and *F. Huper*, Angew. Chem. internat. Edn., 1970, **9,** 891).

A fully saturated alkaloid **skytanthine,** $C_{11}H_{21}N$, has been isolated from dried branches of *Skytanthus acutus.* It contains an *N*-methyl group and, as it gives racemic actinidine on dehydrogenation, was assigned the structure shown below, which is in good agreement with its observed chemical and spectroscopic properties (*C. Djerassi et al.*, Chem. Ind., 1961, 210; Tetrahedron 1962, **18,** 183; *C. G. Casinovi et al.*, Chem. Ind., 1961, 253; Gazz., 1961, **91,** 1037):

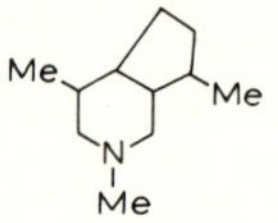

Skytanthine

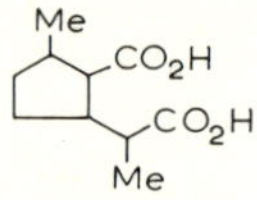

(XCVII)

The various stereoisomers of skytanthine have been synthesised from α-, β-, γ- and δ-nepetalinic acids, stereoisomers of structure XCVII, by reduction to the corresponding diol, tosylation, and condensation with methylamine. Since the absolute stereochemistry of the nepetalinic acids is known it is possible to distinguish between the isomers of skytanthine, and α-, (XCVIII), β-, (XCIX), and δ-skytanthine, (C), have been shown to occur in natural skytanthine (*E. J. Eisenbraum et al.*, Chem. Ind., 1962, 1242; *Casinovi et al.*, Gazz., 1962, **92,** 479).

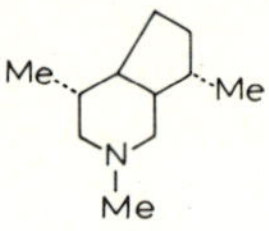

α- (XCVIII)

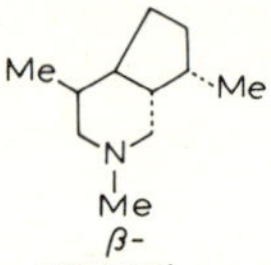

β- (XCIX)

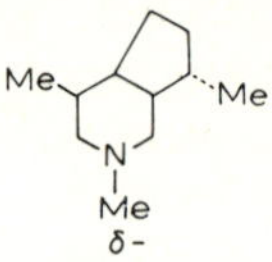

δ- (C)

More recently it has been reported that although α-, and β-skytanthine are present in dried branches, they are absent from original extracts of fresh samples of *Skytanthus acutus* and they are thus transformation products of the genuine alkaloids (*H. H. Appel* and *M. P. Streeter*, Scientia, [Valparaiso], 1970, **36,** 105; C.A., 1971, **74,** 121343a).

Two minor alkaloids of *Skytanthus acutus,* **hydroxyskytanthine I,** and **hydroxyskytanthine II,** $C_{11}H_{21}NO$, have been shown to have the following structures from a detailed analysis of their n.m.r. spectra (*G. Adolphen et al.*, Tetrahedron, 1967, **23,** 3147):

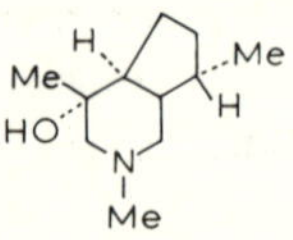

Hydroxyskytanthine I

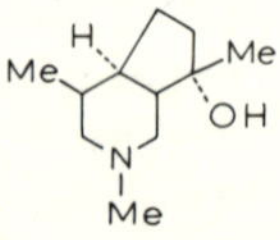

Hydroxyskytanthine II

Dehydration of hydroxyskytanthine I gives **dehydroxyskytanthine,** $C_{11}H_{19}N$, which had previously been isolated from the same species and shown to have either structure CI or CII (*Casinovi et al.*, Chem. Ind., 1963, 984); the latter is thus correct:

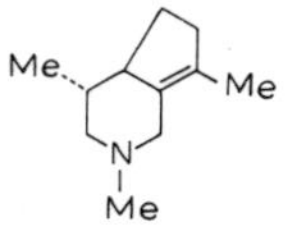

(CI) 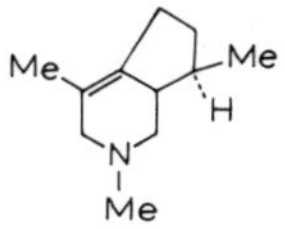 (CII)

Tecomanine, $C_{11}H_{17}NO$, isolated from *Tecoma stans* has a u.v. (λ_{max} 226 nm, logε 4.10) and i.r. spectrum (1700, 1620 cm^{-1}) typical of an α,β-unsaturated ketone, probably contained in a five-membered ring. The n.m.r. spectrum indicates the presence of one vinylic proton (singlet, δ 5.95), an *N*-methyl group (singlet, δ 2.75) and two secondary *C*-methyl groups as doublets centred at δ 1.12 and δ 1.07. Successive reduction of the double bond and the ketone group followed by dehydrogenation yields ($\pm$)-actinidine; the following structure has therefore been assigned to tecomanine (*G. Jones, H. M. Fales* and *W. C. Wildman*, Tetrahedron Letters, 1963, 397):

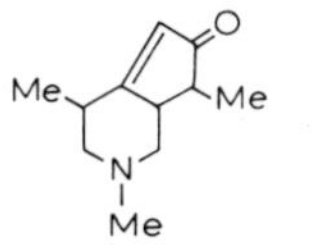

Tecomanine

Tecostanine, $C_{11}H_{21}NO$, is a primary alcohol obtained from *Tecoma stans*. The n.m.r. spectrum shows the presence of an *N*-methyl and a *C*-methyl group and indicates that the two protons of the primary alcohol group are adjacent to a tertiary carbon atom. Removal of the hydroxy group by formation of the tosylate and subsequent reduction gives desoxytecostanine, $C_{11}H_{21}N$. On dehydrogenation this derivative gives ($\pm$)-actinidine and, therefore, tecostanine must be CIII or CIV. The mass spectrum shows that the hydroxymethyl group is attached to the piperidine

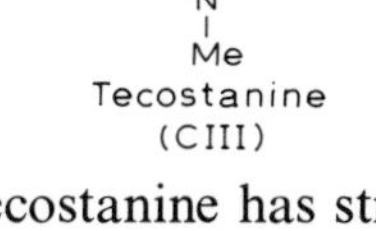

Tecostanine (CIII) (CIV)

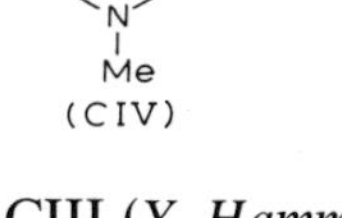

ring and hence tecostanine has structure CIII (*Y. Hammouda, M. Plat* and *J. LeMen*, Bull. Soc. chim. Fr., 1963, 2802).

The properties of the monoterpenoid pyridine and piperidine alkaloids are summarised in Table 10.

TABLE 10

THE MONOTERPENOID PYRIDINE AND PIPERIDINE ALKALOIDS

Alkaloid	*m.p. (°C)*	$[\alpha]_D$	*Picrate m.p. (°C)*	*Methiodide m.p. (°C)*
Actinidine	100–103/9mm[a]	−7.2°	143	
N-*β*-(4-Hydroxyphenyl)ethyl-actinidium chloride	201–203	+50.5°	151–152	
Tecostidine		+4°	153	
N-*β*-(4-Hydroxyphenyl)ethyl-tecostidinium chloride	220–222	+19.3°	65–70	
Valerianine	134	−10.5°		
Skytanthine	54–62/15mm[a]	+42°	135–136	296–298
α-Skytanthine		+79°		237–239
β-Skytanthine	88/10mm[a]	+16.4°		293–295
δ-Skytanthine		+9°		303–305
Hydroxyskytanthine I	94–95	+35.8°		
Hydroxyskytanthine II	119–120	−38.5°		
Dehydroxyskytanthine			127	
Tecomanine	125/0.1mm[a]	−175°	179.5–180.5	240–242
Tecostanine	82	±2°		245

[a] b.p.

Nupharamine, $C_{15}H_{25}NO_2$, is an optically active alkaloid isolated from *Rhizama nupharis.* Dehydrogenation of nupharamine gives an aromatic compound which yields pyridine-2,3,6-tricarboxylic acid on oxidation with permanganate. Successive reduction and Hofmann degradation of nupharamine lead to 1-(3-tetrahydrofuryl)-4,8-dimethyl-8-hydroxynonane (CV), whereas oxidation with permanganate after the first Hofmann degradation gives 1-(3-tetrahydrofuryl)-pentan-4-one, (CVI) (*Y. Arata* and *T. Ohashi,* Yakugaku Zasshi, 1957, **77,** 790; C.A., 1957, **51,** 17756b; *Ohashi, ibid.,* 1959, **79,** 729; C.A., 1959, **53,** 22031f). This information led to the suggestion of

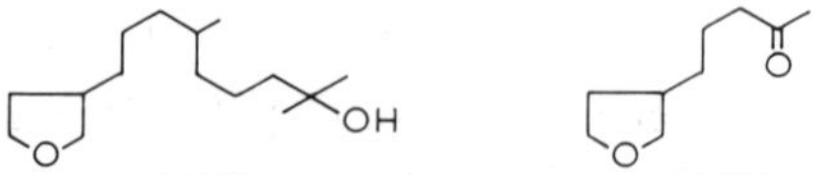

OH

(CV)

(CVI)

the following structure for nupharamine, and this has been confirmed by several syntheses of nupharamine derivatives. Mild catalytic reduction of

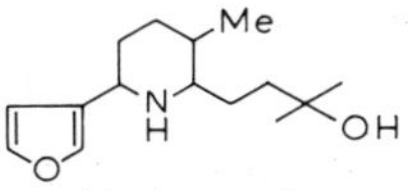
Me
N
H
OH
O

Nupharamine

nupharamine gives deoxynupharamine which has been synthesised as shown in Scheme 10:

Scheme 10

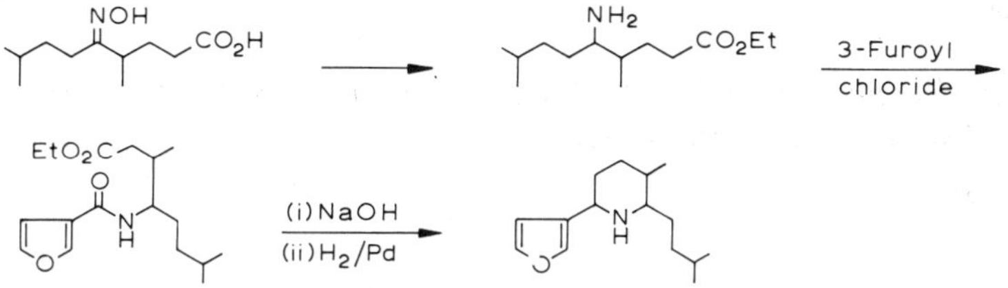
NOH
CO_2H
NH_2
CO_2Et
3-Furoyl
chloride
EtO_2C
O
N
H
(i) NaOH
(ii) H_2/Pd
N
H

The synthetic product was shown to have the same i.r. spectrum as that derived from natural sources (*Arata et al., ibid.*, 1962, **82,** 782; C.A., 1963, **58,** 2432g). *N*-Methylnupharamine has been synthesised as shown in Scheme 11.

Scheme 11

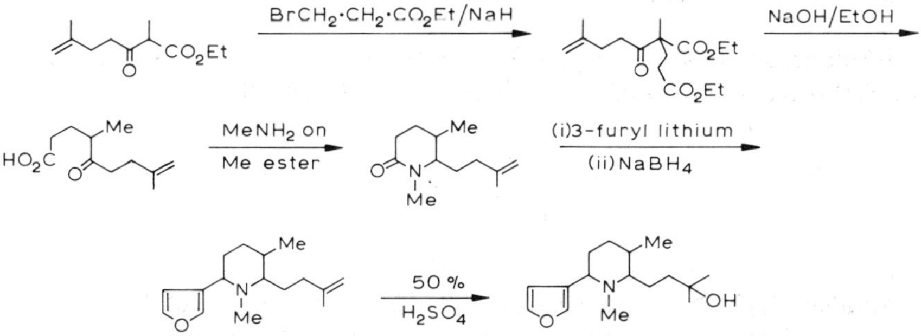
$BrCH_2 \cdot CH_2 \cdot CO_2Et$/NaH
CO_2Et
O
CO_2Et
CO_2Et
NaOH/EtOH
Me
HO_2C
O
$MeNH_2$ on
Me ester
Me
O
N
Me
(i)3-furyl lithium
(ii)$NaBH_4$
Me
N
Me
50 %
H_2SO_4
Me
N
Me
OH

(CVII)

Examination of the various synthetic isomers of CVII showed that the isomer CVIII was identical with (−)-*N*-methylnupharamine derived from natural nupharamine (*S. Matsutani, I. Kawasski* and *T. Kaneko*, Bull. Soc. Chem. Japan, 1967, **40,** 2629):

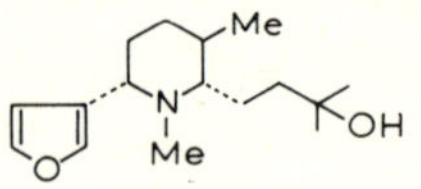

(CVIII) 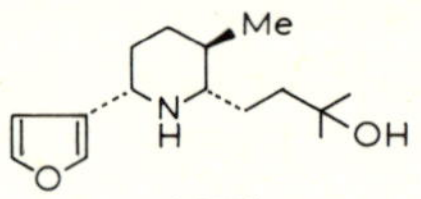 (CIX)

This is in agreement with the configuration (2*R*,3*S*,6*R*) (CIX) assigned earlier to nupharamine from a spectroscopic comparison with the known (−)-deoxynupharidine (*idem, ibid.*, 1963, **36**, 1474).

Nupharamine has b.p. 145–150°/2mm, $[\alpha]_D^{22}$ −35.04°; it forms a *picrolonate*, m.p. 167.5–168°; N-*methylnupharamine* has b.p. 120°/0.05 mm and forms a *methiodide*, m.p. 168–169°.

Anhydronupharamine, $C_{15}H_{23}NO$, has been isolated from *Nuphar japonicum* and shown to have the structure:

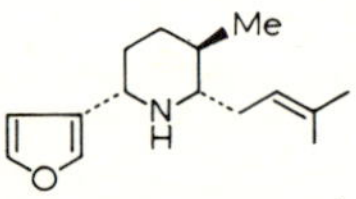

Anhydronupharamine

The action of dilute aqueous hydrochloric acid eventually gives (−)-nupharamine.

Anhydronupharamine has b.p. 108–109°/4mm and forms a *picrolonate*, m.p. 203.5° (*Arata et al.*, Yakugaku Zasshi, 1957, **87**, 1094; C.A., 1968, **68**, 29907r).

3-Epinupharamine, $C_{15}H_{23}NO_2$, has been isolated from *Nuphar luteus* (*C. F. Wong* and *R. T. LaLonde*, Phytochem., 1970, **9**, 1851):

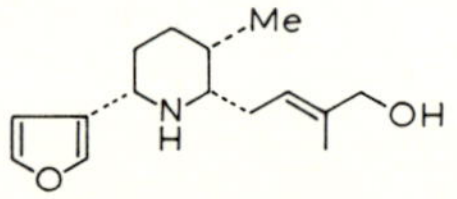

Epinupharamine

Anibine, $C_{11}H_9NO_3$, has been extracted from the sawdust of the rosewood species *Aniba duckei* and *Aniba rosaedora*. It is optically inactive and contains no active hydrogen atoms; functional group analysis shows the presence of one methoxy group and the absence of *N*-methyl or *C*-methyl groups. The i.r. spectrum indicates the presence of a lactone system and, possibly, an enol ether (1580, 1650 and 1730 cm^{-1}). The action of ethanolic alkaline solution on anibine gives a dipotassium salt, $C_{10}H_7K_2NO_4$, which undergoes decarboxylation in acid to give the known diketone CX:

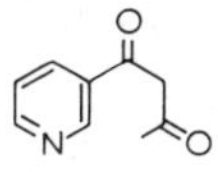

(CX)

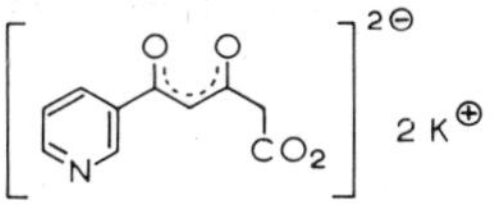
2⊖
CO2
2 K⊕

(CXI)

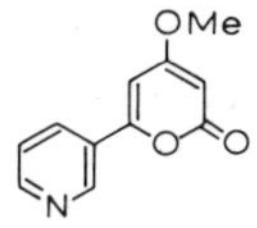
OMe

Anibine
(CXII)

The absence of a *C*-methyl group in anibine requires that the dipotassium saponification product must be CXI; anibine is therefore CXII (*Djerassi et al.*, J. Amer. chem. Soc., 1957, **79,** 4507). Anibine has been synthesised as shown in Scheme 12 (*E. Ziegler* and *E. Noelken*, Monatsh., 1958, **89,** 391):

Scheme 12

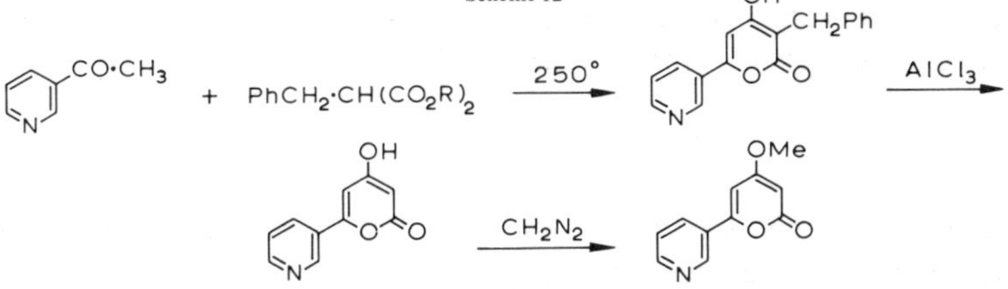
CO·CH3
+ PhCH2·CH(CO2R)2
250°
OH
CH2Ph
AlCl3
OH
CH2N2
OMe
(CXII)

Anibine has m.p. 179–180°; *picrate*, m.p. 199–201°; *hydrochloride*, m.p. 205–230°; *methiodide*, m.p. 233–236°.

The structure and absolute configuration of two alkaloids isolated from the South African arrow poison frog, *Dentrobates histrionicus*, have been elucidated by X-ray crystallography. **Histrionicotoxin,** $C_{19}H_{25}NO$, is (2p*R*,6*S*,7p*S*,8α*S*)-7-(*cis*)-1-buten-3-ynyl)-8-hydroxy-2-(*cis*-2-penten-4-ynyl)-azaspiro[5.5]undecane (CXIII), while in **dihydroisohistrionicotoxin,** $C_{19}H_{27}NO$, (CXIV), the 2-pentenynyl side chain is replaced by an allenic

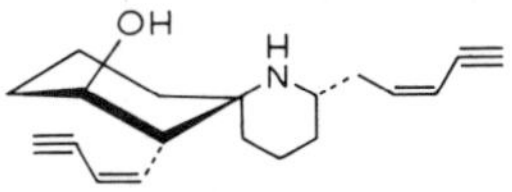
OH
H
N

Histrionicotoxin
(CXIII)

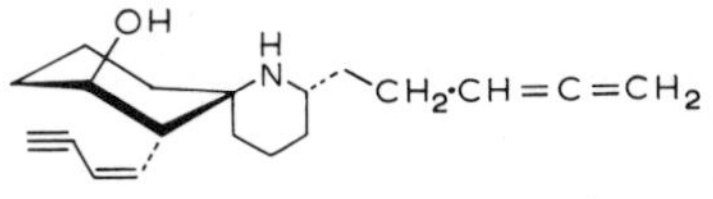
OH
H
N
CH2·CH=C=CH2

Dihydroisohistrionicotoxin
(CXIV)

2-(3,4-pentadienyl) substituent (*B. Witkop et al.*, Proc. nat. Acad. Sci. U.S., 1971, **68,** 1870).

Nigrifactin, $C_{12}H_{17}N$, is an optically inactive alkaloid produced by *Streptomyces* strain No. FFD-101. It may be handled in dilute acidic solution or in the form of a salt, but the free base is unstable and rapidly polymerises.

Nigrifactin hydrochloride displays a characteristic u.v. absorption which shows a considerable hypsochromic shift on basification (300 nm above pH 9.2, 354 nm in neutral or acidic solution). The i.r. spectrum of nigrifactin indicates the presence of C=C (1610 cm^{-1}), and C=N (1645 cm^{-1}) bonds, but the absence of any N–H bonds, while the n.m.r. spectrum shows the presence of six vinylic protons. Nigrifactin is readily reduced to an octahydro derivative which is a secondary amine no longer showing any evidence of unsaturation. The mass spectrum of this reduced derivative shows a peak at *m/e* 84 characteristic of piperidine-ring containing compounds and dehydrogenation gives a pyridine derivative which was shown to be 2-*n*-heptyl pyridine by comparison with an authentic sample. From these data it was deduced that nigrifactin has the following structure (*T. Terashima et al.*, Tetrahedron Letters, 1969, 2535):

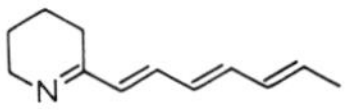

Nigrifactin has been synthesised by an aldol condensation between the piperideine CXV and the aldehyde CXVI, followed by acid-catalysed dehydration of the product CXVII (*H. W. Gschwend, ibid.*, 1970, 2711).

Nigrifactin forms a *picrate*, m.p. 175.5–176°.

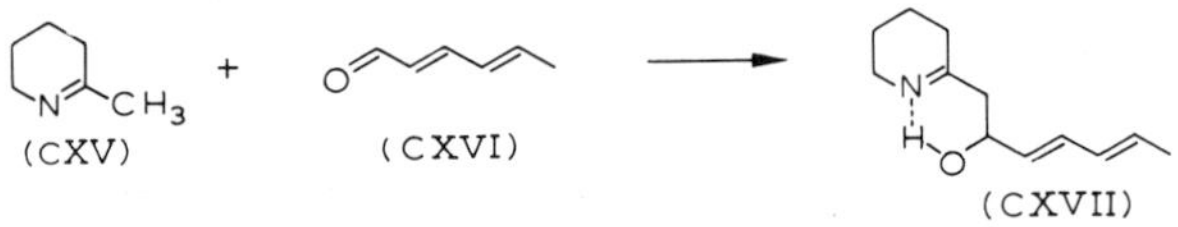

(CXV) (CXVI) (CXVII)

Three new alkaloids, **lythranine, lythranidine** and **lythramine** have been isolated from *Lythrum anceps* (*E. Fujita et al.*, Chem. pharm. Bull. Japan, 1970, **18,** 2216). They were assigned their basic skeletal forms as the result of a chemical and spectroscopic study (*E. Fujita* and *K. Fuji*, J. chem. Soc. C, 971, 1651), while the absolute stereochemistry was determined by X-ray analysis of a bromo derivative of lythranine (*R. J. McClure Jr.* and *G. A. Sim*, J. chem. Soc., Perkin II, 1972, 2073):

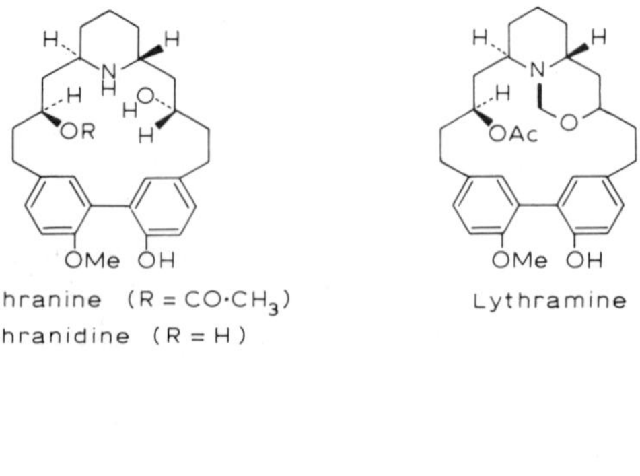

Lythranine ($R = CO{\cdot}CH_3$)
Lythranidine (R = H)

Lythramine

Halfordine, $C_{19}H_{20}N_2O_4$, **halfordinol,** $C_{14}H_{10}N_2O_2$, and **halfordinone,** $C_{19}H_{18}N_2O_3$, have all been isolated from *Halfordia scleroxyla* (see also the subsequent chapter on oxazole alkaloids). Halfordine is a weak tertiary base, halfordinol its hydrolysis product and halfordinone the product of dehydration. Halfordine shows neither NH nor CO bands in the i.r. spectrum, but two bands at 3602 and 3587 cm^{-1} are assignable to hydroxy groups. In addition to halfordinol, strong acidic hydrolysis of halfordine gives a variety of volatile fragments including propan-2-ol and 1-hydroxy-3-methylbutan-2-one. Oxidation of halfordine with periodate gives acetone; halfordine thus contains the side chain $-OCH_2 \cdot CHOH \cdot C(OH)Me_2$, which is consistent with mass spectroscopic data.

Oxidation of *O*-methylhalfordinol with permanganate gives anisic acid, whereas halfordine itself gives 4-carboxyphenoxyacetic acid and nicotinamide. This suggests the partial structures CXVIII and CXIX for halfordine, leaving only one carbon atom unaccounted for. Vigorous catalytic reduction

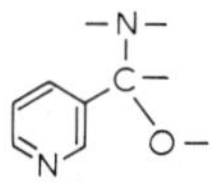

(CXVIII) 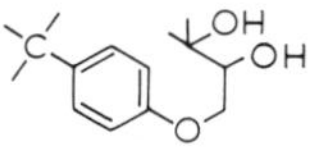 (CXIX)

of halfordinol methochloride gives the amide CXX; the structure CXXI or CXXII can therefore be assigned to halfordine.

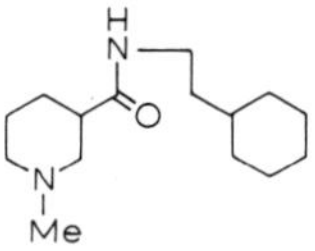

(CXX)

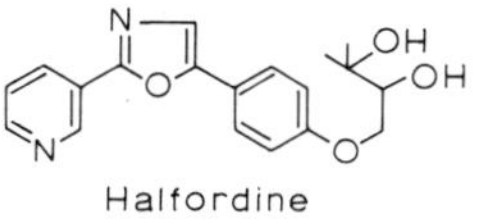

Halfordine (CXXI)

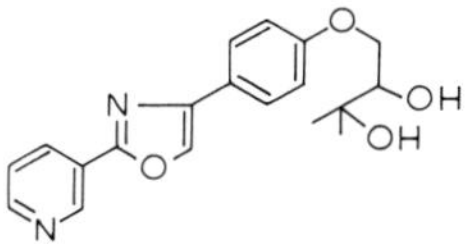

(CXXII)

The mass spectroscopic fragmentation pattern of halfordinol shows a peak at *m/e* 121 which was assigned to the ion CXXIII, and hence structure CX

(CXXIII)

above must be correct for halfordine: halfordinol and halfordinone have the following structures:

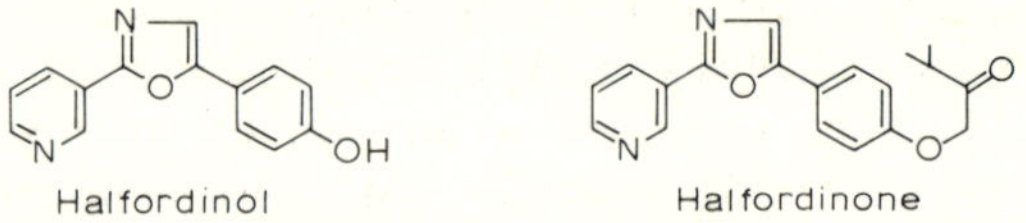

Halfordinol Halfordinone

These structures were later shown to be in accord with more detailed mass spectroscopic and n.m.r. results (*W. D. Crow* and *J. H. Hodgkin*, Tetrahedron Letters, 1963, 85; Aust. J. Chem., 1964, **17**, 119; *Crow, Hodgkin* and *J. S. Shannon, ibid.*, 1965, **18**, 1433).

Halfordine, has m.p. 163° and forms an N-*methyl picrate*, m.p. 143° or 198° depending on the method of crystallisation, a *perchlorate*, m.p. 148° or 206°, and a *chloride*, m.p. 210°. *Halfordinol* has m.p. 255–256° and *halfordinone*, m.p. 132–133°.

RW47, $C_9H_{11}NO$, is a minor alkaloid extracted from *Rauwolfia verticillata*. The i.r. spectrum shows a hydroxy band and an absorption at 1600 cm^{-1}, which, in conjunction with the u.v. spectrum (λ_{max} 258 nm, logε 3.65, shoulder at 267 nm), is attributable to a pyridine ring. This is also consistent with the mass spectrum, which shows a series of peaks (*m/e* 106, 79, 77, 65, 63 and 51) typical for a single pyridine ring. The n.m.r. spectrum indicates a 3,4-disubstituted pyridine (AB doublets at δ 8.18 and 7.09, $J=5.5$ Hz; singlet δ 8.21), a *C*-methyl group (doublet at δ 1.32, $J=7$ Hz), a hydroxy group (singlet at δ 4.96) with one hydrogen on the adjacent carbon (multiplet centred at δ 4.50), and three benzylic protons (multiplet centred at δ 2.72–3.36). From this it was deduced that RW47 has either structure CXXIV or CXXV; the former was considered to be the more likely because of its similarity to the actinidine and skytanthine alkaloids. A detailed analysis of

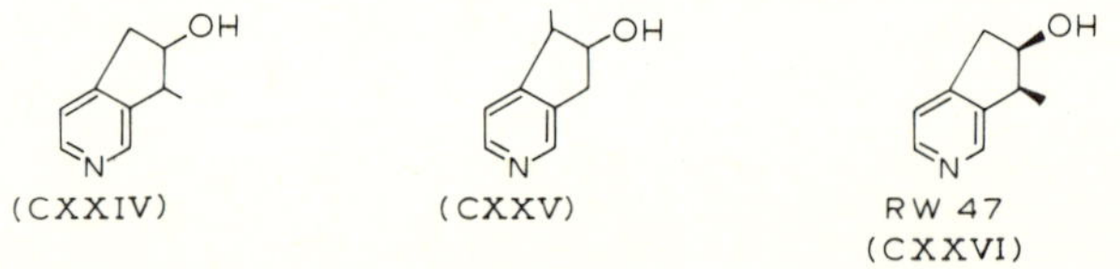

(CXXIV) (CXXV) RW 47 (CXXVI)

the n.m.r. couplings led to the suggested stereochemical formula CXXV (*H. R. Arthur et al.*, Phytochemistry, 1966, **5**, 977; Austral. J. Chem., 1967, **20,** 2505).

RW47 has m.p. 130–132°, $[\alpha]_D$ +27° (chloroform).

An alkaloid, $C_8H_{15}NO_3$, containing a secondary amine function, has been isolated from the fungus *Rhizoctonia leguminicola*. As it shows no sign of aromaticity or unsaturation it is evidently bicyclic; it shows strong hydroxyl absorption in the i.r. spectrum (3350 cm^{-1}) and its reaction with periodate indicates the presence of a *vic*-glycol moiety. The structure CXXVII was assigned on the basis of n.m.r. spin decoupling experiments. The single

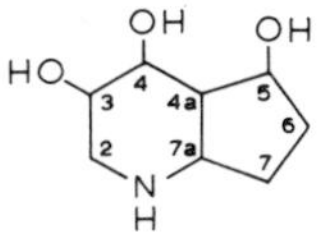

(CXXVII)

proton H-4a (δ 2.05, multiplet) is coupled to H-4 (δ 4.60, doublet of doublets, $J=3$ Hz), to H-5 (δ 4.19, multiplet, $J=6.5$ Hz) and to H-7a (δ 3.05, multiplet, $J=9$ Hz), and this latter coupling indicates that the fusion of the two rings is *trans*, H-4a and H-7a being axial. Thus, as $J_{4,4a}=3$ Hz, H-4 must be equatorial.

The secondary methanol proton H-5 is coupled to a high field proton (δ 2.27, multiplet). H-7a is also coupled to one or more high field protons at δ 1.71 (multiplet). H-3 (δ 4.50, multiplet) is coupled to H-4 ($J=4.5$ Hz) and is thus axial. H-3 is also coupled to H-2ax (δ 2.57, doublet of doublets, $J=7$ Hz) and to H-2eq (δ 3.23, doublet of doublets, $J=2$ Hz); H-2ax and H-2eq are in turn coupled to each other ($J=12$ Hz). This leads to structure formula CXXVIII with undetermined stereochemistry at C-5. This structure

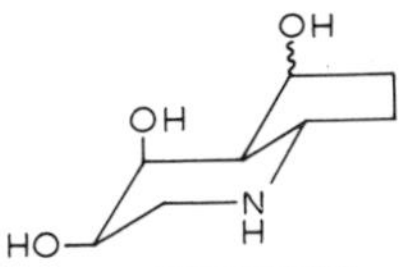

(CXXVIII)

is consistent with all other observations (*F. P. Guegerich et al.*, J. Amer. chem. Soc., 1973, **95,** 2055).

Julocrotine, $C_{18}H_{24}N_2O_3$, is isolated from *Julocroton montevidensis* and has been assigned the following structure on the basis of its chemical and spectroscopic properties:

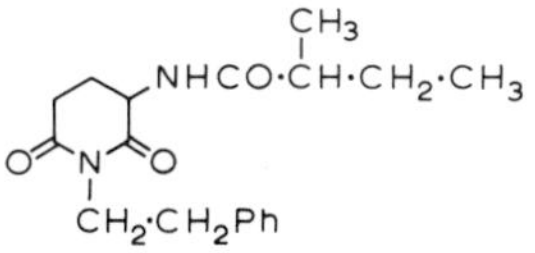

Julocrotine is a crystalline solid, m.p. 108–109°, $[\alpha]_D$ −9° (chloroform), −50° (methanol) (*Djerassi et al.*, Tetrahedron Letters, 1959, 8; J. org. Chem., 1961, **26,** 1184).

Chapter 31

The Quinoline Alkaloids

MALCOLM SAINSBURY

Compounds containing the quinoline ring system are produced by animals, higher plants, fungi and micro-organisms. Reference to the isolation of kynurenic acid (4-hydroxyquinoline-2-carboxylic acid) (C.C.C., Vol. IV F, p. 340), from the urine of dogs, was made over one hundred and fifty years ago (*J. Pelletier* and *J. B. Caventou*, Ann. Chim. Phys., 1820, [ii], **15**, 289, 337); while the therapeutic properties of *Cinchona* bark have been recognized in Western medicine for more than three centuries.

The quinoline alkaloids from higher plants are best classified according to their biogenetic derivation from either anthranilic acid or tryptophan. Since all of the various quinoline derivatives produced by plants of the *Rutacea* family seem to be related to anthranilic acid, they may be considered under a single heading; whereas the *Cinchona* alkaloids, together with the alkaloids of *Calycanthus*, *Camptothecea* and *Melodinus* plants, are closely related to the indole alkaloids with which they frequently occur. These alkaloids arise in nature through the intermediacy of tryptophan.

Quinolines from fungi and micro-organisms are biosynthesised from anthranilic acid but, in view of the nature of the parent systems and for ease of discussion, these are discussed as a separate group.

1. Quinoline alkaloids of the Rutacea family

The plant family Rutacea is subdivided into some one hundred and fifty genera comprising over one thousand individual species; within the members of the family alkaloids are abundant and molecules based upon the quinoline, isoquinoline, pyrrolidine, oxazole, and imidazole ring systems are common.

Currently (1975) over 130 quinoline alkaloids of established structure have been isolated from rutaceous plants (see Table 1). In this review these natural products are classified as simple quinolines, 2- and 4-quinolones, furoquinolines, pyranoquinolines and benzoquinolines (acridones).

TABLE 1

STRUCTURES OF THE RUTACEOUS ALKALOIDS

(a) THE SIMPLE QUINOLINES

Alkaloid name	*Botanical source (genus)*	*M.p. (°C)*	$[\alpha]_D^{\circ}$	*Structure (unless noted)* 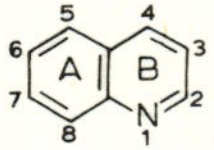 2	3	4	7	8
Quinoline	*Camellia*[1] *Citrus*[2] *Galipea (Cusparia)*[3], *Peganum*[4]	b.p. 237		H	H	H	H	H
Quinaldine	*Galipea*[3], *Peganum*[4]	b.p. 247		Me	H	H	H	H
2-Amylquinoline	*Galipea*[3]	oil		Amyl	H	H	H	H
2-Amyl-4-methoxy-quinoline	*Galipea*[3]	oil		Amyl	H	OMe	H	H
Orixine	*Orixa*[5]	153	+2.2	OMe	a	OMe	$-OCH_2O-$	
Orixinone	*Orixa*[6]	103–104		OMe	b	OMe	$-OCH_2O-$	
4-Methoxy-2-phenyl-quinoline	*Lunasia*[7]	66–67		Ph	H	OMe	H	H
Dubamine	*Haplophyllum*[8,9]	96–97		MDP	H	H	H	H
Graveolinine	*Ruta*[10]	115–116		MDP	H	OMe	H	H
Cusparine	*Galipea*[11,12]	90–92/110		$(CH_2)_2$MDP	H	H	H	H
Cuspareine	*Galipea*[13]	54	−20	$(CH_2)_2$MDP	H	H	H	H B sat. –NMe
Galipine	*Galipea*[14]	114		$(CH_2)_2$MDP	H	OMe	H	H
Galipoline	*Galipea*[14]	193		$(CH_2)_2$MDP	H	OH	H	H
Lunamarinium	*Lunasia*[15]	—		MDP	H	OMe	OMe	H $-\overset{\oplus}{N}Me$

a = $CH_2 \cdot CH(OH) \cdot C(OH)Me_2$; b = $CH_2 \cdot CO \cdot CHMe_2$; MDP = 3,4-methylenedioxyphenyl.

References

1 *R. Yamamoto* and *H. Ito*, Bull. agric. chem. Soc. Japan, 1937, **13**, 74.
2 *T. Ohta, T. Miyazaki* and *T. Yamakawa*, Tokyo, Yakka. Daigaku. Kenkyu. Nempo, 1959, **9** 237; Chem. Zentr., 1962, 15220.
3 *E. Späth* and *J. Pikl*, Ber., 1929, **62**, 2244; Monatsh., 1930, **55**, 352.
4 *B. Kh. Zharekeev et al.*, Khim. priod. Soedin, 1974, 264; C.A., 1974, **80**, 13681v.
5 *M. Tersaka, K. Narahashi* and *Y. Tomikawa*, Chem. Pharm. Bull., Tokyo, 1960, **8**, 1142; *Narahashi, ibid.*, 1962, **10**, 792.
6 *W. J. Donnelly* and *M. F. Grundon*, J. chem. Soc., Perkin I, 1972, 2116.
7 *S. Goodwin, A. F. Smith* and *E. C. Horning*, J. Amer. chem. Soc., 1957, **79**, 2239.
8 *G. P. Sidyakin, V. I. Pastukhova* and *S. Yu. Yunusov*, Uzbeksk. Khim. Zhur., 1962, **6**, 56; C.A., 1963, **58**, 4608 h.
9 *Sidyakin et al.*, Zhur. obshcheĭ. Khim., 1962, **32**, 4091; C.A., 1963, **58**, 14010 g; Khim. Prir. Soedin, 1973, 548.
10 *A. Chatterjee* and *A. Deb*, Chem. and Ind., 1962, 1982; *H. R. Arthur* and *H. T. Cheung*, Austral. J. Chem., 1960, **13**, 510.
11 *Späth* and *O. Brunner*, Ber., 1924, **57**, 1243.
12 *H. Beckurts, P. Nehring* and *A. Lachwitz*, Arch. Pharm., 1891, **229**, 591.
13 *J. Tröger* and *H. Runne*, *ibid.*, 1911, **249**, 174.
14 *Späth* and *G. Papaioanou*, Monatsh., 1929, **52**, 129.
15 *J. R. Price*, Austral. J. Chem., 1959, **12**, 458.

(b) 2-QUINOLONES

Alkaloid name	*Botanical source (genus)*	*M.p. (°C)*	$[\alpha]_D^{\circ}$	*Structure (unless noted)*	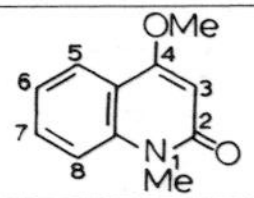		
				3	*7*	*8*	
1-Methyl-2-quinolone	*Galipea (Cusparia)*[3]	74		H	H	H	C_4–H
Casimiroine	*Casimiroa*[16]	199–200		H	–OCH_2O–		
Edulitine (Robustinine)	*Casimiroa*[17] *Haplophyllum*[18]	235–236		H	H	OMe	NH
Folifidine	*Haplophyllum*[19,52]	226–227		H	H	OH	
Folimine	*Haplophyllum*[20,53]	139–140		H	H	OMe	
Halfordamine	*Halfordia*[21,22]	240		H	H	OMe	NH,C_6–OMe
Hydroxylunacridine	*Lunasia*[23], *Orixa*[44]	100–102	+31.5(A)	a	H	OMe	
Hydroxylunidine	*Lunasia*[23], *Ptelea*[24,49]	124–125	+28	a	–OCH_2O–		
Edulinine	*Casimiroa*[17,25] *Citrus*[26] *Eriostemnon*[27] *Ruta*[203]	140–142		a	H	H	
Lunacridine	*Lunasia*[28]	85–86	+30(A)	b	H	OMe	
Lunidine	*Lunasia*[28]	65–67	+28(A)	b	–OCH_2O–		
Atenine (Fagara base)	*Fagara*[29], *Ravenia*[30]	133		c	H	H	
—	*Ptelea*[31]	159–162		c	–OCH_2O–		
Ptelecortine	*Ptelea*[32]	126–128		c	—	OMe	C_6,C_7–OCH_2O–
Ptelefoline, methyl ether	*Ptelea*[32,49]	oil	−13.7(C)	c	H	OMe	C_6–OMe
—	*Haplophyllum*[36]	114–115		c	H	H	NH,C_4–Oc
Preskimmianine	*Dictamnus*[46,86]	151–152		c	OMe	OMe	NH

Isoptelefolidine	*Ptelea*[24]	145–146		d	OMe	OMe	
Ptelefructine	*Ptelea*[24]	146–149		d	—	OMe	C_6,C_7-OCH_2
Ptelefolidine	*Ptelea*[24]	118–119		d	$-OCH_2O-$		
Ptelefoline	*Ptelea*[24,32]	133–135		d	H	OMe	C_6-OMe
Pteleoline	*Ptelea*[32]	oil	+3.7(C)	$CH_2\cdot CHMe\cdot CO_2Me$	$-OCH_2O-$		
Lunidonine	*Lunasia*[28,49]	118–119		$CH_2\cdot CO\cdot CHMe_2$	$-OCH_2O-$		
—	*Ailanthus*[33]	oil		$CH_2\cdot CH_2\cdot CMe=CH_2$	H	H	
Ravenoline	*Ravenia*[34]	144	+6(C)	$CH_2\cdot CH_2\cdot CMe=CH_2$	H	H	C_4-OH
Ravenine	*Ravenia*[30,34]	120–121		H	H	H	C_4-Oc
Bucharaine (Bucharine)	*Haplophyllum*[21,48]	151–152		H	H	H	NH,C_4-Oe
Bucharidine	*Haplophyllum*[35]	251–252		H	H	H	NH,C_4-Of
Flindersine	*Flindersia*[37–39] *Haplophyllum*[36,181] *Geijera*[36a]	186–188		$-CH=CH\cdot CMe-OC_4$	H	H	NH
Foliasidine	*Haplophyllum*[40,41]	141–142	+42(A)	H	H	Oa	
Oricine	*Oricia*[42]	150–152		$-CH=CH\cdot CMe_2-O-$	OMe	H	C_6-OMe
Haplamine	*Haplophyllum*[43]	—		$-CH=CH\cdot CMe_2-O-$	H	H	C_6-OMe
4-Methoxy-1-methyl-2-quinolone	*Hesperthusa*[44] *Zanthoxylum*[45]	100–100.5		H	H	H	
4,7,8-Trimethoxy-1-methyl-2-quinolone	*Spathelia*[51]	143–148		H	OMe	OMe	
3-Isopentenyloxy-7,8-methylenedioxy-4-methoxy-2-quinolone	*Ptelea*[31] *Ptelea*[31]	159–161		b	$-O-CH_2-O-$		
N-Methylflindersine	*Ptelea*[47]	83–85		$-CH=CH\cdot CMe_2O-C_4$	H	H	
3-(3′,3′-Dimethylallyl)-4,6,8-trimethoxy-1-methyl-2-quinolone	*Ptelea*[47]	69–71		c	H	OMe	C_6-OMe

Table 1(b) *(continued)*

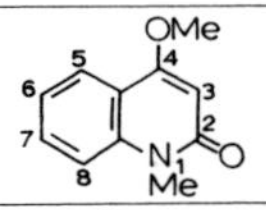

Alkaloid name	*Botanical source (genus)*	*M.p. (°C)*	*$[\alpha]_D^\circ$*	*Structure (unless noted)* 3	7 8	
6-Methoxylunidonine	*Ptelea*[49]	123.5		$-CH_2 \cdot CO \cdot CHMe_2$	$-O-CH_2-O-$	C_6-OMe
6-Methoxylunidine	*Ptelea*[49]	145–147		b	$-O-CH_2-O-$	C_6-OMe
6-Methoxyhydroxy-lunidine	*Ptelea*[49]	181–183		a	$-O-CH_2-O-$	C_6-OMe
Nororixine	*Orixine*[5]	199–200		a	$-O-CH_2-O-$	NH

8 a $= CH_2 \cdot CH(OH) \cdot C(OH)Me_2$; b $= CH_2 \cdot CH(OH) \cdot CHMe_2$; c $= CH_2 \cdot CH = CMe_2$; d $= CH_2 \cdot CH(OH) \cdot CMe = CH_2$; e $= CH_2 \cdot CH = CMe \cdot CH_2 \cdot CH_2 \cdot CH(OH) \cdot C(OH)Me_2$; f $=$ $(CH_2)_2$ … OH, Me, Me, Me

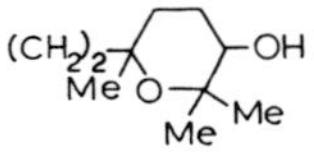

References

16 *F. A. Kincl et al.*, J. chem. Soc., 1956, 4163, 4170.
17 *N. P. Troube, J. W. Murphy* and *A. E. Cross*, Tetrahedron, 1967, **23,** 2061.
18 *I. M. Fakhrutdinova, G. P. Sidyakin* and *S. Yu. Yunusov*, Khim. Prir. Soedin, 1965, **2,** 107; C.A., 1965, **63,** 8423e; *ibid.*, 1967, **3,** 257; C.A., 1967, **67,** 108803m; *T. Kappa, H. Schmidt* and *E. Ziegler*, Z. Naturforsch., 1970, **25b,** 328.
19 *F. A. Stedt* and *K. K. Chen*, J. Amer. pharm. Assoc. Sci. Ed., 1943, **32,** 107.
20 *D. M. Razzakova, I. A. Bessonova* and *Yunusov*, Khim. Prir. Soedin., 1972, **8,** 133; C.A., 1972, **77,** 725637y.
21 *S. M. Sharafutdinova* and *Yunusov*, Khim. Prir. Soedin., 1968, **4,** 264; C.A., 1969, **70,** 47648c.
22 *R. Storer* and *D. W. Young*, Tetrahedron Letters, 1972, 1555; *P. Venturella et al.*, Chem. Ind. (London), 1972, 887; Gazz., 1974, **104,** 297.
23 *S. Goodwin et al.*, J. Amer. chem. Soc., 1959, **81,** 6209.
24 *J. Reisch et al.*, Tetrahedron Letters, 1970, 1945, 3365.
25 *J. Iriarte et al.*, J. chem. Soc., 1956, 4170.
26 *S. R. Johns, J. A. Lamberton* and *A. A. Sioumis*, Austral. J. Chem., 1967, **20,** 1975; 1968, **21,** 1897; 1970, **23,** 419.
27 *E. Lassak* and *J. T. Pinhey, ibid.*, 1969, **22,** 2175.
28 *A. Rüegger* and *D. Stauffacher*, Helv., 1963, **46,** 2329.
29 *I. T. Eshiet* and *D. A. H. Taylor*, Chem. Comm., 1966, 467; J. chem., Soc. C, 1968, 481.
30 *S. K. Talapatra et al.*, Tetrahedron Letters, 1969, 4789.
31 *D. L. Dreyer*, Phytochemistry, 1969, **8,** 1013.
32 *I. Novák et al.*, Tetrahedron Letters, 1972, 449; Herba Hung., 1970, **9,** 23; C.A., 1971, **75,** 77090k.
33 *F. Bohlmann* and *V. S. Bhaskar Rao*, Ber., 1969, **102,** 1774.
34 *B. D. Paul* and *P. K. Bose*, J. Indian Chem. Soc., 1968, **45,** 552; 1969, **46,** 678.
35 *Z. Sh. Faizudinov et al.*, Khim. Prir. Soedin., 1970, **6,** 239; C.A., 1970, **73,** 131179v.
36 *D. Lavie et al.*, Tetrahedron, 1968, **24,** 3011.
36a *D. L. Dreyer* and *A. Lee*, Phytochemistry, 1972, **11,** 763.
37 *H. Matthes* and *E. Schreiber*, Ber., deut. pharm. Ges., 1914, **24,** 385.
38 *R. F. C. Brown et al.*, Austral. J. Chem., 1954, **7,** 348.
39 *Brown, G. K. Hughes* and *E. Ritchie, ibid.*, 1956, **9,** 277; Chem. and Ind., 1955, 1385.
40 *V. I. Pastukhova, G. P. Sidyakin* and *Yunusov*, Dokl. Akad. Nauk Uz.S.S.R., 1964, **21,** 31; C.A., 1965, **62,** 11864c; Khim. Prir. Soedin., 1965, **1,** 27; C.A., 1965, **63,** 8425b.
41 *Z. Sh. Faizudinov, Bessonova* and *Yunusov*, Khim. Prir. Soedin., 1967, **3,** 257; C.A., 1967, **67,** 108803m.
42 *M. O. Abe*, Phytochemistry, 1971, **10,** 3328.
43 *V. I. Akhmezhanova, Bessonova* and *Yunusov*, Khim. Prir. Soedin., 1974, **10,** 109; C.A., 1974, **80,** 121152m.
44 *M. N. S. Nayar, C. V. Sutar* and *M. K. Bhan*, Phytochemistry, 1971, **10,** 2843.
45 *J. L. Pousset, R. R. Paris* and *A. Cavé, ibid.*, 1974, **13,** 1257.
46 *Storer* and *Young*, Tetrahedron, 1973, **29,** 1217.
47 *J. Reich et al.*, Phytochemistry, 1975, **14,** 1678.
48 *Bessonova et al.*, Khim. Prir. Soedin, 1974, **10,** 358.
49 *L. A. Mitscher et al.*, Lloydia, 1975, **38,** 109, 117.
50 *K. Yamamoto* and *Y. Konno*, Chem. Zentr., 1968, **35,** 1412.
51 *Storer et al.*, Phytochemistry, 1973, **29,** 1721.
52 *S. K. Banerjee et al.*, Tetrahedron, 1961, **16,** 251.
53 *Bessonova* and *Yunusov*, Khim. Prir. Soedin, 1973, **9,** 206; 1974, **10,** 52; C.A., 1973, **79,** 32147a; 1974, **80,** 121, 152m.

(c) 4-QUINOLONES *

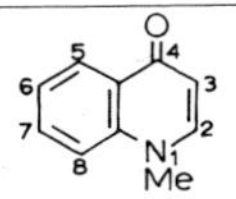

Alkaloid name	*Botanical source (genus)*	*M.p. (°C)*	*Structure (unless noted)* 2	3	7	8	
—	*Acronychia*[54], *Platydesma*[55]	178–179	Me	H	H	H	
—	*Boronia*[56]	112	$(CH_2)_2Me$	H	H	H	N–CH_2OAc
Evocarpine	*Evodia*[57]	34–38	a	H	H	H	a = $(CH_2)_7 \cdot CH{=}CH(CH_2)_3CH_3$
—	*Vespris*[58]	126	b	H	H	H	b = $(CH_2)_9 \cdot CO \cdot CH_3$
—	*Vespris*[58]	~50	c	H	H	H	c = $(CH_2)_8 \cdot CH_2OH$
—	*Vespris*[58]	103	d	H	H	H	d = $CH_2 \cdot CH_2 \cdot CH{=}CH \cdot CH_2 \cdot CH{=}CH \cdot CH_2 \cdot CH_3$
—	*Balfourodendron*[66], *Haplophyllum*[65]	144–145	Ph	H	H	H	
Eduleine	*Casimiroa*[16], *Lunasia*[59]	201	Ph	H	OMe	H	
Eduline	*Casimiroa*[16,60], *Skimmia*[67]	187–188	Ph	H	H	H	6–OMe
Japonin	*Orixa*[61]	143	Ph	OMe	H	H	6–OMe
—	*Lunasia*[62]	174–176	Ph	H	H	H	5–OH
Graveoline (Rutamine)	*Ruta*[12,14,172]	205–206	MDP	H	H	H	
Lunamarine	*Lunasia*[19,23]	245–246	MDP	H	OMe	H	
Lunasia Base-1	*Lunasia*[10]	230–233	MDP	H	H	H	6–OMe
—	*Ruta*[13,64]	224	$(CH_2)_4MDP$	H	H	H	NH
Pilokeanine	*Platydesma*[55]	—	H	e	H	OMe	e = $CH_2 \cdot CH(OH) \cdot CHMe_2$
Folimidine	*Haplophyllum*[65]	225	f	H	H	H	f = 3,4–$C_6H_3(OMe)(OH)$
Acutine	*Haploplyllum*[53]	122–123	g	H	H	H	NH g = $-(CH_2)_3CH{=}CHEt$

MDP = 3,4-methylenedioxyphenyl.

*Structures related to or identical with the pseudanes (p. 221) have been isolated from *Evodia*[57], *Vespis*[58], *Haplophyllum* (*R. M. Rassakowa et al.*, Khim. priod. Soedin., 1973, 206) *Ptelea* and *Ruta* (*Reich et al.*, Phytochemistry, 1975, **14**, 840) species.

References
54 *J. A. Lamberton*, Austral. J. Chem., 1966, **19,** 1995.
55 *F. Werny* and *P. J. Scheuer*, Tetrahedron, 1963, **19,** 1293.
56 *A. M. Duffield* and *P. R. Jefferies*, Austral. J. Chem., 1963, **16,** 123, 282, 292.
57 *R. Tschesche* and *W. Werner*, Tetrahedron, 1967, **23,** 1873.
58 *C. Kan Fan et al.*, Phytochemistry, 1970, **9,** 1283.
59 *R. Johnstone, J. R. Price* and *A. R. Todd*, Austral. J. Chem., 1958, **11,** 562.
60 *H. C. Beyerman* and *R. W. Rooda*, K. Ned. Akad. Wetenschap. Proc., 1960, **B63,** 427, 432.
61 *Ha-hug-Kê, M. Luckner* and *J. Reisch*, Phytochemistry, 1970, **9,** 2199.
62 *N. K. Hart et al.*, Austral. J. Chem., 1968, **21,** 1389.
63 *J. Reisch et al.*, Naturwiss., 1967, **54,** 517.
64 *A. Z. Gulubov, Z. Bozhkova* and *T. O. Sunguryan*, C.A., 1971, **74,** 61600v.
65 *D. M. Razzakova, I. A. Bessonova* and *S. Yu. Yunusov*, Khim. Prir. Soedin., 1972, 133, 755; C.A., 1973, **78,** 84605x.
66 *H. Rapoport* and *K. Holden*, J. Amer. chem. Soc., 1959, **81,** 3738; 1960, **82,** 4395; J. org. Chem., 1961, **26,** 3585.

(d) FUROQUINOLINES

Alkaloid name	*Botanical source (genus)*
Dictamnine	*Aegle*[67], *Balfourodendron*[68], *Boenninghausenia*[69], *Casimiroa*[23], *Chlorilaena*[75], *Decatropis*[76], *Dictamnus*[70,71,77], *Evodia*[72,73,171], *Fagara*[78], *Flindersia*[39], *Halfordia*[74], *Haplophyllum*[79,80], *Helietta*[101], *Hortia*[81], *Glycomis*[102], *Medicosma*[88], *Monnieria*[182], *Orixa*[82], *Pheblalium*[83], *Pitavia*[106], *Ruta*[84,172], *Skimmia*[85], *Zanthoxylum*[45,105]
Skimmianine (β-Fagarine)	*Acronychia*[87], *Aegle*[89,107], *Balfourodendron*[68], *Boronia*[56], *Casimiroa*[23], *Chloroxylon*[90], *Choisya*[91], *Citrus*[2], *Decatropis*[76], *Dictamnus*[70,77], *Eriostemnon*[56], *Esenbeckia*[92], *Evodia*[94], *Fagara*[78,95,96,108], *Flindersia*[39], *Geijera*[94], *Glycomis*[94], *Haplophyllum*[79,97,98,116,130], *Hortia*[81], *Lunasia*[23], *Melicope*[99,135], *Monnieria*[100], *Murraya*[109], *Orixa*[5,82], *Phebalium*[83], *Poncirus*[2], *Ptelea*[31,49,125], *Ruta*[10,110,111,131], *Skimmia*[85,89], *Teclea*[132], *Thamnosma*[112], *Toddalia*[113], *Vespris*[114], *Zanthoxylum*[115,133,134]
γ-Fagarine (Aegelenine, Haplophine)	*Aegle*[117], *Casimiroa*[23], *Dictamnus*[77,118], *Fagara*[78,119], *Geijera*[94], *Glycomis*[5], *Haplophyllum*[18], *Hortia*[81], *Melicope*[135], *Phebalium*[83], *Pitavia*[1], *Ravenia*[30,136,153], *Ruta*[84,121,172], *Thamnosma*[112], *Zanthoxylum*[112,133]
7-Isopentenyloxy-γ-fagarine	*Ptelea*[31]
Kokusaginine	*Acronychia*[87], *Balfourodendron*[68], *Evodia*[72,73,123,124,171], *Flindersia*[126], *Helietla*[101], *Glycomis*[127], *Melicope*[135], *Orixa*[5,61,128], *Phebalium*[83], *Platydesma*[55], *Ptelea*[31,125], *Poncirus*[2], *Ruta*[10,129,171], *Teclea*[132], *Vespris*[58,114]
Kokusagine	*Evodia*[137], *Lunasia*[5,61], *Orixa*[138,139]
Pteleine (6-Methoxydictamnine)	*Dictamnus*[77], *Medicosma*[88], *Platydesma*[55], *Ptelea*[41,125]
Haplopine	*Aegle*[142], *Haplophyllum*[18,41], *Zanthoxylum*[105]
Acronycidine	*Acronychia*[99], *Melicope*[99]
Maculine	*Esenbeckia*[92], *Flindersia*[126], *Ptelea*[31], *Teclea*[132], *Vespris*[141]
Maculosidine	*Balfourodendron*[66,68], *Eriostemnon*[56,143,144], *Esenbeckia*[93], *Flindersia*[126], *Ptelea*[125]
Pteleatinium ion	*Ptelea*[49]
O-Methylpteleofolium ion	*Ptelea*[152]
Evolitrine	*Cusparia*[66,145], *Evodia*[72,73,146], *Orixa*[5], *Phebalium*[83]
Evolitrine, 7-*O*-demethyl	*Evodia*[26], *Poncirus*[2]
Robustine	*Dictamnus*[77], *Haplophyllum*[18], *Zanthoxylum*[105]
Flindersiamine	*Balfourodendron*[66,68], *Esenbeckia*[92], *Flindersia*[126,147,148], *Teclea*[149], *Vespris*[100]
Lunasine	*Lunasia*[150]
Lunacrinium	*Lunasia*[15]
Lunacrinium, hydroxy	*Lunasia*[15]

M.p. (°C)	*[α]°*	*Structure (unless noted)*			
		2	*7*	*8*	
132	—	H	H	H	
177	—	H	OMe	OMe	
142–143	—	H	H	OMe	
101–103	—	H	H	OMe	7-$OCH_2CH{=}CMe_2$
171	—	H	OMe	H	6-OMe
195–197	—	H	–OCH_2O–		
134–135	—	H	H	H	6-OMe
203–204	—	H	OH	OMe	
137–138	—	H	OMe	OMe	5-OMe
196–197	—	H	—	H	6,7-OCH_2O–
134–136	—	H	H	OMe	6-OMe
267–270 $(Cl^{\ominus})$	—	$C(OH)Me_2$	H	OH	$\overset{\oplus}{N}Me$
191–194$(OH^{\ominus})$ [90–93$(Cl^{\ominus})$]	—	$C({=}CH_2)Me$	H	OMe	$\overset{\oplus}{N}Me$
114–115	—	H	OMe	H	
240–242	—	H	OH	H	
147–148	—	H	H	OH	
206–207	—	H	—	OMe	6,7-OCH_2O–
—	—	$CHMe_2$	H	OMe	$\overset{\oplus}{N}Me$, 2,3-H_2
—	—	$CHMe_2$	H	OMe	$\overset{\oplus}{N}Me$
—	—	$C(OH)Me_2$	H	OMe	$\overset{\oplus}{N}Me$

Structure diagram: furo[2,3-b]quinoline ring system (rings A, B, C) with OMe at position 4; positions numbered 1–8.

(continued)

Table 1(d) *(continued)*

Alkaloid name	*Botanical source (genus)*
Balfourodinium-(D)	*Balfourodendron*[66], *Choisya*[151]
Luninium, *O*-methyl	*Lunasia*[150]
Platydesmine, Metho salt	*Skimmia*[85], *Ptelea*[153]
Ribalinium	*Balfourodendron*[173], *Ruta*[174]
O-Methylhydroxyluninium ion	*Ptelea*[49]
Evellerine	*Evodia*[29], *Haplophyllum*[26]
Evodine	*Evodia*[175]
Evoxine	*Choisya*[26], *Evodia*[94,176], *Haplophyllum*[183]
Anhydroevoxine	*Evodia*[177], *Ptelea*[3]
Methylevoxine	*Haplophyllum*[188]
—	*Evodia*[177]
Halfordinine	*Melicope*[135]
Evolatine	*Evodia*[94,124]
Evoxoidine	*Evodia*[176]
Haplophyllidine	*Haplophyllum*[179]
Medicosmine	*Medicosma*[37,88]
Choisyine	*Choisya*[26,91,175]
Platydesmine	*Geijera*[94], *Platydesma*[55], *Skimmia*[85], *Zanthoxylum*[104,123] (racemate in *Melicope perspicuinerva* Merr. et Perry)[135]
Dubinine	*Haplophyllum*[41,180]
Haplophydine	*Haplophyllum*[185]
Glycoperine	*Haplophyllum*[185]
Dubinidine	*Haplophyllum*[9,103,110,180]
Perforine	*Haplophyllum*[178]
Anhydroperforine	*Haplophyllum*[186]
Maculosine	*Flindersia*[126]
Isomaculosidine	*Dictamnus*[46], *Ptelea*[49]
Foliminine	*Haplophyllum*[53]
Confusameline	*Melicope*[189]

M.p. (°C)	*[α]°*	*Structure (unless noted)* 2	7	8	
—	9(A,25°)	C(OH)Me$_2$	H	OMe	$\overset{\oplus}{N}$Me
208–209 (ClO$_4^{\ominus}$)	—	CHMe$_2$	–OCH$_2$O–		$\overset{\oplus}{N}$Me, 2,3-H$_2$
—	—	C(OH)Me$_2$	H	H	$\overset{\oplus}{N}$Me, 2,3-H$_2$
—	—	C(OH)Me$_2$	H	H	NMe, 2,3-H$_2$,6-OH
133–135 (OH$^{\ominus}$)	−17.5 (M)	C(OH)Me$_2$	–OCH$_2$O–		$\overset{\oplus}{N}$Me, 2,3-H$_2$
—	—	H	Oa	H	a = CH$_2$·CH=C(Me)$_2$
153–154	−3(23°)	H	—	OMe	7-OCH$_2$·CH(OH)·CH=CH$_2$Me
154–155	15(A)	H	—	OMe	7-OCH$_2$·CH(OH)·C(OH)Me$_2$
141–144	—	H	—	OMe	7-OCH$_2$·CH–CMe$_2$ O
55–56	—	H	—	OMe	7-OCH$_2$-CH(OH)C(OMe)Me$_2$
145–147	—	H	—	H	7-OCH$_2$·CH–CHMe$_2$ O
150–152	—	H	OMe	OMe	6-OMe
201–202	17(A)	H	—	H	6-OMe, 7-OCH$_2$·CH(OH)·C(OH)Me$_2$
136–137	—	H	—	OMe	7-OCH$_2$·CO·CHMe$_2$
110–111	—	H	b	OMe	b = CH$_2$·CH$_2$·CH(OH)Me
139–140	—	H	H	H	5,6-CH=CHC(Me)$_2$O–
188–189	—	H	OMe	H	5,6-CH$_2$·CH[C(OH)Me$_2$]O–
137–138	47(C)	C(OH)Me$_2$	H	H	2,3-H$_2$
185–186	−73(Ac)	Me	H	H	2,3-H$_2$, 3-CH(OH)CH$_2$OAc
—	—	H	H	Oa	
—	—	H	c	OMe	c = C$_6$H$_{12}$O$_4$
132–133	−63(A)	Me	H	H	2,3-H$_2$, 3-CH(OH)·CH$_2$OH
—	—	H	H	—	C$_8$(OMe)–CH$_2$·CH$_2$·C(OH)Me$_2$, ring A saturated
—	—	H	C$_7$–O–C(Me)$_2$CH$_2$CH$_2$–C$_8$	C$_8$–OMe	ring A saturated
229–230	36(Py,25°)	H	—	H	6,7-OCH$_2$O–, 4-OCH$_2$·CH(OH)·C(OH)Me$_2$
170–172		H	H	OMe	6-OMe, *N*-Me
106–108	—	H	—	—	7,8-CH$_2$·CH$_2$·CMe$_2$–O–
—	—	H	OH	H	

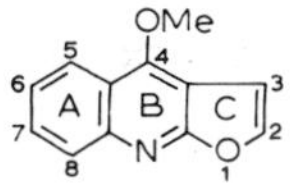

References
67 *A. Chatterjee* and *S. K. Roy*, J. Indian Chem. Soc., 1959, **36,** 267.
68 *O. O. Orazi* and *R. A. Corral*, Anales Asoc. quim. Argentina, 1963, **51,** 174; C.A., 1964, **61,** 959f.
69 *T. Ohta* and *T. Miyazaki*, J. pharm. Soc. Japan, 1958, **78,** 1067.
70 *W. Renner*, Naturwiss., 1961, **48,** 53.
71 *H. Thoms*, Ber. deut. pharm. Ges., 1923, **33,** 68.
72 *J. Rondest et al.*, Phytochemistry, 1968, **7,** 1019.
73 *R. G. Cooke* and *H. K. Hayes*, Austral. J. Chem., 1954, **7,** 273.
74 *W. D. Crow* and *J. H. Hodgkin*, *ibid.*, 1968, **21,** 3075.
75 *J. R. Cannon* and *C. D. Shilkin*, *ibid.*, 1971, **24,** 2181.
76 *X. A. Dominguez et al.*, Phytochemistry, 1971, **10,** 2554.
77 *I. A. Bessonova et al.*, Nauk. Gruez. S.S.R., 1971, **64**, 85; C.A., 1972, **76**, 56582c; Khim. Prir Soedin., 1971, **7**, 675; C.A., 1972, **76**, 124107c.
78 *I. A. Benages et al.*, Phytochemistry, 1974, **13,** 2891.
79 *D. Kurbanov*, *G. P. Sidyakin* and *S. Yu. Yunusov*, Khim. Prir. Soedin., 1967, **3,** 67; C.A., 1967, **67,** 22065 k.
80 *M. F. Grundon* and *T. R. Chamberlain*, J. chem. Soc., C, 1971, 910.
81 *I. J. Pachter et al.*, J. Amer. chem. Soc., 1960, **82,** 5187.
82 *T. Obata*, J. pharm. Soc. Japan, 1939, **59,** 136.
83 *R. C. Cambie*, New Zealand J. Sci., 1959, **2,** 230.
84 *G. Schneider*, Planta Med., 1965, **13,** 425.
85 *D. R. Boyd* and *Grundon*, J. chem. Soc., C, 1970, 556.
86 *R. Storer* and *D. W. Young*, Tetrahedron Letters, 1972, 2199.
87 *J. A. Lamberton* and *J. R. Price*, Austral. J. Chem., 1953, **6,** 66.
88 *Lamberton et al.*, *ibid.*, 1968, **21,** 2357.
89 *Chatterjee* and *S. Bose*, J. Indian chem. Soc., 1952, **29,** 425; 1953, **30,** 33.
90 *A. Mockerjee* and *P. K. Bose*, *ibid.*, 1946, **23,** 1.
91 *V. I. Frolova* and *A. D. Kuzovkov*, J. gen. Chem. U.S.S.R. (Eng. Transl.), 1963, **33,** 121.
92 *J. C. Vitagliano* and *J. Comin*, Anales Asoc. quim. Argentina, 1970, **58,** 59; C.A., 1970, **74,** 34576k.
93 *D. L. Dreyer et al.*, Phytochemistry, 1972, **11,** 705.
94 *S. R. Johns* and *Lamberton*, Austral. J. Chem., 1966, **19,** 895, 1991.
95 *R. Goto*, J. pharm. Soc. Japan, 1941, **61,** 91; *F. G. Torto et al.*, Ghana J. Sci., 1969, **9,** 3.
96 *J. M. Calderwood*, *N. Finkelstein* and *F. Fish*, Phytochemistry, 1970, **9,** 675; *Fish* and *P. G. Waterman*, J. Pharm. Pharmacol., 1971, **23,** 67; *idem*, Phytochemistry, 1972, **11**, 1866.
97 *S. A. Sultancy*, *V. I. Pastukhova* and *Yunusov*, Khim. Prir. Soedin., 1967, **3,** 355; C.A., 1968, **68,** 19546h.
98 *Yunusov* and *G. P. Sidyakin*, Zhur. obshcheĭ Khim., 1952, **22,** 1055; C.A., 1953, **47,** 8084f.
99 *G. K. Hughes et al.*, Nature, 1948, **162,** 233; *Price*, Austral. J. Chem., 1949, **A2,** 249; *F. N. Lahey* and *W. C. Thomas*, *ibid.*, p. 423.
100 *R. Rouffiac*, *I. Fouraste* and *E. Stanislas*, Planta Med., 1969, **17,** 361.
101 *C. A. Mammarella* and *J. Comin*, Anales Asoc. quim. Argentina, 1971, **59,** 579.
102 *D. P. Chakrabory* and *B. K. Barman*, C.A., 1965, **62,** 4066h.
103 *Yunusov* and *Sidyakin*, Zhur. obshcheĭ Khim. 1955, 2009; C.A., 1956, **50,** 9435i.
104 *S. Najjar et al.*, Phytochemistry, 1975, **14,** 2309.
105 *H. Ishii et al.*, J. Pharm. Soc. Japan, 1974, **94,** 322.
106 *H. N. Millan* and *O. M. Silva*, C.A., 1970, **72,** 107836r.
107 *A. Shoeb*, *R. S. Kapil* and *S. P. Popli*, Phytochemistry, 1973, **12,** 2071.
108 *Fish* and *Waterman*, *ibid.*, 1971, **10,** 3322; *ibid.*, 1972, **11,** 3007; J. Pharm. Pharmacol., 1971, **23,** 132S, *L. Fonzes* and *F. Winternitz*, Compt. rend., 1968, **266,** 930.
109 *T. Ohta*, *H.-Y. Hsu* and *C. Noda*, Tokyo, Yakka. Daigaku. Kenkyu. Nekpo, 1959, **9,** 244.
110 *D. Basu* and *S. C. Basa*, J. org. Chem., 1972, **37,** 3035.
111 *E. F. L. J. Anet*, *Hughes* and *E. Ritchie*, Austral. J. Chem., 1961, **14,** 173.
112 *Dreyer*, Tetrahedron, 1966, **22,** 2923.
113 *J. Reisch et al.*, Pharmazie, 1969, **11,** 699.

114 *T. R. Govindachari* and *V. N. Sundarajan*, J. Sci. Ind. Res., India, 1961, **20B**, 298.
115 *Ishii* and *T. Komaki*, Yak. Zasshi, 1966, **86,** 631; C.A., 1966, **65,** 12564b.
116 *M. Ionescu* and *I. Mester*, Phytochemistry, 1970, **9,** 1137.
117 *K.K. Chakravarty*, J. Indian chem. Soc., 1944, **21,** 401.
118 *Hā-Huy-Kê* and *M. Luckner*, Pharmazie, 1966, **21,** 771.
119 *V. Deulofeu*, *R. Labriola* and *J. de Langhe*, J. Amer. chem. Soc., 1942, **64,** 2326.
120 *H. R. Arthur* and *L. Y. S. Loh*, J. chem. Soc., 1961, 4360.
121 *G. Schneider*, Naturwiss., 1965, **52,** 347; Arch. Pharm., 1967, **300,** 953; *A. G. Gonzalez et al.*, Anales de Quim., 1974, **70,** 281; C.A., 1974, 81, 117848k.
122 *Ishii* and *K. Harada*, Yak. Zasshi., 1961, **81,** 238; *Ishii et al.*, *ibid.*, 1972, **92,** 118.
123 *J. A. Diment*, *Ritchie* and *W. C. Taylor*, Austral. J. Chem., 1967, **20,** 565, 1719.
124 *R. J. Gell*, *Hughes* and *Ritchie*, *ibid.*, 1955, **8,** 114.
125 *Frolova*, *Kuzovkov* and *P. N. Kilal'chich*, Zhur. obshcheĭ, Khim., 1964, 34, 3499; C.A., 1965, **62,** 2800f.
126 *R. F. C. Brown et al.*, Austral. J. Chem., 1954, **7,** 181.
127 *A. W. McKenzie* and *Price*, Austral. J. Sci. Res., 1952, **A5,** 579.
128 *M. Terasaka*, J. pharm. Soc. Japan, 1933, **53,** 1046.
129 *T. Ohta et al.*, Chem. and Pharm. Bull., Tokyo, 1960, **8,** 377.
130 *Yunusov et al.*, Khim. Prir. Soedin., 1972, **8,** 133, 343; *E. F. Nesmelova* and *Sidyakin*, *ibid.*, 1973, **9,** 548.
131 *A. A. Gonzalez et al.*, Anales de Quim., 1974, **70,** 60; C.A. 1974, **80,** 130487b.
132 *J. Vaquette*, *J. L. Pousset* and *A. Cavé*, Planta Med. Phytotherap., 1974, **8,** 72.
133 *N. Weber*, Chem. Ber., 1973, **106,** 3769; *N. Decaudin* and *N. Kunesch*, Phytochemistry, 1974, **13,** 505.
134 *Fish et al.*, *ibid.*, 1975, **14,** 2094.
135 *S. T. Murphy*, *E. Ritchie* and *W. C. Taylor*, Austral. J. Chem., 1974, **27,** 187.
136 *K. C. Das* and *P. K. Bose*, Trans. Bose Res. Inst., Calcutta, 1963, **26,** 129.
137 *Hughes* and *K. C. McNeill*, Austral. J. Sci. Res., 1949, **A2,** 429.
138 *W. J. Donnelly* and *Grundon*, J. chem. Soc., Perkin I, 1972, 2116.
139 *M. Terasaka*, *Ohta* and *K. Narahashi*, J. pharm. Soc. Japan, 1953, **73,** 773.
140 *Ritchie*, *Taylor* and *D. V. Willcocks*, Austral. J. Chem., 1960, **13,** 426.
141 *A. K. Ganguly et al.*, Indian J. Chem., 1966, **4,** 334.
142 *D. Basu* and *R. Sen*, Phytochemistry, 1974, **13,** 2329.
143 *A. M. Duffield*, *P. R. Jefferies* and *P. H. Lucich*, Austral. J. Chem., 1962, **15,** 812.
144 *Duffield et al.*, Tetrahedron, 1963, **19,** 593.
145 *Rapoport* and *H. T. G. Hiem*, J. org. Chem., 1960, **25,** 2251.
146 *T. Sato* and *M. Ohta*, Chem. Zent., 1961, **132,** 9042.
147 *F. A. L. Anet et al.*, Austral. J. Sci. Res., 1952, **A5,** 412.
148 *J. R. Cannon et al.*, *ibid.*, 1952, **A5,** 420.
149 *R. R. Paris* and *A. Stambouli*, Compt. rend., 1959, **248,** 3736.
150 *N. K. Hart* and *Price*, Austral. J. Chem., 1966, **19,** 2187.
151 *R. Garestier* and *M. Rideau*, C.R. Acad. Sci. Ser. D., 1972, **274,** 354.
152 *K. Szendrei et al.*, Phytochemistry, 1973, **12,** 2552
153 *L. Mitschler*, *M. S. Bathala* and *J. L. Beal*, J. chem. Soc., D, 1971, 1040.
154 *K. H. Pegel* and *W. G. Wright*, J. chem. Soc., C, 1969, 2327.
155 *F. Dallacker* and *G. Adolphen*, Ann. Chem., 1966, **134,** 691.
156 *G. K. Hughes*, *K. G. Neill* and *E. Ritchie*, Austral. J. Sci. Res., 1950, **A3,** 497.
157 *Hughes* and *Neill*, *ibid.*, 1949, **A2,** 401, 429.
158 *F. D. Popp* and *D. P. Chakraborty*, J. pharm. Sci., 1964, **53,** 968.
159 *I. J. Pachter*, *D. E. Zacharias* and *O. Robeiro*, J. org. Chem., 1959, **24,** 1285.
160 *P. L. MacDonald* and *A. V. Robertson*, Austral. J. Chem., 1966, **19,** 275.
161 *T. R. Govindachari*, *B. R. Pai* and *P. S. Subramaiam*, Tetrahedron, 1966, **22,** 3245.
162 *Govindachari et al.*, *ibid.*, 1970, **26,** 2905.
163 *J. A. Diment*, *Ritchie* and *W. C. Taylor*, Austral. J. Chem., 1967, **20,** 1719.
164 *S. Johne*, *H. Bernasch* and *D. Gröger*, Pharmazie, 1970, **25,** 777.
165 *A. N. Fraser* and *J. R. Lewis*, J. Chem. Soc., Chem. Comm., 1973, 615.

166 *B. Maryse, M. Koch* and *M. Plat*, Phytochemistry, 1974, **13,** 301.
167 *Fraser* and *Lewis*, J. Chem. Soc., Perkin I, 1973, 1173.
168 *A. Cavé et al.*, Planta Med. Phytotherap. 1971, **5,** 327.
169 *S. K. Talapatra, B. C. Maiti* and *B. Talapatra*, Tetrahedron Letters, 1971, 2683.
170 *J. Vaquette et al.*, Planta Med. Phytotherap., 1974, **8,** 57.
171 *Reisch et al.*, Phytochemistry, 1972, **11,** 2121, 2359.
172 *T. N. Vasudevan* and *M. Luckner*, Pharmazie, 1968, **23,** 520; *I. Novák et al.*, *ibid.*, 1965, **20,** 655.
173 *R. A. Corral* and *O. O. Orazi*, Tetrahedron, 1965, **21,** 909.
174 *K. Szendrei et al.*, Pharmazie, 1968, **9,** 519.
175 *R. H. Prager et al.*, Austral. J. Chem., 1962, **15,** 301.
176 *Hughes, Neill* and *Ritchie*, Austral. J. Sci. Res., 1952, **A5,** 401.
177 *D. L. Dreyer*, J. org. Chem., 1970, **35,** 2420.
178 *Z. Sh. Faizutdinova, I. A. Bessonova* and *Yunusov*, Khim. Prior. Soedin., 1968, **4,** 360; C.A., 1969, **71,** 3516a.
179 *T. T. Shakirov, G. P. Sidyakin* and *Yunusov*, Doklady Akad. Nauk U.S.S.R., 1961, **8,** 47; C.A., 1962, **57,** 9902.
180 *Bessonova* and *Yunusov*, Khim. Prir. Soedin., 1969, **5,** 29; C.A., 1969, **71,** 3523z.
181 *Yunusov et al.*, *ibid.*, 1974, **7,** 266; C.A., 1974, **81,** 60896d.
182 *I. Fouraste, J. Gleye* and *E. Stanislas*, Planta Med. Phytotherap., 1973, **7,** 216; C.A., 1974, **80,** 143009x.
183 *A. G. Gonzalez et al.*, Anales de Quim., 1972, **68,** 1133; C.A., 1973, **78,** 82069b.
184 *Bessonova et al.*, Khim. Prir. Soedin., 1974, **10,** 677.
185 *Abdullaev et al.*, *ibid.*, 1974, **10,** 681, 684.
186 *Bessonova et al.*, *ibid.*, 1974, **10,** 682.
187 *Szendrei et al.*, Lloydia, 1973, **36,** 337.
188 *Akhmedzhan et al.*, Khim. Prir. Soedin, 1975, **11,** 272.
189 *T. H. Yang et al.*, J. Pharm. Soc. Japan, 1971, **91,** 782; C.A., 1971, **75,** 95382m.

(e) FUROQUINOLONES

Alkaloid name	*Botanical source (genus)*	*M.p. (°C)*	*[α]°*	*Structure (unless noted)* 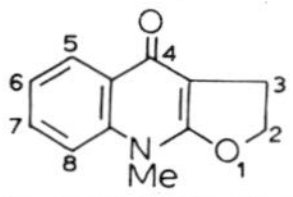 2	7	8	
Ifflaiamine	*Flindersia*[80,190–192]	116–120	−6(Me,25°)	Me	H	H	3-Me_2
Spectabiline (Lemobiline)	*Flindersia*[193], *Ravenia*[30,34]	99(hydrate)	−6(C)	$CHMe_2$	H	H	3-Me
Lunacrine	*Lunasia*[194]	117–118	−51(A,20°)	$CHMe_2$	H	OMe	
Lunine	*Lunasia*[23,59]	222–235	−38(C)	$CHMe_2$	−OCH_2O−		
Lunine, hydroxy	*Lunasia*[23], *Ptelea*[49,195]	228–230	−6(A)	$C(OH)Me_2$	−OCH_2O−		
Lunacrine, hydroxy	*Lunasia*[194]	201–203		$C(OH)Me_2$	H	OMe	
Ptelefolone	*Ptelea*[195]	70–71	+2(C)	$C(=CH_2)Me$	H	OMe	6-OMe
Acrophylline	*Acronychia*[196]	120		—	OMe	H	*N*-$CH_2CH{=}CMe_2$ Δ^2
Acrophyllidine	*Acronychia*[196]	177		—	OMe	H	*N*-$CH_2CH_2C(OH)Me_2$, Δ^2
Ribaline	*Balfourodendron*[173]	223–224 (decomp.)	+68(27°,A)	$C(OH)Me_2$	H	H	6-OH
Isomaculosidine	*Dictamnus*[197], *Ptelea*[49]	170–172		—	H	OMe	6-OMe
Bucharamine (as the acetonide derivative)	*Haplophyllum*[48]	—		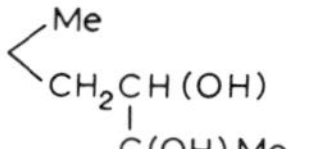 Me, $CH_2CH(OH)C(OH)Me_2$	H	H	3-Me, NH
Isodictamnine	*Dictamnus*[77]	188		—	H	H	Δ^2
		187–188	−5.4	–	H	H	3-OH, 9-OMe
Ptelofolidone	*Ptelea*[198]	152–154		$C(=CH_2)Me$	−OCH_2O−		

References
190 *R. Tschesche et al.*, Tetrahedron, 1964, **20,** 1435.
191 *J. Reisch et al.*, Acta Pharmac. Suec., 1967, 265; C.A., 1968, **68,** 39861k.
192 *J. Bosson et al.*, Austral. J. Chem., 1963, **16,** 480.
193 *T. R. Chamberlain* and *Grundon*, J. chem. Soc., C, 1971, 910.
194 *S. Goodwin* and *E. C. Horning*, J. Amer. chem. Soc., 1959, **81,** 1908.
195 *J. Reisch et al.*, Tetrahedron Letters, 1969, 3803; 1970, 3365; *K. Szendrei et al.*, Lloydia, 1973, **36,** 333.
196 *F. N. Lahey, M. McCamish* and *T. McEwan*, Austral. J. Chem., 1969, **22,** 447.
197 *R. Storer* and *D. W. Young*, Tetrahedron, 1973, **29,** 1217; *M. Gellert et al.*, Herba Hung., 1971, **10,** 123.
198 *Reisch et al.*, Phytochemistry, 1975, **14,** 2722.

				Structure (unless noted)		
				3	9	
Khaplofoline	*Haplophyllum*[41,199]	272–274		H	H	NH
Folifine	*Haplophyllum*[41]	255–256	14(Me)	OH	H	
(+)-Ribalinidine	*Balfourodendron*[200,201], *Ruta*[43]	257–258	15	OH	H	7-OH
(−)-Ribalinidine	*Balfourodendron*[200]	257–258	−15	OH	H	7-OH
Ribalinine	*Balfourodendron*[200]	223–224 (decomp.)	(±)	OH	H	
(+)-Isobalfourodine	*Balfourodendron*[66]	204–205	15(A)	OH	OMe	
(−)-Isobalfourodine (Lunasia base II)	*Lunasia*[23,60,66], *Ptelea*[187]	201–203	−14(A)	OH	OMe	
Neohydroxylunine	*Ptelea*[49]	228–230	—	OH	—	8-OCH_2O-9
Pteleforin	*Ptelea*[198]	93–96				8-OCH_2O-9, 5-OMe, N-de Me ring B aromatic

Key to solvents used in optical rotation measurements: A = ethanol; C = chloroform; Me = methanol. 112176x.

References

199 *D. Kurbanov*, *Bessonova* and *Yunosov*, Khim. Prir. Soedin., 1968, **4,** 58; C.A., 1968, **68,** 112176x.
200 *R. A. Corral* and *O. O. Orazi*, Tetrahedron Letters, 1967, 583.
201 *Corral*, *Orazi* and *I. A. Benages*, *ibid*., 1968, 545; Tetrahedron, 1973, **29,** 205.
202 *W. Scharlemann*, Z. Naturforsch., 1972, **27b**, 806.
203 *D. Boulanger*, *B. K. Bailey* and *W. Steck*, Phytochemistry, 1973, **12,** 2399.

(g) ACRIDONES

Alkaloid name	*Botanical source (genus)*	*M.p. (°C)*	*Structure (unless noted)* 1	2	3	4
			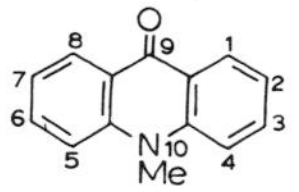			
N-Methylacridone	*Thamnosma*[112]	202–203	H	H	H	H
Acridone,1-OH, *N*-Me	*Ruta*[202]					
Acridone,1-OH, 3-OMe *N*-Me	*Ruta*[202]					
Acridone,4-OH,2,3-diOMe, *N*-Me	*Acronychia*[99,176], *Evodia*[124], *Fagara*[108]	176–177	OH	OMe	OMe	H
Acridone, 1,2,3-triOMe, *N*-Me	*Evodia*[94,123,124]	169–170	OMe	OMe	OMe	H
Acridone, 1,3-diOMe, *N*-Me	*Acronychia*[87,111], *Evodia*[177], *Vespris*[58,141]	163–164				
Arborinine	*Acronychia*[99], *Evodia*[123], *Fagara*[108], *Glycomis*[52], *Lemonia*[169], *Monnieria*[168], *Ravenia*[136], *Ruta*[172], *Teclea*[154], *Vespris*[141]	185–186	OH	OMe	OMe	H
Melicopicine	*Acronychia*[99], *Melicope*[99], *Teclea*[170]	133–134	OMe	OMe	OMe	OMe
Melicopidine	*Acronychia*[99,155], *Baurella*[166], *Evodia*[123,124,156,157], *Melicope*[99,155]	121–122	OMe	$-OCH_2O-$		OMe

Melicopine	*Acronychia*[99,155], *Melicope*[99]	179	OMe	OMe	$-OCH_2O-$		
Evoxanthine	*Balfourodendron*[156], *Evodia*[94,123,124,157,176], *Teclea*[149,154,158,170], *Vespris*[58,141]	217–218 (225)	OMe	$-OCH_2O-$		H	
Evoprenine	*Evodia*[123,163,164]	143	OH	OMe		Oa	H
Evoxanthidine	*Evodia*[155]	312–313	OMe	$-OCH_2O-$		H	*N*-de-Me
Xanthevodine	*Acronychia*[54,159], *Evodia*[155,157,76,177]	213–214	OMe	$-OCH_2O-$		OMe	*N*-de-Me
Xanthoxoline	*Evodia*[157,176]	257–267	OH	OMe	OMe	H	*N*-de-Me
Tecleanthine	*Teclea*[154,170]	158	OMe	$-OCH_2O-$		H	5-OMe
Acronycine	*Acronychia*[99,160], *Baurella*[166]	175	OMe	H	$-OC(Me)_2CH=CH-$		
Acronycine, de-*N*-Me	*Glycomis*[161]	268–270	OMe	H	$-OC(Me)_2CH=CH-$		*N*-de-Me
Acronycine, nor	*Glycomis*[161]	198–200	OH	H	$-OC(Me)_2CH=CH-$		
Acronycine, nor, de-*N*-Me	*Glycomis*[161]	245–246	OH	H	$-OC(Me)_2CH=CH-$		*N*-de-Me
Rutecridone	*Ruta*[191,202]	161–162	OMe	H	$-OCH(CMe=CH_2)CH_2-$		*N*-de-Me
Atalaphylline	*Atlantia*[162]	246	OH	a	OH	a	*N*-de-Me, 5-OH
Atalaphylline, *N*-Me	*Atlantia*[162]	192–193	OH	a	OH	a	5-OH
Atalaphylline, *N*-Me, bicyclo	*Atlantia*[110]	185	c		c		

(continued)

Table 1(g) *(continued)*

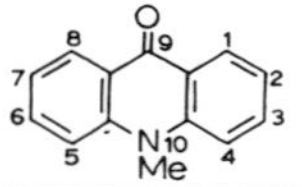

Alkaloid name	*Botanical source (genus)*	*M.p. (°C)*	*Structure (unless noted)* 1	2	3	4	
Atalanine	*Atlantia*[165]	216.5–217.5	OH	H	$-OCH(CHMe_2)CH(OH)-$		5-b
Ataline	*Atlantia*[165]	209–210	OH	H	$-OC(Me)_2CH{=}CH-$		5-b
–	*Atlantia*[167]	190–191.5	OH	a	$-OC(Me)_2CH{=}CH-$		5-OH
–	*Atlantia*[167]	252–254	OH	H	$-OC(Me)_2CH{=}CH-$		5-OH
1,5,6-Trimethoxy-2,3--methylenedioxy-10--methylacridone	*Teclea*[170]	—	OMe	$-OCH_2O-$		H	5-OMe, 6-OMe
6-Methoxytecleanthine	*Teclea*[170]	168	OMe	$-OCH_2O-$		H	6-OMe
1,3,5-Trimethoxy-10--methylacridone	*Teclea*[170]	141	OMe	H	OMe	H	5-OMe
Gravacridondiol	*Ruta*[171]	224–227	OMe	H	$-OCH(R)CH_2-$		$R = C(OH)MeCH_2OH$
Gravacridondiol monomethyl ether	*Ruta*[171]	219–221	OMe	H	$-OCH(R)CH_2-$		$R = C(OH)MeCH_2OMe$
Gravacrindon chlorine	*Ruta*[171]	254–257	OMe	H	$-OCH(R)CH_2-$		$R = C(Cl)MeCH_2OH$
Gravacrindonolchlorine	*Ruta*[171]	223–227	OMe	H	$-OCH(R)CH_2-$		$R = C(Cl)CH_2OH \cdot CH_2OH$
1-Hydroxy-3-methoxy--10-methylacridone	*Fagara*[108]	175–176	OH	H	OMe	H	

$a = CH_2CH{=}C(Me)_2$, b = [structure: O, OH, OMe, N, Me, OH, ⊖O], $c = -OC(Me)_2CH_2 \cdot CH_2-$

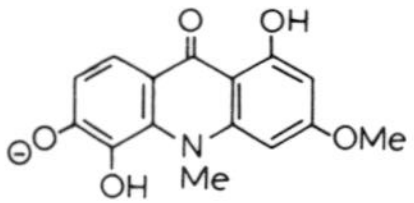

(a) Simple quinolines

A number of simple quinolines, including the parent structure have been isolated from rutaceous plants. Most of these alkaloids are substituted at position 2 of the heterocycle; thus, for example, angostura bark, obtained from *Galipea officinalis* Hancock (syn. *Cusparia trifoliata* Engl.), which has a questionable reputation as a febrifuge, contains quinoline, quinaldine, 2-*n*-amylquinoline, 4-methoxy-2-*n*-amylquinoline, cusparine, cuspareine, galipine and galipoline; see, *inter alia, W. G. Körner* and *C. Bohringer*, Gazz., 1883, **13,** 363; Ber., 1883, **16**, 2305; *H. Beckurts et al.*, Arch. Pharm., 1891, **229**, 591; 1905, **243**, 470; *J. Tröger et al., ibid.*, 1910, **248**, 1; 1911, **249**, 174; 1912, **250**, 494; *E. Späth et al.*, Ber., 1924, **57**, 1687; Monatsh., 1929, **52**, 129; Ber., 1929, **62**, 2244; Monatsh., 1930, **55**, 352.

Cusparine, $C_{19}H_{17}O_3N$, m.p. 91–92°, is a tertiary base containing a methoxyl group and a methylenedioxy group, but no *N*-methyl group. Distillation of the alkaloid with zinc dust affords pyridine, and fusion with potassium hydroxide protocatechuic acid. The methiodide of the alkaloid yields on heating **isocusparine**, which contains an *N*-methyl group and no methoxyl group (*Tröger* and *W. Müller*, Arch. Pharm., 1914, **252**, 459). On heating alone, cusparine is demethylated to pyrocusparine, $C_{18}H_{15}O_3N$, which contains no methoxyl group and exhibits cryptophenolic properties. Oxidation of cusparine with nitric acid gives an acid $C_{10}H_7O_3N$, almost certainly kynurenic acid, which yields quinoline on distillation with zinc dust (*Tröger* and *W. Beck*, *ibid.*, 1913, **251**, 246). From these and other studies the structure, 4-methoxy-2-[2′-(3″,4″-methylenedioxyphenyl)ethyl]quinoline, was assigned to cusparine by *Späth* and *O. Brunner* (Ber., 1924, **57**, 1243), and subsequently proved correct by synthesis: 4-methoxyquinaldine and piperonal condensed to give 4-methoxy-2-(3′,4′-methylenedioxybenzylidene)quinaldine (dehydrocusparine), catalytic hydrogenation of which afforded cusparine.

Cusparine hydrochloride has m.p. 193–194°, the *oxalate* m.p. 155–158° (decomp.) and the *methiodide* m.p. 186°.

Galipine, $C_{20}H_{21}O_3N$, m.p. 113.5°, resembles cusparine very closely in its chemical behaviour, but contains three methoxyl groups and no methylenedioxy group. On oxidation with chromic acid or potassium permanganate the base yields veratric acid, together with kynurenic acid methyl ether, demethylation of which affords kynurenic acid. *Späth* and *Brunner (loc. cit.)* proposed, on this evidence, that galipine was the 3′,4′-dimethoxyanalogue of cusparine, proved to be correct by its synthesis through condensation of 4-methoxyquinaldine and veratraldehyde, followed by reduction (*Späth* and *H. Eberstaller*, Ber., 1924, **57**, 1687). Galipine methiodide is isomerised by heat or alkali to isogalipine, analogous to isocusparine:

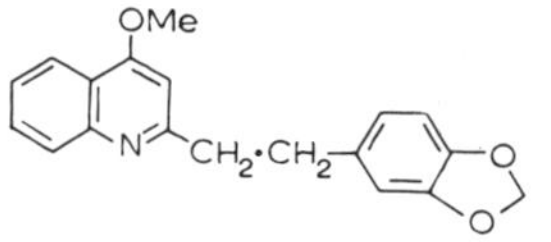

Cusparine

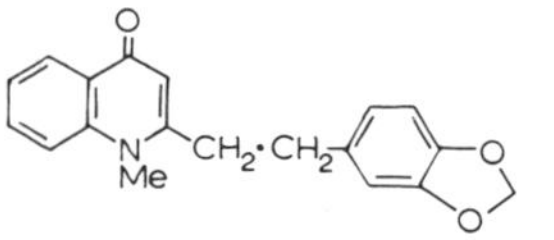

Isocusparine

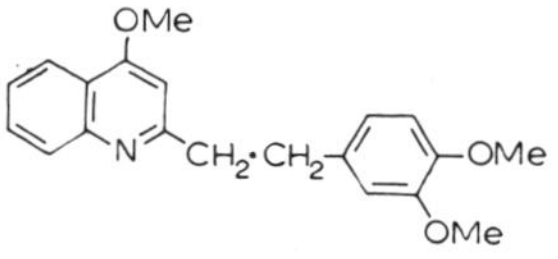

Galipine

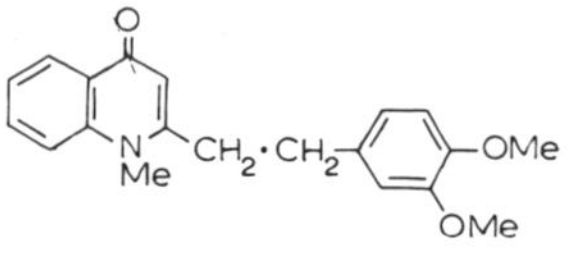

Isogalipine

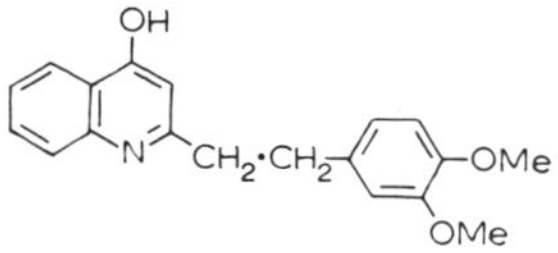

Galipoline

Galipoline, $C_{19}H_{19}O_3N$, m.p. 193°, contains two methoxyl groups and is phenolic. On methylation with diazomethane it is converted into galipine; it is therefore an *O*-demethylgalipine. A decision between the three possible structures was effected by synthesis. Two were synthesised by condensation of 4-methoxyquinaldine with, respectively, vanillin and isovanillin, followed by reduction; neither product was identical with galipoline, which must therefore be the 4-hydroxyquinoline derivative. Confirmation has been provided by condensation of 4-chloroquinaldine with veratraldehyde; the product on treatment with sodium benzyloxide afforded the corresponding 4-benzyloxy-base, which on hydrogenation and removal of the benzyloxy group by hydrolysis with acid gave a product identical with galipoline (*Späth* and *G. Papaioanou*, Monatsh., 1929, **52,** 129).

Cuspareine, $C_{20}H_{25}O_2N$, contains two methoxyl groups, and is an *N*-methyl tertiary base. On distillation with zinc dust quinoline is formed, and the base is not converted into an isomeric compound when its methiodide is heated (contrast cusparine and galipine). Oxidation with potassium permanganate affords veratric acid and a neutral oil which on dehydrogenation yields *N*-methylcarbostyril; on gentle oxidation cuspareine yields isogalipine. Cuspareine has the structure shown and the (±)-base has been synthesised (*Tröger* and *H. Runne*, Arch. Pharm., 1911, **249**, 174; *J. Schläger* and *W. Leeb*, Monatsh., 1950, **81**, 714) by the sequence outlined below:

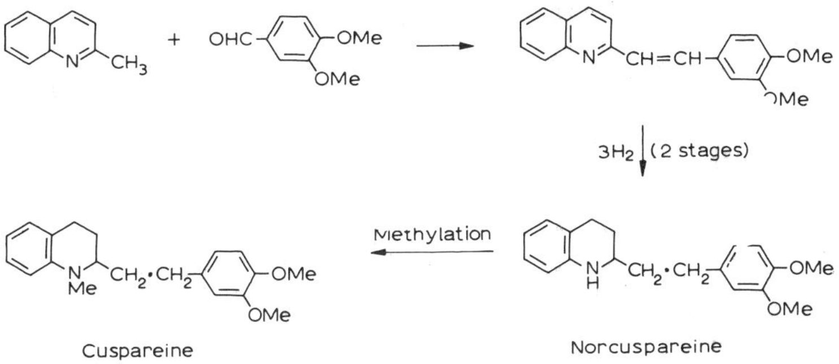

Cuspareine

Norcuspareine

(−)-*Cuspareine* has m.p. 54–56°, $[\alpha]_D$ −20.4° (ethanol); *methiodide*, m.p. 156° (decomp.); (±)-*base*, m.p. 32–35°.

Galipoidine, $C_{19}H_{15}O_4N$, m.p. 233°, is of unknown constitution (*Tröger* and *Runne*, *loc. cit.*).

(*b*) *Simple quinolones*

In addition to the quinolines described above, angostura bark also contains 1-methyl-2-quinolone. The co-occurrence of 2- and 4-quinolones is a common feature in many plants of the Rutaceae; fortunately spectroscopy allows an easy distinction to be made between these two types of alkaloids: in the ultra-violet spectra, for example, both 2- and 4-quinolones show an intense band at λ_{max} 239–258 nm and a band at λ_{max} 322–331 nm of medium intensity. 2-Quinolones, however, show an additional band λ_{max} 263–298 nm which is normally absent in the spectra of 4-quinolones.

The greater basicity of the 4-quinolone structure is reflected in a shift in the absorption maxima when acid is added; this effect is not observed with 2-quinolones (*H. Rapoport* and *K. Holden*, J. Amer. chem. Soc., 1960, **82**, 4395; *E. A. Clark* and *M. F. Grundon*, J. chem. Soc., 1964, 4190; see also *A. W. Sangster* and *K. L. Stuart*, Chem. Reviews, 1965, **65**, 69). The integrated intensity of the carbonyl band in the infra-red spectra of 2- and 4-quinolones is also diagnostic (*N. J. McCorkindale*, Tetrahedron, 1961, **14**, 223), being greater for 2-quinolones. The position of this band, normally 1630–1620 cm^{-1} for 2-quinolones and 1660–1650 cm^{-1} for the 4-isomer, also serves as a point of distinction, but has proved unreliable in a number of examples. N.m.r. spectroscopy also differentiates between the two structural types since the C-5 proton gives rise to a signal at unusually low chemical shift due to the deshielding influence of the C-4 carbonyl group. In the spectrum of eduline, for example, the C-5 proton is characteristically deshielded by the adjacent carbonyl group and thus gives rise to a doublet

($J = 3$ Hz, *meta* coupling) at δ 7.9. Protons at C-7 and C-8 give rise to multiplets situated at δ 7.6–7.2 (*D. R. Boyd* and *Grundon*, J. chem. Soc., C, 1970, 556):

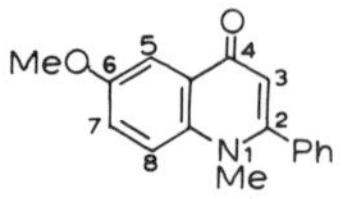

Eduline

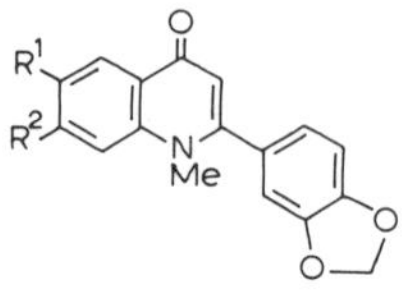

(I, $R^1 = R^2 =$ H)
Graveoline
(II, R^1 = OMe, R^2 = H)
Lunasia-1
(III, R^1 = H, R^2 = OMe)
Lunamarine

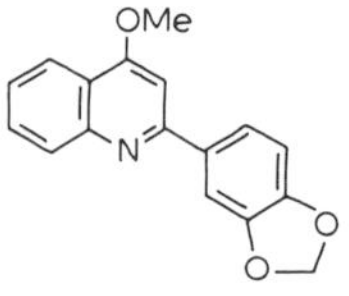

Graveolinine

All of the known 4-quinolones, with the possible exception of pilokeanine (see below), are substituted at C-2, and the majority are *N*-methylated. The C-2 substituent may be a simple alkyl function, but is frequently phenyl or 3,4-methylenedioxyphenyl. **Lunasia-1** (II) and its isomer **lunamarine** (III) isolated from *Lunasia* and *Balfourodendron* species (*S. Goodwin et al.*, J. Amer. chem. Soc., 1959, **81**, 6209; *H. C. Beyerman* and *R. W. Rooda*, Koninkl. Ned. Akad. Wetenschap. Proc., 1959, **B62**, 187; 1960, **B63**, 154; *F. A. Steldt* and *K. K. Chen*, J. Amer. pharm. Assoc., Sci. Ed., 1943, **32**, 107) are representative of this latter type. 2-Aryl-*N*-methyl-4-quinolones are isomeric with 2-aryl-4-methoxyquinolines with which they sometimes co-occur. *Ruta graveolens* L., for example, elaborates both the 4-quinolone, **graveoline** (I) and the quinoline. **graveolinine**. Details of the bioxynthesis of graveoline in *Ruta angustifolia* Pers have been published (*M. Blaschke-Cobet* and *M. Luckner*, Phytochemistry, 1973, **12**, 2392); it is now known that the two oxygen atoms of the methylenedioxy group are introduced by mixed function oxygenation involving an N.I.H. shift.

The *in vitro* isomerization of **galipine** to **isogalipine** has already been mentioned (p. 193); this is a general reaction. By heating 4-methoxyquinolines with methyl iodide in a sealed tube conversion to the corresponding 1-methyl-4-quinolone is effected.

The plant *Vespris ampody* H. Perr (*C. Kan Fan et al.*, *ibid.*, 1970, **9**, 1283) contains the three quinolones IV, V and VI; these structural allocations rest mainly upon spectroscopic evidence, principally mass spectrometry. A study of the mass fragmentation of some model 2-alkyl-4-quinolones has been made (*J. Reisch*, *R. Pagnucco* and *N. Jantos*, *ibid.*, 1968, **7**, 997). In five of the eight compounds studied the molecular ion peak is also the base peak in the spectrum. When, however, a large alkyl group is present in the 2-position a number of modes of fragmentation become evident and in the spectra of IV, V and VI an ion, *m/e* 159, represented as VII, or its equivalent, is the base peak. The ion VII is thought to arise by fragmentation of the alkyl side chain with concomitant hydrogen transfer:

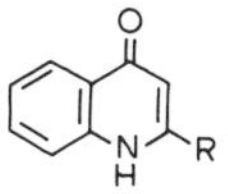

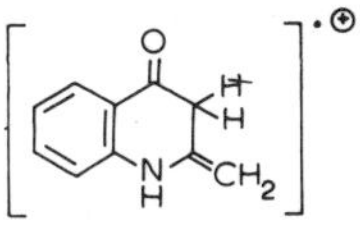

(IV, R = $-CH_2\cdot CH_2\cdot CH=CH\cdot CH_2\cdot CH=CH\cdot CH_2\cdot CH_3$)
(V, R = $-(CH_2)_8\cdot CH_2OH$)
(VI, R = $-(CH_2)_9\cdot CO\cdot CH_3$)

(VII)

1,2-**Dimethyl-4-quinolone** found in *Acronychia* species, is also present in *Platydesma campanulata* Mann (*F. Werny* and *P. J. Scheuer*, Tetrahedron, 1963, **19**, 1293). This plant which is known as pilo-kea in Hawaii, contains a number of other alkaloids including an oily compound, **pilokeanine**; this material, a secondary alcohol, was characterised as the *picrate*, m.p. 216° (decomp.). The ultraviolet spectrum of pilokeanine, λ_{max} 337, 325, 292, 281 sh, 248, 242, 231 sh, 215 (logε 4.39, 4.34, 3.75, 3.66, 4.72, 4.74, 4.63, 4.68) nm, suggests an 8-methoxy-1-methyl-4-quinolone and the following structure is proposed for the alkaloid:

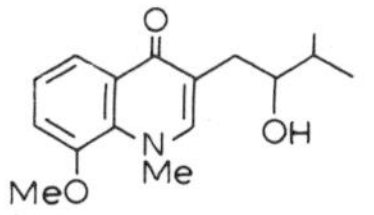

Pilokeanine

Folimidine, m.p. 225°, has been isolated from *Haplophyllum foliosum* Vved (*D. M. Razzakova, I. A. Bessonova* and *S. Yu. Yunusov*, Khim. Prir. Soedin., 1972, 755; C.A., 1973, **78,** 84605x) and an alkaloid, **acutine,** m.p. 122–123° in *H. actifolium* (*D. M. Gulyamova, Bessonova* and *Yunusov, ibid.*, 1971, 850; C.A., 1972, **76,** 138169n) is VIII:

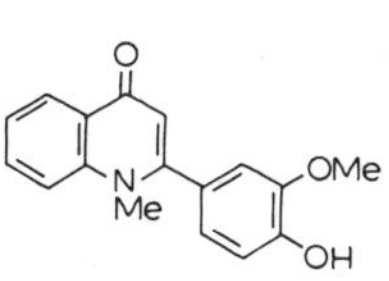

Folimidine

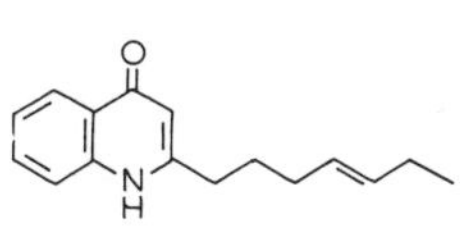

Acutine
(VIII)

While pilokeanine is currently the sole example of a simple 4-quinolone bearing a modified pentenyl substituent, many examples of pentenyl-substituted 2-quinolones are known. Thus, for example, the leaves and stems of *Haplophyllum tuberculatrum* Juss contains the unnamed **alkaloid** $C_{19}H_{23}O_2N$, (IX), λ_{max} 227, 272, 324, 338 (ε 41,300, 32,200, 8,500, 5,800) nm (*D. Lavie et al.*, Tetrahedron, 1968, **24,** 3011). All twenty-three protons of the molecular formula are accounted for by the following interpretation of the 60 MHz ^{1}H-n.m.r. spectrum ($CDCl_3$): the presence of four vinylic methyl groups is disclosed by singlets at δ 1.70 (6*H*), 1.80 and 1.88 ($2 \times 3H$). Two doublets each due to two protons centred at δ 3.50 and δ 4.57 arise from the methylene protons of the prenyl side chains and a complex between δ 5.18–5.82 corresponds to the signals of the two vinylic protons. In the aromatic region a complex multiplet is assigned to the 6, 7 and 8 protons and a double doublet centred at δ 7.75 is due to the C_5-H. The NH group gives rise to a signal at δ 11.7, removable by deuteration:

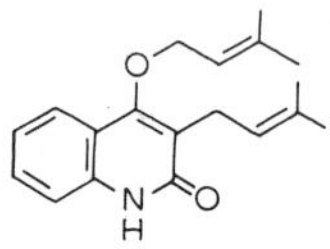

(IX)

The alkaloid **hafordamine** isolated from *Hafordia kendack* Guill. (*W. D. Crow* and *J. H. Hodgkin*, Austral. J. Chem., 1968, **21**, 3075) is 4,6,8-trimethoxy-2-quinolone (XII) (*R. Storer* and *D. W. Young*, Tetrahedron Letters, 1972, 155; *P. Venturella, A. Bellino* and *F. Piozzi*, Chem. and Ind., London, 1972, 887; Gazz., 1974, **104**, 297) as confirmed by synthesis.

2,4-Dimethoxyaniline when reacted with excess diethyl malonate gives the *mono-ethyl ester mono-dimethoxy anilide* (X), m.p. 106–108°, which is cyclised to the quinolone XI, m.p. 288–291°, by heating with polyphosphoric acid. Treatment of this product with diazomethane completes the sequence to hafordamine (XII):

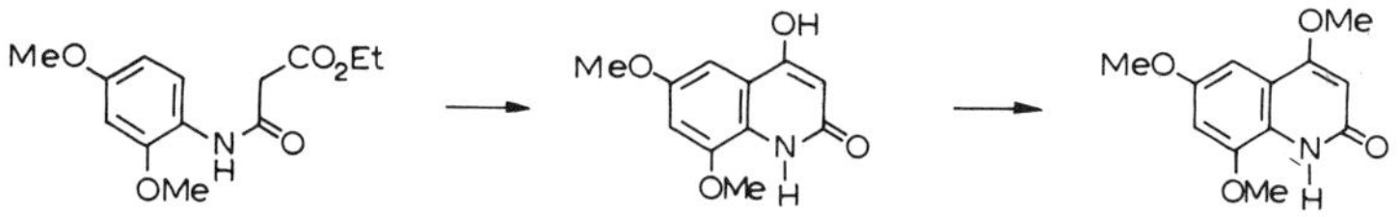

(X) (XI) Hafordamine (XII)

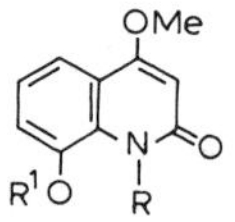

(XIII, R = R¹ = H) Edulitine
(XIV, R = Me, R¹ = H) Folidine
(XV, R = R¹ = Me) Folimine

The genus *Haplophyllum* is an abundant source of 2-quinolone alkaloids; unprenylated examples include **edulitine** (XIII) (*I. M. Fakhrutdinova, G. P. Sidyakin* and *S. Yu. Yunusov*, C.A., 1965, **63**, 8423e), also found in *Casimiroa edulis* Llave et Lex (*N. P. Toube, J. W. Murphy* and *A. E. Cross*, Tetrahedron, 1967, **23**, 2061), **folidine** (XIV) (*Z. Sh. Faizudinov, I. A. Bessonova* and *Yunusov*, C.A., 1967, **67**, 108803m) and **folimine** (XV) (*D. M. Razzakova, Bessonova* and *Yunusov*, *ibid.*, 1972, **77**, 72563y), whereas **foliosidine**, m.p. 141–142°, $[\alpha]_D^{EtOH}$ 41.6°, from *Haplophyllum foliosum* Vved (*V. I. Pastukhova, Sidyakin* and *Yunusov*, *ibid.*, 1965, **62**, 11864c) has a related structure with a hydroxylated prenyl group linked through oxygen at position C-8. While isopentyl derived substituents are common, the alkaloids **bucharaine** (**bucharine** or **bukharidine**) and **bucharidine** isolated from *H. bucharicum* Litv. represent structures which probably arise biogenetically through the agency of geranyl pyrophosphate, or its equivalent. The structure of bucharaine was determined largely by spectroscopic methods (*S. M. Sharafutdinova* and *Yunusov*, *ibid.*, 1969, **70,** 47648c). On heating, bucharaine forms bucharidine (*Faizudinov et al.*, *ibid.*, 1970, **73**, 131179v).

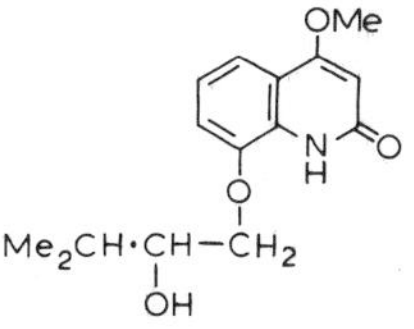

Foliosidine

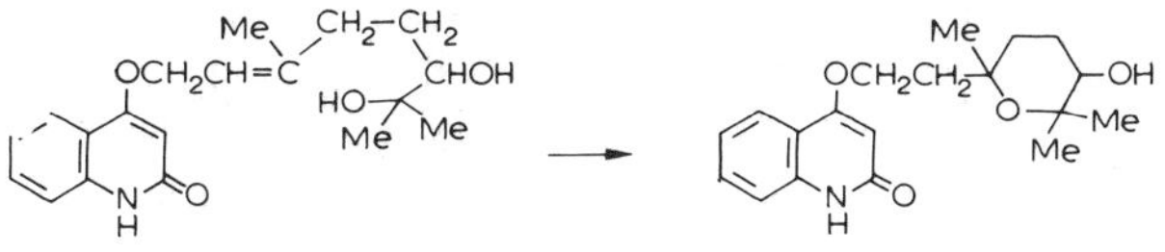

Bucharaine

Bucharidine

Oricine (XVI), m.p. 150–152°, is present in *Oricia suaveolens* (*M. O. Abe*, Phytochemistry, 1971, **10,** 3328) and **haplamine**, which has the related structure XVII, has been isolated from *Haplophyllum perforatum* (*V. I. Akhmezhanova*, *Bessonova* and *Yunusov*, Khim. Prir. Soedin., 1974, 109; C.A., 1974, **80,** 121152m):

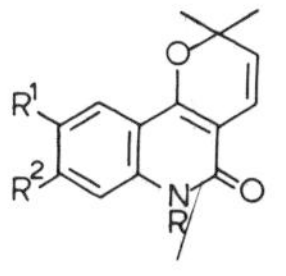

(XVI ; R = Me ; R^1 = R^2= OMe)
(XVII ; R = R^2= H ; R^1 = OMe)

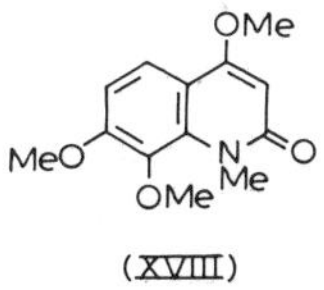

(XVIII)

4,7,8-**Trimethoxy**-1-**methyl**-2-**quinolone** (XVIII), m.p. 143–148°, is present in *Spathelia sorbifolia* L. (*R. Storer et al.*, Tetrahedron, 1973, **29,** 1721), while 4-**methoxy**-1-**methyl**-2-**quinolone**, m.p. 100–100.5° (*hydrate*, m.p. 68–69°), has been obtained from *Hesperethusa crenulata* M. Roem (*M. N. S. Nayar*, *C. V. Sutar* and *M. K. Bhan*, Phytochemistry, 1971, **10,** 2843).

3-Substituted-2,4-dihydroxyquinolines and their 1-methyl derivatives can be prepared conveniently by heating aromatic amines and substituted malonic esters in diphenyl ether (*H. Rapoport* and *K. G. Holden*, J. Amer. chem. Soc., 1959, **81,** 3738; *R. H. Baker*, *G. R. Lappin* and *B. Riegel*, *ibid.*, 1946, **68,** 1284; *M. F. Grundon*, *N. J. McCorkindale* and *M. N. Rodger*, J. chem. Soc., 1955, 4284). If diethyl 3-methylbut-2-enylmalonate (XIX) is used 3-γ,γ-dimethylallylquinoline derivatives may be synthesised. Thus the 2-quinolone XX may be prepared by the action of XIX upon *N*-methyl-2-methoxyaniline (*E. H. Clarke* and *Grundon*, *ibid.*, 1964, 438):

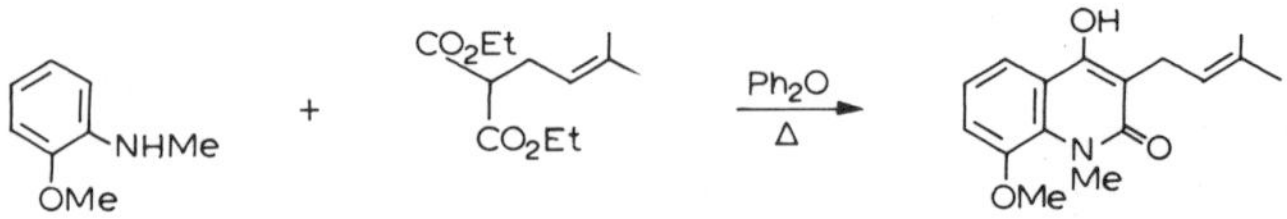

(XIX) (XX)

Hydroboration of XX, followed by treatment with alkaline hydrogen peroxide and then with diazomethane, affords racemic **lunacridine**. The laevorotatory antipode of this structure occurs in *Lunasia* species *(Goodwin et al., loc. cit.)*:

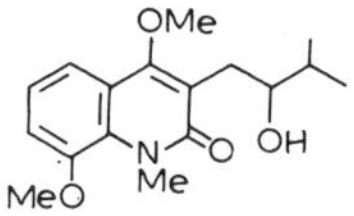

Lunacridine

Accompanying lunacridine in the bark of *Lunasia amara (Blanco* var. *repanda)* Lanterb. et K. Schum. Lanterb. is the homologue **lunidine** and the related diols hydroxylunidine and hydroxylunacridine. Yet another 2-quinolone, **lunidonine**, is also present (*A. Rüegger* and *D. Stauffacher*, Helv., 1963, **46,** 2329), this last molecule is a ketone, since it may be obtained by oxidation of the secondary alcohol lunidine.

Hydroxylunacridine $[\alpha]_D^{25}$ +31.5° is the enantiomorph of **balfourolone**, isolated from extracts of *Balfourodendron riedelianum* Engl. (*B. eburneum* Mello) a South American shrub (*Rapoport* and *Holden*, J. Amer. chem. Soc., 1960, **82,** 4395). Balfourolone is not a true alkaloid however, being an artefact of the quaternary furoquinoline ion *O*-methylbalfourodinium. On treatment with base, during the isolation procedure, the latter substance readily undergoes ring opening to give balfourolone.

While there is no reason to doubt that hydroxylunacridine and hydroxylunidine are natural products, it is interesting to note that they co-occur in *L. amara* with lunacrine and lunine, which are furoquinolines (see next section), lunacridine and *O*-methyllunium ion.

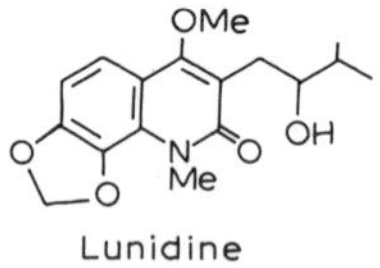

Lunidine

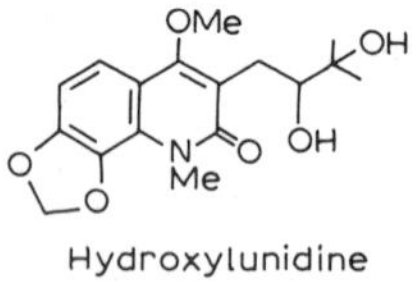

Hydroxylunidine

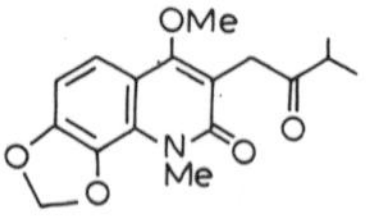

Lunidonine

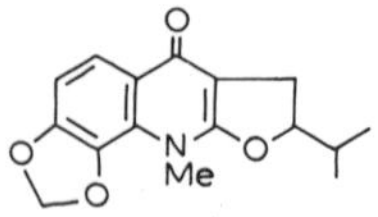

Lunine

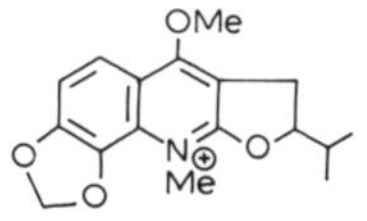

O-Methyllunium ion

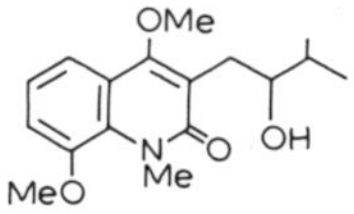

Lunacridine

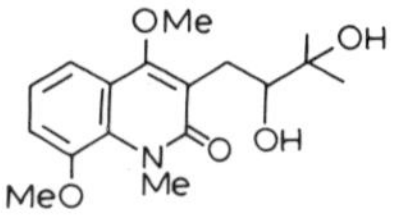

Hydroxylunacridine (+)
Balfourolone (−)

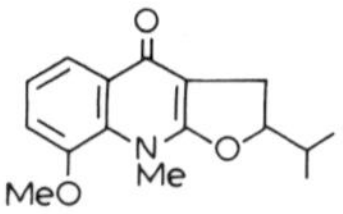

Lunacrine

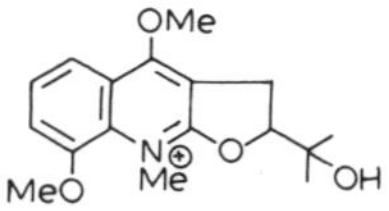

O-Methylbalfourodinium ion

Orixine, an alkaloid of *Orixa japonica* Thunb. (*M. Terasaka et al.*, Chem. pharm. Bull. Japan, 1960, **8,** 523, 1142) is isomeric with hydroxylunidine and recently (*W. J. Donelly* and *Grundon*, J. chem. Soc., Perkin I, 1972, 2116) the corresponding ketone, **orixinone,** has been isolated. A partial asymmetric synthesis of orixine has been described (*R. M. Bowman* and *Grundon*, J. chem. Soc., C, 1967, 2368):

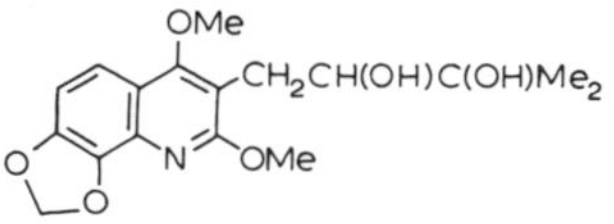

Orixine

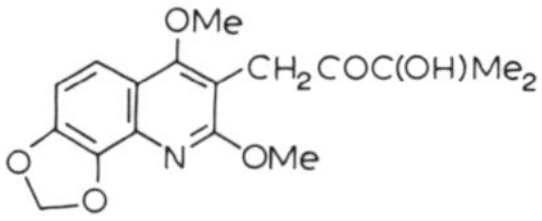

Orixinone

(c) Furoquinolines and furoquinolones

(−)-**Lunacrine** is found in the leaves of *L. amara* (*S. Goodwin* and *E. C. Horning*, 1959, *loc. cit.)* whereas the racemate has been isolated from the bark of *Lunasia* species (*H. C. Beyerman* and *R. W. Rooda*, Proc. Kon. Ned. Akad. Wetenschap., 1959, **62B**, 187).

Spectroscopic studies confirm the [2,3-*b*]dihydrofuroquinoline nature of lunacrine; thus in the ultraviolet spectrum the long wavelength maxima at λ 313–326 nm merge into a band centred at λ_{max} 300 nm when acid is added indicating a 4-quinolone unit of structure, a conclusion substantiated by the low field position of the quartet (τ 1.89, J_{AX}=7 Hz, J_{BX}=2 Hz), due to the C-5 proton, in the ^{1}H-n.m.r. spectrum. The remaining signals of this latter spectrum may be summarized as shown in XXI (*E. A. Clarke* and *Grundon*, J. chem. Soc., 1964, 438):

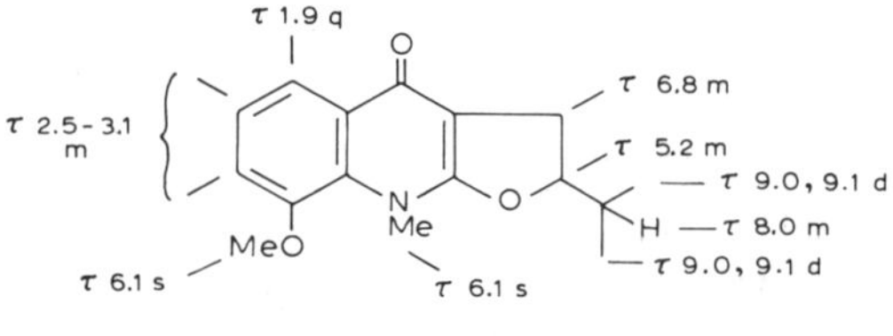

(XXI)

When (+)-**lunacridine** perchlorate is heated, a salt (XXII) is formed which, when treated with lithium bromide gives (+)-lunacrine. If (−)-lunacrine is heated with aqueous alkali, hydrolysis to an alkali soluble product occurs and this, with diazomethane, gives (+)-lunacridine. In addition to the fusion procedure described above, (+)-lunacridine may also be converted into (+)-lunacrine, the enantiomorph of the natural alkaloid, by heating with α-toluenesulphonyl chloride or, more simply, by treating with strong acid. Clearly, these interconversions involve stereochemical inversion, and this is rationalized by the sequence shown below which incorporates as the critical step an intramolecular S_N2 cyclization of lunacridine.

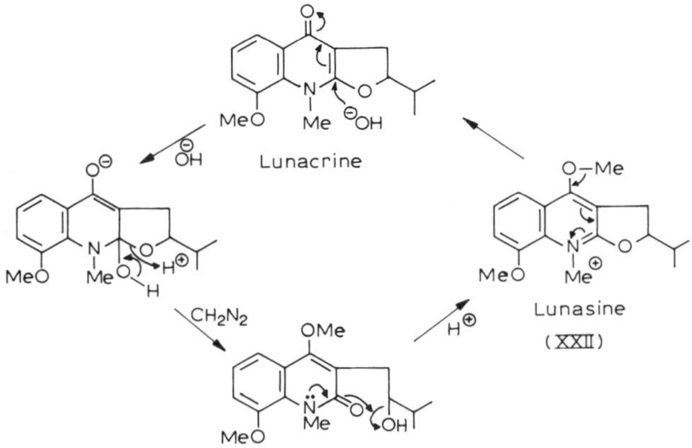

Lunacrine

Lunasine

(XXII)

Lunacridine

J. R. Price (Austral. J. Chem., 1959, **12,** 458) has shown that the salt XXII corresponds to the alkaloid **lunasine** isolated from *L. quercifolia.* In addition to the 4-quinolone lunacrine, a product of acid treatment of lunacridine is the angular 2-quinolone XXIII. It is assumed that this arises by demethylation of the oxygen atom at C-4 in lunacridine and then cyclization at this position·

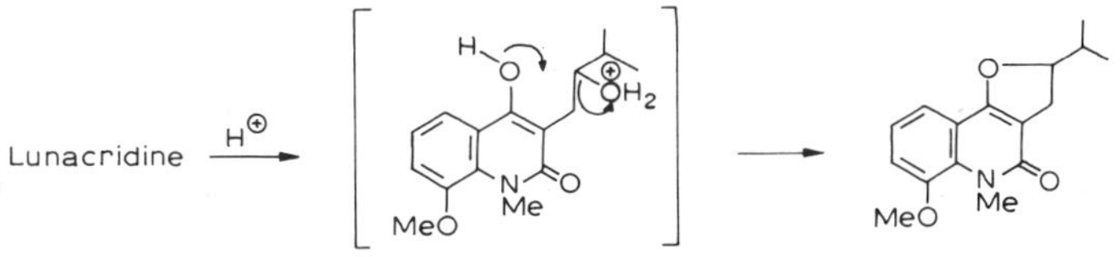

Lunacridine $\xrightarrow{H^{\oplus}}$ [intermediate] ⟶ (XXIII)

Dictamnine (XXIV, X=Y=Z=H), **skimmianine** (XXIV, X=H, Y=Z=OMe), **γ-fagarine** (XXIV, X=Y=H, Z=OMe) and **kokusaginine** (XXIV, X=Y=OMe, Z=H) represent four of the commonest furoquinoline alkaloids:

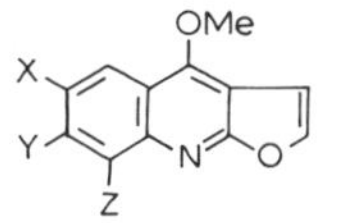

(XXIV)

Skimmianine is found in *Fagara capensis* Thunb. (*F. Fish* and *P. G. Waterman*, J. Pharm. Pharmacol., 1971, **23**, 67), in *F. chalybea* (*idem*, Phytochemistry, 1972, **11**, 1866), in *F. xanthoxyloides* Lam. (*F. G. Torto et al.*, Ghana J. Sci., 1969, **9**, 3) and in species from many other genera of the Rutacea (see Table 1). The same alkaloid, accompanied by γ-fagarine occurs in *Zanthoxylum tsihanimposa* H. perr (*N. Decaudain* and *N. Kunesch*, Phytochemistry, 1974, **13**, 505) and, together with kokusaginine, (±)-platydesmine and halfordinine (6,7,8-trimethoxydictamnine), in *Melicope perspicuinerva* Merr. et Perry (*S. T. Murphy*, *E. Ritchie* and *W. C. Taylor*, Austral. J. Chem. 1974, **27**, 187). Skimmianine and other furoquinolines have been isolated from *Haplophyllum* species (*K. Ubaidullaev*, *I. A. Bessonova* and *S. Yu. Yunusov*, Khim. Prir. Soedin., 1972, **8**, 343; C.A., 1973, **78**, 2010n; *D. M. Razzakoba*, *Bessonova* and *Yunusov*, *ibid*., 1972, **8**, 133; C.A., 1972, **77**, 72563x).

Decatropis bicolor (Zucc.) Radlk. is a source of skimmianine and dictamnine (*X. A. Dominguez*, *D. Butruille* and *J. Wapinsky*, Phytochemistry, 1971, **10**, 2554) and these compounds co-occur with 4-methoxy-1-methyl-2-quinolone in *Zanthoxylum decaryi* (*J. Vaquette et al.*, *ibid*., 1974, **13**, 1257). The leaves of *Teclea unifoliata* Baill. contain maculine, skimmianine and kokusaginine (*Vaquette*, *J. L. Pousset* and *A. Cavé*, Plantes Médicinales et Phytothérapie, 1974, **8**, 72).

Dictamnine, robustine and haplopine are obtained from *Zanthoxylum arnottianum* Maxim (*H. Ishii et al*., J. pharm. Soc. Japan, 1974, **94**, 322).

In the ^{1}H-n.m.r. spectra the protons of the furan system, present in these structures, give rise to an AX quartet δ 7.53–7.58 (C-2) and δ 6.93–7.02 (C-3) (*T. T. Batterham* and *J. A. Lamberton*, Austral. J. Chem., 1965, **18**, 859); in the mass spectrum an intense molecular ion peak is observed. Peaks due to the loss of carbon monoxide and methyl radical are characteristic and an $M^{\oplus} - 43$ ion is common.

When heated with methyl iodide in a sealed tube, isomerization to the corresponding *N*-methyl-4-quinolone occurs. Dictamnine, for example, forms **isodictamnine**, m.p. 188° (*Y. Asahina*, *T. Ohta* and *M. Inubuse*, Ber., 1930, **63**, 2045). This reaction occurs *via* the quaternary salt XXV:

(XXV) Isodictamnine

Hydrogenolysis of dictamnine over platinum catalyst in acetic acid solution affords 3-ethyl-4-hydroxy-2-quinolone. In contrast, catalytic hydrogenation using palladium in ethanol gives the dihydrofuro[2,3-*b*]quinoline (*Lamberton* and *J. R. Price*, Austral. J. Chem., 1953, **6**, 66; *Ohta*, J. pharm. Soc. Japan, 1953, **73**, 63; *Ohta*, *T. Miyazaki* and *Y. Mori*, *ibid*., 1954, **74**, 708.

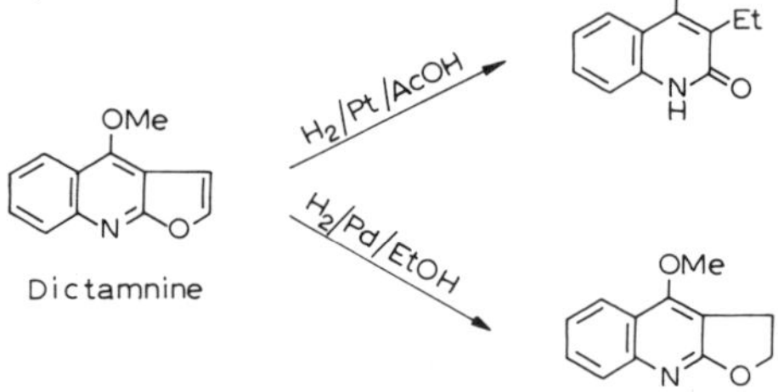

$H_2/Pt/AcOH$ $H_2/Pd/EtOH$

Dictamnine

Most furoquinolines have been synthesised, the preparation of dictamnine being typical: ethyl 2-anilino-4-hydroxyfuran-3-carboxylate (XXVI) in boiling diphenyl ether gives the quinol XXVII which with diazomethane yields a mixture of the *O*- and *N*-methyl derivatives XXVIIIa and XXVIIIb. If XXVIIIa is heated with phosphoryl chloride containing a trace of water the 3-chlorofuroquinoline results. Finally, hydrogenolysis over palladium on calcium carbonate affords dictamnine. Completion of the same reaction sequence with the *N*-methyl isomer XXVIIIb gives isodictamnine (*H. Tuppy* and *F. Böhn*, Angew. Chem., 1956, **68**, 388):

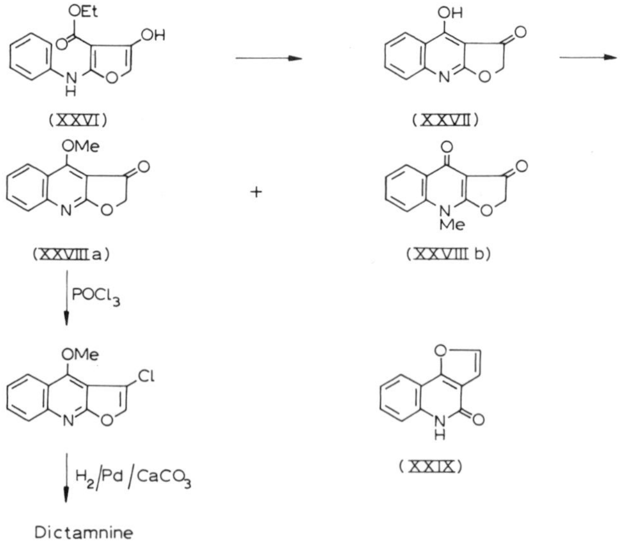

(XXVI) (XXVII)

(XXVIIIa) + (XXVIII b)

$POCl_3$

$H_2/Pd/CaCO_3$

Dictamnine

(XXIX)

If dictamnine is demethylated under rigorous conditions, by heating with hydrogen bromide in acetic acid, rearrangement to the angular compound XXIX is observed.

A number of other syntheses of dictamnine have been reported (*cf. Grundon* and *N. J. McCorkindale*, Chem. and Ind., London, 1956, 1091; J. chem. Soc., 1957, 2177; *T. Sato* and *M. Ohta*, J. chem. Soc. Japan, 1956, **77,** 1630; Bull. chem. Soc. Japan, 1957, **30**, 708; Chem. and pharm. Bull., Tokyo, 1957, **5**, 87; *R. G. Cooke* and *H. F. Hayes*, Austral. J. Chem., 1958, **11**, 225; *T. Ohta* and *Y. Mori*, Yakugaku Zasshi, 1962, **82**, 549).

An interesting route to the dihydro-alkaloids has been developed (*Y. Kuwayama*, Chem. and pharm. Bull., Tokyo, 1961, **9,** 719; Yakugaku Zasshi, 1962, **82**, 703; *Kuwayama* and *Y. Matsuda*, *ibid.*, 1965, **85**, 731; *Kuwayama et al.*, *ibid.*, 1968, **88**, 1050); α-acetobutyrolactone is reacted with magnesium in ethanol containing a few drops of carbon tetrachloride, *o*-nitrobenzoyl chloride is then added furnishing the benzoyl derivative XXX. Deacetylation in ethanolic ammonium hydroxide solution and treatment of the product with diazomethane yields the enol ether. Cyclization to the tricyclic system is achieved by reduction to the corresponding amine in an acidic medium:

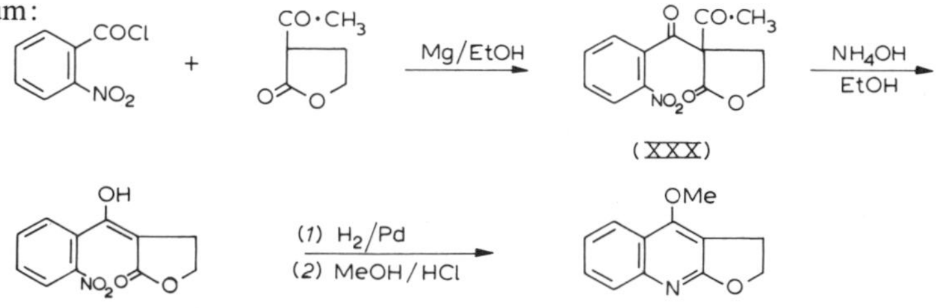

(XXX)

Dihydrodictamnine

Dihydro derivatives of evolitrine, skimmianine, haplopine, kokusaginine, maculine and pteleine have been prepared similarly. Dehydrogenation to the parent alkaloid is normally effected by the use of *N*-bromosuccinimide.

All the above synthetic methods are, however, somewhat lengthy and yields of final product are variable. A more efficient route to the furoquinolines has recently been devised (*J. F. Collins et al.*, Chem. Comm., 1972, 1029; J. chem. Soc., Perkin I, 1973, 94). The alkaloids dictamnine, γ-fagarine and skimmianine have been synthesised (48–66% yield) from 4-methoxy-3-(3-methylbut-2-enyl)-2-quinolones by ozonolysis, or by reaction with osmium tetroxide–periodate, followed by cyclization of the resultant aldehydes with polyphosphoric acid. The necessary isoprenylated starting materials were prepared by the alkylation of 2,4-dimethoxyquinolines and treatment with hydrogen chloride. The method is illustrated by the synthesis of dictamnine:

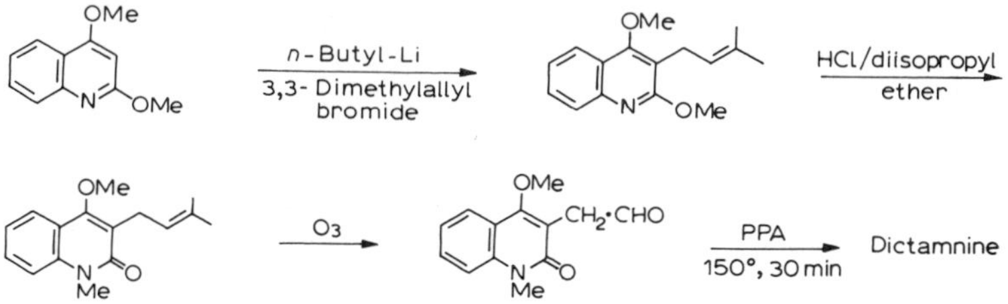

The alkaloid **ifflaiamine** from the wood of *Flindersia ifflaiana* F. Muell (*J. Bosson et al.*, Austral. J. Chem., 1963, **16**, 480) belongs to the rather rare group of natural 1,1-dimethylallyl derivatives. The structure of this molecule was established by spectroscopic methods (*E. N. Marvell, D. R. Anderson* and *J. Ong*, J. org. Chem., 1962, **27**, 1109; *A. von Habich et al.*, Helv., 1962, **45**, 1943) and by synthesis (*T. R. Chamberlain* and *M. F. Grundon*, Tetrahedron Letters, 1967, 3547; J. chem. Soc., C, 1971, 91).

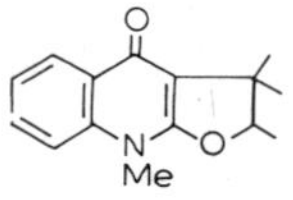

Ifflaiamine

When the 3,3-dimethylallyl ether, **ravenine** (XXXI) is heated at 130–140° for five hours three products XXXIII, XXXIV and XXXV are obtained. The formation of these products may be rationalized as follows:

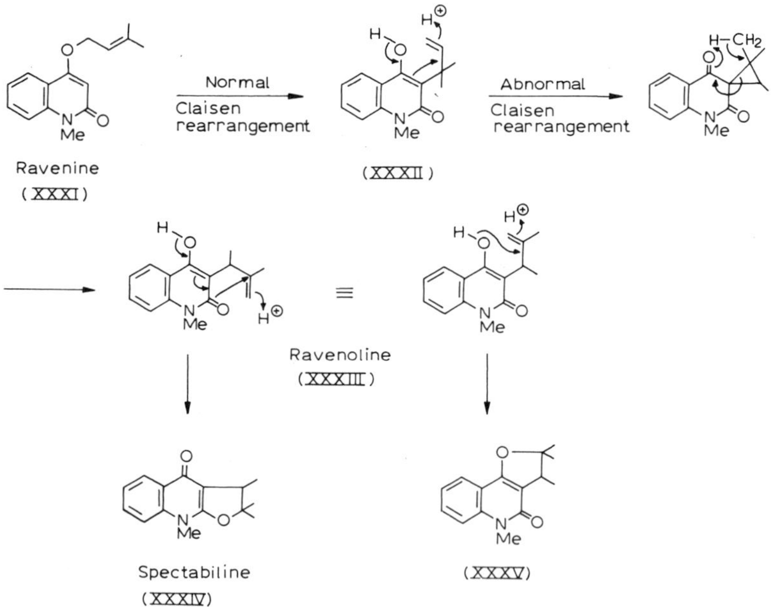

Ravenine (XXXI)

Normal Claisen rearrangement

(XXXII)

Abnormal Claisen rearrangement

Ravenoline (XXXIII)

Spectabiline (XXXIV)

(XXXV)

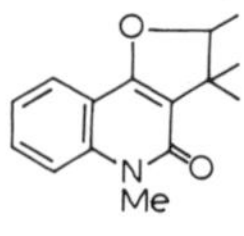

(XXXVI)

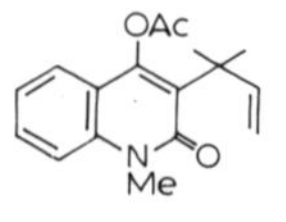

(XXXVII)

Compound XXXIII results from an abnormal Claisen rearrangement, and the other products XXXIV and XXXV are formed from it by attack of the 2- and 4-oxygen atoms respectively upon the isolated double bond of the side chain. The two modes of cyclization appear to proceed at about the same rate.

If the rearrangement is conducted for a longer period (20 h), 3% of the angular molecule XXXVI is formed, which is considered to arise from the normal Claisen product XXXII, thus suggesting that the abnormal sequence is reversible (*cf. R. M. Roberts* and *R. G. Landolt*, J. org. Chem., 1966, **31**, 2699). Ifflaiamine was isolated in only 6% yield when the ether XXXI was heated in the presence of anhydrous sodium carbonate, the major product (87%) being XXXVI. Small quantities of XXXIII, XXXIV and XXXV were also formed. This result lends support to the suggested mechanism for the formation of these "abnormal" molecules, since the presence of base would clearly inhibit the transfer of a hydrogen atom from the acidic 4-hydroxyl group of the intermediate XXXII to the side chain. If the ether XXXI is heated in *N*-methylpiperidine, containing acetic anhydride, XXXVII is formed; when this is treated with hydrogen bromide in acetic acid the natural product ifflaiamine is isolated in 56% yield.

Probably it is of biogenetic significance that the ether XXXI corresponds to the alkaloid **ravenine** from the rutaceous plant *Ravenia spectabilis* Engl. (*D. B. Paul* and *P. K. Bose*, J. Indian chem. Soc., 1968, **45,** 552; 1969, **46**, 678). **Ravenoline** (XXXIII) is also a constituent of the same plant. Furthermore, structure XXXIV corresponds to the alkaloid **spectabiline** again present in *R. spectabilis* (*S. B. Talapatra et al.*, Tetrahedron Letters, 1969, 4789).

A review of chemosystematics within the Rutacea family is now available (*Fish* and *Waterman*, Taxon, 1973, **22,** 177).

For further general synthetic studies see, for example, *N. S. Narasinham* and *R. S. Mali* (Tetrahedron Letters, 1973, 843); *R. Storer* and *D. W. Young* (Tetrahedron, 1973, **29**, 1215); *T. Sekika* (Bull. chem. Soc., Japan, 1973, **46**, 577); *P. Shanmugan, P. Lakshimarayama* and *R. Palaniappan* (Monatsh., 1973, **104**, 633); *Shanmugan* and *Lakshimarayama* (Z. Naturforsch., 1972, **27b**, 474); *V. N. Ramachandran et al.* (Indian J. Chem., 1973, **11**, 1088); *Grundon* and *K. J. James* (Tetrahedron Letters, 1971, 4727); *R. A. Corral, O. O. Orazi* and *I. A. Benages* (Tetrahedron, 1973, **29,** 205); *Bowman, Grundon* and *James* (J. Chem. Soc., Perkin I, 1973, 1055); *L. Maat et al.* (Rec. Trav. chim., 1973, **92**, 1399).

Foliminine, m.p. 107–108° (*I. A. Bessonova* and *S. Yu. Yunusov*, Khim. Prir. Soedin., 1974, **10,** 52), occurs in *Haplophyllum foliosum* Vved. where it is accompanied by **foliosine, folimidine** and **folimine**.

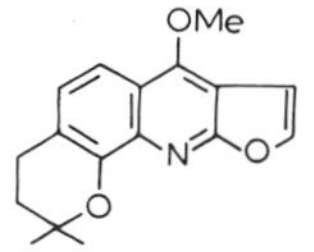

OMe N O O

Foliminine

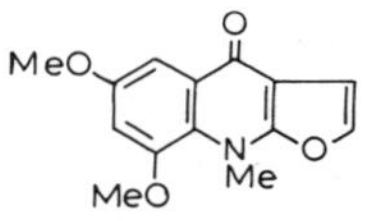

O MeO N O Me MeO

Isomaculosidine

The quaternary salt **pteleatinium chloride** is present in *Ptelea trifoliata* L. (*L. Mitscher, M. S. Bathala* and *J. L. Beal*, J. chem. Soc., D, 1971, 1040). **Evoxine** and its monoacetate occur in *Haplophyllum hispanicum* (*A. G. Gonzalez, R. M. Ordonez* and *F. R. Luis*, Anales de Quim., 1972, **68,** 1133; C.A., 1973, **78**, 82069b) and **haplopine** is found in *Aegle marmelos* Correa (*D. Basu* and *R. Sen*, Phytochemistry, 1974, **13**, 2329).

Preskimmianine is accompanied by a new alkaloid, **isomaculosidine**, m.p. 170–172°, in *Dictamnus alba* (*R. Storer* and *D. W. Young*, Tetrahedron, 1973, **29**, 1217).

Flindersine, the major alkaloid of *Flindersia australis* R Br. wood, has been synthesised from the epoxide XXXVIII by treatment with sodium hydroxide in dimethyl sulphoxide. The reaction takes place *via* the intermediate allylic alcohol (XXXIX) followed by an electrocyclization reaction of its dehydration product. Demethylation with aqueous hydrobromic acid completes the sequence to the alkaloid (*R. M. Bowman, M. F. Grundon* and *K. J. James*, Chem. Comm., 1970, 666):

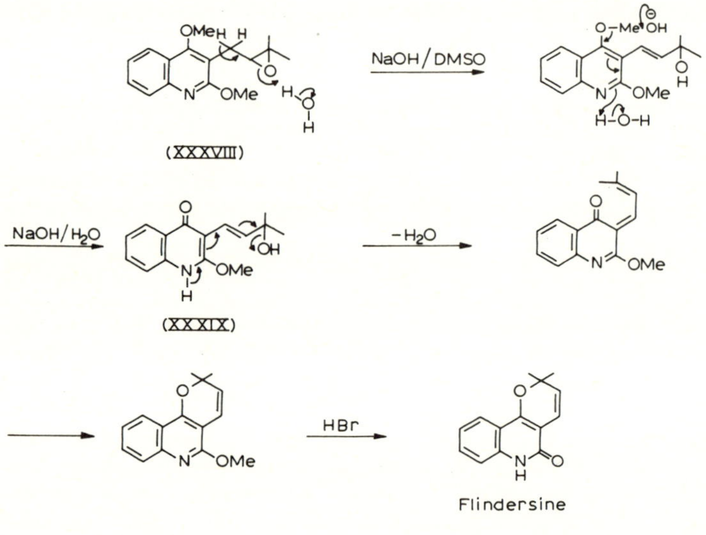
OMe
H H
N
OMe
H
O
H
(XXXVIII)
NaOH/DMSO
O–Me
OH
N
OMe
H–O–H
NaOH/H₂O
O
OH
N
OMe
H
(XXXIX)
–H₂O
O
N
OMe
O
N
OMe
HBr
O
N
H
O
Flindersine

The absolute configuration of a number of furano- and pyrano-quinoline alkaloids has been established (*J. F. Collins* and *Grundon*, *ibid.*, 1969, 1078; J. chem. Soc., Perkin I, 1973, 161).

Perforine and **haplofilidine** from *Haplophyllum perforatum* Kar. et Ker. (*Z. Sh. Faizutdinova, Bessonova* and *Yunusov*, Khim. Prir. Soedin., 1967, **3**, 356; C.A., 1968, **68**, 69160y) represent an unusual variant upon the normal furoquinoline theme, since the carbocyclic ring is non-aromatic in both molecules:

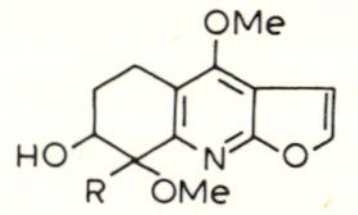

R = $-(CH_2)_2C(OH)Me_2$ Perforine
R = $-CH_2{\cdot}CH{=}CMe_2$ Haplofilidine

(d) Pyranoquinolines

The pyrano[2,3-*b*]quinoline group is far less common than the related furo-[2,3-*b*]quinolines and, so far, these alkaloids are restricted to the genera *Balfourodendron*, *Haplophyllum*, *Lunasia*, *Ptelea* and *Ruta*. Indeed, structural diversity is less than might be supposed since **lunacrinol (lunasia II),** $[\alpha]_D^{25°}$ $-15°$ (EtOH), from *L. amara* (*S. Goodwin et al.*, J. Amer. chem. Soc., 1959, **81,** 6209; *H. C. Beyerman* and *R. W. Rooda*, Proc. Kon. Ned. Akad. Wetenschap., 1959, **62B**, 187; 1960, **63B**, 154) and **isobalfourodine**, $[\alpha]_D^{25°}$ $+14°$ (EtOH), from *B. riedelianum* Engl. (*B. eburneum* Mello) (*H. Rapoport* and *K. G. Holden*, J. Amer. chem. Soc., 1959, **81,** 3738; 1960, **82,** 4395) are the enantiomers of structure XL.

Lunacrinol when treated with dimethyl sulphate gives the quaternary salt, which is readily decomposed by alkali to racemic hydroxylunacridine:

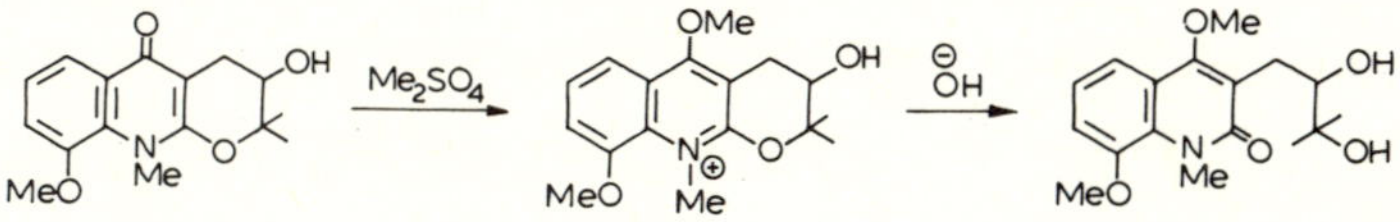

(XL) Hydroxylunacridine

A synthesis of (±)-isobalfourodine has been achieved (*R. M. Bowman* and *M. F. Grundon*, Chem. Comm., 1965, 216) commencing from the 3-dimethylallyl-4,8-dimethoxyquinolone; this when treated with peroxylauric acid gave a mixture of the furo- and pyrano-quinolines (XLI and XLII). Heating the pyranoquinoline with methyl iodide furnished (±)-isobalfourodine.

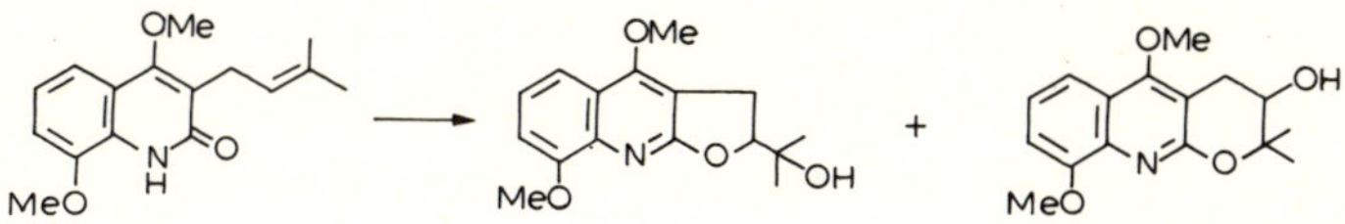

(XLI) (XLII)

In a similar way the tertiary alcohol XLI gave (±)-balfourodine. The alkaloid **khaplofoline** from *Haplophyllum foliosum* (Vved) (*I. M. Fakhrutdinova, G. P. Sidyakin* and *S. Yu. Yunusov*, Uzbeksk. Khim. Zhur., 1963, **7**, (4), 41; C.A., 1963, **59**, 15, 331) has been prepared by an analogous route (*Bowman* and *Grundon*, J. chem. Soc., C, 1966, 1084):

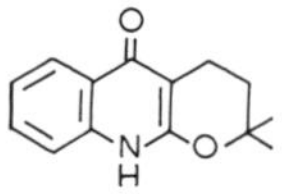

Khaplofoline

N.m.r. spectroscopy allows a clear distinction to be made between the pyrano- and furo-quinolines of this series; in particular, the pyrano-compounds produce a singlet at $\sim 8.6\,\tau$ ($>CMe_2$) whilst the methyl resonances ($-CHMe_2$) of the furo-analogues appear as a multiplet in this region (*E. A. Clarke* and *Grundon*, J. chem. Soc., 1964, 4190).

(e) Acridone alkaloids

Some thirty alkaloids derived from 9-acridone have been isolated from rutaceous plants; in the main structural diversity stems from the substitution of *O*-methyl, methylenedioxy or *O*-prenyl functions at the 1-, 2-, 3- or 4-positions.

The chemistry of this group has been reviewed [*J. E. Saxton*, in "The Acridines", ed. *R. M. Acheson*, Wiley, Interscience, New York, 2nd. Edn., 1973, p. 379 (Vol. 9 in "The Chemistry of Heterocyclic Compounds", general eds. *A. Weissberger* and *E. C. Taylor*)].

2,3,4-**Trimethoxy**-10-**methylacridone** has been found in the bark of *Evodia alata* F. Muell (*R. J. Gell, G. K. Hughes* and *E. Ritchie*, Austral. J. Chem., 1955, **8,** 114). The compound was identified by comparison with a synthetic specimen (*Hughes* and *Ritchie*, Austral. J. sci. Res., 1951, **A4**, 423); 4-**hydroxy**-2,3-**dimethoxy**-10-**methylacridone** is also a constituent of the plant. This molecule is present in plants from the genera *Acronychia* (*F. N. Lahey, M. McCamish* and *T. McEwan*, Austral. J. Chem., 1969, **22,** 447; *Hughes, K. G. Neill* and *Ritchie, ibid.*, 1952, **A5**, 401) and *Fagara* (*F. Fish* and *P. G. Waterman*, Phytochemistry, 1972, **11**, 3007):

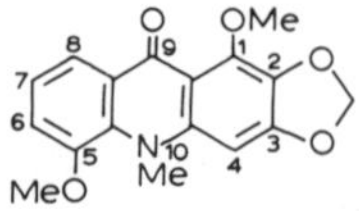

Tecleanthine

Melicopine, melicopidine and **melicopicine** have been isolated from *Melicope fareana* F. Muell (*Hughes et al.*, Nature, 1948, **162**, 223; *J. R. Price*, Austral. J. sci. Res., 1949, **A2**, 249; *W. D. Crow* and *Price*, *ibid.*, p. 255; *Crow*, *ibid.*, p. 264; *Price*, *ibid.*, p. 272; *Crow* and *Price*, *ibid.*, p. 282) and from *Melicope* and *Acronychia* species (*J. A. Lamberton* and *Price*, Austral. J. Chem., 1953, **6**, 66; *Lamberton*, *ibid.*, 1966, **19**, 1995):

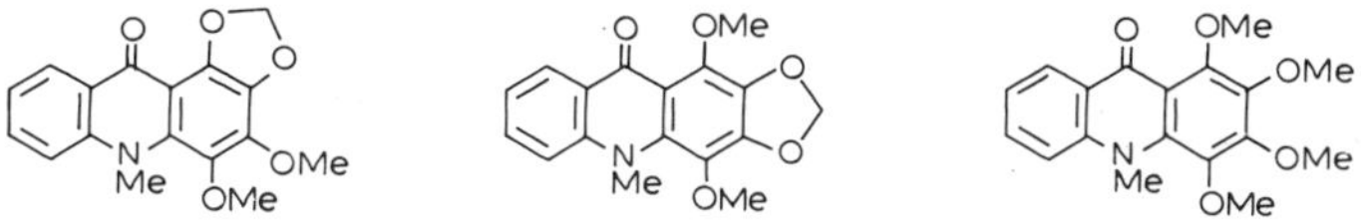

Melicopine Melicopidine Melicopicine

All three alkaloids yield, on oxidation with nitric acid, 1-methyl-4-quinolone-3-carboxylic acid and all show ultraviolet absorption spectra characteristic of an acridone system. Melicopine, melicopidine and melicopicine are accompanied in the bark of the Australian scrub ash *Acronychia baueri* Schott by **xanthevodine** and also by **acronycine.** Xanthevodine is also found in *Evodia* species (*F. Dallacker* and *G. Adolphen*, Ann., 1966, **691**, 134) where it occurs with **xanthoxoline**; this latter compound has been synthesised (*M. Ionescu* and *M. Vlassa*, Rev. Roum. Chim., 1971, **16**, 743; C.A., 1971, **75**, 49372k):

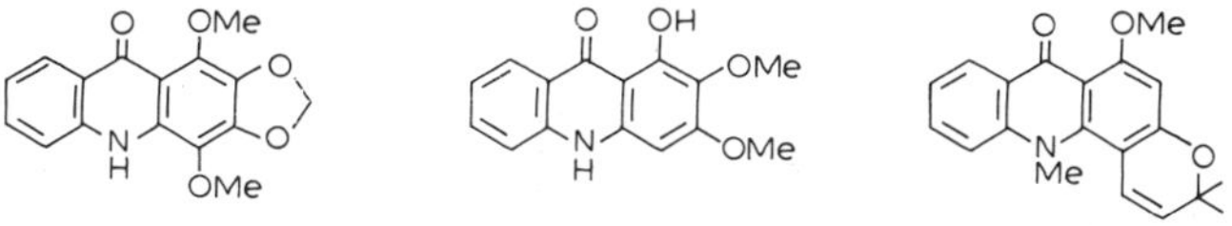

Xanthevodine Xanthoxoline Acronycine

Melicopicine, **tecleanthine** and **evoxanthine** accompany two novel alkaloids 6-**methoxytecleanthine,** m.p. 168°, and 1,3,5-**trimethoxy**-10-**methylacridone,** m.p. 141°, in the bark of *Teclea boiviniana* Baill. (*J. Vaquette et al.*, Plantes Médicinales et Phytothérapie, 1974, **8**, 57).

Acronycine which has a tetracyclic structure (*Lahey* and *W. C. Thomas*, Austral. J. sci. Res., 1949, **A2**, 423; *P. L. MacDonald* and *A. V. Robertson*, Austral. J. Chem., 1966, **19**, 275), shows antitumour activity (*G. V. Svoboda*, Lloydia, 1966, **29**, 206). The alkaloid has been synthesised (*J. R. Beck et al.*, J. Amer. chem. Soc., 1967, **89**, 3934): 3,4-dihydro-5,7-dimethoxy-2-quinolone if treated with 2-iodobenzoic acid and cuprous iodide forms the *N*-(2′-carboxyphenyl)-2-quinolone XLIII, which when reacted successively with polyphosphoric acid and methanolic hydrogen chloride affords the acridone XLIV (R = Me). Selective demethylation of the methoxyl function at C-1, using boron trichloride, yields XLIV (R = H) which upon alkylation with methyl-lithium in tetrahydrofuran affords the tertiary alcohol XLV. Demethylation and cyclisation of this material by fusion with pyridine hydrochloride gives XLVI. This with methyl iodide produces nordihydroacronycine. Dehydrogenation with 2,3-dichloro-5,6-dicyanobenzoquinone and methylation with dimethyl sulphate complete the synthetic sequence to acronycine:

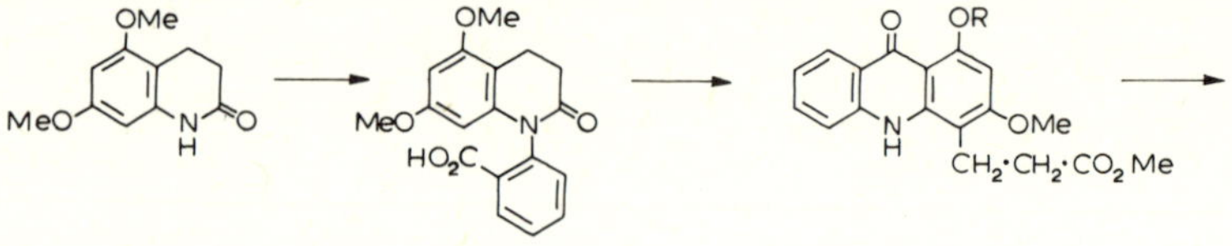

(XLIII) (XLIV)

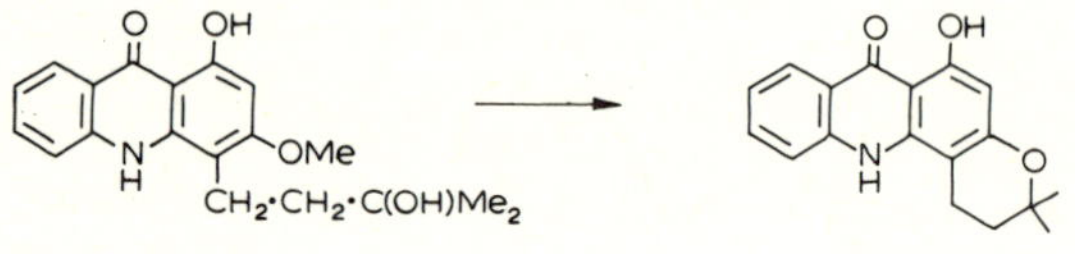

(XLV) (XLVI)

A simpler synthesis starting from 1,3-dihydroxy-9-acridone has been devised (*J. Hlubecek, Ritchie* and *W. C. Taylor*, Chem. and Ind., 1969, **50**, 1809); when treated with 3-chloro-3-methylbutyne in presence of base this compound affords XLVII, which may be de-alkylated, by heating with *N,N*-dimethylaniline, to XLVIIa; methylation of the latter gives acronycine:

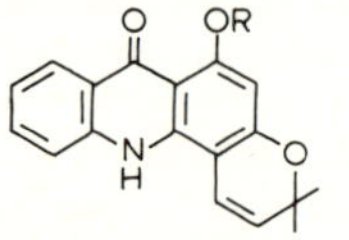

(XLVII, R = $CMe_2 \cdot C \equiv CH$)
(XLVIIa, R = H)

More recent syntheses of acronycine and de-*N*-methylnoracronycine have been reported (*W. M. Bandaranayake et al.*, J. chem. Soc., Perkin I, 1974, 998; see also *I. H. Bowen et al.*, *ibid.*, 1972, 2524; *Ionescu*, *Vlasca* and *I. Gola*, J. prakt. Chem. 1972, **314,** 441).

Rutacridone, $[\alpha]_{589}^{21}$ $-43°$, from *Ruta graveolens* L. (*J. Reisch et al.*, Acta Pharm. Suecica, 1967, **4**, 265; *W. Scharleman*, Z. Naturforsch., 1972, **27b**, 806), is the homologue of acronycine and has been allocated the following structure although the linear representation is not completely ruled out:

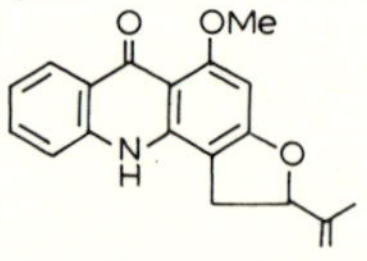

Rutacridone 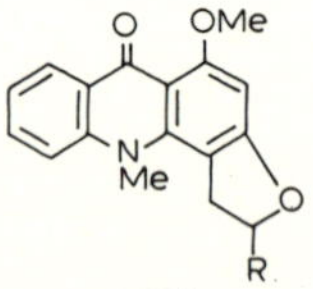 (XLVIII)

Rutacridone is accompanied in *Ruta graveolens* by **gravacridondiol** (XLVIII, R=C(OH)(Me)CH_2OH), m.p. 224–227°, **gravacridondiol monomethyl ether** (XLVIII, R=C(OH)(Me)CH_2OMe), m.p. 219–221°, **gravacridonchlorine** (XLVIII, R=C(Cl)(Me)CH_2OH), m.p. 254–257°, and **gravacridonolchlorine** (XLVIII, R=C(Cl)(CH_2OH)CH_2OH), m.p. 223–227° (*Reisch et al.*, Phytochemistry, 1972, **11**, 2121, 2359).

Although a number of *O*-prenylacridone alkaloids are known (see Table 1), **atalaphylline** (XLIX, R = H) and *N*-**methylatalaphylline** (XLIX, R = Me) were the first two open chain *C*-isoprenoid acridones to be discovered (*T. R. Govindachari et al.*, Tetrahedron, 1970, **26**, 2905):

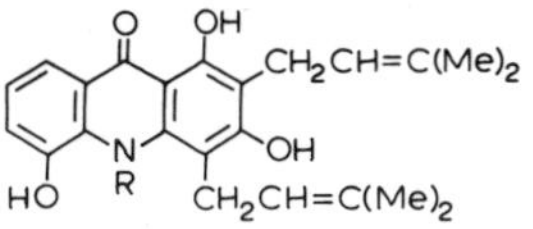

(XLIX) 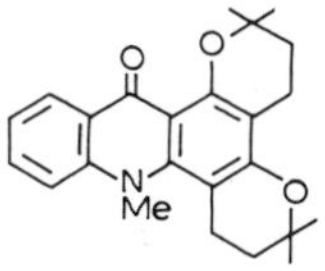 (L)

The related pentacyclic alkaloid bicyclo-*N*-methylatalphylline (L) has recently been isolated from *Atlantia monophylla* Correa (*D. Basu* and *S. C. Basa*, J. org. Chem., 1972, **37**, 3055).

The same source provides two new acridone *alkaloids* (LI), m.p. 252–254°, and (LII), m.p. 190–191.5° (*A. W. Fraser* and *J. R. Lewis*, J. chem. Soc., Perkin I, 1973, 1173):

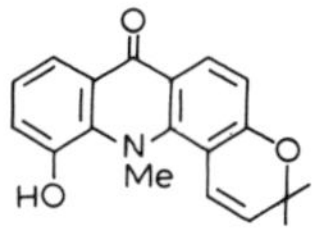

(LI) 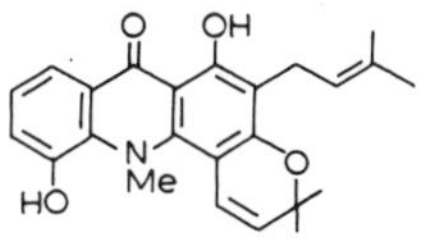 (LII)

The isolation of LI as a natural product is interesting since it has been suggested that this compound is a metabolite of acronycine in rat bile (*G. Guroff et al.*, Science, 1967, **157**, 1524).

The isolation of two dimeric alkaloids **atalanine**, m.p. 216.5–217.5°, and **ataline**, m.p. 209–210°, have been reported from *Atlantia ceylanica*; the assignment of a bi-acridone structure for these compounds is based on chemical and spectroscopic evidence (*Fraser* and *Lewis*, J. chem. Soc., D, 1973, 615):

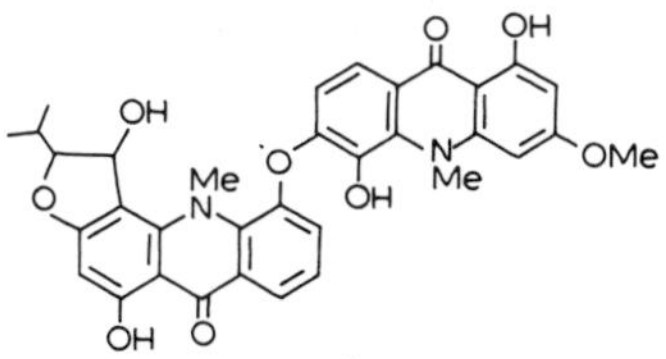

Atalanine

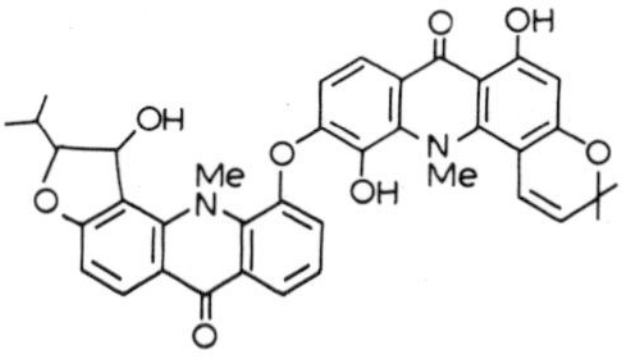

Ataline

(f) Biosynthesis of the rutaceous quinoline alkaloids

It is suggested (*I. Monkovič, I. D. Spencer* and *A. O. Plunkett*, Canad. J. Chem., 1967, **45**, 1935; *D. Gröger*, Lloydia, 1969, **32**, 221), that in the biosynthesis anthranilic acid is incorporated as a discrete unit into all the rutaceous quinoline alkaloids; tryptophan is not involved. Thus, for example, when isotopically labelled [$^{14}CO_2H$,^{15}N]anthranilic acid is fed to the plant *Dictamnus albus* L dictamnine, containing the same ^{14}C:^{15}N ratio as the substrate, may be isolated. In addition [^{3}H]anthranilic acid is incorporated into skimmianine by the plant *Skimmia japonica* Thunb. (*M. Matsuo, M. Yamazaki* and *Y. Kasida*, Biochem. biophys. Res. Commun., 1966, **23**, 679) and into acridone alkaloids in plants of the genera *Glycomis* (*D. Gröger* and *S. Johne*, Z. Naturforsch., 1968, **23b**, 1072; *Johne, H. Bernasch* and *Gröger*, Pharmazie, 1970, **25**, 777), *Evodia* and *Melicope* (*R. H. Prager* and *H. M. Thredgold*, Austral. J. Chem., 1969, **22**, 2627). Evidence has been presented (*Lewis, Bowen* and *P. Gupta*, J. chem. Soc., D, 1970, 1625) to suggest that a 2-aminobenzophenone intermediate is involved in the biosynthesis of the acridones (Scheme 1) and that the additional necessary carbon atoms are provided by "acetate":

Scheme 1

Biosynthesis of the acridone alkaloids

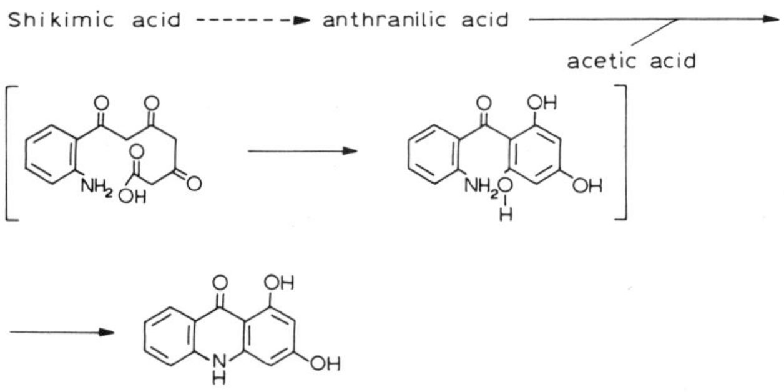

Indeed, the very frequent occurrence of 2,4-dioxygenation observed in the quinoline alkaloids prompts the speculation that "acetate" provides C-2 and C-3 of the quinoline ring of most of the rutaceous alkaloids and, in support of this view, it has been shown that [2-^{14}C]acetic acid gives rise to skimminine specifically labelled at the β-position of the nitrogen heterocycle when fed to plants producing this alkaloid (*Matsuo, Yamazaki* and *Kasida, loc. cit.*). Other related experiments support the general role of "acetate" in the biosynthesis of these alkaloids (*M. Luckner*, Pharmaz. Prax., 1965, 155; Pharmazie, 1963, **18**, 93; *S. Ghosal*, Sci. and Cult., 1964, **30**, 142; *Fish* and *Waterman*, *loc. cit.*). Only in the 2-phenylquinolines, is it thought probable that anthranilic acid combines with a phenylpropane derivative; formed from shikimic acid, to give a quinol-4-one-3-carboxylic acid, which subsequently undergoes decarboxylation and subsequent modification (Scheme 2). Acetic acid is presumed not to be involved (*cf. Luckner* in "Biosynthese der Alkaloide", ed. *K. Mothes* and *H. R. Schütte*, VEB Deutscher Verlag der Wissenschaften, Berlin, 1969):

Scheme 2

Biosynthesis of 2-phenylquinolines

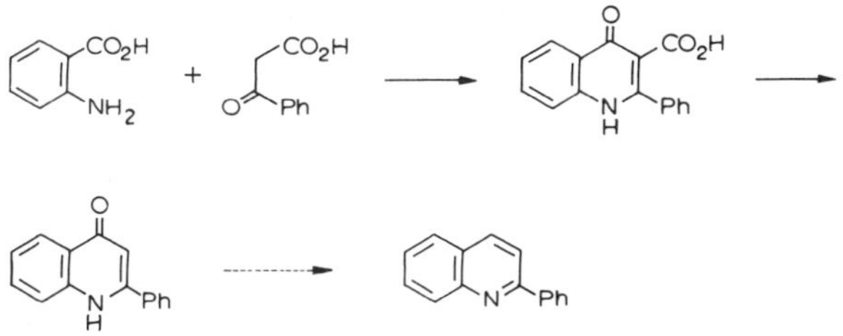

Similarly, the biosynthesis of molecules containing 2-alkyl substituents terminated by aryl groups, such as galipine [LIV, R = $-(CH_2)_2 \cdot C_6H_3(OMe)_2$] from angostura bark (p. 193), probably involves the chain extension of a phenylpropane-carboxylic acid by acetate (malonate) prior to condensation with activated anthranilic acid. In the case of 2-alkylquinolones, for example, 2-oxoundecylquinol-4-one (LIII, R = $-(CH_2)_9CO \cdot CH_3$) from *Vespris ampody* (p. 196), it seems that anthranilic acid combines with an appropriately functionalized long chain unit biosynthesised entirely from acetic and malonic acids.

Scheme 3

Biosynthesis of 2-alkylquinoline derivatives

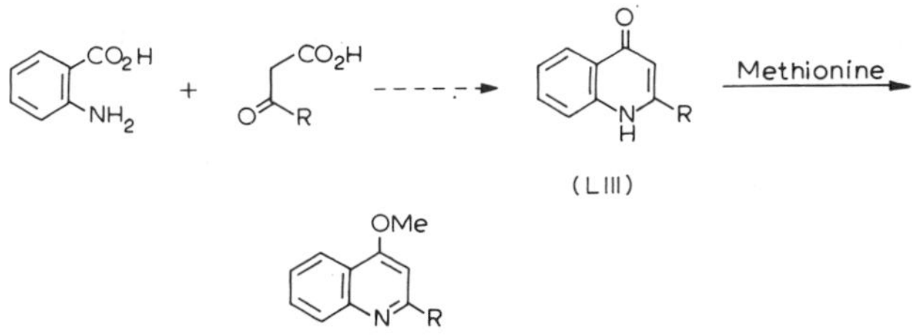

Methionine

(LIII)

(LIV)

Many alkaloids listed in Table 1 contain prenyl side chains (C_5), often modified into furan or pyran rings. It is most probable that these are derived from mevalonic acid through the intermediacy of γ,γ-dimethylallyl pyrophosphate (*cf. W. D. Ollis* and *I. O. Sutherland* in "Chemistry of Natural Phenolic Compounds", ed. *Ollis*, Pergamon Press, New York, 1961). The most likely biosynthetic route to a molecule such as dictamnine can be summarized as shown (Scheme 4) (*M. Cobet* and *Luckner*, Phytochemistry, 1970, **10**, 1031). In order to test this hypothesis [3-^{14}C]-2,4-dihydroxyquinoline was fed to *Skimmia japonica* whereupon radioactive dictamnine and the metho salt of platydesmine could be isolated (*J. F. Collins* and *M. F. Grundon*, Chem. Comm., 1969, 621). When the isotopically labelled dimethylallylquinolone (LV) was administered to the plant the same products were obtained specifically labelled as indicated by the asterisks in the formulae:

Scheme 4
Biosynthesis of furoquinoline alkaloids

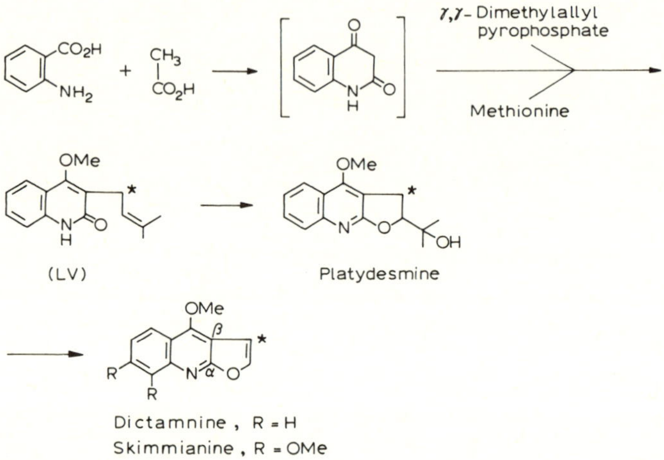

γ,γ-Dimethylallyl pyrophosphate

Methionine

(LV)

Platydesmine

Dictamnine, R = H
Skimmianine, R = OMe

The furan ring of the alkaloid skimmianine is also derived from mevalonate (*A. O. Colonna* and *E. G. Gros*, *ibid.*, 1970, 674).

More recently it has been shown that (+)-platydesmine is incorporated by *S. japonica* into both dictamnine and skimmianine suggesting that the hydroxylation of the homocyclic ring occurs very late in the biosynthetic sequence (*Grundon* and *K. J. James*, *ibid.*, 1971, 1311; see also *D. Boulanger et al.*, Phytochemistry, 1973, **12**, 2399).

It is interesting to note that *in vitro* platydesmine may be converted into dictamnine by the action of lead acetate and iodine (*R. Storer* and *D. W. Young*, Tetrahedron Letters, 1972, 2199).

For further studies of the biosynthesis of furoquinolines see *R. M. Bowman*, *J. F. Collins* and *M. F. Grundon*, J. chem. Soc., Perkin I, 1973, 626; *Collins et al.*, *ibid.*, 1974, 2177; 1975, 302; *D. Boulanger*, *B. K. Bailey* and *W. Steck*, Phytochemistry, 1973, **12**, 2399. It has been shown that dictamnine is converted into skimmianine in the plant *Choisya ternata* and furthermore that the 4-methoxyl group of a 3-prenyl-2-quinolone precursor is retained (*Grundon*, *D. M. Harrison* and *C. G. Spyropoulos*, J. chem. Soc., D, 1974, 51; see also *Grundon et al.*, J. Chem. Soc., Perkin I, 1974, 2181).

2. Miscellaneous alkaloids of probable biogenetic derivation from anthranilic acid

(a) Echinorine and echinine

The seeds of several *Echinops* species (Compositae) contain the alkaloid

echinorine (I) (*P. Schröder* and *M. Luckner*, Arch. Pharm., 1968, **301**, 39). When treated with aqueous alkali, echinorine yields N-*methyl-4-quinolone* (II), m.p. 152°, whereas with ammonia 1-methyl-4-iminoquinoline (III) (*monoperchlorate*, m.p. 242–244°) is formed. Both of these substances, named **echinopsine** and **echinopsidine**, respectively, have been claimed previously as alkaloids in their own right (*M. Greshoff*, Rec. Trav. chim., 1900, **19,** 360; *B. Avramova*, Farmatsiya, Sofia, 1964, **14** (6), 29; C.A., 1965, **63**, 5943b); they should now be regarded as artefacts of echinorine:

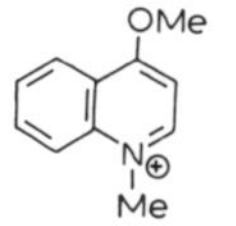

OMe; N⊕; Me

Echinorine (I)

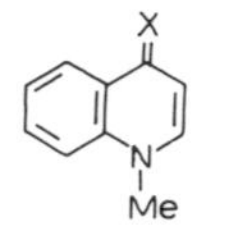

X; N; Me

(II, X = O)
(III, X = NH)

The seeds of *Echinops ritro* L. also contain an oily compound **echinine**, shown spectroscopically to be 1,2-dihydro-4-hydroxy-3-methoxy-1-methylquinoline (*W. Doepke* and *G. Fritsch*, Pharmazie, 1969, **24**, 782):

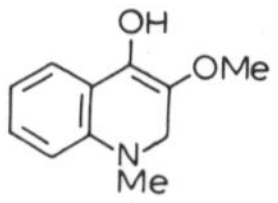

OH; OMe; N; Me

Echinine

So far, no crystalline derivatives have been described. The biosynthesis of echinorine in *E. ritro* has been studied; anthranilic acid and "acetate" are involved (*Schröder* and *Luckner*, *ibid.*, 1966, **21**, 642; Planta Med., 1968, **16**, 99). Scheme 1 outlines the *in vivo* formation of the alkaloid:

Scheme 1

Biosynthesis of echinorine

CO_2H; NH_2 + CH_3–CO_2H → [O; NH_2; CO_2H] —Methionine→ OMe; N⊕; Me

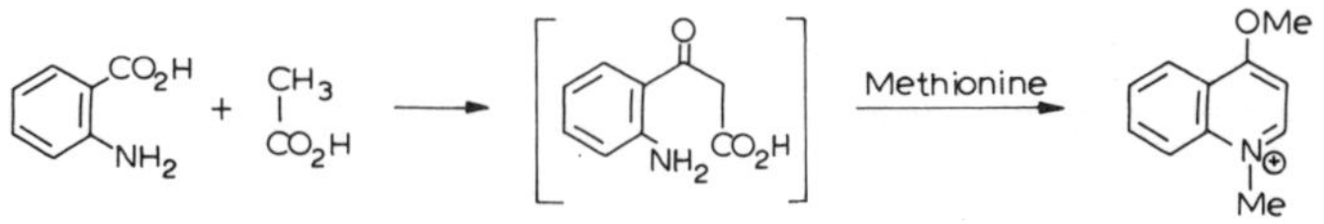

(*b*) *The alkaloids of* Macrorungia longistrobus

The South Africa shrub *Macrorungia longistrobus* C. B. Clarke (Acanthaceae) contains a number of alkaloids of which **macrorine** (IV), **normacrorine** (V), **isomacrorine** and **macrorungine** were the first to be isolated (*H. Jordaan* and *V. P. Joynt*, J. chem. Soc. Suppl., 1964, 5966; J. chem. Soc., 1965, 3001).

The n.m.r. spectra of all the alkaloids, with the exception of normacrorine, indicate the presence of one *N*-methyl group per molecule, but apart from this the spectra show only signals in the aromatic region. *Macrorine*, m.p. 160°, and *isomacrorine*, m.p. 110° form the same *methiodide* (VI), m.p. 196°, where methylation of normacrorine with dimethyl sulphate affords a 2:1 mixture of macrorine and isomacrorine.

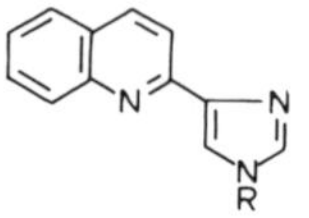

(IV, R = Me) Macrorine
(V, R = H) Normacrorine

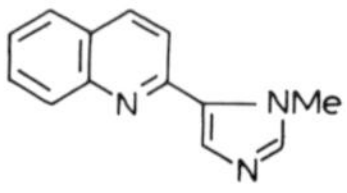

Isomacrorine

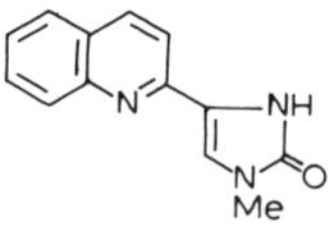

Macrorungine

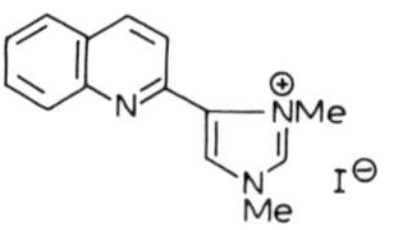

(VI)

The pale yellow alkaloid *macrorungine*, m.p. 267–270°(decomp.), contains an amide function and, on reduction with lithium tetrahydridoaluminate, yields macrorine.

Since macrorungine is stable to both acid and alkaline hydrolysis the carbonyl group is located between the two nitrogen atoms of the imidazole ring. The respective sites of the *N*-methyl functions in macrorine and its isomer have been determined by chemical degradation, and normacrorine has been synthesised by the following route: 2-acetylquinoline oxime was transformed by a Neber rearrangement into 2-ω-amino-acetylquinoline hydrochloride (VII) treatment of which with potassium thiocyanate, followed by desulphurization with nitric acid gave normacrorine:

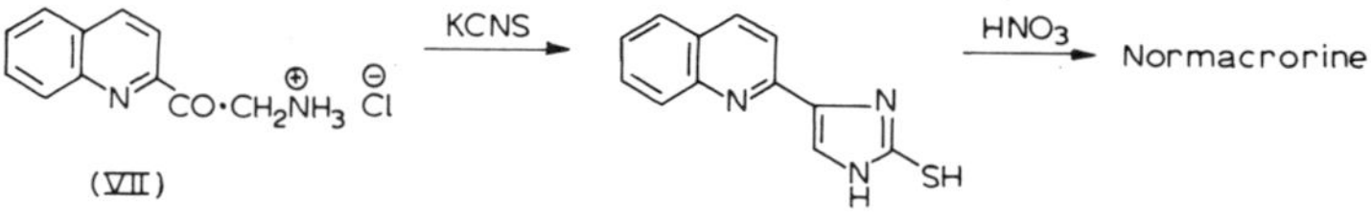

(VII) $\xrightarrow{KCNS}$ $\xrightarrow{HNO_3}$ Normacrorine

Three new alkaloids **longistrobine**, m.p. 145–148°, **isolongistrobine**, m.p. 132–136° and **dehydroisolongistrobine**, m.p. 131°, were isolated subsequently from *M. longistrobus* (*R. R. Arndt, S. H. Eggers* and *A. Jordan*, Tetrahedron, 1969, **25**, 2767). On distillation with zinc dust longistrobine and isolongistrobine yield macrorine and isomacrorine, respectively. The same products are formed when dehydrolongistrobine and dehydroisolongistrobine are subjected to acid hydrolysis. Jones' oxidation of isolongistrobine affords dehydroisolongistrobine identical with the natural product, whereas longistrobine yields dehydrolongistrobine which has not so far been found in nature.

Originally it was thought that all three alkaloids contained the quinoline nucleus; isolongistrobine, for example, was considered to have structure VIII:

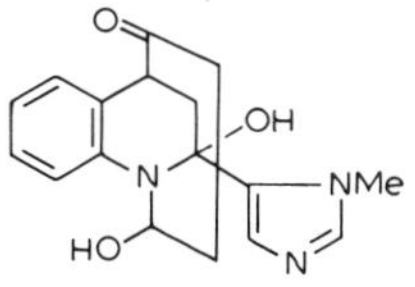

(VIII)

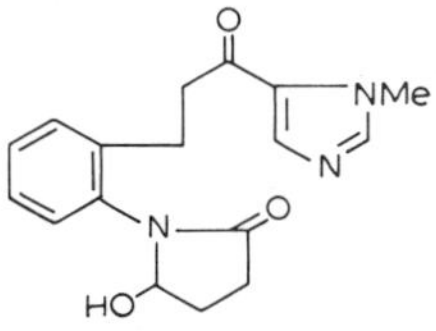

Isolongistrobine
(IX)

However, *M. A. Wuonola* and *R. B. Woodward* (J. Amer. chem. Soc., 1973, **95**, 284) have pointed out that the position of the carbonyl band in the *infrared* spectrum of isolongistrobine, 1660–1680 cm^{-1}, is lower than that which would be anticipated for structure VIII. Moreover, no hydroxyl bands appear in the *infrared* spectrum of dehydroisolongistrobine. These workers propose the structural alternative IX for isolongistrobine, whereas dehydroisolongistrobine is XV. The latter conclusion was verified by the following synthesis: methyl 1-methyl-2-imidazole-5-carboxylate was treated with the Grignard reagent from methyl phenyl sulphone and the product, X, alkylated with 2-nitrobenzyl bromide, in the presence of potassium tertiary butoxide, to yield XI. This material, on reduction with lithium amalgam, gave a mixture of the tetrahydroquinoline (XII), which can be dehydrogenated to isomacrorine, and XIII. When treated with β-methoxycarbonylpropionyl chloride, the latter molecule affords the amide XIV; oxidation of this with chromic oxide in pyridine and subsequent heating gives dehydroisolongistrobine (XV):

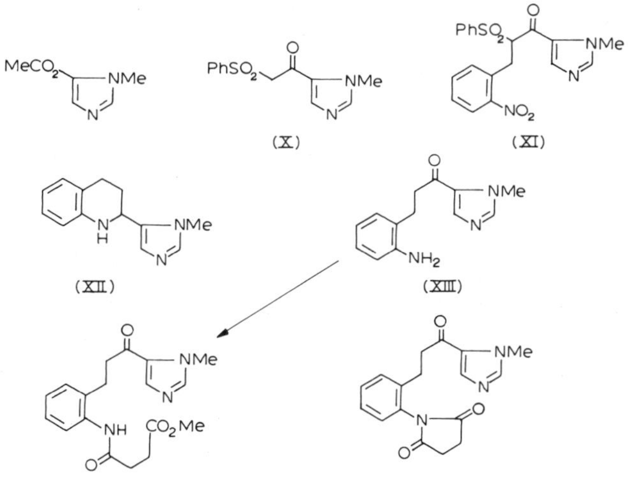

(X)

(XI)

(XII)

(XIII)

(XIV)

Dehydroisolongistrobine
(XV)

(c) Cryptolepine

This alkaloid ($C_{16}H_{12}N_2$) is obtained from the root extracts of the African plants *Cryptolepis triangularis* N.E. Br. and *C. sanguinolenta* (Lindl.) Schlecter (Asclepiadaceae) (*E. Clinquart*, Bull. Acad. roy. Med. Belg., 1929, **9**, 627; C.A., 1930, **24**, 1139; *E. Delvaux*, J. Pharm. Belg., 1931, **13**, 955, 973) and is reported to have marked hypertensive properties (*Raymond-Hamet*, Compt. rend. Soc. biol., 1938, **207**, 1016).

Cryptolepine crystallizes as deep purple needles, m.p. 175–178°, from aqueous ethanol and forms a series of well characterized salts with acids (*hydrochloride*, m.p. 263–265°; *nitrate*, m.p. 261–262°; *perchlorate*, m.p. 331–333°; *picrate*, m.p. 265–267°). It has been shown to have the tetracyclic structure shown below (*E. Gillert, Raymond-Hamet* and *E. Schlitter*, Helv., 1951, **34**, 642) and the chemistry of this alkaloid has been reviewed (*J. E. Saxton* in "The Alkaloids", ed. *R. H. F. Manske*, Vol. VIII, Academic Press, New York, London, 1965, p. 19).

Interestingly, this structure was synthesised (*F. Fichter et al.*, Ber., 1906, **39**, 3932; 1907, **40**, 3478; 1910, **43**, 3489) some twenty years before the alkaloid was first isolated:

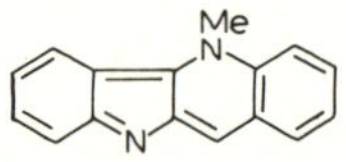

Me
N
N

Cryptolepine

Biosynthetically the molecule probably forms by the combination of two molecules of *N*-methylanthranilic acid (*Luckner* in "Biosynthese der Alkaloide", eds. *K. Mothes* and *H. R. Schütte*, VEB Deutscher Verlag der Wissenschaften, Berlin, 1969, p. 538).

3. Miscellaneous alkaloids containing the quinoline ring system

Fabianine. An extract of the terminal branchlets of the South American plant *Fabiana imbricata* Ruiz. and Par., family Solanacea, has been used to treat renal and urinary infections in cattle (*H. Kunz-Krause*, Arch. Pharm., 1899, **237**, 1). During an attempt to isolate the active principle(s) of this plant the alkaloid *fabianine*, $C_{14}H_{21}NO$, was obtained as a colourless mobile liquid (*picrate*, m.p. 114–116°) (*O. E. Edwards* and *N. F. Elmore*, Canad. J. Chem., 1962, **40**, 256).

In the *infrared* spectrum of the alkaloid a band at 3280 cm^{-1} shows the presence of a hydroxyl group, and bands at 1596 and 1577 cm^{-1}, together with the *ultraviolet* spectral data (λ_{max} 213, 273 nm, log ε 3.9, 3.8) suggest the presence of a pyridine unit of structure in the alkaloid. This conclusion is confirmed by an analysis of the ^1H-

n.m.r. spectrum (summarized in the formula) and leads to the structural allocation (I) for this natural product.

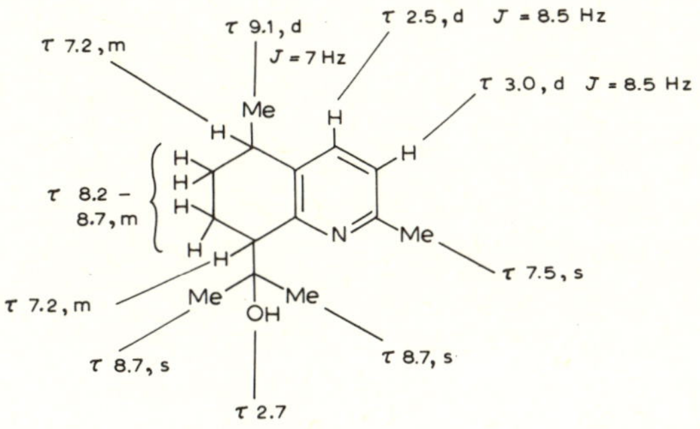

Fabianine

(I)

When fabianine is heated with sodium hydroxide in ethylene glycol solution a retro-aldol reaction occurs affording acetone and 2,5-*dimethyl*-5,6,7,8-*tetrahydroquinoline* (II), b.p. 68–73°, 4×10^{-2} mm. This latter substance is also present in the extracts of *F. imbricata*. On dehydrogenation fabianine gives 2,5-dimethylquinoline. Fabianine has recently been synthesised (*P. Shimizu et al.*, Recherches, 1974, **19**, 241).

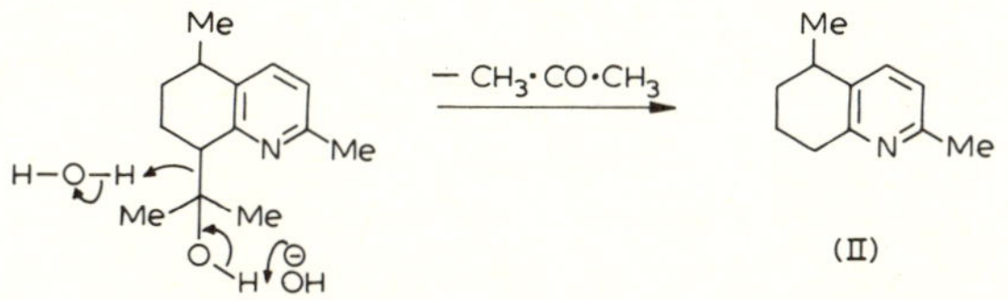

(II)

(a) Quinolines from fungi and micro-organisms

(i) Pseudanes

A number of 2-*n*-alkyl-4-hydroxyquinolines (see Table 2), collectively known as pseudanes, have been isolated from cultures of the bacterium *Pseudomonas aeruginosa* (Schroeter) Migula (*J. W. Cornforth* and *A. T. James*, Biochem. J., 1956, **63**, 124; *M. Hashimoto* and *K. Hattori*, Chem. and pharm. Bull., Tokyo, 1967, **15**, 718; *E. E. Hays et al.*, J. biol. Chem., 1945, **159,** 725; *J. W. Lightbown*, Nature, 1950, **166**, 356; J. gen. Microbiol., 1954, **11**, 477; *R. Takeda*, J. Fermentation Technol., 1959, **37**, 59; *I. C. Wells*, J. biol. Chem., 1952, **196**, 331).

TABLE 2*

THE PSEUDANES

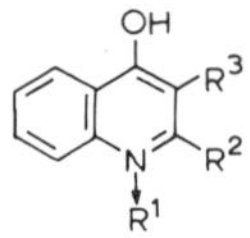

*Name**	*M.p. (°C)*	R^1	R^2	R^3
Pseudane-VII	146–147	—	$-(CH_2)_6CH_3$	H
Pseudane-IX	139	—	$-(CH_2)_8CH_3$	H
Δ^1-Pseudane-IX	153–155	—	$-CH=CH\cdot(CH_2)_6CH_3$	H
Pseudane-VII *N*-oxide	158–160	O	$-(CH_2)_6CH_3$	H
Pseudane-VIII *N*-oxide		O	$-(CH_2)_7CH_3$	H
Pseudane-IX *N*-oxide	148–149	O	$-(CH_2)_8CH_3$	H
Pseudane-XI *N*-oxide		O	$-(CH_2)_{10}CH_3$	H
3-Hydroxypseudane-VII		—	$-(CH_2)_6CH_3$	OH
3-Methyl-Δ^2-pseudane-VII		—	$-CH_2CH=CH-(CH_2)_3CH_3$	CH_3

*The Roman numeral refers to the number of carbon atoms in the C-2 side chain. Pseudanes are formed also by higher plants of the Rutaceae family, see p. 178.

Biosynthetically these molecules probably arise through the combination of anthranilic acid with an acetate derived fatty acid derivative *e.g.* III. Decarboxylation of the 3-carboxylic acid function may then afford the pseudane structure. Support for the following scheme is provided by radiotracer studies (see *M. Luckner*, in "Biosynthese der Alkaloide", eds. *K. Mathes* and *H. R. Schütte*, VEB Deutscher Verlag der Wissenschaften, Berlin, 1969, p. 530, and references cited therein):

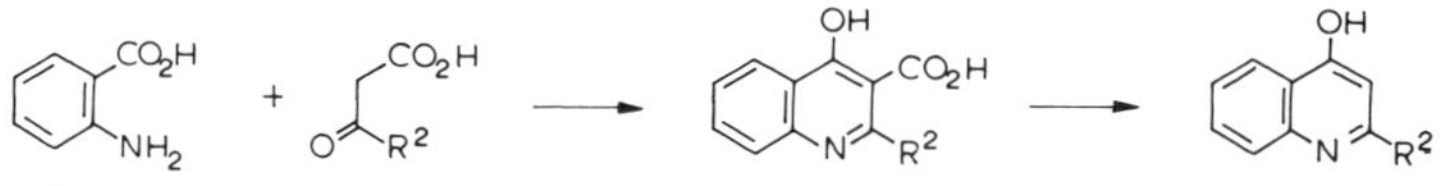

(III)

(ii) Viridicatine, viridicatol, cyclopenine and cyclopenol

A. Bracken, A. Pocker and *H. Raistrick* (Biochem. J., 1954, **57**, 587) described the characterization of two nitrogen-containing metabolites from cultures of the mould *Penicillium cyclopium* Westling. One of these products had been previously isolated from *P. viridicatum* Westling (*K. G. Cunningham* and *G. G. Freeman, ibid.*, 1953, **53**, 328) and called **viridicatine,** $C_{15}H_{11}NO_2$, m.p. 269°. The second product was named **cyclopenine**.

During a later study (*J. H. Birkinshaw et al.*, J. Biochem., 1963, **89**, 196) it was discovered that the substance cyclopenine was, in reality, a mixture of two closely related compounds. For one of these ($C_{17}H_{14}N_2O_3$, m.p. 183–184°, $[\alpha]^{20}_{5461}$ $-291°$) the name cyclopenine has been retained, whereas the other molecule ($C_{17}H_{14}N_2O_4$, m.p. 215° (decomp.), $[\alpha]^{20}_{5461}$ $-309°$) is now called **cyclopenol.**

A fourth compound ($C_{15}H_{11}NO_3$, m.p. 280°), given the name **viridicatol**, was also isolated (*Luckner* and *Mothes*, Tetrahedron Letters, 1962, 1035; Arch. Pharm., 1963, **296**, 18), and this molecule together with its 3-O-*methyl ether*, m.p. 257°, has also been found in *P. puberulum* (*D. J. Austen* and *M. B. Meyers*, J. chem. Soc., 1964, 1197).

Viridicatine is shown to be 2,3-dihydroxy-4-phenylquinoline (IV) by degradative studies and by an unambiguous synthesis (*Cunningham* and *Freeman*, *loc. cit.; B. Eistert* and *H. Selzer*, Z. Naturforsch., 1962, **17b**, 3):

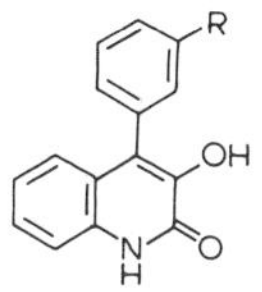

(IV, R = H) Viridicatine
(V, R = OH) Viridicatol

The *ultraviolet* spectrum of viridicatol is very similar to that of viridicatine ($\lambda_{max.}^{MeOH}$ 226, 284, 304 sh; 316, 329 sh nm), and like the latter molecule it gives an intense green colour with ethanolic ferric chloride solution. Both compounds show *infrared* absorption bands at ν 3200, 1650 and 1635 cm^{-1}. Viridicatol forms a *diacetate*, m.p. 195–196°.

Degradation of viridicatol, by oxidation with alkaline hydrogen peroxide, affords 2-amino-3′-hydroxybenzophenone (VI, R = OH); viridicatine under similar conditions yields 2-aminobenzophenone (VI, R = H). Consequently, viridicatol is 2,3-dihydroxy-4-(3-hydroxyphenyl)quinoline (V). This conclusion has been confirmed by synthesis (*Luckner* and *Y. S. Mohammed*, Tetrahedron Letters, 1964, 1987).

In the *infrared* spectrum of cyclopenine, bands at ν 3090 (NH), 1700 (CO), 1630 (CONH), 990 and 885 cm^{-1} are observed. On methylation with diazomethane, cyclopenine yields a monomethyl derivative and upon hydrolysis with dilute acid affords viridicatine, methylamine, benzoic acid and carbon dioxide. Oxidation with hydrogen peroxide in glacial acetic acid affords 3-*N*-methyl-2,4-quinazolinedione (VII), anthranilic acid, benzoic acid and benzaldehyde. From this evidence cyclopenine is adjudged to be VIII (R = H) and cyclopenol, which gives viridicatol on acid hydrolysis, is the phenol VIII (R = OH) (*idem, ibid.*, 1963, 1953).

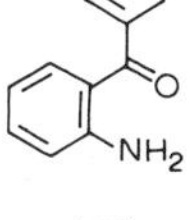

(VI)

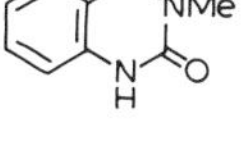

(VII)

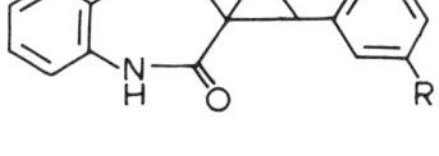

Cyclopenine
(VIII)

The benzodiazepine structures of the metabolites cyclopenine and cyclopenol are further supported by mass spectrometry (*Luckner et al.*, Tetrahedron, 1969, **25**, 2575) and by synthesis (*H. Smith*, *P. Wegfahrt* and *H. Rapoport*, J. Amer. chem. Soc., 1968, **90**,

1668; *Rapoport et al.*, J. org. Chem., 1969, **34**, 1359; *J. D. White et al.*, Tetrahedron, 1970, **26**, 233).

It has been shown that cyclopenase, an enzyme from the mycelium of *P. viridicatum*, catalyses the transformation of cyclopenine and cyclopenol into the quinolines viridicatine and viridicatol, respectively (*Luckner*, *K. Winter* and *J. Reisch*, European J. Biochem., 1969, **7**, 380). Since cyclopenine and cyclopenol are biosynthesised from anthranilic acid and phenylalanine (*L. Nover* and *Luckner*, Abhandlungen der Deutschen Akad. Wissenschaften, Berlin, 1971, 525; 1972, 535), the suggestion has been made that the biosynthesis of viridicatine and viridicatol can be summarized as shown in the following scheme (where E represents an electron receptor site on the assembling enzyme)*:

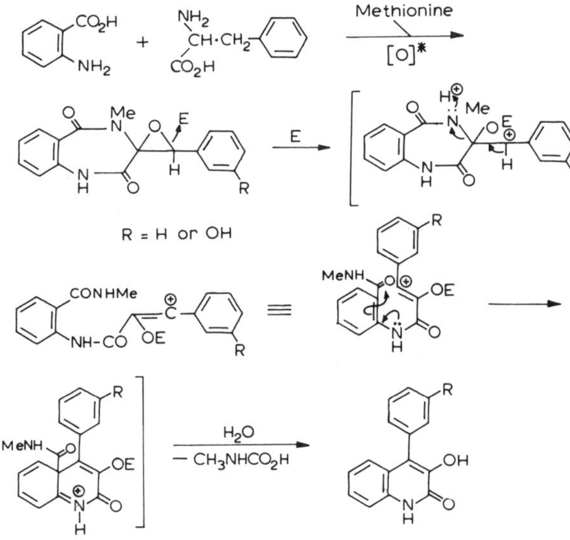

Methionine

[O]*

R = H or OH

H_2O

$- CH_3NHCO_2H$

The epoxide and aromatic oxygen atoms are introduced by molecular oxygen (*L. Nover* and *M. Luckner*, F.E.B.S. Letters, 1969, **3**, 292).

Viridicatin is metabolised by *Penicillium crustosum* (*M. Taniguchi* and *Y. Satomura*, Agr. biol. Chem., Tokyo, 1970, **34**, 506) and further details of the biosynthesis of cyclopenine and cyclopenol have been discussed by *Nover* and *Luckner* (European J. Biochem., 1969, **10**, 265).

Nitramin represents a new decahydroquinoline alkaloid; it is present in extracts of *Nitraria schoberi* (*N. Yu. Novgorodova*, *S. K. Maekh* and *S. Yu. Yunusov*, Khim. Prir. Soedin., 1973, 1969; C.A., 1973, **79**, 32150w):

---

* Although some aspects of this sequence seem mechanistically rather speculative, the general conclusions are substantiated by experiment (see *J. Framm et al.*, European J. Biochem., 1973, **37**, 78).

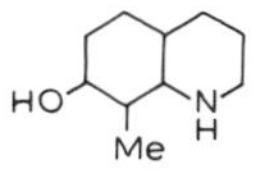

Nitramin

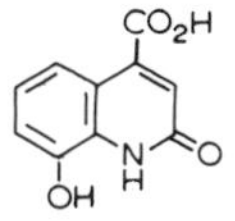

Zeanic acid

Zeanic acid, from corn steep liquor, is a natural plant growth regulator (*H. Fukumi et al.*, J. pharm. Soc., Japan, 1974, **94**, 768).

4. Quinoline alkaloids derived from tryptophan

(a) The cinchona alkaloids

(i) Occurrence

The cinchona alkaloids occur in the bark of *Cinchona* and *Remijia* species (Family Rubiaceae)*, which are indigenous to the eastern tropical slopes of the Andes. Cinchona trees are cultivated commercially in India, Ceylon and the East Indies, and, more recently, in Central America and parts of S. America. Cinchona preparations have been used medicinally in S. America for many centuries, and were introduced into Europe in the early 17th century. Until quite recent times these preparations have been the sole treatment for malaria. The treatment is now in competition with synthetic antimalarial drugs, the development of which was accelerated by the Japanese invasion of the East Indies in 1942, which cut off about nine-tenths of the world's supply of cinchona bark.

Investigations on the alkaloidal content of cinchona bark began as long ago as 1810, and since then almost 30 bases have been isolated from the source. The principal alkaloids are quinine, quinidine, cinchonine and cinchonidine. Numerous processes for extraction of the alkaloids from the bark have been described (see, *inter alia, C. Béguin*, Pharm. Acta Helv., 1939, **14**, 109; *L. Rosenthaler* and *O. N. Jalcindag, ibid.*, 1941, **16**, 149; *J. Büchi et al., ibid.*, 1942, **17**, 1, 73; *I. M. Heilbron et al.*, B.P. 558,320/1943), and chromatographic, spectroscopic and polarographic methods for the detection and determination of the alkaloids have been worked out (see, *P. von Vácha et al.*, Planta Med., 1964, **12**, 406; *M. Vincze-Vermes* and *Z. Vincze*, Acta Pharm. Hung., 1973, **43**, 49; C.A., 1973, **79**, 5481).

---

* Although primarily found in Rubiaceous plants of the genera *Cinchona, Remijia* and *Ladenbergia*, these alkaloids also occur in plants of the families Annonacea and Loganiacea (see *M. Luckner* in "Biosynthese der Alkaloide", eds. *K. Mothes* and *H. R. Schütte*, VEB Deutscher Verlag der Wissenschaften, Berlin, 1969, p. 516). The leaves of *Olea europaea* L and *Ligustrum vulgare* L *(Oleaceae)* also contain cinchona alkaloids (*G. von Schneider* and *W. Kleinert*, Planta Med., 1972, **22** (2), 109).

(ii) Individual quinoline alkaloids and related compounds

(*1*) **Cinchonine.** The alkaloid cinchonine, $C_{19}H_{22}ON_2$, is present in most cinchona barks, and may be purified *via* its crystalline sulphate. It is a ditertiary base, but does not contain an *N*-methyl group, and is unsaturated and optically active. The oxygen atom is present as a secondary alcoholic group, since cinchonine forms a mono-acetate, and on careful oxidation with chromic acid is converted into the corresponding ketone, **cinchoninone***, $C_{19}H_{20}ON_2$. Cinchonine contains one double bond, since on hydrogenation it takes up one molecular equivalent of hydrogen with the formation of the saturated **dihydrocinchonine,** $C_{19}H_{24}ON_2$. Further, the double bond is part of a vinyl group, since on ozonolysis cinchonine yields formaldehyde and an aldehyde, **cinchoninal**, $C_{18}H_{20}O_2N_2$, and on mild permanganate oxidation formic acid and a carboxylic acid, **cinchotenine,** $C_{18}H_{20}O_3N_2$, are obtained (*L. Seekles*, Rec. Trav. chim., 1923, **42**, 69; *Z. H. Skraup et al.*, Ann., 1879, **197**, 235; 1879, **199**, 348; Monatsh., 1895, **16**, 159; Ber., 1895, **28**, 12).

Reactions and structure. When fused with caustic potash, cinchonine yields quinoline, this reaction being the first recorded formation of quinoline (*C. Gerhardt*, Ann., 1842, **44**, 279). Lepidine (4-methylquinoline), 3-ethylpyridine and β-collidine (3-ethyl-4-methylpyridine) are formed simultaneously.

More vigorous oxidation of cinchonine with chromic acid leads to a mixture of products, arising from the splitting of the molecule into two fragments. Amongst these products are cinchoninic acid (quinoline-4-carboxylic acid), cinchomeronic acid (pyridine-3,4-dicarboxylic acid) and pyridine-2,3,4-tricarboxylic acid.

A further important oxidation product is **meroquinene**, $C_9H_{15}O_2N$ (*W. Koenigs*, Ber., 1894, **27**, 1501). It is an unsaturated iminocarboxylic acid, which forms a nitrosamine, and which on further oxidation yields **cincholoiponic acid**, $C_8H_{13}O_4N$, and **loiponic acid**, $C_7H_{11}O_4N$, together with formic acid. Meroquinene on reduction forms a saturated dihydro derivative, **cincholoipon**, $C_9H_{17}O_2N$; since cincholoiponic acid is a dicarboxylic acid, and is saturated, meroquinene must contain a vinyl group, which is presumably the same group present in cinchonine, unaffected by the oxidation. When heated with hydrochloric acid at 250–260° meroquinene is transformed into β-collidine. Cincholoiponic acid is an iminodicarboxylic acid, and is also formed by the direct oxidation of cinchonine. On heating with sulphuric acid it yields γ-picoline and carbon dioxide, and on oxidation it yields loiponic acid, which is the lower homologous iminodicarboxylic acid. These transformations are best explained on the basis of structures I, II and III for meroquinene, cincholoiponic acid and loiponic acid, respectively (*Koenigs*, *ibid.*, 1897, **30**, 1326; 1902, **35**, 1357; *P. Rabe* and *R. Pasternack*, *ibid.*, 1916, **49**, 2753). *Koenigs* (*loc. cit.*, 1897) drew attention to the fact that loiponic acid

* Strictly speaking the oxidation product is cinchonidinone, derived from cinchonidine and epimeric with cinchoninone. See section on stereochemistry, p. 234 *et seq.*

was isomeric with hexahydrocinchomeronic acid (piperidine-3,4-dicarboxylic acid), which is of known structure, being formed by the oxidation of isoquinoline, followed by hydrogenation.

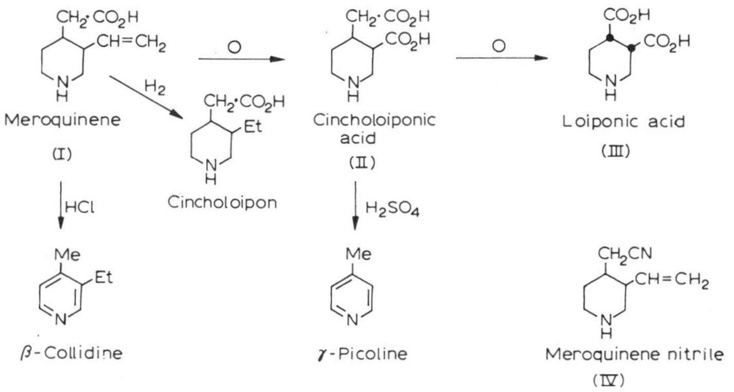

Meroquinene (I) — Cincholoipon — Cincholoiponic acid (II) — Loiponic acid (III)

β-Collidine — γ-Picoline — Meroquinene nitrile (IV)

This acid is obtained as a mixture of *cis*- and *trans*-forms, converted by the action of alkali into the more stable *trans*-form. Loiponic acid is also converted into this *trans*-acid on treatment with alkali, so that the compound is correctly formulated as *cis*-piperidine-3,4-dicarboxylic acid (III). The constitution (II) for cincholoiponic acid has been confirmed also by synthesis (*A. Wohl* and *M. S. Losanitsch*, Ber., 1907, **40**, 4698; *Wohl* and *R. Maag*, *ibid.*, 1909, **42**, 627), the racemic acids obtained being resolved with brucine.

From these and other observations, it appeared that cinchonine was composed of quinoline and 3-vinylpiperidine units linked in the 4-positions, though the nitrogen atom in the piperidine part must be involved in the linkage in some way, since cinchonine is not a secondary base. *L. Pasteur* had observed over a century ago that when cinchonine is treated with acid, it is isomerised to **cinchotoxine** (Compt. rend., 1853, **37**, 110; *W. von Miller* and *G. Rohde*, Ber., 1894, **27**, 1187, 1280; 1895, **28**, 1056; 1900, **33**, 3214). Cinchotoxine is a ketone, and one of the nitrogen atoms is secondary. With amyl nitrite cinchotoxine affords an α-hydroxyimino derivative, which when subjected to the Beckmann rearrangement with phosphorus pentachloride yields cinchoninic (quinoline-4-carboxylic) acid and **meroquinene nitrile** (3-vinylpiperidyl-4-acetonitrile) (IV) (*Rabe*, Ann., 1906, **350**, 180). It became apparent that the degradations outlined were in most cases preceded by an isomerisation of cinchonine to cinchotoxine, and it was postulated (*von Miller* and *Rohde, loc. cit.*) that in cinchonine the quinoline system is linked with a bicyclic system in which the nitrogen atom occupies a bridgehead position, as in partial structure V. In particular. the structure VI for cinchonine was advanced by *Rabe* in 1908 (Ber., 1908, **41**, 62; see also Ann., 1909, **365**, 353; 1910, **373**, 85) as being the best explanation of the transformations described (see Scheme 1, p. 228).

Scheme 1
Transformations of cinchonine

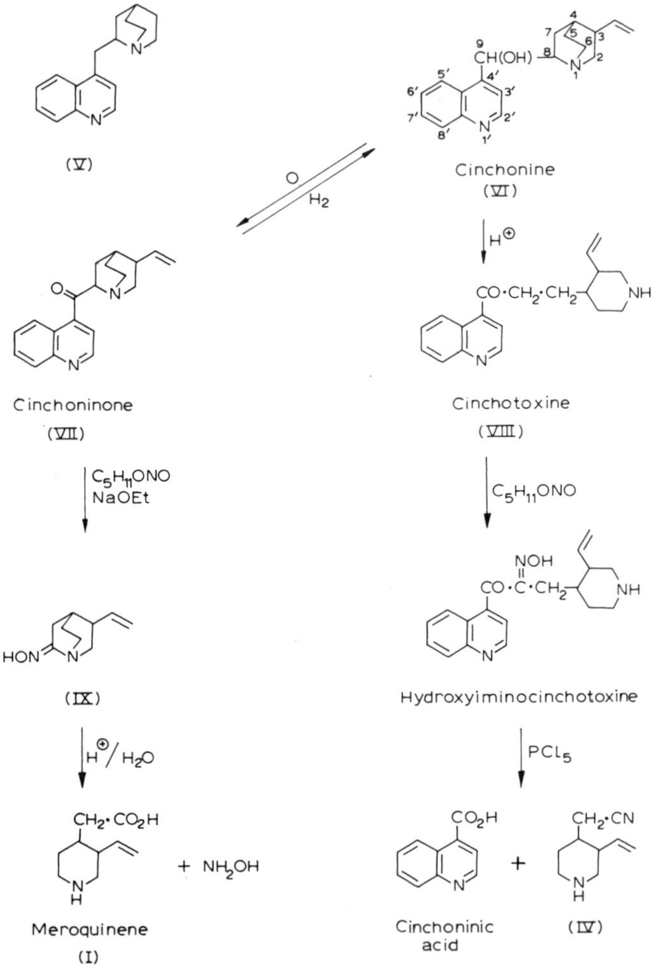

(V)

O

H_2

Cinchonine
(VI)

$H^{\oplus}$

Cinchoninone
(VII)

$CO \cdot CH_2 \cdot CH_2$ NH

Cinchotoxine
(VIII)

$C_5H_{11}ONO$
NaOEt

$C_5H_{11}ONO$

NOH
$CO \cdot C \cdot CH_2$ NH

HON N

(IX)

Hydroxyiminocinchotoxine

$H^{\oplus}/H_2O$

PCl_5

$CH_2 \cdot CO_2H$ + NH_2OH

Meroquinene
(I)

CO_2H + $CH_2 \cdot CN$

Cinchoninic acid

(IV)

Cinchoninone is therefore VII, and the isomerisation of cinchonine to cinchotoxine (VIII) is explained by the partial ring-cleavage of the tricyclic (quinuclidine) system, a C–N bond being ruptured. Cinchotoxine may be converted into cinchoninone (and thence to cinchonine) by sodium hypobromite, *via* an *N*-bromo derivative, which cyclises *via* the reactive CH_2 group in VIII, with elimination of hydrogen bromide. The presence of the quinuclidine system in cinchoninone has been demonstrated by cleavage of the compound with amyl nitrite and sodium ethoxide to β-vinyl-α'-quinuclidone oxime (IX), which on acid hydrolysis yields hydroxylamine and meroquinene (I) (*Rabe et al.*, Ber., 1907, **40**, 3655; 1908, **41**, 62; Ann., 1909, **364**, 330).

Other transformations which are explicable on the basis of structure VI for cinchonine include the behaviour of the base on chlorination and dehydrochlorination. When cinchonine is treated with phosphorus pentachloride it yields cinchonine chloride (VI; OH replaced by Cl), which is dehydrochlorinated by alcoholic potash to **cinchenine** (cinchene), $C_{19}H_{20}N_2$ (X). This unsaturated base is cleaved hydrolytically by phosphoric acid to lepidine and meroquinene (*Koenigs et al.*, Ber., 1884, **17**, 1986; 1890, **23**, 2669; 1894, **27**, 900). Prolonged heating of the anhydro-base (X) with acids affords **apocinchenine** (XI) and ammonia. Apocinchenine is a phenolic base which yields cinchoninic acid on oxidation. Its structure has been proved by degradation and by synthesis of its methyl and ethyl ethers (*J. Kenner* and *F. S. Statham*, J. chem. Soc., 1935, 299). For observations on the mechanism of the formation of apocinchenine, see *R. B. Turner* and *R. B. Woodward*, "The Alkaloids, Chemistry and Physiology", eds. *R. H. F. Manske* and *H. L. Holmes*, Academic Press, London, Vol. III, 1953, pp. 13–15.

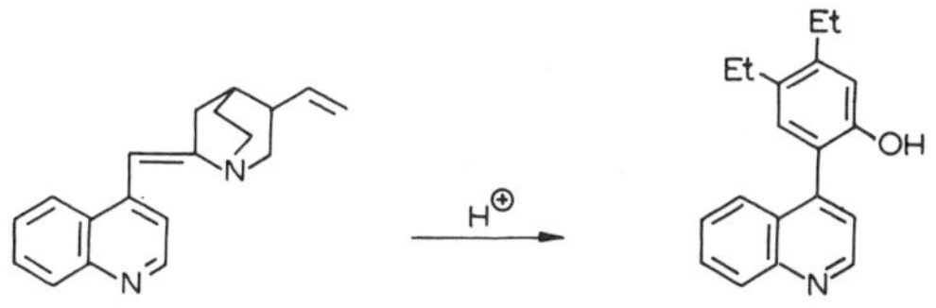

Cinchenine (X) Apocinchenine (XI)

Some further transformations of cinchonine are represented in the following scheme:

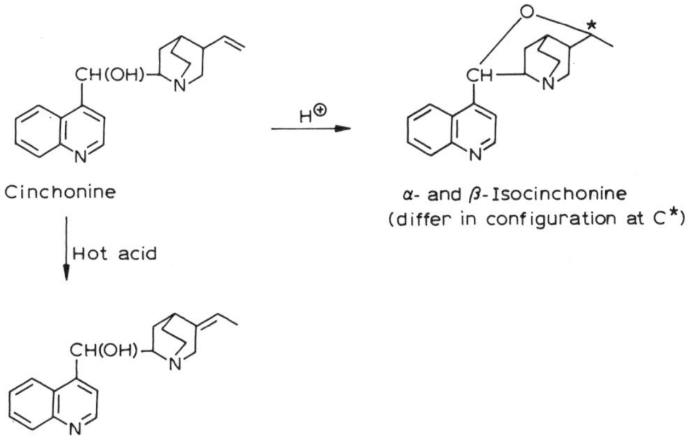

Cinchonine

α- and β-Isocinchonine
(differ in configuration at C*)

Apocinchonine
(Apocinchonidine, β-Cinchonidine)

Amongst the products formed by the acid-catalysed isomerisation of cinchonine are α- and **β-isocinchonines**, which differ in configuration at C*. These compounds are seven-membered cyclic ethers formed by interaction between the secondary alcohol group and the vinyl group in the alkaloid (*inter alia, E. Jungfleisch* and *U. Léger*, Ann.

chim., 1920, [ix], **14**, 59; *Léger*, Bull. Soc. chim. Fr., 1938, [v], **5**, 183). With hot acid cinchonine is converted into **apocinchonine**, formed by shift of the vinyl double bond to the neighbouring exocyclic position. The possibility of geometrical isomerism about the new alkenic bond has been substantiated by the isolation of **apocinchonidine** and **β-cinchonidine**, geometrical isomerides, as transformation products of cinchonidine, a steroisomer of cinchonine (see pp. 234–239) (*O. Hesse*, Ann., 1880, **205**, 314; 1888, **243**, 149; *G. Neumann*, Monatsh., 1892, **13**, 651; *Léger*, Bull. Soc. chim. Fr., 1919, [iv], **25**, 571).

Selective methods have recently been developed for the catalytic reduction of cinchonine (*J. Suszko* and *B. Golankiewicz*, Rocz. Chem., 1968, **42**, 477; C.A., 1968, **69**, 44084j; see also C.A., 1969, **71**, 50289n), and it has been demonstrated that cinchonine, cinchonidine, quinine and quinidine undergo photochemical reduction to the corresponding 9-deoxy compounds when the aqueous hydrochloric acid solutions are irradiated with broad spectrum ultraviolet radiation (*V. I. Sternberg, E. T. Travecedo* and *W. E. Musa*, Tetrahedron Letters, 1969, 2031; Tetrahedron, 1970, **35**, 4131).

(*2*) **Cinchonidine,** $C_{19}H_{22}ON_2$, occurs in most varieties of cinchona bark, and especially in *C. succirubra.* It is stereoisomeric with cinchonine (VI), the two alkaloids differing in configuration at $C_{(8)}$ and $C_{(9)}$ (see stereochemistry of cinchona alkaloids, pp. 235–236).

(*3*) **Quinine,** $C_{20}H_{24}O_2N_2$, the most important cinchona alkaloid, was first obtained, though probably in an impure state, as long ago as 1792. The base is optically active though the specific rotation varies markedly with solvent and pH.

Reactions and structure. Quinine shows chemical properties very closely related to those of cinchonine. Quinine contains a methoxyl group, and is, in fact, a methoxycinchonidine. The methoxyl group is located in the quinoline component of the quinine molecule, since on oxidation with chromic acid quinine yields quininic acid, 6-methoxyquinoline-4-carboxylic acid (XII) in place of cinchoninic acid, the products obtained by breakdown of the quinuclidine component being the same as from cinchonine. Quininic acid has been synthesised by the methods given in C.C.C., Vol. IV F, p. 342 and also by *Rabe et al.* (Ber., 1931, **64**, 2487) by the following route:

MeO–C₆H₄–NH₂ + CH₃–CO–CH₂–CO₂Et → MeO–C₆H₄–NH–CO–CH₂–CO–CH₃ —H_2SO_4→ 6-MeO-4-CH₃-2-OH-quinoline —(a) PCl_5; (b) 2H→ 6-MeO-4-CH₃-quinoline —PhCHO→ 6-MeO-4-(CH=CHPh)-quinoline —O→ 6-MeO-4-CO_2H-quinoline

Quininic acid
(XII)

This synthesis has been further explored by subsequent workers (*A. D. Ainley* and *H. King*, Proc. roy. Soc., 1938, **125B**, 60; *R. C. Elderfield et al.*, J. org. Chem., 1946, **11**, 803).

Quinine is therefore represented by structure XIII. On mild oxidation it affords quininone* (XIII; CHOH replaced by CO); with acids it undergoes the "toxine" rearrangement to quinotoxine (XIV) and on permanganate oxidation it yields formic acid and **quitenine,** the analogue of cinchotenine (p. 226). It forms a series of transformation products exactly analogous to those formed by cinchonine.

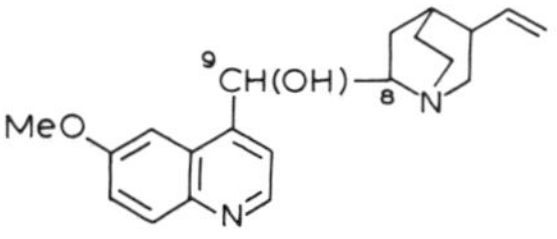

Quinine
(XIII)

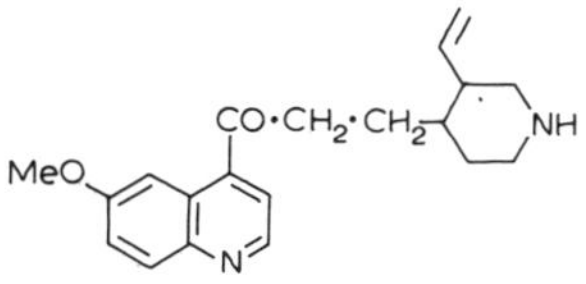

Quinotoxine
(XIV)

The constitution of quinine has been confirmed by a wealth of synthetic studies (see p. 239 *et seq.*) and also by intensive spectroscopic analysis. A discussion of the possible fragmentation modes in the mass spectrometry of quinine and its derivatives, for example, has been given (*G. Spiteller* and *M. Spiteller-Friedman*, Tetrahedron Letters, 1963, 153) (the base peak in the spectrum normally occurs at *m/e* 136) and the natural abundance ^{13}C-n.m.r. spectrum of quinine has also been interpreted (*W. O. Cram, Jr., W. C. Wildman* and *J. D. Roberts*, J. Amer. chem. Soc., 1971, **93**, 990; see also *E. Wenkert et al.*, Accounts of Chemical Research, 1974, **7**, 46).

(*4*) **Quinidine,** $C_{20}H_{24}O_2N_2$, is present in small quantity in most cinchona barks. It bears a similar relation to quinine as cinchonine does to cinchonidine, differing from quinine only in the configuration at $C_{(8)}$ and $C_{(9)}$ in formula XIII (see stereochemistry of cinchona alkaloids, p. 234).

(*5*) **Quinotoxine** (quinicine), $C_{20}H_{24}O_2N_2$, (XIV), already described as the isomerisation product of quinine, is also found in cinchona bark (*von Miller, Rohde* and *E. Fussenegger*, Ber., 1900, **33**, 3214).

(*6*) **Epiquinine** and **epiquinidine** are diastereoisomers of quinine found in the mother-liquors after removal of the major alkaloids (*Rabe et al.*, Ann., 1932, **492**, 242; J. pr. Chem., 1939, **154**, 66; Ber., 1908, **41**, 67 and subsequent papers). They may be artifacts.

(7) **Dihydrocinchonine** (cinchotine), **dihydrocinchonidine** (cinchamidine), **dihydroquinine** and **dihydroquinidine** are all minor constituents of extracts of cinchona bark. They are all related to the respective primary base, the vinyl group having been reduced to ethyl. It is possible they are artifacts.

---

*Owing to epimerization the oxidation product is really quinidinone. See section on stereochemistry, p. 234.

(*8*) **Cupreine,** $C_{19}H_{22}O_2N_2$, occurs in cuprea bark (*Remijia pedunculata*, a plant closely related to the cinchonas) (*B. H. Paul* and *A. J. Cownley*, Pharm. J., 1881, **12**, 497; 1884, **15**, 221, 401; *Hesse*, Ber., 1882, **15**, 857; 1883, **16**, 60; *D. Howard* and *J. Hodgkin*, Pharm. J., 1881, **12**, 528; J. chem. Soc., 1882, **41**, 66), and also in *Cinchona succiruba* (*K. A. Khaleque, M. Aminuddin* and *S. Azim-Uk-Khuda*, Sci. Res., Dacca, Pakistan, 1965, **2**, 1).

Cupreine is a phenolic base which on methylation affords quinine; it is therefore *O*-demethylquinine (XIII, HO in place of MeO).

(*9*) **Dihydrocupreine,** a degradation product of quinine, has been established as an alkaloid in *Timonius kaniensis* Val. (Family *Rubiaceae*). It occurs as a glassy non-crystalline solid, $[\alpha]_D$ $-143°$ (EtOH), *picrate*, m.p. 250–252° (*S. R. Johns* and *J. A. Lamberton*, Austral. J. Chem., 1970, **23**, 211):

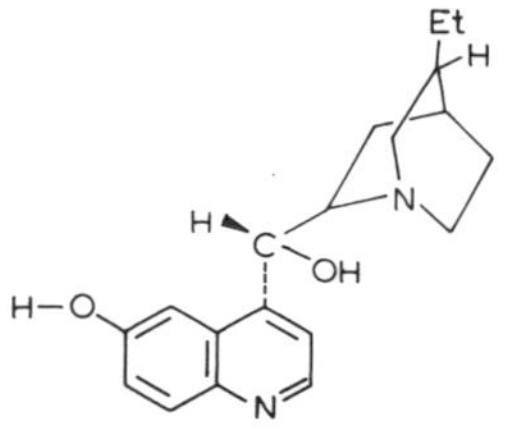

Dihydrocuprein

(iii) Cinchona alkaloids derived from indole

In addition to the quinoline alkaloids described so far, the cinchona alkaloids comprise also a series of indole derivatives, which are important in connection with the biochemical synthesis of the quinoline series (see p. 244). For this reason a brief description of some of them is included again in this chapter.

Cinchonamine, $C_{19}H_{24}ON_2$, is a minor cinchona alkaloid, first isolated by *A. Arnaud* (Compt. rend., 1881, **93**, 593; 1883, **97**, 174). It gives colour reactions characteristic of the indole alkaloids (*Raymond-Hamet, ibid.*, 1941, **212**, 135; 1945, **220**, 670; 1945, **221**, 307), and behaves as a monoacidic base, containing a primary alcohol group and a basic tertiary nitrogen atom, which does not carry a methyl group. On oxidation with chromic acid cinchonamine behaves differently from the cinchona bases already considered in that a quinuclidine carboxylic acid is obtained. This compound was assigned structure XV on the grounds that on decarboxylation it yielded 3-vinylquinuclidine, which on catalytic hydrogenation afforded 3-ethylquinuclidine, identical with a synthetic product (*V. Prelog et al.*, Ann., 1940, **545**, 247). On treatment with acetic anhydride cinchonamine affords diacetylallocinchonamine, which on oxidation yields 3-β-acetoxyethylindole-2-carbaldehyde (XVI); **diacetylallocinchonamine** is therefore XVII, and cinchonamine XIX (*R. Goutarel et al.*, Helv., 1950, **33**, 150; *W. I. Taylor, ibid.*,

p. 164). On dehydrogenation with selenium, cinchonamine affords **dehydrocinchonamine** (XVIII), a change which recalls the transformation of quinuclidine into γ-ethylpyridine (*Prelog* and *K. Balenovic*, Ber., 1941, **74**, 1508).

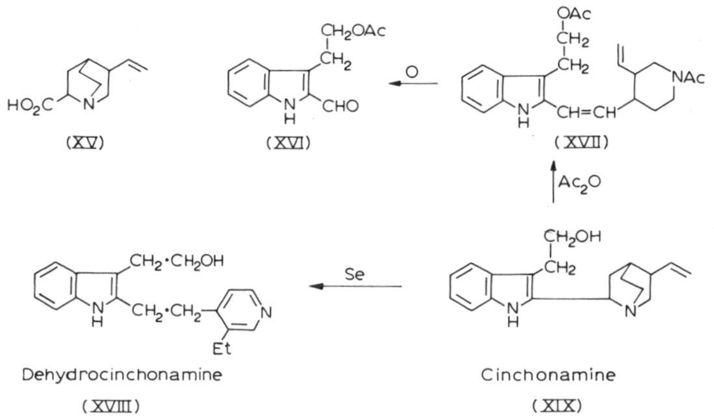

(XV) (XVI) (XVII)

Dehydrocinchonamine (XVIII) Cinchonamine (XIX)

There is some evidence to suggest that cinchonamine is an intermediate in the biosynthesis of the quinoline alkaloids (see p. 246). The reverse process, the conversion of cinchonine (and cinchonidine) to dihydrocinchonamine, has been achieved in the laboratory (*E. Ochiai* and *M. Ishikawa*, Pharm. Bull., Tokyo, 1957, **5**, 498; *Ochiai, Ishikawa* and *Y. Oka*, *ibid.*, 1959, **7**, 744). Cinchonamine has been synthesised (*Ch'en Ch'ang-pai, R. P. Evstigneeva* and *N. Preobrazhenskii*, Proc. Acad. Sci. U.S.S.R. Chem. Sect. (Eng. Transl.), 1958, **123**, 927).

Quinamine, $C_{19}H_{24}O_2N_2$, was first isolated from cinchona bark by *O. Hesse* (Ber., 1872, **5**, 265; 1877, **10**, 2152; Ann., 1873, **166**, 217; 1877, **199**, 333; 1881, **207**, 288). The alkaloid resembles cinchonamine in its reactions, yields 2,3-dimethylindole on distillation with zinc dust and shows indole colour reactions (*K. S. Kirby*, J. chem. Soc., 1945, 528). The same quinuclidinecarboxylic acid (XV), obtained from cinchonamine, is also formed by the oxidation of quinamine with chromic acid (*T. A. Henry, Kirby* and *G. E. Shaw*, *ibid.*, p. 524). On reduction with lithium tetrahydridoaluminate, quinamine affords cinchonamine (*Goutarel et al., loc. cit.*), and conversely, cinchonamine is converted into quinamine by peracetic acid (*B. Witkop*, J. Amer. chem. Soc., 1950, **72**, 3211). Quinamine does not contain an aromatic indole nucleus, on the basis of diazo-coupling reactions. One oxygen atom is present as a tertiary hydroxyl group; the other must be ethereal in character, since quinamine is not an *N*-oxide, nor does it show a carbonyl band in its infrared absorption spectrum. The immediate lithium tetrahydridoaluminate reduction product of quinamine is a dihydroxydihydroindole, which readily loses water to form an aromatic indole (cinchonamine) (*C. C. J. Culvenor et al.*, J. chem. Soc., 1950, 1485). Structure XX (*Witkop, loc. cit.*) appears to explain the properties and reactions of quinamine most satisfactorily:

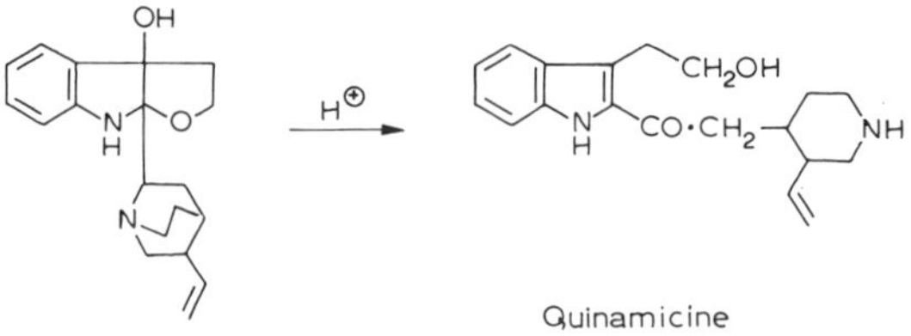

Quinamine
(XX)

Quinamicine
(XXI)

A change similar to the "toxine" rearrangement occurs when quinamine is treated with dilute acids, the product being **quinamicine** (XXI) *(Henry, Kirby* and *Shaw, loc. cit.).* When quinamine is heated with acetic anhydride, acetylapoquinamine is formed, hydrolysis of which affords **apoquinamine** (XXII); the mechanism of the formation of this apobase has been discussed by *R. B. Turner* and *R. B. Woodward* ("The Alkaloids, Chemistry and Physiology", eds. *R. H. F. Manske* and *H. L. Holmes*, Academic Press, London, 1953, Vol. III, pp. 53, 54):

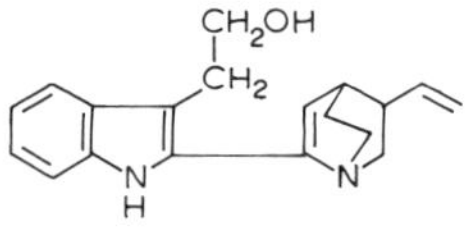

Apoquinamine
(XXII)

Conquinamine, $C_{19}H_{24}O_2N_2$, is a minor cinchona base, isolated by *Hesse* (Ber., 1877, **10**, 2152). It resembles quinamine very closely, and can be transformed into apoquinamine and quinamicine under appropriate conditions. In fact the natural product conquinamine is identical with the artefact epiquinamine which is obtained from quinamine by treatment with base. Quinamine can be partly regenerated from epiquinamine under the same conditions and it is clear that the two substances differ only in configuration at $C_{(8)}$ (see p. 239).

For further details of cinchona alkaloids derived from indole see *V. Boekelheide* and *Prelog* in "Progress in Organic Chemistry", ed. *J. W. Cook*, Butterworths, London, 1955, Vol. 3, Chapter 5, and also *W. I. Taylor* in "The Alkaloids", ed. *Manske* and *Holmes*, Academic Press, London, Vol. XI, Chapter 3; Vol. XIV, Chapter 5.

(iv) Stereochemistry of the cinchona bases

Each of the major alkaloids has five chiral centres, at positions $C_{(3)}$, $C_{(4)}$, $C_{(8)}$, $C_{(9)}$ and $N_{(1)}$ in the adjoining formula (R^1=4-quinolyl or 6-methoxy-4-quinolyl, R^2=vinyl or ethyl). $C_{(4)}$ and $N_{(1)}$, of course, being bridge-head positions are stereochemically interrelated.

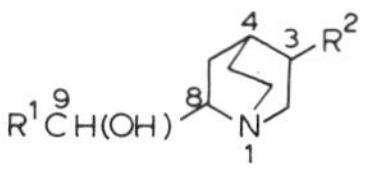

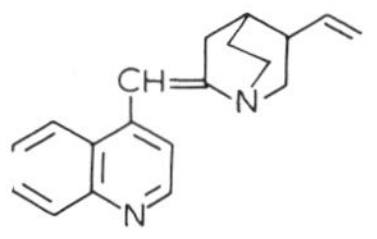

R¹ C H(OH) 9 8 4 3 R² N 1

CH= N N

Quinene

(XXIII)

Configuration at $C_{(3)}$, $C_{(4)}$

Cinchonine and cinchonidine both yield cinchotoxine (VIII) on treatment with acids; it follows that in these two bases the configurations at $C_{(3)}$ and $C_{(4)}$ must be the same. For a similar reason quinine and quinidine must also have identical configurations at these two centres. Confirmation is provided by the facts that cinchonine and cinchonidine both yield cinchenine (X) and quinine and quinidine yield quinene (XXIII), on dehydration. All four alkaloids yield meroquinene (I) on oxidative degradation; since the methods used do not involve inversion at $C_{(3)}$ or $C_{(4)}$, all the bases must therefore have the same configurations in these positions. Further oxidation of meroquinene yields cincholoiponic and loiponic acids, and as has already been shown these acids are *cis* in configuration; the four alkaloids must then have the hydrogen atoms at $C_{(3)}$ and $C_{(4)}$ *cis* to each other. *Prelog* and *E. Zalán* (Helv., 1944, **27**, 535) proved conclusively the presence of a *cis*-arrangement at $C_{(3)}$–$C_{(4)}$ by degrading cinchonine to optically inactive 1,2-diethylcyclohexane. All of the cinchona quinoline alkaloids may thus be represented as (XXIV) where the stereochemistry at $C_{(8)}$ and $C_{(9)}$ is uncommitted:

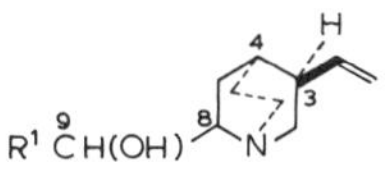

R¹ C H(OH) 9 8 4 H 3 N

(XXIV)

H CH₂ 8 N N

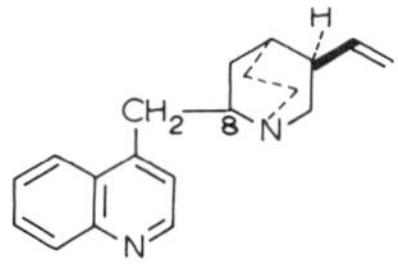

(XXV)

At the time this work was concluded the absolute configuration at $C_{(3)}$, implied in this formula, was not known and thus the antipodal arrangement was equally possible. By good fortune the original arbitrary assignment is now known to be correct (see p. 238).

Configuration at $C_{(8)}$

Cinchonine can be converted into deoxycinchonine, and cinchonidine into deoxycinchonidine (*W. Koenigs*, Ber., 1896, **29**, 372; 1895, **28**, 3143), both deoxy-bases being represented by XXV. The deoxy-bases are isomeric, not identical. It follows that cinchonine and cinchonidine differ in the steric arrangement at $C_{(8)}$. Quinine and quinidine differ similarly. Cinchonine and quinidine, but not cinchonidine and quinine, are readily convertible into iso-bases which have a cyclic ether structure (see p. 229). The ether bridge in these compounds can be formed only when the vinyl group at $C_{(3)}$ and the >CHOH group at $C_{(8)}$ are in the *cis* relationship and it follows that in cinchonine and quinidine the hydrogen atoms at $C_{(8)}$ and $C_{(3)}$, and thus at $C_{(4)}$, are all *cis*-orientated. In cinchonidine and quinine the hydrogen atoms at $C_{(8)}$ and $C_{(3)}$ are *trans*:

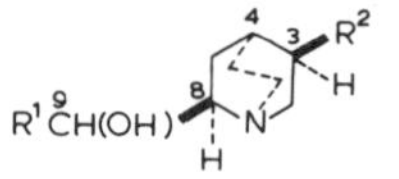

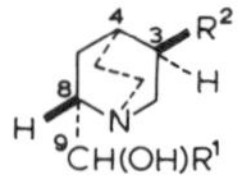

Cinchonine	Cinchonidine
Quinidine	Quinine
Epicinchonine	Epicinchonidine
Epiquinidine	Epiquinine

Configuration at $C_{(9)}$

As this centre is not incorporated in a rigid system the determination of its configuration is more difficult. An early attempt to settle this question made use of the principle of optical superposition (*H. King* and *A. D. Palmer*, J. chem. Soc., 1922, **121**, 2577). As shown above, all four bases have the same configurations at $C_{(3)}$ and $C_{(4)}$; this means that there are, in both the cinchonine and quinine series, four possible isomers (and their mirror-images). These isomers are cinchonine, cinchonidine, epicinchonine and epicinchonidine, and a similarly named group in the quinine series, the normal members being the natural alkaloids. The specific rotations of the eight bases are as follows:

Cinchonine	$+224°$	Quinine	$-158°$
Cinchonidine	$-111°$	Quinidine	$+254°$
Epicinchonine	$+120°$	Epiquinine	$+43°$
Epicinchonidine	$+63°$	Epiquinidine	$+102°$

These figures show that in the most strongly dextrorotatory bases the partial rotatory contributions of $C_{(8)}$ and $C_{(9)}$ are both positive in cinchonine and quinidine, and both negative in cinchonidine and quinine. It was argued that cinchonine and cinchonidine must differ in configuration at $C_{(9)}$, since they already differ at $C_{(8)}$, and so also must quinine and quinidine. The structure of quinine is thus methoxycinchonidine whereas quinidine corresponds to methoxycinchonine.

The configuration of each base at $C_{(9)}$ may be deduced by a correlation of the alkaloids with (−)-ephedrine (XXVI) and (+)-ψ-ephedrine (XXVII), formulated as Fischer projections*. Epiquinine is a stronger base than quinine (*P. Rabe*, Ber., 1943, **76**, 251; Ann., 1949, **561**, 132), and ψ-ephedrine is more strongly basic than ephedrine; by analogy quinine is probably related to ephedrine in configuration, and epiquinine to ψ-ephedrine. Inspection of models shows that this difference in basic strengths may be due to differing distances between the hydroxyl group and the nitrogen atom in the two epimerides; in epiquinine and ψ-ephedrine these two groups are sufficiently near to permit intramolecular hydrogen bonding, with consequent stabilisation of the conjugate acid form and increase in basic strength (*Prelog* and *O. Hafliger*, Helv., 1950, **33**, 2021).

* These projection formulae represent the absolute configurations of the ephedrines, but this fact is not significant to the argument.

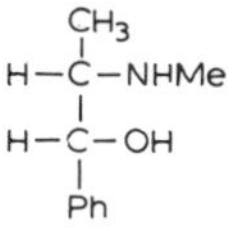

Ephedrine (XXVI) ψ-Ephedrine (XXVII) Quinine (XXVIII) Epiquinine (XXIX)

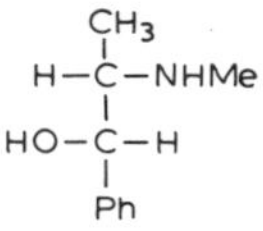

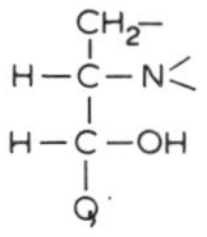

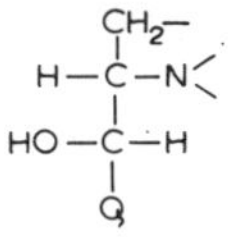

In quinine and ephedrine the bulky quinolyl and phenyl groups, respectively repel the quinuclidyl and methyl groups, and thus inhibit the formation of a hydrogen bond. The configuration of the $C_{(8)}$–$C_{(9)}$ systems in quinine and epiquinine may therefore be written as (XXVIII) and (XXIX) respectively.

In support of this conclusion it has been observed that the $C_{(9)}$ epimers form chelates with copper sulphate, whereas the natural bases do not (*Z. Földi*, *T. Földi* and *A. Földi*, Chem. and Ind., 1957, 465; Acta chim. Acad. Sci. Hung., 1958, **16**, 185). Some additional evidence in favour of the above assignments is provided by chemical studies (*Y. Kobayashi*, Chem. pharm. Bull., Tokyo, 1959, **7**, 209). At one stage, there were some misgivings concerning the interpretation of the early work (see for example *W. E. Doering*, *G. Cortes* and *L. H. Knox*, J. Amer. chem. Soc., 1947, **69**, 1700) and a case for the reversal of the stereochemical assignments was presented (*G. G. Lyle* and *W. Gaffield*, Tetrahedron Letters, 1963, 1371; Tetrahedron, 1967, **23**, 51) based upon ORD studies. The original assignments, however, are correct (*Prelog*, Tetrahedron Letters, 1964, 2037) and the relative configurations of the alkaloids are settled by X-ray structural analysis of quinine sulphate and selenate (*H. Mendel*, Proc. Kon. Nederl. Akad. Wetenschappen, 1955, **58B**, 132). *Lyle* and *L. K. Keefer* (Tetrahedron, 1967, **23**, 3253) have also shown that if the quaternary salts of the dihydro derivatives of quinine, quinidine, cinchonine and cinchonidine and their respective epimers are converted stereospecifically into the corresponding rigid oxiranes (XXX)*, by known procedures (*Rabe*, *K. Dussel* and *R. Teske-Guttman*, Ann., 1949, **561**, 159), then the configurations of the parent structures may be determined by the size of the coupling constant of the signals due to the oxirane ring protons in the ^{1}H-n.m.r. spectra.

Dihydrocinchonine benzochloride, for example, gives specifically the *trans*-oxirane in which the coupling constant $J_{8,9} = 2$ Hz*. On the other hand epidihydroquinidine benzochloride forms the *cis*-oxirane $J_{8,9} = 4$ Hz (also see *Lyle* and *Keefer*, J. org. Chem., 1966, **31**, 3921).

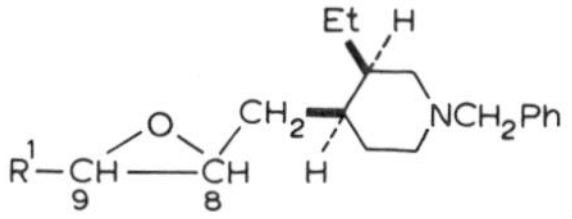

(XXX)

* The formation of oxiranes from quaternary salts of β-aminoethanols involves *retention* of configuration at the carbinol centre, but *inversion* at the carbon atom to which the ammonium function was originally bonded (*A. C. Cope* and *E. R. Turnball*, Organic Reactions, 1960, **11**, 352, 389). The numbering system adopted here relates the centres back to the cinchona parent.

Thus the relative configurations of the centres at $C_{(8)}$ and $C_{(9)}$ in the alkaloids is adjudged to be *erythro* (see Table 3).

TABLE 3

CONFIGURATIONS OF THE CINCHONA ALKALOIDS

Alkaloid	*Relative stereochemistry of $C_{(8)}$, $C_{(9)}$ system*	*Configuration at $C_{(9)}$*
Quinine	*erythro*	*R*
Epiquinine	*threo*	*S*
Quinidine	*erythro*	*S*
Epiquinidine	*threo*	*R*
Cinchonidine	*erythro*	*R*
Epicinchonidine	*threo*	*S*
Cinchonine	*erythro*	*S*
Epicinchonine	*threo*	*R*

The key to the absolute stereochemical configuration of the cinchona bases is the degradation of cinchonine to (−)-4-ethyl-3-methylhexane, which has been correlated with glyceraldehyde and is now known to have the absolute configuration XXXI (*Prelog* and *Zalán*, *loc. cit.*, p. 545; *Prelog* and *Häfliger*, *loc. cit.*).

$$\begin{array}{c} CHEt_2 \\ | \\ H-C-Me \\ | \\ Et \end{array}$$

(XXXI)

This enables the absolute configuration of all the alkaloids at $C_{(3)}$ to be fixed, and as the configurations at all the other asymmetric centres relative to $C_{(3)}$ are known, with the aid of models the following absolute configurations for the bases may be written:

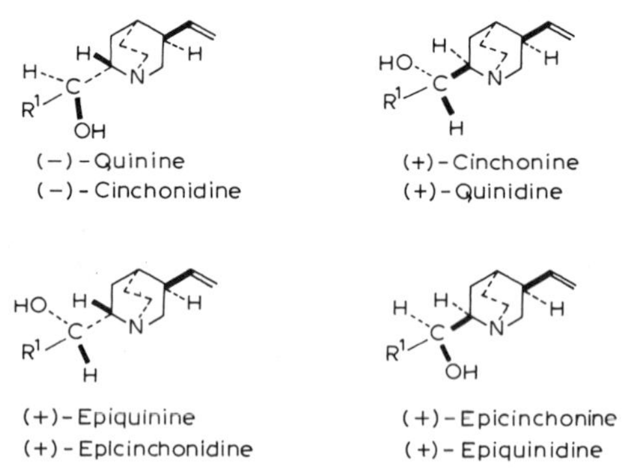

(R = 4-quinolyl or 6-methoxy-4-quinolyl)

The various stereoisomerides are interconvertible by the action of alkali; quinine, for example, affords a mixture of quinine, epiquinine, quinidine and epiquinidine (*Rabe et al.*, Ann., 1932, **492**, 258; J. pr. Chem., 1939, **154**, 66).

The interconversion of cinchonine and cinchonidine into dihydrocinchonamine (*Ochiai et al., loc. cit.*) demonstrates the stereochemical relationship between the quinoline and indole bases, but since both quinolines gave the same dihydrocinchonamine, no definite conclusion could be reached concerning the stereochemistry at $C_{(8)}$. Subsequently, *Y. K. Sawa* and *H. Matsumura* (Tetrahedron, 1970, **26**, 2919) synthesised dihydrocinchonamine from quinine so establishing the configuration of this centre. Cinchonamine is represented therefore by structure XXXII:

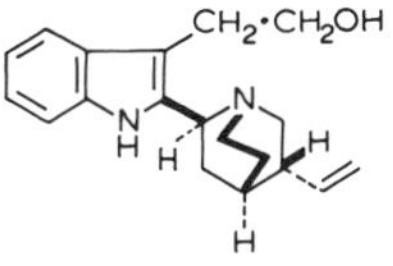

Cinchonamine

(XXXII)

(v) Physical properties of cinchona bases

The physical properties of cinchona bases and their derivatives are summarised in Table 4.

(vi) Synthetical studies in the cinchona series

The structures of the cinchona alkaloids have represented a challenge to organic chemists for nearly a century. Early work centred upon the synthesis of the degradation products of the alkaloids, some of these studies have already been mentioned. A further example is provided by the partial synthesis of dihydrocinchonine (VI, $-CH{=}CH_2$ replaced by Et). *P. Rabe* and *K. Kindler* (Ber., 1918, **51**, 1360) condensed ethyl cinchoninate (XXXIII) with the ethyl ester of *N*-benzoylhomocincholoipon (XXXIV), which had been been obtained earlier by oxidation of *N*-benzoyldihydrocinchotoxine (*A. Kaufmann et al., ibid.*, 1916, **49**, 2302). De-ethoxycarbonylation and hydrolysis of the product afforded **dihydrocinchotoxine** (XXXV):

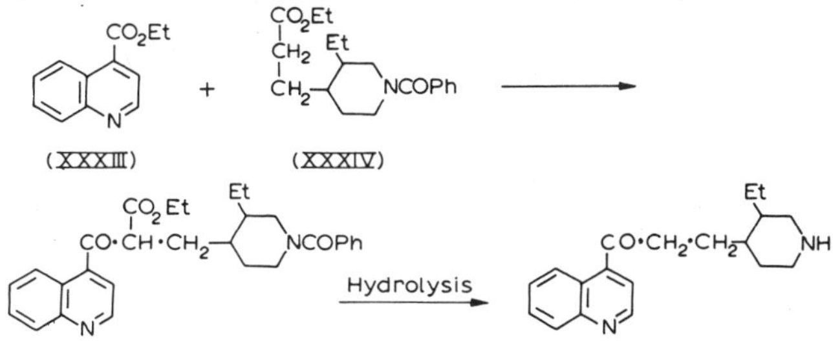

(XXXIII) (XXXIV)

Hydrolysis

Dihydrocinchotoxine

(XXXV)

TABLE 4

CINCHONA ALKALOIDS AND RELATED COMPOUNDS

Base	*M.p. (°C)*	$[\alpha]_D^{\circ}$	*Hydrochloride* $B \cdot 2HCl$, *m.p. (°C)*	*Sulphate* $B_2 \cdot H_2SO_4$, *m.p. (°C)*	*"Thalleioquin" reaction*[a]	*Fluorescence in dilute sulphuric acid*
Cinchonine	264	+263.7[b]	217–218 (mono)	200	−	−
Cinchonidine	204.5	−178[b]	242 (mono)	205	−	−
Quinine	174.5–175	−284.5[b]	158–160 (mono)	205 (dihydrate)	+	+
Quinidine	173.5	+334.2[b]	258–259 (mono)		+	+
Epiquinine	oil	+43.3[c]	196 (decomp.)			
Epiquinidine	113	+102.4[c]	195–196 (decomp.)			
Dihydroquinine	173.5	−235.7[b]	206–208 (mono)		+	+
Dihydroquinidine	169.5	+299[b]	273–274 (mono)		+	+
Dihydrocinchonine	267	+225.8[b]	220–221 (mono)	195		
Dihydrocinchonidine	232	−95.8[c]	202–203 (mono)			
Cupreine	198	−175.5[c]		257	+	
Cinchonamine	194	+121.1[c]	208[d]			
Quinamine	185–186	+104.5[c]	166–167 (mono)	175–176[e]		
Cinchoninone	126–127	+71 to +76	245–247			
Quininone	107–108	+73.8	210–212			
Cinchotenine	197	+209[f]	213–214[g]			
Quitenine	286 (decomp.)	−298[f]			+	+
Quinotoxine	oil	+38.6[hh]	179–180 (mono)			
Cinchotoxine	58–59	+49.6				
Meroquinene	223–224 (decomp.)	+27.5[i]	165[j]			
Cincholoiponic acid[k]	221–222	+30.1[i]				
Loiponic acid	259 (decomp.)					
Cinchenine	123–125					
Quinenine	81–82				+	
Apocinchenine	209–210					
Apoquinenine	246					
Apocinchonine	216–218	+150				
Apoquinine	184	−214.8				
α-Isocinchonine	130.4	+53.1				
β-Isocinchonine	130.7	−61.6				
Apoquinamine	115–117	−32.9[b]				
Quinamicine	203–205	−17.5[l]	275–276[d]			
Conquinamine	121	+204.6[c]				

[a] An acid solution of the alkaloid is treated with bromine water, followed by ammonia in excess; a deep green colour is produced. [b] In Ṅ/10

Dihydrocinchotoxine on treatment with sodium hypobromite yields dihydrocinchoninone (*Rabe*, Z. angew. Chem., 1913, **26**, 543; *cf. idem*, Ber., 1911, **44**, 2088). Reduction of dihydrocinchoninone with aluminium and alkali, or sodium isopropoxide in toluene (*R. B. Woodward et al.*, J. Amer. chem. Soc., 1945, **67**, 1425) affords a mixture of four racemic bases from which (±)-**dihydrocinchonine** and (±)-**dihydrocinchonidine** have been isolated. *E. Koenigs* and *W. Ottmann* (Ber., 1921, **54**, 1343) synthesised (±)-homocincholoipon (XXXVI) from *β*-collidine, available synthetically (*L. Ruzicka*, Helv., 1919, **2**, 338), by the following route:

Me, Et, N — $CCl_3{\cdot}CHO$ → $CH_2{\cdot}CH(OH){\cdot}CCl_3$, Et, N — NaOH → $CH{=}CH{\cdot}CO_2H$, Et, N — 8H → $CH_2{\cdot}CH_2{\cdot}CO_2H$, Et, NH

β-Collidine (XXXVI)

The product was a mixture of (±)-*cis*- and (±)-*trans*-homocincholoipon, which were separated. The ethyl ester of the former was resolved with (+)-tartaric acid by *Rabe et al.* (Ber., 1931, **64**, 2487); this resolution completes the synthesis of (+)-dihydrocinchonine.

Dihydroquinine has been synthesised by the same route as employed for dihydrocinchonine (*Rabe* and *Kindler*, *ibid.*, 1919, **52**, 1842); this synthesis was later extended to the synthesis of (−)-dihydroquinine by starting with (+)-homocincholoipon ethyl ester. **Dihydroquinidine** was also formed (*Rabe et al., loc. cit.*; Ann., 1932, **492**, 242; Ber., 1939, **72**, 263). The total synthesis of **quinine** was achieved by *Woodward* and *W. von E. Doering* (J. Amer. chem. Soc., 1944, **66**, 849; 1945, **67**, 860). *M. Prostenik* and *Prelog* (Helv., 1943, **26**, 1965) had achieved a partial synthesis of **quinotoxine** by condensation of ethyl quininate with *N*-benzoylhomomeroquinene ethyl ester, the latter being derived by Beckmann rearrangement of *N*-benzoylcinchotoxine oxime. *Woodward* and *Doering (loc. cit.)* synthesised homomeroquinene (XXXVIII), and thereby completed the total synthesis of quinine, by the following route, shown in outline. The starting point was 7-hydroxyisoquinoline, obtained from 3-hydroxybenzaldehyde and aminoacetal *via* a *Pomeranz–Fritsch* synthesis:

HO, N — CH_2O / piperidine → HO, N, $CH_2NC_5H_{10}$ — 2 H → HO, N, Me — 4 H / AcOH →

HO, NAc, Me — 6 H / Ni → HO, NAc, Me (Mixture of isomers) — CrO_3 → O, NAc, Me *cis*-form (XXXVII a) + O, NAc, Me *trans*-form (XXXVII b)

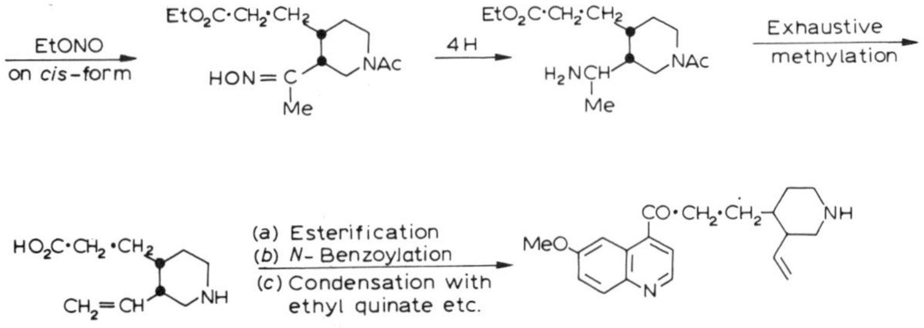

EtONO on *cis*-form → $EtO_2C{\cdot}CH_2{\cdot}CH_2$ … HON=C(Me) … NAc → 4H → $EtO_2C{\cdot}CH_2{\cdot}CH_2$ … $H_2NCH(Me)$ … NAc → Exhaustive methylation →

$HO_2C{\cdot}CH_2{\cdot}CH_2$ … $CH_2{=}CH$ … NH

(±)-Homomeroquinene

(XXXVIII)

(a) Esterification
(b) *N*-Benzoylation
(c) Condensation with ethyl quinate etc.

MeO … $CO{\cdot}CH_2{\cdot}CH_2$ … NH

(±)-Quinotoxine

(XXXIX) ≡ (XIV)

(+)-Quinotoxine was obtained by resolving the (±)-form with (+)-tartaric acid. The two oxodecahydroisoquinolines (XXXVIIa) and (XXXVIIb) were separated by taking advantage of the fact that the *cis*-ketone forms a crystalline hydrate. Thenceforward the synthesis was continued with this isomer. Quinotoxine had already been converted into quinine by the action of sodium hypobromite, following by reduction (*Rabe* and *Kindler*, Ber., 1918, **51**, 466).

More recently a stereoselective total synthesis of quinine has been accomplished (*M. Uskoković, J. Gutzwiller* and *T. Henderson*, J. Amer. chem. Soc., 1970, **92**, 203, 204). This commences with a four step preparation of (±)-*N*-benzoylmeroquinene (XL, R = H). The natural 3(*R*),4(*S*)-enantiomer is depicted. The methyl ester (XL, R = Me) of the racemate was then reacted with the lithium salt of 6-methoxylepidine (XLI) to give the ketone (XLII):

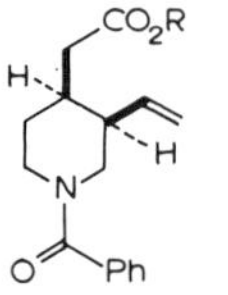

CO_2R H H N O Ph

(XL)

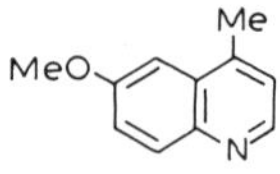

Me MeO N

(XLI)

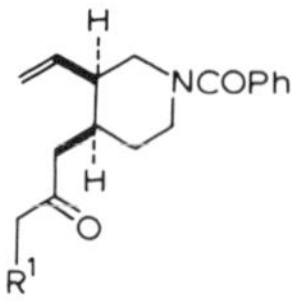

H NCOPh H O R^1

(XLII)

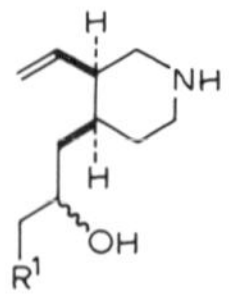

H NH H OH R^1

(XLIII)

(R^1 = 6-methoxylepidyl)

Reduction of this product with diisobutylaluminium hydride gave the alcohol (XLIII) as a 3:2 mixture of $C_{(8)}$ epimers. Resolution with dibenzoyl-(+)-tartaric acid afforded a 3:2 mixture of these epimers with the natural 3(*R*),4(*S*) configuration.

Cyclization to a mixture of deoxyquinine and deoxyquinidine (XLIV) was effected by heating the epimeric alcohols with acetic acid in benzene and finally base catalysed hydroxylation with molecular oxygen gave predominantly the *erythro* products quinine (XLV) and quinidine (XLVI), contaminated with only small quantities of the *threo* products epiquinine and epiquinidine:

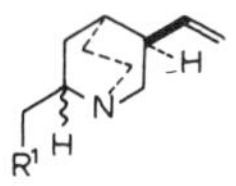

(XLIV) 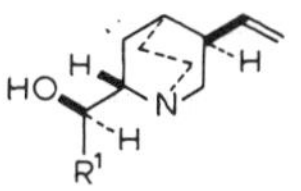 (XLV) 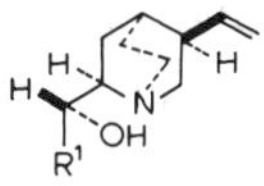 (XLVI)

Separation of the individual components was achieved by a combination of crystallization and chromatography. (See also *M. Gates, B. Sagavanam* and *W. L. Schreiber*, J. Amer. chem. Soc., 1970, **92**, 205). An alternative and improved preparation of *N*-benzoylmeroquinene has subsequently been described (*M. R. Uskoković et al., ibid.*, 1971, **93**, 5902) and another synthesis of quinine and quinidine from this same starting material is reported (*G. Grethe et al., ibid.*, 1971, **93**, 5904). In this (XL, R = H) was converted into the correctly functionalized quinuclidine derivative (XLVII) prior to combination with 6-methoxy-4-quinolyllithium. From this reaction a mixture of bases containing 20% quinine and 23% quinidine was obtained directly (see also *Grethe et al.*, Helv., 1973, **56**, 1485; *J. Gutzwiller* and *Uskoković, ibid.*, p. 1494).

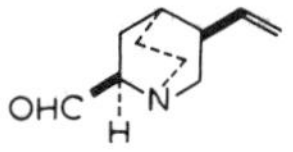

(XLVII)

Stereoselective synthesis of 9-epiquinine and 9-epiquinidine are described by the same group (Helv., 1972, **55**, 1044), and yet another synthesis of quinine and quinidine is reported by an American group (*E. C. Taylor* and *S. F. Martin*, J. Amer. chem. Soc., 1972, **94**, 6218) by a route which differs from the preceding methods in the manner by which the necessary intermediates deoxyquinine and deoxyquinidines (XLIV) are prepared. Details of these approaches to quinine and its stereomers and reviewed by *Uskoković* and

Grethe in "The Alkaloids", ed. *R. H. F. Manske*, Academic Press, London, Vol. XIV, 1973, p. 183.

(vii) Biosynthesis of cinchona alkaloids

The quinoline alkaloids of *Cinchona* are found principally in the bark, only trace amounts are present in the leaves, but for the co-occurring indole alkaloids the reverse is true and structures such as cinchonamine (XIX) etc. are most abundant in the leaves. This naturally leads to the view that the quinoline alkaloids are derived biosynthetically from indolic precursors (*R. Goutarel et al.*, Helv., 1950, **33**, 150). Support for this conclusion is provided by the specific incorporation of tryptophan into quinine (*N. Kowanko* and *E. Leete*, J. Amer. chem. Soc., 1962, **84,** 4919). Furthermore, the relationship of the C_9 unit of quinine (XLVIII) to that of the C_{9-10} unit of the *Corynanthé* alkaloids, shown by thickened lines in XLIX, is best interpreted if the quinoline bases are formed by the late modification of corynantheal or a related structure. The co-occurrence of cinchonamine with the quinolines is, itself, extremely suggestive of a common biosynthetic pathway.

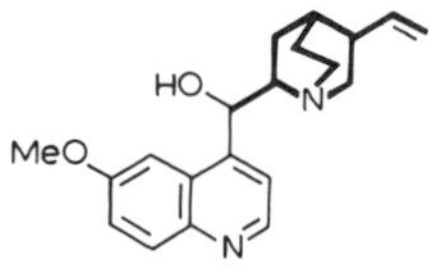

(XLVIII) 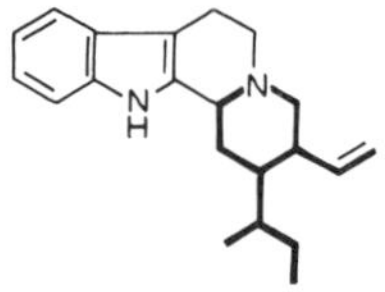 (XLIX)

The precursor of the C_{9-10} unit in the indole series is known to be geraniol, in a sequence involving loganin (*A. R. Battersby et al.*, Chem. Comm., 1966, 890; *idem, ibid.*, 1968, 133; *P. Loew* and *D. Arigoni, ibid.*, p. 137), secologanin and vinicoside (*Battersby, A. R. Burnett* and *P. G. Parsons*, Chem. Comm., 1968, 1280, 1282; J. chem. Soc., C, 1969, 1187, 1193).

Tracer experiments (*E. Leete* and *J. N. Wemple*, J. Amer. chem. Soc., 1966, **88**, 4743; 1969, **91**, 2698; *Battersby et al.*, Chem. Comm., 1966, 888) show that geraniol is incorporated in a predictable manner into quinine, and, as further direct evidence of a common biosynthetic route to both types of molecules, both loganin (*Battersby* and *E. S. Hall, ibid.*, 1970, 194) and sweroside (*H. Inouye, S. Ueda* and *Y. Takeda*, Tetrahedron Letters, 1969, 407) are known to be incorporated specifically into quinine in *Cinchona* species.

The biosynthesis of quinine and related alkaloids may thus be summarized by the following scheme:

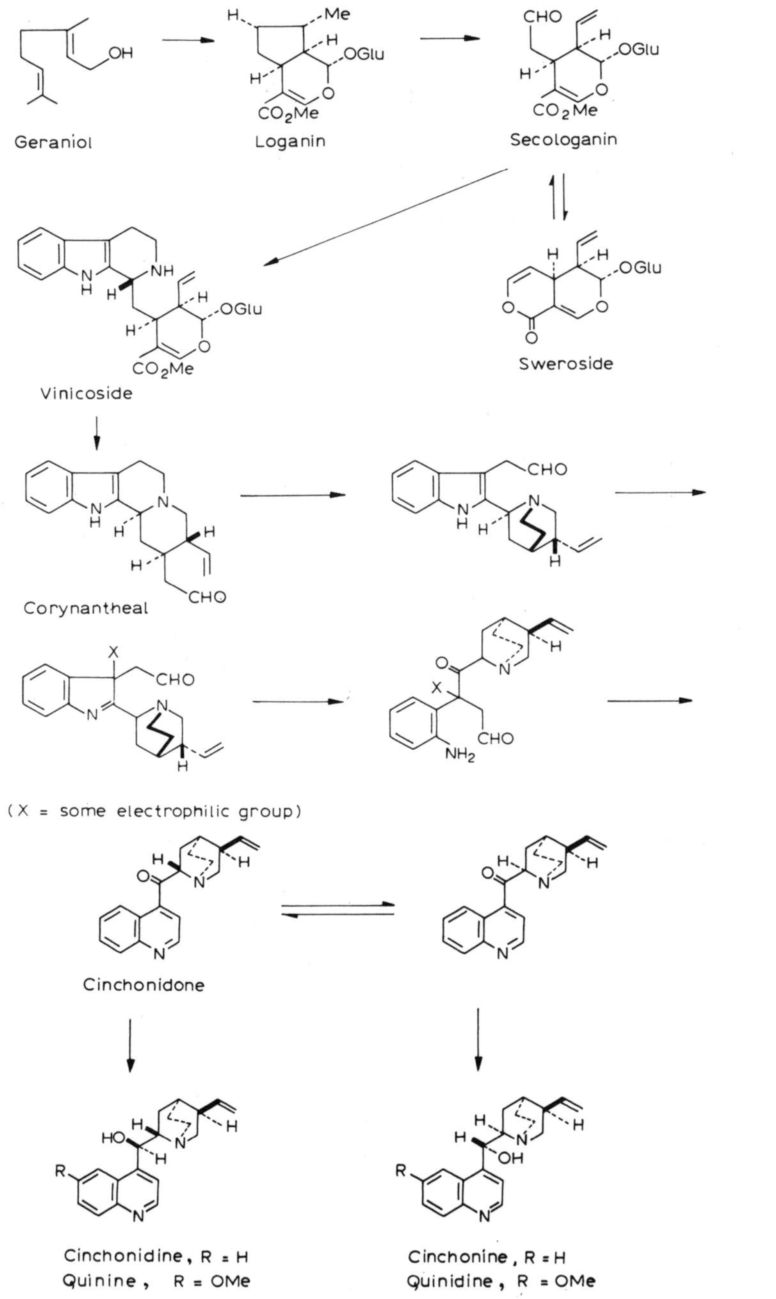

Geraniol

Loganin

Secologanin

Vinicoside

Sweroside

Corynantheal

(X = some electrophilic group)

Cinchonidone

Cinchonidine, R = H
Quinine, R = OMe

Cinchonine, R = H
Quinidine, R = OMe

It has been shown that corynantheal serves as an efficient precursor of quinine, cinchonidine and cinchonine (*Battersby* and *R. J. Parry*, Chem. Comm., 1971, 30) and, through dilution analyses (*idem*, *ibid*., p. 31), the presence of cinchonidone was proven in *Cinchona legeriana*. Moreover, when [11-^3H]-cinchonidone was administered to *C. ledgeriana* shoots activity was incorporated efficiently into cinchonidine and cinchonine, indicating that the methoxyl function is introduced at a very late stage in the biosynthetic sequence. A low incorporation into quinoline was also observed.

(b) Camptothecin

The plant *Camptothecea accuminata* Decne family Nyssaceae, elaborates the alkaloid **camptothecin** ($C_{20}H_{16}O_4N_2$), m.p. 264–267°, $[\alpha]_D^{25}$ +31.3°. Spectroscopic analysis, including X-ray crystallography, indicates the pentacyclic structure L* for this compound (*M. E. Wall et al*., J. Amer. chem. Soc., 1966, **88**, 3888):

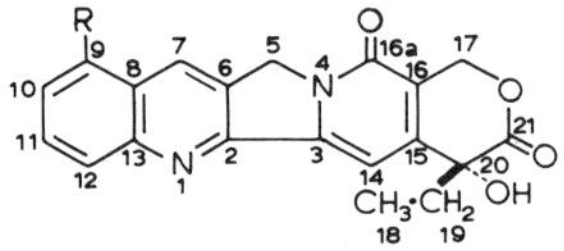

(L, R = H) Camptothecin
(LI, R = OH)
(LII, R = OMe)

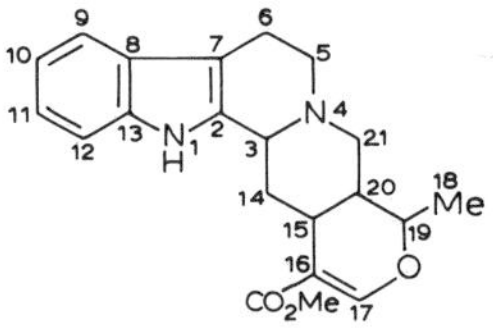

Ajmalicine
(LIII)

Camptothecin is also present in *Mappia foetida* Miers (Fam. Olacaceae), where it is accompanied by minor amounts of 9-*methoxycamptothecin* (LII), m.p. 254–255° (*T. R. Govindachari* and *N. Viswanathan*, Indian J. Chem., 1972, **10**, 453). The last compound, together with 9-*hydroxycamptothecin* (LI) m.p. 268–270°, occurs in *C. accuminata* (*M. C. Wani* and *Wall*, J. org. Chem., 1969, **34**, 1364).

* The numbering system shown for camptothecin emphasises its probable biogenetic relationship to indole alkaloids such as ajmalicine (LIII).

The unique structure of camptothecin allied to the early, but as yet unconfirmed, reports of useful antitumour and antileukaemic activity* has provided a powerful stimulus to efforts directed at the synthesis of the molecule (*M. Shamma* and *N. Ludvik*, Tetrahedron, 1969, **25**, 2275; *J. A. Kepler et al.*, J. org. Chem., 1969, **34**, 3853; *Shamma* and *L. Novak*, Coll. Czech. chem. Comm., 1970, **35**, 3280; *T. Kametani et al.*, Chem. and Ind., 1970, 1323; Tetrahedron, 1970, **26**, 5753; *E. Winterfeldt* and *H. Rudunz*, J. chem. Soc., D, 1971, 374; *T. K. Liao, W. H. Nyberg* and *C. C. Cheng*, J. heterocyclic Chem., 1971, **8**, 373; *Wani et al.*, J. chem. Soc., D, 1970, 404).

Eventually these labours have been rewarded and a number of successful routes to racemic camptothecin have been reported (*G. Stork* and *A. G. Schultz*, J. Amer. chem. Soc., 1971, **93**, 4074; *R. Volkman et al., ibid.*, 1971, **93**, 5576; *T. Sugasawa, T. Toyoda* and *K. Sasakura*, Tetrahedron Letters, 1972, 5109; Chem. Pharm. Bull., Tokyo, 1974, **22**, 771; *C. Tang* and *H. Rapoport*, J. Amer. chem. Soc., 1972, **94**, 8615; *A. I. Meyer et al.*, J. org. Chem., 1973, **38**, 1974; *A. S. Kende et al.*, Tetrahedron Letters, 1973, 1307; *M. Shamma et al.*, Tetrahedron, 1973, **29**, 1949; *E. J. Carey et al.*, J. org. Chem., 1975, **40**, 2140).

It seems probable that camptothecin is biogenetically derived from indole precursors which stem from the combination of tryptamine with a C_{10} unit of terpenoid origin (*Shamma*, Experientia, 1968, **24**, 107); an outline biosynthetic route was suggested:

Tryptamine + (monoterpenoid unit of appropriate functionality) ----→

----→ Camptothecin

In the absence of biochemical experiments, the justification for this scheme rested primarily upon circumstantial structural evidence and the demonstration that substituted indoles can be auto-oxidized in high yield to quinolones (*Winterfeldt*, Ann., 1971, **745**, 23). From these considerations a "biogenetically modelled" total synthesis of camptothecin has been developed (*M. Boch et al.*, Ber., 1972, **105**, 2126). The tetrahydro-β-carboline derivative

* The high toxicity of this compound probably rules out its application to the human situation.

LIV was reacted with the anion of di-*tert*-butyl malonate to give LV. Autooxidation and rearrangement of LV in dimethylformamide solution containing sodium hydride gave the 4-quinolone, LVI, in 75% yield. Reaction of LVI with thionyl chloride, followed by dehalogenation of the product LVII and reduction with di-isobutylaluminium hydride at −70° gave the tetracyclic alcohol LVIII, cyclisation of this with trifluoroacetic acid yielded de-ethyldesoxycamptothecin (LIX), which when alkylated with ethyl iodide and sodium iodide and reacted with oxygen, in the presence of cupric ions, formed (±)-camptothecin:

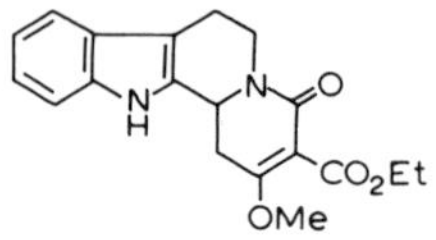

(LIV) 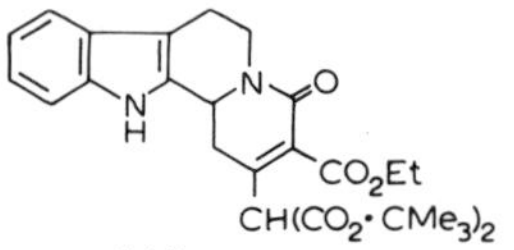 (LV)

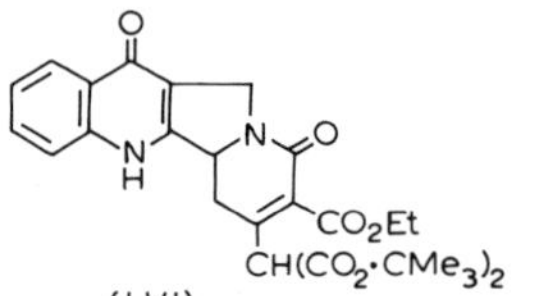

(LVI) 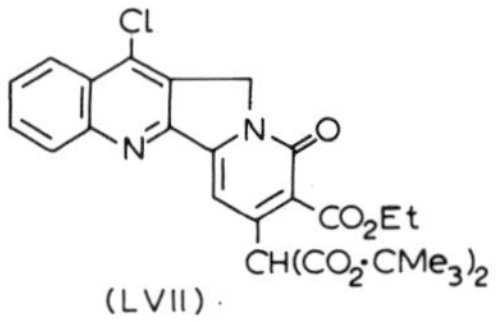 (LVII)

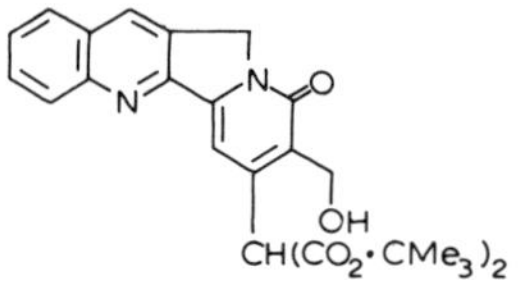

(LVIII) 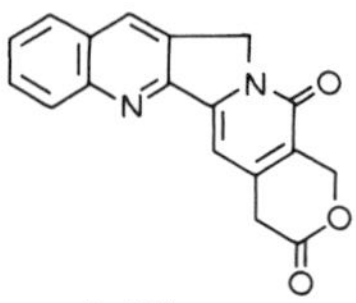 (LIX)

Recently it has been established that the biosynthesis of camptothecin proceeds *via* isovincoside lactam (strictosamide) (*C. R. Hutchinson et al.*, J. Amer. chem. Soc., 1974, **96**, 5609), thus confirming earlier speculations.

Mappicine, m.p. 251–252°, from *Mappia foetida* Miers is clearly related to camptothecin and this inter-relationship has been confirmed by synthesis (*T. R. Govindachari, K. R. Ravindranath* and *N. Viswanathan*, J. chem. Soc., Perkin I, 1974, 1215; *T. Kametani et al.*, Heterocycles, 1975, **3**, 167).

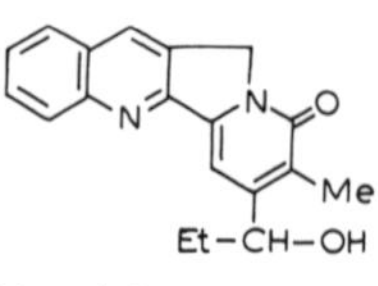

Mappicine

(c) Calycanthine, calycanthidine, folicanthine chimonanthine and mesochimonanthine

(i) Calycanthine

The alkaloid **calycanthine** ($C_{22}H_{26}N_4$), $[\alpha]_D^{EtOH}$ +684°, B·HCl, m.p. 216–217°, B·2HCl, m.p. 221–222° is the major alkaloid of the seeds of Calycanthaceous plants* (*E. Eccles*, Proc. Amer. pharm. Assoc., 1888, **84**, 382; Pharm. J., 1888, **18**, 822; *E. Späth* and *W. Stroh*, Ber., 1925, **58**, 2131; *R. H. F. Manske* and *L. Marion*, Canad. J. Res., 1939, **B17**, 293; *Manske*, J. Amer. chem. Soc., 1929, **51**, 1836).

Normally calycanthine crystallises as the *hydrate*, m.p. 216–218°, but may also be obtained in the anhydrous form which melts at 245°. The alkaloid is dibasic and contains two tertiary methylamino and two secondary amino groups. Destructive distillation over zinc or sodium hydroxide, or dehydrogenation with selenium, gives rise to a number of compounds of which skatole, β-ethylindole, lepidine, quinoline, *N*-methyltryptamine, β-carboline and the base calycanine ($C_{16}H_{10}N_2$) have been characterized (*Marion* and *Manske*, Canad. J. Res., 1938, **B16**, 432; *G. Barger*, *J. Madinaveitia* and *P. Streuli*, J. chem. Soc., 1939, 510). The structure of **calycanine** (*Sir Robert Robinson* and *H. J. Teuber*, Chem. Ind., 1954, 783) was proven by synthesis (*R. B. Woodward et al.*, Proc. chem. Soc., 1960, 76; *K. W. Gopinath*, *Govindachari* and *S. Rajappa*, Tetrahedron, 1960, **8**, 291; *T. Kobayashi* and *R. Kikumoto*, *ibid.*, 1962, **18**, 813; *V. M. Clark* and *A. Cox*, *ibid.*, 1966, **22**, 3421):

N N NMe N

Calycanine (LX)

Oxidation of calycanthine with silver acetate affords a further base which was shown to be LX (*E. Späth et al.*, Monatsh. Chem., 1948, **79**, 11; *K. Eiter*, *ibid.*, 1948, **79**, 17; *Eiter* and *M. Nagy*, *ibid.*, 1949, **80**, 607).

The constitution of the parent alkaloid, a point of controversy for many years, was finally settled by X-ray crystallographic analysis of the dihydrobromide (*T. A. Hamor et al.*, Proc. chem. Soc., 1960, 78; *Hamor* and *J. M. Robertson*, J. chem. Soc., 1962, 194) and by chemical methods (*Woodward et al., loc. cit.*). Calycanthine is a compact three dimensional unit in which the planes of the benzene rings are inclined at 61° to each other. The molecule has been synthesised (see below):

---

* The family Calycanthacea, the taxonomic status of which is uncertain, comprises only two genera *Calycanthus* L. (in N. America) and *Chimonanthus* Lindl. (*Meratia* Loisel) (in Asia). Calycanthine also co-occurs with pyrrolizidine alkaloids in *Bhesa archboldiana* (Merrill and Perry) Ding. Hon., family Celastraceae (*C. C. Culvenor et al.*, Austral. J. Chem., 1970, **23**, 1279).

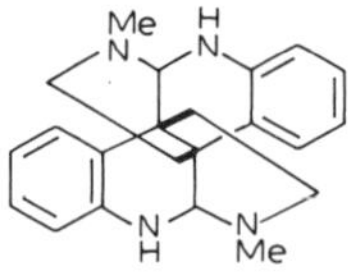

Calycanthine

(ii) Folicanthine and chimonanthine

Folicanthine, $C_{24}H_{30}N_4$, co-occurs with calycanthine in *Calycanthus floridus* L. (*K. Eiter* and *O. Svierak*, Monatsh. Chem., 1951, **82**, 186) and *C. occidentalis* Hook and Arn. (*H. F. Hodson* and *G. F. Smith*, J. chem. Soc., 1957, 1877).

It is a saturated tertiary base, $[\alpha]_D^{MeOH}$ −364°, m.p. 118–119°, which forms a *dipicrate*, m.p. 179°, and a di-iodide, m.p. 220°. Chemical (*Hodson, R. Robinson* and *Smith*, Proc. chem. Soc., 1961, 465; *Hodson et al.*, Chem. Ind., 1958, 1551), X-ray crystallographic (*I. J. Grant et al.*, Proc. chem. Soc., 1962, 148) and synthetic studies (*T. Hino* and *S. Yamada*, Tetrahedron Letters, 1963, 1757) confirm structure LXI for folicanthine, whereas **chimonanthine**, m.p. 188–189°, $[\alpha]_D$ −329°, from the leaves of *Chimonanthus fragrans* Lindl. (*Meratia praecox* Rehd et Wils) *(Hodson et al.*, 1961, *loc. cit.)* is LXII.

Mesochimonanthine (LXIV), m.p. 198–200°, accompanies chimonanthine in this plant (*E. S. Hall, F. McCapra* and *A. I. Scott*, Tetrahedron, 1967, **23**, 4131). The structures LXII, LXIII and LXIV are supported by synthesis (*Hall et al., loc. cit.; J. B. Hendrickson, R. Göschke* and *R. Rees*, Tetrahedron, 1964, **20**, 565). The absolute configurations of chimonanthine and folicanthine follow from circular dichroism measurements (*S. F. Mason* and *G. W. Vane*, J. chem. Soc., B, 1966, 370):

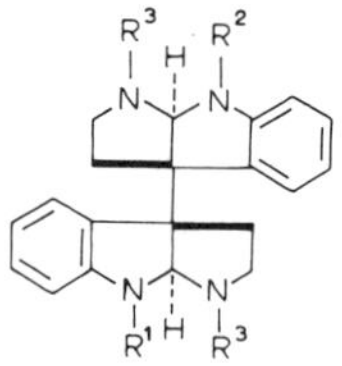

(LXI, $R^1 = R^2 = R^3 = Me$) Folicanthine
(LXII, $R^1 = R^2 = H$; $R^3 = Me$) Chimonanthine
(LXIII, $R^1 = H$; $R^2 = R^3 = Me$) Calycanthidine

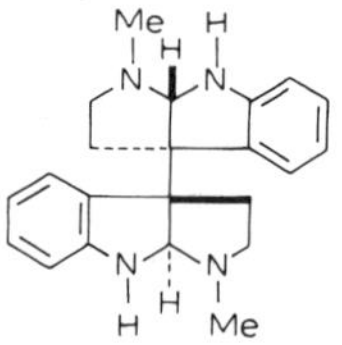

Mesochimonanthine (LXIV)

(iii) Calycanthidine

Calycanthidine, $C_{23}H_{28}N_4$, m.p. 142°, $[\alpha]_D^{20}$ −285° (MeOH), is also found in *Calycanthus* species (*G. Barger, A. Jacob* and *Madinaveitia*, Rec. Trav. chim., 1938, **57,** 548). *N*-Methylation of calycanthidine yields folicanthine, consequently structure LXIII is allocated to this molecule.

(iv) Biosynthesis of the calycanthaceous alkaloids

Feeding experiments have shown that [2′-^{14}C]-tryptophan is a precursor of calycanthine in *Calycanthus floridus* and also of folicanthine and chimonanthine in *Chimonanthus fragrans* (*H. R. Schütte* and *B. Maier*, Arch. Pharm., 1965, **298,** 459; *D. G. O'Donovan* and *M. F. Keogh*, J. chem. Soc., 1966, 1570).

The probable biosynthetic pathway from tryptophan to these alkaloids is outlined below, where the asterisk indicates the position of the radioactive label (see *K. Mothes* and *Schütte* in "Biosynthese der Alkaloide", *loc. cit.*, pp. 510–550 and references cited therein).

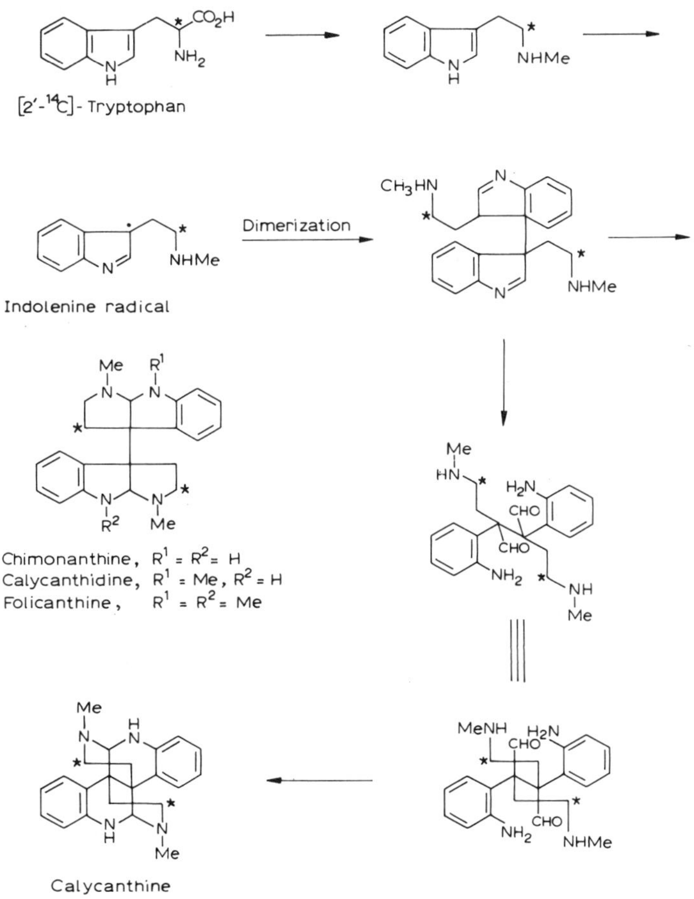

[2′-^{14}C]-Tryptophan

Dimerization

Indolenine radical

Chimonanthine, $R^1 = R^2 = H$
Calycanthidine, $R^1 = Me$, $R^2 = H$
Folicanthine, $R^1 = R^2 = Me$

Calycanthine

(d) 2-Quinolone alkaloids of Melodinus scandens

(i) Meloscine and epimeloscine

The genus *Melodinus* of the Apocynaceae family elaborates a number of aspidospermine type alkaloids (see *B. Gilbert* in "The Alkaloids", ed. *R. H. F. Manske*, Academic Press, London, 1968, Vol. XI, Chapter 9, p. 242). In addition to indolic derivatives, a number of alkaloids containing the 3,4-dihydro-2-quinolone ring system are known, thus **meloscine** (LXV) and its epimer (LXVI) are found in the plant *Melodinus scandens* Forst. (*K. Bernauer, G. Englert* and *W. Vetter*, Experientia, 1965, **21**, 374; Helv., 1969, **52**, 1886).

The structural assignments LXV and LXVI were deduced by a combination of spectroscopic methods, in particular, ^{1}H-n.m.r. and mass spectrometry. **Epimeloscine** isomerizes to meloscine in the presence of strong base and since models indicate the *cis*-BC fused structure LXV to be the more stable, this representation corresponds to meloscine:

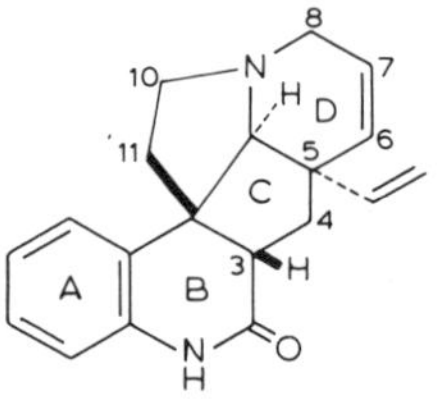

Meloscine
(LXV)

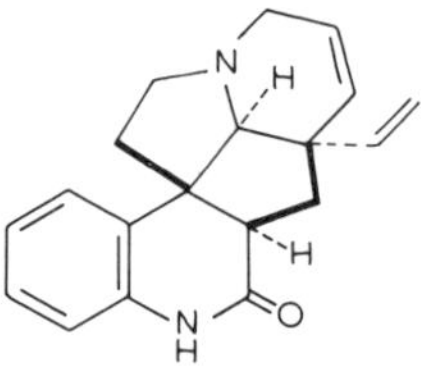

Epimeloscine
(LXVI)

The stereochemical assignments indicated in the formulae were deduced by comparison of the o.r.d. curves of tetrahydromeloscine ($-CH_2 \cdot CH_3$ instead of $-CH{=}CH_2$ in LXV) with those of Aspidospermidine alkaloids of known structure (see *W. Klyne et al., ibid.*, 1966, **49**, 833) and by an X-ray crystallographic analysis of meloscine-(N_{β})-methobromide (*W. E. Oberhänsli, ibid.*, 1969, **52**, 1905).

The $N_{(\beta)}$-oxide derivative of epimeloscine has proved to be a natural product in its own right (*H. Mehri, M. Plat* and *P. Potier*, Ann. Pharm. Fr., 1971, **29**, 291).

(ii) Scandine and meloscandonine

A further two alkaloids containing a quinolone system have been isolated from *M. scandens* (Table 5), **scandine*** (LXVII), which contains a methoxycarbonyl group and on treatment with acid hydrolyses to meloscine, and **meloscandonine** (LXVIII) (*M. Plat et al.*, Tetrahedron Letters, 1970, 3395· B. *K. Bernauer et al.*, Helv., 1969, **52**, 1886):

* Previously called "Alkaloid-X".

Scandine
(LXVII)

Meloscandonine
(LXVIII)

Biosynthesis. It seems likely that the meloscine alkaloids derive from a preformed tabersonine-like base, *e.g.* LXIX, which undergoes an oxidative ring expansion reaction to form scandine; the remaining alkaloids could then arise from scandine in obvious ways.

(LXIX) → → Scandine

TABLE 5

PHYSICAL DATA OF *MELODINUS* ALKALOIDS

Alkaloid	$[\alpha]_D^{25°}$	*M.p. (°C)*	λ_{max} *(log) nm*
Meloscine	+133.8	188–190	253 (4.04)
Epimeloscine	+238	amorph. (*picrate* m.p. 210–215)	253 (4.01)
Meloscandonine	+72($CHCl_3$)*	–	254 (3.92)
Scandine	+254	188–192	257.5 (3.94)

*$[\alpha]_{578}^{20°}$.

(iii) Perloline and perlolidine

Perloline is an alkaloid occurring in several species of grass, one of the richer sources being perennial rye grass *Lolium perenne* L. It seems that hybrd species of grasses may be developed in which the concentration of this alkaloid is reduced, a need prompted by reports that it causes digestive inhibition in ruminants (*R. C. Buckner et al.*, Crop Sci., 1973, **13**, 666). The alkaloid $C_{20}H_{18}N_2O_4$, *methyl ether*, m.p. 248–252°, was first isolated in New Zealand during 1940 (*J. A. D. Jeffreys*, J. chem. Soc., 1964, 4504 and references cited). After studying the ultraviolet spectrum, which is pH dependent, *W. S. Metcalf* (New Zealand J. Sci. Technol., 1954, **35B**, 473) concluded that perloline contained a dimethoxyphenyl group which was lost on oxidative degradation to the simpler base **perlolidine** ($C_{12}H_8N_2O$), m.p. 338° (decomp.), also present in the grass, and that perlolidine was a mono- (or bis-)-aza-anthracene. However, the spectrum of perlolidine recorded in a neutral buffer contains a band at λ_{max} 349 nm which suggests (*G. Ferguson, Jeffreys* and *G. A. Sim*, J. chem. Soc., B, 1966, 454) an angular rather than linear assembly. X-Ray crystallographic analysis of perloline mercuri-chloride indicates this alkaloid to be the first example of a naturally occurring benzo[*c*-2,7]naphthyridine derivative. It has the structure:

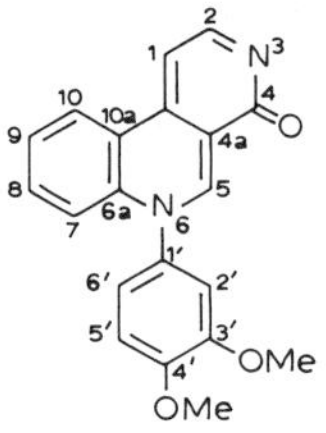

Perloline

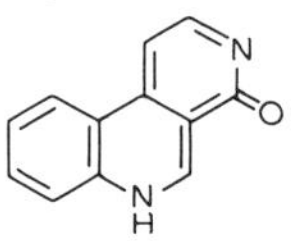

Perlolidine

The structure for perlolidine has been confirmed by synthesis (*M. A. Akhtar et al.*, J. chem. Soc., C, 1967, 859; *R. N. Seelye* and *D. W. Stanton*, Tetrahedron Letters, 1966, 2633).

From a structural point of view, it would seem that the biosynthesis of perloline requires the elaboration of tryptamine with some as yet unknown precursor or precursors.

(iv) Perhydroquinolines

Lobinaline, m.p. 108–110°, $[\alpha]_D^{26}$ +38°, was originally isolated from *Lobelia cardinalis* L. by *Manske* (Canad. J. Res., 1938, **163**, 445) and more recently by *Gupta* and *Spenser* (Canad. J. Chem., 1971, **49**, 384) from *L. inflata.* Related alkaloids, **Syphilobine-A**, $C_{28}H_{22}N_2O_2$, m.p. 192–194°, optically inactive, and **Syphilobine-F,** $C_{28}H_{26}N_2O_3$, m.p. 222–223°, $[\alpha]_D$ −1.6° (pyridine) occur in *L. syphilitica.* Syphilobine-A is a dimethoxy derivative of lobinaline (*R. Tschesche et al.*, Tetrahedron, 1964, **20**, 2885). Lobinaline is formed *in vivo* by the dimerisation of an α-phenacylpiperidine unit, the piperidine ring of which is derived from lysine and the PhC_2− side-chain from phenylalanine (*Gupta* and *Spenser, loc. cit.*).

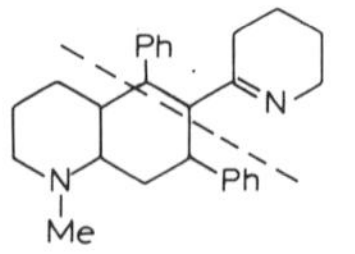

Lobinaline

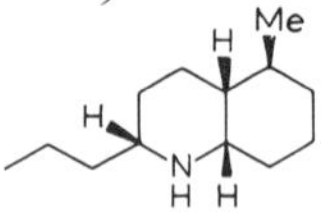

Pumiliotoxin -C[1]

Pumiliotoxin-C[1] is one of the toxic principals of the skin of the highly coloured Panamanian frog *Dendrobates pumilio.* Isolated as the *hydrochloride*, m.p. 230–240°, its structure was determined by X-ray crystallography (*J. N. Daly et al.*, Ann., 1969, **729,** 198). The racemate has been synthesised in three steps from cyclohexane-1,3-dione (*G. Habermehl* and *W. Kissing*, Ber., 1974, **107**, 2326).

Chapter 32

The Acridine Alkaloids

B. P. SWANN and A. McKILLOP

The acridine group of alkaloids (often also called the acridone alkaloids) has a short history. The first alkaloids in this family were isolated from trees of the Australian rain forests (order *Rutaceae*) (*G. K. Hughes et al.*, Nature, 1948, **162,** 223). They have since been found to be fairly widely distributed across the *Rutaceae* spp. and frequently occur together with furanoquinoline and β-carboline derivatives. All are polyoxygenated 9-acridones, and only a few examples are known in which these oxygen substituents are found in both benzene rings. An excellent summary of the early history of these alkaloids is available (*J. R. Price*, "Acridine Alkaloids", in "The Alkaloids", Vol. 2, Ed. *R. H. F. Manske* and *H. L. Holmes*, Academic Press, New York, 1952, p. 353) and recently a full review has appeared (*J. E. Saxton*, "The Acridine Alkaloids" in "The Chemistry of Heterocyclic Compounds", Ed. *A. Weissberger* and *E. C. Taylor*, Vol. 9, 2nd Edn., 1973, p. 379).

Considerable work on these alkaloids has been carried out in recent years, largely because of the pronounced antitumor activity shown by acronycine, a member of this group. The alkaloids are all crystalline, very weak bases, particularly those bearing an *N*-methyl substituent, and salt formation is frequently very difficult or impossible. Hydrochloride formation can often result in concomitant loss of the C-1 methoxyl group, and this reaction, which occurs readily on heating, gives the corresponding noracridone. Since nearly all common extraction procedures involve treatment with acid at some stage it is difficult to evaluate whether some of the noracridones which have been reported actually exist in the plant or are artifacts of the extraction procedure (see *inter alia: Price, loc. cit.*, p. 356; *J. Reisch et al.*, Pharmazie, 1971, **4,** 208).

Spectral characteristics

Ultraviolet spectra. Acridone alkaloids generally show two intense absorption bands at 380–420 nm (logε 3.5–4.0) and at 240–280 nm (logε 4.4–4.8) (*Reisch et al., ibid.*, p. 209; *R. D. Brown* and *F. N. Lahey*, Austral. J. sci. Res., 1950, **3**, 593, 615; *A. W. Sangster* and *K. L. Stuart*, Chem. Reviews, 1965, **65**, 69; *A. Albert*, "The Acridines", 2nd Edn., Edward Arnold, London, 1966, p. 184). In addition a band at 300 nm which is only an inflection in 9-acridone is greatly increased by hydroxyl or methoxyl substituents in the benzene rings (*Brown* and *Lahey, loc. cit.*). The long wave band undergoes a hypsochromic shift with alkoxyl groups in the 1-, 3-, or 4- positions. The spectra are pH-dependent and acidification generally causes a bathochromic shift in the main short wave band and a hypsochromic shift in the long wave band. The presence of a C-1 hydroxyl function and a C-3 methoxyl function gives a visual red shift to the spectra because of a different tautomeric form ("Nor-effect", *Brown* and *Lahey, loc. cit.*). This is generally of the order of 7–10 nm and results in a change of the alkaloid from yellow to red in colour:

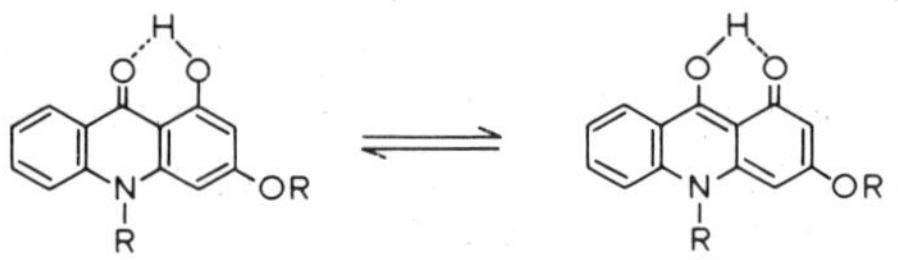

Fluorescence spectra. All of the acridones show intense fluorescence spectra which are very characteristic of the oxygenation pattern (*Reisch et al., loc. cit.*, p. 210).

Infrared spectra. The two bands that are useful for diagnostic purposes are the N–H stretch of *N*-demethylated acridones at 3,300–3,100 cm^{-1} (*T. R. Govindachari et al.*, Tetrahedron, 1966, **22**, 3245) and the strong intramolecular hydrogen bond of a C-1 hydroxyl function (*J. Hlubucek, E. Ritchie* and *W. C. Taylor*, Austral. J. Chem. 1970, **23**, 1881; *R. J. Gell, Hughes* and *Ritchie, ibid.*, 1955, **8**, 114). Details of other bands are summarised by *Reisch* (*Reisch et al., loc. cit.*, p. 212).

Mass spectra. The mass spectra of the acridone alkaloids are again very dependent on the oxygenation pattern and full details are available for most of the alkaloids (*J. H. Bowie et al.*, Austral. J. Chem., 1967, 20, 1179; *Reisch et al., loc. cit.*, p. 213).

Nuclear magnetic resonance spectra. The C-1 hydroxyl proton often occurs as low as 14–15δ and other characteristic shifts are seen depending on the oxygenation pattern (*Reisch et al., loc. cit.*, p. 212 and refs. therein). A Nuclear Overhauser Effect has been observed in acronycine (I) between the *N*-methyl group and H_a proton of the chromene ring (*Hlubucek, Ritchie* and *Taylor, loc. cit.*).

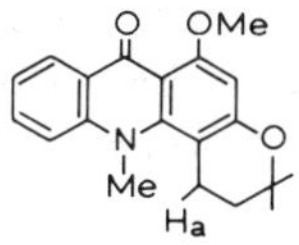

Acronycine
(I)

The acridone alkaloids may be conveniently divided into two major groups. The first of these comprises alkaloids which are oxygenated in ring A or ring B of the acridone system, and they may or may not also be methylated at nitrogen. The second group of compounds is distinguished by the presence of C_5 side chains in ring A. The properties of the compounds in each group are summarised in Tables 1 and 2.

Two dimeric acridone alkaloids, **atalanine,** m.p. 216.5–217.5°, and **ataline**, m.p. 209–210°, have recently been isolated (*A. W. Fraser* and *J. R. Lewis*, Chem. Comm., 1973, 615) and their structures assigned as II and III, respectively:

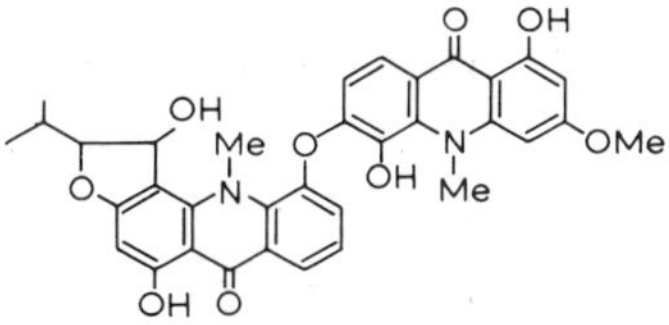

Atalanine
(II)

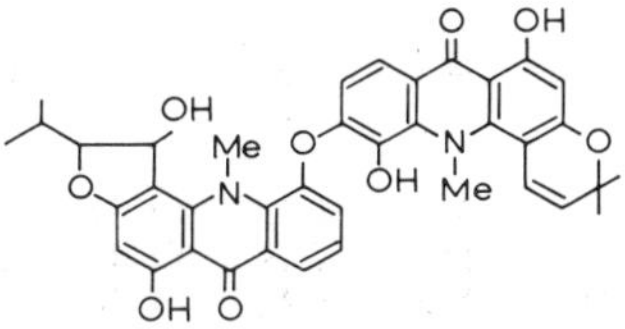

Ataline
(III)

Reactions of acridone alkaloids

As described earlier, when 1-methoxy-substituted acridone alkaloids are heated with hydrochloric acid, rapid demethylation of the ether occurs and the corresponding 1-hydroxy compounds are formed. The hydroxyl group in these alkaloids is strongly hydrogen bonded, and hence does not behave as a typical phenolic hydroxyl group. The noralkaloids are more highly coloured than the methyl ethers, they are even weaker bases than the parent acridones, and exhibit the properties of cryptophenols. Thus, they are insoluble in alkali and are frequently not reactive to diazomethane. They can, however, be methylated using potassium carbonate and methyl iodide, and give green colours with ferric chloride.

The highly oxygenated ring A in many of the alkaloids gives rise to a variety of unusual reactions. Oxidation with nitric acid of melicopine, melicopicine, melicopidine, evoxanthine, noracronycine and dihydroacronycine gives 1-*methyl*-4-*quinolone*-3-*carboxylic acid*, m.p. 294–296° (*Hughes* and

K. G. Neill, Austral. J. sci. Research, 1949, **2,** 429; *L. J. Drummond* and *Lahey*, *ibid.*, 1949, **2,** 630). Under milder oxidation conditions, melicopicine can be converted into the two quinones IV and V (*W. D. Crow*, *ibid.*, 1949, **2,** 264):

TABLE 1

RINGS A AND B OXYGENATED ACRIDONE ALKALOIDS

Name	R^1	R^2	R^3	R^4	R^5	R^6	*M.p. (°C)*	*Derivatives M.p. (°C)*	*Re*
N-Methylacridone	Me	H	H	H	H	H	203		1
1-Hydroxy-*N*-methyl-acridone	Me	OH	H	H	H	H	192–194		2
1,3-Dimethoxy-*N*-methylacridone	Me	OMe	H	OMe	H	H	158–160	Hydrochloride, 135–136; picrate, 203–205	3
Arborinine	Me	OH	OMe	OMe	H	H	176–177		4,
Xanthoxoline	H	OH	OMe	OMe	H	H	265–267		4
1,2,3-Trimethoxy-*N*-methylacridone	Me	OMe	OMe	OMe	H	H	168–170		6
Evoxanthine	Me	OMe	$O–CH_2–O$		H	H	217–218		4
Evoxanthidine	H	OMe	$O–CH_2–O$		H	H	312–313		4
Melicopicine	Me	OMe	OMe	OMe	OMe	H	133–134		7
Melicopidine	Me	OMe	$O–CH_2–O$		OMe	H	121–122	Hudrochloride, 88–90 (decomp.); picrate, 134–135	7
Xanthevodine	H	OMe	$O–CH_2–O$		OMe	H	213–214	Hydrochloride, 137 (decomp.); picrate, 145–146	7
Melicopine	Me	OMe	OMe	$O–CH_2–O$		H	178–179		7
Tecleanthine	Me	OMe	$O–CH_2–O$		H	OMe	158	Picrate, 171	8

References
1 *D. L. Dreyer*, Tetrahedron, 1966, **22,** 2923.
2 *J. Reisch et al.*, Experientia, 1971, **27,** 1005.
3 *G. K. Hughes, K. G. Neill* and *E. Ritchie*, Austral. J. sci. Research, 1952, **5,** 401.
4 *J. A. Lamberton* and *J. R. Price*, *ibid.*, 1952, **5,** 401.
5 *S. C. Pakrashi et al.*, Chem. and Ind., 1961, 464.
6 *A. K. Ganguly et al.*, Indian J. Chem., 1966, **4,** 334.
7 *Price*, Austral. J. sci. Research, 1949, **2,** 249.
8 *K. H. Pegel* and *W. G. Wright*, J. chem. Soc. C, 1969, 2327.

TABLE 2

ACRIDONE ALKALOIDS WITH A C_5 SIDE CHAIN

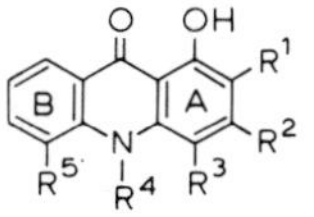

ompound	R^1	R^2	R^3	R^4	R^5	M.p. (°C)	Ref.
voprenine	OMe	$OCH_2CH{=}CMe_2$	H	Me	H	143	1
talaphylline	$CH_2CH{=}CMe_2$	OH	$CH_2CH{=}CMe_2$	H	OH	246	2
-Methylatala-phylline	$CH_2CH{=}CMe_2$	OH	$CH_2CH{=}CMe_2$	Me	OH	192–193	2

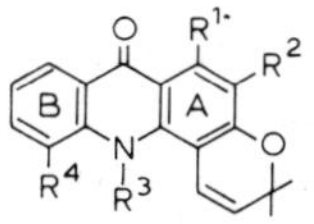

ompound	R^1	R^2	R^3	R^4	M.p. (°C)	Derivatives M.p. (°C)	Ref.
es-*N*-methyl-noracronycine	OH	H	H	H	246		3
es-*N*-methyl-acronycine	OMe	H	H	H	268–270 (decomp.)	Hydrochloride, 137–141; picrate, 222–224 (decomp.)	3
oracronycine	OH	H	Me	H	307		3
cronycine	OMe	H	Me	H	175–176	Hydrochloride, 125–130(decomp.); picrate, 150–154	4
	OH	H	Me	OH	252–254		5
	OH	$CH_2CH{=}CMe_2$	Me	OH	190–191		5

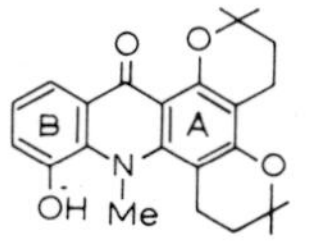

Compound	M.p. (°C)	Ref.
-Methylbicyclo-atalaphylline	185	6

eferences

J. A. Diment, E. Ritchie and W. C. Taylor, Austral. J. Chem., 1967, **20,** 1719.
T. R. Govindachari et al., Tetrahedron, 1970, **26,** 2905.
Idem, ibid., 1966, **22,** 3245.
F. N. Lahey and W. C. Thomas, Austral. J. sci. Research, 1949, **2,** 423.
A. W. Fraser and J. R. Lewis, J. chem. Soc., Perkin I, 1973, 1173.
D. Basu and S. C. Basa, J. org. Chem., 1972, **37,** 3035.

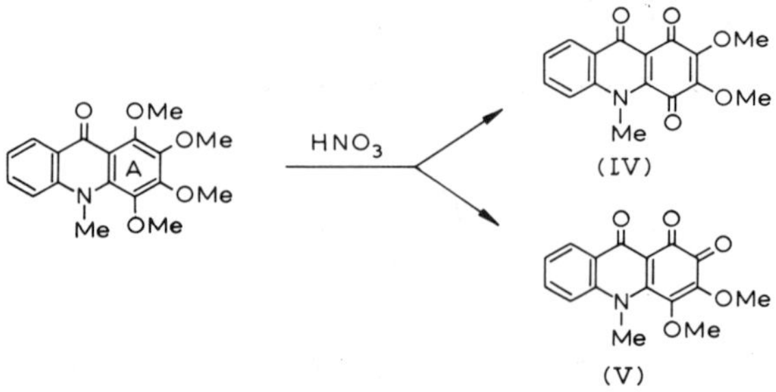

Compounds which contain a methylenedioxy bridge such as evoxanthine, melicopine and melicopidine are hydrolysed by alcoholic potash to monohydric phenols (*Crow* and *Price*, *ibid.*, 1949, **2,** 255). Reactions with bromine in various hydroxylic and non-hydroxylic solvents give rise to many products, the nature of which depends on the oxygenation pattern. In particular, addition across the electron-rich double bonds is a predominant reaction in methanol (*R. H. Prager* and *H. M. Thredgold*, Austral. J. Chem., 1968, **21,** 229; 1969, **22,** 1477, 1493, 1503).

Those alkaloids which have been synthesised have generally been prepared by variations of the procedure outlined in Scheme I for tecleanthine. Condensation of the appropriately substituted alkoxyiodobenzene with the appropriately substituted anthranilic acid, using copper as catalyst, gives the diphenylaminecarboxylic acid VI. Treatment of VI with phosphoryl chloride results in cyclisation to the 9-chloroacridine derivative VII, hydrolysis of which with aqueous acid and subsequent methylation gives tecleanthine (VIII) (*V. N. Ramachandran et al.*, Indian J. Chem., 1972, **10**, 14).

Scheme I

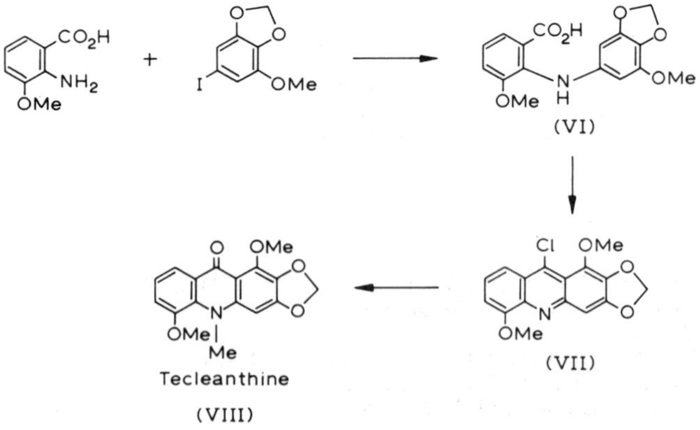

The following alkaloids have been synthesised using this general procedure: melicopicine (*Hughes, Neill* and *Ritchie*, Austral. J. sci. Research, 1950, **3,** 497), arborinine (*Gell, Hughes* and *Ritchie, loc. cit.*), 1,3-dimethoxy-*N*-methylacridone (*M. Ionescu* and *I. Mester*, Rev. Roum. Chim., 1969, **14,** 789), 1,2,3-trimethoxy-*N*-methylacridone and xanthoxoline (*Tonescu* and *M. Vlassa, ibid.*, 1971, **16,** 743), xanthevodine and melicopidine (*F. Dallacker G. Adolphen*, Tetrahedron Letters, 1965, 2023; Ann., 1966, **691,** 138), evoxanthine and evoxanthidine (*idem, ibid.*, 1966, **691,** 134).

The acronycine group of alkaloids

The acronycine group of the acridone alkaloids is a small group of compounds derived formally from 1,3-dihydroxyacridone and one or more isoprene units. Interest is situated in the title compound acronycine, which possesses broad spectrum antitumor activity (*G. H. Svoboda*, Lloydia, 1966, **29,** 206; *Svoboda et al.*, J. pharm. Sci., 1966, **55,** 758). The alkaloid was originally isolated in 1949 from the Australian rain-forest tree *Acronychia baueri* (*Lahey* and *W. C. Thomas*, Austral. J. sci. Research, 1949, **2,** 423). Investigations showed that acronycine gave a red hydrochloride with 10% hydrochloric acid which, on being heated above 130°, gave noracronycine, later isolated as a genuine alkaloid from *Glycosmis pentaphylla* (*Govindachari et al.*, Tetrahedron, 1966, **22,** 3245). This latter compound could be reconverted into acronycine by treatment with dimethyl sulphate, and contained one *N*-methyl but no *O*-methyl groups. Acronycine could be hydrogenated over Raney nickel to *dihydroacronycine*, $C_{20}H_{21}NO_3$, m.p. 140–141.5°. Further degradative studies showed acronycine to possess one of the two structures IX or X (*Brown et al.*, Austral. J. sci. Research, 1949, **2,** 622; *L. J. Drummond* and *Lahey, ibid.*, 1949, **2,** 630):

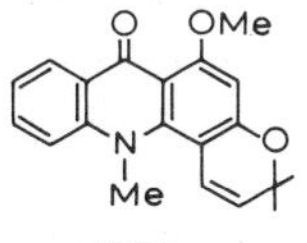

(IX) 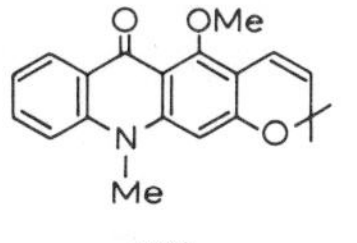 (X) 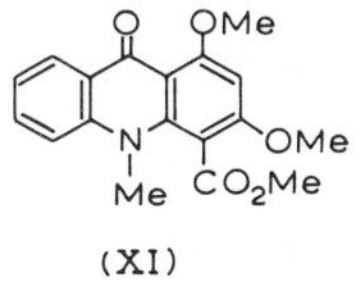 (XI)

The correct structure was first deduced as IX in 1966 (*P. L. Macdonald* and *A. V. Robertson*, Austral. J. Chem., 1966, **19,** 275), when it was shown that the product obtained by the stepwise ozonolysis, methylation, oxidation and esterification of acronycine was 1,3-dimethoxy-4-methoxycarbonyl-*N*-methylacridone (X), and not the 2-methoxycarbonyl isomer. Structure IX was also confirmed by careful examination of the n.m.r. spectrum

(*Govindachari et al.*, Tetrahedron, 1966, **22,** 3245). Isomer X has been synthesised and shown not to be identical with acronycine (*C. S. Oh* and *C. V. Greco*, J. hetero. Chem., 1970, **7,** 261). Acronycine was originally synthesised by a group at Lilly, who used the route shown in Scheme II to prepare noracronycine (*J. R. Beck et al.*, J. Amer. chem. Soc., 1967, **89,** 3934; 1968, **90,** 4706).

Scheme II

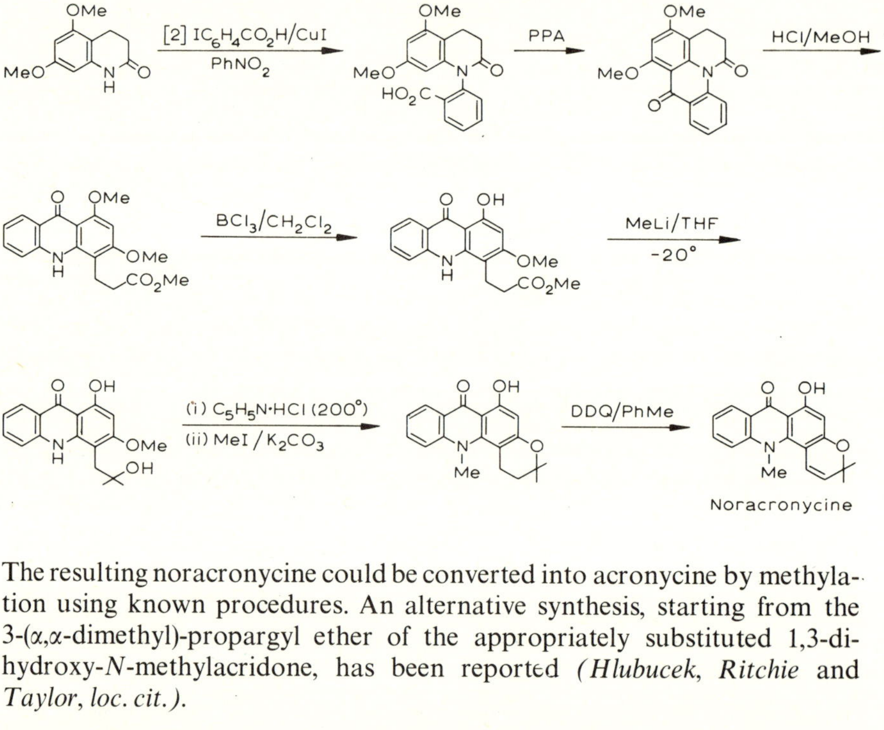

The resulting noracronycine could be converted into acronycine by methylation using known procedures. An alternative synthesis, starting from the 3-(α,α-dimethyl)-propargyl ether of the appropriately substituted 1,3-dihydroxy-*N*-methylacridone, has been reported (*Hlubucek, Ritchie* and *Taylor, loc. cit.*).

In addition to the three closely related alkaloids noracronycine, des-*N*-methylacronycine and des-*N*-methylnoracronycine the four other known isoprene-containing acridone alkaloids bear a close resemblance to acronycine. Treatment of atalaphylline (XII) with formic acid at 80–100° gives *bicycloatalaphylline* (XIII), m.p. 251–253°, which can be methylated with diazomethane to give O-*methylbicycloatalaphylline*, m.p. 225–227° (XIV) (*Govindachari et al.*, Tetrahedron, 1970, **26,** 2905). *N*-Methylation gives an identical product to that formed by reaction of diazomethane with the alkaloid *N*-methylbicycloatalaphylline (*D. Basu* and *S. C. Basa*, J. org. chem.

1972, 37, 3035). Similar cyclisations could account for the formation of the C_5 skeleton in acronycine, possibly *via* a precursor such as evoprenine (see Table 2).

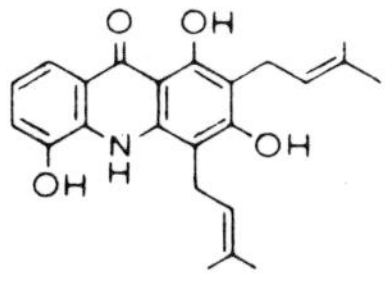

Atalaphylline

(XII)

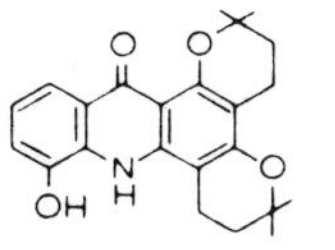

Bicycloatalaphylline

(XIII)

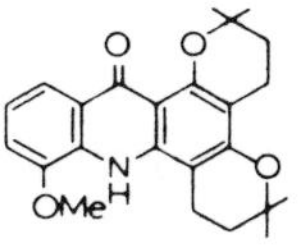

(XIV)

Chapter 33

The Alkaloids of the Morphine Group

K. W. BENTLEY

The alkaloids related to morphine have been subjected to more detailed study than those of any other group of comparable size, partly because of the very important pharmacological properties of the major alkaloids as analgesics and drugs of addiction, and partly because of the ease with which some of them, thebaine in particular, undergo a wide variety of molecular rearrangements. For many years only six bases in the group were known, but recently the number of structurally identified alkaloids, including those of partly modified structure and ring homologues, has increased rapidly and at the time of this review stands as forty one.

The following sub-groups may be identified and treated separately for the purposes of discussion:

1. The morphine–thebaine group consisting of bases containing a 4,5-oxide bridge, exemplified by morphine (I, R = H) and codeine (I, R = Me).

2. The morphinandienone or sinomenine–salutaridine group, bearing no such oxide bridge and exemplified by sinoacutine (II).

3. The hasubanonine group, in which the nitrogen atom is linked to C-14 instead of C-9, exemplified by hasubanonine (III).

4. The homomorphine group, comprising ring homologues of bases related to I and II with an additional carbon atom between C-9 and C-10.

One bimolecular alkaloid, cancentrine, is known in which a ring-C modified codeine comprises half of the molecule. Bases of groups 1 and 3 belong only to the series of absolute stereochemistry shown, whereas those of groups 2 and 4 are found in both enantiomeric series.

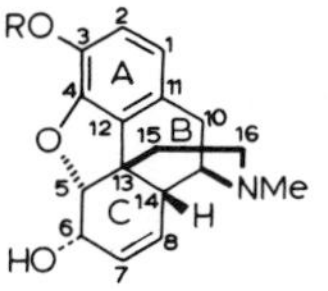

Morphine (R = H)
(I)

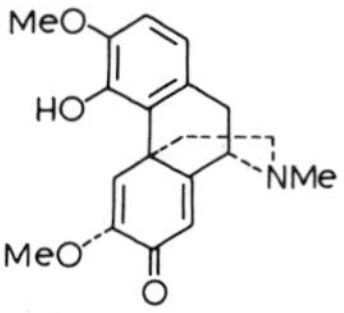

Sinoacutine
(II)

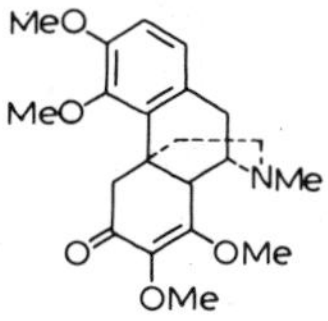

Hasubanonine
(III)

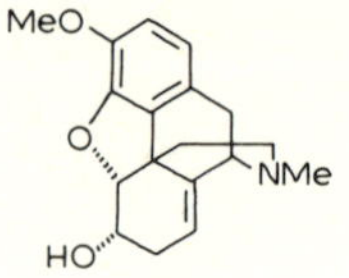

Neopine
(IV)

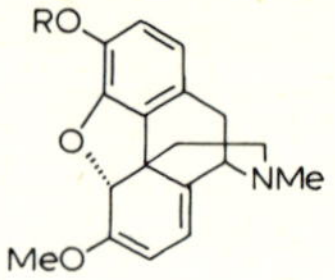

(V)

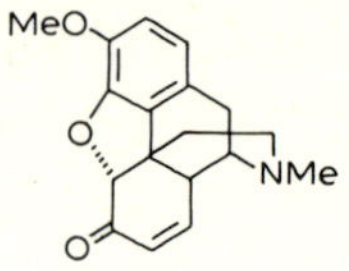

Codeinone
(VI)

The group has been reviewed in great detail up to 1952 by *K. W. Bentley*, "The Chemistry of the Morphine Alkaloids", The Clarendon Press, Oxford, 1954, and from 1952 to 1970 by *G. Stork* and by *Bentley*, "The Alkaloids", edited by *R. H. F. Manske*, Academic Press, New York, Volume VI, 1960, pp. 219–245 *(Stork)* and Vol. XIII, 1971, pp. 1–163 *(Bentley)*.

1. The morphine–thebaine group

This group comprises the six very closely related alkaloids morphine (I, R = H)*, codeine (I, R = Me), neopine (IV), thebaine (V, R = Me), 16-hydroxythebaine and oripavine (V, R = H), all of which occur in *Papaver* species. The relationships between the alkaloids have been established by interconversions. The physical constants of the bases and some of their salts are given in Table 1.

TABLE 1

MORPHINE–THEBAINE ALKALOIDS

	m.p. (°C)	$[\alpha]_D$ *(°)*	*m.p. (°C) HCl*	*m.p. (°C) HBr*	*m.p. (°C)* $HClO_4$	*m.p. (°C) Picrate*	*m.p. (°C) MeI*
Morphine	247–248	−130.9	200		150	163–165	286
Codeine	156–157	−111	264	190–192		196–197	270
Neopine	127	−28.1		282–283			
Thebaine	193	−218.6				217	224
16-Hydroxy-thebaine	126–128						
Oripavine	200–201	−211.8	244–245				207–208

* The numbering is based on that of phenanthrene.

Occurrence. Morphine, codeine, neopine and thebaine all occur in opium, the quantities in a good grade being approximately 10%, 0.2–0.8%, 0.005% and 0.2–0.8%, respectively, though they vary between wide limits. The seeds of *Papaver somiferum* have been reported to be alkaloid-free, (*M. Kerbosch*, Arch. Pharm., 1910, **248**, 536; *A. Heiduschka*, Schweiz. Apoth. Ztg., 1919, **57**, 447) but patents have been granted covering the isolation of morphine and codeine from them (*B. A. Klyachkina et al.*, Russ. P., 53,168, 53,224/1933). Thebaine also occurs in *Papaver orientale*, but only in the growing season; during periods of withering and rest only the aporphine alkaloid isothebaine can be isolated (*J. Gadamer*, Arch. Pharm., 1911, **249**, 39; *W. Klee*, *ibid.*, 1914, **252**, 211). Oripavine, which has not been detected in opium, occurs in *Papaver orientale* alongside thebaine (*S. Junusov et al.*, Ber., 1935, **68**, 2158) and also in *Papaver bracteatum* (*V. V. Kiselev* and *R. R. Konovalova*, J. gen. Chem. U.S.S.R., 1948, **18**, 142). Recently a very rich source of thebaine has been reported (*N. Sharghi* and *I. Lalenzari*, Nature, 1967, **213**, 1244) in certain strains of *Papaver bracteatum* the dried latex from which yielded 25% by weight of thebaine; further work has confirmed that this strain could be a valuable commercial source of alkaloids of this group.

Isolation. For many years morphine was extracted from opium by the "Gregory process". In this, concentrated aqueous calcium chloride is added to a concentrated aqueous extract of opium, when calcium meconate, lactate and sulphate are precipitated and may be removed by filtration. Concentration of the filtrate then affords the "Gregory salt", which is a mixture of the hydrochlorides of morphine and codeine; this is dissolved in water and the morphine precipitated by ammonia, the more soluble codeine being recovered from the aqueous solution by extraction with benzene. Thebaine may be subsequently recovered as its sparingly soluble salicylate or bitartrate. Neopine, which is present in opium in only very minute amounts, may be isolated from the final mother liquors of processing after all other alkaloids have been removed (*J. J. Dobbie* and *A. Lauder*, J. chem. Soc., 1911, **99**, 34), and this remains the only source of the base. The method of isolation has been improved and neopine may be separated from codeine through the sulphate (*A. H. Homeyer* and *W. L. Shilling*, J. org. Chem., 1947, **12**, 356). For other methods of extraction see references cited by *Bentley (op. cit.)*. A recent review of modern methods is given by *A. Barbier* (Ann. Pharm. Fr., 1947, **5**, 121; C.A., 1948, **42**, 1023). Modern commercial processes use liquid–liquid extraction methods, the precise details of which are not published, and very efficient recoveries of morphine from opium and from poppy capsule are known to be achieved. Extraction of thebaine from the latex of *P. bracteatum* mentioned above is simply achieved by hot water followed by basification.

Commercial preparation of codeine. Most of the codeine used commercially is prepared by the methylation of morphine, and for this the following processes have been used (references, *Bentley*, *loc. cit.*): methylation with (a) dimethyl sulphate or sulphite and alkali or sodium alkoxides, (b) sodium methoxide and trimethyl phosphate, methyl nitrate or methyl esters of organic sulphonic acids, (c) diazomethane and (d) aryltrimethylammonium hydroxides. Methylation of morphine *N*-oxide followed by the reduction of the product has also been used; this method together with (c) and (d) above, gives no quaternary salts. The most satisfactory method is (d), developed by *W. Rodionov* (Bull. Soc. chim. Fr., 1926 [iv], **39**, 305); for details see *Bentley (op. cit.)*, *R. E. Lutz* and *L. F. Small* ("The Chemistry of the Opium Alkaloids", U.S. Treasury Dept., Pub. Health Suppl. No. 103, 1931) and *H. L. Holmes* ("The Alkaloids", Edited by *Manske* and *Holmes*, Academic Press, New York, Vol. II, 1952, Ch. 8), and gives yields >98%.

Detection and estimation. Colour tests for the detection of and references to methods for the estimation of these alkaloids are given by *Bentley (op. cit.)*.

(a) Structure

Morphine, codeine and thebaine have the compositions $C_{17}H_{19}O_3N$ (*M. A. Laurent*, Ann. Chim. Phys., 1847, [iii], **19**, 359), $C_{18}H_{21}O_3N$ (*T. Anderson*, Trans. roy. Soc. Edinburgh, 1850, **20**, 57) and $C_{19}H_{21}O_3N$ (*idem*, J. pr. Chem., 1852, **57**, 358), respectively. Morphine is easily recognised as a phenol; methylation of the phenolic group yields codeine (*E. Grimaux*, Compt. rend., 1881, **92**, 1140) which is therefore morphine methyl ether. Morphine forms a diacetyl derivative (heroin) (*O. Hesse*, Ann., 1883, 222, 203; *C. R. A. Wright*, J. chem. Soc., 1874, **27**, 1031) and codeine a monoacetyl derivative *(Wright, loc. cit.)* so both bases must contain an alcoholic group; this is confirmed by the production of the so-called halogen-morphides and -codides (p. 283) on treatment of the alkaloids with phosphorus halides, thionyl halides and halogen hydracids. The oxidation of codeine with chromic acid (*F. Ach* and *L. Knorr*, Ber., 1903, **36**, 3067; *S. P. Findlay* and *Small*, J. Amer. chem. Soc., 1950, **72**, 3247) and by the Öppenauer process (*Findlay* and *Small*, *ibid.*, 1951, **73**, 4001) gives a ketone, codeinone, (VI), showing that the alcoholic group is secondary; the ketone is also obtained by the hydrolysis of thebaine with dilute sulphuric acid (*Knorr* and *H. Horlein*, Ber., 1906, **39**, 1409), indicating that thebaine is the methyl ether of an enolic form of codeinone. The relationship between the three alkaloids is thus clear, and most of the evidence on which the structure for morphine is based is derived from work on codeine and thebaine. Oripavine has the composition $C_{18}H_{19}O_3N$ and is a phenol giving thebaine on methylation, and neopine gives dihydrocodeine on reduction. 16-Hydroxythebaine, which has been recovered from opium using thin layer chromatographic techniques, has had its structure assigned on the basis of spectroscopic and mass spectrometric studies (*E. Brochmann-Hanssen et al.*, J. org. Chem., 1972, **37**, 1881).

The characteristic feature of these alkaloids is the ease with which they undergo deep-seated degradations leading to the extrusion of a nitrogen-containing side-chain with the production of phenanthrene derivatives or suffer rearrangement with aromatisation to derivatives of phenanthrene or biphenyl as a result of migration of that same side-chain. The implications of these peculiarities were only recognised by *J. M. Gulland* and *R. Robinson* (J. chem. Soc., 1923, **123**, 980; Mem. Proc. Manchester lit. phil. Soc., 1925, **69**, 79) and the key reactions that led them to propose the correct structures for these alkaloids may be summarised as follows:

(i) The degradation of thebaine methiodide and codeinone methiodide to acetylthebaol (VII, R = Me) and to 4,6-diacetoxy-3-methoxyphenanthrene (VII, R = Ac) on heating with sodium acetate and acetic anhydride, demonstrating the positions of the oxygen substituents on the phenanthrene skeleton (*R. Pschorr, C. Seydel* and *W. Stohrer*, Ber., 1902, **35**, 4400).

(ii) Hofmann degradation of codeine methiodide to an olefin, α-codeimethine (VIII) (*Grimaux*, Compt. rend., 1881, 93, 591) isomerisation of this to β-codeimethine (IX) (*L. Knorr* and *S. Smiles*, Ber., 1902, **35**, 3009) and further degradation of the methiodides of these bases to methylmorphenol (X) by alkalis (*E. Vongerichten* and *O. Dittmer*, *ibid.*, 1906, **39**, 1718) and to acetylmethylmorphol (XII) by acetic anhydride (*Hesse, loc. cit.; Pschorr* and *C. Sumuleanu*, Ber., 1900, **33**, 1810):

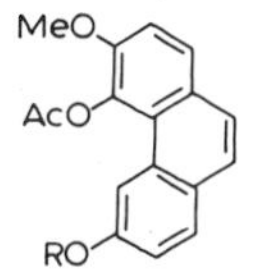

MeO, AcO, RO
(VII)

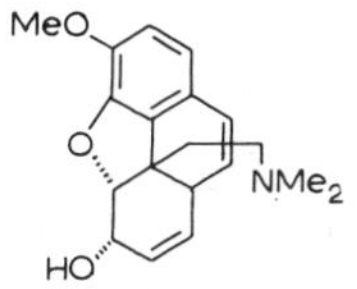

MeO, O, NMe$_2$, HO
α - Codeimethine
(VIII)

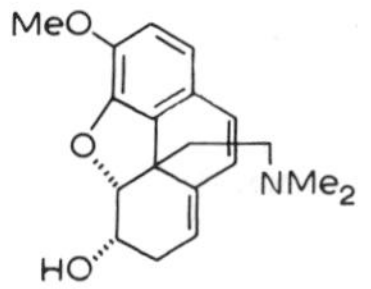

MeO, O, NMe$_2$, HO
β - Codeimethine
(IX)

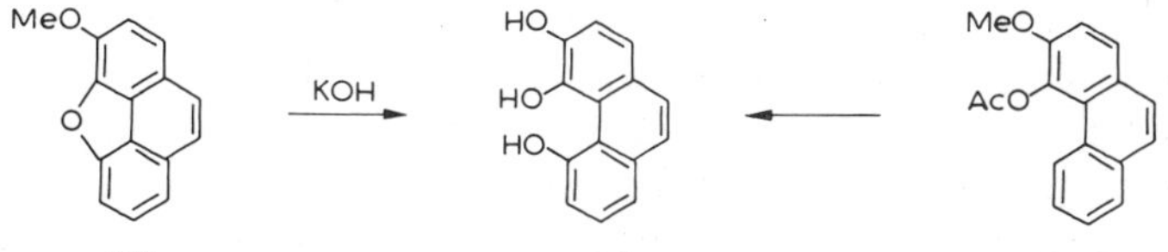

MeO, O
(X)

KOH →

HO, HO, HO
(XI)

←

MeO, AcO
(XII)

(iii) The hydrolysis of methylmorphenol (X) to 3,4,5-trihydroxyphenanthrene (XI) by fused potassium hydroxide (*Vongerichten* and *Dittmer*, *loc.*

cit.), and the synthesis of 3,4,5- and 3,4,6-trimethoxyphenanthrene and 3,4-dimethoxyphenanthrene for comparison with material prepared from XI, VII and XII.

(iv) The dehydration and rearrangement of morphine to apomorphine (XIII) (*Pschorr et al.*, Ber., 1906, **39**, 3124; 1907, **40**, 1998, 2001) and the rearrangement of thebaine to morphothebaine (XIV) (*Knorr* and *Pschorr*, Ber., 1905, **38**, 3153) by acids.

(v) The rearrangement of thebaine and of codeinone to thebenine (XV) on heating with dilute hydrochloric acid (*Hesse*, Ann., 1870, **153**, 47).

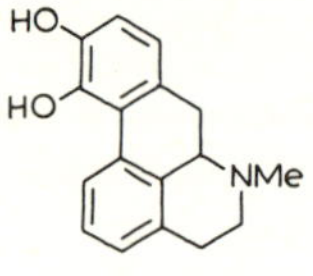

Apomorphine (XIII) Morphothebaine (XIV) Thebenine (XV)

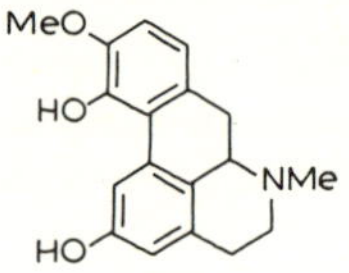

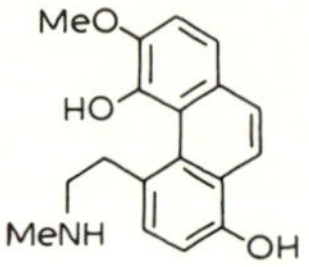

Gulland and *Robinson* reasoned that the nitrogen-containing side-chain in these alkaloids must be so linked to a partially hydrogenated phenanthrene skeleton that aromatisation to a phenanthrene or biphenyl system could not take place without extrusion or migration of that chain. The nitrogen end of the side-chain was known to be attached at C-9 or C-10 (see below), only the former being seriously considered following work on apomorphine (XIII) and morphothebaine (XIV), hence it was deduced that the carbon end of the chain must be located at one of the two possible angular positions C-13 or C-14 so that its extrusion is an essential part of any aromatisation process. Of these two possibilities the former (C-13) was chosen since this alone led to a satisfactory mechanism for the rearrangement of thebaine to thebenine (see formulae CXLII–CLIII, p. 295).

Earlier formulae advanced by *Knorr* and *Horlein* (Ber., 1907, **40**, 3301) and modified by *H. Weiland* and *E. Koraleck* (Ann., 1923, **433**, 267) based on a C-5 attachment of the side-chain failed to account for universal extrusion or migration of the chain on aromatisation, gave no explanation of the failure of β-codeinethine (IX) to aromatise to a derivative of naphthalene and necessitated a mechanism for the thebenine transformation that involved an inexplicable fission of the nitrogen-containing ring and an allylic shift of the C-6 hydroxyl group under highly improbable conditions. The only supporting evidence for a C-5 placing of the side-chain was the apparent location of the chain at that position in thebenine.

Still earlier formulae based on a C-8 attachment of the side-chain (*Pschorr* and *H. Einbeck*, Ber., 1907, **40**, 1980), following work on apomorphine, became untenable when it was shown that morphine and codeine could be converted with an allylic shift of the C-6 hydroxyl group, through the halogenocodides, into isomeric bases (see below) bearing –CH(OH)– at this position. Two isomers of codeine, ψ-codeine and allo-ψ-codeine, prepared in this way, on oxidation gave the ketone ψ-codeinone (XVI) degradable to 4,8-diacetoxy-3-methoxyphenanthrene (*Knorr* and *Horlein, loc. cit.*).

Evidence concerning the attachment of the nitrogen end of the side-chain came from a study of the degradation of hydroxycodeine. This base, obtained by the oxidation of codeine with chromic acid (*Ach* and *Knorr*, *loc. cit.*), bears an additional hydroxyl group (now known to be located at C-10, *H. Rapoport* and *G. W. Stevenson*, J. Amer. chem. Soc., 1964, **76**, 1796) which was shown to be at C-9 or C-10 since hydroxycodeine on subjection to the same

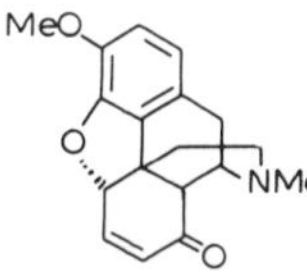

ψ-Codeinone (XVI)

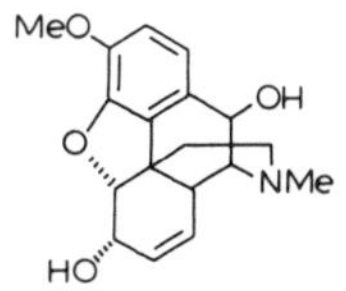

10-Hydroxycodeine (XVII)

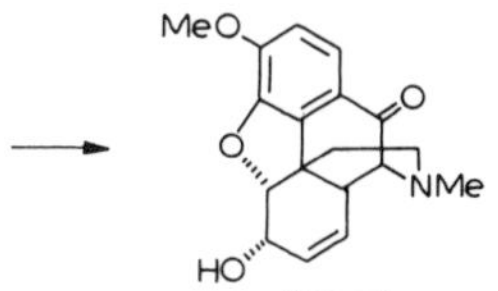

(XVIII)

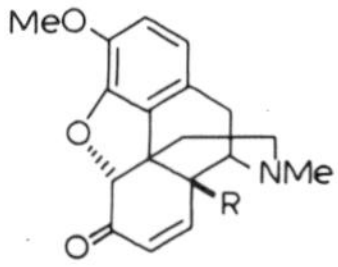

(XIX)

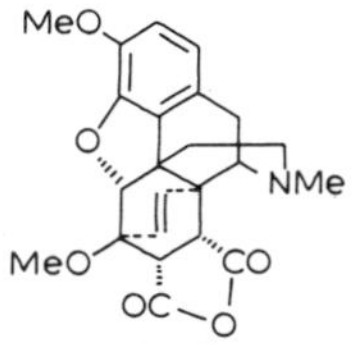

(XX)

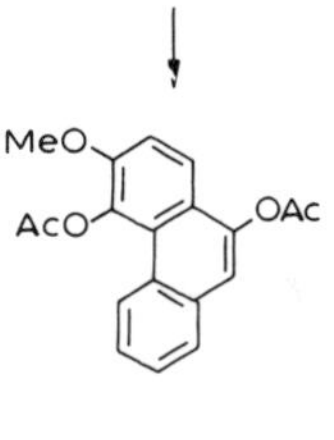

(XXI)

sequence of reactions that degraded codeine (I, R = Me) to acetylmethylmorphol (XII) (Hofmann degradation followed by acetolysis) gave an acetoxyacetylmethylmorphol (XXI) which gave acetylmethylmorphol-9,10-quinone on oxidation with chromic acid, with loss of the extra acetoxy group. In the intermediate methine base of this sequence, XVIII, the new oxygen function appeared as an oxo group, showing that a 9,10 double bond must be introduced during Hofmann degradation, and hence the nitrogen atom must be attached at C-9 or C-10 (XVII) (*Pschorr* and *Einbeck*, *loc. cit.; Knorr* and *Horlein*, Ber., 1907, **40**, 2042). No positive proof of the attachment of the nitrogen to C-9 was obtained until the synthesis of morphine was achieved, though rational mechanisms for the conversion of thebaine into thebenine and phenyldihydrothebaine required such a linkage.

The formulation of thebaine as the conjugated diene (V, R = Me) and not the alternative $\Delta^{5,7}$- or $\Delta^{5,8}$-diene is in accord with its ready conversion into 14-bromocodeinone (XIX, R = Br) and 14-hydroxycodeinone (XIX, R = OH) by treatment in acetic acid with bromine and hydrogen peroxide, respectively (*M. Freund*, *ibid*., 1906, **39**, 844; *Freund* and *E. Speyer*, J. prakt. Chem., 1916, **94**, 135) and with the addition of dienophils to give adducts, such as that with maleic anhydride (XX), that are not enol ethers (*W. Sandermann*, Ber., 1938, **71**, 648; *C. Schöpf, K. von Gottberg* and *W. Petri*, Ann., 1938, **536**, 216).

Direct evidence of the attachment of the side-chain at C-13 or C-5 was sought by examination of the Beckmann transformation of dihydrocodeinone oxime (XXII). The Beckmann transformation of ketoximes usually gives amides, but α-hydroxyoximes and their ethers generally suffer carbon–carbon fission to give nitriles and carbaldehydes or ketones. Such a course for the reaction with dihydrocodeinone oxime would give a carbaldehyde with a C-13 linkage of the side-chain (XXII–XXIV) or a ketone with a C-5 linkage (XXV). The production of a nitrile was observed, but positive identification of the product as an aldehyde or ketone could not be achieved. It was assigned the structure XXIV after the conversion of its *O*-methyl ether into an oxime and further Beckmann transformation of this to a product, XXVII, that retained all of its atoms when subjected to Hofmann degradation. The other possible products of this sequence derived from the ketone XXV, namely the lactam XXVI or an isomer, would lose *N*,*N*-dimethylethylenediamine or dimethylaminopropionic acid under the alkaline conditions of the Hofmann degradation (*Schöpf*, Ann., 1927, **452**, 411):

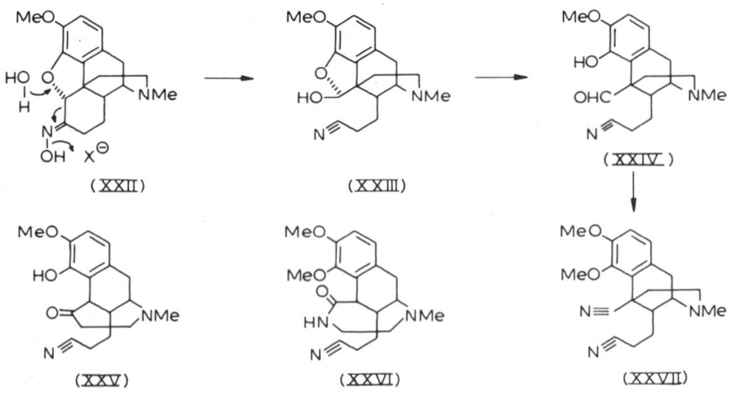

(XXII) (XXIII) (XXIV)

(XXV) (XXVI) (XXVII)

These reactions remained the basis on which the structures of the morphine alkaloids rested until the total synthesis of derivatives of codeine was achieved in 1949, and the validity of the Gulland and Robinson structures rested for many years on their unique ability to explain all of the reactions and complex rearrangements of the bases in this group.

(b) Total synthesis of morphine

The first total synthesis of the carbon–nitrogen skeleton of morphine was achieved by *R. Grewe.* The benzyloctahydroisoquinoline (XXVIII, $R = R^1 = H$) (Naturwiss., 1946, **33**, 333; *Grewe et al.*, Ber., 1948, **81**, 279; Chem. Ber., 1951, **84**, 527), was heated with phosphoric acid, when it underwent cyclisation to *N*-methylmorphinan (XXIX, $R = R^1 = H$) in which the arrangement of groups at the asymmetric centres is the same as that in morphine. The name morphinan is given to the unsubstituted skeleton. In a similar manner the base XXVIII ($R = R^1 = OMe$) was cyclised by heating with concentrated hydrochloric acid, during which process partial demethylation occurred, the product being (±)-tetrahydrodeoxycodeine (XXIX; $R = OMe$, $R^1 = OH$), identical with the racemate prepared by mixing equal amounts of (−)- and (+)-tetrahydrodeoxycodeine prepared from the thebaine and sinomenine series, respectively (*Grewe, A. Mondon* and *E. Nolte*, Ann., 1949, **564**, 161). This synthesis has been extended to provide a synthesis of dihydrothebainone (XXXII) from the phenolic product XXXIII obtained by reduction of the isoquinoline derivative XXX with sodium and liquid ammonia (*G. O. Morton, R. O. Waite* and *J. Shavel*, Tetrahedron Letters, 1967, 4055) though the overall yield is very poor. Any synthesis of dihydrothebainone can now be regarded as a total synthesis of morphine following the work of *M. Gates*, who achieved the first total synthesis in the following way.

After preliminary work by which a route to the C-14 epimer of *N*-methylmorphinan (XXIX, $R = R^1 = H$) was devised (*Gates et al.*, J. Amer. chem. Soc., 1948, **70**, 2261; 1950, **72**, 228, 114; Experientia, 1949, **5**, 235), 4-cyanomethyl-5,6-dimethoxy-1,2-naphthaquinone (XXXI) was condensed with butadiene and the product XXXIV catalytically reduced to the cyclic amide XXXV. The cyclisation during the reduction is exactly analogous to the reductive cyclisation of 2-cyanoacetophenone to 3-methylisoindol-1-one. Wolff–Kishner reduction of XXXV gave the amide, which was reduced with lithium tetrahydridoaluminate to the secondary amine, which was then methylated by Eschweiler's method to the *N*-methyl compound XXXVI.

This base was shown to be identical with β-Δ^6-dihydrodeoxycodeine methyl ether, the configuration at C-14 being epimeric with that found in morphine (*Gates* and *G. Tschudi*, *ibid*., 1950, **72**, 4839). The inversion at C-14 necessary for the conversion of the base XXXVI into morphine was accomplished in the following way (*idem*, *ibid*., 1952, **74**, 1109). Hydration of the double bond, followed by partial demethylation and Öppenauer oxidation led to β-dihydrothebainone (XXXIX), the C-14 epimer of XXXII, (identical with material prepared from thebaine by *Small* and *G. L. Browning*, J. org. Chem., 1939, **3**, 618) and bromination of this followed by treatment of the tribromo compound XXXVIII with dinitrophenylhydrazine gave the dinitrophenylhydrazone of 1-bromo-Δ^7-thebainone (XXXVII), inversion at C-14 occurring with remarkable ease in the hydrazone. Cleavage of the hydrazone with acetone and acid then afforded 1-bromo-Δ^7-thebainone (XL), which was reduced to dihydrothebainone (XXXII). Treatment of β-dihydrothebainone with three equivalents of bromine, 2,4-dinitrophenylhydrazine, acetone and acid gave, *via* XLII, 1-bromocodeinone (XLI), which on reduction gave codeine (I, R = Me) and demethylation of this with pyridine hydrochloride afforded morphine (I, R = H):

(XXVIII) $\xrightarrow{H^{\oplus}}$ *N*-Methylmorphinan (R = R^1 = H) (XXIX)

(XXX) $\xrightarrow{Na/NH_3,\ Bu^tOH}$ (XXXIII) $\xrightarrow{H^{\oplus}}$ Dihydrothebainone (XXXII)

(XXXI) → (XXXIV) $\xrightarrow{H_2,\ CuCrO_3}$ (XXXV) $\xrightarrow{1.\ W/K;\ 2.\ LiAlH_4;\ 3.\ HCHO}$ (XXXVI)

3 stages

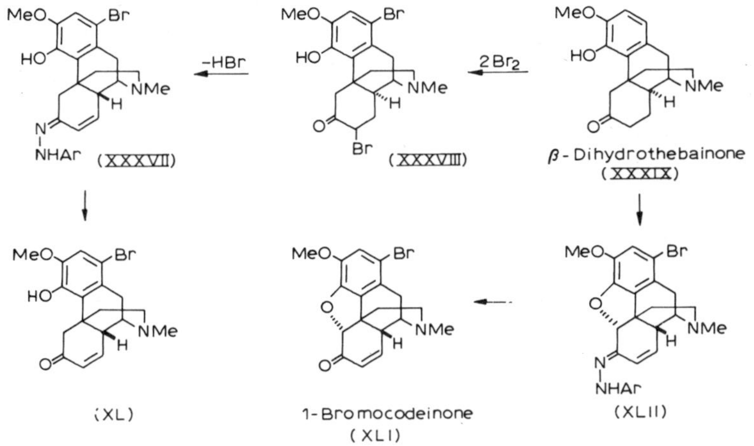

A further synthesis of dihydrothebainone, involving no epimerisation, has been achieved by *D. Ginsburg* and his co-workers (J. chem. Soc., 1951, 936; 1953, 1524, 2664; 1954, 3052). 2-(2,3-Dimethoxyphenyl)cyclohex-2-ene-1-one was converted in three steps into the dione XLIII, which was then transformed by the sequence shown in formulae XLIII–XLVIII into the oxime XLVIII. Hydrolysis and reduction of this gave nordihydrothebainol, which was oxidised and *N*-methylated by formaldehyde and formic acid to dihydrothebainone (XXXII):

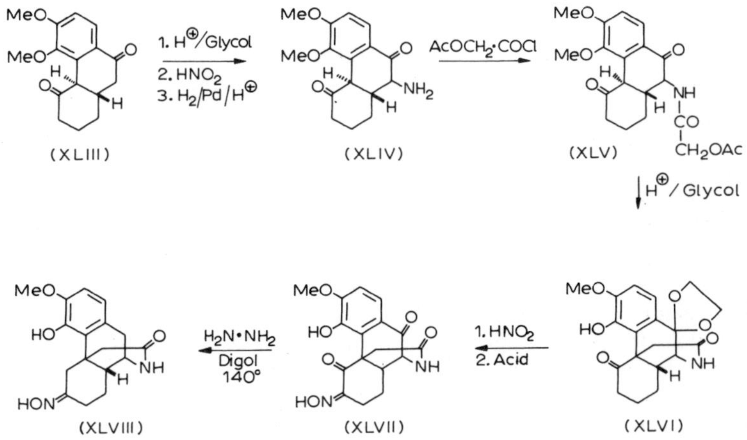

The formal synthesis of thebaine has been accomplished from dihydrocodeinone *via* its enol ether, dihydrothebaine (XLIX), which reacts with methyl hypobromite (bromine and methanol) to give the bromo-acetal L, and dehydrobromination of this gives codeinone dimethyl acetal (LI), which loses methanol to give thebaine (V, R = Me) on warming with toluene-4-sulphonic acid in chloroform (*Rapoport, H. N. Reist* and *C. H. Lowell*, J. Amer. chem. Soc., 1956, **78**, 5128). Since neopine can be prepared from thebaine (see below) a formal synthesis of this alkaloid can also be regarded as accomplished:

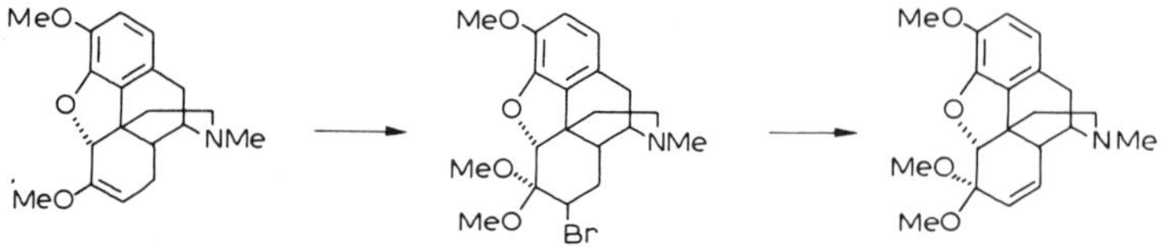

(XLIX) (L) (LI)

(c) Stereochemistry

(i) Morphine alkaloids

The stereochemistry of the morphine alkaloids has been the subject of theoretical discussion by *L. F. Fieser* and *M. Fieser* ("Natural Products Related to Phenanthrene", Amer. chem. Soc. Monograph No. 70, Rheinhold, New York, 3rd. Edn., 1949), *Stork (loc. cit.)* and *K. W. Bentley* and *H. M. E. Cardwell* (J. chem. Soc., 1955, 3252), experimental investigation by *H. Rapoport et al.* (J. Org. Chem., 1950, **15**, 1350; J. Amer. Chem. Soc., 1952, **74**, 2630; 1953, **75**, 5329) and by *J. Kalvoda et al.* (Helv., 1955, **38**, 1847) and crystallographic study (*M. McKay* and *D. C. Hodgkin*, J. chem. Soc., 1955, 3261; *J. Fridrichsons, M. F. Mackay* and *A. M. Mathieson*, Tetrahedron, 1970, **26**, 1969; *J. H. van den Hende* and *R. Nelson*, J. Amer. chem. Soc., 1967, **89**, 2901).

The correct configuration has been deduced as follows:

14-Hydroxydihydrocodeinone (LII) on exhaustive methylation affords the cyclic ether "dihydrohydroxycodeone" (LIII) (*C. Schöpf* and *F. Borkowsky*, Ann., 1927, **452**, 211), from which it was deduced that the side-chain at C-13 and the hydroxyl group at C-14 are in the *cis* relationship, as in LII. Assuming that the hydroxyl group at C-14 in 14-hydroxydihydrocodeinone has the same orientation as the hydrogen atom at C-14 in dihydrocodeinone, the latter base must have a *cis* arrangement of the side-chain and the hydrogen atom. The C-14:H and C-9:N bonds must also be *cis*, as a *trans* linking of the side-chain is impossible. The arrangements of the C-5:O bond *trans* to the side-chain an C-13 results in a strainless oxide ring, but a *cis* arrangement introduces a high degree of strain into this ring. Catalytic hydrogenation of dihydrocodeinone affords exclusively dihydrocodeine, and *Fieser* and *Fieser* on the basis of the hydrogenation of the carbonyl group from the less hindered side allotted a *cis* disposition of the C-5:O and C-6:OH bonds. The resulting structure for codeine is LIV:

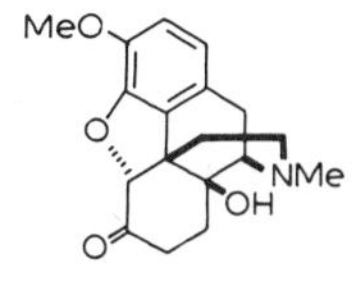

14-Hydroxycodeinone (LII) (LIII) 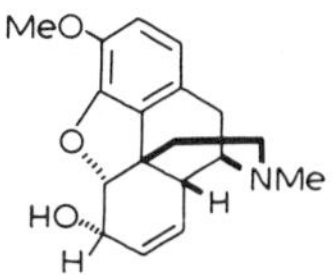 Codeine (LIV)

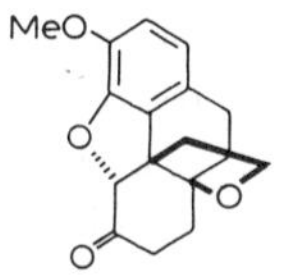

(*1*) The arrangement at C-5 and C-6 (*Rapoport* and *G. B. Payne*, J. org. Chem., 1950, **15**, 1350). Ozonolysis of dihydrocodeine (LV) affords α-ozodihydromorphilactonate and this may be reduced with lithium tetrahydridoaluminate to tetrahydromorphitetrol. A stereoisomer of this tetral, tetrahydro-α-isomorphitetrol, may be prepared from dihydroisocodeine (LVI). Oxidation of tetrahydromorphitetrol with lead tetra-acetate proceeds three time faster than the oxidation of the α-iso compound, indicating that in the former, the vicinal hydroxyl groups are *cis* to each other, and in the latter *trans*. Accordingly, the C-5:O and C-6:OH bonds in codeine and morphine have the *cis* arrangement:

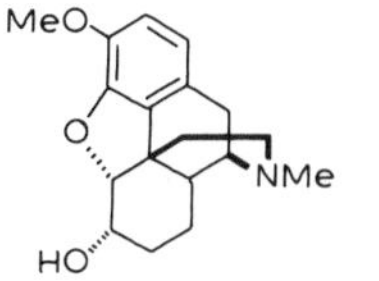

(LV) (LVI = C-6 epimer) (LVII) 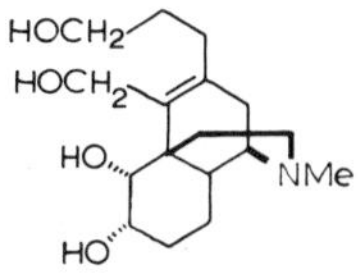 (LVIII)

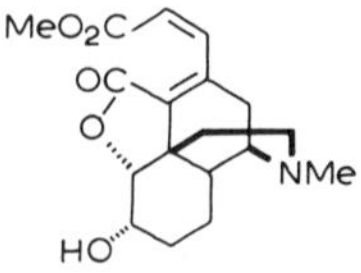

(*2*) The arrangement at C-6 and C-13 (*Rapoport* and *Payne*, J. Amer. chem. Soc. 1952, **74**, 2630). The relative stereochemistry of these two centres was inferred from a study of the exhaustive methylation of dihydrocodeine (LV) and its C-6 epimer dihydroisocodeine (LVI). Hofmann degradation of tetrahydrocodeimethine (from dihydrocodeine) is accompanied by some methylation of the C-6 hydroxyl at the expense of some quaternary salt, and the degradation of tetrahydroisocodeimethine also furnishes a cyclic ether 6-codiran. Since the cyclic ether system must be *cis* fused to the hydrophenanthrene system, the 6-codiran product must have structure LX and dihydroisocodeine must be LIX in view of the previously determined relationship of oxygen functions at C-5 and C-6. However, a strain-free model of the alternative structure LXI can also be constructed, and it is possible to rationalize the production of this from an isomeric structure for dihydroisocodeine (LXII), the low yield of codiran on this basis being accounted for by postulating an abnormal reaction in which $-\overset{\oplus}{N}Me_3$ is replaced by $-OH$ with the resulting ethanol side-chain performing a substitution with inversion at C-6. These results therefore do not allow of an unambiguous allocation of configuration to γ-tetrahydrocodeimethine and codeine.

(*3*) The arrangement at C-13 and C-14 (*Rapoport* and *J. B. Lavigne*, *ibid*., 1953, **75**, 5329). Exhaustive methylation of dihydrothebainone (XXXII) with reduction of the intermediate methine base yields the cyclic ether thebenone (LXIII), the bis-hydroxyimino derivative of which (LXIV) can be degraded through the dinitrile LXV and amido acid LXVI to the cyclic imide LXVII. By contrast the C-14 epimer of thebenone, *i.e.* β-thebenone prepared from β-dihydrothebainone (XXXIX), can be degraded only as far as the acid isomeric with LXVI. It can be argued that cyclization to the imide would only be possible with a *cis* arrangement of $-CONH_2$ and $CH_2 \cdot CO_2H$ and, hence, that thebenone and dihydrothebainone must have the defined structures XXXII and LXIII, respectively. However, a strain-free model of LXVIII can also be constructed, and hence, uncertainty remains.

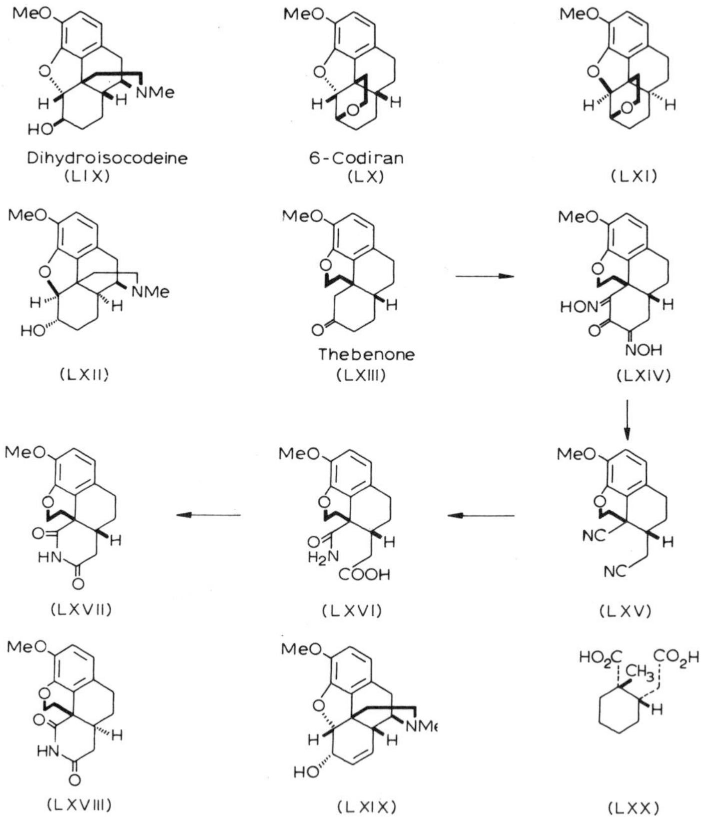

Dihydroisocodeine (LIX)

6-Codiran (LX)

(LXI)

(LXII)

Thebenone (LXIII)

(LXIV)

(LXVII)

(LXVI)

(LXV)

(LXVIII)

(LXIX)

(LXX)

The whole stereochemical solution has, however, been put beyond doubt by X-ray crystallographic studies which first confirmed the stereochemistry represented by I (*McKay* and *Hodgkin*, *loc. cit.*) and later the absolute stereochemistry given above

(*Fridrichsons et al., loc. cit.; van den Hende* and *Nelson, loc. cit.*) and chemically by the degradation of thebaine to an acid having the absolute stereochemistry shown in LXX (*Kalvoda et al., loc. cit.*).

(*ii*) *B/C*-trans-*morphine and its derivatives*

The first derivatives of morphine having a B/C-*trans* fusion to be obtained were β-thebainone-A and β-dihydrothebainone (XXXIX) (*Small* and *Browning, loc. cit.*) and of these β-thebainone-A is less stable than thebainone-A (XLI, unbrominated), its C-14 epimer (*Gates* and *Tschudi*, J. Amer. chem. Soc., 1956, **78**, 1380). Bromination of dihydrothebainone (XXXII) proceeds with the formation of the 1,7-dibromo compound and treatment of this with alkali results in S_N2' expulsion of the 7-bromine by phenoxide ion, with closure of the 4,5-oxide bridge and formation of 1-bromodihydrocodeinone in good yield (*Schöpf et al.*, Ann., 1930, **483**, 157; 1932, **492**, 213; *Gates* and *Tschudi, loc. cit.; Gates* and *M. S. Shepard*, J. Amer. chem. Soc., 1962, **84**, 4125). Treatment of β-dihydrothebainone (XXXIX) in the same way gives only a poor yield of 1-bromo-B/C-*trans*-dihydrocodeinone (*Gates* and *Shepard, loc. cit.*). The examination of models shows that the least strained configuration for this ketone has a *cis* arrangement of the 4,5-oxide bridge and C-13 side-chain (LXXI) and that there is considerable strain in the furan ring of the alternative LXXII in which the stereochemical arrangement at C-5 and C-13 is the same as in morphine itself. The examination of models and plausible reaction mechanisms also leads to the structure LXXI, the poor yield in the process being attributable to the need for epimerisation at C-7 to give the axial bromide to meet the *cis* requirements of the S_N2' reaction (*K. W. Bentley*, "The Alkaloids", *loc. cit.*):

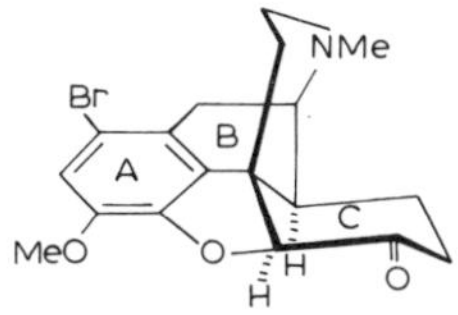

1-Bromo-B/C-*trans*-dihydrocodeinone
(LXXI)

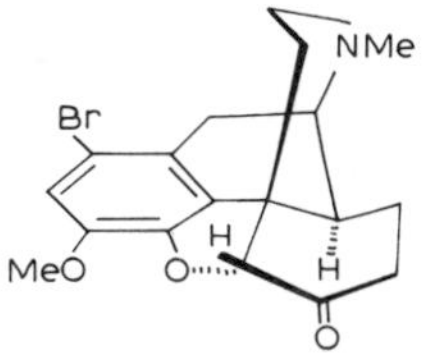

(LXXII)

Other preparations of B/C-*trans*-morphine derivatives, culminating in the production of *trans*-morphine itself, by processes that leave no doubt about the stereochemistry at C-5, have been reported. Δ^8-Deoxycodeine (LXXIII) on treatment with diborane, followed by oxidation of the initial product, gave a mixture of dihydro-ψ-codeine (LXXIV) and B/C-*trans*-dihydroallo-ψ-codeine (LXXV) which was shown to have the B/C-*trans* arrangement by reduction of its toluenesulphonyl ester to B/C-*trans*-tetrahydrodeoxycodeine, isomeric with the base XXIX ($R = OMe$, $R^1 = OH$) (p. 276) prepared by catalytic hydrogenation of LXXIII (*H. Kugita* and *M. Takeda*, Chem. and Ind., 1964, 2099; Chem. Pharm. Bull., Japan, 1965, **13**, 1422):

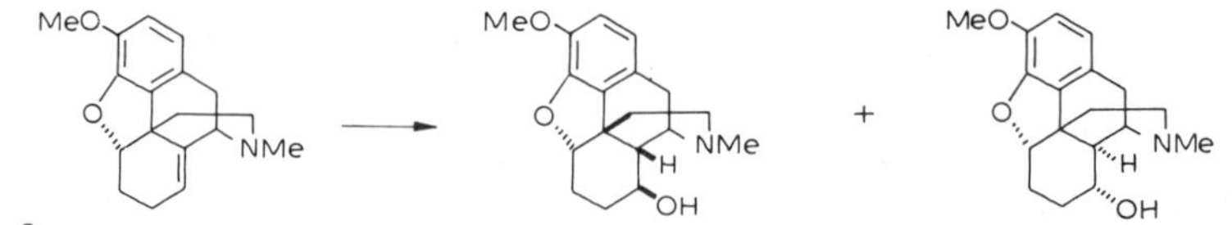

Δ^8- Deoxycodeine (LXXIII)

Dihydro-ψ-codeine (LXXIV)

B/C-*trans*-dihydroallo-ψ-codeine (LXXV)

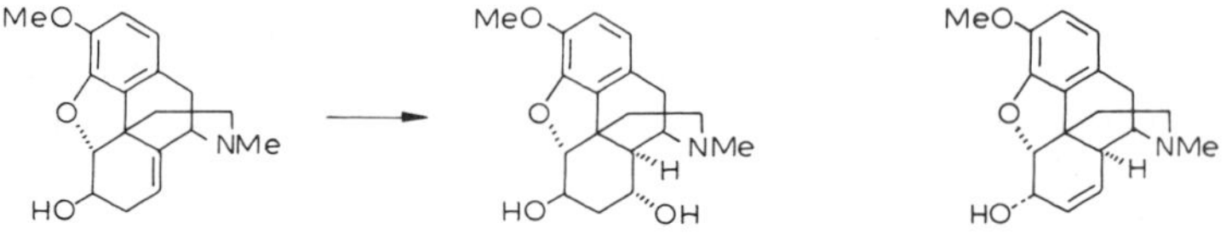

Isoneopine (LXXVI)

(LXXVII)

B/C-*trans*-codeine (LXXVIII)

A direct approach to *trans*-codeine from neopine (IV) was not possible since the hydroxyl group hinders attack of the double bond by diborane on the appropriate face, but isoneopine (LXXVI) readily affords the *trans*-diol LXXVII, the ditoluenesulphonyl ester of which on treatment with potassium acetate and dimethylformamide suffers loss of toluenesulphonic acid and S_N2 displacement at C-6 to give, after hydrolysis, B/C-*trans*-**codeine**, (LXXVIII), m.p. 98–102°; $[\alpha]_D$ +61°; *hydrobromide*, m.p. 233–235°; *picrate*, m.p. 244–245°, the overall yield from isoneopine being 38%. Demethylation of *trans*-codeine with pyridine hydrochloride yields B/C-*trans*-**morphine**, m.p. +1MeOH, 108–109°; $[\alpha]_D$ +80°; *hydrochloride*, m.p. 263–265°; picrate, m.p 217–220° (*Kugita et al.*, Tetrahedron, 1969, **25**, 1851; J. med. Chem., 1970, **13**, 973). Reduction of *trans*-codeine in this way, followed by Öppenauer oxidation gave *trans*-dihydrocodeinone identical with the base prepared from Gates's 1-bromo-*trans*-dihydrocodeinone (LXXI or LXXII), but the correlation is not stereochemically unambiguous:

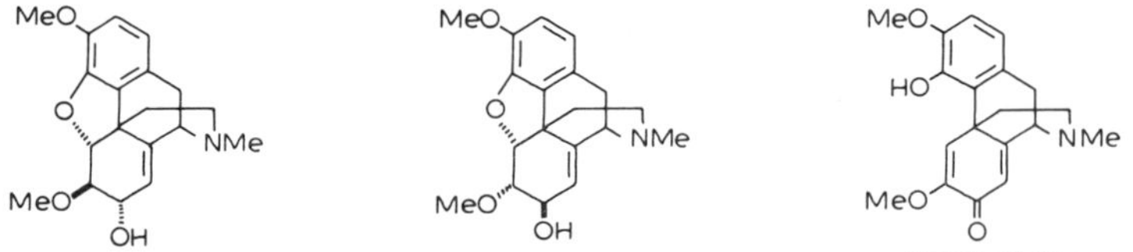

(LXXIX)

(LXXX)

Salutaridine (LXXXI)

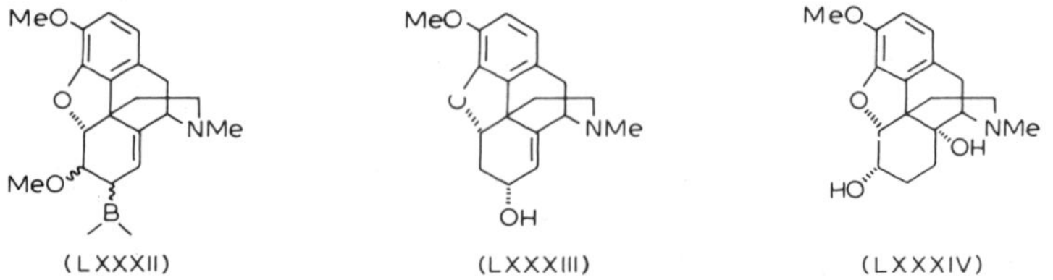

(LXXXII)

(LXXXIII)

(LXXXIV)

Other bases in the morphine series will react with diborane, for example, thebaine with one equivalent gives the isomeric alcohols LXXIX and LXXX, both of which can be oxidised to the alkaloid salutaridine (LXXXI), whereas with an excess of the reagent the products are LXXIX, LXXVII, LXXXIII and LXXXIV, the last three of which doubtless arise from the $\Delta^{6,8}$-diene resulting from elimination in the intermediate LXXXII (*idem, loc. cit.*), analogous eliminations having been observed in similarly constituted alkylboranes (*D. J. Pasto* and *C. C. Cumbo*, J. Amer. chem. Soc., 1964, **86**, 4343).

(d) Reactions and derivatives of the morphine–thebaine group of alkaloids

(i) The halogeno-morphides and -codides

The replacement of the alcoholic hydroxyl group of morphine and codeine by halogen yields the halogeno-morphides and -codides. This reaction may be effected with concentrated hydrochloric acid, phosphorus halides and oxyhalides and thionyl halides (*A. Matthiessen* and *C. R. A. Wright*, Proc. roy. Soc., 1869, **18**, 83; *E. Vongerichten*, Ann., 1881, **210**, 105; *F. H. Lees*, J. chem. Soc., 1907, **91**, 1408; *E. Speyer* and *H. Rosenfeld*, Ber., 1925, **58**, 1113). The iodo compounds may be obtained from the bromo compounds by treatment with potassium iodide (*L. F. Small* and *F. L. Cohen*, J. Amer. chem. Soc., 1931, **53**, 2214).

Two isomers of the chloro compounds exist; α-**chlorocodide**, m.p. 151–153°, $[\alpha]_D$ −383°, has the structure LXXXV (*Small et al.*, J. org. Chem., 1940, **5**, 334) and on heating with concentrated hydrochloric acid it is converted into β-**chlorocodide**, m.p. 152–153°, $[\alpha]_D$ −10°, (LXXXVI) by an α,γ-shift of halogen; the point of attachment of the halogen in this compound has not been rigorously proved, but is inferred from the reactions of the compound and from the mode of its formation (*G. Stork*, "The Alkaloids", Ed. *R. H. F. Manske* and *H. L. Holmes*, Academic Press, New York, 1952, Vol. II, Ch. 8). The bromo and iodo compounds are believed to have structures analogous to the β-chloro compounds:

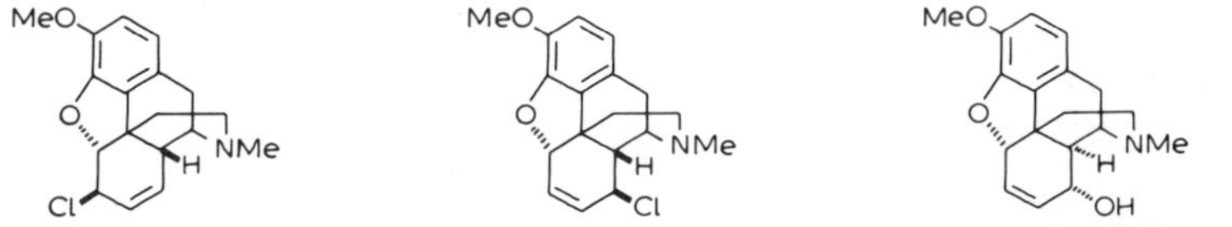

α-Chlorocodide (LXXXV)

β-Chlorocodide (LXXXVI)

Allo-ψ-codeine (LXXXVII)

The reactions can be rationalised on the basis of the stereochemistry of structure I by postulating the formation of α-chlorocodide by S_N2 displacement of hydroxyl by attack from the less hindered side of the molecule. Hydrolysis of the halide by the S_N2

mechanism would, however, involve attack on the more hinderd side, and accordingly the hydrolysis takes place by the S_N2' process to give **allo-ψ-codeine** (LXXXVII), m.p. 116–117°, $[\alpha]_D$ −235°, with the hydroxyl at C-8 and its C-8 epimer **ψ-codeine**, m.p. 181–182°, $[\alpha]_D$ −96.8°. By the same token isocodeine cannot be converted into α-chlorocodide, but by an S_N2' reaction gives β-chlorocodide, which is obtained directly from allo-ψ-codeine. This halide on hydrolysis affords mainly isocodeine by the S_N2' process, and, since α-chloride is readily transformed into the β-compound, some *iso-codeine*, the C-6 epimer of codeine, m.p. 171–172°, $[\alpha]_D$ −150.6°, accompanies ψ-codeine and allo-ψ-codeine during the hydrolysis of this isomer (*Stork, loc. cit.*; *Stork* and *F. H. Clarke*, J. Amer. chem. Soc., 1956, **78**, 4619). Replacement of the halogen atoms by precisely analogous reactions using amines can also occur.

The preparation of *B/C*-trans-*ψ-codeine*, m.p. 130–131°, $[\alpha]_D$ −24.4°, and trans-*allo-ψ-codeine*, m.p. 159–160.5°, $[\alpha]_D$ −89.7°, has also been reported (*Kugita, Takeda* and *Inoue, loc. cit.*).

(ii) Exhaustive methylation of codeine

Codeine and its isomers readily suffer Hofmann degradation of their quaternary salts to give the olefinic codeimethines.

Codeine and its C-6 epimer give **α-codeimethine** (VIII, p. 271), m.p. 118.5°, $[\alpha]_D$ −212°, and *α-codeimethine*, m.p. 167–169°, $[\alpha]_D$ +64.3°, which can be isomerised by heat or alkalis to the more highly conjugated **β-codeimethine** (IX) (p. 271) m.p. 134–135°, $[\alpha]_D$ +438°, and **δ-codeimethine**, m.p. 111–113°, $[\alpha]_D$ +256°, and the epimeric pair ψ and allo-ψ-codeine give **ε-codeimethine** (LXXXVIII), m.p. 129–130°, $[\alpha]_D$ −120°, and its epimer **ξ-codeimethine**, (oil, *perchlorate*, m.p. 117–118°, $[\alpha]_D$ −154°) in which further conjugation cannot occur *(Bentley, loc. cit.)*.

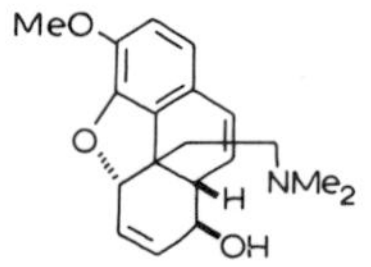

ε- Codeimethine
(LXXXVIII)

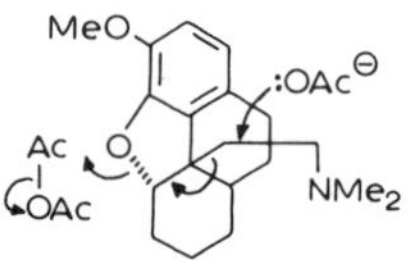

(LXXXIX)

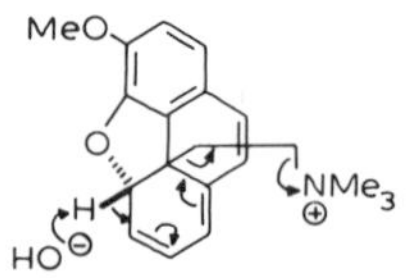

(XC)

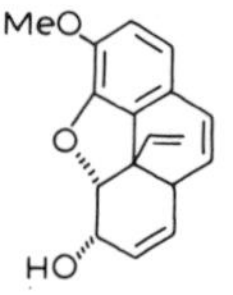

(XCI)

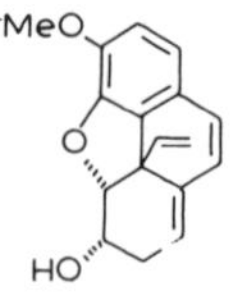

(XCII)

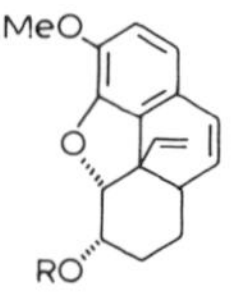

(XCIII)

Further degradation of these methines is dependent upon the conditions. Heating with acetic anhydride and sodium acetate gives acetylmethylmorphol (XII) (p. 271), which presumably arises by dehydration and displacement of side-chain and oxygen bridge as in LXXXIX, and further Hofmann degradation proceeds with dehydration and extrusion of the side-chain in XC to give methylmorphenol (X) (p. 271). Simple degradation to angular vinyl compounds can, however, be achieved by pyrolysis of the amine oxides (*Bentley*, *J. C. Ball* and *J. P. Ringe*, J. chem. Soc., 1956, 1963), the products being the *olefins* XCI (oil, $[\alpha]_D$ $-125°$; *dinitrobenzoate*, m.p. 165°) and XCII, m.p. 107–108°, $[\alpha]_D$ $+434°$, from α- and β-codeimethine, respectively.

The α-, β-, and δ-methines can be reduced to tetrahydro compounds and the ε- and ξ-methines can also be reduced with opening of the oxide bridge, being allylic ethers. The dihydromethines can be prepared from dihydrocodeine and its three isomers. All of the reduced methines can be degraded further by pyrolysis of the amine oxides, or by Hofmann degradation of their quaternary salts, without aromatisation. Hofmann degradation results in methylation of the C-6 hydroxyl group at the expense of quaternary salt in a proportion of the material, for example dihydrocodeinemethine gives both the alcohol (XCIII, R=H) and the ether (XCIII, R=Me) (*H. Rapoport*, J. org. Chem., 1948, **13**, 714) and γ-tetrahydrocodeimethine corresponding products and the cyclic ether codiran (LX) (p. 280) (*Rapoport* and *Payne*, J. Amer. chem. Soc., 1952, **74**, 2630).

(iii) The deoxy compounds

The deoxycodeines and deoxymorphines are bases in which the alcoholic oxygen atom has been removed; they and their reduction products can be prepared from codeine and morphine derivatives, especially the halogeno compounds, by reduction processes. Two series of derivatives, phenolic and non-phenolic, may be distinguished.

Deoxycodeine-E (Δ^7-*deoxycodeine*) (XCIV) may be prepared by the reduction of *p*-toluenesulphonylcodeine with lithium tetrahydridoaluminate; it has m.p. 84° and $[\alpha]_D$ $-68°$ (*Rapoport* and *R. M. Bonner*, J. Amer. chem. Soc., 1951, **73**, 2872; *P. Karrer* and *C. Widmark*, Helv., 1951, **34**, 34); the Δ^8-*isomer* (**deoxycodeine-D**), m.p. 61.5–62°, $[\alpha]_D$ $-21.1°$, may be prepared in the same way from neopine (*Rapoport* and *Bonner*, *loc. cit.*) and also from 8-chlorodihydrocodide by prolonged boiling with sodium in cyclohexanol (*L. F. Small* and *J. E. Mallonee*, J. org. Chem., 1940, **5**, 350).

Deoxycodeine-C (Δ^6-*deoxycodeine*) (XCV), m.p. 105–106°, $[\alpha]_D$ $-199.4°$, results from the action of sodium methoxide on 6-chlorodihydrocodide (*Small* and *Cohen*, *loc. cit.*). A phenolic isomer, $\Delta^{5,7}$-*deoxycodeine* (**deoxycodeine-A**) (XCVI), m.p. 157°, $[\alpha]_D$ $+119°$, may be prepared by the reduction of the halogenocodides with zinc dust/alcohol (*L. Knorr* and *R. Waentig*, Ber., 1907, **40**, 3860; *Knorr* and *H. Horlein*, *ibid.*, 1907; **40**, 4883) and by the reduction of these bases by methylmagnesium iodide (*Small* and *Cohen*, *loc. cit.*).

The phenolic bases, **dihydrodeoxycodeine-B** (Δ^5-*dihydrodeoxycodeine*) (XCVII), m.p. 128–131° and 170–173°, $[\alpha]_D$ −106.9°, **dihydrodeoxycodeine-C** (Δ^6-*dihydrodeoxycodeine*), m.p. 109–111°, $[\alpha]_D$ +5.6° and *tetrahydrodeoxycodeine*, m.p. 123–124° and 157–158°, $[\alpha]_D$ −72.3° (XXIX, R=OMe, R^1=OH) (p. 276) and the non-phenolic **dihydrodeoxycodeine-D** (XCVIII), m.p. 106–107°, $[\alpha]_D$ −82.5°, result from the reduction of the halogenocodides under a variety of conditions (*Small et al., loc. cit.*; *Knorr* and *Horlein*, Ber., 1907, **40**, 376; *Small* and *Cohen*, J. Amer. chem. Soc., 1931, **53**, 2227; *Mosettig et al., loc. cit.*; *Bentley, loc. cit.*). The last of these can also be prepared by the Clemmensen and electrolytic reductions of dihydrothebainone (*E. Speyer* and *S. Siebert*, Ber., 1921, **54**, 1519; *E. Ochiai*, J. pharm. Soc. Japan, 1926, **538**, 99); the optical antipode has been obtained from sinomenine (*K. Goto* and *H. Sudzuki*, Bull. chem. Soc. Japan, 1929, **4**, 244) and the racemate has been synthesised (*R. Grewe et al.*, Ann., 1949, **564**, 161). The phenolic $\Delta^{8,(14)}$-**dihydrodeoxycodeine-E** (XCIX) has been prepared by the electrolytic reduction of 14-bromocodeinone (XIX, R=Br) (p. 273) (*Speyer* and *K. Sarre*, Ber., 1924, **57**, 1404; *Small* and *Cohen, loc. cit.*; *U. Weiss, T. Rull* and *R. B. Bradley*, J. org. chem., 1968, **33**, 3000):

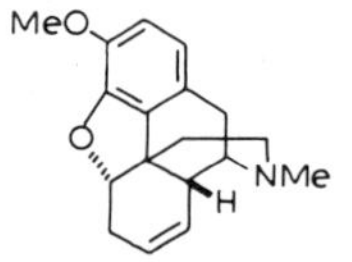

Deoxycodeine-E (XCIV)

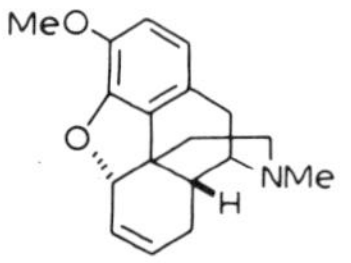

Deoxycodeine-C (XCV)

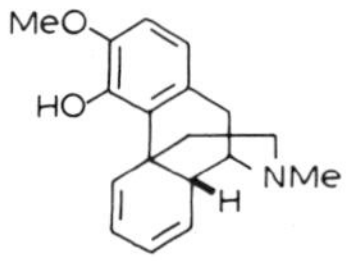

Deoxycodeine-A (XCVI)

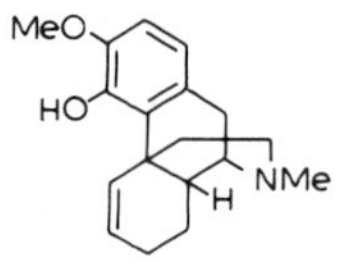

Dihydrodeoxycodeine-B (XCVII)

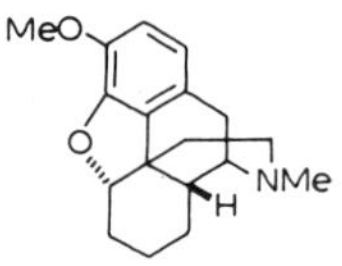

Dihydrodeoxycodeine-D (XCVIII)

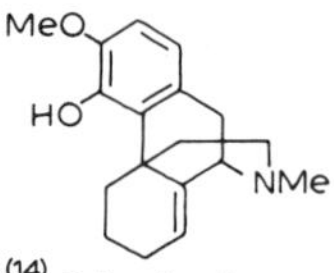

$\Delta^{8,(14)}$-Dihydrodeoxycodeine-E (XCIX)

Δ^7-, Δ^8- and $\Delta^{5,7}$-deoxymorphine, dihydrodeoxymorphine and tetrahydrodeoxymorphine have also been prepared (*Bentley, loc. cit.*).

(iv) The reduction of thebaine

Thebaine (V; R=Me) (p. 268) contains a highly reactive doubly unsaturated phenol ether system, $Ar-\overset{1}{O}-\overset{2}{C}H-\overset{3}{C}(OMe)=\overset{4}{C}H-\overset{5}{C}H=\overset{6}{C}-$, that can suffer reduction in all possible ways, by chemical and catalytic processes.

Lithium tetrahydridoaluminate causes 1,2-reduction to the phenolic conjugated diene, *β*-**dihydrothebaine** (C), m.p. 171–172°, $[\alpha]_D$ +307° (*H. Schmid* and *P. Karrer*, Helv. 1951, **34**, 19; *G. Stork*, J. Amer. chem. Soc., 1952, **74**, 768) and sodium and liquid ammonia bring about 1,4-reduction to the unconjugated diene, *dihydrothebaine-φ*(CI), m.p. 154°, $[\alpha]_D$ +31.4° (*Bentley, R. Robinson* and *A. E. Wain,* J. chem. Soc., 1952, 958). Catalytic reduction under a variety of conditions affords non-phenolic *dihydrothebaine* (XLIX), m.p. 162–163°, $[\alpha]_D$ −267°, (*M. Freund et al.*, Ber., 1920, **53**, 2250) by 5,6-reduction; neopine methyl ether (VI, OH=OMe) (*L. F. Small,* J. org. Chem., 1955, **20**, 953) by 3,4-reduction; *tetrahydrothebaine* (CIII), m.p. 83°, $[\alpha]_D$ −153.4° (*C. Schöpf* and *L. Winterhalder*, Ann., 1927, **452**, 232) by successive 3,6- and 4,5-reduction; *dihydrothebainone* Δ^6-*enol methyl ether* (CIV), m.p. 165–166°, $[\alpha]_D$ −118.4° (*Small* and *Browning, loc. cit.*) by successive 1,6- and 2,5-reduction and *dihydrothebainol 6-methyl ether* (CV), m.p. 141–142°, $[\alpha]_D$ −23.4° (*Small* and *Browning, loc. cit.*) by 1,2-, 3,6- and 4,5-reduction. The Δ^5-*enol methyl ether* of dihydrothebainone, m.p. 145–146°, $[\alpha]_D$ −61.5°, isomeric with CIV, can be prepared by the catalytic reduction of the diene CI, or by the reduction of dihydrothebaine (XLIX) with sodium and liquid ammonia (*Bentley, Robinson* and *Wain, loc. cit.*). Both of these enol ethers are very rapidly hydrolysed to dihydrothebainone (XXXII) by acids, and this ketone is always obtained when thebaine is catalytically reduced in acid solution:

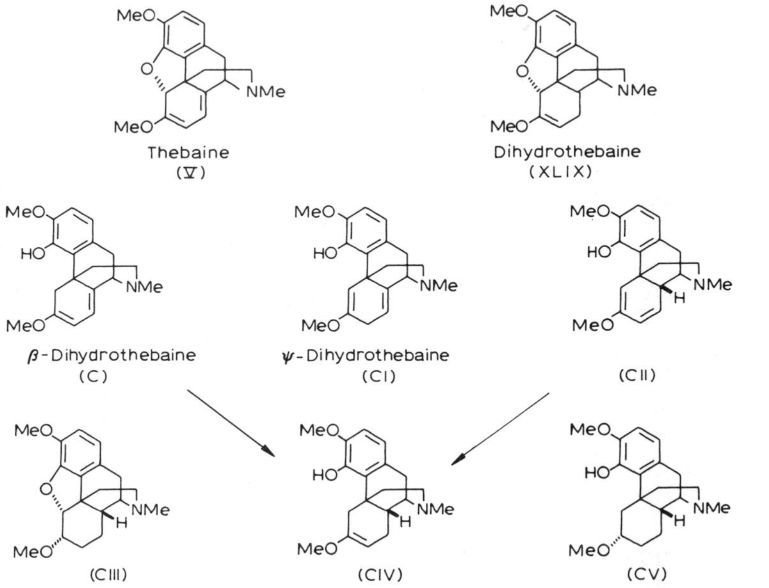

Thebaine (V)

Dihydrothebaine (XLIX)

β-Dihydrothebaine (C)

ψ-Dihydrothebaine (CI)

(CII)

(CIII)

(CIV)

(CV)

An *isomer* of the dienes C and CI, of structure CII, m.p. 154–156°, $[\alpha]_D$ +9.6°, is obtainable by the treatment of codeine methyl ether with sodium ethoxide. This reaction involves a *trans* elimination of the C-6 hydrogen and C-5 oxygen substituents since isocodeine methyl ether is unaffected under the same conditions (*Small* and *Browning, loc. cit.*). Both dienes C and CII can be reduced to the enol ether CIV.

The unconjugated diene CI methiodide can be degraded to the *methine base* CVI, m.p. 99°, which can be further degraded to the N-free *product* CVII, m.p. 88° (*Bentley, Robinson* and *Wain, loc. cit.*), though the methine methyl ether is reported to extrude the side-chain on further degradation to give 3,4,6-trimethoxyphenanthrene (*Freund*, Ber., 1905, **38**, 3234). The diene methiodide is also unstable to acids, suffering ring opening as well as hydrolysis, giving *β*-thebainone-A-methine (CVIII), the $\Delta^{8,9}$ isomer of which can be obtained by the action of acids on the methine base CVI (*Bentley, Robinson* and *Wain, loc. cit.*):

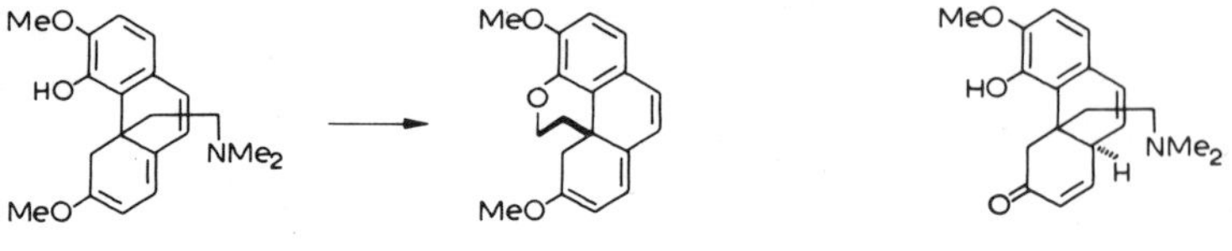

(CVI) (CVII) (CVIII)

Dihydrothebaine (XLIX) can be hydrolysed by hot hydrochloric acid to the related ketone dihydrocodeinone, from which it may be prepared by methylation with potassium *tert*-butoxide and dimethyl sulphate (*Rapoport, Reist* and *Lowell, loc. cit.*). Exhaustive methylation of the base proceeds normally to give a methine base and a nitrogen-free product (*Freund et al., loc. cit.*; *R. S. Cahn*, J. chem. Soc., 1932, 702; *Bentley, Ball* and *Ring, loc. cit.*). Dihydrothebaine and the closely related dihydrocodeinone enol acetate readily react with Grignard reagents, undergoing 1,2- and 1,4-addition to the allylic ether system and giving phenolic 5- and 7-alkyl enol ethers, hydrolysed under the conditions of isolation to 5- and 7-alkyldihydrothebainones, CIX and CX. These bases can be cyclised like dihydrothebainone itself, by treatment with bromine and alkali, to substituted dihydrocodeinones, which can be reduced (removal of 1-bromine) and demethylated to substituted dihydromorphinones (*Small et al.*, J. Amer. chem. Soc., 1936, **58**, 1547; J. org. Chem., 1938, **3**, 204) of which the most important is the drug **metopon**, *5-methyldihydromorphinone* (CXI), m.p. 243–245°, $[\alpha]_D$ −140.7° (*G. Stork* and *L. Bauer*, J. Amer. chem. Soc., 1953, **75**, 4373):

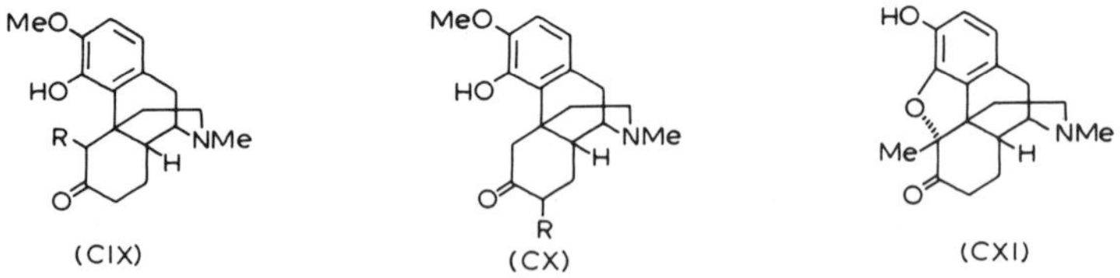

(CIX) (CX) (CXI)

(v) The thebainones

Reductive scission of the 4,5-oxide bridge and hydrolysis of the enol ether system of thebaine affords the phenolic ketones known as thebainones of which five isomers are known.

Thebainone-A (Δ^7-*thebainone*, CXII), m.p. 151–152°, $[\alpha]_D$ −42.5°, can be prepared by hydrolysis of the enol ether CII (*Small* and *Browning*, *loc. cit.*), by the reduction of thebaine with stannous chloride and hydrochloric acid under carefully controlled conditions (*Schöpf* and *H. Hirsch*, Ann., 1931, **489**, 224) and by the catalytic rearrangement of codeine over noble metals (*U. Weiss* and *N. Weiner*, J. org. Chem., 1949, **14**, 194). It is also a minor product of the hydrolysis of dihydrothebaine-ϕ (CI) and β-dihydrothebaine, the major product of which is β-thebainone-A, the C-14 epimer of CXII, with a *trans* fusion of rings B and C (*Small* and *Browning*, *loc. cit.*; *Bentley*, *Robinson* and *Wain*, *loc. cit.*; *M. Gates* and *R. Helg*, J. Amer. chem. Soc., 1953, **75**, 379). Of these two isomers CXII is the more stable (*Gates* and *Tschudi*, *ibid.*, 1956, **78**, 1380), the less stable isomer being formed in the hydrolysis of the dienes C and CI by a kinetically determined process, and is obtained from the B/C *trans* isomer in good yield by epimerisation in acetic acid (*Gates* and *Helg*, *loc. cit.*):

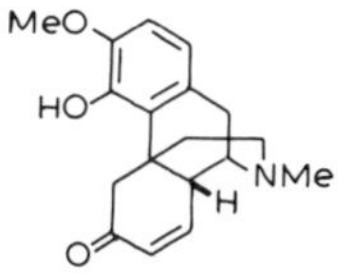

Thebainone-A (CXII) 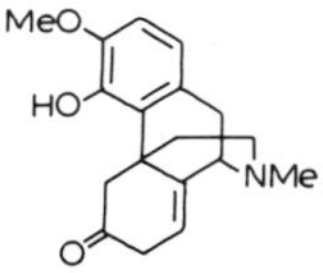 Thebainone-B (CXIII) 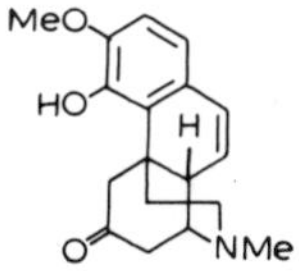 Thebainone-C (CXIV)

The β,γ-unsaturated ketone (CXIII), **thebainone-B**, is obtained as the *hydrobromide*, m.p. 205–207°, $[\alpha]_D$ +15.8°, by the hydrolysis of dihydrothebaine-ϕ (CI) with alcoholic hydrobromic acid (*Bentley* and *Wain*, J. chem. Soc., 1952, 967), and is stable only in the form of its salts; it is readily isomerised to β-thebainone-A.

Thebainone-C ("α-*thebainone*", CXIV), m.p. 185–186°, $[\alpha]_D$ +158.5°, is obtained by the hydrolysis of dihydrothebaine-ϕ (CI) with warm sulphurous acid and is doubtless formed *via* a normethine analogous to CVIII, with epimerisation at C-14 prior to addition of the amine to the enone system. The base has a B/C *cis* fusion but a model of the B/C *trans* isomer (unknown) is equally free of strain (*Bentley* and *H. M. E. Cardwell*), ibid., 1955, 3245).

The fifth isomer of thebainone, metathebainone, is the product of a rearrangement and is described in a later section.

Dihydrothebainone (XXXII) (p. 276), m.p. +$\frac{1}{2}$EtOH, 148–152°, $[\alpha]_D$ −80.1°, may be obtained by the reduction of thebainone-A (CXII) and thebainone-B (CXIII), by the hydrolysis of the enol ether CIV and its Δ^5-isomer, and by the reduction of thebaine under a variety of conditions in acid solution (*Freund et al.*, *loc. cit.*; *A. Skita et al.*, Ber., 1921, **54**, 1560; *H. Wieland* and *M. Kotake*, Ann., 1925, **444**, 69). Its epimer, β-*dihydrothebainone* (XXXIX), oil, *perchlorate*, m.p. 254–255°, $[\alpha]_D$ −32.5°, can be prepared by the hydrogenation of β-thebainone-A (*Small* and *Browning*, *loc. cit.*). Both these ketones on exhaustive methylation give methine bases which can be reduced and degraded further to thebenone (LXIII) and its C-14 epimer (*Wieland* and *Kotake*, *loc. cit.*; *Small* and *Browning*, *loc. cit.*).

(vi) The thiocodides

When bromo or β-chlorocodide (LXXXVI) is heated with ethanethiol S_N2' replacement of halogen by SEt occurs, the product being **α-ethylthiocodide** (CXV) (two forms, m.p. 77–79° and 88–89°, $[\alpha]_D$ −344.6°). Under similar conditions α-chlorocodide (LXXXV) gives **δ-ethylthiocodide** (CXVI) (*perchlorate*, m.p. 223–224°, $[\alpha]_D$ +54.3°). When the α-isomer is heated with sodium ethoxide it is converted into an analogue of the diene CII, obtainable in the same way from codeine methyl ether, namely **β-ethylthiocodide** (CXVII) (m.p. 148°, $[\alpha]_D$ −45.5°) the sulphoxide of which was erroneously given the name γ-ethylthiocodide (*R. Pschorr* and *A. Rollett*, Ann., 1910, **373**, 1; *D. E. Morris* and *L. F. Small*, J. Amer. chem. Soc., 1934, **56**, 2159).

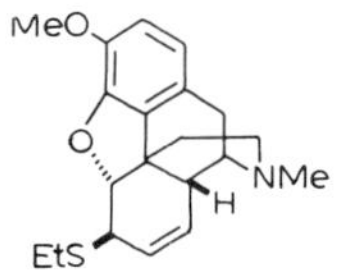

α-Ethylthiocodide (CXV)

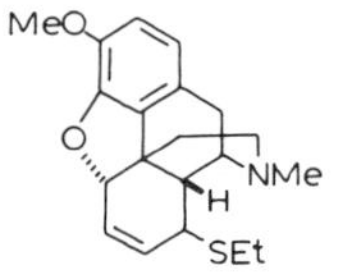

δ-Ethylthiocodide (CXVI)

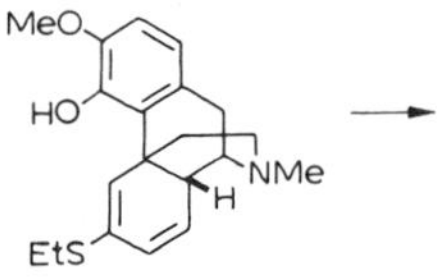

β-Ethylthiocodide (CXVII)

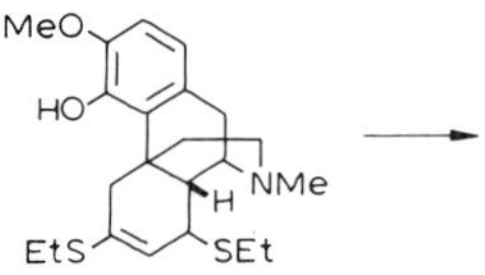

(CXVIII)

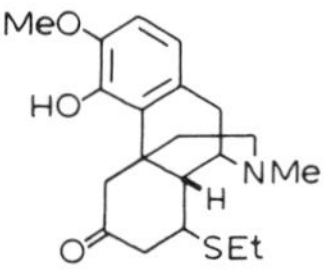

(CXIX)

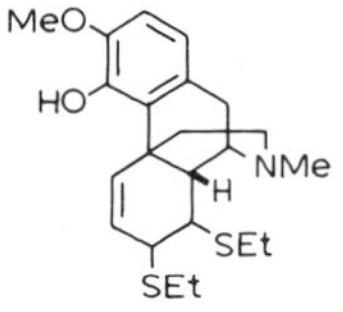

(CXX)

The hydrolysis of β-ethylthiocodide yields thebainone-A (CXII), and the liberated ethanethiol adds to unchanged codide to give **dihydro-β-diethyldithiocodide** (CXVIII), hydrolysis of which gives 8-ethylthiodihydrothebainone (CXIX), also obtainable from thebainone-A and ethanethiol. δ-Ethylthiocodide, being an allylic phenyl ether, can readily be reduced to phenolic bases and will add ethanethiol to give the base CXX isomeric with CXVIII (*Small* and *Morris*, *loc. cit.*). Deoxycodeine-C (XCV) will add ethanethiol in the same way to give a 7-ethylthiodihydro compound.

(vii) 14-Bromo- and 14-hydroxy-codeinone

Treatment of thebaine with bromine in acetic acid, or with *N*-bromosuccinimide gives 14-*bromocodeinone* (XIX, R = Br), m.p. 156–157°, which can be catalytically reduced to *neopinone* (CXXI), m.p. 125°, $[\alpha]_D$ −9.8°. This can be reduced to neopine (IV) and can be isomerised to codeinone (VI) on charcoal (*H. Conroy*, J. Amer. chem. Soc., 1955, **77**, 5960). Treatment of neopinone with alkali causes dimerisation by Michael addition of the enolate ion to codeinone generated in solution. 14-Bromocodeinone on treatment with Claisen's alkali suffers S_N2' displacement of bromide ion by hydroxyl and gives the enolised diketone (CXXIII, R = H) which can be methylated to the alkaloid salutaridine (CXXIII, R = Me) (*D. E. Rearick* and *M. Gates*, Tetrahedron Letters, 1970, 507) the reaction involving oxide bridge displacement in the intermediate 7-hydroxy ketone CXXII:

(CXXI) (CXXII) (CXXIII)

14-Hydroxycodeinone (XIX, R = OH) is obtained from thebaine and hydrogen peroxide in acetic acid (*Freund* and *Speyer*, *loc. cit.*) and from 14-bromocodeinone by the action of hydroxylamine (*Freund*, Ber., 1906, **39**, 844). Unlike codeinone it is stable to rearrangement and aromatising degradations as a consequence of the angular substituent at C-14. The enol acetate of 14-acetoxydihydrocodeinone will react with Grignard reagents in the same way as dihydrocodeinone enol acetate and dihydrothebaine, giving 14-hydroxy-analogues of the ketones CIX and CX (*R. E. Lutz* and *Small*, J. org. Chem., 1939, **4**, 220).

14-*Nitrocodeinone*, m.p. 172.5–173°, may be prepared by the action of tetranitromethane on thebaine in methanol, followed by hydrolysis of the resulting dimethyl acetal; it can be reduced to 14-aminocodeinone (*R. M. Allen* and *G. W. Kirby*, Chem. Comm., 1970, 1346).

(viii) Migration of nitrogen from C-9 to C-14

The reduction of 14-bromocodeinone (XIX, R = Br) with excess of sodium tetrahydridoborate gives neopine (IV) and its C-6 epimer isoneopine (*via* neopinone CXXI), and an isomer of codeine, **indolinocodeine** (CXXVI, R = H), m.p. 129–131°, $[\alpha]_D$ +23.7°, formed from 14-bromocodeine (CXXIV) *via* a carbonium ion and the aziridinium salt CXXV (*S. Okuda et al.*, J. org. Chem., 1962, **27**, 4121). The solvolysis of 14-bromocodeine doubtless proceeds through the same salt and yields the bases CCXVI, R = OH, OAc and OMe (*K. Abe et al.*, Chem. pharm. Bull., Japan, 1969, **17**, 1847, 1917).

The structure of CXXVI, R = H follows from its spectra, from the degradation of its methiodide to β-codeimethine (IX) and from its reduction and oxidation to an isomer of dihydrocodeinone that can be further reduced to a phenolic ketone CXXIX isomeric with, but different from both B/C *cis* (XXXII) and B/C *trans* (XXXIX) dihydrothebainone. Confirmation of the structure also follows from the conversion of the base into an isomer of tetrahydrodeoxycodeine (XXIX, R = OMe, R^1 = OH) and its C-14 epimer. The base CXXVI, R = OAc, can be converted through the 6-chloro compound into the alcohol CXXVIII, and the methanesulphonate of this can be solvolysed with rearrangement to 7-methoxydeoxyneopine (CXXVII) (*Abe et al., loc. cit.*):

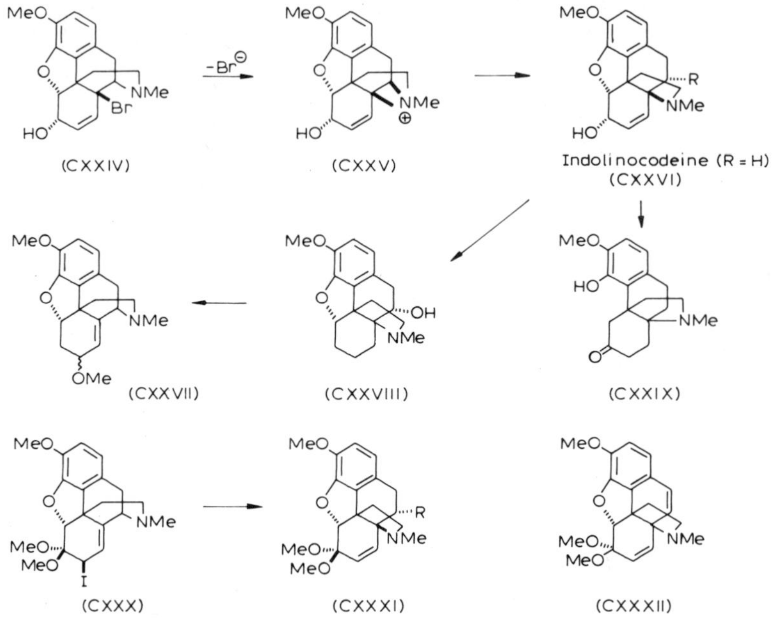

(CXXIV) (CXXV) Indolinocodeine (R = H) (CXXVI)

(CXXVII) (CXXVIII) (CXXIX)

(CXXX) (CXXXI) (CXXXII)

Iodination of thebaine in methanol gives the 7-iodoacetal CXXX and solvolysis of this proceeds with migration of nitrogen to give the bases CXXXI (R = OH, OAc, OMe, ONO, N_3, NC, NMe_2) and degradation to the base CXXXII; the aziridinium salt intermediate in these processes has been isolated as the perchlorate (*Allen* and *Kirby*, J. chem. Soc. Perkin I, 1973, 363).

(ix) Diels–Alder reactions of thebaine

Thebaine contains a highly reactive diene system and very readily adds unhindered dienophils giving, for example, adducts with maleic anhydride (XX), benzoquinone (CXXXIII), methyl vinyl ketone (CXXXIV), acrolein, ethyl acrylate, acrylonitrile and nitrosobenzene (CXXXV). The reactions are under electronic and stereochemical control and unsymmetrical dienes only add in the direction shown and give almost exclusively 7 α-substituents. (*Bentley et al.*, J. org. Chem., 1958, **23**, 1720; J. Amer. chem. Soc., 1967, **89**, 3267; Chem. Comm., 1969, 1411; J. chem. Soc. Perkin I, 1972, 870; *O. Hromatka* and *G. Sengtschmid*, Monatsh., 1971, **102**, 1022; *R. Giger et al.*, Tetrahedron, 1973, **29**, 2387; *R. Rubinstein et al., ibid.*, 1974, **30**, 1201).

The adduct with nitrosobenzene CXXXV is readily hydrolysed to 14-phenylhydroxylaminocodeinone (CXXXVIII) from which 14-anilinodihydrocodeinone can be prepared by reduction. The adduct with methyl vinyl ketone reacts readily with Grignard reagents to give a series of alcohols CXXXVII (R = Me) most of which are powerful analgesics and these may be demethylated with alkalis to the corresponding phenols CXXXVIII (R = H), which are very much more potent analgesics (see below p. 302) (*Bentley et al.*, J. Amer. chem. Soc., 1967, **89**, 3273, 3281). The ketone CXXXIV can be converted, through the dimethyl acetal or enol ether and a Vilsmeier reagent, into the methoxyenal CXXXVI, from which a variety of heterocyclic compounds can be produced (*J. J. Brown* and *R. A. Hardy*, Belg. P., 724,258/1969; C.A., 1970, **72**, 21816x).

Codeinone (VI) can be converted by treatment with secondary amines into a series of dienamines which also undergo Diels–Alder reactions with dienophils to give bases such as the ketone CXXXIX (*Brown et al.*, U.S.P., 3,318,885/1967; C.A., 1967, **67**, 82295v). Thebaine will react with acetylenic dienophils, *e.g.* with ethyl propiolate, to give adducts such as CXL, which are unstable on heating when they are converted into benzazocines CXLI (*H. Rapoport* and *P. Sheldrick*, J. Amer. chem. Soc., 1963, **85**, 1636); similar arrangements occur with 7,8-unsaturated bases derived from the ketone CXXXIV:

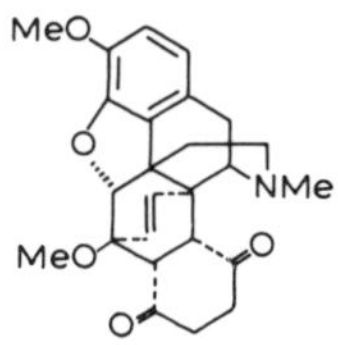

(CXXXIII) 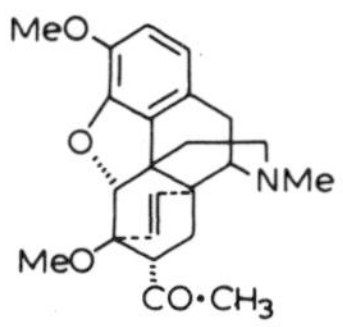 (CXXXIV) 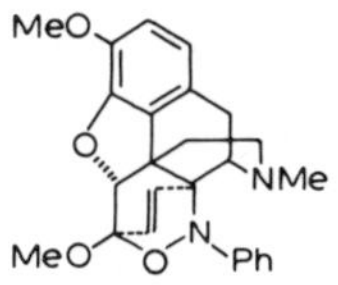 (CXXXV)

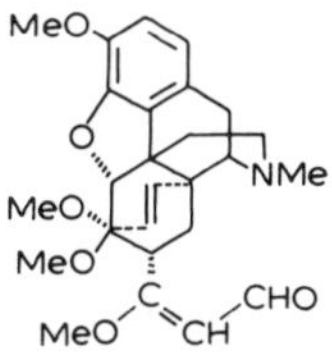

(CXXXVI) 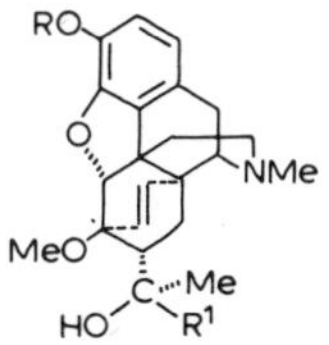 (CXXXVII) 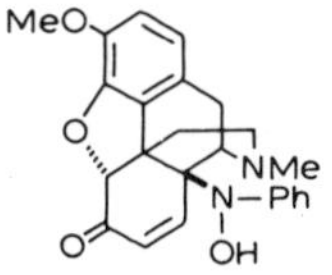 (CXXXVIII)

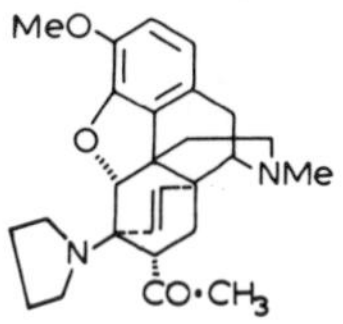

(CXXXIX) 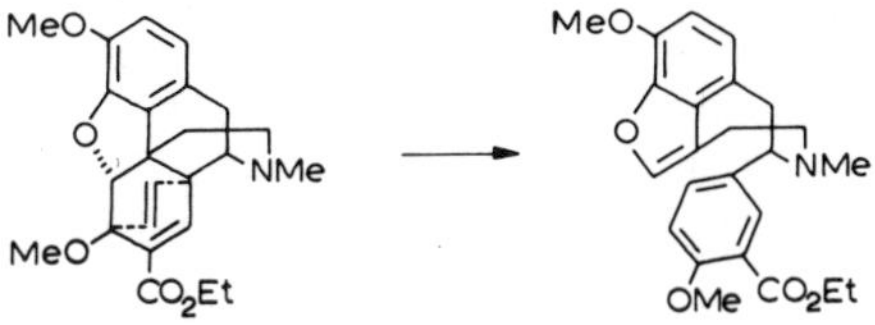 (CXL) (CXLI)

Bases of the series CXXXIV and CXXXVII undergo a wide variety of acid- and base-catalysed rearrangements (see below).

(x) Molecular rearrangements involving the C-13 side-chain

The alkaloids of the morphine group, and their derivatives very rapidly suffer molecular rearrangement with migration or loss of the side-chain. The extrusion of the side-chain in the acetolysis and Hofmann degradation of the codeimethines and thebaine has been considered above. Most of the other rearrangements are observed with thebaine, though codeinone and neopinone, into which thebaine may be converted in acid solution, behave similarly, and are initiated by attack of the cyclic ether oxygen atom by a proton or other positive ion with opening of the 4,5-ether bridge. The resulting electron deficit at C-5 in thebaine/codeinone is then made good by migration of the side-chain from C-13 to C-14, or, in derivatives of the alkaloids, by migration of some other group, as in CXXXVII, or by introduction of an electron pair from an external source. These rearrangements may be summarised as follows.

In concentrated hydrochloric acid thebaine dissolves to give an orange halochromic solution, which contains the protonated dienone CXLIV, formed *via* CXLIII, (*R. Robinson*, Nature, 1947, **160**, 815; *W. Fleischhacker et al.*, Monatsh., 1968, **99**, 300; *R. T. Channon et al.*, Chem. Comm., 1969, 92). This solution on reduction, chemically or catalytically, gives the enone *metathebainone* CXLVII, m.p. $+1\frac{1}{2}$ H_2O 88–90°, +MeOH 115–118°, $[\alpha]_D$ −419°, isomeric with thebainones CXII and CXIII; (*R. Pschorr et al.*, Ber., 1905, **38**, 3160; *C. Schöpf* and *F. Borkowsky*, Ann., 1927, **458**, 148; *Cahn*, J. chem. Soc., 1933, 1038), the structure of which has been proved by degradation to 14-ethyl-3,4-dimethoxy-6-oxo-6,7,8,9,10,14-hexahydrophenanthrene identical with material prepared by synthesis (*Bentley et al.*, Tetrahedron, 1965, **21**, 255).

In concentrated acid the nitrogen atom of the dienone CXLIV remains fully protonated and further rearrangement on heating, following protonation of the carbonyl group, proceeds with dienone–phenol rearrangement CXLVI to morphothebaine (CXLV) (*L. Knorr* and *R. Pschorr*, Ber., 1905, **38**, 3153), the structure of which has been proved by degradation to 3,4,6-trimethoxy-8-vinylphenanthrene and conversion of this through the 8-carboxylic acid into 3,4,6,8-tetramethoxyphenanthrene (*Pschorr* and *Rettberg*, Ann., 1910, **373**, 51) and by synthesis of morphothebaine dimethyl ether (*J. M. Gulland* and *C. J. Virden*, J. chem. Soc., 1928, 921).

The conversion of morphine into apomorphine (XIII) proceeds by a similar mechanism after dehydration of the alkaloid, the rearranged carbonium ion CLI suffering further migration of the side-chain from C-14 to C-8.

In dilute acid solution thebaine is converted into the largely unprotonated dienone CXLVIII, in which an unshared pair of electrons is available on the nitrogen atom, and further protonation of the carbonyl group results in production of the quaternised Schiff base CXLIX, which is very rapidly hydrolysed to the amino-carbaldehyde CL.

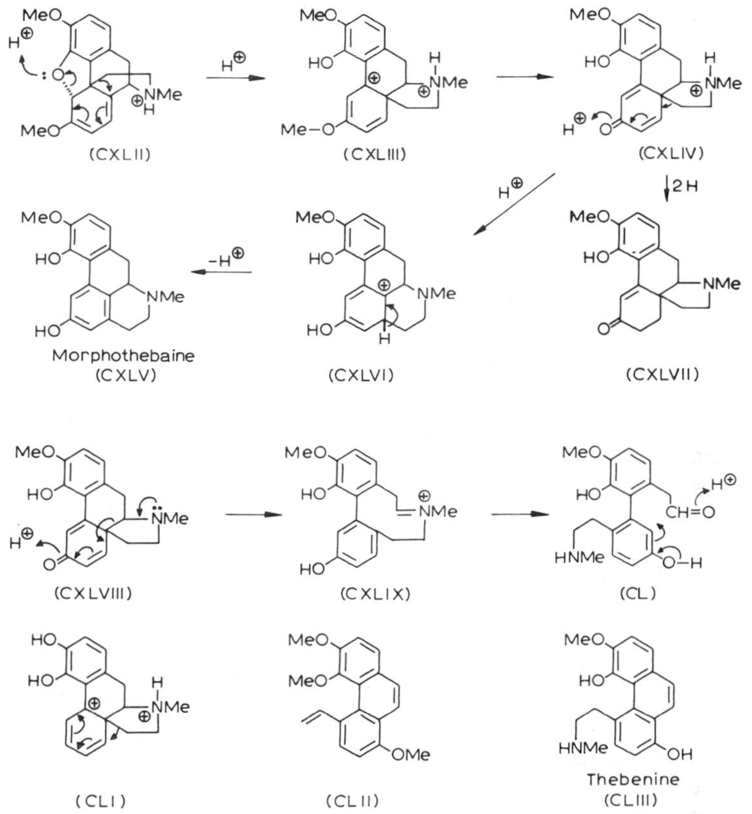

(CXLII) (CXLIII) (CXLIV)

Morphothebaine (CXLV) (CXLVI) (CXLVII)

(CXLVIII) (CXLIX) (CL)

(CLI) (CLII) Thebenine (CLIII)

This aldehyde then suffers cyclodehydration to the secondary base thebenine (CLIII), the whole rearrangement from thebaine being complete in 90 seconds in boiling 2*N* hydrochloric acid. It may be noted that this rearrangement cannot be satisfactorily formulated if the side-chain is initially attached at C-5 or C-14, or if the nitrogen atom is attached to C-10, and this mechanism was an important keystone in the arguments of *Gulland* and *Robinson (loc. cit.)* culminating in their proposal of the correct structures of morphine and thebaine. The structure of thebenine was determined by degradation to the olefin CLII and reduction of this to 5-ethyl-3,4,8-trimethoxyphenanthrene, identical with material prepared by synthesis (*Gulland* and *Virdan*, *loc. cit.*).

The reaction between thebaine (CLIV) and Grignard reagents follows a course basically similar to that of the thebenine transformation, but, as the conditions are neutral and anhydrous, hydrolysis of the C-6 methoxyl

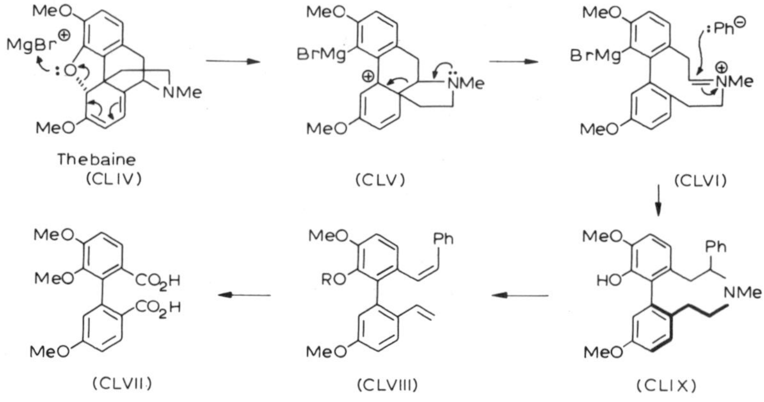

Thebaine (CLIV) (CLV) (CLVI)

(CLVII) (CLVIII) (CLIX)

group and the Schiff base quaternary salt does not occur. Instead the Grignard reagent brings an electron pair to the Schiff base salt, which is thus reduced to a tertiary base, the reaction with phenylmagnesium bromide being represented in CLIV⟶CLIX (*Robinson*, *loc. cit.*; *Bentley* and *Robinson*, J. chem. Soc., 1952, 947). The end-product, phenyldihydrothebaine (CLIV), exists in four stereoisomeric forms. One of the centres of dissymmetry is the asymmetric carbon atom (originally C-9) attached to nitrogen, which is configured in both senses by the entry of the phenyl group, though one configuration predominates ($\sim$90%) as a result of steric hindrance to attack by $Ph^{\ominus}$ on one side of the double bond in CLVI. The second centre of dissymmetry is the biphenyl system, which is generated in one configuration only from the alkaloid CLIX, on which the nine-membered ring places a restraint to rotation overcome by heating above 200° C (*L. F. Small et al.*, J. org. Chem., 1939, **3**, 509; 1947, **12**, 847).

Phenyldihydrothebaine shows none of the properties of an olefin or an enol ether. Catalytic reduction only occurs under forcing conditions and then results in hydrogenolytic cleavage of the nitrogen-containing ring. Cleavage of the methoxyl groups occurs only with concentrated hydrobromic acid and affords a trihydric phenol readily methylated to phenyldihydrothebaine dimethyl ether. These properties are sufficient to show that the near-aromatic ring of thebaine has become fully aromatic during the Grignard reaction with migration of the side-chain from C-13. If this is accepted then the nitrogen-free product CLVIII (R = H), obtained by two successive Hofmann degradations, and therefore containing two double bonds generated by those degradations, can contain no asymmetric carbon atom and must owe its optical activity to restricted rotation in a hindered biphenyl system *(Robinson, loc. cit.)*. The two aromatic nuclei cannot be coplanar in such a hindered biphenyl and hence none of the compounds of this series shows biphenyl absorption in the ultraviolet, a fact regarded by *Small* (J. org. Chem., 1947, **12**, 847) as an insuperable objection to structures based on the aromatisation of ring C.

The structure CLIX for phenyldihydrothebaine was confirmed by oxidation of the base to 4-methoxyphthalic acid and of the nitrogen-free product CLVIII (R = Me) to 5,6,5′-trimethoxydiphenic acid (CLVII) (*Bentley* and *Robinson, loc. cit.*). The parent "dihydrothebaine" from which phenyldihydrothebaine is derived, *i.e.* the base CLIX (Ph = H) has been prepared by reduction of the iminium salt CLVI generated from thebaine and anhydrous magnesium iodide (*Bentley*, J. Amer. chem. Soc., 1967, **89**, 2464).

ψ-Codeinone (XVI) reacts with methylmagnesium iodide to give methyldihydro-ψ-codeinone, which shows no properties of an olefin or a ketone (*R. E. Lutz* and *Small*, *ibid.*, 1935, **57**, 2651), and doubtless has the structure of CLX.

The mechanistic processes set out in formulae CXLII–CXLIX, and CLIV–CLIX have been the basis of an elegant synthesis of the alkaloid **protostephanine** (CLXII) from the dienone CLXI (*A. R. Battersby et al.*, Chem. Comm., 1968, 1214):

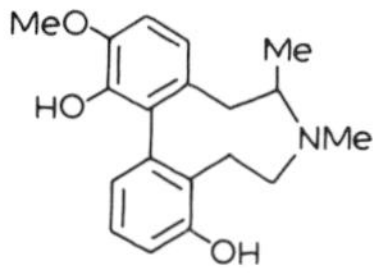

MeO, HO, Me, NMe, OH

(CLX)

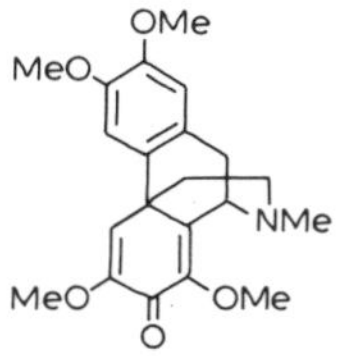

OMe, MeO, NMe, MeO, OMe, O

(CLXI)

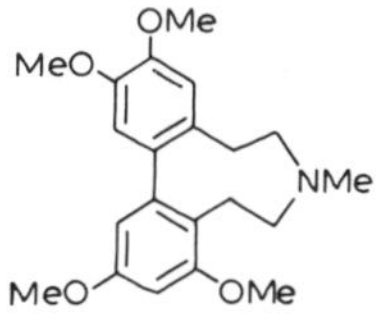

OMe, MeO, NMe, MeO, OMe

Protostephanine

(CLXII)

(xi) Molecular rearrangements not involving the C-13 side-chain

Opening of the 4,5-oxide bridge without migration of the C-13 side-chain occurs if an electron pair can be brought to C-5 by an external reagent or by migration of some other group. External sources provide such an electron pair in the reactions of dihydrothebaine (CLXIII, A = OMe), dihydrocodeinone enol acetate (CLXIII, A = OAc) and deoxycodeine-C (CLXIII, A = H) with Grignard reagents (*L. F. Small et al.*, J. Amer. chem. Soc., 1936, **58**, 192, 1547; J. org. Chem., 1938, **3**, 204), in the conversion of δ-ethylthiocodide (CXVI) (p. 290) into the base CXX (p. 290) and in the reduction of dihydrocodeinone with ethanethiol (see CLXIV) *via* the thio compound CLXV to dihydrothebainone (XXXII) (p. 276) (*T. D. Perrine* and *Small*. J. org. Chem., 1952, **17**, 1540):

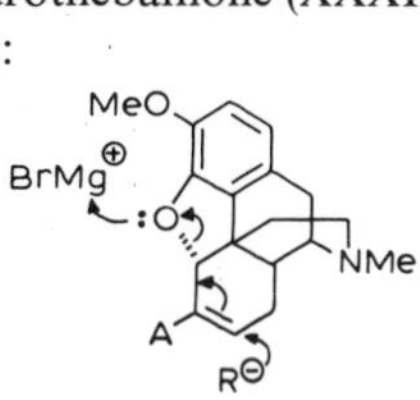

MeO, BrMg⊕, :O, NMe, A, R⊖

(CLXIII)

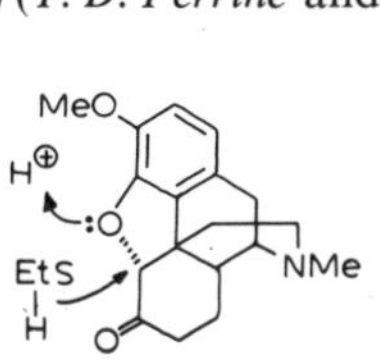

MeO, H⊕, :O, EtS, H, NMe, O

(CLXIV)

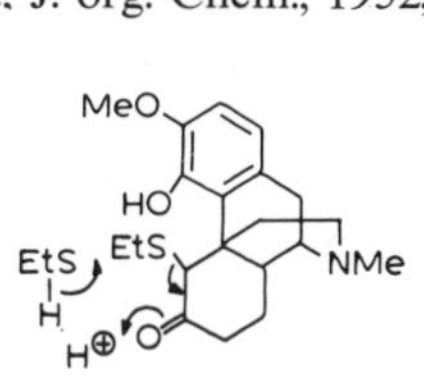

MeO, HO, EtS, EtS, H, NMe, H⊕, O

(CLXV)

An electron pair can be made available to C-5 from elsewhere in the molecule in derivatives of the Diels–Alder adducts of thebaine. The quinol CLXVI, obtained from the adduct CXXXIII, rearranged by acids as shown, with migration of the quinol nucleus to C-5 and the production of flavothebaone (CLXVII) (*C. Schöpf et al.*, Ann., 1938, **536**, 216; *Bentley et al.*, J. org. Chem., 1956, **21**, 1348; 1957, **22**, 409, 418; 1958, **23**, 941). A similar rearrangement occurs when the alcohol CLXX is heated with mineral acid, and this proceeds through the alkenylcodeinone CLXVIII, obtainable from CLXX and formic acid, which can be separately converted into the phenol CLXXI (*idem*, J. Amer. chem. Soc., 1967, **89**, 3293). A related rearrangement of the ketone CLXIX, in the presence of base, occurs with displacement of the oxide bridge and attack at C-6 by the phenate anion so generated, the product being an isomer CLXXIII of the original ketone, which is very readily hydrolysed to the diketone CLXXIV. A methyl ether of the intermediate phenol CLXXII can be obtained by the addition of a methylating agent during the reaction. The ketone CLXXIII has been converted into the olefin CLXXI by the successive action of methyl-lithium and warm hydrochloric acid (*idem*, J. org. Chem., 1958, **23**, 1725; J. Amer. chem. Soc., 1967, **89**, 3312).

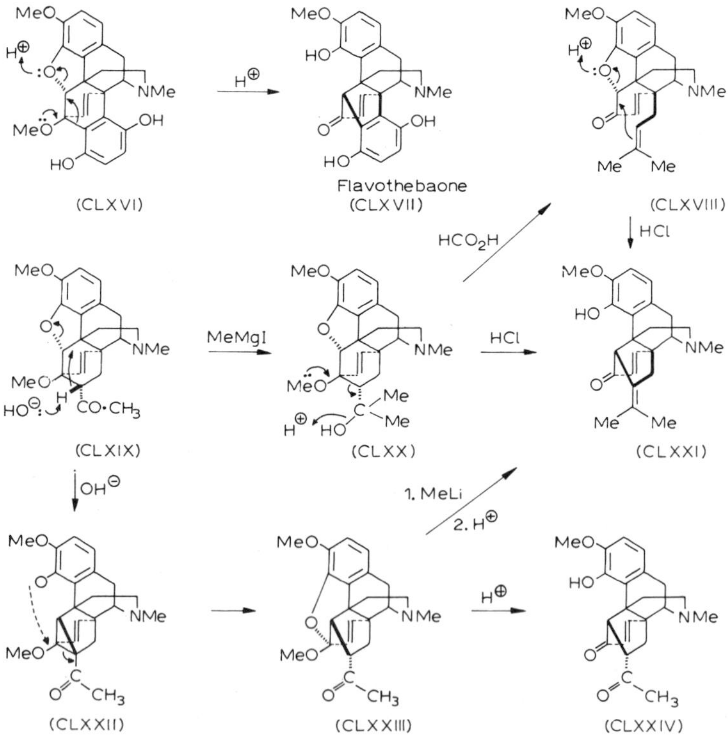

(CLXVI)

Flavothebaone
(CLXVII)

(CLXVIII)

(CLXIX)

(CLXX)

(CLXXI)

(CLXXII)

(CLXXIII)

(CLXXIV)

Other acid-catalysed rearrangements of CLXIX and CLXVII and their homologues, involving only the bridged ring-C and C-7 substituent, in which the oxide bridge remains intact, have also been observed (*idem*, *ibid*., 1967, **89**, 3293, 3303; J. chem. Soc. C., 1969, 2229, 2239; *Bentley*, "The Alkaloids," Ed. *R. H. F. Manske*, Academic Press, New York, Vol. XIII, 1960).

(xii) Flavothebaone

The structure of flavothebaone was deduced from the reactions of the product of Hofmann degradation of its trimethyl ether. The methine base (CLXXV) is transformed by heating with alkalis into a ψ-methine (CLXXVII) by dealdolisation of the enone system and loss of the resulting angular formyl group (CLXXVI):

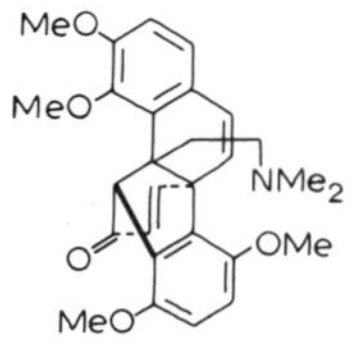

(CLXXV) 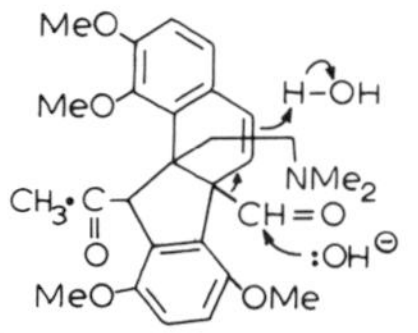 (CLXXVI) 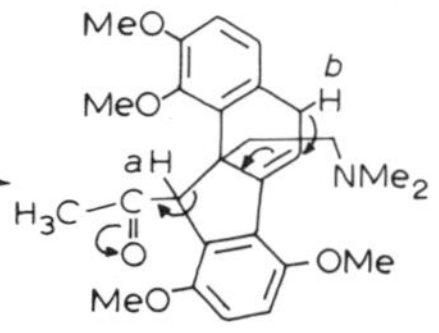 (CLXXVII)

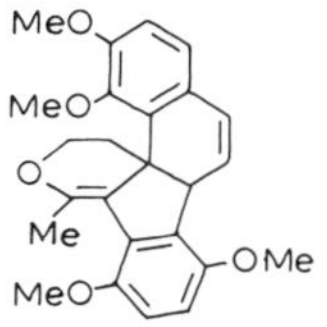

(CLXXVIII) 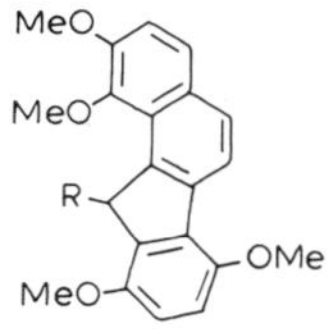 (CLXXIX) 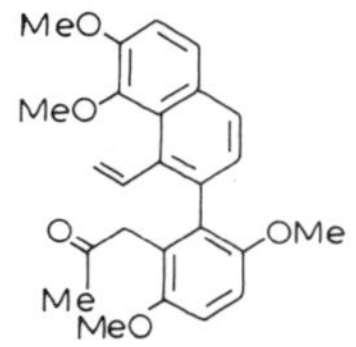 (CLXXX)

When the ψ-methine quaternary salt is subjected to Hofmann degradation two protons are readily removable. Removal of proton *a* results in elimination of ethylene and trimethylamine in a process exactly analogous to the conversion of α-codeimethine into methylmorphenol (X) (p. 271), the primary product being the ketone CLXXIX (R = Ac), part of which is hydrolysed to the benzfluorene CLXXIX (R = H). An alternative pathway following removal of this proton involves displacement of trimethylamine by the enolate ion, the product being the enol ether CLXXVIII. Removal of the proton *b* can be followed by the bond shifts shown which, together with normal Hofmann elimination, leads to the phenylnaphthalene CLXXX, which is the major product of the reaction (*Bentley et al.*, J. org. Chem., 1957, **22**, 409; J. chem. Soc. Perkin I, 1974, 682). Of these the structure of the benzfluorene (CLXXIX, R = H) has been confirmed by synthesis (*idem*, J. chem. Soc., 1963, 4055).

Normal Hofmann degradation of the methines CLXXV and CLXXVII cannot be accomplished, but as with codeimethines unrearranged angular vinyl compounds can be obtained by pyrolysis of the related amine oxides (*idem*, J. org. Chem., 1957, **22**, 422). The ψ-methine oxime CLXXXI loses acetonitrile under the conditions of the Beckmann transformation, giving the base CLXXXII, which can be aromatised with migration of the side-chain to CLXXIX ($R = CH_2 \cdot CH_2NMe_2$) by heating with formic acid (*Bentley* and *J. P. Ringe*, *ibid*., 1957, **22**, 424):

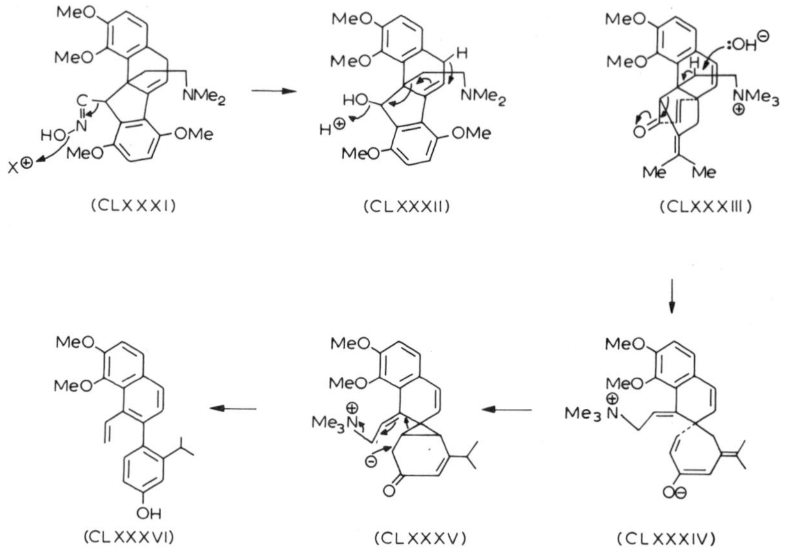

(CLXXXI) (CLXXXII) (CLXXXIII)

(CLXXXVI) (CLXXXV) (CLXXXIV)

A rearrangement similar to that resulting in the naphthalene derivative CLXXX occurs on degradation of the methohydroxide CLXXXIII, which gives the phenylnaphthalene CLXXXVI, likely intermediates being the salts CLXXXIV and CLXXXV (*Bentley et al*., J. chem. Soc. C., 1969, 2225). The degradation of CLXXVII to CLXXX could, of course, well proceed by an initial step analogous to that represented in CLXXXIII.

(xiii) Reaction of thebaine with metal carbonyls

Thebaine reacts with tricarbonyliron to give a complex CLXXXVII from which the iron may be removed by ferric chloride in acetone. The methoxydiene system is thereby stabilised and reactions such as Hofmann degradation and reductive fission of the 4,5-oxide bridge in acids, impossible with thebaine, can be easily accomplished. Treatment of the complex with hydrofluoroboric acid affords the mesomeric cation CLXXXVIII, which on boiling with water or methanol rearranges as shown to the iminium salt CLXXXIX, and this can be reduced with sodium tetrahydridoborate to the tertiary base and converted into a pseudo-base by sodium bicarbonate (*A. J. Birch et al*., Chem. Comm., 1968, 531; Australian J. Chem., 1969, **22**, 971).

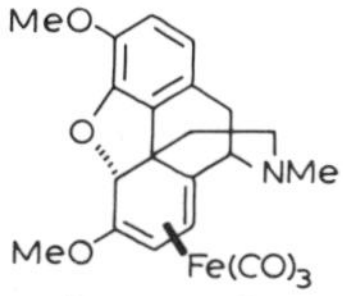

(CLXXXVII) 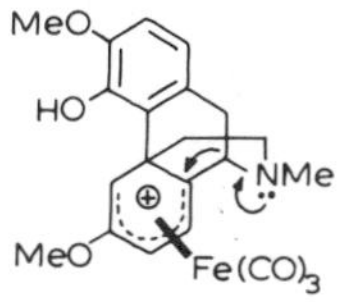 (CLXXXVIII) 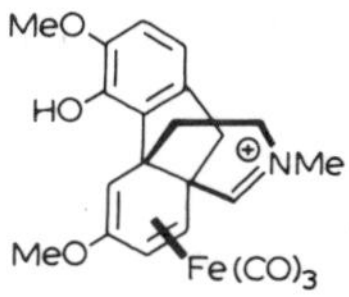 (CLXXXIX)

(xiv) ψ-Morphine

Oxidation of morphine with 2,2′-coupling of two molecules by a radical pairing process to give **ψ-morphine** can be effected by a variety of reagents of which the best is alkaline potassium ferricyanide (*K. W. Bentley*, "The Chemistry of The Morphine Alkaloids", Clarendon Press, Oxford, 1954). The structure of the base has been confirmed by the preparation of 1,1′-dibromotetrahydro-ψ-morphine by the oxidation of 1-bromodihydromorphine (*Bentley* and *S. F. Dyke*, J. chem. Soc., 1959, 2574). The properties of the base are generally similar to those of morphine and acetolysis of the bis-quaternary salts proceeds with aromatisation to 3,3′,4,4′-tetra-acetoxy-2,2′-biphenanthryl (*K. Goto* and *Z. Kitasato*, Ann., 1930, **481**, 81; J. chem. Soc. Japan, 1933, **54**, 178). Isomer of ψ-morphine have been prepared by the oxidation of α,β and γ-isomorphine, and bimolecular bases have also been obtained by the oxidation of dihydrodeoxymorphine-D, thebainone-A, dihydrothebainone and metathebainone *(Bentley, op. cit.)*.

(e) The relationship between structure and analgesic activity

Morphine is an extremely effective analgesic but its clinical use is restricted because of its addiction potential and by the development of tolerance to its action. Attempts have been made to obtain analgesics without these disadvantages by modification of the molecule and by the synthesis of the compounds bearing some resemblance to parts of the morphine structure. The following generalisations may be made, though there are exceptions to most of them.

(a) Analgesic activity is increased by *O*-acetylation, by methylation, oxidation or elimination of the C-6 hydroxyl group, reduction of the 7,8-double bond, introduction of a hydroxyl or esterified hydroxyl group at C-14 or a methyl group at C-6 and by the replacement of the *N*-methyl group by certain groups such as β-phenylethyl and β-furylethyl.

(b) Activity is decreased by methylation of the phenolic hydroxyl group, by the replacement of NMe by NH, NEt, NPr, NBu and most other groups, quaternisation of the nitrogen atom, fission of the nitrogen-containing ring and opening of the 4,5-oxygen bridge with the production of a C-4 hydroxy group. Elimination of the oxide bridge to give 4-deoxy compounds generally results in the retention and in some cases enhancement of activity.

(c) Extremely high activity is found in many derivatives of the Diels–Alder adducts of thebaine. The ketone CXXXIV is as potent as morphine and the series of tertiary alcohols (CXXXVII, R = Me) derived from this contains compounds up to 500 times as potent as the alkaloid ($R^1 = CH_2 \cdot CH_2Ph$). Demethylation of these alcohols results in phenols of even higher activity, maximum potencies about 5,000–10,000 times that of morphine being found in the bases CXXXVII, R = H, $R^1 = Pr^n$, Bu^n, Am^n and Am^i.

2. The sinomenine – salutaridine group

The bases of this sub-group differ from those of the morphine–thebaine group in that they do not contain the 4,5-oxygen bridge. At the time of this review thirteen alkaloids are known and are listed, with their main sources, in Table 2 (*K. W. Bentley*, "The Alkaloids", Clarendon Press, Oxford, 1954, Vol. XIII).

TABLE 2

THE SINOMENINE – SALUTARIDINE GROUP OF ALKALOIDS

	m.p. (°C)	*[α]$_D$(°)*	*Source*
Sinomenine	161/182	−70.7	*Sinomenium acutum*
Isosinomenine	196–197	+94	*Sinomenium acutum*
Salutaridine	197–198	+111	*Croton salutaris; P. orientale*
Sinoacutine	197–198	−112	*S. acutum; C. balsamifera*
Norsinoacutine	140–142	−107	*C. balsamifera*
Dihydrosalutaridine	198–203	−76.1	*C. linearis*
Dihydronorsalutaridine	208–212	−69.1	*C. linearis*
Amurine	213–215	+10	*P. feddei; P. radicatum; P. nudicaule*
Nudaurine	201–202	−52	*P. nudicaule*
Flavinantine	130–132	−14.5	*C. flavens*
Flavinine	130–132	−6	*C. flavens*
Pallidine		−32	*Corydalis pallida*
Delavaine	145–150	−240.7	*Stephania delavayi*

Sinomenine (I) may be isolated from the ground roots and stems of the plants by heating with 95% ethyl alcohol for several days. After removal of the solvent the residue is treated with lead acetate and finally the basic material is taken up in hydrochloric acid and the other organic materials removed by extraction with ether. The alkaloid hydrochloride crystallises from the aqueous solution when this is concentrated. The base, when liberated from the hydrochloride, may be recrystallised from benzene, when it is obtained as needles, m.p. 161–162° with solidification and remelting at 182°.

Sinomenine has the composition $C_{19}H_{23}NO_4$ and contains two methoxyl groups, one phenolic hydroxyl group and an α,β-unsaturated ketone system. Hofmann degradation proceeds very easily to give, under mild conditions, the achro*methine* (II) m.p. 179°, $[\alpha]_D$ +72.6°, which is very easily isomerised to the roseo*methine* (III) m.p. 163°, $[\alpha]_D$ +135.7°, and the violeo*methine* (VI), m.p. 172–173°, $[\alpha]_D$ +437.8° (*K. Goto* and *H. Shishido*, Bull., chem. Soc. Japan, 1931, **6**, 79). Degradation of sinomenine methiodide by heating with benzoic anhydride yields dibenzoylsinomenol (V, R = COPh), which can be hydrolysed to sinomenol (V, R = H) also obtainable by further Hofmann degradation of the methines II, III and VI. Ethylation of sinomenol gives 4,6-diethoxy-3,7-dimethoxyphenanthrene (V, R = Et) identical with material prepared by synthesis, thus demonstrating the position of the phenolic hydroxyl and carbonyl groups (*Goto* and *H. Sudzuki*, *ibid*., 1929, **4**, 163). The loss of the nitrogen-containing side-chain during the aromatisation reactions is characteristic of the morphine group:

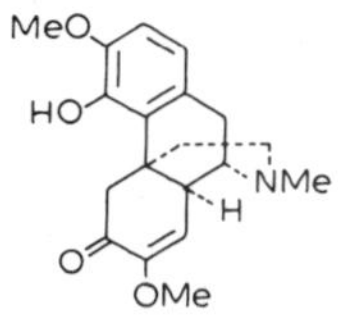

Sinomenine (I)

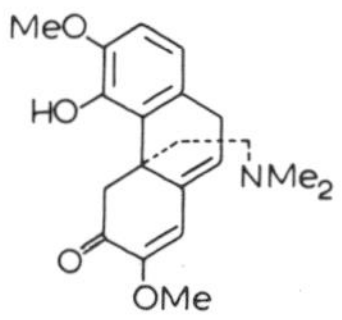

(II)

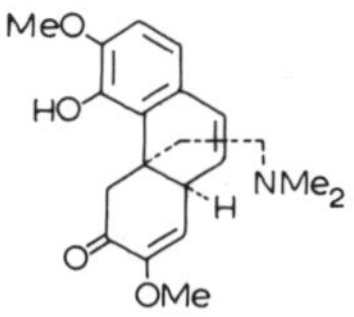

(III)

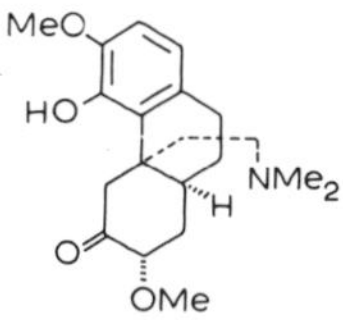

(IV)

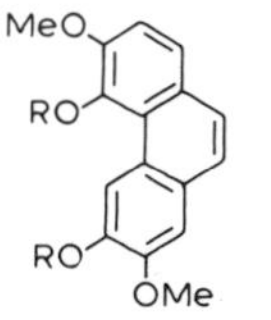

(V)

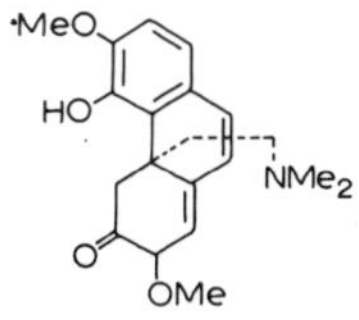

(VI)

Catalytic reduction of sinomenine yields *dihydrosinomenine* (IV), m.p. 198°, $[\alpha]_D$ +193.6°, which, being an α-methoxyketone can be further reduced with loss of the C-7 methoxyl group to the optical antipode of dihydrothebainone (XXXII) (p. 276) and, under the conditions of the Clemmensen reduction, to (+)-tetrahydrodeoxycodeine (the antipode of the base XXIX, R = OMe, R^1 = OH). Both of these bases form racemates with the corresponding bases prepared from thebaine, showing that sinomenine belongs to a series enantiomorphic with that to which morphine and thebaine belong and these reactions clearly indicate the gross structure of the alkaloid (I), which is 7-methoxy-(+)-thebainone (*Goto et al*., *ibid*., 1929, **4**, 244; *H. Kondo* and *E. Ochiai*, Ann., 1929, **470**, 224; Ber., 1930, **63**, 646).

Demethoxydihydrosinomenine, which is (+)-dihydrothebainone, has been converted, by methods based on those used by *Gates* in his synthesis of morphine (see p. 275), into (+)-thebainone-A, (+)-codeinone, (+)-codeine and (+)-morphine. (+)-Dihydrocodeinone has been converted into (+)-dihydrothebaine and (+)-codeinone has been rearranged to (+)-metathebainone, (+)-morphothebaine and thebenine, and (+)-codeine has been converted through (+)-α-chlorocodide into enantiomorphs of iso-, ψ- and allo-ψ-codeine (*Goto et al.*, Proc. Japan Acad., 1954, **30**, 66, 769; 1957, **33**, 477; 1958, **34**, 60; 1959, **35**, 472; 1960, **36**, 282; *Goto*, "Sinomenine", Kitasato Institute, Tokyo, 1964).

The treatment of sinomenine with bromine and then with alkalis results, as with derivatives of thebainone, in nuclear bromination and closure of the 4,5-oxide bridge, and gives 1-bromosinomeneine which is 1-bromo-7-methoxy-(+)-codeinone (VII) (*C. Schöpf* and *T. Pfeiffer*, Ann., 1930, **483**, 157).

Both sinomenine and 1-bromosinomeneine contain an enol ether system that can be hydrolysed with the generation of a carbonyl group, the products of hydrolysis being sinomeninone (VIII) and 1-bromosinomenine ketone (VII), respectively (*Goto* and *Sudzuki*, Bull. chem. Soc. Japan, 1929, **4**, 271). (−)-Sinomeninone can be prepared by the hydrolysis of 7-bromodihydrocodeinone dimethyl acetal* (L) (see p. 278), which presumably proceeds through the enolised 7-hydroxy ketone X, since the 7-hydroxy acetal behaves in the same way (*W. Fleischhacker* and *H. Markut*, Monatsh., 1971, **102**, 643).

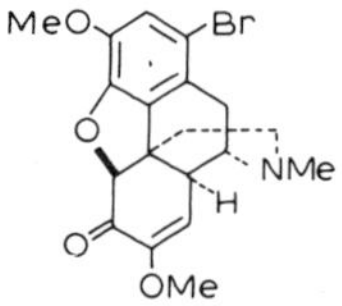

(VII) 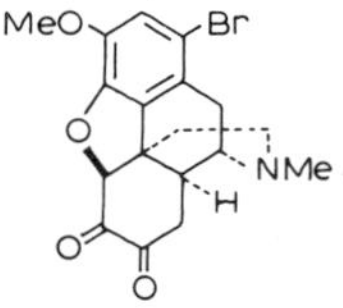 (VIII) 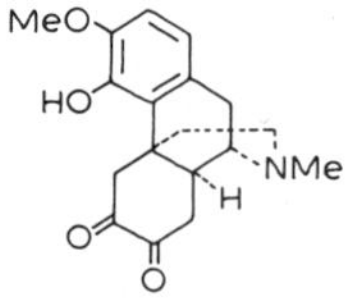 (IX)

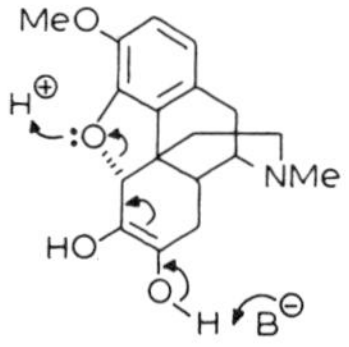

(X) 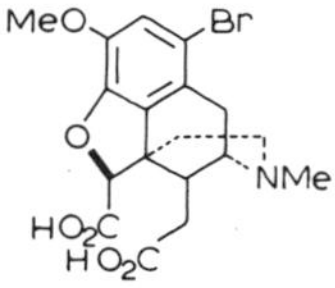 (XI) 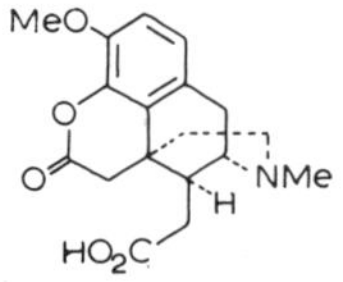 (XII)

The two carbonyl groups of sinomeninone differ in activity, that at C-7 being the more easily enolised and methylation of (−)-sinomeninone, prepared from 7-bromodihydrocodeinone dimethyl acetal, with methyl sulphate and sodium *tert*-butoxide gives the optical antipode of sinomenine methyl ether (*Goto* and *I. Yamamoto*, Proc. Japan Acad., 1958, **34**, 619).

**N.B.* Designated also as 7-bromodihydrocodeinone dimethyl ketal; the term ketal is not now accepted in the I.U.P.A.C. Rules (see Rule C331).

The two α-diketones VIII and IX can be oxidised with hydrogen peroxide to 1-bromosinomeneinic acid (XI) and sinomeninic acid (XII) (*Goto* and *Shishido*, Ann., 1933, **501**, 304). Benzilic acid transformation of 1-bromosinomeneine ketone (VIII) with alkalis yields 1-bromosinomenilic acid (XIII) which gives 1-bromosinomenilone (XIV, R = Br) on warming above 35° with 20% oleum. Catalytic reduction of the bromo-ketone gives sinomenilone (XIV, R = H), but reduction with sodium amalgam results in opening of the 4,5-oxide bridge and the production of 1-bromohydrosinomenilone (XV, R = Br) reducible catalytically to dihydrosinomenilone (XV, R = H). Enantiomorphs of these bases have been prepared from thebaine (*Goto et al., ibid.*, 1932, **495**, 122; 1932, **499**, 169; 1933, **501**, 304):

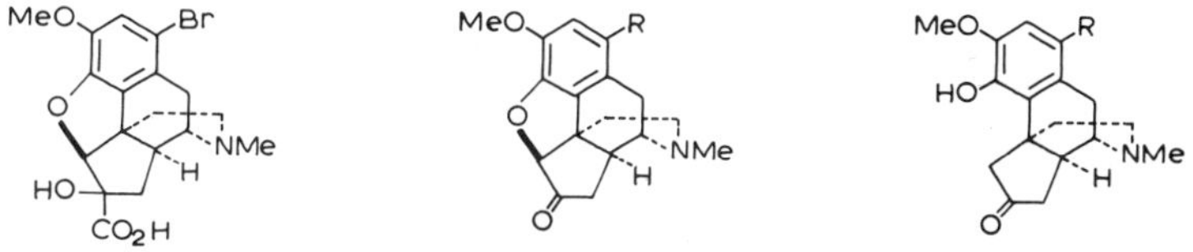

MeO, Br, O, NMe, H, HO, CO_2H

(XIII)

MeO, R, O, NMe, H, O

(XIV)

MeO, R, HO, NMe, H, O

(XV)

Oxidation of sinomenine with silver nitrate, alkaline potassium ferricyanide, or similar reagents gives 1,1′-disinomenine, which exists in two diastereoisomeric forms probably as a consequence of restriction of rotation about the biphenyl linkage. Both bases yield the same disinomenol derivative on acetolysis. (*Goto* and *Sudzuki*, Bull. chem. Soc. Japan, 1929, **4**, 107, 129; *Kondo* and *Ochiai*, J. pharm. Soc. Japan, 1927, **549**, 923).

Isosinomenine, which is the enantiomorph of dihydrosalutaridine (XIX) has been alleged to accompany sinomenine in *S. acutum* but this is doubtful and, since treatment of sinomenine with methanolic hydrogen chloride gives isosinomenine *via* the common α-diketone (IX), it is probably an artefact (*Y. Sawa et al.*, Tetrahedron, 1961, **15**, 144, 154; 1964, **20**, 2255.

Salutaridine (XVII), $C_{19}H_{21}NO_4$, contains a phenolic hydroxyl group and a dienone system. It is known to be the first alkaloid of the morphine group to appear in *Papaver* species in which it arises by internal oxidative coupling of reticuline (XVI), $R = R^1 = H$). Its production in this way, which has been duplicated in the laboratory, indicates that it has the structure XVII and this has been confirmed by its preparation from dihydrothebaine-ϕ (CI) (p. 287) by oxidation (of the *O*-acetyl ester) successively with selenium dioxide and manganese dioxide (*D. H. R. Barton et al.*, J. chem. Soc. 1965, 2423), from 14-bromocodeinone as previously described (p. 290) and by diazotisation of the benzyl ether of the amine XVI ($R = NH_2$, $R^1 = Me$) (*T. Kametani et al.*, J. chem. Soc. C, 1969, 2030). The absolute stereochemistry of the base follows from its conversion through the salutaridinols into thebaine (see below), and from the reduction of its methyl ether with sodium and ammonia to a benzylisoquinoline that gives (−)-laudanosine on methylation (*Barton et al.*, *ibid.*, 1967, 128):

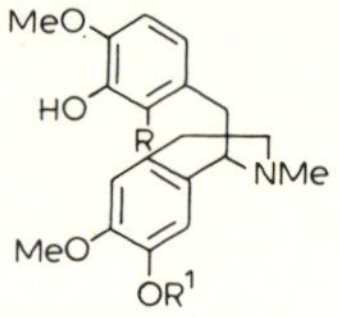

(XVI)

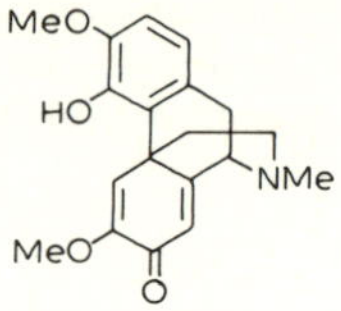

Salutaridine
(XVII)

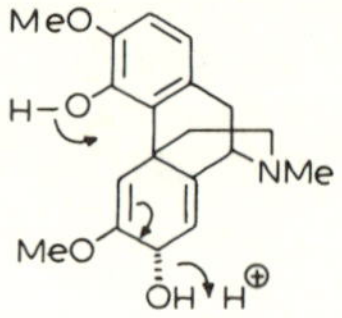

(XVIII)

The chemistry of the alkaloid has not been widely studied but reduction with sodium tetrahydridoborate affords a mixture of C-7 epimeric secondary alcohols, the salutaridinols, both of which give thebaine on treatment with acid at pH 4, by allylic expulsion of hydroxyl, which may be represented as a concerted S_N2' process, though, since only the epimer XVIII has the correct stereochemistry for such a process, it may proceed through a carbonium ion. Only the C-7 epimer of XVIII is efficiently converted into thebaine in the poppy and this may well proceed by S_N2' displacement in a (phosphate?) ester of XVIII generated by S_N2 replacement through the intervention of an enzyme system (*idem*, *ibid.*, 1965, 2423).

Norsinoacutine, $C_{18}H_{19}NO_4$, is a secondary base that on methylation with formaldehyde and formic acid gives **sinoacutine**, which is the antipode of salutaridine (*A. R. Battersby* and *T. H. Brown*, Chem. Comm., 1966, 170; *J.-H. Hsu et al.*, Sci. Sinica, 1964, **13**, 2016; C. A., 1965, **62**, 9183). Norsinoacutine has been reduced with sodium tetrahydridoborate to a mixture of norsinoacutinols which on treatment with mineral acid yield the enantiomorph of dehydronormetathebainone, the secondary base related to the enedione CXLIV (*K. L. Stuart et al.*, J. chem. Soc. C, 1969, 1681). This rearrangement is precisely similar to the first stage of the rearrangement of nudaurine presented below, the loss of a proton from an analogue of the carbonium ion XXVI being the same process as that represented in formulae CXLIII–CXLIV (p. 295).

Dihydrosalutaridine (XIX) **and dihydronorsalutaridine** obtained from *Croton linearis*, have the compositions $C_{19}H_{23}NO_4$ and $C_{18}H_{21}NO_4$ and are related as tertiary and secondary bases, dihydronorsalutaridine giving dihydrosalutaridine methiodide on complete methylation. The structure of dihydrosalutaridine (XIX) was deduced from spectral studies and a demonstration that it is the enantiomorph of isosinomenine, thus confirming the B/C *cis* fusion (*L. J. Haynes et al.*, Chem. Comm., 1967, 15). Catalytic reduction of dihydrosalutaridine gives dihydrosalutaridinol and tetrahydrosalutaridinol (XX), which are assumed to be the products of addition of hydrogen on the least hindered side of the molecule.

Amurine (XXI), $C_{19}H_{19}NO_4$, obtained from certain *Papaver* species, was initially assigned a proaporphine structure, but on the basis of its rearrangement to the phenanthrene XXIV by dilute hydrochloric acid has been reformulated as the salutaridine analogue XXII. The base XXIV is an analogue of thebenine (CLIII) (p. 295) and is doubtless formed by a rearrangement as shown in formulae XXI–XXIV exactly analogous to the rearrangement of thebaine under comparable conditions (*H. Flentje et al.*, Naturwiss., 1965, **52**, 259):

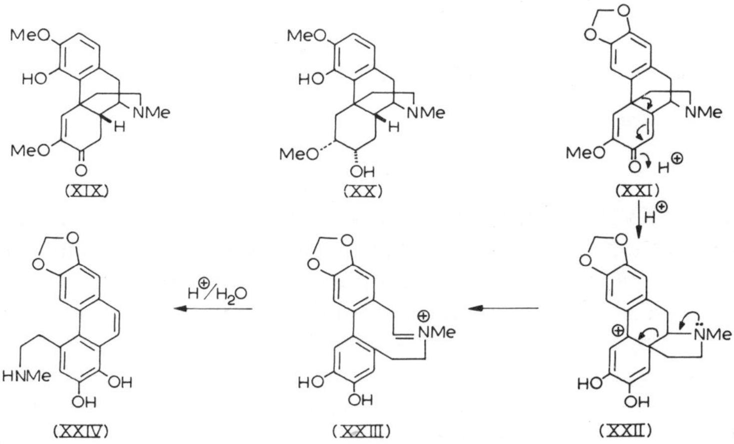

(XIX) (XX) (XXI)

(XXIV) (XXIII) (XXII)

[It may be noted that acid-catalysed rearrangement of 4-*O*-methylsalutaridine methoperchlorate leads to 7-hydroxy-4-*O*-methyldehydrometathebainone methoperchlorate, derived from the dienone CXLIV (p. 295), and finally to 7-hydroxy-4-*O*-methylmorpholthebaine methoperchlorate (*G. Heinisch* and *F. Vieboeck*, Monatsh., 1971, **102**, 77).] The structure XXI for amurine has been confirmed by the synthesis of the alkaloid by heating the diazonium salt obtained from 1-(2-amino-4,5-methylenedioxybenzyl)-6,7-dimethoxy-2-methyltetrahydroisoquinoline, a process similar to that involved in the synthesis of salutaridine from the amine XVI ($R = NH_2$, $R^1 = Me$) (*Kametani et al.*, J. chem. Soc. C, 1969, 801), and through a benzyne intermediate derived from 1-(2-bromo-4,5-methylenedioxybenzyl)-7-hydroxy-6-methoxy-2-methyltetrahydroisoquinoline (*idem, ibid.*, 1971, 2712).

Nudaurine (XXV), has the composition $C_{19}H_{21}NO_4$ and contains a secondary alcoholic hydroxyl group. On heating with dilute hydrochloric acid it is converted into a deoxy analogue XXVII of the base XIV prepared by the rearrangement of amurine. On this basis and its spectroscopic properties it has been assigned the structure of amurinol (XXV) and its optical properties and other studies show that it belongs to the same stereochemical series as salutaridine and that the disposition of the hydroxy group is as shown (*Haynes et al., loc. cit.* and *Flentje et al., loc. cit.*; *Barton et al.*, Ber., 1967, **100**, 2457). Since it is in a lower state of oxidation than amurine aromatisation can only take place with loss of an oxygen function:

NMe MeO OH H⊕ → ⊕ NMe HO —Thebenine cyclisation→ HNMe OH

Nudarine (XXV) (XXVI) (XXVII)

Pallidine (XXVIII), which is isomeric with salutaridine and sinoacutine, has the composition $C_{19}H_{21}NO_4$ and has been assigned the structure XXVIII on the basis of spectroscopic and optical studies (*Kametani et al.*, Chem. Comm., 1969, 301). It is the enantiomorph of isosalutaridine which has been synthesised by the oxidation of **reticuline** (XVI, $R = R^1 = H$) with manganese dioxide (*B. Franck et al.*, Angew. Chem., 1967, **79**, 980, 1066) and with potassium ferricyanide (*Kametani et al.*, J. chem. Soc. C, 1969, 2034), and hence belongs to the sinoacutine series.

Flavinantine, (XXXIX), $C_{19}H_{21}NO_4$, is another alkaloid isomeric with salutaridine and shown by its optical properties to belong to the same absolute stereochemical series. Spectroscopic studies show that it must have the structure XXIX or be the enantiomorph of pallidine (XXVIII), but since it is not identical with isosalutaridine, prepared by the oxidation of reticuline, though the *O*-methyl ethers of the two alkaloids are identical, it must have the constitution XXIX and this has been confirmed by synthesis of racemic flavinantine by thermal cyclisation of the diazonium salt obtained from 1-(2-amino-4-benzyloxy-5-methoxybenzyl)-6,7-dimethoxy-2-methyltetrahydroisoquinoline (*C. Chambers* and *Stuart*, Chem. Comm., 1968, 328; *Kametani et al.*, J. chem. Soc. C, 1969, 520, 1063, 2034; 1972, 2446). Flavinantine methyl ether occurs naturally in *Nemuaron vieillardii.*

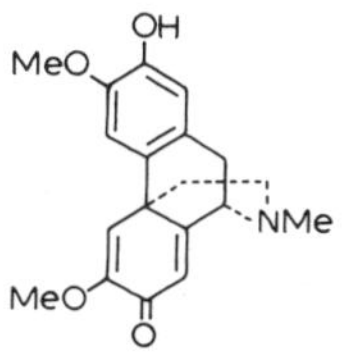

Pallidine (XXVIII) Flavinantine (XXIX) Delavaine (XXX)

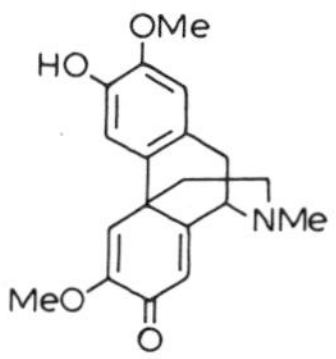

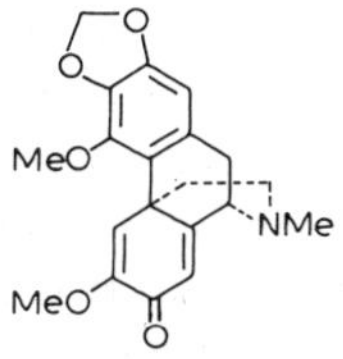

Flavinine, $C_{18}H_{19}NO_4$, isomeric with norsinoacutine, has the properties of a secondary base and, since it can be methylated to *N*-methylflavinine methiodide, identical with flavinantine methiodide, must have the structure XXIX (NMe = NH) (*Kametani et al.*, Chem. Comm., 1969, 1301).

Delavaine (XXX), only recently discovered, has been assigned the structure XXX (?) or that of the 2-methoxy-3,4-methylenedioxy isomer, on the basis of n.m.r. spectroscopic studies (*I. I. Fadeeva et al.*, Khim. prirod. Soedinenii, 1970, **6**, 140; C. A., 1970, **73**, 45639p). A contemporary publication, however, assigns it to the hasubanonine group with the structure LIX (p. 318) (*T. L. Il'inskaya et al.*, Postep. Dziedzinie Leku Rosl. Pr. Ref. Dosw. Wygloszne Symp., 1970, 98; C.A., 1973, **78**, 58647t).

3. The hasubanonine group

The bases of this sub-group differ from those of the sinomenine–salutaridine sub-group in the linkage of the carbon–nitrogen side-chain which forms a bridge between C-13 and C-14 rather than C-13 and C-9. The alkaloids have been related to bases in the indolinocodeine series (see p. 291) obtained from 14-bromocodeinone. They resemble the sinomenine–salutaridine bases in not containing a 4,5-oxide bridge. Included also are three alkaloids in which the aromatic nucleus has undergone fission and ring contraction, the other parts of the molecule and general substitution pattern being unchanged. At the time of this review fifteen alkaloids of the sub-group have been identified, and are listed with their physical properties and main sources in Table 3 (see also *K. W. Bentley*, "The Alkaloids", Clarendon Press, Oxford, 1954, Vol. XIII).

Hasubanonine (I), $C_{21}H_{27}NO_5$, is a non-phenolic base containing an α,β-unsaturated ketone system the double bond of which must be fully substituted as it resists hydrogenation. The methine base on heating with acetic anhydride suffers loss of the nitrogen-containing side-chain with aromatisation to acetylhasubanol, which can be hydrolysed and ethylated to 6-ethoxy-3,4,8-trimethoxyphenanthrene (*H. Kondo et al.*, Ann. Rept. ITSUU

Lab., 1950, **1**, 50; and *H. Matsumara*, J. pharm. Soc. Japan, 1963, **83**, 991). These reactions indicated that the alkaloid belonged to the morphine group, but the n.m.r. spectrum of hasubanonine lacks signals attributable to the proton at C-9 in the system –CH–N– characteristic of all cyclic bases in the morphine series except those related to indolinocodeine (CXXVI) (p. 292). The structure I for the alkaloid was deduced as follows.

Hasubanonine was reduced to a mixture of epimeric allylic alcohols II both of which gave hasubanonine on re-oxidation, and these on heating with hydrobromic acid lost methanol to give the enone III. Clemmensen reduction of this enone gave an isomer of tetrahydrodeoxycodeine (VI) which was shown to be the antipode of, and form a racemate with, a base of that undoubted structure prepared from indolinocodeine (CXXVI) by conversion into dihydroindolinocodeinone, Wolff–Kishner reduction to the Δ^5-olefinic phenol, reduction and *O*-methylation (*M. Tomita et al.*, Tetrahedron Letters, 1964, 2937; Chem. Pharm. Bull., Japan, 1965, **13**, 538). Hofmann degradation of hasubanonine methiodide in methanol gives the acetal methine base IV (R = Me) and in ethanol the homologue (R = Et), which are readily hydrolysed to the corresponding diketone, and aromatisation of this during acetolysis with loss of the C-7 oxygen function must proceed through the isomerised form V (*T. Ibuka et al.*, J. pharm. Soc. Japan, 1967, **87**, 1014):

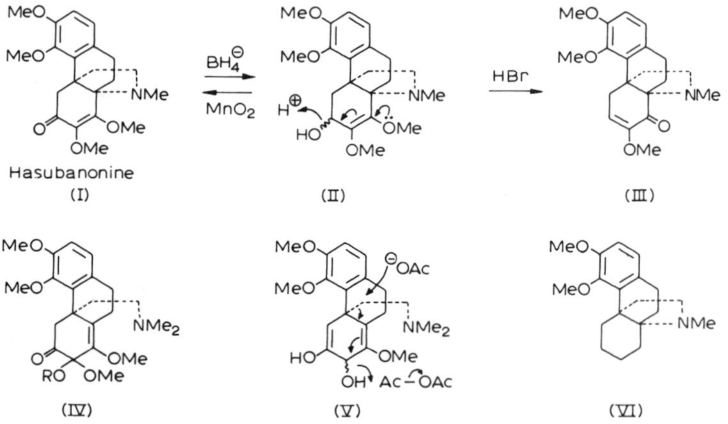

MeO, MeO, NMe, O, OMe, OMe
Hasubanonine
(I)

$BH_4^{\ominus}$ / MnO_2

(II)

HBr

(III)

(IV) (V) (VI)

A total synthesis of (±)-hasubanonine has been achieved from the diketone VII, obtainable from the 7-ketone by oxidation with lead tetra-acetate and boron trifluoride to the 8-acetoxy ketone, bromination at C-6, dehydrobromination and hydrolysis. Further bromination of the diketone followed by methylation gave the enol ether VIII which was hydrolysed to the methoxy-*β*-diketone IX and methylation of this with

TABLE 3

THE HASUBUNONINE GROUP OF ALKALOIDS

	m.p. (°C)	*[α]$_D$(°)*	*Source*
Hasubanonine	116–117	−219.5	*Stephania japonica*
Homostephanoline	233	−247.8	*S. hernandifolia*
Aknadinine	70	−283	*S. hernandifolia; S. sasaki*
Aknadicine	156	−200	*S. hernandifolia*
Aknadilactam		−212	*S. sasaki*
Hernandolinol			*S. hernandifolia*
Cepharamine	186–187	−248	*S. cepharantha*
Metaphanine	232	−41.1	*S. japonica*
Prometaphanine	207*	−32*	*S. japonica*
Stephabyssine	178–180	−58.9	*S. abyssinica*
Stephaboline	186–188	+34.7	*S. abyssinica*
Prostephabyssine	196–198*	−105*	*S. abyssinica*
Miersine	222		*S. japonica*
Stephasunoline	233	+121.1	*S. japonica*
Stephamiersine	165	+33	*S. japonica*
Epistephamiersine	98	+64.1	*S. japonica*
Oxostephamiersine	298	+88.3	*S. japonica*
Delavaine(?)	145–150	−240.7	*S. delavayi*
Oxodelavaine			*S. delavayi*
Stephavanine	229–230	+30	*S. abyssinica*
Stephisoferuline	133–135**	+48	*S. hernandifolia*
Acutumine	238	−206	*Sinomenium acutum; Menispermum dauricum*
Acutumidine	239–241	−212	*S. acutum; M. dauricum*
Acutuminine	175–177	−110	*M. dauricum*

* Methiodide.
**Chloroform solvate.

diazomethane gave aknadilactam (XII) 16-oxohasubanonine and their Δ^6-6,7-dimethoxy-8-oxo isomers. 16-Oxohasubanonine yields hasubanonine when reduced with lithium tetrahydridoaluminate and re-oxidised with manganese dioxide (*idem*, Tetrahedron Letters, 1970, 4811):

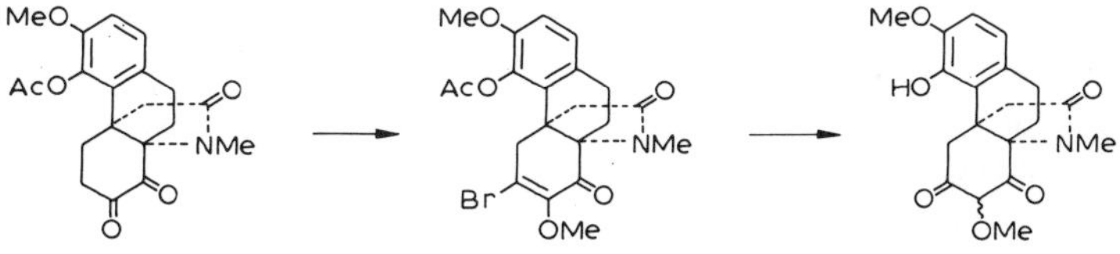

MeO AcO O NMe O O (VII) → MeO AcO O NMe Br O OMe (VIII) → MeO HO O NMe O O OMe (IX)

Homostephanoline and **aknadinine** are isomeric phenolic bases of composition $C_{20}H_{25}NO_5$ and give hasubanonine on *O*-methylation. Homostephanoline has been identified as the phenol X by *O*-ethylation and conversion through the methine base into *O*-ethylhomostephanol, the methyl ether of which is identical with 3-ethoxy-4,6,8-trimethoxyphenanthrene (*Y. Watanabe et al.*, J. pharm. Soc. Japan, 1965, **85**, 584). Aknadinine has been shown to be the isomeric phenol XI (R = Me) by n.m.r. spectroscopic and X-ray crystallographic studies (*S. M. Kupchan et al.*, J. org. Chem., 1968, **33**, 4529); it has been independently isolated and identified under the name **hernandoline** (*I. I. Fadeeva et al.*, Khim. prirod. Soedinenii, 1967, **3**, 106; Farmatsiya, Moscow, 1970, **19**, 28). The related secondary alcohol has also been reported to occur naturally in *Stephania hernandifolia* as the alkaloid **hernandolinol** (*idem*, Khim. prirod. Soedinenii, 1970, **6**, 492):

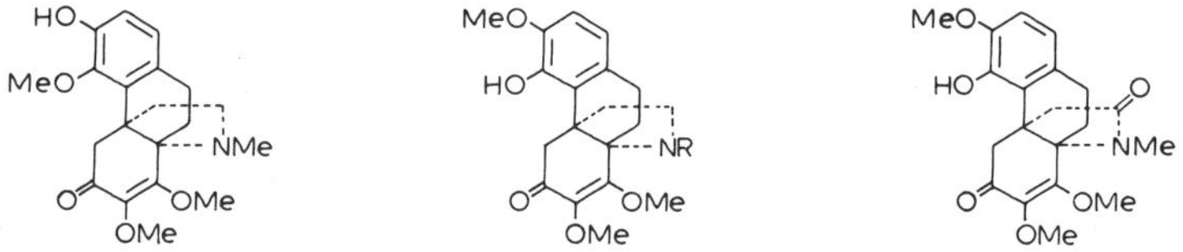

Homostephanoline (X) | Aknadicine (XI) | Aknadilactam (XII)

Aknadicine, $C_{19}H_{23}NO_5$, is a secondary base giving aknadinine on methylation with formaldehyde and formic acid and hence has the structure XI (R = H) (*B. K. Moza et al.*, Chem. and Ind., 1969, 1178). **Aknadilactam** is a γ-lactam and can be prepared by the oxidation of *O*-acetylaknadinine followed by mild hydrolysis with sodium carbonate and thus has the structure XII (*J. Kunitomo et al.*, Tetrahedron Letters, 1969, 3287).

Cepharamine. The structure of this alkaloid, $C_{19}H_{23}NO_4$, which is a phenol containing a free *para* position has been determined by reduction to a pair of epimeric alkaloids XIV, which were hydrolysed to the hydroxyketone XV and then reduced to tetrahydrodeoxyindolinocodeine (the 4-hydroxy analogue of the base XVI) and by its n.m.r. spectrum which is consistent only with the orientation of substituents shown in XIII or its Δ^5-6-methoxy-7-oxo isomer (*Tomita* and *M. Kozuka*, *ibid.*, 1966, 6229).

The structure XIII has been confirmed by synthesis of (±)-cepharamine from 1-cyanomethyl-7,8-dimethoxy-2-oxo-1,2,3,4-tetrahydronaphthalene, which with methyl vinyl ketone and base gave the lactam XVIII (R = Me, R^1 = H). This was *N*-methylated, partially *O*-demethylated and acetylated to the lactam XVIII (R = Ac, R^1 = Me), which was then brominated and hydrolysed to the α-diketone XVII:

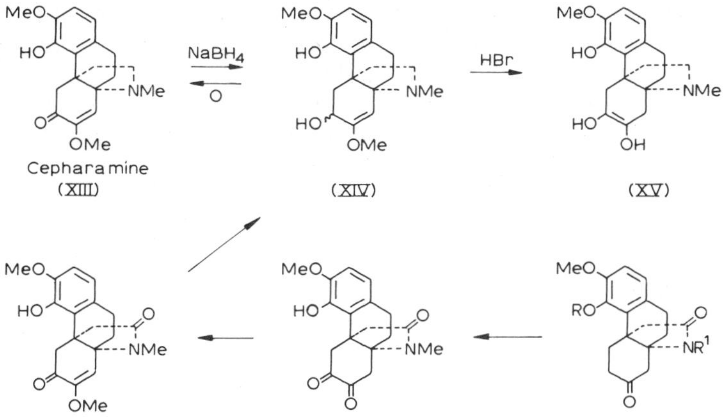

Cepharamine (XIII) (XIV) (XV)

(XVI) (XVII) (XVIII)

This diketone was converted into the enol methyl ether XVI, which was reduced with lithium tetrahydridoaluminate, and the product, XIV, oxidised to (±)-cepharamine (*Y. Inubushi et al.*, Tetrahedron Letters, 1970, 1611). The keto-lactam XVIII ($R=R^1=Me$) has been synthesised by a slightly different route (*S. L. Keely et al.*, Tetrahedron, 1970, **26**, 4729).

Metaphanine (XXII), $C_{19}H_{23}NO_5$, contains two methoxyl groups, one ketonic carbonyl group and one hydroxyl group. The fifth oxygen atom is non-functional and must be present in an ether system and, since no reducible double bonds are present, the molecule must be pentacyclic. Acetolysis of metaphanine gives 7,8-diacetoxy-3,4-dimethoxyphenanthrene (identified by conversion into the 7,8-diethoxy compound). Reduction of the carbonyl group gives dihydrometaphanine, acetolysis of which gives 8-acetoxy-3,4-dimethoxyphenanthrene, indicating that the carbonyl group is located at C-7. The complete structure of the alkaloid was deduced from n.m.r. spectroscopic studies and the results of Huang Minlon reduction and of benzilic acid rearrangements (*K. Takeda*, Ann. Rept. ITSUU Lab., 1960, **11**, 64; *Tomita et al.*, Tetrahedron Letters, 1964, 3605; Chem. Pharm. Bull., Japan, 1965, **13**, 695, 704; *Ibuka*, J. pharm. Soc. Japan, 1965, **85**, 579).

Huang–Minlon reduction of the base XXII gives dehydrodeoxometaphanine-A (XX), -B (XXI) and -D (XIX) and the first two of these have been converted as shown into the base XXIII, previously prepared from hasubanonine, and the enantiomorph of tetrahydrodeoxyindolinocodeine. The base XIX is also obtainable by desulphurisation of metaphaninedithiol acetal. Benzylic acid transformation must proceed through the diketone form XXV and gives the potassium salt XXVI which can be cyclised to the hydroxy-δ-lactone, and hydrogenolysed to the hydroxy-acid XXVII, and this on reduction and oxidation with periodic acid gives the ketone XXX. Hofmann degradation of metaphanine methiodide must also involve benzilic acid transformation since the product undoubtedly has the structure XXVIII, which could clearly arise from the methiodide of the transformation product XXVI as shown in XXIX (*H. L. de Waal et al.*, Tetrahedron Letters, 1966, 6169). (±)-Metaphanine has been synthesised from the 8-acetoxy derivative of the ketone XXVIII (R = Ac, R[1] = Me). This on acetalisation, oxidation, hydrolysis and methylation gave the ketone XXXI which was reduced and converted into the tetrahydropyranyl ether XXXI. Oxidation and hydrolysis of the ether XXXII gave the lactam XXXIII and this on reduction and hydrolysis gave (±)-metaphanine (XXII) (*Ibuka et al.*, Tetrahedron Letters, 1972, 1393):

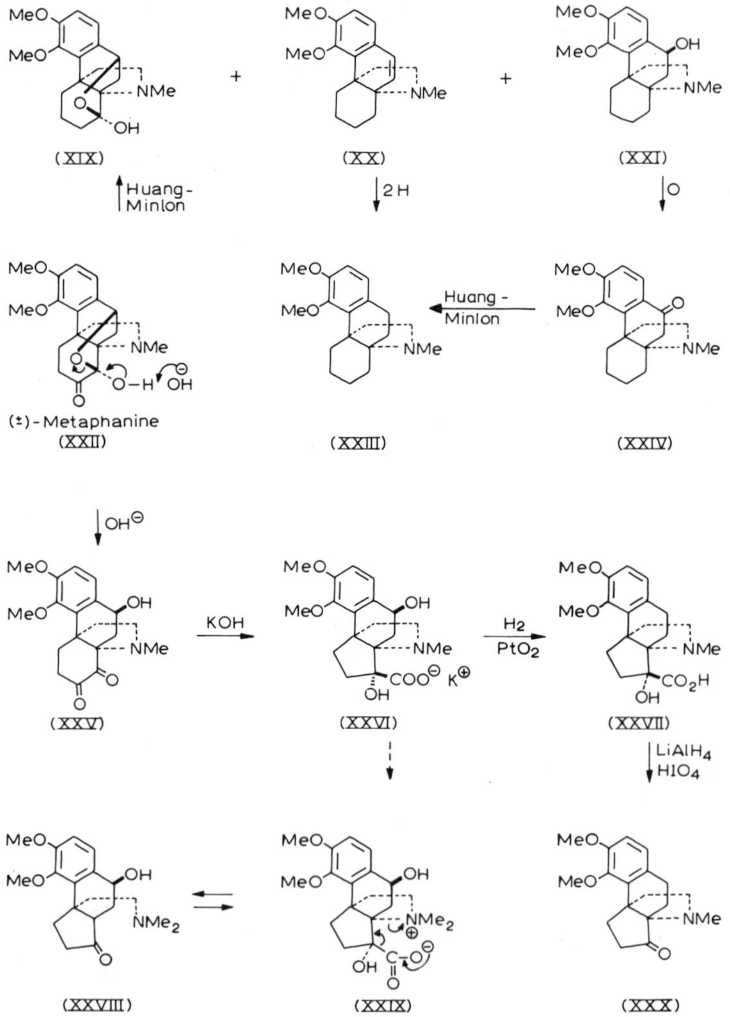

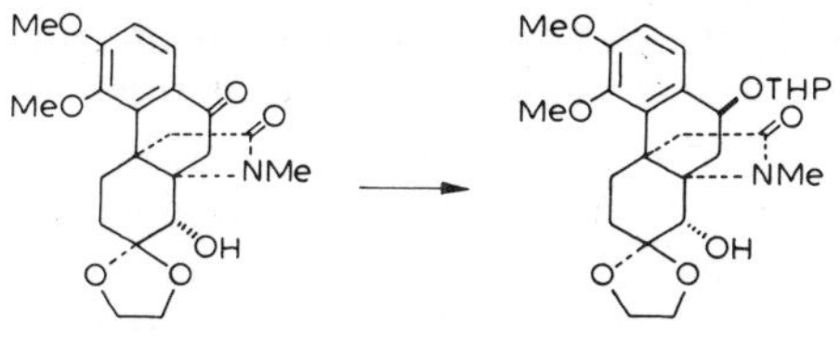

(XXXI) → (XXXII) 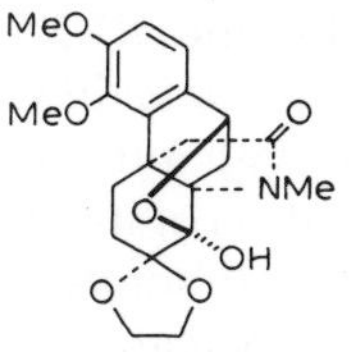 (XXXIII)

Prometaphanine, $C_{20}H_{25}NO_5$, is an enol ether that gives metaphanine on hydrolysis with dilute acid. Since it gives 8-acetoxy-3,4,7-trimethoxyphenanthrene on acetolysis it must contain a methoxyl group at C-7 and n.m.r. spectroscopic studies show that it exists as an equilibrium mixture of the ketone XXXIV and the hemiacetal XXXV, the position of equilibrium depending on the solvent. It can be reduced to dihydroprometaphanine (XXXVI) and oxidised to the diketone XXXVII (*Tomita et al.*, Tetrahedron Letters, 1964, 3167; J. pharm. Soc. Japan, 1967, **87**, 381).

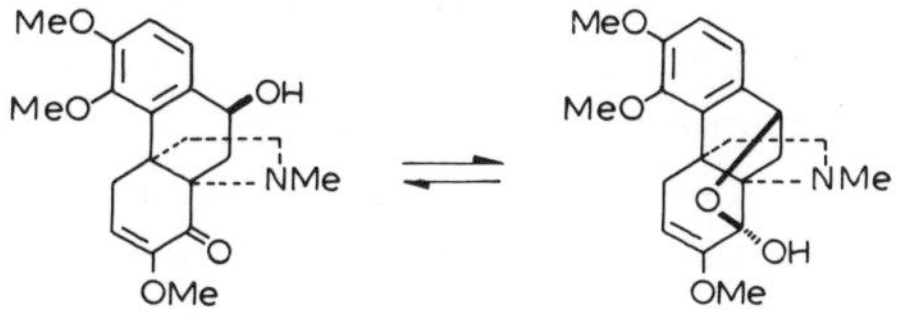

(XXXIV) ⇌ (XXXV)
Prometaphanine

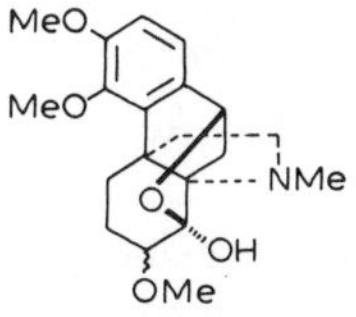

(XXXVI)

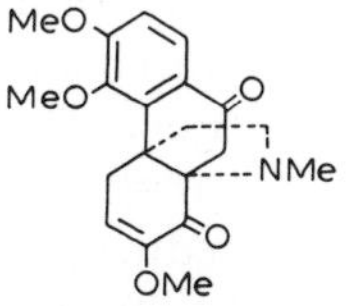

(XXXVII)

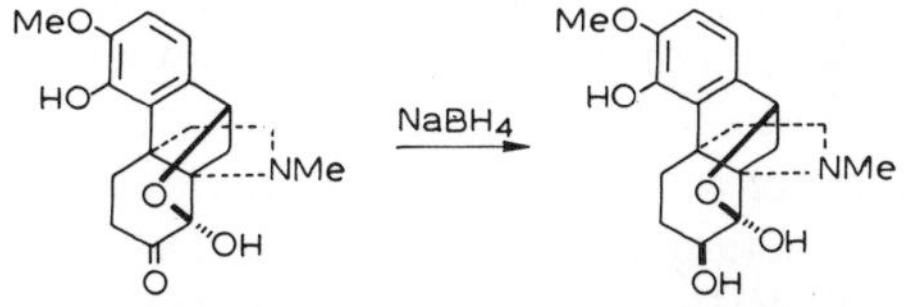

Stephabyssine
(XXXVIII)

NaBH4 →

Stephaboline
(XXXIX)

HCl ↗

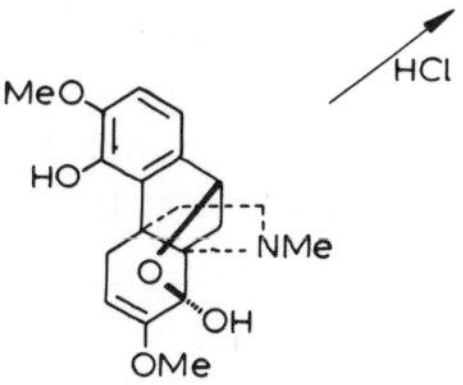

Prostephabyssine
(XL)

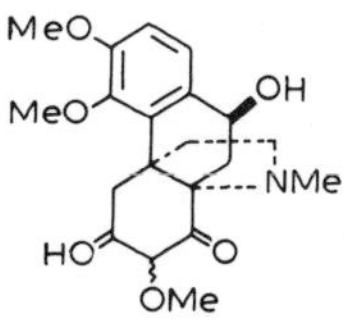

Miersine
(XLI)

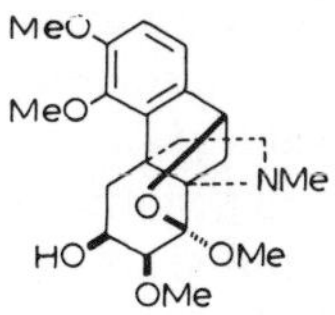

Stephasunoline
(XLII)

Stephabyssine (XXXVIII), **stephaboline** (XXXIX) and **prostephabyssine** (XL) from *Stephania abyssinica*, are closely related to metaphanine. All are phenols with a free *para* position (giving a blue colour with Gibbs reagent). Methylation of stephabyssine gives metaphanine and reduction with sodium tetrahydridoborate gives stephaboline. Prostephabyssine is the enol methyl ether of stephabyssine into which it is converted by hydrolysis with acids (*Kupchan et al.*, J. org. Chem., 1973, **38**, 151).

The mass spectral fragmentation pattern of **miersine** (XLI), $C_{20}H_{29}NO_6$, is characteristic of a hasubanan alkaloid and, since it is easily converted into a hemiacetal XLII (R = Me?), can be dehydrated with bases and occurs naturally with metaphanine and prometaphanine, it has been assigned the structure XLI (*A. R. Battersby et al.*, reported by *C. W. Thorner*, Phytochem., 1970, **9**, 157). Such a structure, being a β-hydroxyketone, would be very easily dehydrated to prometaphanine.

The structure XLII has been assigned to **stephasunoline** on the basis of its production by the mild acid hydrolysis of dihydroepistephamiersine (see below). This structure was also assigned to an alkaloid from *S. hernandifolia* given the name "hernandine" (*T. N. Il'inskaya*, Khim. prirod. Soedinenii, 1971, **7**, 180; C.A., 1971, **75**, 36408b), but as this name has also been given to an aporphine alkaloid obtained from *Hernandia bivalvis* (*K. S. Sohn et al.*, Tetrahedron Letters 1966, 5279) it should not be used in this context.

Stephamiersine and **epistephamiersine** are epimeric α-methoxy-ketones, which can be equilibrated in methanolic sodium hydroxide. Only one oxygen atom is unaccounted for in four methoxyl groups and one carbonyl, and this is shown to be present in a mixed acetal system (as in metaphanine) by the reduction of epistephamiersine with sodium tetrahydridoborate and mild hydrolysis of the resulting alcohol with acid to stephasunoline (XLII). Acetolysis of the two alkaloids yields 1,3-diacetoxy-2,5,6-trimethoxy- and 1,2,3-triacetoxy-5,6-dimethoxyphenanthrene. Oxidation of stephamiersine gives the lactam **oxostephamiersine** (XLV) which can be reduced to the alcohol XLVIII, convertible by hydrochloric acid into the unsaturated ketone XLVII, and this can be reduced to the 16-oxo analogue of the base III (p. 310) obtainable in a similar manner from 16-oxohasubanonine *via* the alcohol XLVI. These reactions serve unambiguously to determine the structures of these alkaloids (*M. Matsui et al.*, *ibid.*, 1973, 4263).

A recent publication assigns the structure XLIX to **delavaine** and L to 16-**oxodelavaine** (*Il'inskaya et al.*, Postep Dziedzinie Leku Rosl. Pr. Ref. Dosw. Wygloszone Symp., 1970, 98; C. A., 1973, **78**, 58647t) although the morphinandienone structure XXX (p. 276) has also been suggested.

Stephisoferuline (hernandifoline) (LI), $C_{29}H_{32}NO_9$, can be hydrolysed by bases to 3-hydroxy-4-methoxycinnamic acid and the alcohol *stephuline*,

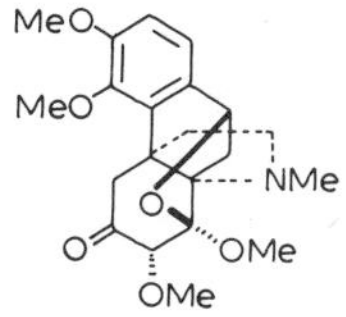

(XLIII) 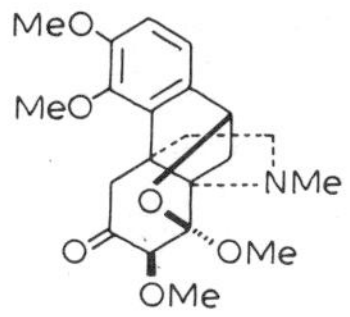 (XLIV) 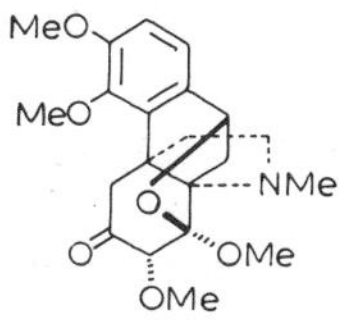 Oxostephamiersine (XLV)

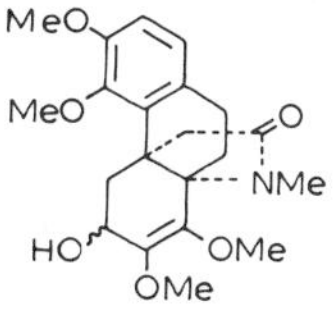

(XLVI) 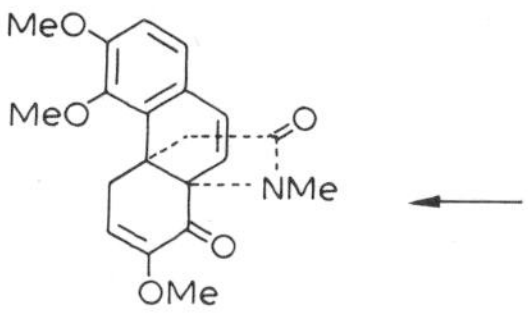 (XLVII) 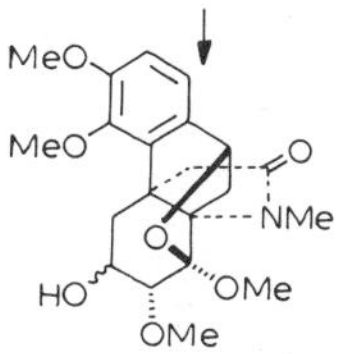 (XLVIII)

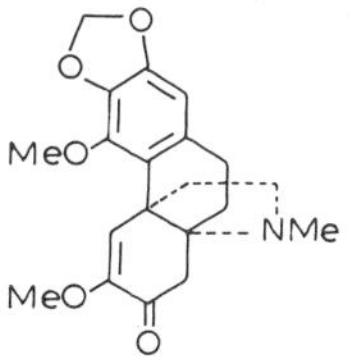

Delavine (XLIX) (L)

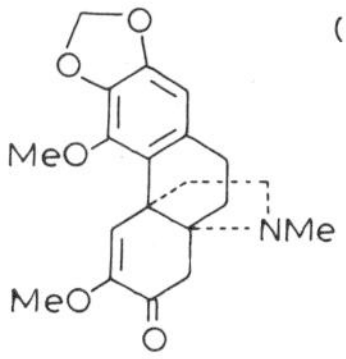

m.p. 223–225°, $[\alpha]_D$ +93°, which can be acetylated and dehydrated by acid to the olefinic ketone LII. Reduction of the 9,10 double bond of LII followed by treatment with acetone dimethyl acetal and *p*-toluenesulphonic acid affords the rearranged triacetate LIII, which can be prepared from aknadicine (XI, R=H) (p. 312) by reduction with sodium tetrahydridoborate, hydrolysis and acetylation. Since aknadicine has been converted into aknadinine the structure of which has been confirmed by an X-ray crystallographic study, stephisoferuline can be assigned the structure LI (*Kupchan* and *M. I. Suffness*, Tetrahedron Letters, 1970, 4975). **Hernandifoline** appears to have the same structure (*D. A. Fesenko et al.*, Khim. prirod. Soedinenii, 1971, **7**, 158; C. A., 1971, **75**, 49369q):

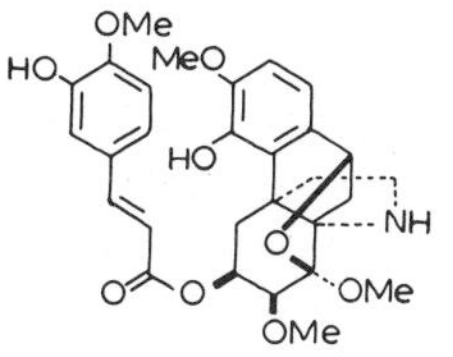

Stephisoferuline (LI) (LII) 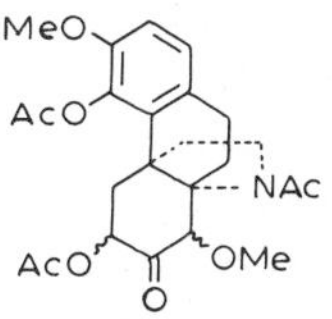 (LIII)

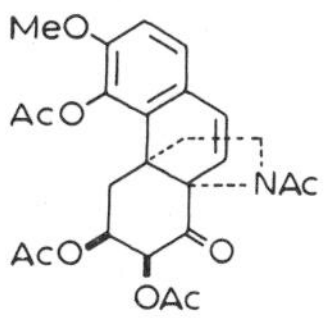

Stephavanine (LIV), $C_{26}H_{26}NO_9$, can be hydrolysed by bases to vanillic acid and the alcohol stephine (LV) and this can be oxidised to the ketone 6-dehydrostephine, which is converted by bases into isodehydrostephine LVI. As the substitution pattern of the alkaloid is different from that of others in the hasubanonine sub-group chemical conversion into materal of proved constitution could not be effected and the structure and absolute configuration of stephavanine was determined by X-ray crystallography (*Kupchan et al.*, J. Amer. chem. Soc., 1970, **92**, 5756):

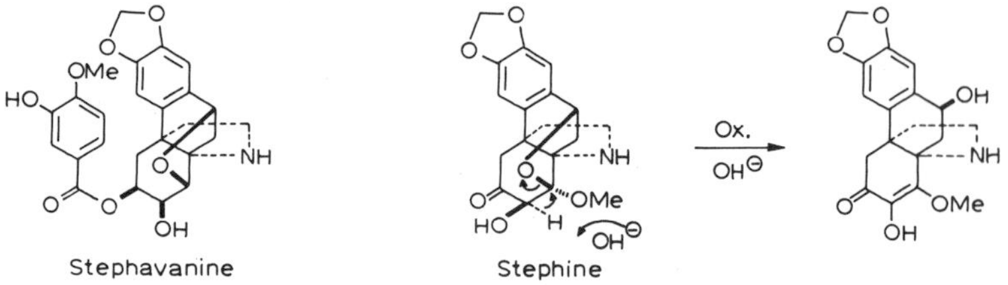

Stephavanine
(LIV)

Stephine
(LV)

Ox.
$OH^{\ominus}$

(LVI)

Acutumine has been assigned the structure LVII (R = Me) after an X-ray crystallographic study and since **acutumidine** is a secondary base giving acutumine on *N*-methylation its structure must be LVII (R = H) (*Tomita et al.*, Tetrahedron Letters, 1967, 2421, 2425; *K. Goto et al.*, Proc. Japan Acad., 1966, 42, 1181). **Acutuminine** has been assigned the structure of deoxy-acutumine (LVII, R = Me, OH = H) on the basis of spectroscopic studies (*Y. Okamoto*, Tetrahedron Letters, 1969, 1933).

Acutumine can be oxidised by manganese dioxide to the related enedione and the *O*-acetate can be reduced with loss of a methoxy group to the base LX. Reduction with zinc and acetic acid yields the carbinolamine LIX, presumably *via* the secondary base LVIII and heating with zinc and acetic anhydride results in aromatisation and extrusion of the side-chain with the production of the diacetates LXI (R = OMe) and LXI (R = H). These nitrogen-free compounds can be hydrolysed and methylated to the corresponding phenol ethers, which can be oxidised to the indanones LXXII (R = OMe) (identical with material prepared by synthesis) and LXXII (R = H):

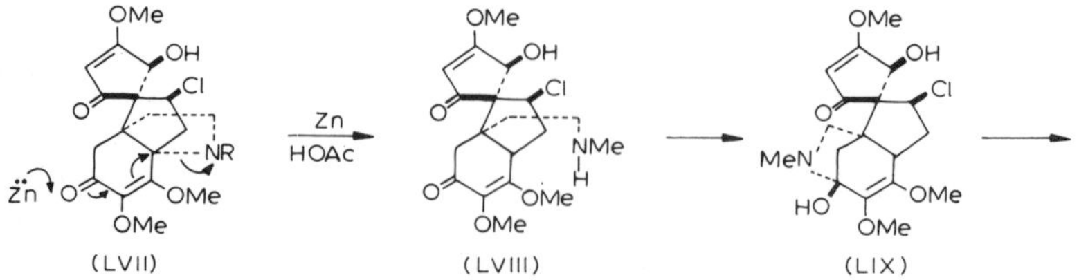

(LVII)

Zn
HOAc

(LVIII)

(LIX)

4. The homomorphine alkaloids

A small group of alkaloids has recently been discovered composed of bases of the morphine group having a seven-membered ring B, and clearly derived from phenethylisoquinoline alkaloids in the same way that bases of the morphine group are derived by oxidation from benzylisoquinolines. At the time of this review seven alkaloids have been identified, and these are listed with their physical properties and sources in Table 4.

TABLE 4

THE HOMOMORPHINE ALKALOIDS

	m.p. (°C)	*[α]D(°)*	*Source*
Androcymbine	199–201	−260	*Androcymbium melanthioides*
(+)-Kreysiginine	149	+89	*Kreysigia multiflora*
(−)-Kreysiginine	149	−89	*Colchicum cornigerum*
Alkaloid CC-2	172–174	+40	*C. cornigerum*
Alkaloid CC-3b	210–212	+25	*C. cornigerum*
Alkaloid CC-10	209–212	+308	*C. cornigerum*
Alkaloid CC-20	210–212	+324	*C. cornigerum*

Androcymbine has the composition $C_{21}H_{25}NO_5$, is phenolic and contains three methoxyl groups and a cross-conjugated dienone system spectroscopically similar to that in salutaridine (XVII, p. 306). Its methyl ether can be oxidised to 3,4,5-trimethoxyphthalic acid and reduced with sodium and liquid ammonia to the phenethylisoquinoline (II, R=Me, R^1=H). These results indicate that androcymbine has the structure I, the absolute configuration being deduced from optical rotatory dispersion studies (*A. R. Battersby et al.*, Chem. Comm. 1965, 228). The alkaloid and its methyl ether have been synthesised by photolytic cyclisation of the bromo phenols

(II, R = H, R^1 = Br) and (II, R = Me, R^1 = Br), respectively, in the presence of sodium iodide; in the absence of this salt cyclisation took a different course to give the homoaporphine alkaloid multifloramine (*T. Kametani* and *M. Kiozume*, J. chem. Soc. C, 1971, 3976; *Kametani et al.*, J. org. Chem., 1971, **36**, 3733):

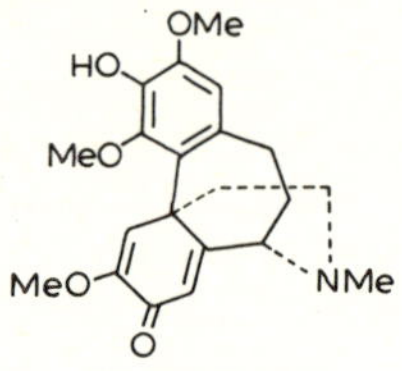

Androcymbine (I)

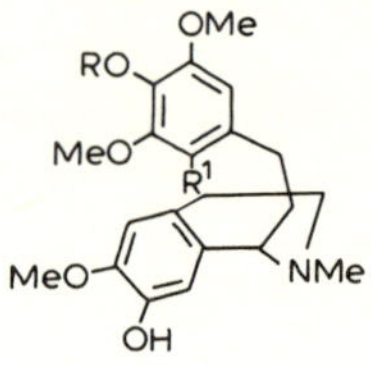

(II)

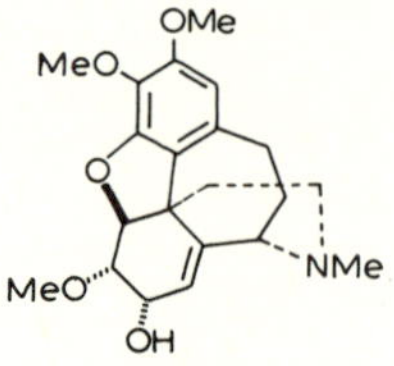

Kreysiginine (III)

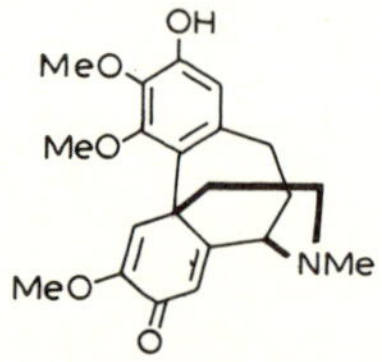

Alkaloid CC-10 (IV)

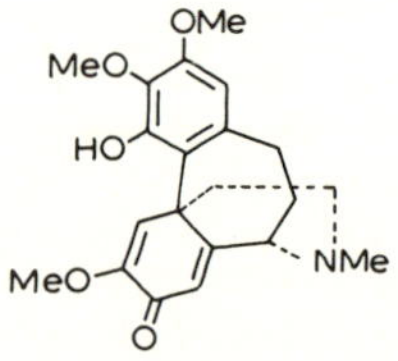

(V)

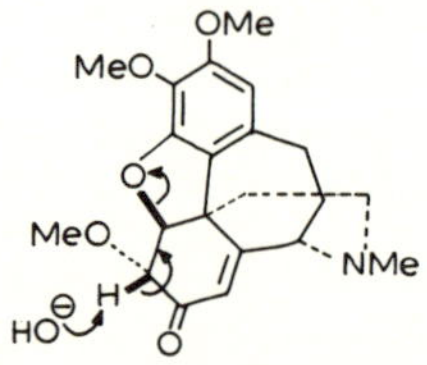

(VI)

Kreysiginine, a non-phenolic base, $C_{21}H_{27}NO_5$, contains one hydroxyl group, three methoxyl groups and one reducible double bond and can be oxidised to an α,β-unsaturated ketone (VI) which is isomerised by base to a phenolic dienone V, isomeric with androcymbine and giving *O*-methylandrocymbine on methylation. Kreysiginine thus has the structure III, which has been confirmed by an X-ray crystallographic study (*Battersby et al.*, Chem. Comm., 1968, 695; *J. Fridrichsons et al.*, Tetrahedron, 1970, **26**, 1969).

Alkaloid CC-10 is isomeric with androcymbine and the dienone V derived from kreysiginine, but different from both, though it yields the antipode of *O*-methylandrocymbine on methylation. It must therefore have the structure IV (*Battersby et al.*, J. chem. Soc. C, 1971, 3514).

Alkaloids CC-20 (VII), **CC-3b** (VIII) and **CC-2** (IX), isolated with alkaloid CC-10 and (−)-kreysiginine from *Colchicum cornigerum*, have been shown by spectroscopic studies to have the structures indicated, and the structure of alkaloid CC-2 has been confirmed by an X-ray crystallographic study of its *O*-acetyl ester *(idem, loc. cit.)*:

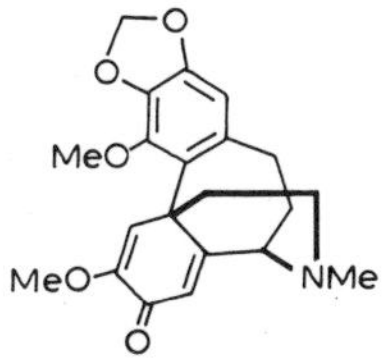

Alkaloid CC-20 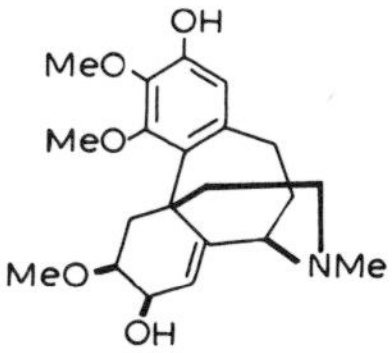 Alkaloid CC-3b 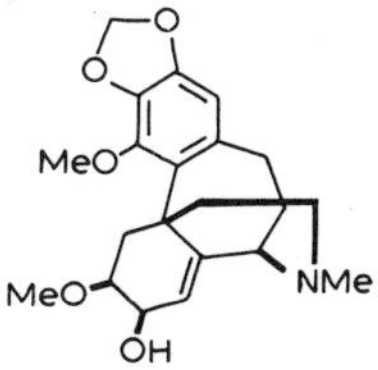 Alkaloid CC-2

Cancentrine (X) is the only bimolecular alkaloid so far reported representing the union of a morphine-like base and a base of a different series. *Cancentrine*, m.p. 238°, isolated from *Dicentra canadensis*, has the composition $C_{36}H_{34}N_2O_7$ and contains three aromatic methoxyl groups, one phenolic hydroxyl and one *N*-methyl group. Hofmann degradation affords a *methine base* XI, m.p. 230°, which can be methylated with diazomethane and reduced catalytically to the *base* XII, m.p. 189°, which was proved to have the structure and absolute configuration shown by an X-ray crystallographic study. The linkage of the nitrogen atom as shown in X and the position of the phenolic hydroxyl group were inferred from n.m.r. spectroscopic studies of the alkaloid and codeine (*G. R. Clarke et al.*, J. Amer. chem. Soc., 1970, **92**, 4998):

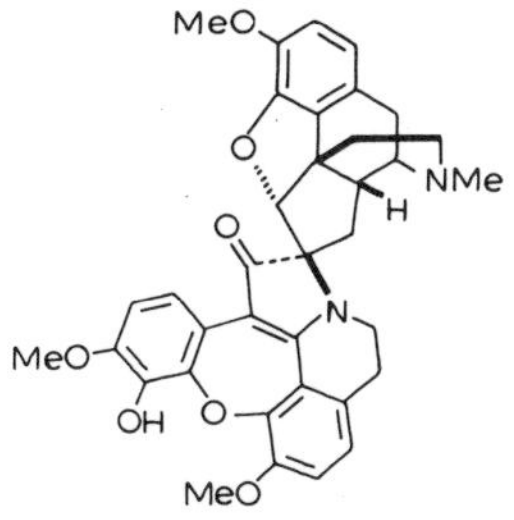

Cancentrine
(X)

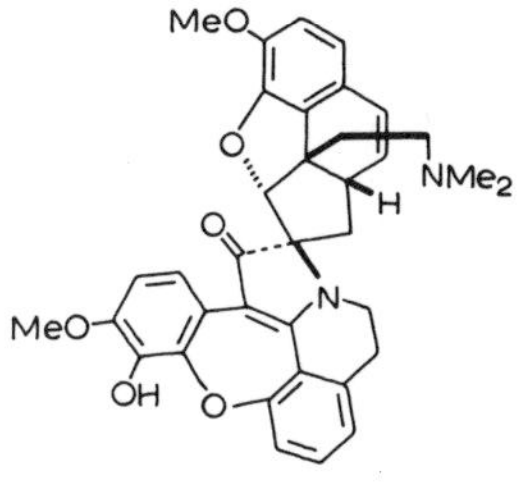

(XI)

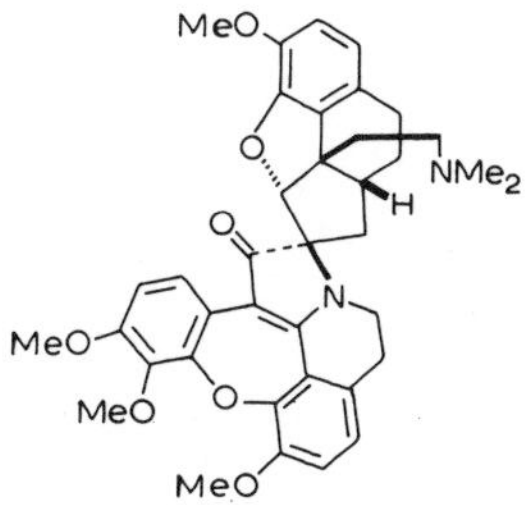

Chapter 34

Diterpenoid Alkaloids

A. R. PINDER

Alkaloids with a diterpenoid or closely related carbon skeleton have been known for many years, particularly because of their pharmacological properties, and in some cases highly toxic character. They occur chiefly in the plant families *Ranunculaceae*, *Garryaceae*, *Compositae* and *Rosaceae*, and for the purpose of discussion are conveniently divided into two main groups*:

1. Alkaloids having a C_{20} skeleton, divisible further into three groups: (*a*) *Garrya* alkaloids, (*b*) atisine alkaloids, and (*c*) alkaloids with a modified atisane skeleton.
2. Alkaloids generally regarded as diterpenoid, having a skeleton related to the diterpenes but containing only 19 (occasionally 18) carbon atoms, and divisible into four groups: (*a*) lycoctonine-type alkaloids, (*b*) aconitine-type alkaloids, (*c*) lactonic alkaloids, and (*d*) *Daphniphyllum* alkaloids.

Diterpene alkaloids have been reviewed in several places: see, for example *S. W. Pelletier* and *L. H. Keith*, in "Chemistry of the Alkaloids", Van Nostrand–Reinhold, New York, 1970; *idem*, in "The Alkaloids", ed. *R. H. F. Manske*, Vol. XII, Chapters 1 and 2, Academic Press, New York, 1970; *Pelletier*, Tetrahedron, 1961, **14**, 76; Experientia, 1974, **20**, 1; Quart. Reviews chem. Soc., 1967, **21**, 525; *E. S. Stern*, in "The Alkaloids", ed. *Manske*, Vol. 7, Chapter 22, Academic Press, New York, 1960; *K. Wiesner* and *Z. Valenta*, Fortschr. Chem. org. Naturst., 1958, **16**, 26; *H.-G. Boit*, "Ergebnisse der Alkaloid-Chemie bis 1960", pp. 851–1009, Akademie-Verlag, Berlin, 1961; *Pelletier* and *W. S. Page*, in M.T.P. International Review of Science, Vol. 9, Chapter 9, Butterworths, London, 1973.

* The author is indebted to Professor *S. W. Pelletier*, University of Georgia, for his helpful co-operation during the preparation of this chapter. A.R.P.

1. C_{20}-diterpene alkaloids

(a) The Garrya *alkaloids*

The *Garrya* alkaloids have the carbon framework of the parent hydrocarbon kaurane (I) (see *C.C.C.*, Vol. IIC, p. 380), which conforms to the isoprene rule. In these bases carbon atoms 19 and 20 are linked *via* the nitrogen atom of methylamine, ethylamine or 2-aminoethanol, leading to the general skeleton (II), known as the veatchine skeleton.

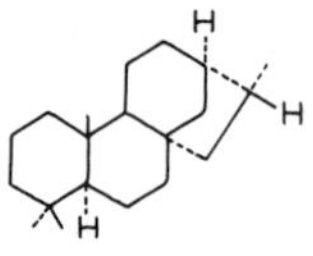

(I) Kaurane

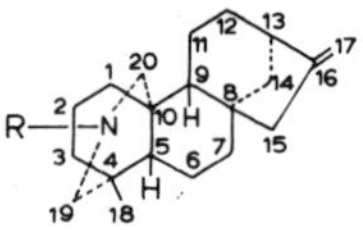

(II) Veatchine skeleton

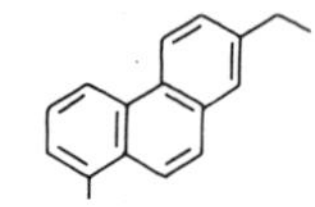

(III)

The first isolation of pure alkaloids from *Garrya* species was described by *J. F. Oneto* (J. Amer. pharm. Assoc., 1946, **35**, 204), who separated veatchine and garryine from *G. veatchii* Kellogg bark, though alkaloids had been detected in *G. fremontii* and *G. racemosa* much earlier (*Ross*, Amer. J. Pharm., 1877, [iv], **7**, 585).

Garryine, $C_{22}H_{33}NO_2$, has more recently been separated from *G. veatchii* bark by countercurrent distribution (*Wiesner* and collaborators, Canad. J. Chem., 1952, **30**, 608). The same is true of the isomeric base **veatchine**, and these two bases are conveniently discussed together because (*a*) veatchine is converted quantitatively into garryine by alcoholic alkali, (*b*) either base on reduction with lithium tetrahydridoaluminate yields the same dihydroveatchine, and on catalytic hydrogenation each affords tetrahydroveatchine, and (*c*) the i.r. spectra of the alkaloids are almost indistinguishable. Veatchine is a distinctly stronger base (pK_a 11.5) than garryine (pK_a 8.7), and yields a basic *O*-acetate, whilst dihydroveatchine yields a basic *O,O*-diacetate. Evidently the reduction of the alkaloids involves the rupture of a cyclic ether group; this view is confirmed by the fact that dihydroveatchine contains two active hydrogens, whereas garryine and veatchine contain only one (*Wiesner et al.*, Ber., 1953, **86**, 800; *M. F. Bartlett, W. I. Taylor* and *Wiesner*, Chem. and Ind., 1953, 173).

Dehydrogenation of both alkaloids by means of selenium yields 7-ethyl-1-methylphenanthrene (III) and 7-ethyl-1-methyl-3-azaphenanthrene (IV), the products being identified by spectroscopy and synthesis (*Wiesner et al.*, *ibid.*, 1954, 132; *Bartlett* and *Wiesner*, *ibid.*, p. 542). On the basis of these results and others, of biogenetic considerations, and of a possible relationship between the *Garrya* alkaloids and the well-known tetracyclic diterpene phyllocladene (V) (see *C.C.C.*, Vol. IIC, p. 380), structure VI has been assigned to dihydroveatchine.

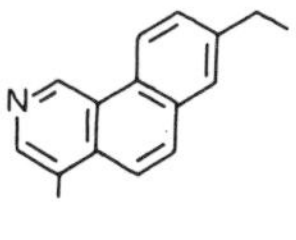

(IV)

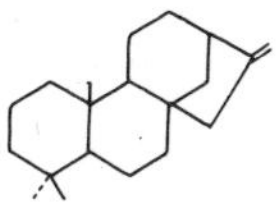

(V) Phyllocladene

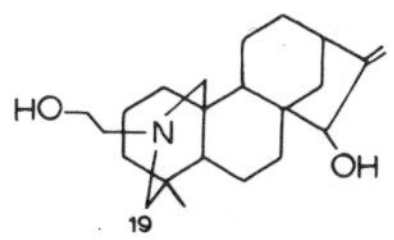

(VI) Dihydroveatchine

Garryine and veatchine are consequently formulated as VII and VIII respectively, each containing an oxazolidine ring cleaved by hydrogenolysis (*Wiesner et al.*, J. Amer. chem. Soc., 1954, **76**, 6068). These respective assignments emerge from the following observations. Firstly, there is a marked difference in basic strength between the two alkaloids and an argument has been presented reconciling this difference with the respective structures. Next, a parallelism exists between garryine and veatchine on the

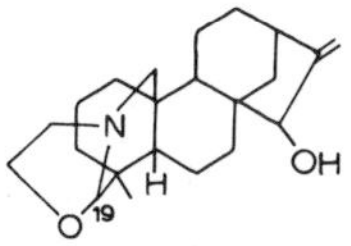

(IX)

(VII) Garryine

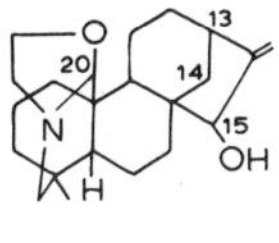

(VIII) Veatchine Garryfoline

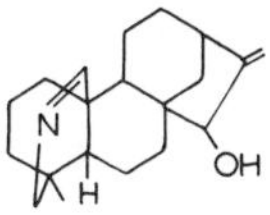

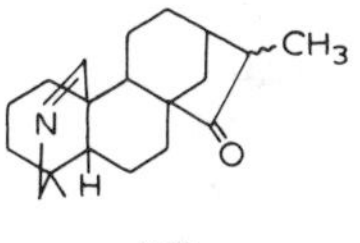

(X)

one hand, and isoatisine and atisine (see pp. 334 *et seq.*) on the other (*Wiesner* and *J. A. Edwards*, Experientia, 1955, **11**, 255), which supports the assignments. Finally, extensive degradative studies on the bases are in agreement. Amongst these is the behaviour of garryine and veatchine on pyrolysis with selenium, which yields two isomeric pyro-bases, $C_{20}H_{29}NO$, formulated as IX and X, the former an alcohol and the latter a ketone. Reduction of the former afforded a dihydro-alcohol (IX; no heterocyclic double bond) which is also a secondary amine and reacts with ethylene chlorohydrin to give dihydroveatchine (VI). Oxidation of dihydroveatchine with osmium tetroxide gave garryine (VII), with cyclisation at the less hindered

19-position. The presence of an oxazolidine unit in both garryine and veatchine is thus demonstrated.

Oxidation of veatchine with permanganate yields, under mild conditions, two isomeric lactams, oxoveatchine-A and -B, which are respectively γ- and δ-lactams, on i.r. evidence. They are a consequence of oxidation at the two carbon atoms α to the nitrogen, and are formulated as XI and XII. Garryine

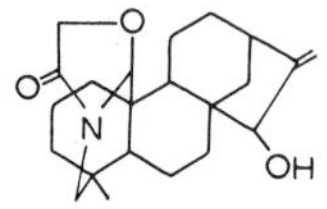

(XI) Oxoveatchine-A 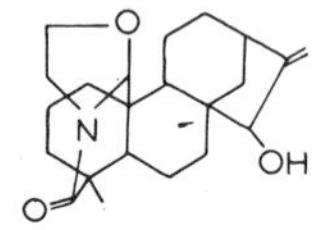 (XII) Oxoveatchine-B 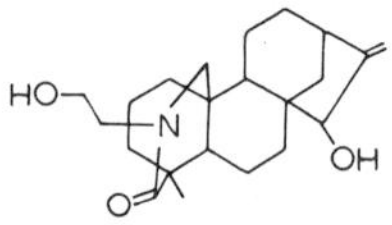 (XIII)

on similar treatment gives oxogarryine (XIII), a δ-lactam, only, presumably by oxidation of the intermediate hydrate formed by hydrolytic cleavage of the oxazolidine ring. Reduction of oxogarryine with lithium tetrahydridoaluminate furnishes dihydroveatchine (VI). Under more vigorous conditions oxidation of veatchine yields two lactam dicarboxylic acids, the result of cleavage of the 5-membered carbocyclic ring, which are formulated as XIV and XV:

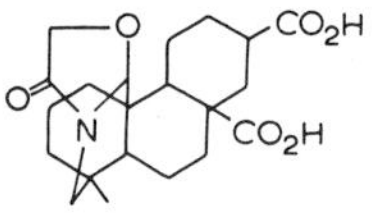

(XIV) 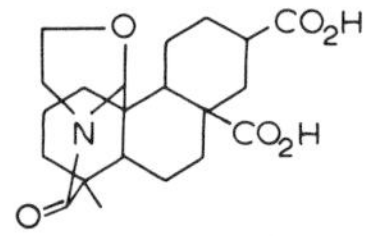 (XV) 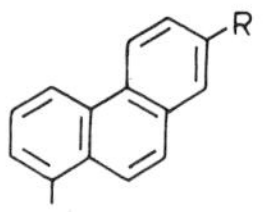 (XVI; R = Me) (XVII; R = CO_2H)

These acids yield anhydrides readily, and are, on i.r. evidence, respectively γ- and δ-lactams; the anhydrides are derived from glutaric anhydride (ν_{CO} 1800, 1762 and 1800, 1775 cm^{-1} respectively). Each acid forms a dimethyl ester with diazomethane, but these on hydrolysis yield monomethyl esters only; one of the carboxyl groups is evidently hindered (angular) (*Bartlett et al.*, Chem. and Ind., 1953, 323). Acid XV affords pimanthrene (1,7-dimethylphenanthrene) (XVI) and 1-methylphenanthrene-7-carboxylic acid (XVII) on dehydrogenation; the glycol obtained by reduction of the dimethyl ester of XV with lithium tetrahydridoaluminate also give pimanthrene on dehydrogenation, confirming the relative positions of the methyl and non-angular CO_2H groups in XV.

Basic strength of garryine and veatchine. As already intimated veatchine is a much stronger base than garryine. In hydroxylic solvents the alkaloids exist in part as iminium hydroxides, the latter being more strongly basic than the normal tertiary amine structure (*Pelletier* and *W. A. Jacobs*, Chem. and Ind., 1955, 1385; *O. E. Edwards* and *T. Singh*, Canad. J. Chem., 1954, **32**, 465).

$$\text{N-C}_{20}\text{H-O (oxazolidine)} + H_2O \rightleftharpoons \overset{\oplus}{N}{=}C_{20}\ {}^{\ominus}OH \ \ \text{-OH}$$

In veatchine, therefore, there will be a preponderance of the iminium hydroxide structure in hydroxylic solvents; it seems plausible that the reason for this is one of steric interference involving the hydrogen at $C_{(20)}$ and those at $C_{(13)}$ and $C_{(14)}$ in the tertiary amine structure VIII of veatchine, as revealed by models. There is serious steric repulsion when $C_{(20)}$ is tetrahedral. In the iminium hydroxide structure $C_{(20)}$ is trigonal and the strain is relieved. On the other hand, in garryine (VII) this steric factor does not operate to the same extent, so that the above equilibrium in this case lies much further to the left, and garryine is a weaker base. A similar consideration explains why veatchine is readily isomerised to garryine. In solution the isomerisation proceeds *via* the iminium structure by prototropy. The steric factors discussed above are responsible for garryine's having a lower free energy than veatchine, and consequently the normal base (veatchine) tends to be transformed into the "iso-base" (garryine). The correctness of this view is supported by some experiments on model oxazolidines (*N. J. Leonard*, *K. Conrow* and *R. R. Sauers*, J. Amer. chem. Soc., 1958, **80**, 5185). This isomerisation is shown by other alkaloids in this group (see below).

Synthesis. The total synthesis of the racemic forms of veatchine and garryine has been described by *W. Nagata* and co-workers, *ibid.*, 1964, **86**, 929; 1967, **89**, 1499), along the following lines. The tricyclic ketone XVIII was converted in several steps into the dimesylate XIX (*Nagata et al.*, *ibid.*, 1963, **85**, 2342), which was converted into the cyclopentene XX by heating with collidine. This compound was hydroborated under steric control, using bis-2-butyl-3-methylborane to give, after oxidation and hydrolysis, diol XXI, accompanied by a small amount of the isomeric diol resulting from the other possible direction of hydroboration. The former was rearranged, by conversion to its monobrosylate and subsequent treatment with base, to ketone XXII. A Wittig reaction with the latter afforded olefin XXIII. Birch reduction was next used to demesylate the nitrogen atom, whereupon reaction with ethyl chloroformate gave XXIV. Allylic bromination of the latter, followed by epoxidation and debromination with zinc, led to allylic alcohol XXV with some of the isomeric endocyclic allylic alcohol. The former was hydrolysed ($NCO_2Et \rightarrow NH$), then reacted with ethylene chlorohydrin, to give (±)-dihydroveatchine (VI), the "natural" epimeride of which had already been converted into garryine and then veatchine (see above). Other syntheses have been described by *S. Masamune* (J. Amer. chem. Soc., 1964, **86**, 288, 290, 291), *Valenta*, *Wiesner* and *C. M. Wong* (Tetrahedron Letters, 1964, 2437), *Wiesner et al.* (Coll. Czech. chem. Comm., 1966, **31**, 602; Tetrahedron Letters, 1968, 6279), and by *T. Matsumoto et al.*, (*ibid.*, p. 1127). Another involves conversion of an intermediate encountered in the

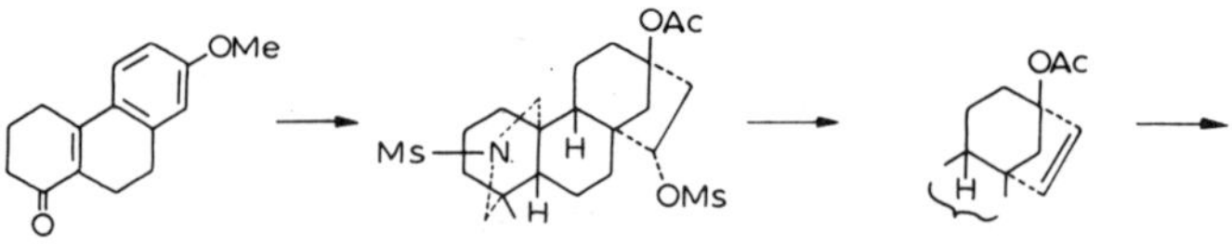

(XVIII) (XIX) (XX)

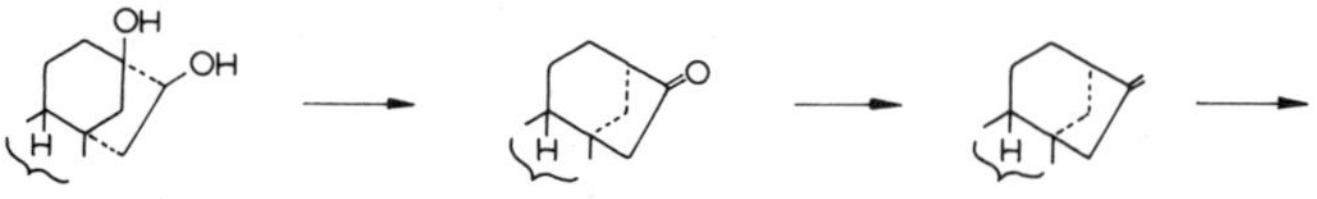

(XXI) (XXII) (XXIII)

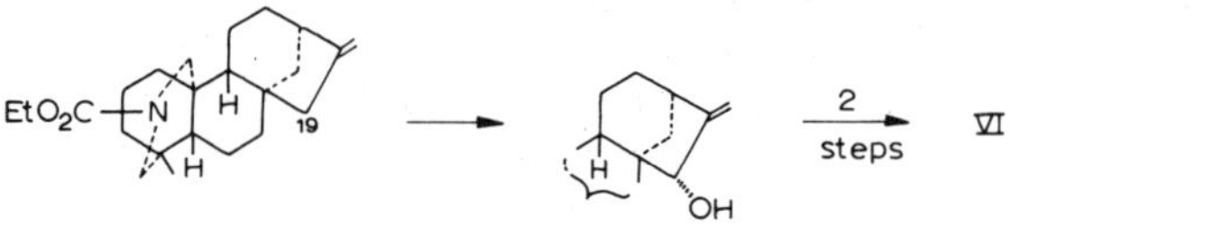

(XXIV) (XXV) 2 steps VI

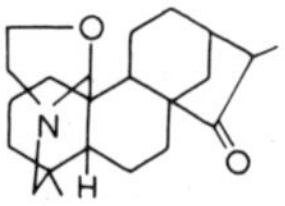

(XXVa) Cuauchichicine

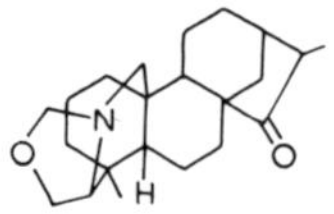

(XXVI) Isocuauchichicine

synthesis of kaur-16-en-19-oic acid into a tetracyclic lactam, itself an intermediate in the *Wiesner (loc. cit.)* synthesis of veatchine and garryine (*K. Mori*, *K. Saeki* and *M. Matsui*, Agric. and biol. Chem., Japan, 1971, **35**, 956). For other examples of synthetic endeavours in this area see *O. E. Edwards*, Specialist Periodical Reports, Chemical Society, Alkaloids, Vol. 1, 1971, p. 357.

Cuauchichicine, $C_{22}H_{33}NO_2$, is found in *Garrya laurifolia* and is an isomer of veatchine and garryine. It is a cyclopentanone, is saturated and is a tertiary base. A Kuhn–Roth estimation reaveals the presence of two *C*-methyl groups; no hydroxyl group or $C{=}CH_2$ group is present. Pyrolysis of the alkaloid yields a pyro-base identical with one of those from veatchine. This pyro-base on reduction with lithium tetrahydridoaluminate, followed by reaction with ethylene chlorohydrin, affords tetrahydroepiveatchine, also obtainable from cuauchichicine by similar reduction. The basic strength of cuauchichicine is comparable with that of veatchine rather than garryine, and the alkaloid is consequently formulated as XXVa (*C. Djerassi et al.*, J. Amer. chem. Soc., 1954, **76**, 5889; 1955, **77**, 4801, 6633).

Cuauchichicine is isomerised by base to **isocuauchichicine**, a transformation exactly analogous to the veatchine–garryine rearrangement *(vide supra)*. Isocuauchichicine, which is an artifact, is consequently formulated as XXVI.

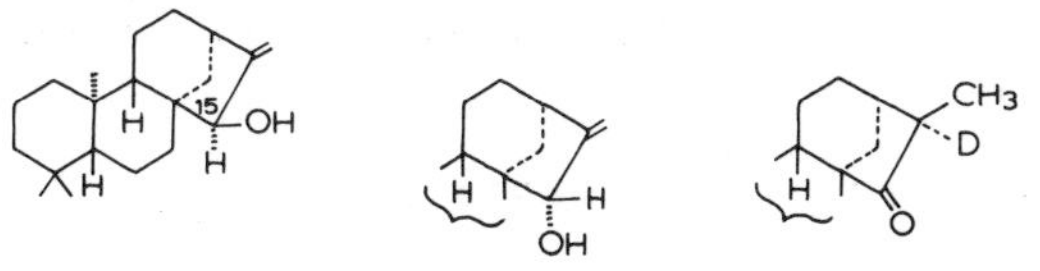

(XXVII) (XXVIII) (XXIX)

Garryfoline, isomeric with cuauchichicine, is also found in *G. laurifolia*. It is a tertiary base containing a *C*-methyl group, a secondary alcoholic group, and an exocyclic methylene group, formaldehyde and a cyclic ketone being formed on ozonolysis. Treatment with acid causes it to be isomerised to cuauchichicine (XXVa), and reduction with lithium tetrahydridoaluminate yields dihydrogarryfoline, in turn isomerised by acids into dihydrocuauchichicine. The latter is ketonic and yields on the same type of reduction tetrahydroepiveatchine. Garryfoline is thus to be identified with 15-epiveatchine (see VIII); it is clearly related to veatchine because its pK_a value (11.8) approximates to that of that base (11.5) *(idem, loc. cit.)*.

(i) Stereochemistry of the Garrya *alkaloids*

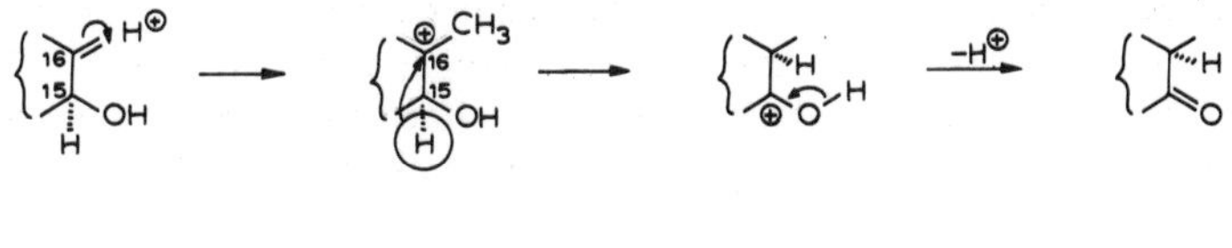

The allylic alcohol–ketone isomerisation exemplified by the garryfoline–cuauchichicine change described above gives a clue concerning the stereochemistry of the *Garrya* bases. Veatchine, it should be noted, is stable to mineral acid, and it must be concluded that this difference in behaviour is due to the difference in configuration of the >CHOH group at position 15. Working with the epimeric kaur-16-en-15-ols as models, *M. F. Barnes* and *J. MacMillan* (J. chem. Soc. C, 1967, 361) have shown that the 15β-ol (XXVII) is rapidly rearranged by acid to the expected α-methyl ketone, whereas the 15α-epimer (XXVIII) is stable under the same conditions, the configurations of the two alcohols having been settled unequivocally by their methods of synthesis. The mechanism of the rearrangement was established to involve a 15,16-hydride shift, since 15-deuterokaur-16-en-15β-ol (XXVII, C····D at position 15) yielded on acid treatment 16*R*,16-deuterokauran-15-one (XXIX). The strict requirement of an *exo*-15-hydrogen (or deuterium) for rearrangement is explained by the fact that a 16-carbonium ion is generated by attack of a proton on the exocyclic methylene group. In this species only an *exo*-15 C–H bond is parallel to the vacant *p*-orbital of this ion; this view is corroborated by n.m.r. spectral study. It would seem reasonable to extend this mechanism to those *Garrya* and related alkaloids which rearrange under the same conditions. It follows, then, that garryfoline, which rearranges readily to cuauchichicine (see above), has a 15α hydrogen atom (and therefore a 15β hydroxyl group), whilst veatchine, stable to acid, has the epimeric configuration at this position. The relative configurations of *Garrya* alkaloids, on the basis of this rearrangement, are expressed in Table 1.

TABLE 1

GARRYA ALKALOIDS AND RELATED COMPOUNDS

Alkaloid or related compound	*M.p. (°C)*	$[\alpha]_D^{CHCl_3}$ *(°)*	*Oxygen of* $OCH_2 \cdot CH_2N<$ *attached to position*	*Substituents*	
				at position 15	*at position 16*
Veatchine	119–120	−69	20	····OH	$=CH_2$
Garryine	74–82*	−84	19	····OH	$=CH_2$
Garryfoline	130–133	−60	20	–OH	$=CH_2$
Isogarryfoline	140–144	−57	19	–OH	$=CH_2$
Cuauchichicine	152–155	−71	20	=O	$\sim CH_3$
Isocuauchichicine	134–136	−84	19	=O	$\sim CH_3$
Napelline	166	−42.7**	—	–OH	$=CH_2$
Isonapelline	121		—	=O	$\sim CH_3$
Songorine	212		—	–OH	$=CH_2$
Isosongorine	203		—	=O	$\sim CH_3$
Lucidusculine	170		—	–OAc	$=CH_2$
Songoramine	211–212		—	–OH	$=CH_2$
Anopterine	222–223	−12	—	H_2	$=CH_2$

* Hydrate.
**Hydrobromide in water.

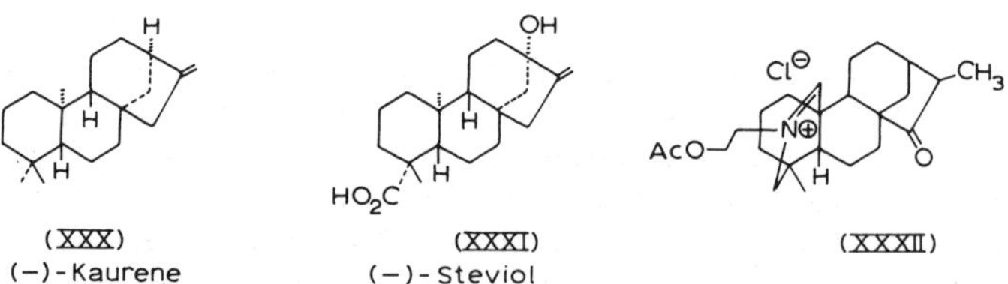

(XXX) (−)-Kaurene (XXXI) (−)-Steviol (XXXII)

Other aspects of the stereochemistry of the *Garrya* bases are bound up with a stereochemical correlation with atisine, another diterpene alkaloid (see p. 334), and with the diterpenes (−)-kaurene (XXX) and (−)-steviol (XXXI), both of established absolute configuration (*C.C.C.*, Vol. IIC, pp. 461, 462). Garryfoline (VIII) is convertible into an acetate iminium chloride XXXII, which by Hofmann degradation of the nitrogen side-chain affords the corresponding azomethine. Treatment of the latter with nitrous acid yielded a hemiacetal XXXIII, which on Wolff–Kishner reduction suffered attack on both 15-keto and masked 19-aldehyde functions, leading to alcohol XXXIV. Oxidation

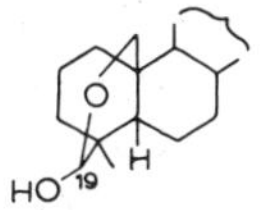

(XXXIII)

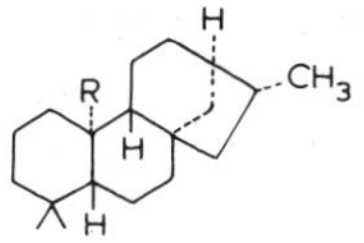

(XXXIV; R = CH_2OH
(XXXV; R = CH_3)

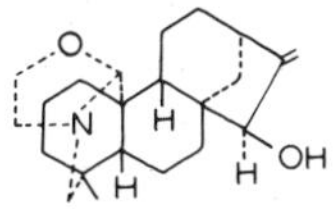

(XXXVI)
(−)-Garryfoline

of the latter (CrO_3) afforded the corresponding carbaldehyde, which on further Wolff–Kishner reduction afforded XXXV, a hydrocarbon which proved to be identical with (−)-"β"-dihydrokaurene, the minor hydrogenation product of (−)-kaurene (XXX), itself identical with "stevane-B", a degradation product of (−)-steviol (XXXI). The hydrocarbon is of known absolute configuration XXXV, and it follows that natural (−)-garryfoline must be represented by XXXVI. The absolute configurations of related alkaloids follow either from similar correlations or from correlation with garryfoline. Further, veatchine and garryfoline have been converted into the 17-nor-16-ketones XXXVII and XXXVIII, respectively; these compounds showed a positive Cotton effect in their o.r.d. curves with amplitude closely similar to that shown by ketone XXXIX obtained from (+)-phyllocladene (XL), of known absolute configuration.

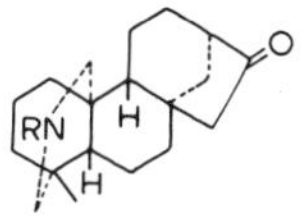

(XXXVII; R = $AcOCH_2 \cdot CH_2$)
(XXXVIII; R = Ac)

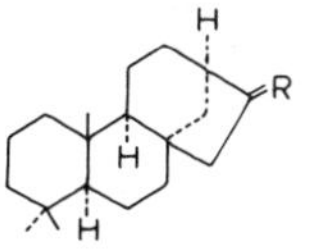

(XXXIX; R = O)
(XL; R = CH_2) (+)-Phyllocladene

With the reasonable assumption that the configuration at position 9 will not affect the sign of the Cotton effect, it follows that the absolute configurations of garryfoline at positions 8 and 13 are the same as in phyllocladene (*H. Vorbrueggen* and *Djerassi*, Tetrahedron Letters, 1961, 119; J. Amer. chem. Soc., 1962, **84**, 2990; *Djerassi et al.*, *ibid.*, 1961, **83**, 3163, 3720).

Napelline, $C_{22}H_{33}NO_3$, is a trihydric alcohol occurring in *Aconitum napellus* L. It is an *N*-ethyl tertiary base and contains *C*-methyl and exocyclic $C{=}CH_2$ groups (*Wiesner et al.*, Chem. and Ind., 1957, 173; *A. D. Kuzovkov*, J. gen. Chem. U.S.S.R., 1953, **23**, 521). Its structure is conveniently considered alongside that of **songorine,** $C_{22}H_{31}NO_3$, another alkaloid from the same source, which contains two hydroxyl groups and a cyclohexanone carbonyl group, and the other groups listed for napelline. It yields napelline on reduction with lithium tetrahydridoaluminate, and is therefore napelline in which one of the hydroxyl functions has been oxidised to a ketone. Napelline is readily isomerised by acids to **isonapelline**, a cyclopentanone, and thus resembles garryfoline (p. 329); likewise, songorine is isomerised to **isoson-**

gorine. Isonapelline yields 7-ethyl-1-methyl-3-azaphenanthrene, and songorine yields 7-ethyl-1,9-dimethylphenanthrene, on dehydrogenation with selenium, both these results pointing to a veatchine-type skeleton for the compounds. The following observations indicate that napelline, songorine, isonapelline and isosongorine are to be formulated as XLI, XLII, XLIII and XLIV, respectively (*T. Okamoto et al.*, Chem. pharm. Bull., Tokyo, 1965, **13**, 1270; *T. Sugasawa, ibid.*, 1961, **9**, 889, 897; *Wiesner, S. Ito* and *Valenta*, Experientia, 1958, **14**, 167):

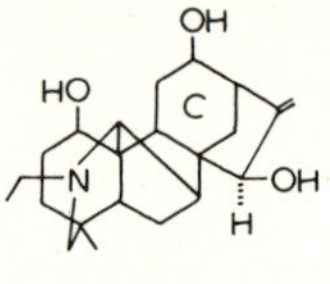

(XLI) Napelline

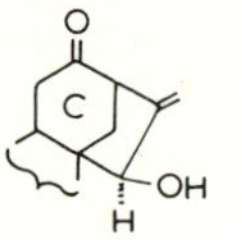

(XLII) Songorine

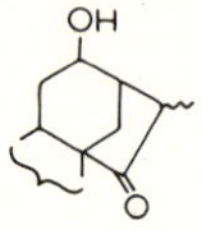

(XLIII) Isonapelline

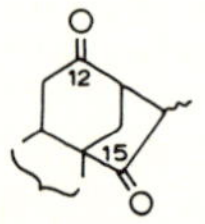

(XLIV) Isosongorine

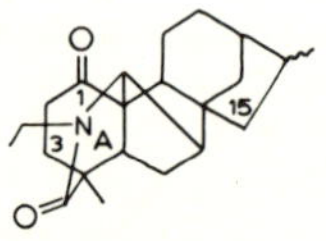

(XLV)

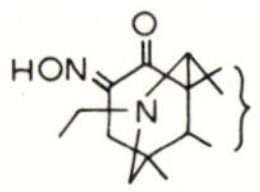

(XLVI)

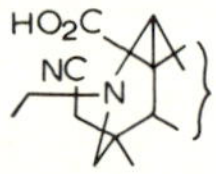

(XLVII)

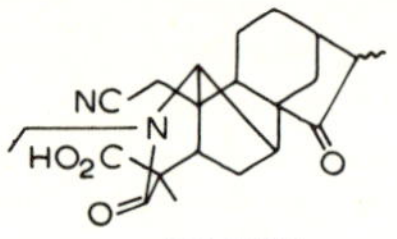

(XLVIII)

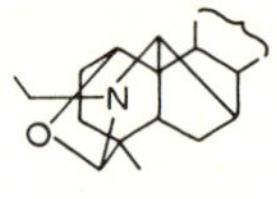

(XLIX)

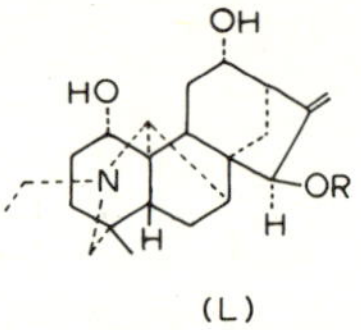

(L) Napelline (R = H) LuciduscuLine (R = Ac)

Isosongorine (XLIV) on Huang-Minlon reduction gave a 12,15-bisdeoxy derivative which on oxidation (CrO_3) afforded a keto-δ-lactam (XLV) ($\nu_{C=O}$ 1635 and 1707 cm^{-1}); the latter formed a monobenzylidene derivative and exchanged approximately two deuterium atoms with NaOD in MeOD–D_2O. The oxo group in the lactam is thus flanked by a CH_2 group and the C=O group in ring A must be at position 1 or 3. Songorine (XLII) can likewise be reduced to 12-deoxysongorine and this on rearrangement with acid gave isodeoxysongorine (XLIV; CH at position 12), which on similar oxidation gave a dioxolactam (XLV; C=O at position 15). Treatment of the latter with isoamyl nitrite yielded a hydroxyimino derivative XLVI; benzenesulphonyl chloride and alkali cleaved this to a nitrile-carboxylic acid XLVII. This acid resisted attempts to decarboxylate it thermally or under acid conditions. If the ring A C=O group in the dioxolactam had been at position 3, the nitrile-carboxylic acid would have been XLVIII, which is a malonamide and which would have been expected, from analogies, to suffer thermal or acid-catalyzed decarboxylation smoothly. Consequently the ring A hydroxyl group in napelline, songorine and the iso-bases must be placed at position 1. Oxidation of dihydronapelline with silver oxide affords a methanolamine ether formulated as partial structure XLIX; models show that the formation of this derivative requires the hydroxyl at $C_{(1)}$ to be equatorial, and ring A to be a boat. The $C_{(20)}$–$C_{(7)}$ linkage present in these compounds is revealed by the formation of 7-ethyl-1,9-dimethylphenanthrene as a dehydrogenation product.

Lucidusculine, $C_{24}H_{35}NO_4$, occurring in *Aconitum lucidusculum*, contains an acetoxy group, three secondary hydroxyl groups, an exocyclic C=CH_2 group and an *N*-ethyl tertiary nitrogen atom. Its structure has been established partly by the observation that hydrolysis of the base yields napelline and acetic acid, and by X-ray diffraction analysis of its hydriodide. The absolute configuration of lucidusculine is represented by L (R=Ac), and napelline is therefore L (R=H). The absolute configurations of the other members of this subgroup follow (*A. Yoshino et al.*, Chem. pharm. Bull., Tokyo, 1965, **13**, 1270; *Yoshino* and *Y. Iitaka*, Acta Cryst., 1966, **21**, 57).

Songoramine, $C_{22}H_{29}NO_3$, is an alkaloid from *Aconitum karakolicum* and *A. songoricum*, the molecular formula of which is that of a dehydrosongorine. Spectral study revealed the presence of an *N*-ethyl group, a carbonyl group and an exocyclic C=CH_2 group, probably β,γ to the C=O group. The relative positions of the last two groups are associated with a bathochromic shift of the nπ^* band and moderate enhancement of $\varepsilon_{max.}$ in the u.v. spectrum ($\lambda_{max.}$ 295 nm, log $\varepsilon_{max.}$ 2.60), and had been noted amongst diterpene alkaloids earlier (*O. E. Edwards* and *L. Marion*, Canad. J. Chem., 1954, **32**, 195). The presence of a methanolamine ether group was shown by reductive ring opening and immonium salt formation. The fully reduced alkaloid proved to be identical dihydrosongorine, and oxidation (Ag_2O) of songorine (XLII) gave songoramine. Songoramine must therefore be assigned structure LI (*M. S. Yunusov et al.*, Khim. prirod. Soedin., 1970, 101; C.A., 1970, **73**, 131178u).

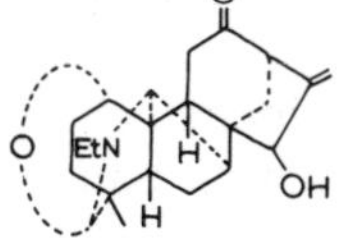

(LI)
Songoramine

Anopterine, $C_{31}H_{43}NO_7$, is the first diterpenoid alkaloid to be encountered in the *Escalloniaceae* family, being the major base found in the bark and leaves of *Anopterus macleayanus* and *A. glandulosus*, both of Australian origin. Anopterine is an ester-alkaloid yielding on hydrolysis anopteryl alcohol, $C_{21}H_{31}NO_5$ and tiglic acid. The former forms a tetra-acetyl derivative; an anomalous scattering X-ray diffraction analysis of the methiodide of this ester showed it to have structure and absolute configuration LIa, an azomethine. Chemical and spectroscopic studies on the alcohol, combined with the above result, pointed to structure and absolute configuration LIb for anopteryl alcohol, the azomethine structure for the methiodide being explained by an oxidation (atmospheric?) during a slow reaction with methyl iodide, or possibly:

$$>\ddot{N}-\underset{H}{\underset{|}{C}}H- \quad CH_3-I \longrightarrow >\overset{\oplus}{N}=CH- \;+\; CH_4$$

The locations of the tigloyloxy groups in anopterine (LIc) were settled by n.m.r. spectral analysis on the alkaloid (*J. A. Lamberton et al.*, Tetrahedron Letters, 1972, 2727).

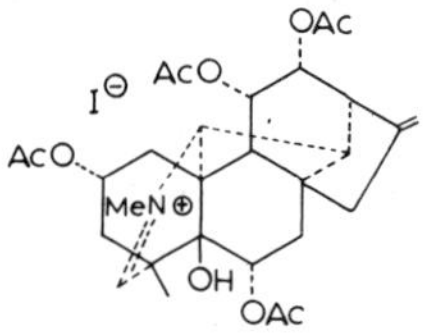

(LIa)

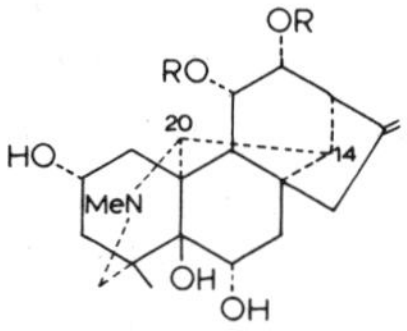

(LIb; R = H) Anopteryl alcohol

(LIc; R = *cis*- $-\underset{\underset{O}{\|}}{C}-\overset{\overset{CH_3}{|}}{C}=CH\cdot CH_3$) Anopterine

The alkaloid is unique in that it has a $C_{(20)}$–$C_{(14)}$ linkage (encountered in atisine alkaloids, but not previously in the veatchine group).

(b) The atisine alkaloids

A close parallel exists between these bases and those of the foregoing subgroup; realisation of this relationship simplified greatly structural studies in both areas.

Atisine, $C_{22}H_{33}NO_2$, is the principal alkaloid of *Aconitum heterophyllum.* It is a tertiary *N*-ethyl base which on n.m.r. and i.r. spectral evidence contains hydroxyl (secondary), exocyclic $C{=}CH_2$ and *C*-methyl groups, and is isomeric with the *Garrya* alkaloids, which it resembles in its chemical behaviour. For example, it is a strong base (pK_a 12.8), easily isomerised to isoatisine (pK_a 10.3) (*W. A. Jacobs* and *L. C. Craig*, J. biol. Chem., 1942, **143**, 589; 1943, **147**, 567; *C. F. Huebner* and *Jacobs*, *ibid.*, 1947, **170**, 515; 1948, **174**, 1001; *S. W. Pelletier* and *Jacobs*, Chem. and Ind., 1955, 1385).

Complex hydride reduction yields dihydroatisine, analogous to dihydroveatchine. Further reduction affords tetrahydroatisine, which contains the group $>NCH_2CH_2OH$. Atisine contains one reactive hydrogen atom (*O. E. Edwards* and *T. Singh*, Canad. J. Chem., 1954, **32**, 465; 1955, **33**, 448).

A notable difference between atisine and the *Garrya* alkaloids is the result of dehydrogenation with selenium. Atisine yields 1-methylphenanthrene, 6-ethyl-1-methylphenanthrene and 6-ethyl-1-methyl-3-azaphenanthrene (LII) (*Jacobs* and *Craig, loc. cit.*; *Huebner* and *Jacobs*, J. biol. Chem., 1947, **170**, 203; *D. M. Locke* and *Pelletier*, J. Amer. chem. Soc., 1958, **80**, 2588; 1959, **81**, 2246). It was conjectured, on this basis, that atisine differed from the *Garrya* bases in the mode of attachment of ring D. Careful consider-

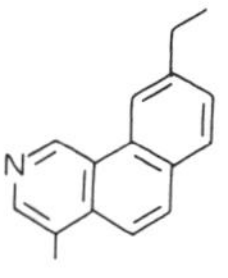

(LII) 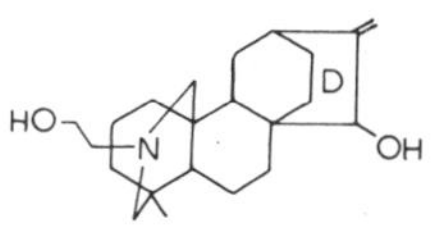 (LIII) 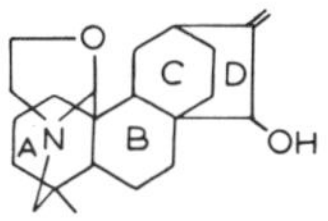 (LIV) Atisine (LV) Isoatisine

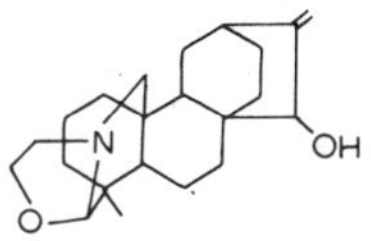

ation of these and other factors led to the proposal of structures LIII, LIV and LV for dihydroatisine, atisine and isoatisine respectively (*K. Wiesner et al.*, Chem and Ind., 1954, 132; *Pelletier* and *Jacobs*, J. Amer. chem. Soc., 1954, **76**, 4496).

The positive positions of the $=CH_2$ and $-OH$ groups in atisine were apparent from the ease of acid-catalysed isomerisation of both atisine and isoatisine to 2-methyl ketones, behaviour analogous to that of garryfoline (p. 329). Evidence for the exact placing of these groups is concerned with the azomethine alcohol LVI, derived from atisine (*idem, ibid.*, 1956, **78**, 4139, 4144; *D. Dvornik* and *Edwards*, Chem. and Ind., 1956, 248; Canad. J. Chem., 1957, **35**, 860), which was converted in several steps into the oxo acid LVII (*idem*, Chem. and Ind., 1958, 623; Canad. J. Chem., 1964, **42**, 137).

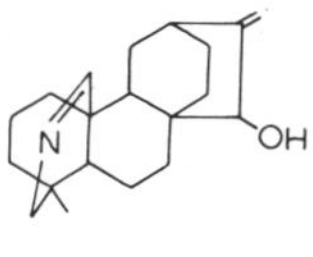

(LVI) 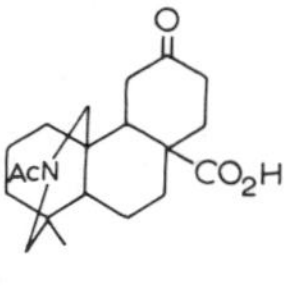 (LVII) 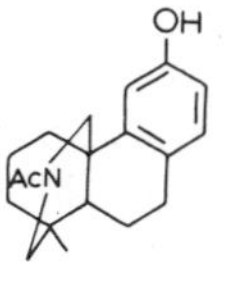 (LVIII)

Bromination–dehydrobromination of the latter yielded the phenol LVIII and a ketonic γ-lactone with $\nu_{C=O}$ 1796, 1734, 1639 cm^{-1}, the latter being formulated as LIX, with union of the lactone ring at either position 11 or 13. Formation of this lactone reveals a 1,4-relationship between the keto and carboxylic acid functions in LVII, and shows the presence of a bicyclo[2.2.2]octane system in atisine, when taken in conjunction with dehydrogenation evidence, and another sequence of transformations (*Pelletier*, Chem. and Ind., 1958, 1116; *Pelletier* and *P. C. Parthasarathy*, J. Amer. chem. Soc., 1965, **87**, 777).

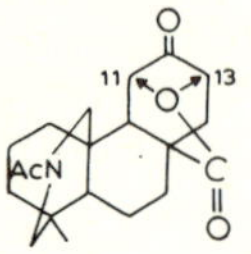

(LIX) 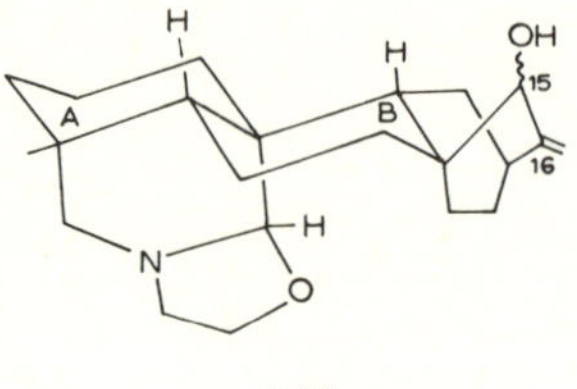 (LX) Atisine (β-OH) (LXI)

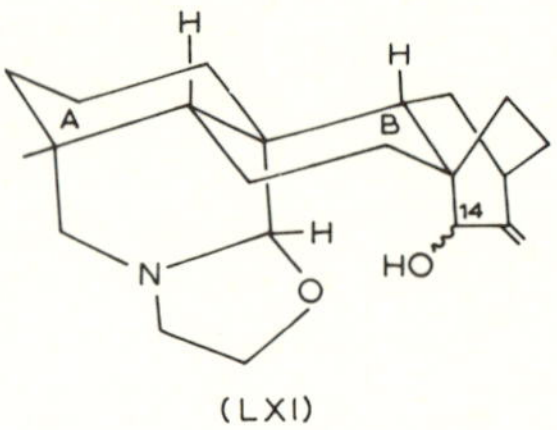

Evidence for the presence of an oxazolidine unit in atisine and its congeners includes the observation that atisine on oxidation yields a pair of lactam-dicarboxylic acids (*cf.* veatchine). Likewise, oxidation of isoatisine yields an oxoatisinedicarboxylic acid (*cf.* oxidation of garryine to oxogarryine) (*Jacobs*, J. org. Chem., 1951, **16**, 1593; *Pelletier* and *Jacobs*, J. Amer. chem. Soc., 1956, **78**, 4139). Other evidence is concerned with the ease of isomerisation of atisine by base to isoatisine (*cf.* the veatchine→garryine rearrangement), along with similar isomerisations observed with atisine derivatives. Thirdly, it has been possible to regenerate atisine from the azomethine alcohol LVI by treatment with ethylene chlorohydrin (*Dvornik* and *Edwards*, *loc. cit.*), and isoatisine has been reconverted into atisine (*Edwards* and *Singh*, Canad. J. Chem., 1954, **32**, 465; 1955, **33**, 448; *Pelletier* and *K. Kawazu*, Chem. and Ind., 1963, 1879; *Pelletier*, *Kawazu* and *K. V. Gopinath*, J. Amer. chem. Soc., 1965, **87**, 5229).

Stereochemistry of atisine. Conformational considerations of the structure of ajaconine (another alkaloid of the atisine group; see p. 339) indicate that rings A and B in atisine are *trans*-fused (*A. J. Solo* and *Pelletier*, Chem. and Ind., 1960, 1108). This allows two possible partial configurations LX and LXI to be written for atisine: in the first the allyl alcohol group is placed on the branch of the bicyclo[2.2.2]octane which is *trans* to the nitrogen atom, and in the second, it is placed on the *cis* branch. It was observed that reduction of two ketones LXII and LXIII with tetrahydridoborate gave rise in each case to two epimeric alcohols, all of which yielded readily an *O*-acetate. The hydroxyl groups therein are consequently unhindered, and it would seem reasonable to place the allylic alcohol unit on the *trans* branch of the bicyclooctane system, as in LX. The formation of a pair of epimers in each case supports this view, since reduction of an oxo-group at $C_{(14)}$ in LXI would be expected to be

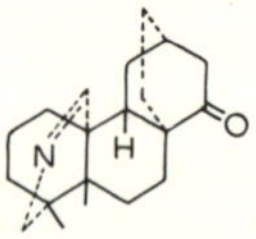

(LXII) 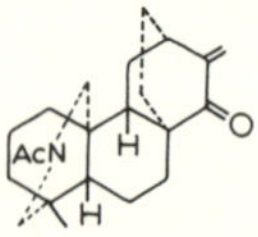 (LXIII) 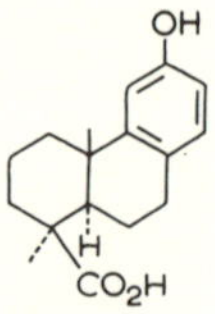 (LXIV)

stereoselective because of severe steric crowding in this area. pK_a measurements agree with this formulation, it being possible to present a plausible argument relating basic strength to interaction between nitrogen and hydroxyl functions, more pronounced in LXI than in LX (*Dvornik* and *Edwards*, *ibid.*, 1958, 623; Canad. J. Chem., 1964, **42**, 137). The results of detailed experimental studies demonstrate that the relative stereochemistry of atisine is represented by LX, in which the hydroxyl group is 15β (*idem*, Tetrahedron, 1961, **14**, 54; *Pelletier*, *ibid.*, p. 76). The absolute configuration of atisine rests in part on a correlation between atisine and veatchine (*Pelletier*, J. Amer. chem. Soc., 1960, **82**, 2398; *Pelletier* and *Locke*, *ibid.*, 1965, **87**, 761), pointing to expression LX (15β-OH) for natural atisine. Independent evidence for this assignment has been provided by *J. W. ApSimon* and *Edwards* (Canad. J. Chem., 1962, **40**, 896), who synthesised from podocarpic acid (LXIV) (*C.C.C.* Vol. IIC, p. 379), of known absolute configuration, a phenol LXV, which proved to be the enantiomer of phenol LVIII described above, derived from atisine:

OH
AcN
H

(LXV)

Several partial syntheses of atisine have been described (for a summary see *Pelletier* and *Keith*, in "The Alkaloids", ed. *Manske*, Vol. XII, pp. 188–202, Academic Press, New York, 1970). A stereospecific total synthesis of racemic atisine takes the following course (*W. Nagata et al.*, J. Amer. chem. Soc., 1963, **85**, 2342; 1967, **89**, 1483), shown in outline:

OMe → HCN → CN, OMe, H → (a) $Ph_3P{=}CHOMe$ (b) $H_3O^{\oplus}$ (c) Methylation → CN, OMe, OHC, H

(a) Hydrolysis (b) Ethylation (c) $LiAlH_4$ → HN, OMe, H → (a) Birch reduction (b) Mesylation (c) Hydrolysis → MsN, H, H

HCN → H, CN → (a) Acetalisation (b) LiMe (c) Hydrolysis (d) Base → OH, H → (a) Acetylation (b) $NaBH_4$ (c) Mesylation

OAc, H, H, OMs → $^{\ominus}OH$ → H → (a) Acetalisation (b) BH_3, H_2O_2 (c) Hydrolysis (d) Mesylation → H, $CH_2{\cdot}CH_2OMs$

(continued)

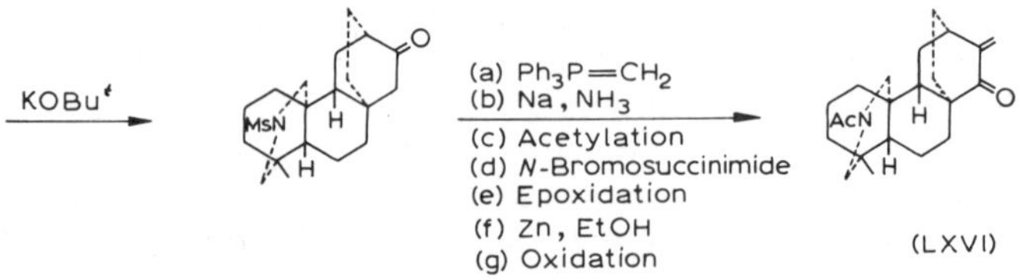

KOBu^t →

(a) $Ph_3P{=}CH_2$
(b) Na, NH_3
(c) Acetylation
(d) *N*-Bromosuccinimide
(e) Epoxidation
(f) Zn, EtOH
(g) Oxidation

(LXVI)

The final product LXVI has previously been converted into atisine (*Pelletier* and *Parthasarathy*, Tetrahedron Letters, 1963, 205).

Atidine, $C_{22}H_{33}NO_3$, accompanies atisine in *A. heterophyllum.* It is a tertiary base containing two hydroxyl groups, is ketonic (the carbonyl group being part of a six- or higher-membered ring) and contains an exocyclic $C{=}CH_2$ group, a *C*-methyl group and an ethanolimino ($>NCH_2 \cdot CH_2OH$) group, most of these features being discernible by i.r. spectroscopy, and confirmed by n.m.r. spectral study.

Neither of the hydroxyl groups is tertiary, since an *O,O*-diacetate is formed without difficulty. Oxidation of tetrahydroatidine with lead tetra-acetate gave glyoxal, though this behaviour is not diagnostic for the presence of an ethanolimino group (being shown also by some *N*-ethyl compounds). However, n.m.r. spectroscopy reveals the presence of the group. Reduction of atidine with tetrahydridoborate yields a triol. These and other properties suggested that atidine belongs to the atisine family, and pointed to its being a ketodihydroatisine; this view was confirmed by the formation of dihydroatisine by Huang-Minlon reduction of atidine. Atidine is formulated as LXVII, the keto group being placed at position 7 as a consequence of a correlation between the alkaloid and ajaconine (see below), another member of this sub-group.

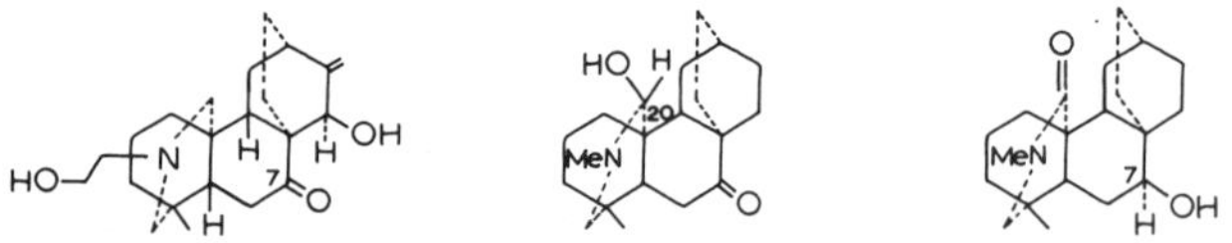

(LXVII) Atidine (LXVIII) (LXIX)

Ajaconine on reduction with tetrahydridoborate yields a triol, convertible to a triacetate. These two derivatives proved to be identical, respectively, with a pair of compounds obtained from dihydroatidine. This correlation shows that the C=O group in atidine and the potential hydroxyl group in ajaconine are located in the same position; the latter group is shown to be at position 7 on independent evidence (see below) (*Pelletier*, Chem. and Ind., 1956, 1016; 1957, 1670; J. Amer. chem. Soc., 1965, **87**, 799; *Dvornik* and *Edwards*, Tetrahedron, 1961, **14**, 54; Chem. and Ind., 1957, 952). The absolute configuration of atidine, also represented by LXVII, follows from that of atisine, already discussed (p. 336).

Ajaconine, $C_{22}H_{33}NO_3$, is the first reported case of the occurrence of an atisine alkaloid in a *Delphinium* sp., actually in *D. ajacis* L. It was early recognized as being related to atisine by its conversion to an azomethine base already encountered as a transformation product of that alkaloid. Secondly, reduction of atidine yields a mixture of epimeric alcohols one of which is identical with dihydroajaconine.

Ajaconine contains a methanolamine ether grouping; it forms immonium salts, and the ether linkage can be cleaved reductively, by tetrahydridoborate. The latter reaction generates a secondary alcohol in which the hydroxyl group is to be placed at position 7 on the following grounds. The methanolamine LXVIII, obtainable from ajaconine, on treatment with base yields, along with the expected iso-base, a hydroxylactam LXIX, the structure being supported by compelling spectral evidence. This compound is a consequence of an intramolecular Cannizzaro-type hydride-transfer reaction between positions 20 and 7. Sequential oxidation and reduction of LXIX yields its epimeride [LXIX; inverted at $C_{(7)}$]; this configuration of the hydroxyl group (7α) must clearly be that corresponding to ajaconine, which can be formulated as LXX. This structure also depicts the absolute configuration of the alkaloid, derived from correlations with atisine and atidine (*Dvornik* and *Edwards, loc. cit.*; Proc. chem. Soc., 1958, 305; *Pelletier, loc. cit.*).

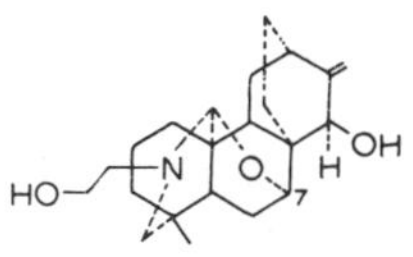

(LXX)

Ajaconine

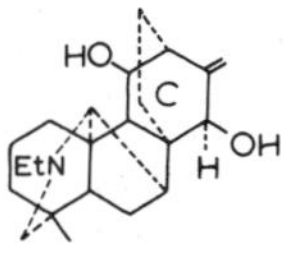

(LXXI)

Denudatine

Denudatine, $C_{22}H_{33}NO_2$, is an atisine-type alkaloid found in *D. denudatum* (*M. Götz* and *Wiesner*, Tetrahedron Letters, 1969, 4369). An atisine skeleton was suspected because the alkaloid rearranged to a methyl ketone readily. A hydroxyl group was placed in ring C as shown as a consequence of the following transformations, carried out on the derived lactam:

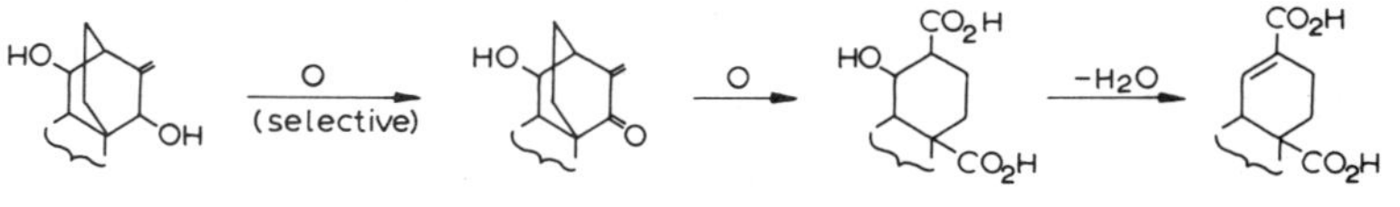

λ_{max} 216 nm

The structure and stereochemistry of the base were revealed by an X-ray diffraction analysis, pointing to expression LXXI (*F. Brisse, ibid.*, 1969, 4373); a similar study on denudatine methiodide confirmed the assignment (*L. H. Wright et al.*, Chem. Comm., 1970, 359).

Spiradine-F and **Spiradine-G.** These are two bases found in *Spiraea japonica*, spiradine-F being the monoacetate of spiradine-G. Spiradine-G, $C_{22}H_{31}NO_3$, has a molecular formula suggesting an atisine derivative.

An exocyclic $C{=}CH_2$ group was detected by oxidation to a cyclohexanone. Two methanolamine ether groups were diagnosed because reduction with tetrahydridoborate yielded a tetrahydrotriol, whilst catalytic hydrogenation generated a saturated, hexahydrotriol; dehydrogenation with selenium of the latter led to 6-isopropyl-1-methylphenanthrene. N.m.r. spectroscopy established that one of the cyclic ether functions was an oxazolidine ring. The alkaloid was capable of being oxidised to a hydroxylactam, indicating that the oxazolidine ring is in the isoatisine position (see p. 334). The other ether bridge was investigated by oxidation of the alkaloid to the corresponding ketone LXXIII, which was reduced and oxidised to what proved to be an enolic cyclic 1,2-diketone:

(a) Reduction
(b) Oxidation

(LXXII)
Spiradine-G

(LXXIII)

The second ether linkage must therefore terminate at position 7, and LXXII thus becomes an acceptable formulation for spiradine-G. The 6-hydroxyl group in the alkaloid is given the β-configuration because the n.m.r. spectral singlet of the $C_{(4)}$-methyl group suffered an appreciable downfield shift when spiradine-G was acetylated to spiradine-F. The absolute stereochemistry of the base, also represented by LXXII, was revealed by molecular rotation difference studies, and by the o.r.d. curve of ketone LXXIII (*M. Toda* and *Y. Hirata*, Tetrahedron Letters, 1968, 5565). Spiradine-F is thus LXXII (OAc in place of OH).

Atisine alkaloids are summarised in Table 2.

TABLE 2

ATISINE ALKALOIDS

Alkaloid	*M.p. (°C)*	*$[\alpha]_D$ (°) (solvent)*
Atisine	60	unstable in solution
Atidine	182.5–183.5	−47 (chloroform)
Ajaconine	172	−119 (ethanol)
Denudatine	248	+0.15 (ethanol)
Spiradine-F	114–117*	
Spiradine-G	168–170	−137 (methanol)

* Hydrochloride.

(c) Alkaloids with a modified atisane skeleton

A subgroup of diterpene alkaloids is characterised by the presence of an atisine framework with additional ring fusions. The members are mainly found in *Aconitum* spp. native to India and Japan.

Hetisine, $C_{20}H_{27}NO_3$, is a minor base of *A. heterophyllum* and of *Delphinium cardinale* (*M. H. Benn*, Canad. J. Chem., 1966, **44**, 1; *W. A. Jacobs* and *L. C. Craig*, J. biol. Chem., 1942, **143**, 605; *Jacobs* and *C. F. Huebner*, *ibid.*, 1947, **170**, 189). It contains one olefinic bond, three active hydrogen atoms, and a tertiary, basic nitrogen atom (not *N*-methyl). The exocyclic methylene character of the unsaturation was revealed by i.r. absorption study, and confirmed by formaldehyde formation on ozonolysis. A Kuhn–Roth analysis showed the presence of one *C*-methyl group in the alkaloid and two in dihydrohetisine, both results being confirmed by n.m.r. spectroscopy. Dihydrohetisine shows no near u.v. absorption and hence must be heptacyclic (*A. J. Solo* and *S. W. Pelletier*, J. Amer. chem. Soc., 1959, **81**, 4439; J. org. Chem., 1962, **27**, 2702). Hetisine forms both di- and tri-acetates, and does not react with periodate, lead tetra-acetate nor acetone (to form an acetonide); evidently three esterifiable hydroxyl groups are present, which are neither vicinal nor 1,3-diaxial in relationship. Dehydrogenation of hetisine yields a mixture of hydrocarbons amongst which pimanthrene (1,7-dimethylphenanthrene) was identified. Aside from these preliminary studies, the main evidence for formulating hetisine as LXXIV is an X-ray diffraction study on hetisine hydrobromide (*M. Przybylska*, Canad. J. Chem., 1962, **40**, 566; Acta Cryst., 1963, **16**, 871).

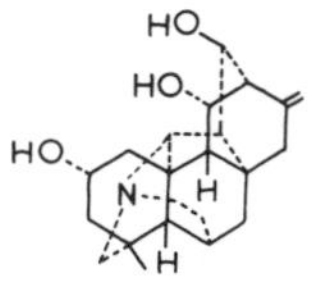

(LXXIV) Hetisine (LXXV) (LXXVI)

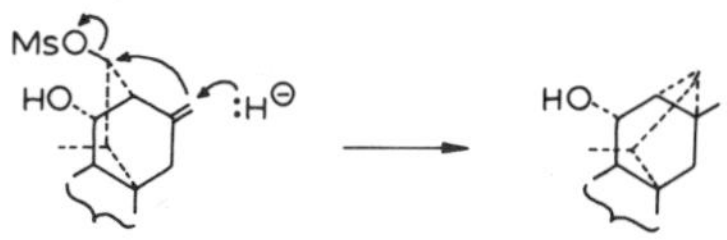

Hetisine diacetate on oxidation yields a ketone which on hydrolysis affords a dihydroxyketone, **hetisinone**; this compound (LXXIV; C=O at position 2), occurs in *D. cardinale* and *D. denudatum*. Its structure has been corroborated by chemical and spectroscopic study (*Pelletier*, *L. H. Keith* and *P. C. Parthasarathy*, J. Amer. chem. Soc., 1967, **89**, 4146; *Benn*, *loc. cit.*; *Pelletier et al.*, Canad. J. Chem., 1968, **46**, 2635; Tetrahedron, 1961, **14**, 76). The absolute configuration of both bases has not been proved rigorously; that depicted in LXXIV is based on analogy with atisine and the *Garrya* alkaloids.

Structure LXXIV for hetisine permits an explanation for an earlier observation that the *methiodide*, m.p. 325°, obtained by heating hetisine with methyl iodide at 100° (sealed tube), underwent Hofmann degradation to yield a Hofmann base $C_{21}H_{29}NO_3$. This compound possessed only one olefinic linkage, and no C=O group (*Jacobs* and *Huebner*, J. biol. Chem., 1947, **170**, 189). Later work revealed that milder conditions (methyl iodide in methanol at reflux) furnished a different methiodide, which underwent normal Hofmann degradation to a product containing a new double bond. It is apparent that the first methiodide is a rearranged salt, and it is suggested that the over-all reaction took the following course (*Wiesner*, *Z. Valenta* and *L. G. Humber*, Tetrahedron Letters, 1962, 621), leading to a hemiacetal:

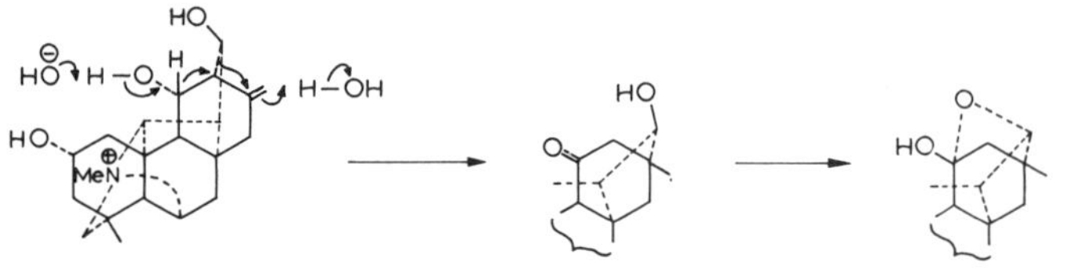

Hetisine methohydroxide

Hemiacetal

Another unusual reaction was the behaviour of the monomesylate LXXV on reduction with lithium tetrahydridoaluminate. The product showed in its n.m.r. spectrum a signal (singlet) due to a deshielded methyl group, and though no mention of high-field cyclopropyl proton signals was made, structure LXXVI was advanced for it, corroborated by an X-ray diffraction analysis of the methiodide (*H. E. Wright*, *M. G. Newton* and *Pelletier*, Chem. Comm., 1969, 507).

Ignavine, $C_{27}H_{33}NO_6$, occurs in the roots of *A. sanyoense*, *A. tasiromontanum* and *A. japonicum*; it is an ester-alkaloid which on hydrolysis yields benzoic acid and a basic product anhydroignavinol, formulated as $C_{20}H_{27}NO_4$ on mass spectrometric evidence (*Pelletier*, *S. W. Page* and *Newton*, Tetrahedron Letters, 1970, 4825). **Anhydroignavinol** results as a consequence of both hydrolysis and dehydration; attempts to prepare ignavine methiodide are likewise accompanied by dehydration (*E. Ochiai* and co-workers, J. pharm. Soc. Japan, 1952, **72**, 816, 1605; 1956, **76**, 550).

Ignavine's six oxygen atoms are accounted for as an ester group and possibly four hydroxyl groups; of the latter one is adjacent to the benzoyloxy group, because anhydroignavinol is oxidisable by periodate, whilst ignavine is not. The alkaloid likewise forms diacyl derivatives which are also anhydro-compounds, *viz.* derivatives of anhydroignavinol (*idem*, Chem. pharm. Bull., Tokyo, 1953, **1**, 60). An exocyclic methylene group present in ignavine is part of an allylic alcohol unit of the atisine–*Garrya* alkaloid type, since ignavine is readily isomerised to a methyl ketone and oxidised to a conjugated enone (*idem*, *ibid*., 1959, **7**, 550). An X-ray diffraction analysis of anhydroignavinol methiodide establishes the structure of the base as LXXVII (*Pelletier*, *Page* and *Newton*, *loc. cit.*). Since anhydroignavinol contains neither an

(LXXVII; R = H) Anhydroignavinol (LXXIX)
(LXXVIII; R = PhCO) Ignavine (monohydrate?)

ether linkage nor an additional double bond it seems likely that ignavine is a hydrate containing one molecule of water of crystallisation, to be formulated as LXXVIII.

Hypognavine, $C_{27}H_{31}NO_5$, found in varieties of *A. sanyoense*, is similar to ignavine in its chemical behaviour.

For example, it is a benzoyl ester, is a tertiary amine (not *N*-methyl), and contains an exocyclic C=CH_2 group and two esterifiable hydroxyl groups. On hydrolysis it yields an alkamine **hypognavinol,** $C_{20}H_{27}NO_4$ which, however, is formed by simple hydrolysis unaccompanied by dehydration. Likewise, hypognavinol can be acylated or converted into its methiodide to yield the expected products, without simultaneous dehydration. Hypognavine undergoes acid-catalysed rearrangement to a methyl ketone, and it may be oxidised to a conjugated enone; clearly, an allylic secondary alcohol system of the type found in atisine and the *Garrya* alkaloids is present. Hypognavinol on dehydrogenation yields several alkylphenanthrenes, including 1,8-dimethyl-, 1,7-dimethyl-6-*n*-propyl- and 3-ethyl-1,8-dimethyl-; the last two are also obtained from anhydroignavinol, which suggests that ignavine and hypognavine have the same carbon skeleton. Partial structure (LXXIX) was assigned to hypognavinol on these and other grounds (*Ochiai et al.*, Pharm. Bull., Japan, 1953, **1**, 152; *S. Sakai*, J. pharm. Soc., Japan, 1956, **76**, 1054; Chem. pharm. Bull., Tokyo, 1958, **6**, 448; *Ochiai et al., ibid.*, p. 327). A significant point was the finding that hypognavinol undergoes a first stage of Hofmann degradation with ease, yielding the des-base LXXX accompanied by the ring D isomerised structure LXXXI. However, a second stage of this degradation is resisted; on the basis of structure LXXX it is seen that the only hydrogen β to the nitrogen is at $C_{(14)}$, but elimination of this would lead to a 14,20 double bond which would violate Bredt's rule.

(LXXX) (LXXXI) (LXXXII) Hypognavinol

The size of the hetero-ring in LXXIX was revealed by the usual procedure of oxidation to a lactam (*Sakai, loc. cit.* and *ibid.*, 1959, **7**, 50,55). Of the three unplaced hydroxyl groups in LXXIX two are readily acylated, and one is not; the tertiary nature of the latter was confirmed by oxidation studies. Detailed investigations have revealed that there is in hypognavinol a 1,2-glycol grouping, and that this is located in ring A. In hypognavine one of these glycol hydroxyls is esterified with benzoic acid (no reaction with periodate). The precise location of the glycol system was settled by oxidation of the Hofmann base LXX under mild conditions, which yielded a methanolamine ether reducible to the original des-base by tetrahydridoborate. The ether forms salts showing i.r. bands characteristic of an immonium group $>C{=}\overset{\oplus}{N}<$ (~1680 cm $^{-1}$); these salts regenerate the des-base on basification. Models show that the most favourable position from which a hydroxyl group could indulge in such ether formation is 2α (axial) (ether formation *via* $C_{(20)}$ would be forbidden because of violation of Bredt's rule on salt formation). These transformations are pictured as follows *(Sakai, loc. cit.)*:

(LXXX) $\underset{NaBH_4}{\overset{Ag_2O}{\rightleftharpoons}}$ [structure: 20, 2, MeN, O, OH, }2(OH)] $\xrightarrow{HX}$ [structure: $X^{\ominus}$, HO, MeN$^{\oplus}$, OH, }2(OH)]

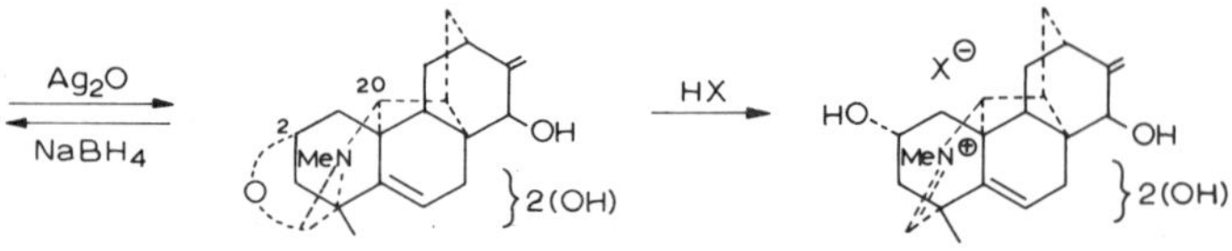

The position of the second glycolic hydroxyl group was settled by an X-ray diffraction analysis of hypognaviol methiodide, which also revealed the location of the fourth hydroxylic function, and enabled hypognavinol to be formulated in detail as LXXXII (*Pelletier*, *Page* and *Newton*, Tetrahedron Letters, 1971, 795). The position of the benzoyl group in hypognavine is as yet unsettled.

Kobusine, $C_{20}H_{27}NO_2$, is a constituent alkaloid of *A. kamtschaticum (fischeri)*, *A. sachalinense*, *A. lucidusculum* and *A. yesoensis.* It has two secondary hydroxyl groups, both in cyclohexane rings, and one is allylic, of the type characteristic of *Garrya* and atisine bases. Dehydrogenation of the alkaloid yielded 1,7-dimethyl-6-*n*-propylphenanthrene (*cf.* ignavine and hypognavine). Structure LXXXIII for kobusine has been proposed as a consequence of detailed oxidation and reduction studies (*M. Natsume*, Chem. pharm. Bull., Tokyo, 1959, **7**, 539; *T. Okamoto, ibid.*, p. 44).

The allylic hydroxyl is 15β; the location of the other was narrowed to either position 11 or 13 by a sequence leading to lactone formation between this group and position 16. An X-ray diffraction analysis on kobusine methiodide showed that in fact the second hydroxyl group is 11β, and kobusine is to be formulated as LXXXIII (*Pelletier et al.*, Chem. Comm., 1970, 98). The cleavage of the $C_{(14)}$–$C_{(20)}$ bond in the alkaloid by a sequence of reactions has been described (*Okamoto et al.*, Chem. pharm. Bull., Tokyo, 1975, **23**, 3030).

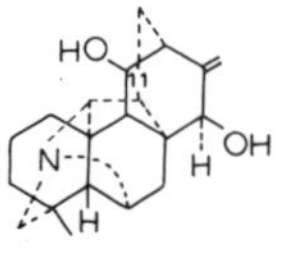

[structures: HO, 11, N, H, OH; HO, N, H, 6, OH, OH; HO, 11, RO, 2, N, H, OH]

(LXXXIII) Kobusine

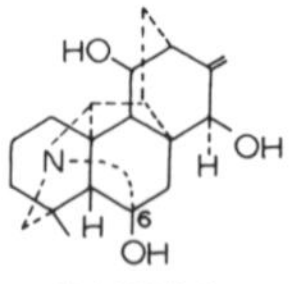

(LXXXIV) Pseudokobusine

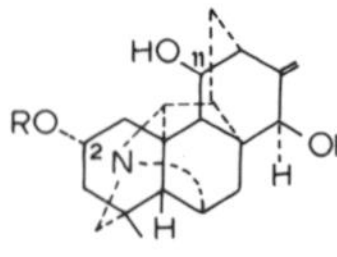

(LXXXV; R = H) Isohypognavinol
(LXXXVI; R = PhCO) Isohypognavine

Pseudokobusine, $C_{20}H_{27}NO_3$, is found in *A. yesoensis* and *A. lucidusculum*, and is very closely related to kobusine. It contains the usual features characteristic of an alkaloid of the kobusine type, together with three acylable hydroxyl groups, including the allylic one. Oxidation experiments show two of the hydroxyls, including the allylic one, are secondary and in six-membered rings, whilst the third is tertiary *(Natsume, loc. cit.)*.

Dehydrogenation of the alkaloid was attended by results essentially the same as those for kobusine, suggesting that pseudokobusine is a hydroxylated kobusine, the additional hydroxyl being tertiary. It became apparent that this group is part of a masked aminoketone unit. This was revealed by reduction of pseudokobusine to its dihydro-derivative and then oxidation to dihydropseudokobusindione, which on acetylation yielded an *N*-acetylseco derivative containing a third oxo-group in a six-membered ring and no hydroxyl:

Since this third hydroxyl is tertiary and gives rise to a cyclohexanone on acetylation it must be placed at position 6 and consequently have the β-configuration. Pseudokobusine must thus be formulated as LXXXIV (*idem*, Chem. pharm. Bull., Tokyo, 1962, **10**, 879, 883). This structure has been confirmed by several correlations between kobusine and pseudokobusine *(Natsume, loc. cit.)*.

Isohypognavine, $C_{27}H_{31}NO_4$, occurs in *A. majimai* and *A. japonicum*; its name is unfortunate since it is not an isomer of hypognavine. It is the benzoate of an alkamine **isohypognavinol**, which contains three unhindered hydroxyl groups, one of which is allylic and of the type encountered in typical *Garrya* and atisine bases. Investigations of a type similar to those already described for hetisine, ignavine and other alkaloids of this subgroup point to structure LXXXV for isohypognavinol, and LXXXVI for isohypognavine.

A correlation with kobusine (LXXXIII) allows one hydroxyl group to be placed at $C_{(11)}$. The third hydroxyl group must be at position 2 because reduction of isohypognavinol (LXXXV) with sodium in propanol yields a deoxy-base LXXXVII, the ethiodide of which underwent Hofmann degradation to a des-base, reducible to a

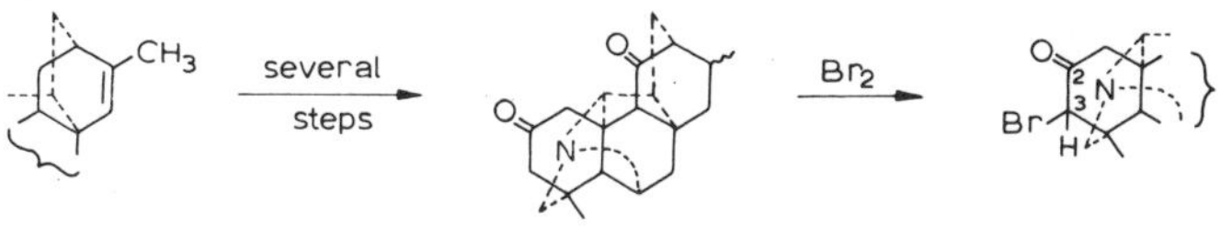

(LXXXVII) (LXXXVIII) (LXXXIX)

tetrahydro-base, which on oxidation gave a diketone LXXXVIII. This diketone afforded a monobromo derivative, the n.m.r. spectrum of which showed a C*H*Br signal at δ4.01 ppm (singlet). This observation, accompanied by the deshielding of the *C*-methyl group from δ1.14 to 1.31 ppm, permits structure LXXXIX to be assigned to the bromoketone, and thus the third hydroxyl group of isohypognavinol is at $C_{(2)}$. The benzoyloxy group of isohypognavine is placed at $C_{(2)}$ largely by structure analogy with ignavine and **vakognavine** (see below) (*S. Junussov*, J. gen. Chem. U.S.S.R., 1948, **18**, 515; *Ochiai et al.*, J. pharm. Soc. Japan, 1955, **75**, 638; 1956, **76**, 550; *Okamoto, Natsume* and *S. Kamata*, Chem. pharm. Bull., Tokyo, 1964, **12**, 1124).

The **spiradines** constitute a small group of bases found in *Spiraea japonica*, several of which are closely inter-related. **Spiradine-A** and **spiradine-B** are respectively a ketone and its secondary alcoholic reduction product; **spiradine-C** on treatment with alcoholic potash yields spiradine-B. Spectroscopic and chemical investigations showed that spiradine-A contains a keto group in a six-membered ring, an exocyclic $C{=}CH_2$ group, a tertiary hydroxyl group, a *C*-methyl group and a tertiary basic nitrogen atom. An X-ray diffraction analysis of its methiodide showed that the alkaloid is represented by XC, which also depicts its absolute configuration (*G. Goto et al.*, Tetrahedron Letters, 1968, 1369).

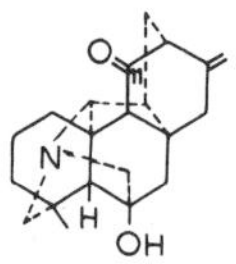

(XC) Spiradine-A (XCI) 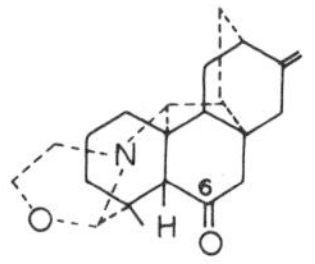 (XCII) Spiradine-D

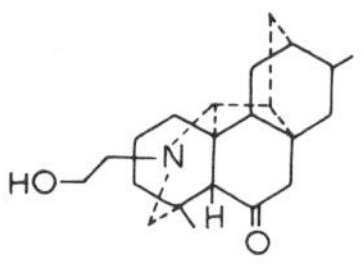

Spiradine-B and -C are thus respectively XC ($\sim\sim$OH at position 11) and XC ($\sim\sim$OAc at position 11).

Interesting aspects of spiradine-A include its pK_a value of 8.35, whilst that of spiridine-B is 9.54 and spiradine-C 9.02 (all in 50% aqueous methanol). Since four bonds separate $C_{(11)}$ from the nitrogen atom the effect must be largely *trans*-spatial in character. Next, acetylation of spiradine-A yielded an *N*-acetylketone, demonstrating that the hydroxyl group is part of a masked aminoketone unit (*cf.* pseudokobusine, p. 345). Finally, treatment of the same base with methyl iodide furnished an *N*-methylketone in a related manner:

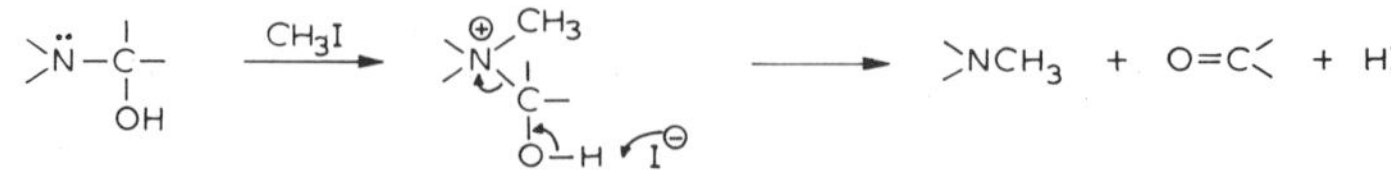

$$>\ddot{N}-\overset{|}{\underset{OH}{C}}- \xrightarrow{CH_3I} >\overset{\oplus}{N}(CH_3)-\overset{|}{C}-O-H \;\; I^{\ominus} \longrightarrow >NCH_3 + O{=}C< + HI$$

Further methylation with CH_3I–Ag_2O generated an ether:

$$>N:\ (CH_3)\ \rightarrow\ >C{=}O\quad CH_3{-}I\ \xrightarrow{Ag_2O}\ >\overset{\oplus}{N}(CH_3){-}\overset{|}{\underset{|}{C}}{-}OCH_3\ +\ I^{\ominus}$$

Spiradine-D contains a carbonyl group, an exocyclic $C{=}CH_2$ group and an ether linkage. Reduction with tetrahydridoborate yields a hydroxydihydro-base formed without loss of oxygen.

This observation, the two additional carbon atoms relative to spiradine-A, and an n.m.r. spectral signal at δ 4.22 ppm (singlet, N–C*H*–O) suggested an oxazolidine ring was present. On the assumption of a diterpene alkaloid framework, the C=O group was believed to be in rings A or B because of the pK_a (9.35) and because of the formation of normal quaternary ammonium salts rather than immonium salts. A correlation with spiradine-A (XC) was effected as follows: the latter was subjected to Wolff–Kishner reduction and the product reacted with 2-chloroethanol and reduced sequentially. The final product XCI was also obtained from spiradine-D by reduction with tetrahydridoborate followed by hydrogenation (*Goto* and *Y. Hirata*, Tetrahedron Letters, 1968, 2989). Spiradine-D may consequently be formulated as XCII. The inertness of the 6-oxo group in XCII towards reduction with tetrahydridoborate is unexpected.

Miyaconitine, $C_{23}H_{29}NO_6$, occurs in *A. miyabei.* On the basis of its molecular formula and functional groups it was recognised as a hexacyclic tertiary base containing acetoxy, *N*-methyl and secondary and tertiary OH groups, along with an exocyclic $C{=}CH_2$ group and two oxo groups, both in six-membered rings. Structure XCIII was proposed on chemical and spectroscopic evidence (*Y. Ichinohe et al., ibid.*, 1970, 2323). An X-ray diffraction analysis of the hydrobromide confirmed this structural assignment, and revealed the relative and absolute stereochemistry depicted in XCIII (*H. Shimanouchi, Y. Sasada* and *T. Takeda, ibid.*, p. 2327).

(XCIII; R = H,···OH) Miyaconitine
(XCIV; R = O) Miyaconitinone

(XCV)
Hetidine

The carbonyl at position 6 in the alkaloid has an unusually low i.r. frequency (1678 cm^{-1}) due in part to hydrogen bonding with the neighbouring hydroxyl group, and in part to trans-annular interaction with the nitrogen atom. Salts of miyaconitine do not show this i.r. band, since they are formed as follows:

$$H_3C{-}\overset{|}{\underset{|}{N}}:\ \rightarrow\ \overset{|}{\underset{|}{C}}{=}\overset{\oplus}{O}{-}H\ \longrightarrow\ H_3C{-}\overset{|\oplus}{\underset{|}{N}}{-}\overset{|}{\underset{|}{C}}{-}OH$$

Miyaconitinone, $C_{23}H_{27}NO_6$, from the same source, is related to miyaconitine in that it is the corresponding 6,7-diketone XCIV *(Ichinohe et al., loc. cit.)*.

Hetidine, $C_{21}H_{27}NO_4$, is a minor base of *A. heterophyllum* (*Pelletier, R. Aneja* and *K. W. Gopinath*, Phytochemistry, 1968, **7**, 625). Owing to paucity of material it was possible only to make spectral measurements, which revealed functional groups, and to effect an X-ray diffraction analysis of hetidine hydriodide, which enabled structure XCV (relative configuration) to be assigned to the base (*Pelletier et al.*, Chem. Comm., 1970, 393). Hetidine hydriodide is analogous in structure to miyaconitine salts (see above). Hetidine shows unusually short wavelength ($\lambda_{max.}$ 209 nm, ε 5,200) absorption in the u.v. region, presumably owing to *trans*-spatial carbonyl interaction.

Vakognavine, $C_{34}H_{37}NO_{10}$, is found in *A. salmatum*. Spectral investigations revealed benzoate, tertiary *C*-methyl, *N*-methyl, formyl and three acetoxyl groups (*N. Singh* and *S. S. Jaswal*, Tetrahedron Letters, 1968, 2219). The formyl group is absent in salts of the alkaloid. An X-ray diffraction analysis of vakognavine hydriodide has shown the cation to be of structure XCVI, and hence the free base is XCVII (*Wright et al.*, American Crystallog. Assocn. Summer Meeting, August, 1971, Abstract C7).

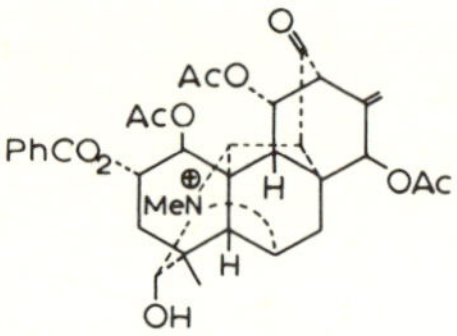

(XCVI)

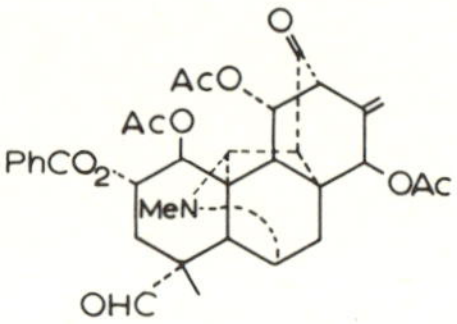

(XCVII)
Vakognavine

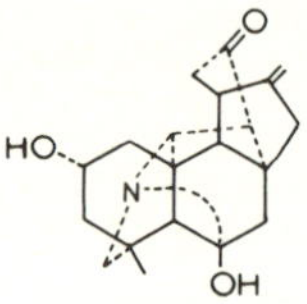

(XCVIII)
Delnudine

Delnudine, $C_{20}H_{25}NO_3$, is a delphinium alkaloid, found in *Delphinium denudatum* (*M. Götz* and *Wiesner*, Tetrahedron Letters, 1969, 5335). Its molecular formula and the presence of carbonyl and exocyclic methylene groups indicated a heptacyclic structure. The u.v. spectral absorption ($\lambda_{max.}$ 300 nm, log ε 1.76) revealed that the keto group and $C{=}CH_2$ were in close proximity. A methanol unit is present, since on acetylation an *N*-acetyl ketone results, and a secondary alcohol is in the neighbourhood of the nitrogen atom because of a fall in the pK_a value of two units when it is oxidised to the ketone. Structure XCVIII is consistent with this and other evidence; it has been corroborated by an X-ray diffraction analysis of delnudine hydrochloride (*K. B. Birnbaum, ibid.*, p. 5245).

Staphisine, $C_{43}H_{60}N_2O_2$ (on mass spectral evidence, *m/e* 636.4648 for parent peak), is a constituent alkaloid of *D. staphisagria* (*Jacobs* and *Craig*,

J. biol. Chem., 1941, **141**, 67). Chemical studies on it were hindered by its lability and by its tendency to undergo complex rearrangements on degradation. Spectral study (u.v., i.r. and n.m.r.) pointed to the presence of two NMe groups, a methoxyl group, a cyclopropane system and a conjugated diene grouping ($\lambda_{max.}^{EtOH}$ 268 nm, ε 17,300), the latter being confirmed by hydrogenation to saturated tetrahydrostaphisine. A single-crystal X-ray analysis of staphisine monomethiodide revealed XCIX to be the structure of the alkaloid, which is seen to have a dimeric structure composed of two molecules of the atisine type (*Pelletier et al.*, J. Amer. chem. Soc., 1972, **94**, 1754).

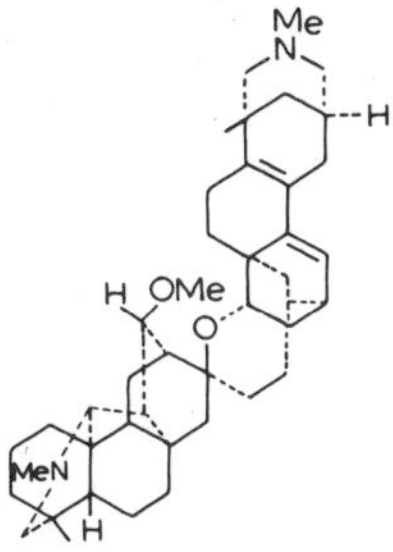

(XCIX)
Staphisine

Modified atisane alkaloids are collected in Table 3.

TABLE 3

ALKALOIDS WITH A MODIFIED ATISANE SKELETON

Alkaloid	*M.p. (°C)*	*[α]$_D$ (°) (solvent)*
Hetisine	145 (decomp.)*	+13.7 (EtOH)
Hetisinone	268–270	+18
Ignavine	172–174	+85.3 (EtOH)
Anhydroignavinol	302–304	
Hypognavine	239–241	+127.1 (MeOH)
Hypognavinol	307–308	+67.7 (MeOH)
Kobusine	267–267.5	+104.4 (MeOH)
Pseudokobusine	271	
Isohypognavine	135	
Isohypognavinol	292–295	
Spiradine-A	281	
Spiradine-B	260	
Spiradine-C	249	
Spiradine-D	135	
Miyaconitine	218 (decomp.)	−87.8 ($CHCl_3$)
Miyaconitinone	285 (decomp.)	−27.6 (AcOH)
Hetidine	218–221	
Vakognavine		
Delnudine	235–237	
Staphisine	200–208	−152 to −162 (benzene)

* Complete fusion at 253–256°.

2. C_{19}-diterpene alkaloids

This family of diterpenoid bases includes a number of highly toxic ester bases, heavily substituted by hydroxyl and methoxyl groups. The lethal dose for humans is about 2–5 mg, and because of this high toxicity preparations have been used extensively as arrow poisons. The alkaloids affect the autonomic nervous system, are depressor bases, and are cardiac inhibitors. On the other hand the alkamines obtained by hydrolysis of the ester alkaloids are comparatively non-toxic.

For discussion these alkaloids are conveniently divisible into four groups, two large and two small. These may be described as: *(a)* lycoctonine-type alkaloids, *(b)* aconitine-type alkaloids, *(c)* lactonic alkaloids and *(d)* *Daphniphyllum* alkaloids. Groups *(a)* and *(b)* are very closely related, having the same carbon skeleton but different oxygen-substitution patterns. X-ray diffraction analyses have featured prominently in structural studies in these areas, sometimes because of paucity of material for examination, and also because of complicated structures involved. Other alkaloids have been correlated with members of established constitution and configuration. The various forms of spectroscopy have been used extensively in structure elucidation.

(a) Lycoctonine-type alkaloids

The key compound in this subgroup is **lycoctonine** itself, $C_{25}H_{41}NO_7$, which is the parent aminoalcohol of a number of ester-alkaloids found in *Aconitum* and *Delphinium* genera. Naturally occurring esters of lycoctonine have been known for many years (for early investigations see *E. S. Stern*, in "The Alkaloids", ed. *R. H. F. Manske*, Vol. IV, Chapter 37, Academic Press, New York, 1954). The acids involved in the esters are anthranilic, *N*-acetylanthranilic, *N*-succinylanthranilic and *N*-methylsuccinylanthranilic acids, and other related acids. The esters are very readily hydrolysed to lycoctonine, and it is likely that lycoctonine itself is not a natural product. In spite of extensive efforts employing chemical and later physicochemical methods to attack the problem of the structure of lycoctonine, the question was not solved until an X-ray diffraction analysis was effected on deoxymethylenelycoctonine hydriodide (*M. Przybylska* and *L. Marion*, Canad. J. Chem., 1956, **34**, 185). This analysis, however, did not reveal the absolute stereochemistry of lycoctonine: a later study of a similar nature finally settled this point (*idem, ibid.*, 1959, **37**, 1843), and permitted lycoctonine to be formulated as I (*O. E. Edwards*, *Marion* and *D. K. R. Stewart*, *ibid.*, 1956, **34**, 1315).

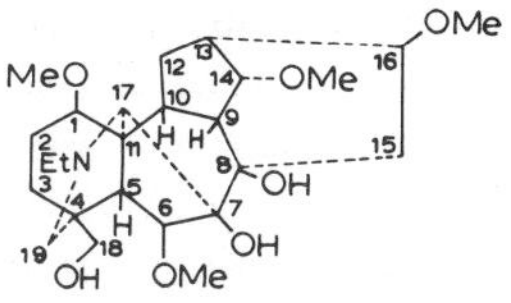

(I)
Lycoctonine

Early investigations revealed that lycoctonine contained a primary alcohol group (oxidation to an aldehyde, lycoctonal, containing the same number of carbon atoms). There are also present a 1,2-ditertiary glycol grouping (periodate oxidation), four methoxyl groups and an *N*-ethyl group. Mild oxidation (Ag_2O) of lycoctonine yielded a hydroxylycoctonine, which forms anhydronium salts and is formulated as II, the anhydronium salt cation being III (*Z. Valenta*, Chem. and Ind., 1959, 633; *Edwards, M. Los* and *Marion*, Canad. J. Chem., 1959, **37**, 1996; *Valenta* and *I. G. Wright*, Tetrahedron, 1960, **9**, 284).

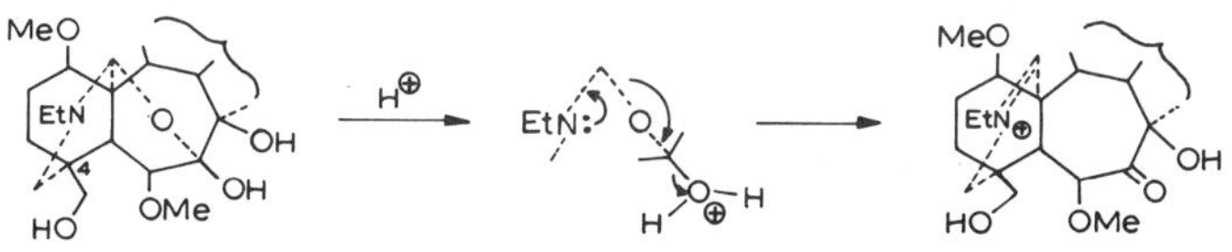

(II) (III)

Hydroxylycoctonine (II) has been a key substance in the study of the chemistry of lycoctonine. Reduction of II with tetrahydridoborate yielded a triol, the reaction being accompanied by elimination of H_2O. Oxidation of the triol with periodate cleaved the seven-membered ring to give a keto-aldehyde which underwent spontaneous hemiacetal formation with the $-CH_2OH$ group at position 4 to give IV. On hydrogenation of hydroxylycoctonine isolycoctonine is obtained; as with the tetrahydridoborate reduction above, a molecule of water is eliminated simultaneously and isolycoctonine (V) is a cycloheptanone. This compound has been subjected to a series of degradations and transformations, the results of which are consonant with structure V (*Edwards, Los* and *Marion, loc. cit.*).

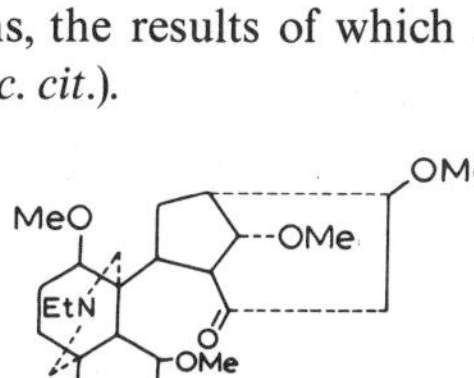

(IV) 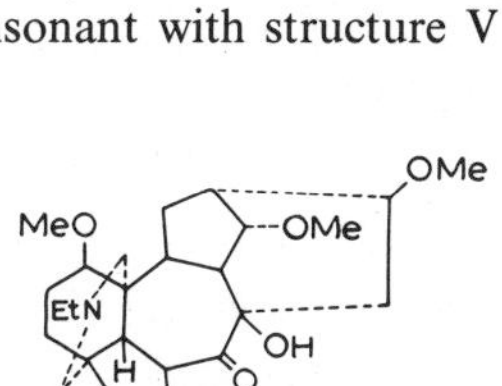 (V)

Ester-alkaloids derived from lycoctonine. Difficulties have been encountered in settling the location of the acyloxy groups in some ester-alkaloids derived from lycoctonine, which has three hydroxyl groups. The primary alcohol group would be expected to be the one esterified, for steric reasons: this proves to be the case in the known natural bases. The acids involved are anthranilic acid and its *N*-acyl derivatives. **Anthraniloyllycoctonine,** occurring in *Delphinium barbeyi, D. ajacis, D. consolida, D. brownii* and other spp., has structure VI (R = NH_2). Other alkaloids based on lycoctonine are listed in Table 4. The hydrolysis of **elatine** yields not lycoctonine but its 7,8-methylenedioxy derivative (**elatidine**), demethylenation of which yielded lycoctonine. Similarly, demethylenation of elatine by heating with phloroglucinol and acid yielded anthraniloyllycoctonine.

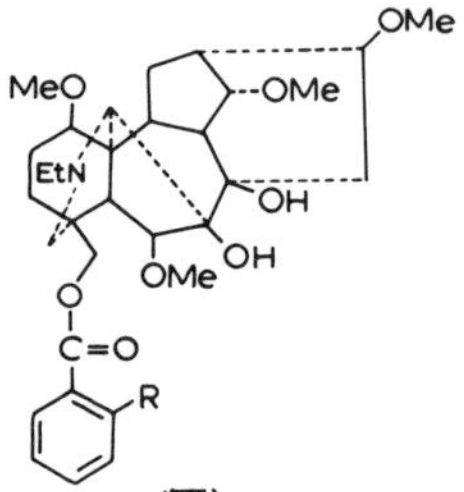

(VI)
Anthraniloyllycoctonine (R = NH_2)

Delphatine, $C_{26}H_{43}NO_7$, an alkaloid of *D. biternatum*, is 18-*O*-methyllycoctonine (*M. S. Yunusov* and *S. Yunusov*, Dokl. Akad. Nauk S.S.S.R., 1969, **188**, 1077).

Delpheline, $C_{25}H_{39}NO_6$, occurs in *D. elatum* (*J. A. Goodson*, J. chem. Soc., 1943, 139; 1944, 665), and was investigated in detail by *R. C. Cookson* and *M. E. Trevett* (*ibid.*, 1956, 2689, 3121), who observed it to be very closely akin to lycoctonine.

A correlation with the latter has been achieved by two groups separately. Lycoctonine was successively oxidised, reduced, oxidised and then subjected to a pinacol–pinacolone rearrangement. The methoxyl group at $C_{(6)}$ was hydrogenolysed off by reduction with sodium amalgam–ethanol and finally oxidation with selenium dioxide used to generate the 1,2-diketone VII:

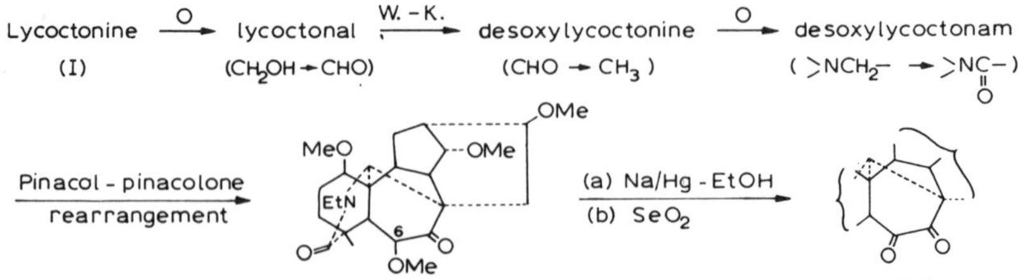

(VII)

TABLE 4

ESTER-ALKALOIDS OF LYCOCTONINE

Alkaloid	*R in structure VI*	*Formula*	*M.p. (°C)*	*Source*	*Ref.*
Anthraniloyl-lycoctonine	$-NH_2$	$C_{32}H_{46}N_2O_8$	135	*D. barbeyi* *D. ajacis* *D. consolida* *D. brownii* and others	1
Ajacine	AcNH–	$C_{34}H_{48}N_2O_9$	154	*D. ajacis*	2
Lycaconitine	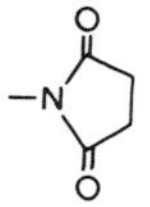	$C_{36}H_{48}N_2O_{10}$	amorphous	*A. lycoctonum* *A. gigas* and others	3
Methyllycacon-itine	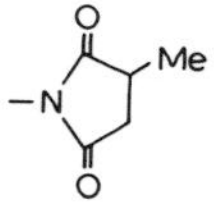	$C_{37}H_{50}N_2O_{10}$	130	*D. elatum* *D. brownii* and others	4
Delsemine	$-NHCO\cdot CHMe\cdot CH_2\cdot CONH_2$ or $-NHCO\cdot CH_2\cdot CHMe\cdot CONH_2$	$C_{37}H_{53}N_3O_{10}$	125	*D. semibarbatum* *D. oreophilum*	5
Elatine*	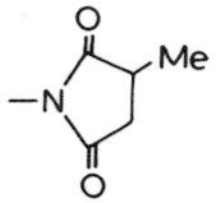	$C_{38}H_{50}N_2O_{10}$	224	*D. elatum*	6
Avadharidine	$-NHCO\cdot CH_2\cdot CH_2\cdot CONH_2$	$C_{36}H_{51}N_3O_{10}$	89	*A. orientale*	6,7

*See text.

References
1 *L. Marion* and *O. E. Edwards*, J. Amer. chem. Soc., 1947, **69**, 2010; *W. B. Cook* and *A. O. Beath, ibid.*, 1952, **74**, 1411.
2 *O. Keller* and *O. Völker*, Arch. Pharm., 1913, **251**, 207; *J. A. Goodson*, J. chem. Soc., 1944, 108; 1945, 245; *M. V. Hunter*, Pharm. J., 1943, **150**, 82; Quart. J. Pharm. Pharmacol., 1944, **17**, 302.
3 *H. Schulze* and *E. Bierling*, Arch. Pharm., 1913, **251**, 8; *H. Suginome* and *K. Ohno*, J. Fac. Sci. Hokkaido Univ., Ser. III, Chem., 1950, **44**, 36.
4 *Manske*, Canad. J. Res., 1938, **16B**, 57; *L. Marion* and *Manske*, *ibid.*, 1946, **24B**, 1; *Goodson*, J. chem. Soc., 1943, 169; *Yunusov* and *N. B. Abubakirov*, J. gen. Chem. U.S.S.R., 1949, **19**, 269; 1951, **21**, 967; 1952, **22**, 1461.
5 *Yunusov* and *Abubakirov*, *loc. cit.*
6 *A. D. Kuzovkov* and *T. F. Platonova*, J. gen. Chem. U.S.S.R., Eng. Transl., 1959, **29**, 2746.
7 *Kuzovkov* and *P. S. Massagetov*, C.A., 1956, **50**, 1852.

The final product VII was identical with dehydrodemethyleneoxodelpheline pinacone, already known as a transformation product of delpheline (*Edwards*, *Marion* and *K. H. Palmer*, Canad. J. Chem., 1958, **36**, 1097). The second correlation (*M. Carmack* and co-workers, J. Amer. chem. Soc., 1958, **80**, 497) involved conversion of delpheline into desoxylycoctonine. These and the earlier investigations enable delpheline to be formulated as VIII. Feeding experiments with methyl-^{14}C-labelled methionine show that the methoxyl methyls could be derived from this source, but not the $-OCH_2O-$ group (*E. J. Herbert* and *G. W. Kirby*, Tetrahedron Letters, 1963, 1505).

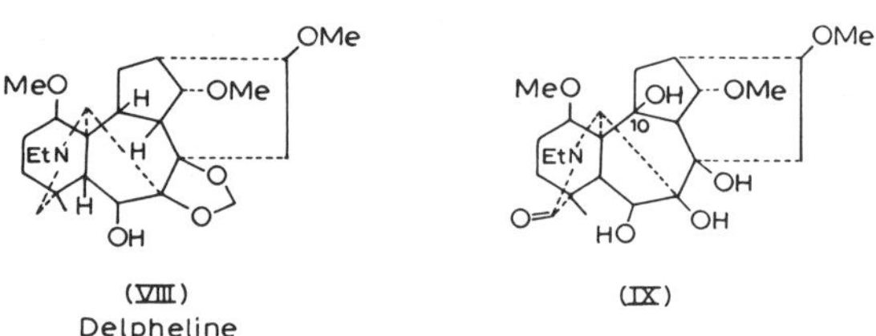

(VIII)
Delpheline

(IX)

Deltaline, $C_{27}H_{41}NO_8$, is a constituent alkaloid of *D. elatum*, *D. occidentale* and *D. barbeyi*, and was formerly called eldeline or delphelatine (*J. F. Couch*, J. chem. Soc., 1936, 684; *Carmack et al.*, *loc. cit.*; *M. S. Rabinovich*, J. gen. Chem. U.S.S.R., Eng. Transl., 1952, **22**, 1702).

Treatment of the alkaloid with thionyl chloride yielded acetylchlorodelpheline, reduction of which with lithium tetrahydridoaluminate (OAc→OH and Cl→H) afforded delpheline (VIII). Deltaline, which was found to contain a methylenedioxy group, three methoxyl groups, an acetoxyl group and a hydroxyl group, must therefore be an acetate ester of delpheline, containing an additional hydroxyl group. The latter was observed to be unreactive, being unattacked by chromic oxide during a lengthy period of exposure, and not being acetylated by heating with acetyl chloride, and was therefore revealed to be tertiary. The stability of this group to lithium tetrahydridoaluminate and catalytic reduction indicated that it was not in the "lycoctonine position". Oxidation of deltaline with permanganate followed by hydrolysis and demethylenation gave the lactam IX. Oxidation of this product with periodate yielded a γ-lactone X, containing an aldehyde group. The fact that this product is a γ-lactone, coupled with the fact that deltamine, the aminoalcohol obtained by hydrolysis of deltaline, is not attacked by periodic acid, settles unequivocally that the "extra" hydroxyl group in deltaline is at position 10; deltaline may thus be formulated as XI and deltamine as XII (*A. D. Kuzovkov* and *T. F. Platonova*, J. gen. Chem. U.S.S.R., Eng. Transl., 1959, **29**, 2746, 3801; 1961, **31**, 1264; *Carmack*, *D. W. Mayo* and *J. P. Ferris*, J. Amer. chem. Soc., 1959, **81**, 4110).

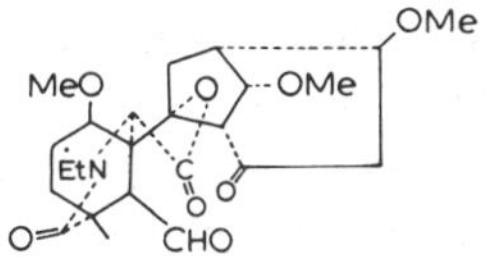

(X)

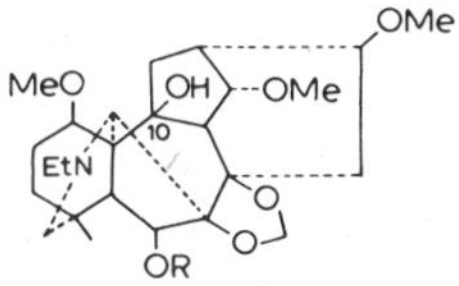

(XI ; R = Ac) Deltaline
(XII ; R = H) Deltamine

Cookson and *Trevett* (*loc. cit.*; *T. A. Crabb* and *Cookson*, Tetrahedron Letters, 1964, 679) have confirmed these structures by n.m.r. spectroscopy.

Delcosine, $C_{24}H_{39}NO_7$, is found in the roots of *D. ajacis*, *D. consolida*, *A. lucidusculum*, *A. japonicum*, *Consolida regalis* and in the seeds of *D. orientale* (*J. A. Goodson*, J. chem. Soc., 1945, 245; *L. Marion* and *O. E. Edwards*, J. Amer. chem. Soc., 1947, **69**, 2010; *W. I. Taylor*, *W. E. Walles* and *Marion*, Canad. J. Chem., 1954, **32**, 780; *H. Suginome et al.*, Proc. Japan Acad., 1946, **22**, 122; Bull. chem. Soc. Japan, 1959, **32**, 354; *S. Furusawa*, *ibid.*, p. 399; *A. Suzuki*, *T. Amiya* and *T. Matsumoto*, *ibid.*, 1961, **34**, 455; *Ochiai et al.*, J. pharm. Soc. Japan, 1956, **76**, 550; Chem. pharm. Bull., Tokyo, 1958, **6**, 730; *V. Skaric* and *Marion*, Canad. J. Chem., 1960, **38**, 2433).

Because of the molecular formula a lycoctonine-type skeleton was assumed. Four hydroxyl groups, three methoxy groups and an *N*-ethyl group were detected; the fact that oxidation of delcosine with chromic acid produced neither acid nor aldehyde was interpreted to mean that the alkaloid has a $-CH_2OMe$ group at $C_{(4)}$ as opposed to the $-CH_2OH$ group of lycoctonine. The easy formation of methanolamine ether linkages in several reactions of delcosine indicated the possible presence of a hydroxyl group at $C_{(1)}$; regulated oxidation of the alkaloid with chromic acid tended to confirm this, by leading to a molecule containing both 5- and 6-membered ring carbonyl groups. The facile methanolamine ether formation requires, further, that the hydroxyl must have the α-configuration. These and other considerations lead to expression XIII for delcosine, methanolamine ether formation occurring *via* the CH_2 group at $C_{(4)}$ to give cyclic ethers of partial structure XIV (*Marion* and *Skaric*, J. Amer. chem. Soc., 1958, **80**, 4434, and *loc. cit.*). An X-ray diffraction analysis of delcosine hydrobromide has shown that the delcosine skeleton is the same as that of lycoctonine and that all

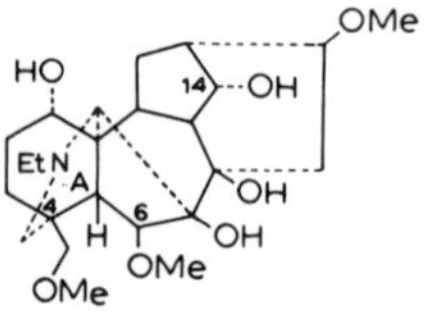

(XIII)
Delcosine

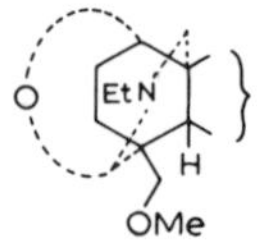

(XIV)

substituents except the hydroxyl group at position 1 have the same orientation as in the latter. Ring A of delcosine is boat-shaped, whereas in lycoctonine it is the normal chair, because of hydrogen bonding between the 1-hydroxyl group and the nitrogen atom (*Marion*, Pure and Applied Chem., 1963, **6**, 621).

Monoacetyldelcosine accompanies delcosine in many of the natural sources listed above. It is formed by acetylation of delcosine with acetic acid using trichloroacetic acid as catalyst and on hydrolysis yields delcosine (*Taylor*, *Walles* and *Marion*, *loc. cit.*). Oxidation of the base (chromic oxide–pyridine) yielded a cyclohexanone derivative, indicating that the acetoxy group must be placed at position 14 and the alkaloid formulated as XIII (OAc replacing OH at position 14).

14-**Dehydrodelcosine** (XIII, C=O at position 14) occurs in *A. japonicum.* Its structure has been established by correlation with delcosine (*Ochiai et al.*, J. pharm. Soc. Japan, 1955, **75**, 638; Chem. pharm. Bull., Tokyo, 1958, **6**, 730; *Skaric* and *Marion*, Canad. J. Chem., 1960, **38**, 2433).

Delsoline, $C_{25}H_{41}NO_7$, occurring in *D. consolida*, is an *O*-methyldelcosine. When delcosine (XIII) is methylated with sodium hydride–methyl iodide at room temperature a regiospecific reaction occurs at $C_{(14)}$, leading to delsoline, which is consequently XIII (OMe in place of OH at position 14) (*F. Sparatore*, *R. Greenhalgh* and *Marion*, Tetrahedron, 1958, **4**, 157; *Marion* and *Skaric*, J. Amer. chem. Soc., 1958, **80**, 4434), the site of the secondary hydroxyl group in the alkaloid having been shown earlier to be in a six-membered ring by oxidation studies.

Browniine, $C_{25}H_{41}NO_7$, occurs in the Canadian plant *D. brownii.* It is isomeric with lycoctonine and shows marked resemblance to that alkaloid.

Oxidation (chromic oxide–pyridine) afforded a ketolactam, the oxo group being in a 5-membered ring. Acetylation gave a monoacetate, indicating the presence of one secondary alcohol group. Oxidation of the acetate ($KMnO_4$) gave oxobrowniine acetate, a lactam, which on hydrolysis furnished oxobrowniine, this same product being formed from the above ketolactam by stereoselective reduction with tetrahydridoborate. Both oxobrowniine and the ketolactam consumed lead tetra-acetate (1 mol.) readily, giving secodi- and secotri-ketones, respectively, indicating that they, and also browniine, contain a pair of vicinal tertiary hydroxyl groups. Thermal elimination of methanol could be effected from the triketone, resulting in an α,β-unsaturated ketone and indicating the presence of a methoxyl group at $C_{(16)}$, and that the two remaining oxygens are part of methoxyl groups, probably in positions 1 and 18 as in lycoctonine (I). Browniine was therefore formulated tentatively as XV, which was confirmed by a correlation with lycoctonine. Oxobrowniine was subjected to a pinacol–pinacolone rearrangement, and the resultant anhydro-oxobrowniine methylated. The product proved to be identical with the methyl ether of lycoctonam (*Benn, M. A. M. Cameron* and *Edwards*, Canad. J. Chem., 1963, **41**, 477):

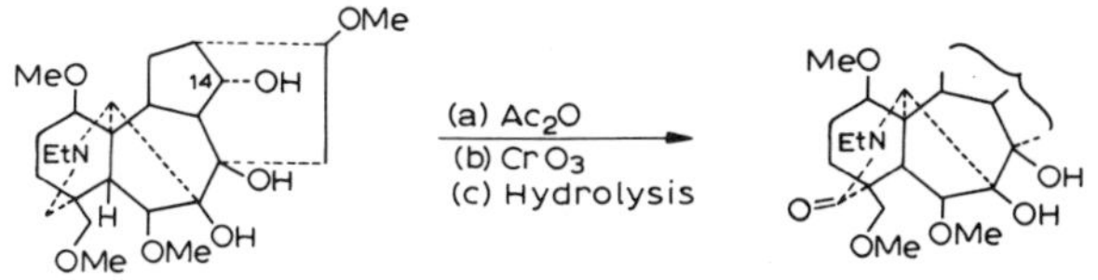

(XV)
Browniine

(a) Ac_2O
(b) CrO_3
(c) Hydrolysis

Oxobrowniine

(a) $H^{\oplus}$ (rearrangement)
(b) Methylation

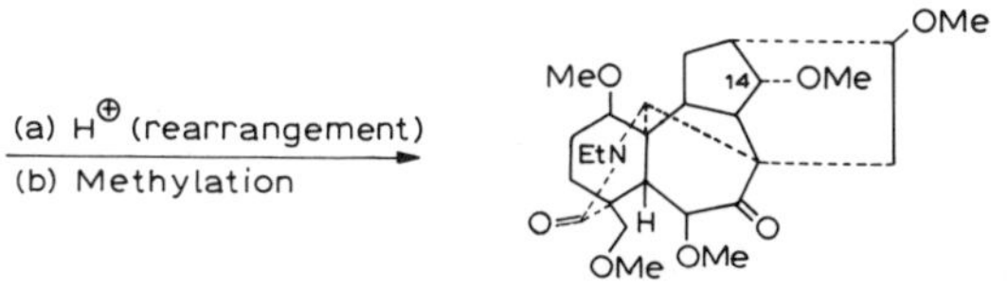

Anhydro-oxobrowniine
(14-methyl ether of anhydrolycoctonam)

14-**Dehydrobrowniine,** $C_{25}H_{39}NO_7$, has been isolated from *D. cardinale.* Reduction with tetrahydroborate yielded browniine stereospecifically, and, conversely, oxidation of browniine with dichromate gave 14-dehydrobrowniine. The latter is therefore (XV, C=O at position 14) (*Benn, ibid.*, 1966, **44**, 1).

Lycoctonine alkaloids are listed in Table 5.

TABLE 5

LYCOCTONINE-TYPE ALKALOIDS

Alkaloid	*M.p. (°C)*	*$[\alpha]_D$ (°) (solvent)*
Lycoctonine	143	+53
Elatidine		
Delphatine	106	+39
Delpheline	227	−27
Deltaline	181	−28
Delcosine	204	+57
Monoacetyldelcosine	193–195	+34 (chloroform)
14-Dehydrodelcosine	212–213.5	
Delsoline	216	+52
Browniine	amorphous, 212*	+25* (ethanol)
14-Dehydrobrowniine	161–163	

* Perchlorate.

(b) Aconitine-type alkaloids

The most important difference between these bases and those related to lycoctonine is that the former do not contain an oxygen function at position 7. There is thus no real or potential 1,2-ditertiary glycol unit in the aconitine-related alkaloids. Another difference relates to the methoxyl group at position 6: all lycoctonine alkaloids have such a hydroxyl, methoxy or acyloxy group, and in stereochemistry it is 6*β*, but such a group is not found in this position in all the aconitine bases, and when it is present it is 6α. Thirdly, many of the aconitine bases have a bridgehead hydroxyl group at position 13; none of the known lycoctonine alkaloids possesses such a group.

Two methods are available for settling the configuration of a methoxyl group at $C_{(6)}$. X-ray diffraction analysis has been used, and a chemical method has been devised. The latter involves brief treatment of the alkaloid with neutral permanganate in aqueous acetone. If the base contains a 6α methoxyl group, oxidation is accompanied to a large extent by *N*-de-ethylation. On the other hand, if the alkaloid has a 6*β* methoxyl group a lactam is the principal product. This difference is explicable in terms of steric hindrance (similar to the familiar 1,3-diaxial interaction) by the 6α methoxyl group of oxidation at $C_{(19)}$ (see formula I, p. 352); contrariwise a 6*β* methoxyl group exerts no such effect (*O. Achmatowicz, Y. Tsuda* and *L. Marion*, Canad. J. Chem., 1965, **43**, 2336).

Aconitine, $C_{34}H_{47}NO_{11}$, has been known as the principal alkaloid of many *Aconitum* spp. for well over a century. It is a physiologically active constituent of several Chinese drugs. Extensive early work carried out on the alkaloid was not very fruitful as far as structure elucidation was concerned, because of the complexity of the structure. The base is tertiary (*N*-ethyl), contains two ester groups (acetoxy and benzoyloxy), four methoxyl groups and three hydroxyl groups. Hydrolysis affords an alkamine, **aconine**. The combined efforts of two groups ultimately led to two possible structures for aconitine (*K. Wiesner, G. Büchi* and collaborators, Tetrahedron Letters, 1959, 15); a choice of one of these was made possible by an X-ray diffraction analysis of a compound derived from aconitine (*Marion* and *M. Przybylska*, Canad. J. Chem., 1959, **37**, 1116, 1843), and pointed to structure XVI for aconitine, in which the configurations of the methoxyl group at $C_{(1)}$ and the hydroxyl at $C_{(3)}$ still remained to be settled (see below under delphinine). As far as the latter group is concerned, chemical and conformational evidence has been presented showing that the hydroxyl group is 3α and equatorial (*F. W. Batchelor, R. F. C. Brown* and *Büchi*, Tetrahedron Letters, 1960, No. 10, 1).

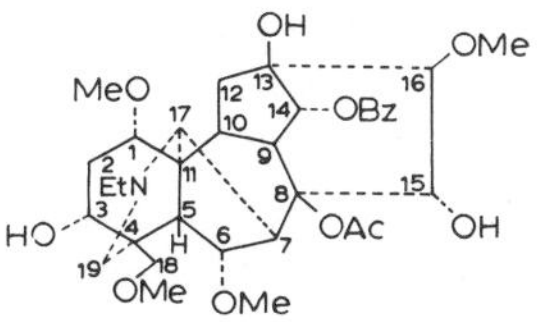

(XVI)
Aconitine

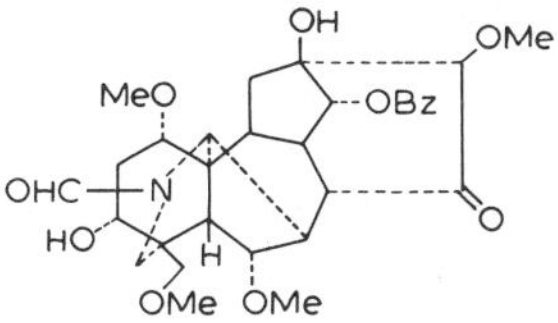

(XVII)
Pyro-oxonitine

A key compound in structural study on aconitine was oxonitine, a permanganate oxidation product of aconitine. Difficulties concerning the formulation of this compound were concerned with the unexpected presence of an *N*-formyl group in the molecule rather than an *N*-acetyl group. The same compound is formed by oxidation of **mesaconitine**, another *Aconitum* base, which is the *N*-methyl analogue of aconitine (see below) (*W. A. Jacobs, R. C. Elderfield* and *L. C. Craig*, J. biol. Chem., 1956, **218**, 439; *Jacobs* and *S. W. Pelletier*, J. Amer. chem. Soc., 1954, **76**, 4048; Chem. and Ind., 1960, 591; *R. B. Turner, J. P. Jeschke* and *M. S. Gibson*, J. Amer. chem. Soc., 1960, **82**, 5182; *R. Majima* and *H. Suginome*, Ber., 1925, **58**, 2048; *S. Morio*, Ann., 1929, **476**, 181). It was demonstrated that this formyl group in oxonitine does not arise from the *N*-ethyl group in aconitine by, for example, conducting the aconitine oxidation in the presence of methanol, when the yield of oxonitine was trebled. Tracer studies confirmed this deduction.

Pyrolysis of oxonitine afforded pyro-oxonitine accompanied by acetic acid, this elimination being characteristic of alkaloids of this group with an acetoxy group at position 8. The immediate result of the elimination is an enol, which rearranges to the ketone pyro-oxonitine (XVII), and from a knowledge of the mechanism of such thermal eliminations it may be deduced that the acetoxy group at $C_{(8)}$ and the hydroxyl group at $C_{(15)}$ in aconitine are *trans*.

The hydrolysis of oxonitine generated a base **oxonine** along with benzoic and acetic acids. Oxonine (XVIII) on oxidation with chromic anhydride–pyridine yielded a 3,14-diketone (an acyloin) which underwent spontaneous base-catalysed rearrangement to a new acyloin XIX.

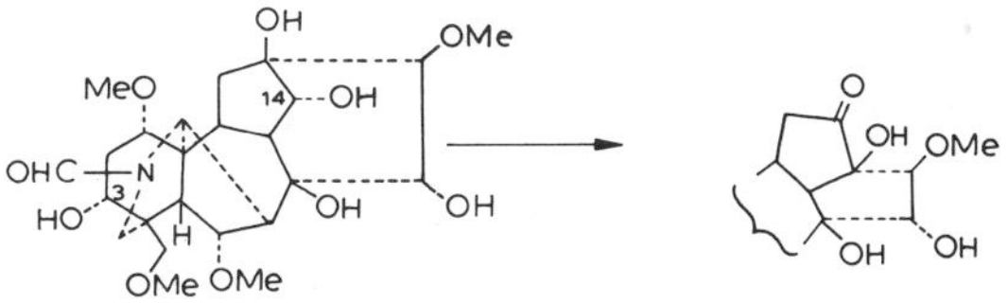

(XVIII)
Oxonine

(XIX)

Pyroaconitine, formed by pyrolytic elimination of acetic acid from aconitine (XVI) afforded on Wolff–Kishner reduction a product showing no carbonyl absorption in its i.r. spectrum, and having one fewer methoxyl group. This product was shown to be XX by spectral study and correlation with **pseudaconitine**, another alkaloid of this group (see p. 366) (*Tsuda* and *Marion*, Canad. J. Chem., 1963, **41**, 1485).

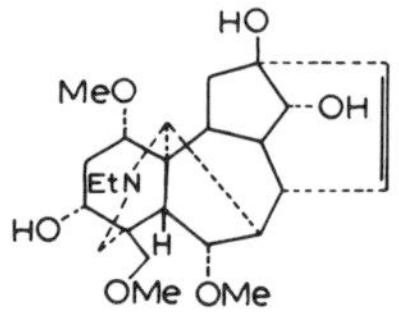

HO, MeO, OH, EtN, HO, H, OMe OMe

(XX)

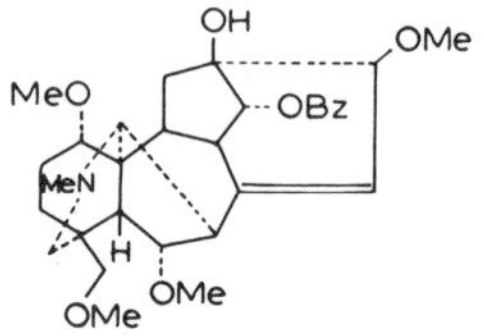

OH, OMe, MeO, OBz, MeN, H, OMe, OMe

(XXI)

The position of the benzoate ester group in aconitine was settled by n.m.r. spectral study: a doublet (1H) at δ 4.9 ppm ($J=4.5$ Hz) appeared in the spectra of aconitine and several closely related derivatives, but was missing in the spectra of the corresponding alkamines. This signal was ascribed to the $\diagdown$C*H*OCOPh proton, and the benzoyloxy group consequently cannot be located at positions 3 or 13; nor can it be at position 15 because of the formation of pyroaconitine on pyrolysis of aconitine. A supporting piece of evidence is the observation that the $O{=}CCH_3$ protons of aconitine and congeners have n.m.r. signals at unusually high field (δ 1.3), explicable in terms of the diamagnetic anisotropy of the aromatic nucleus, which is capable of close approach to the $C_{(8)}$ acetoxy protons. Finally, the benzoyloxy group is assigned the α-configuration because then the $H_{(9)}$–$H_{(14)}$ dihedral angle is $\sim 80°$ and the calculated coupling constant zero (*idem, ibid.*, p. 1634).

For discussion of further reactions of aconitine see *L. H. Keith* and *Pelletier*, in "The Chemistry of the Alkaloids", ed. *Pelletier*, Chapter 18, Van Nostrand–Reinhold, New York, 1970.

Delphinine, $C_{33}H_{45}NO_9$, is the major alkaloid of *D. staphisagria*, and was first isolated well over a century ago. Preliminary studies showed it to contain four methoxyl groups, an acetoxyl group, a benzoyloxy group, a tertiary alcohol group and a tertiary *N*-methyl group (*Jacobs* and *Pelletier*, J. Amer. chem. Soc., 1956, **78**, 3542). Structural studies on delphinine have been very much bound up with those on aconitine, conclusions reached from results on one base being generally of value in connexion with the other.

Delphinine suffers pyrolytic elimination of acetic acid, just as does aconitine, the product in the former case being pyrodelphinine (XXI); however, pyrodelphinine is not a ketone like pyroaconitine, but is an alkene, which fact means that delphinine carries no hydroxyl group at position 15. Pyro-bases containing an *N*-alkyl group show a characteristic u.v. absorption maximum at 235–245 nm (ε 6,000), which vanishes when the base is protonated or the alkyl group replaced by an acyl group.

This spectral behaviour is useful in specific structure diagnosis in this area; it has been explained as being due to an interaction, in the free bases, between the nitrogen lone pair of electrons, the $C_{(17)}$–$C_{(7)}$ bond and the bridgehead 8–15 double bond, which models show to be highly strained:

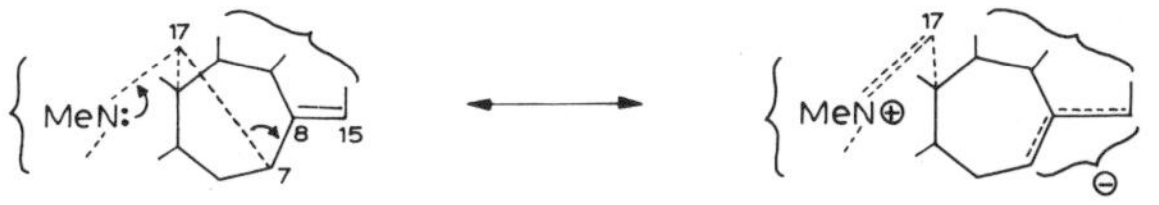

Models show that orbital overlap is possible between the N–$C_{(17)}$ and $C_{(7)}$–$C_{(8)}$ or $C_{(8)}$–$C_{(15)}$ olefinic bonds (*Wiesner et al.*, Tetrahedron Letters, 1960, 17). Such mesomerism will not, of course, be possible if the nitrogen atom has been protonated or is part of an amide.

Pyrodelphinine (XXI) and other pyro-bases lacking a hydroxyl function $C_{(15)}$ in this series suffer acid-catalysed allylic rearrangement with ease:

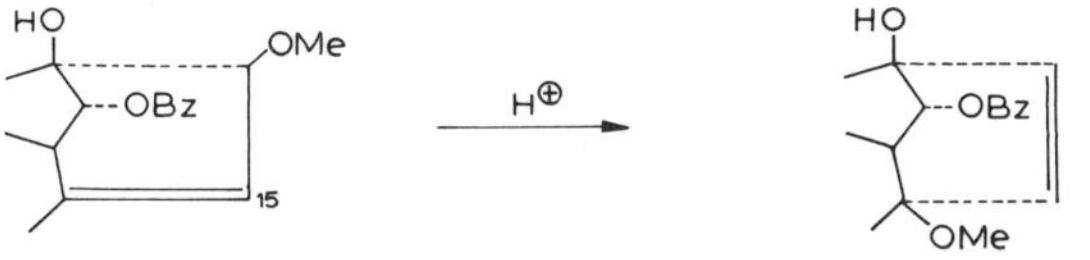

Pyro-base Isopyro-base

The rearrangement is of an intermolecular nature: in acetic acid as solvent an acetoxy-isopyro-base was formed (*Wiesner*, *F. Bickelhaupt* and *D. R. Babin*, Experientia, 1959, **15**, 93).

Hydrolysis of delphinine affords the parent alkamine **delphonine**, which on Hofmann degradation suffers deep-seated rearrangement of the normal Hofmann base. Considerable effort was devoted to the structure of the rearrangement product (*Wiesner et al.*, Tetrahedron, 1960, **9**, 254). Finally, a correlation between aconitine and delphinine was achieved (*Wiesner*, *D. L. Simmons* and *L. R. Fowler*, Tetrahedron Letters, 1959, No. 18, 1). This involved the isopyro-base from α-oxodelphonine (the oxidation product of delphonine; $>NCH_3 \rightarrow >N\text{–}CHO$), which was converted in several stages into a phenolic product; this was methylated and then compared with a product obtained from pyro-oxonitine by a similar route, followed by replacement of the $C_{(3)}$-hydroxyl by chlorine and subsequent hydrogenolysis. As a consequence of these and other detailed studies delphinine may be formulated as XXII.

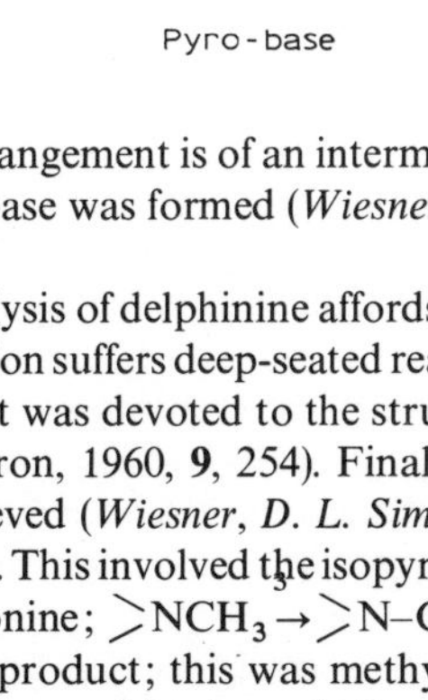

(XXII)
Delphinine

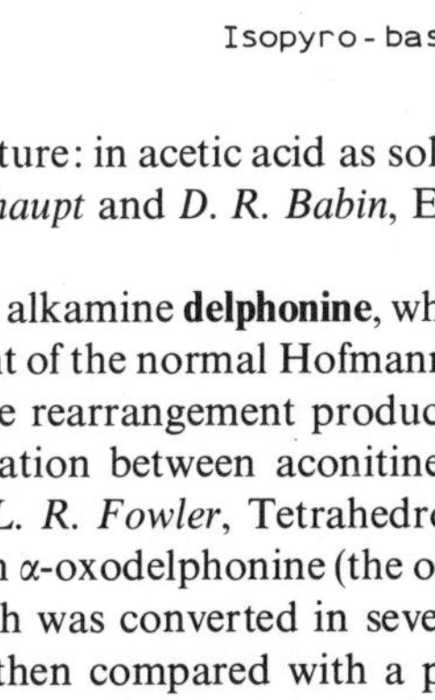

(XXIII ; R^1 = OMe, R^2 = Ac) Alkaloid-A
(XXIV ; R^1 = R^2 = H) Alkaloid-B

In spite of earlier chemical arguments to the contrary, an X-ray diffraction analysis of a derivative of delphinine in which the 1-methoxyl group is still present revealed that it has the α-configuration in delphinine, and therefore, because of correlations, in aconitine and a number of related bases (*G. I. Birnbaum et al., ibid.*, 1971, 867).

Alkaloids of D. bicolor. Two minor alkaloids of this plant have been isolated (*A. J. Jones* and *M. H. Benn, ibid.*, 1972, 4351; Canad. J. Chem., 1973, **51**, 486). The first, known as **alkaloid-A**, $C_{25}H_{39}NO_6$, contains a tertiary methyl, an *N*-ethyl, an acetoxyl, two methoxyl and two hydroxyl groups, on i.r. and n.m.r. spectral evidence. The acetoxyl group was shown to be at $C_{(8)}$ by pyrolysis studies, acetic acid being formed and monitored by n.m.r. spectroscopy. The other, **alkaloid-B**, $C_{22}H_{35}NO_5$, contains a tertiary methyl, an *N*-ethyl, a methoxyl and three hydroxyl groups. A carbon-13 n.m.r. spectral study and comparison of the two bases allows them to be formulated as XXIII and XXIV, respectively.

Jesaconitine, $C_{35}H_{49}NO_{12}$, differs structurally from aconitine only in the nature of one of the ester groups, which is *p*-anisoyloxy rather than benzoyloxy. It occurs in *A. fischeri* tubers, and accompanies aconitine frequently in several other spp. (*K. Makoshi*, Arch. Pharm., 1909, **247**, 243).

Thermal treatment of jesaconitine gave the pyro-base (pyrojesaconitine) (*Majima*, *Suginome* and *Morio*, Ber., 1924, **57**, 1486), and, on the assumption that acetic acid is eliminated, the acetoxy group can be placed at position 8 as in aconitine (XVI). An analysis of the n.m.r. spectrum of jesaconitine made it possible to locate the *p*-anisoyloxy group at position 14, since the proton at this position was responsible for a signal at δ4.62 ppm (doublet), whilst when a hydroxyl group is at position 14 the same signal is found at ~4.2. Next, the acetoxy group protons at $C_{(8)}$ are observed to be markedly shielded (1.36, normally ~2.0); this behaviour has been found to be general for all the aconitine bases with an aroyloxy group at $C_{(14)}$ and an acetoxy group at $C_{(8)}$. Finally, the pyro-reaction has been studied by n.m.r. spectroscopy; as the reaction proceeded the high-field methyl signal of the acetate ester group gradually diminished in intensity and a new signal due to the methyl protons of acetic acid became apparent. Also, the aromatic proton signals remained unchanged during the thermolysis (*Keith* and *Pelletier*, Chem. Comm., 1967, 993; J. org. Chem., 1968, **33**, 2497). Jesaconitine is therefore represented by (XVI; OBz at position 14 replaced by $OCO \cdot C_6H_4OMe$-*p*).

Mesaconitine, $C_{33}H_{45}NO_{11}$, occurs in many *Aconitum* spp., often accompanying aconitine. It is the *N*-methyl analogue of aconitine, a relationship which was easily revealed by the successive de-ethylation and methylation of aconitine, which yielded mesaconitine (*Tsuda*, *Achmatowicz* and *Marion*, Ann., 1964, **680**, 88). Mesaconitine is consequently (XVI; EtN replaced by MeN).

Hypaconitine, $C_{33}H_{45}NO_{10}$, commonly occurs alongside aconitine in *Aconitum* spp. It is formed when mesaconitine is regiospecifically dehydrated with thionyl chloride, to yield a Δ^2-base which on catalytic hydrogenation afforded hypaconitine, which is consequently (XVI; MeN in place of EtN, and CH_2 at position 3) (*Tsuda, Achmatowicz* and *Marion, loc. cit.*).

Deoxyaconitine, $C_{34}H_{47}NO_{10}$, also frequently accompanies aconitine in the same sources. It has structure (XVI; CH_2 at position 3), and is obtainable from aconitine by regiospecific dehydration ($SOCl_2$) followed by hydrogenation (*idem, ibid.*).

Lappaconitine, $C_{32}H_{44}N_2O_8$, is the *N*-acetylanthranilic ester of a trihydroxytrimethoxyalkamine base **lappaconine,** $C_{23}H_{37}NO_6$. It occurs in various *Aconitum* spp., especially *A. ranunculaefolium.* A lycoctonine-type skeleton was presumed.

N.m.r. spectral studies, along with oxidation experiments, enabled the various functional groups to be located (*N. Mollov et al.*, Tetrahedron Letters, 1964, 2711; *Mollov, M. Tada* and *Marion, ibid.*, 1969, 2189). An X-ray diffraction analysis of lappaconine hydrobromide revealed XXV as the structure of the alkamine (*Birnbaum, ibid.*, p. 2193), a structure which is unusual in that carbon atom 18 and the 6-methoxyl group are absent compared with lycoctonine and aconitine bases generally. Lappaconitine is the first bisnor (C_{18}) diterpene alkaloid yet encountered, the position of the *N*-acetylanthraniloyl residue being as yet uncertain.

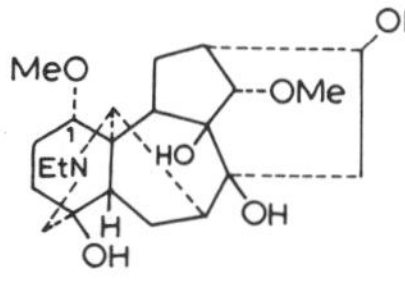

(XXV) Lappaconine

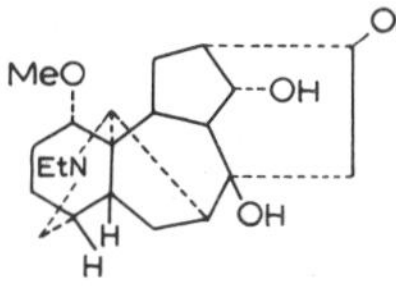

(XXVI) Aconosine

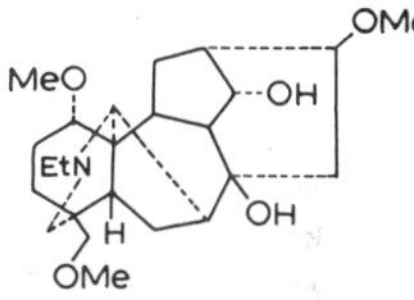

(XXVII) Talatizamine

Lappaconidine, $C_{22}H_{35}NO_6$, an alkaloid of *A. leucostomum*, contains an *N*-ethyl tertiary nitrogen atom, two methoxyl groups and four hydroxyl groups, all capable of being acylated. Methylation yielded tetramethyllappaconine. Lappaconidine and lappaconine consequently differ in that the former has four hydroxyl groups, and when one of these is methylated lappaconine results. Mass spectrometric study allows this hydroxyl group to be placed at $C_{(1)}$, and lappaconidine is therefore (XXV; OMe at position 1 replaced by OH) (*V. A. Tel'nov et al.*, C.A., 1971, **74**, 76573; 1972, **76**, 99877). Some reactions of the base have been described (*idem, ibid.*, 1971, **74**, 42527).

Excelsine, $C_{22}H_{33}NO_6$, occurring in the roots of *A. excelsum*, yields on acetylation a triacetate under mild conditions, but a tetra-acetate on more vigorous treatment. A link-up with lappaconidine (XXV; OH in place of OMe at position 1) was provided by the observation that reduction of excelsine using a Raney nickel catalyst gave this alkaloid. On the basis of these and other chemical and spectral properties excelsine is formulated as XXVIII (*Tel'nov, M. S. Yunusov* and *S. Y. Yunusov, ibid.*, 1971, **74**, 76573; 1973, **78**, 159947).

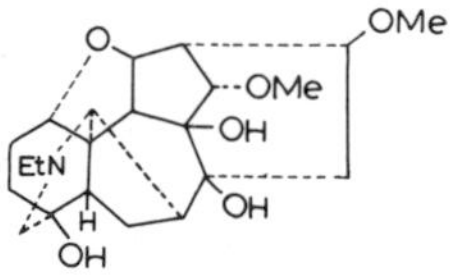

(XXVIII) Excelsine

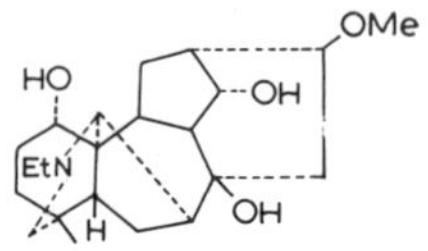

(XXIX) Karacoline

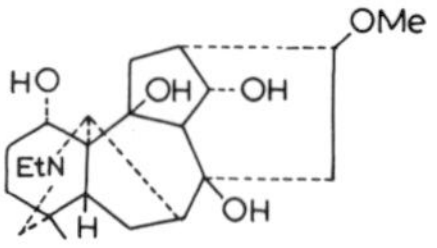

(XXX) Karakolidine

Karacoline, $C_{22}H_{35}NO_4$, found in the roots of *A. karacolicum* of Central Asia, forms a diacetate or a triacetate, according to the conditions. Oxidation with permanganate resulted in an internal methanolamine ether, reconvertible to the original alkaloid on catalytic hydrogenation. Oxidation with chromic acid, on the other hand, afforded a diketone (C=O in 5- and 6-membered rings). As a consequence of these and other investigations a hydroxyl group may be placed at $C_{(1)}$. Mass spectral study, with comparisons, located a methoxyl group at $C_{(16)}$. Karacoline was ultimately formulated as XXIX, as a result of detailed chemical and spectroscopic study (*Yunusov et al., ibid.*, 1973, **79**, 32152).

Karakolidine, $C_{22}H_{35}NO_5$, from the same source, also formed an internal α-methanolamine ether on oxidation with permanganate, behaviour characteristic of a 1α-hydroxyl group. A further resemblance to karacoline was the formation of a diketone with C=O groups in 5- and 6-membered rings on oxidation with chromic oxide. Karakolidine tetra-acetate on pyrolysis yielded, after hydrolysis, pyrokarakolidine, isomerised to isopyrokarakolidine by hot acid. From these chemical transformations and from spectroscopic analyses, structure XXX has been allotted to karakolidine (*Yunusov et al., loc. cit.*).

Aconosine, $C_{22}H_{35}NO_4$, has been isolated from *A. nasutum*, and has been investigated chiefly by i.r. and n.m.r. spectroscopy, which demonstrated that two methoxyl groups, two hydroxyl groups and an *N*-ethyl group were present. Its mass spectrum was characteristic of alkaloids of the lycoctonine group. One of the hydroxyls was shown to be secondary and in a five-membered ring by oxidation. Structure XXVI has been proposed for the base (*Yunusov* and co-workers, C.A., 1972, **77**, 72562).

Talatizamine, $C_{24}H_{39}NO_5$, from *A. talassicum* and *A. variegatum* has as functional groups *N*-ethyl, two hydroxyl (secondary and tertiary) and three methoxyl. Its mass spectrum is typical of alkaloids of the lycoctonine group, and it forms a pyro-base and an isopyro-base, characteristic of these alkaloids with hydroxyl and methoxyl groups at positions 8 and 16, respectively. The methoxyl group in the pyro-base is susceptible to hydrogenolysis by lithium tetrahydridoaluminate, confirming its presence; n.m.r. spectral analysis assigned to it the β-configuration. The same spectral investigation located an α-OH at $C_{(14)}$ and a methoxyl group at $C_{(18)}$. Talatizamine is formulated as XXVII on this and related evidence (*R. A. Konovalova* and *A. P. Orekhov*, Zhur. obshcheĭ Khim., 1940, **10**, 745; *Yunusov et al.*, *ibid.*, 1954, **24**, 2237; *Mollov et al.*, Tetrahedron, 1971, **27**, 819; *Pelletier*, *Keith* and *P. C. Parthasarathy*, J. Amer. chem. Soc., 1967, **89**, 4146). The total synthesis of talatizamine has been reported recently by *Wiesner* and co-workers (*ibid.*, 1974, **96**, 4990).

Cammaconine, $C_{23}H_{37}NO_5$, also found in *A. variegatum*, has been shown by n.m.r. spectral study to contain three hydroxyl groups, two methoxyl groups and an *N*-ethyl group *(Mollov et al., loc. cit.)*. On total methylation it afforded the same product as obtained upon similar treatment of talatizamine, and it was clear as a consequence that cammaconine and talatizamine differ only in their *O*-methylation patterns. Oxidation of cammaconine with chromic anhydride–pyridine afforded two lactam products, one a monoketone (C=O in a 5-membered ring) and the other a diketone (C=O groups in 5- and 6-membered rings), both compounds containing, on spectral evidence, two methoxyl groups, and neither being aldehydic. Consequently cammaconine contains a $-CH_2OMe$ group at $C_{(4)}$ and a secondary hydroxyl at $C_{(14)}$, and because it is different from **isotalatizidine** (XXXIII; OH at $C_{(1)}$ in place of OMe) (see below), the second hydroxyl must be 16β and cammaconine represented by XXXI *(idem, loc. cit.)*:

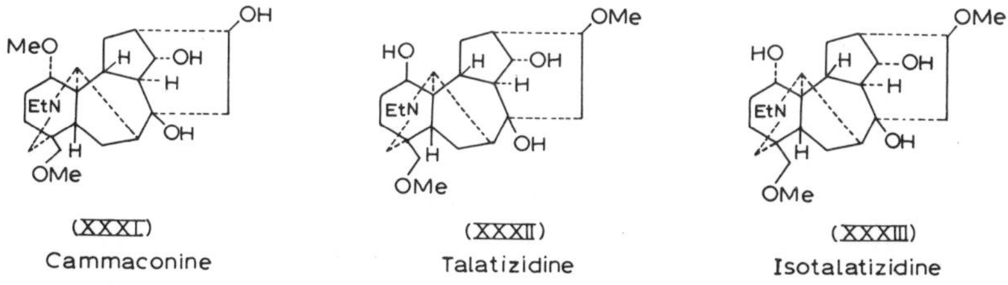

(XXXI) Cammaconine (XXXII) Talatizidine (XXXIII) Isotalatizidine

Talatizidine, $C_{23}H_{37}NO_5$, occurs in the Central Asian *A. talassicum*, along with its $C_{(1)}$-epimeride isotalatizidine (mentioned above) (*M. S. Rabinovich* and *Konovaloza*, Zhur. obshcheĭ Khim., 1942, **12**, 329; *Konovalova* and *Orekhov*, Bull. Soc. chim. Fr., 1940, [v], **7**, 95). A third alkaloid, **condelphine**, $C_{25}H_{39}NO_6$, also belongs to this subgroup: it has been known for a considerable time as a constituent base of *D. confusum.* It is an *O*-acetate of isotalatizidine, since on hydrolysis it yields this alkaloid and acetic acid. Chemical and spectroscopic studies have enabled structures XXXII, XXXIII and XXXIV to be assigned to talatizidine, isotalatizidine and condelphine, respectively (*A. D. Kuzovkov* and *T. F. Platonova*, J. gen. Chem. U.S.S.R., 1961, **31**, 1286; *Pelletier, Keith* and *Parthasarathy, loc. cit.*; Tetrahedron Letters, 1966, 4217). N.m.r. spectroscopy has been particularly helpful, especially in the assignment of configurations and ring conformations.

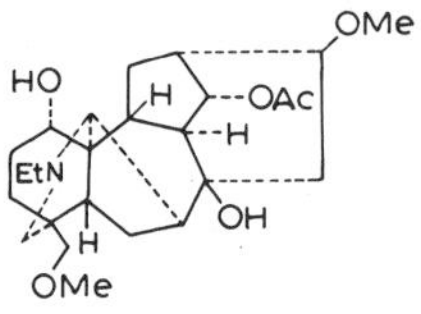

(XXXIV)
Condelphine

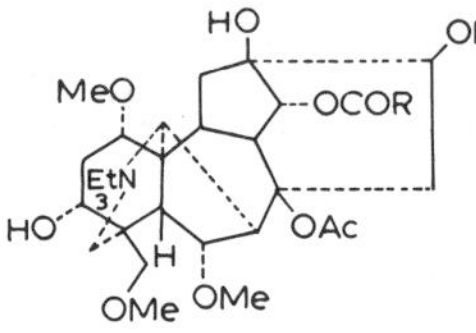

(XXXV; R = C_6H_5) Indaconitine

(XXXVI; R = 3,4-dimethoxyphenyl [OMe, OMe]) Pseudaconitine

Condelphine contains an *N*-ethyl group (tertiary), two methoxyl groups, an acetoxy group and two hydroxyl groups (one secondary, one tertiary), chiefly on n.m.r. spectral evidence. Detailed chemical studies revealed that all three bases belong to the aconitine family of bases. Talatizidine and isotalatizidine yield the same diketone on oxidation, and therefore almost certainly differ only in the configuration at $C_{(1)}$; in condelphine and isotalatizidine intramolecular –O–H···NEt hydrogen bonding is discernible on i.r. spectral study, so that the configuration of the hydroxyl group at $C_{(1)}$ must be α. The presence of a hydroxyl group at position 1 had been proved earlier by chemical and spectroscopic study. The absolute configuration of condelphine (and thence of related compounds) has been settled by an X-ray diffraction analysis of its hydriodide. Structure XXXIV has been shown, by the anomalous dispersion method, to represent the natural base, and, likewise, expressions XXXII and XXXIII represent the absolute configurations of talatizidine and isotalatizidine (*Pelletier* and *D. L. Herald*, unpublished results).

Pseudaconitine, $C_{36}H_{51}NO_{12}$, an alkaloid present in roots of *A. ferox*, of Himalayan origin and other closely related spp., on hydrolysis yields veratric and acetic acids, and an alkamine **pseudaconine. Indaconitine,** $C_{34}H_{47}NO_{10}$, another natural base from *A. chasmanthum*, likewise on hydrolysis affords pseudaconine, and benzoic and acetic acids. *R. E. Gilman* and *Marion* converted indaconitine into delphinine (XXII) by removal of the 3-hydroxyl group and replacement of the *N*-ethyl group by an *N*-methyl (Tetrahedron Letters, 1962, 923). Indaconitine may therefore be formulated as XXXV, in which the configuration of the hydroxyl group at position 3 remains to be settled. Next, the pyrolysis of pseudaconitine yielded a pyro-base identical with the Wolff–Kishner reduction product XX of pyroaconitine (p. 360). This result proves that there is in pseudaconitine a 3α-hydroxyl group, and also in indaconitine. Pseudaconitine is formulated as XXXVI (*Tsuda* and *Marion*, Canad. J. Chem., 1963, **41**, 1485).

Bikhaconitine, $C_{36}H_{51}NO_{11}$, is an alkaloidal constituent of *A. spictatum*. It is another ester-alkaloid which on total hydrolysis affords veratric and acetic acids and an alkamine **bikhaconine**. The alkaloid suffers a pyrolytic elimination of acetic acid to yield pyrobikhaconitine. Six methoxyl groups were detected, two of them aromatic and four aliphatic (n.m.r. spectral evidence). The n.m.r. spectrum also revealed an *N*-ethyl group, an acetoxyl group, three aromatic hydrogens, and a one-proton doublet at δ 4.87; the acetoxy CH_3 signal appeared at high field. Evidently the two ester groups of bikhaconitine are located and oriented as in pseudaconitine (XXXVI). A correlation with the latter alkaloid was achieved by dehydration of pseudaconitine with thionyl chloride (elimination of the 3-hydroxyl group), followed by catalytic hydrogenation to deoxypseudaconitine, which proved to be identical with bikhaconitine (XXXVII) (*Tsuda* and *Marion*, *ibid.*, 1963, **41**, 3055).

Chasmaconitine, $C_{34}H_{47}NO_9$, occurs in the roots of *A. chasmanthum*. Its n.m.r. spectrum showed that four methoxyl groups, an acetoxyl, a benzoyloxy and an *N*-ethyl group were present, and its i.r. spectrum revealed that a single hydroxyl group was also in the molecule.

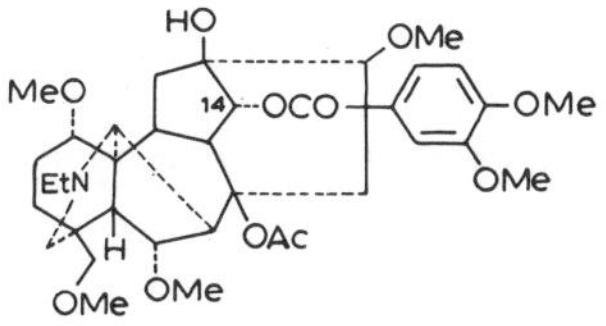

(XXXVII)
Bikhaconitine

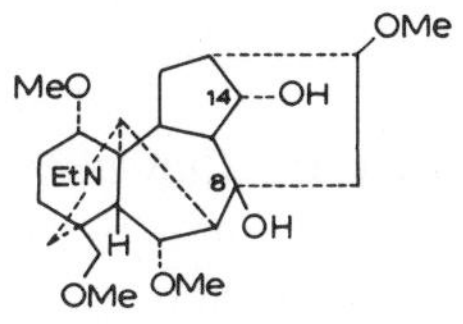

(XXXVIII)
Chasmanine

The acetoxyl methyl group was shielded (δ 1.27 ppm), typical of the influence of an aromatic ester at $C_{(14)}$ on an acetoxy group at $C_{(8)}$. Total hydrolysis of chasmaconitine afforded acetic and benzoic acids and the alkamine bikhaconine, already encountered (see above). Chasmaconitine is consequently deoxyindaconitine (XXXV; CH_2 at position 3) or, regarded in another light, the *N*-ethyl analogue of delphinine (XXII), with which it has been correlated by *N*-demethylation followed by *N*-ethylation (*Achmatowicz* and *Marion*, *ibid.*, 1964, **42**, 154; *Marion et al.*, *ibid.*, 1965, **43**, 825, 1023).

Chasmanthinine, $C_{36}H_{49}NO_9$, from the same source, differs from chasmaconitine only in the nature of the ester group at $C_{(14)}$, which in this base is *trans*-cinnamyloxy. This group was detected by n.m.r. spectroscopy and by hydrolysis, which also furnished bikhaconine. Chasmanthinine is consequently XXXVII (veratroyloxy residue at position 14 replaced by *trans*-cinnamyloxy) (*idem*, *loc. cit.*; *Keith* and *Pelletier*, Chem. Comm., 1967, 993). The structure has been substantiated by n.m.r. spectral analysis.

Chasmanine, $C_{25}H_{41}NO_4$, is also found in *A. chasmanthum*. I.r. and n.m.r. spectral investigations have revealed the following functional groups: four methoxyl, two hydroxyl and a tertiary *N*-ethyl. One of the hydroxyl groups is in a cyclopentanol environment, since oxidation with chromic acid yields a cyclopentanone; the other is tertiary. Chasmanine may be formulated as XXXVIII as a consequence of detailed chemical and spectral study similar to that described for the foregoing alkaloids. An interesting point about this structure is the configuration of the methoxyl group at position 1. A correlation between chasmanine and the lycoctonine-type alkaloid browniine (XV; p. 357), which has a 1β-methoxyl group, has been reported (*O. E. Edwards*, *L. Fonzes* and *Marion*, Canad. J. Chem., 1966, **44**, 583); this suggests that in chasmanine likewise the methoxyl group at $C_{(1)}$ is β. However, a more recent correlation has been achieved between chasmanine and **delphisine**, a newly-isolated alkaloid of *D. staphisagria* (see below). The latter base has a hydroxyl group at $C_{(1)}$, and i.r. spectral studies and X-ray diffraction analysis leave no doubt that this group is α (see below). The correlation with delphisine (outlined below) indicates that chasmanine must have a 1α (equatorial) methoxyl group, as in XXXVIII (*Pelletier*, *Z. Djarmati* and *S. Lajsic*, J. Amer. chem. Soc., 1974, **96**, 7817). X-ray crystallographic analysis of chasmanine 14-α-benzoate hydrochloride demonstrates unequivocally that the methoxyl group is 1α; hence the correlation between chasmanine and browniine referred to above must be erroneous (*Pelletier*, *W. H. De Cauip* and *Djarmati*, Chem. Comm., 1976, 253).

Delphisine, $C_{28}H_{43}NO_8$, an alkaloid of *D. staphisagria* seeds, has been formulated as XXXIX as a result of an X-ray diffraction analysis of its hydrochloride. Spectral investigations support this assignment, which also represents the absolute configuration of the alkaloid (*Pelletier et al.*, J. Amer. chem. Soc., 1974, **96**, 7815). An important revelation from these studies is the α (equatorial) orientation of the 1-hydroxyl group. Delphisine is convertible, by oxidation with chromic acid–aqueous pyridine, into the 1-ketone (dehydrodelphisine), and thence by hydrolysis into a keto-diol (both OAc replaced by OH, and C=O at position 1). Reduction of the keto-diol with tetrahydridoborate afforded two $C_{(1)}$-epimeric triols, separable by chromatography. This lack of stereospecificity was expected since access to the C=O group by the $\overset{\ominus}{B}H_4$ ion is hindered roughly equally from both sides. The more polar of the two epimers proved to be identical with the hydrolysis product of delphisine, and is assigned the 1α configuration both because of its correlation with delphisine and because it shows strong intramolecular hydrogen bonding in its i.r. spectrum (OH and EtN on the same side of ring A).

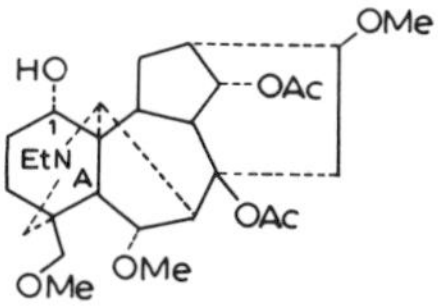

(XXXIX)
Delphisine

The less polar triol is consequently the 1β epimer, both configurations being corroborated by n.m.r. spectral analysis. The more polar (1α) epimer proved to be identical with **neoline** (XL), the alkamine of **neopelline** (see below), which therefore also has a 1α-hydroxyl group. Chasmanine (XXXVIII) has been correlated with neoline by treatment of both bases with methyl iodide–sodium hydride, leading to a common *O*,*O*-dimethylchasmanine (or *O*,*O*,*O*-trimethylneoline). Chasmanine must therefore have a 1α-methoxyl group *(idem, loc. cit.)*. A ^{13}C n.m.r. spectral analysis of chasmanine, neoline and delphisine on the one hand, and lycoctonine and browniine on the other, corroborates this assignment.

Homochasmanine, $C_{26}H_{43}NO_6$, another alkaloid of *A. chasmanthum*, contains, on n.m.r. spectral evidence, an *N*-ethyl and five methoxyl groups. It yields a monoacetate; consequently the sixth oxygen atom must be alcoholic. It was conjectured as a consequence that homochasmanine might be an *O*-methylchasmanine. The above monoacetate shows in its n.m.r. spectrum a triplet signal at δ 4.77 ppm, assigned to the C*H*OAc proton. Clearly, if homochasmanine is an *O*-methylchasmanine, the fifth methoxyl group must be at position 8 in XXXVIII. This formulation has been corroborated by subjecting chasmanine diacetate to reaction with methanol at high temperature and pressure. This reaction is characteristically regiospecific amongst aconitine-type alkaloids, and effects replacement of an 8-acetoxyl group by an 8-methoxyl group. The product was then hydrolysed to yield a compound found to be identical with homochasmanine, which is therefore XXXVIII (OMe at position 8 in place of OH (*Achmatowicz* and *Marion*, Canad. J. Chem., 1965, **43**, 1093).

Neopelline, $C_{33}H_{45}NO_8$, was first encountered as an impurity in crude aconitine isolated from *A. napellus* (*H. Schulze* and *G. Berger*, Arch. Pharm., 1924, **262**, 553). It is an ester-alkaloid which on hydrolysis generates acetic and benzoic acids and an alkamine, **neoline.** Neoline has also been isolated from the same source, but may be an artefact (*W. Freudenberg* and *E. F.*

Rogers, J. Amer. chem. Soc., 1937, **59**, 2572). Neoline contains three alcoholic hydroxyl groups, two secondary and one tertiary. Mild acetylation produces a diacetate in which the secondary hydroxyls have been acetylated. Oxidation of the diacetate ($KMnO_4$), followed by hydrolysis, generated *N*-desethylneoline; if an aconitine-type framework is assumed this behaviour can be rationalised by participation of a 1α-hydroxyl group with the *N*-acetyl group. Triacetylneoline suffers the type of pyrolysis commonly met with in the aconitine series, affording pyrodiacetylneoline, which shows a characteristic olefinic doublet (1H) in its n.m.r. spectrum, enabling a β-methoxyl group to be placed at $C_{(16)}$ and an acetoxyl group at $C_{(8)}$. These and other recognisable features of the chemistry and spectroscopy of neoline permit it to be formulated as XL (*Wiesner* and co-workers, Tetrahedron Letters, 1960, No. 3, 17). The structure of neopelline, which is an acetate-benzoate of neoline, is as yet unsettled.

Aconitine alkaloids are collected in Table 6.

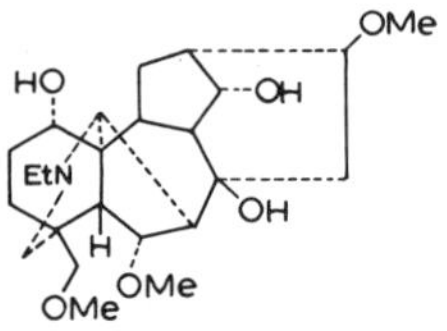

(XL)
Neoline

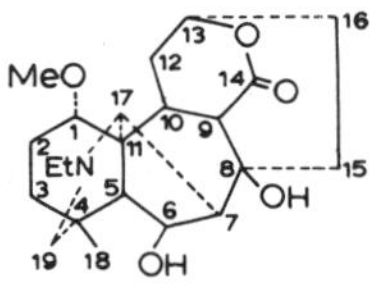

(XLI)
Heteratisine

(*c*) *Lactonic alkaloids*

A small group of diterpenoid bases, generally found in the mother-liquors after extraction of aconitine from *A. heterophyllum*, are lactonic in character and contain a modified lycoctonine skeleton. The best-known member of the group is heteratisine.

Heteratisine, $C_{22}H_{33}NO_5$, is isolable from the weakly basic fraction of *A. heterophyllum.* It may be an artefact, obtained during extraction as a hydrolysis product of its benzoate ester. The structure of heteratisine has been unravelled by both X-ray diffraction analysis of its hydrobromide monohydrate (*M. Przbylska*, Canad. J. Chem., 1963, **41**, 2911) and by chemical and spectroscopic studies, culminating in expression XLI for the alkaloid.

A lactone ring in the alkaloid was detected by its behaviour towards alkali. Two hydroxyl groups were detected by active hydrogen determina-

TABLE 6

ACONITINE-TYPE ALKALOIDS

Alkaloid	*M.p. (°C)*	*[α]D (°) (solvent)*
Aconitine	205	+19
Delphinine	200	+25
Delphonine	78	+38 (EtOH)
Alkaloid-A	>240*	+10 ($CHCl_3$)
Alkaloid-B	190–191	+16 ($CHCl_3$)
Jesaconitine	130	
Deoxyaconitine	178	+16 (EtOH)
Mesaconitine	209	+26
Hypaconitine	198	+22 ($CHCl_3$)
Lappaconitine	227	+34
Lappaconidine	No data published	
Excelsine	No data published	
Karacoline	No data published	
Karakolidine	No data published	
Aconosine	No data published	
Talatizamine	146	
Cammaconine	135–137	
Isotalatizidine	116	
Condelphine	157	+21
Talatizidine	219	−20 (MeOH)
Pseudaconitine	203–205	+24 ($CHCl_3$)
Pseudaconine	93–94	+49 ($CHCl_3$)
Indaconitine	202–203	+18.2 (EtOH)
Bikhaconitine	105–115**	+16 (EtOH)
Bikhaconine	gum	+33.4 (EtOH)
Chasmaconitine	181	+10 (EtOH)
Chasmanthinine	160	+10 (EtOH)
Chasmanine	90–91	+23.6 (EtOH)
Homochasmanine	106	+19 (EtOH)
Neopelline	amorphous	
Neoline	161	+22 (EtOH)
Delphisine	121–122	+7.1 (EtOH)

* Hydriodide; free base amorphous.
**Monohydrate.

tion, a methoxyl group by Zeisel estimation, and an *N*-ethyl group (tertiary) by Herzig–Meyer estimation. The alkaloid appeared to be saturated from its behaviour towards hydrogen in the presence of a catalyst (*W. A. Jacobs et al.*, J. biol. Chem., 1942, **143**, 147, 571; 1947, **170**, 515).

These features in the molecule have been corroborated more recently by i.r. and n.m.r. spectral studies on heteratisine and its simple derivatives (*R. Aneja, D. M. Locke* and *S. W. Pelletier*, Tetrahedron, 1973, **29**, 3297). Treatment of the base with acetic anhydride–pyridine afforded a mono-acetate; on n.m.r. spectral evidence a secondary hydroxyl group has been esterified, and the second hydroxyl group is consequently tertiary. Confirmation is provided by oxidation of heteratisine with chromic oxide to dehydroheteratisine, a hydroxyketone. Dehydrogenation of heteratisine with selenium was not helpful as a degradation procedure, yielding no aromatic hydrocarbons or heterobases, yet the molecular formula and functionality suggested an atisine skeleton was present. An aconane framework was consequently considered. Oxidation of heteratisine acetate with chromic acid–pyridine yielded, as major product, oxoheteratisine acetate, $C_{24}H_{33}NO_7$, a lactam capable of being hydrolysed to oxoheteratisine, a dihydroxylactam. The lactam nature of these two compounds was established by their neutrality and the presence of an *N*-ethyl group in each by n.m.r. spectral analysis. A six-membered lactam structure was supported by i.r. spectroscopy and by analogy with known aconitine-type alkaloids. Further, the quaternary CH_3 signal in heteratisine acetate shifts 0.33 ppm, downfield on lactam formation, suggesting that partial structure XLII is present in oxoheteratisine, in which the methyl group is held in the trigonal plane of the >C=O group. N.m.r. evidence suggests that $C_{(17)}$ is a methine (⩾CH) carbon atom.

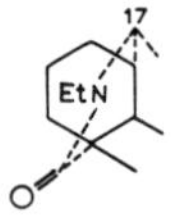

(XLII)

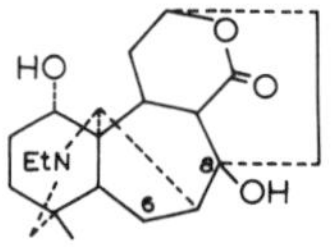

(XLIII)
Heterophylline

A phenanthrenoid-type ring system was rejected because of the behaviour on dehydrogenation with selenium, and consideration was next given to the incorporation of the lactone ring into an aconane framework.

Oxidation of heteratisine with chromic acid in acetic acid afforded dehydroheteratisine, identified as a cyclopentanone by its i.r. absorption; a similar result attended the oxidation of the lactam oxoheteratisine. These two products differ in their u.v. absorption: the dehydro-ketone shows $\lambda_{max.}^{C=O}$ 270 nm, whilst for the keto-lactam the value is 313 nm. This type of behaviour had already been encountered with, for example, dehydrodelpheline and dehydro-oxodelpheline, in the lycoctonine group of alkaloids (see p. 351). Molecular rotation difference measurements emphasize the correspondence between the two pairs and suggest that heteratisine has a 6β-hydroxyl structure analogous to that of delpheline (VIII; p. 354).

A detailed n.m.r. spectral analysis on heteratisine, side-by-side with dehydroheteratisine, led to the suggestion that the lactonic C=O group is attached to $C_{(9)}$, attention being focused on the low-field signal of the $C_{(9)}$-H atom. Its chemical shift requires $C_{(9)}$ to be united with an electron-withdrawing group. Similar analysis revealed that the oxygen atom of the lactone ring is linked to either $C_{(13)}$ or $C_{(16)}$. Heteratisine is not attacked by periodic acid and therefore does not contain a 1,2-glycol grouping; further, the C=O group, the lactonic oxygen atom and the two hydroxyls cannot be located on adjacent carbon atoms. The tertiary hydroxyl group was placed at $C_{(8)}$ as a result of an n.m.r. spectral study of that region of the molecule, particular attention being paid to the signal assigned to the $C_{(9)}$-hydrogen, which appears as a symmetrical doublet and is therefore spin-coupled to only one hydrogen atom. Evidently there is no hydrogen at $C_{(8)}$ and by analogy with numerous other aconitine alkaloids the tertiary hydroxyl is placed there. Careful assignment of signals in the n.m.r. spectrum of heteratisine indicated that the methoxyl group was of the type >CH–OMe. In other alkaloids of this group a methoxyl group is found at positions 1 and/or 16. In heteratisine a methoxyl at $C_{(16)}$ is unlikely because in that position the group would be expected to suffer 1,2-elimination readily with base. Support for placing the group at $C_{(1)}$ includes the fact that heteratisine and its acetate display a C*H*OMe signal at an unexpectedly high field, but in 19-oxodehydroheteratisine the signal is normal. Models reveal that in the latter compound a β-hydrogen at $C_{(1)}$ lies in the deshielding planes of both C=O groups (at positions 6 and 19). These and other studies point to structure XLI for heteratisine, in agreement with the X-ray diffraction analysis result (*Aneja*, *Locke* and *Pelletier*, *loc. cit.*; *Pelletier* and *Aneja*, Tetrahedron Letters, 1965, 215). The absolute stereochemistry depicted in XLI is not proven, but is suggested by analogy with that of other diterpene alkaloids.

Heterophyllidine, $C_{21}H_{31}NO_5$, occurs in small quantity along with heteratisine. It contains a δ-lactone ring, and its structure has been arrived at very largely through mass spectral comparisons with heteratisine (*Pelletier* and *Aneja*, *ibid.*, 1967, 557). Its mass spectrum in the *m/e* 98–300 region is identical with that of heteratisine. The $M^{\oplus}$ ion peak appears at 377, but there is no peak at M−3, and the base peak is at M−17. This suggests that heterophyllidine is *O*-demethylheteratisine (XLI; OMe replaced by OH).

Heterophylline, $C_{21}H_{31}NO_4$, from the same source, has also been investigated by mass spectrometry. There is no peak at B−18 (B=base peak), indicating that one of the hydroxyl groups at position 6 or 8 is missing in the most abundant ion. N.m.r. spectral studies on the alkaloid indicate that there is such a group at $C_{(8)}$ so that heterophylline is XLIII (*Pelletier* and *Aneja*, *loc. cit.*).

Heterophyllisine, $C_{22}H_{33}NO_4$, a third lactonic base from the same source, is formulated as desoxyheteratisine (XLI; H_2 at position 6) on similar grounds (*idem*, *ibid.*).

Lactonic alkaloids are listed in Table 7.

TABLE 7

LACTONIC DITERPENE ALKALOIDS

Alkaloids	*M.p. (°C)*	*[α]D (°) (MeOH)*
Heteratisine	268	+40
Heterophyllidine	270	+42
Heterophylline	222	+10.5
Heterophyllisine	178	+15.5

(d) Daphniphyllum *alkaloids*

The plant *Daphniphyllum macropodum* contains a considerable variety of alkaloids, about thirty having been isolated to date. They are listed in Table 8 (p. 378).

The alkaloids fall into three main skeletal groups, according to the type of nitrogen heterocycle present, which are exemplified by the structures of daphniphylline (XLIV), secodaphniphylline (XLV) and yuzurimine (XLVI). All the *Daphniphyllum* alkaloids contain a 2-azabicyclo[3.3.1]nonane ring system. Investigations on them are largely due to *Y. Hirata* and his collaborators, and most information has been obtained by X-ray diffraction analysis of suitable salts of the bases. The absolute configuration of the alkaloids has emerged from such a study on daphmacrine methiodide (*C. S. Gibbons* and *J. Trotter*, J. chem. Soc. B, 1969, 840).

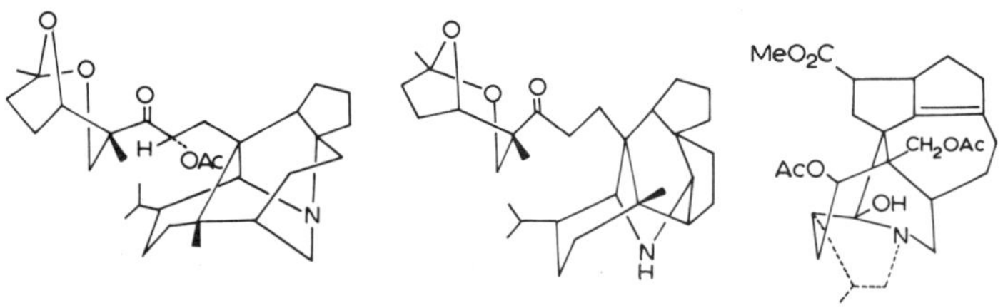

(XLIV) Daphniphylline　(XLV) Secodaphniphylline　(XLVI) Yuzurimine

(i) Daphniphylline subgroup

There are structural differences amongst these bases in the nature of the heterocyclic oxygen unit present, and some are more heavily oxygenated than others. **Daphniphylline** (XLIV) on hydrolysis yields the α-ketol, which on oxidation with periodate affords the acid (XLVII) and the aldehyde (XLVIII) (*H. Irikawa et al.*, Tetrahedron, 1968, **24**, 5691). Sequential reduc-

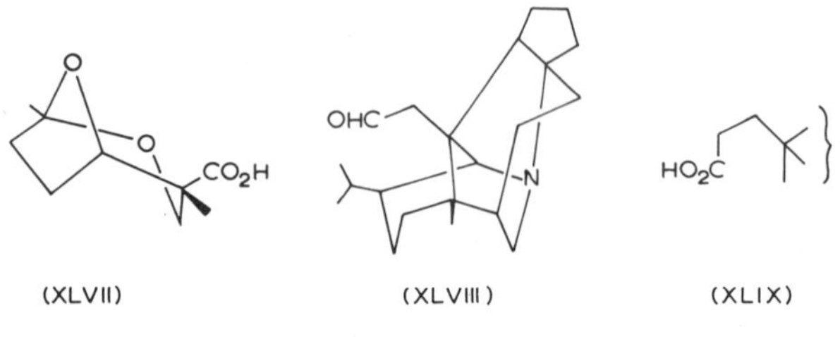

(XLVII)　(XLVIII)　(XLIX)

tion, tosylation, treatment with cyanide ion, and hydrolysis of the latter produced homodaphniphyllic acid (XLIX), the methyl ester of which is another alkaloid of this subgroup (*M. Toda, S. Yamamura* and *Hirata*, Tetrahedron Letters, 1969, 2585; *M. Toda et al.*, Tetrahedron, 1974, **30**, 2683). The mass spectra of members of this subgroup show intense peaks corresponding to two characteristic cleavages, one of which is concerned with the loss of the heterocyclic oxygen unit, and the other with a cleavage within the azabicyclo framework (*Irikawa et al., loc. cit.*).

Daphniphyllidine, $C_{30}H_{47}NO_4$, has been shown on chemical and spectral evidence to be represented by L. It has been correlated with daphniphylline (XLIV), which on hydrolysis to the α-ketol followed by treatment with methanolic sodium methoxide afforded daphniphyllidine, presumably *via* either an enediol intermediate:

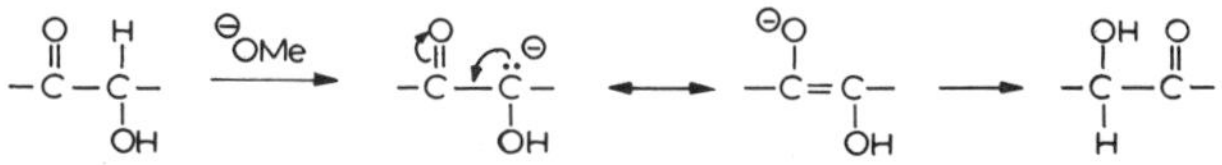

or *via* a 1,2-hydride shift (*Toda et al.*, Tetrahedron Letters, 1973, 797):

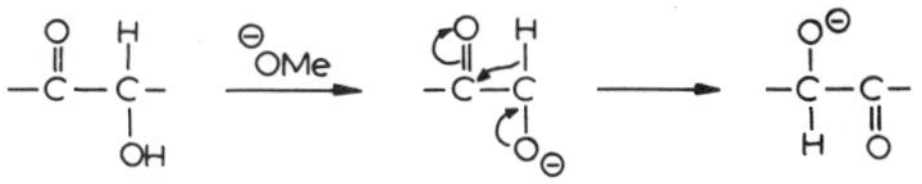

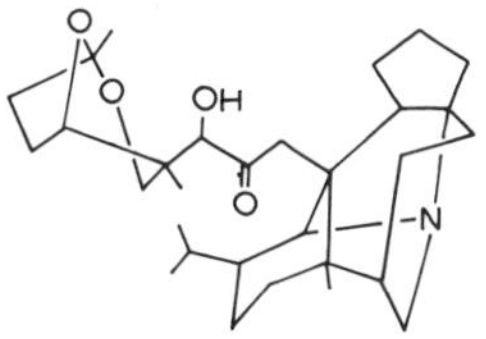

(L)
Daphniphyllidine

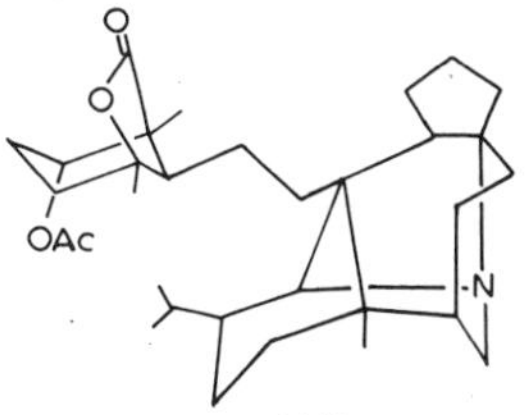

(LI)
Daphmacrine

Daphmacrine, $C_{32}H_{49}NO_4$, is formulated as LI on the basis of X-ray analysis of its methiodide (*Gibbons* and *Trotter, loc. cit.*).

Daphmacropodine, $C_{32}H_{51}NO_4$, (=dihydrodaphmacrine) has been formulated on the basis of its conversion into daphmacrine (LI). On oxidation with Jones reagent daphmacropodine affords daphmacrine, and the former is therefore assigned structure LI, the lactonic C=O being replaced by >C(H)(OH) (*T. Nakano, M. Hasegawa* and *Y. Saeki*, J. org. Chem., 1973, **38**, 2404).

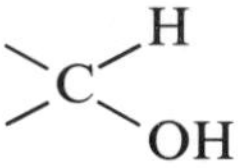

(ii) Secodaphniphylline subgroup

Secodaphniphylline (XLV) resembles daphniphylline (XLIV) in that oxidation gives rise to homosecodaphniphyllic acid, which has a structure analogous to that of homodaphniphyllic acid (XLIX), and of which the methyl ester is also a natural alkaloid (*Toda et al.*, Tetrahedron Letters, 1969, 1821; J. chem. Soc. Japan, 1970, **91**, 103). The two bases have been correlated by conversion into a common degradation product (*Toda, Hirata* and *Yamamura*, Tetrahedron, 1972, **28**, 1477), and their structures and absolute configurations settled unequivocally by an X-ray diffraction analysis of methyl *N*-bromoacetylhomosecodaphniphyllate (*K. Sasaki* and *Hirata*, J. chem. Soc. B, 1971, 1565).

(iii) Yuzurimine subgroup

Macrodaphnine is deoxyyuzurimine *N*-oxide, $C_{27}H_{37}NO_7$. The structures of these bases have generally been elucidated by X-ray diffraction analysis of appropriate salts.

Daphnilactone-A is a minor *Daphniphyllum* base. Spectroscopic study has revealed the presence of an isopropyl group, a tertiary methyl group and a δ-lactone ring. Its structure has been established as LII by X-ray diffraction analysis (*idem*, Tetrahedron Letters, 1972, 1275).

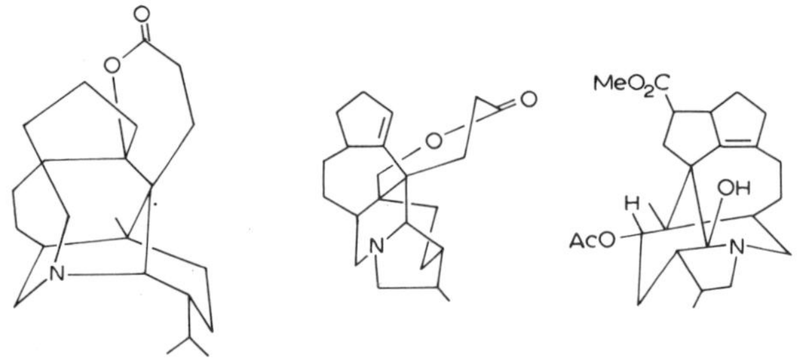

(LII) Daphnilactone-A (LIII) Daphnilactone-B (LIV) Yuzurimine-A

Daphnilactone-B. This alkaloid is a major component from the same source. Spectral study indicated the presence of a secondary methyl group, a vinyl hydrogen and a lactone ring (six- or higher-membered). Its structure LIII has been settled by X-ray diffraction study (*idem*, *ibid*., p. 1891), using the direct phase determination method. Daphnilactone-A and -B may be regarded as constituting a fourth group of *Daphniphyllum* alkaloids, but they are clearly close relatives of yuzurimine (XLVI) (*Toda et al.*, Tetrahedron, 1974, **30**, 2683).

Yuzurimine. The structure of yuzurimine (XLVI) was first revealed by X-ray diffraction analysis of its hydrobromide (*Salurai*, *Salkabe* and *Hirata*, Tetrahedron Letters, 1966, 6309). Subsequent work has shown that this formulation is consonant with the spectral and chemical properties of the alkaloid (*Irikawa*, *Yamamura* and *Hirata*, Tetrahedron, 1972, **28**, 3727). The n.m.r. spectrum of yuzurimine hydrochloride has been analysed carefully; it contains some unusual signals owing to shielding and deshielding of protons.

Yuzurinime-A is a minor *D. macropodum* alkaloid, and has been formulated largely on spectral evidence, particularly as a result of n.m.r. spectral comparison with yuzurimine. It has also been correlated with yuzurimine, and can be assigned structure LIV *(Irikawa et al., loc. cit.)*.

Yuzurimine-B, another minor base, also owes its structure elucidation largely to spectral study and comparisons with yuzurimine. It is formulated as LV *(idem, loc. cit.)*.

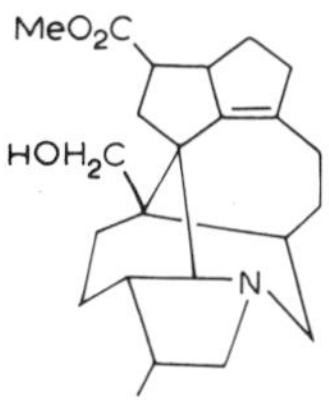

(LV)
Yuzurimine-B

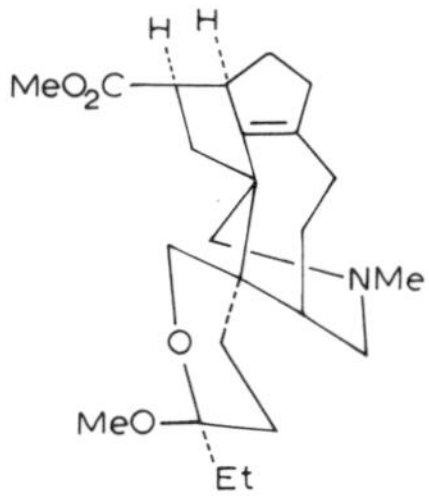

(LVI)
Yuzurine

Yuzurine, $C_{24}H_{37}NO_4$, is found in the bark and leaves of *D. macropodum.* Its mass spectrum is in agreement with the assigned molecular formula ($M^{\oplus}$ 403). Its n.m.r. spectrum revealed the presence of an *N*-methyl group, two methoxyl groups (one part of a CO_2Me group); its methiodide had two *N*-methyl signals, and yuzurine is consequently an *N*-methyl tertiary base. Certain features of the structure were deduced from spectral analyses, but the detailed structure and relative configuration was established by an X-ray diffraction analysis of yuzurine methiodide. This pointed to structure LVI for the alkaloid (*Yamamura et al.*, Tetrahedron Letters, 1974, 2023). Yuzurine is apparently identical with alkaloid A_2 (see Table 8) (*Toda*, *Hirata* and *Yamamura*, Tetrahedron, 1972, **28**, 1477).

For further details concerning *Daphniphyllum* bases see *Yamamura* and *Hirata*, in "The Alkaloids", ed. *Manske*, Vol. XV, Chap. 2, Academic Press, New York, 1975.

TABLE 8

DAPHNIPHYLLUM ALKALOIDS

Alkaloid	*M.p. (°C)*	*Molecular formula*	*Ref.*
Daphniphylline (daphniphyllamine)	240 (B · HCl)	$C_{32}H_{49}NO_5$	1
Codaphniphylline	267 (B · HCl)	$C_{30}H_{47}NO_3$	1
Daphmacrine	> 300 (B · HBr)	$C_{32}H_{49}NO_4$	2
Daphnimacropine	307 (B · MeI)	$C_{31}H_{49}NO_3$	3
Daphmacropodine	215	$C_{32}H_{51}NO_4$	3
Secodaphniphylline	130	$C_{30}H_{47}NO_3$	4
Daphniphyllidine	264	$C_{30}H_{47}NO_4$	4
Yuzurimine (macrodaphnidine)	253 (B · HCl)	$C_{27}H_{37}NO_7$	5
Yuzurimine-A	252 (B · HCl)	$C_{25}H_{35}NO_5$	6
Yuzurimine-B	284 (B · HCl)	$C_{23}H_{33}NO_3$	6
Macrodaphniphyllamine	153	$C_{23}H_{33}NO_4$	3
Macrodaphnine	181.5	$C_{27}H_{37}NO_7$	7
Macrodaphniphyllidine	306 (B · HBr)	$C_{25}H_{35}NO_4$	3
Yuzurimine-C	187	$C_{23}H_{29}NO_5$	4
Yuzurimine-D	195	$C_{24}H_{31}NO_5$	4
Alkaloid A_1	226 (B · MeI)	$C_{23}H_{33}NO_3$	4
Alkaloid A_2 (yuzurine)	230 (B · MeI)	$C_{24}H_{37}NO_4$	4,11
Methyl homodaphniphyllate	234 (B · HCl)	$C_{23}H_{37}NO_2$	8
Methyl homosecodaphniphyllate	103	$C_{23}H_{37}NO_2$	9
Daphnilactone-A	195.5	$C_{23}H_{35}NO_2$	4
Neoyuzurimine	198	—	4
Neodaphniphylline	244	—	4
Daphnilactone-B	92–94	$C_{22}H_{31}NO_2$	10

References
1 *H. Irikawa, N. Sakabe, S. Yamamura* and *Y. Hirata*, Tetrahedron, 1968, **24**, 5691.
2 *T. Nakano, Y. Saeki, C. S. Gibbons* and *J. Trotter*, Chem. Comm., 1968, 600.
3 *Nakano* and *Saeki*, Tetrahedron Letters, 1967, 4791.
4 *M. Toda, Irikawa, Yamamura* and *Hirata*, J. chem. Soc. Japan, 1970, **91**, 103; *Toda, H. Nirva, Hirata* and *Yamamura*, Tetrahedron Letters, 1973, 797.
5 *H. Sakurai, Sakabe* and *Hirata, ibid.*, 1966, 6309.
6 *Sakurai et al., ibid.*, 1967, 2883.
7 *Nakano* and *B. Nilsson, ibid.*, 1969, 2883.
8 *Toda, Yamamura* and *Hirata, ibid.*, 1969, 2585.
9 *Irikawa et al., ibid.*, 1969, 1821.
10 *K. Sasaki* and *Hirata, ibid.*, 1972, 1891.
11 *Yamamura et al., ibid.*, 1974, 2023.

3. Bis-diterpene alkaloids

Three bis-diterpene alkaloids have been isolated from *Delphinium staphisagria*, designated **staphidine,** $C_{42}H_{58}N_2O$, **staphinine**, $C_{42}H_{56}N_2O_2$, and **staphimine**, $C_{41}H_{54}N_2O$. Their structures have been established by ^{1}H- and ^{13}C-n.m.r. spectroscopy to be as represented by LVII, LVIII and LIX, respectively (*S. W. Pelletier et al.*, Tetrahedron Letters, 1976, 1055).

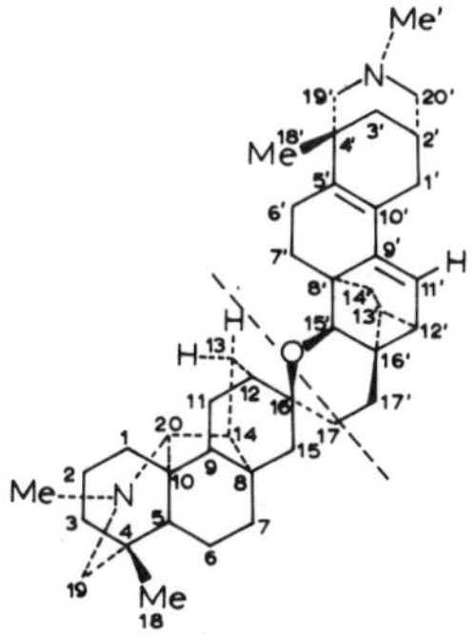

(LVII)
Staphidine

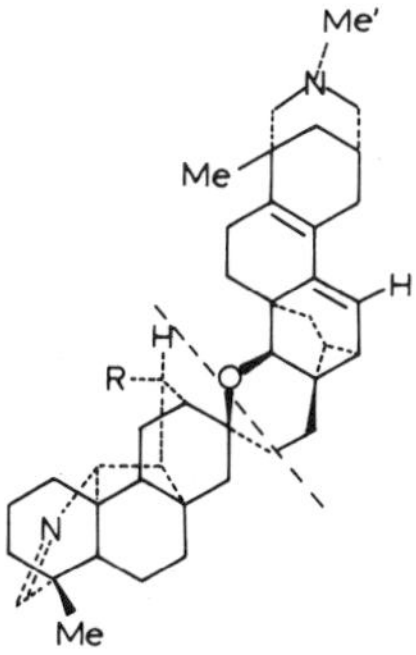

(LVIII; R = –OMe) Staphinine
(LIX; R = –H) Staphinine

Staphidine has m.p. 213–216°, $[\alpha]_D^{24}$ −160° (c=2.0, benzene), $\lambda_{max.}^{EtOH}$ 268 nm (ε 17,300). *Staphinine* is amorphous, $[\alpha]_D^{24}$ −57.5° (c=1.0, benzene), $\lambda_{max.}^{EtOH}$ 268 nm (ε 17,300) and *staphimine* also is amorphous, $[\alpha]_D^{24}$ −58.5° (c=1.0, benzene), u.v. absorption similar to the other two bases.

Chapter 35

Steroidal Alkaloids

A. R. PINDER

Steroidal alkaloids may be defined as natural bases possessing the steroidal carbon framework (basic or modified) in which a nitrogen atom is attached to the polycyclic system or to the side-chain. In the latter event the nitrogen may be part of an additional ring. For discussion the compounds are divided into two main groups:

(*1*) Alkaloids containing a normal steroid skeleton, with a nitrogen atom in the side-chain, attached to the polycyclic system or part of an additional ring; these bases are related to cyclopentenophenanthrene.

(*2*) Alkaloids containing a modified steroid skeleton, with a nitrogen atom in the side-chain, frequently as part of a ring system.

It should be noted that some of these bases do not satisfy structurally the generally accepted definition of an alkaloid (at least one nitrogen atom to be part of a heterocyclic system). Consequently some of them might be better described as steroidal bases. Several reviews of this subject have appeared (see, for example "The Alkaloids," eds., *R. H. F. Manske* and *H. L. Holmes*, Academic Press, New York and London, Vols., III, VII, IX, X and XIV; "Chemistry of the Alkaloids", ed. *S. W. Pelletier*, van Nostrand and Reinhold, New York and London, 1970, Chapter 2 (by *G. Habermehl*); *L. F.* and *M. Fieser*, "Steroids", Reinhold, New York, 1959, Chapter 22; *H. G. Boit*, "Ergebnisse der Alkaloid-Chemie bis 1960", Akademie-Verlag, Berlin, 1961; *R. Tschesche*, Fortschr. Chem. org. Naturst., 1966, **24**, 99). Synthetic endeavours in this area have been reviewed by *G. Adam* (Z. chem., 1963, **3**, 403). Recent developments are summarised in Specialist Periodical Reports, Alkaloids, Vols. 1–6 (Chemical Society, London, 1968–1976).

1. Normal steroidal alkaloids

The normal steroidal bases are best divided into five sub-groups for discussion. They are all derived from the steroidal hydrocarbon 5α-pregnane (allopregnane), and differ from one another principally in the number and positions of amino-groups attached to the carbon framework.

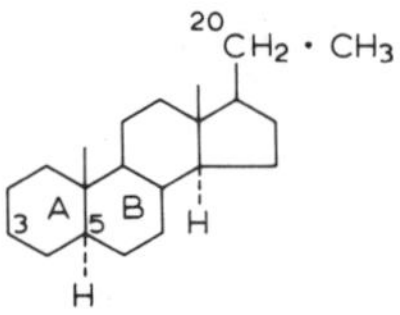

5α-Pregnane (allopregnane)

When there is a hydrogen atom at $C_{(5)}$, it has the α-configuration: that is, rings A and B are *trans* fused. The five subgroups are: (*a*) bases related to 3-amino-5α-pregnane; (*b*) bases related to 20-amino-5α-pregnane; (*c*) bases related to 3,20-diamino-5α-pregnane; (*d*) bases related to 3-aminoconanine; and (*e*) bases related to 20-piperidyl-5α-pregnane. Differences between related bases are often stereochemical.

Two reactions which have been commonly used in structure determination in this area are the Eschweiler–Clarke methylation, whereby primary amines and *N*-methyl secondary amines are fully methylated with formaldehyde and formic acid to *N,N*-dimethyl tertiary bases (*H. T. Clarke, H. B. Gillespie* and *S. Z. Weisshaus*, J. Amer. chem. Soc., 1933, **55**, 4571; *R. N. Icke, B. B. Wisegerver* and *G. A. Alles*, Org. Synth., 1955, Coll. Vol. 3, p. 723), and the Ruschig degradation, by means of which an amino-group is converted into an oxo group via the chloroamine and ketimine (*H. Ruschig et al.*, Chem. Ber., 1955, **88**, 883).

(*a*) Bases related to 3-amino-5α-pregnane

These bases are derived from 3α-amino-5α-pregnane (I) or its 3β-epimeride (II).

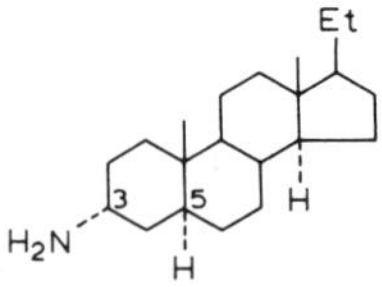

(I) 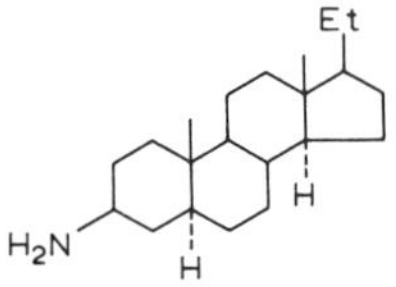 (II) 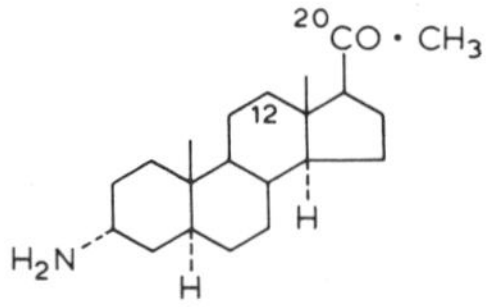 Funtumine (III)

Funtumine, $C_{21}H_{35}NO$, is a constituent alkaloid of *Funtumia latifolia*, a *Holarrhena* species (*M.-M. Janot, Q. Khuong-Huu* and *R. Goutarel*, Compt. rend., 1958, **246**, 3076). It is a methyl ketone and a primary amine. On Eschweiler–Clarke methylation it affords an *N,N*-dimethyl tertiary base which on Wolff–Kishner reduction yields the known 3α-dimethylamino-5α-pregnane (*H. Favre et al.*, J. chem. Soc., 1953, 1115). Ruschig degradation of funtumine furnishes 5α-pregnane-3,20-dione, the formation of which enables the keto-group in the alkaloid to be placed at position 20 (*Janot et al., loc. cit.*; Compt. rend., 1959, **248**, 982). Finally, Wolff–Kishner reduction of the alkaloid itself generates 3α-amino-5α-pregnane *(idem, loc. cit.)*, and it must therefore be formulated as 3α-amino-5α-pregnan-20-one (III). **Bokitamine,** occurring in the leaves of *H. wulfsbergii*, is 12β-hydroxyfuntumine (*S. Nellé et al.*, Compt. rend., 1970, **271C**, 153).

Funtumidine, $C_{21}H_{37}NO$, also present in *F. latifolia*, is the secondary alcohol corresponding to funtumine, into which it is transformed by oxidation with chromic acid *(Janot et al., loc. cit.)*. Conversely, reduction of funtumine with sodium and ethanol affords funtumidine. The configuration of the asymmetric centre at position 20 was established by the observation that Ruschig degradation of funtumidine yields 20α-hydroxy-5α-pregnan-3-one, identical with authentic material, and further by the fact that funtumine on catalytic reduction or reduction with tetrahydridoborate affords the 20β-epimer (isofuntumidine). Ruschig degradation of the latter afforded 20β-hydroxy-5α-pregnan-3-one. A study of the molecular rotation shifts accompanying acetylation of these epimeric keto-alcohols enabled them to be distinguished, and configurations assigned. Funtumidine is therefore 3α-amino-20α-hydroxy-5α-pregnane (*idem, loc. cit.*; Bull. Soc. chim. Fr., 1960, 1669).

Kurchiline, $C_{23}H_{37}NO_2$, occurs in the leaves of *Holarrhena antidysenterica*, and is an *N,N*-dimethyl tertiary base, a methyl ketone and a secondary alcohol, and contains one olefinic linkage, most of these features being detected by i.r. and n.m.r. spectroscopy. Strong intramolecular hydrogen bonding, revealed by the former, indicated the presence of a 1,2-amino-alcohol grouping. The mass spectrum, showing a strong peak at *m/e* 100, as opposed to 84 found with steroids carrying an isolated NMe group at position 3, is in accord with the location of the secondary alcohol group at $C_{(2)}$. On catalytic hydrogenation (uptake H_2), followed by Wolff–Kishner reduction, 3β-dimethylamino-5α-pregnan-2α-ol resulted; this compound was synthesised from the known 2α-acetoxy-5α-pregnan-3-one oxime by reduction with lithium tetrahydridoaluminate, followed by *N*-methylation. Kurchiline must therefore be formulated as IV (*Janot, P. Longevialle* and *Goutarel*, Bull. Soc. chim. Fr., 1964, 2158).

Kurchiline (IV) Kurchiphylline (V) Kurchaline (VI) Holadysine (VII)

Kurchiphylline, isomeric with kurchiline and from the same source, differs from the latter in the location (1748 cm $^{-1}$) of its C=O band in the i.r. spectrum, this frequency being diagnostic of a cyclopentanone. The strong negative optical rotation of the alkaloid, in comparison with known steroid structures, suggested the keto group is at $C_{(16)}$. Finally, Wolff–Kishner reduction affords the same base as does kurchiline. Kurchiphylline is consequently formulated as V (*idem, loc. cit.*, and *ibid.*, 1966, 1212).

Kurchiphyllamine, a congener of the foregoing two bases, is *N*-demethylkurchiphylline, since it is a secondary base which on Eschweiler–Clarke methylation yields kurchiphylline *(Janot et al., loc. cit.)*.

Kurchaline, $C_{23}H_{37}NO_2$, also from *H. antidysenterica*, contains two hydroxyl groups (one being allylic), a dimethylamino group, and two trisubstituted double bonds, on spectral evidence. Largely by comparisons of its n.m.r. spectrum with those of 3,16-dihydroxypregna-5,17(20)-dienes of known structure and stereochemistry, the structure VI has been assigned to it (*Janot et al., loc. cit.*, p. 1212; *W. R. Benn* and *R. M. Dodson*, J. org. Chem., 1964, **29**, 1142).

Holadysine, $C_{22}H_{35}NO$, from the same source, is a secondary keto base, the C=O stretching frequency (1742 cm^{-1}) of which is indicative of a cyclopentanone. The carbonyl group is placed at $C_{(16)}$ on mass spectral evidence, and a trisubstituted double bond present is placed in the "cholesterol position". Structure VII is confirmed by structural correlation with another related alkaloid **holamine** (see below): *N*-acetylholadysine on Wolff–Kishner reduction affords 3α-methylaminopregn-5-ene, identical with a product obtainable from this base *(Janot et al., loc. cit.)*.

Holadysamine, a congener, is isomeric with holadysine. It contains two trisubstituted double bonds, and secondary alcoholic and methylamino functions. N.m.r. evidence has been used to locate the unsaturation at the 5,6 and 16,17 positions, by comparison with steroids of known structure, in particular 3,20-dihydroxypregna-5,6-dienes; it was also possible thus to place the hydroxyl group at $C_{(20)}$. A negative increment in $[M]_D$ on acetylation of this group leads to the assignment to it of the α-configuration, and holadysamine is formulated as VIII *(Janot et al., loc. cit.)*.

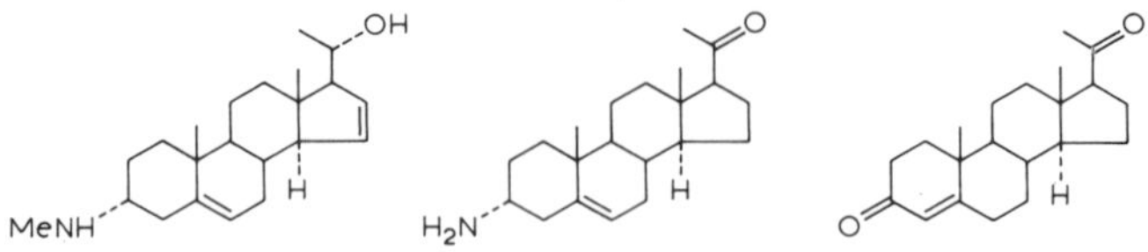

Holadysamine (VIII) Holamine (IX) Progesterone

Holamine, $C_{21}H_{23}NO$, found in *H. africana* leaves, is a ketonic primary amine. A Ruschig reaction (see above) yields progesterone, and hydrogenation furnishes funtumine (III). Eschweiler–Clarke methylation, followed by Wolff–Kishner reduction, gives the known 3α-dimethylaminopregn-5-ene. These observations point to structure IX for the base (*Janot, A. Cavé* and *Goutarel*, Compt. rend., 1960, **251**, 559). This formulation was confirmed by synthesis of the alkaloid from pregnenolone (X, R=H), the tosylate (X, R=Ts) of which reacted with sodium azide (S_N2 displacement) to give the 3α-azide. Reduction with lithium tetrahydridoaluminate, followed by re-oxidation of the $C_{(20)}$ hydroxyl group, afforded holamine (IX) (*Goutarel et al.*, Bull. Soc. chim. Fr., 1962, 646).

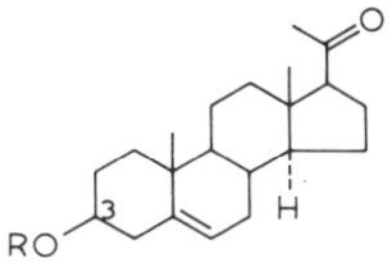

(X)

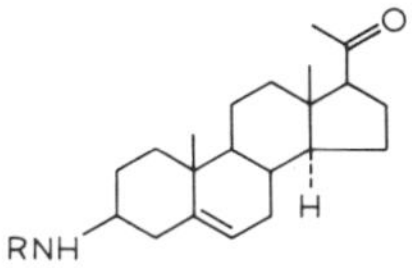

Holaphylline (R = Me)
Holaphyllamine (R = H)
(XI)

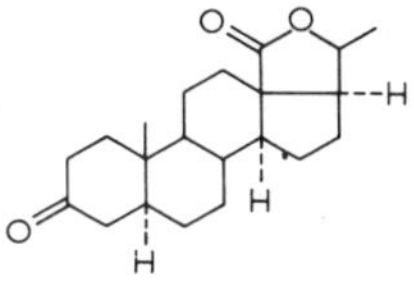

(XII)

Holaphylline, $C_{22}H_{35}NO$, also found in *H. africana*, contains a keto group, an *N*-methyl secondary basic group, and a trisubstituted double bond. Wolff–Kishner reduction gives 3β-methylaminopregn-5-ene, a known compound, and Ruschig degradation affords progesterone. These and related transformations allow holaphylline to be formulated as XI (R=Me) (*Janot, Cavé* and *Goutarel, ibid.*, 1959, 896).

Holaphyllamine, $C_{21}H_{33}NO$, is an amorphous base from the same source. It is the primary amine XI (R=H) corresponding to holaphylline, since Eschweiler–Clarke methylation of both bases leads to the same *N,N*-dimethyl tertiary base *(idem, loc. cit.; Goutarel et al., loc. cit.)*.

Holaphyllidine, a new alkaloid of this group, is believed to be 3β-methylaminopregn-5-en-20β-ol (*Leboeuf, Cavé* and *Goutarel*, Compt. rend., 1964, **259**, 3401). **Holaphyllinol** is its 20α-epimeride (*idem*, Ann. pharm. franç., 1969, **27**, 217); it occurs in *H. floribunda* leaves.

Paravallarine, $C_{22}H_{33}NO_2$, is the chief alkaloid of *Paravallario microphylla.* It is a secondary *N*-methyl base and also contains a γ-lactone moiety (ν_{co} 1766 cm^{-1}), confirmed by reduction with lithium tetrahydridoaluminate, which furnishes a diol, $C_{22}H_{37}NO_2$. Further, hydrolysis yields a hydroxy acid, convertible to a methyl ester. Quantitative hydrogenation reveals the presence of one olefinic bond, yielding a dihydro-base, Ruschig degradation of which converted it into a known lactone XII (*B. Kamber et al.*, Helv.,

1960, **43**, 347). The double bond is placed in the 5,6 position as a consequence of the (+)-shift in optical rotation accompanying hydrogenation; also, Ruschig degradation of the base itself afforded a Δ^4-enone. The methylamino group at $C_{(3)}$ is given the β-configuration because the hydrogenation referred to above yields only the 5α-dihydro base; experience shows that some 5β-isomer would be expected if the 3-substituent were α. Optical rotation values are in agreement with a 3β-methylamino group, and paravallarine is formulated as XIII (*J. Le Men*, Bull. Soc. chim. Fr., 1960, 860). Both 2α-hydroxy-*N*-methylparavallarine (**lantine**) (*A. Cavé, P. Potier* and *Le Men*, Ann. pharm. franç., 1967, **25**, 107) and the 2β-epimeride (*R. M. Bernol-Santos*, C. A., 1971, **74**, 10364p) have been isolated from the stem bark of *Kibatalia gitingensis.*

Paravallaridine, $C_{22}H_{33}NO_3$, is a minor alkaloid of *P. microphylla*, and is closely related to the foregoing base, having in addition to the same functional groups a secondary alcoholic group. This is located in ring D since oxidation leads to a ketone with i.r. absorption at 1745 cm^{-1}. Further, it is placed at $C_{(16)}$ on optical rotatory evidence, the oxidation being accompanied by a marked laevorotatory shift characteristic of 16-ketones. It is possible to convert the alkaloid into *N*-methyldihydroparavallarine, providing a link-up between the alkaloids and permitting structure XIII (R=OH) to be advanced for paravallaridine, in which the α-configuration of the hydroxyl group is assigned on optical rotatory difference grounds (*Huang-Minlon* and *Chung-Tungshun*, Tetrahedron Letters, 1961, 666; *Le Men, C. Kan* and *R. Beugelmans*, Bull. Soc. chim. Fr., 1963, 597).

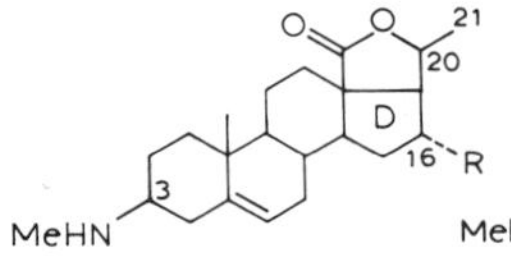

Paravallarine (R = H)
Paravallaridine (R = OH)
(XIII)

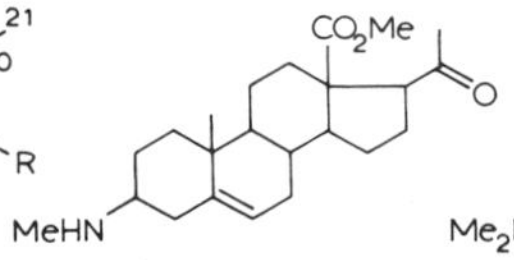

(XIV)

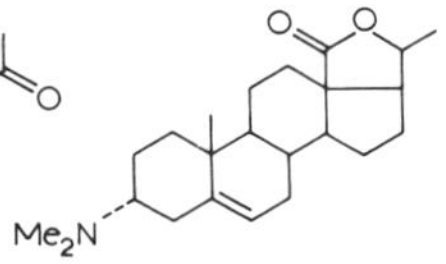

Kibataline
(XV)

20-*epi*-*N*-**Methylparavallarine,** occurring in *Kibatalia gitingensis*, is, as its name implies, an *N*-methylated $C_{(20)}$ epimeride of paravallarine. The epimeric arrangement at $C_{(20)}$ was inferred from a study of the n.m.r. downfield chemical shifts of the $C_{(21)}$ methyl group and the $C_{(20)}$ hydrogen atom, compared with paravallarine. Confirmation of this structure has been provided by a synthesis of the base from paravallarine: alkaline hydrolysis followed by treatment with diazomethane and oxidation yielded the keto-ester (XIV). Reduction of the keto group with sodium tetrahydridoborate, then *N*-methylation with formaldehyde–formic acid, which also induced lactonisation, gave a mixture from which the base could be isolated (*Cavé*, *P. Potier* and *Le Men*, Bull. Soc. chim. Fr., 1965, 2502).

Kibataline, $C_{23}H_{35}NO_2$, found in the same plant, is a tertiary *N,N*-dimethyl base which, spectrally, is almost indistinguishable from *N*-methylparavallarine. In particular, the mass spectra are virtually identical, suggesting that the difference between the two bases is of a stereochemical nature. Reduction of the 5,6 double bond in kibataline affords two isomers, 5α and 5β, pointing strongly to the α-configuration of the basic function at $C_{(3)}$. Structure XV has thus been advanced for kibataline and has been confirmed by partial syntheses of its 5α-dihydro derivative (*Cavé et al., ibid.*, 1964, 2415).

Two alkaloids, **dictyolucidine** and **dictyolucidamine** have a D-homosteroid carbon framework, but are conveniently included here. They are, respectively, $C_{22}H_{39}NO_2$ and $C_{23}H_{41}NO_2$, and occur in the roots of *Dictyophleba lucida.* Both are monoacidic bases, and dictyolucidamine is an *N,N*-dimethyl tertiary base, and is the *N*-methyl derivative of dictyolucidine, and *N*-methyl secondary base. Each is saturated, contains three *C*-methyl groups (Kuhn–Roth analysis), and is a secondary-tertiary 1,2-glycol. Eschweiler–Clarke methylation of dictyolucidine yields an *N,N*-dimethyl derivative in which the two hydroxyl groups have been replaced by a methylenedioxy group; the same product results from dictyolucidamine. Its formation indicates that the hydroxyl groups are favourably disposed stereochemically for bridging (*e.g. cis* in a cyclic system). This compound XVI has been synthesised from 17α-hydroxyprogesterone as follows:

17α-Hydroxyprogesterone —base→ —Me_2NH→ —$NaBH_4$→ 17α + 17β —$HCHO, HCO_2H$ or *cis*→ —Pd—H_2→ (XVI)

The alkaloids must therefore be formulated as XVII and XVIII (*M. Truong-Ho et al.*, Bull. Soc. chim. Fr., 1963, 2332; *Janot et al.*, Compt. rend., 1965, **260**, 6118):

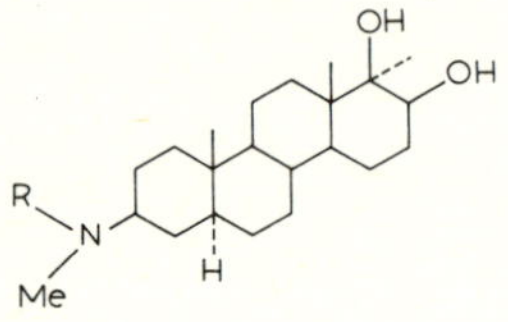

(XVII; R = H) Dictyolucidine
(XVIII); R = Me) Dictyolucidamine

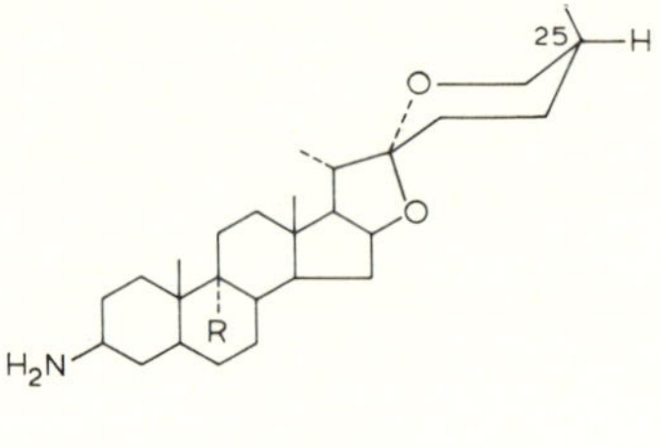

Jurubidine (R = H)
Paniculidine (R = OH)
(XIX)

Jurubidine, $C_{27}H_{45}NO_2$, representative of a novel type of basic sapogenin, is found in *Solanum paniculatum* and *S. tonrum*, where it occurs as its glycoside **jurubine**. Mass and infrared spectroscopic analyses have figured prominently in structural studies, pointing to a 3β-amino-5α-,22α-*O*-spirostane structure (XIX; R=H). The β-orientation of the amino group was settled by molecular rotation difference comparisons with model compounds of known stereochemistry, as well as by o.r.d. measurements (*K. Schreiber, H. Ripperger* and *H. Budzikiewicz*, Tetrahedron Letters, 1965, 3999; Chem. Ber., 1967, **100**, 1725; *Schreiber* and *Ripperger*, *ibid.*, 1966, 5997). Partial synthetic studies have confirmed the structural assignment (*Schreiber et al., loc. cit.; M. E. Wall, T. Perlstein* and *S. G. Levine*, J. Amer. chem. Soc., 1960, **82**, 1444). The sugar unit in jurubine is D-(+)-glucose, which is bonded to the masked OH group at $C_{(26)}$ *(Schreiber et al., loc. cit.).*

Paniculidine, $C_{27}H_{45}NO_3$, occurring as its glycoside **paniculine** in *S. paniculatum*, contains a primary amino group, and its mass spectral fragmentation pattern suggested it was a hydroxyjurubidine, with the OH group at positions 9, 10 or 14. The chemical shifts of the angular methyl groups in the n.m.r. spectrum of the alkaloid were consistent only with the presence of a 9α hydroxyl function, so that paniculidine is 9α-hydroxyjurubidine (XIX, R=OH) (*K. Meyer* and *F. Bernouilli*, Pharm. Acta Helv., 1961, **36**, 80; *Schreiber et al., loc. cit.*). N.m.r. and i.r. spectral studies suggest paniculidine may well be a mixture of $C_{(25)}$-epimerides.

Three new alkaloids have been isolated from the roots of *S. paniculatum*; they are **isojuripidine** (XX), **isojuribidine** (XXI) and **isopaniculidine** (XXII) (*S. Cambiaghi, E. Dradi* and *R. Longo*, Ann. Chim., Italy, 1971, 61, 99):

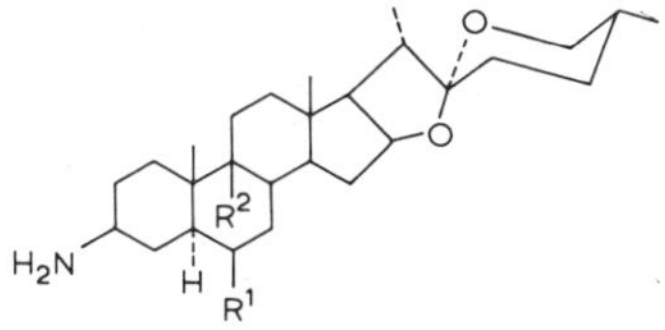

Isojuripidine (XX) (R^1 = OH, R^2 = H)
Isojuribidine (XXI) (R^1 = R^2 = H)
Isopaniculidine (XXII) (R^1 = H, R^2 = OH)

(b) Bases related to 20-amino-5α-pregnane

The 20-aminopregnane alkaloids are derived from the annexed structure. Mass spectrometry has proved a valuable tool for investigation, since the bases break down to yield intense ions of *m/e* $R^1 + R^2 + 42$, as shown (*L. Dolejs et al.*, Coll. Czech. chem. Comm., 1963, **28**, 1584; *W. Vetter et al.*, Bull. Soc. chim. Fr., 1963, 1324).

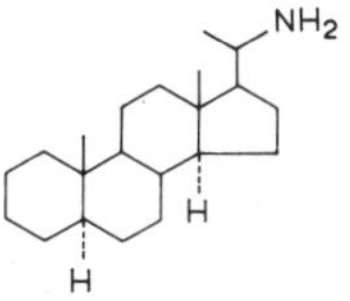

$$CH_3CH{=}\overset{\oplus}{N}R^1R^2$$

$$m/e = R^1 + R^2 + 42$$

20-Amino-5α-pregnane

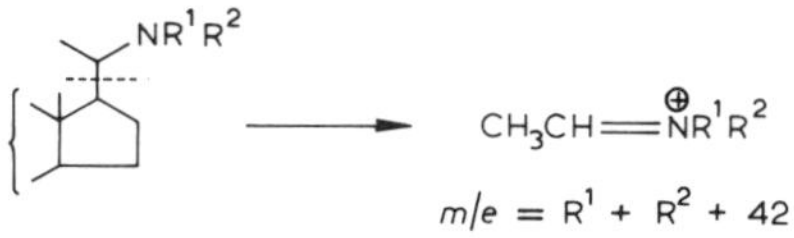

The **funtuphyllamines, A, B** and **C**, are respectively primary ($C_{21}H_{37}NO$), *N*-methyl secondary ($C_{22}H_{39}NO$) and *N,N*-dimethyl tertiary ($C_{23}H_{41}NO$) bases, all found in *Funtumia africana*. Bases B and C also occur in *Malouetia bequaertiana*, B in *M. glandulifera* and C in *Funtumia elastica*. Both A and B on *N*-methylation yield C. Base B proved to be identical with the already known 20α-methylamino-5α-pregnan-3β-ol (*P. Buchschacher et al.*, J. Amer. chem. Soc., 1958, **80**, 2905); this is in accord with the Ruschig degradation of the base to 3β-hydroxy-5α-pregnan-20-one, also known. The three alkaloids must therefore be formulated as XXIII ($R^1 = R^2 =$ H), XXIII ($R^1 =$ H, $R^2 =$ Me) and XXIII ($R^1 = R^2 =$ Me), respectively. Base A has been converted into B by *N*-formylation. followed by reduction with lithium tetrahydridoaluminate and it has also been synthesised by reduction of the oxime of 3β-hydroxy-5α-pregnan-20-one (see above) (*M. M. Janot, Q. Khuong-Huu* and *R. Goutarel*, Compt. rend., 1960, **250**, 2445; *Janot et al.*, Bull. Soc. chim. Fr., 1962, 111).

Funtuphyllamine-A ($R^1 = R^2 = H$)
Funtuphyllamine-B ($R^1 = H, R^2 = Me$)
Funtuphyllamine-C ($R^1 = R^2 = Me$)

(XXIII)

Funtumafrine-B ($R^1 = H, R^2 = Me$)
Funtumafrine-C ($R^1 = R^2 = Me$)

(XXIV)

Funtumafrine-B and **-C,** both present in *F. africana* leaves are respectively the ketonic oxidation products of funtuphyllamine-B and -C, being easily obtained from the latter by oxidation with chromic acid *(idem, loc. cit.)*. They are represented by XXIV ($R^1 = H$, $R^2 = Me$) and XXIV ($R^1 = R^2 = Me$), respectively.

Holafebrine, $C_{21}H_{35}NO$, is found in *Funtumia latifolia, Holarrhena febrifuga* and *Kibatalia arborea.* It is 20α-amino-5-pregnen-3β-ol, because it can be converted into funtuphyllamine-A (see above) by simple catalytic hydrogenation of the 5,6 double bond. Its 3β-D-glucoside occurs in the roots of *Conopharyngia pachysiphon* and is named **conopharyngine** (*Janot et al.*, Bull. Soc. chim. Fr., 1962, 285). Holafebrine is therefore XXIII ($R^1 = R^2 = H$, $\Delta^{5,6}$).

Irehamine, $C_{22}H_{37}NO$, bears a similar relationship to funtuphyllamine-B, into which it is converted by catalytic hydrogenation. It must therefore be assigned structure XXIII ($R^1 = H$, $R^2 = Me$, $\Delta^{5,6}$).

Irehine, $C_{23}H_{39}NO$, like irehamine, has been isolated from *Funtumia elastica* leaves, and also from *Buxus sempervirens* leaves. It is related to funtuphyllamine-C in the same way as irehamine is related to funtuphyllamine-B, and must therefore be formulated as XXIII ($R^1 = R^2 = Me$, $\Delta^{5,6}$) (*M. Truong-Ho et al., ibid.*, 1963, 2332; *Z. Voticky* and *J. Tomko*, Coll. Czech. chem. Comm., 1965, **30**, 348; *Tomko et al.*, Chem. Zvesti, 1964, **18**, 721).

Terminaline, $C_{23}H_{41}NO_2$, found in *Pachysandra terminalis* is a 1,2-glycol, since it is attacked by periodate to yield a dialdehyde. It is also an *N,N*-dimethyl tertiary base, and the basic group is placed at $C_{(20)}$ because of the appearance of an intense peak at m/e 72 (=42+30) in the mass spectrum. The two hydroxyl groups are located at $C_{(3)}$ and $C_{(4)}$, because the dialdehyde oxidation product on reduction ($NaBH_4$) followed by acetylation affords a diacetate identical with that obtainable from the diosphenol XXV, of known structure, by a similar sequence. Finally, the hydroxyl groups are *trans* oriented, because terminaline does not form an acetonide, and further, they are both equatorial because the n.m.r. signal of the $C_{(19)}$ methyl group has a chemical shift ruling out 1,3-diaxial interaction between it and a hydroxyl group at $C_{(4)}$. Terminaline is consequently XXVI (*T. Kikuchi, S. Uyeo* and *T. Nishinaga*, Tetrahedron Letters, 1965, 1933).

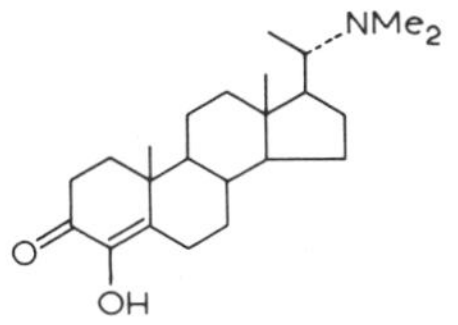

(XXV)

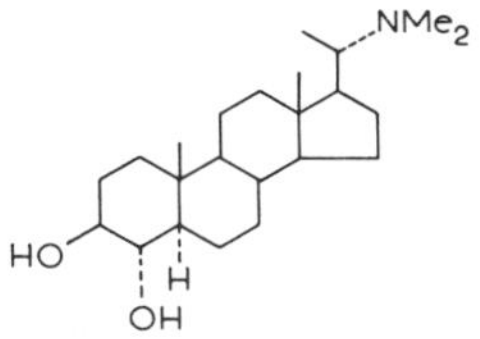

Terminaline
(XXVI)

3α-Methoxy-20α-dimethylaminopregn-5-ene has been found in *Sarcococca pruniformis*. Its structure was established by Emde degradation and catalytic hydrogenation to 3α-methoxy-5α-pregnane, identical with an authentic sample prepared from 5α-pregnane-3-one (*J. M. Kohli, A. Zaman* and *A. R. Kidwai*, Phytochem., 1971, **10**, 442).

(c) Bases related to 3,20-diamino-5α-pregnane

Irehdiamine-A, -B and **-C** are three closely related bases conveniently considered together. The first occurs in *Funtumia elastica* leaves, the second in the same plant and in *Holarrhena antidysenterica*, and the last in *F. latifolia*. Base A is diprimary, B is primary-*N*-methyl secondary, and C is primary-*N,N*-dimethyl tertiary. All three bases on total methylation afford the same ditertiary base, and since both A and B, on Ruschig degradation, yield progesterone (see p. 384), the amino groups of all three bases must be located at positions 3 and 20 in the pregnane skeleton. Each base contains a double bond, placed at $C_{(5)}$–$C_{(6)}$ on molecular rotation difference evidence. Since the stereochemistry of the tetramethyl derivative described above is known to be 3β,20α, it follows that base A must be assigned structure XXVII ($R^1 = R^2 = R^3 = R^4 = H$). Base B, which is a monomethyl derivative of A, must be formulated as XXVII ($R^1 = R^3 = R^4 = H$, $R^2 = Me$) rather than the other possibility because its dihydro-derivative is formed when dihydroholaphylline (see p. 385) is converted into its oxime which is reduced to mixture of 20α- and 20β- diamines. One of these proved to be identical with dihydroirehdiamine-B (*M. Truong-Ho, Q. Khuong-Huu* and *R. Goutarel*, Bull. Soc. chim. Fr., 1963, 594; *V. Černy, L. Labler* and *F. Šorm*, Coll. Czech. chem. Comm., 1959, **24**, 378; Chem. Listy, 1957, **51**, 2351;

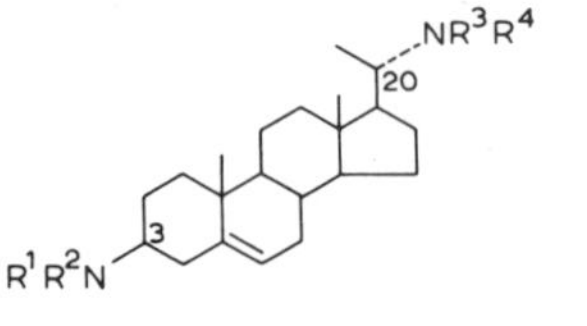

Irehdiamine-A ($R^1 = R^2 = R^3 = R^4 = H$)
Irehdiamine-C ($R^1 = R^2 = Me, R^3 = R^4 = H$)
Irehdiamine-B ($R^1 = R^3 = R^4 = H, R^2 = Me$)
Kurchimine ($R^1 = R^2 = R^4 = H, R^3 = Me$)
(XXVII)

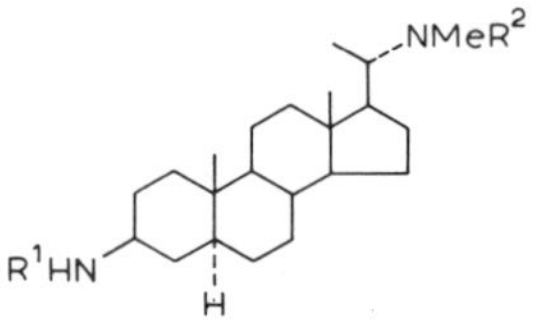

Chonemorphine ($R^1 = H, R^2 = Me$)
Dictyodiamine ($R^1 = R^2 = H$)
Dictyophlebine ($R^1 = R^2 = Me$)
(XXVIII)

Khuong-Huu et al., Bull. Soc. chim. Fr., 1965, 3035). The structure, XXVII ($R^1 = R^2 = R^4 = H$, $R^3 = Me$), referred to above, is **kurchimine**, another base of *H. antidysenterica* (*R. Tschesche*, *ibid.*, 1965, 1219).

Kurchessine, $C_{25}H_{44}N_2$, occurring in *H. antidysenterica* bark, in *Sarcococca pruniformis* leaves, and in *Pachysandra terminalis*, is identical with the total methylation product of the irehdiamines, and is therefore XXVII ($R^1 = R^2 = R^3 = R^4 = Me$) (*Tschesche* and *P. Otto*, Chem. Ber., 1962, **95**, 1144; *Labler* and *Šorm*, Coll. Czech. chem. Comm., 1963, **28**, 2345; *W. F. Knaack* and *T. A. Geissman*, Tetrahedron Letters, 1964, 1381; *A. Chatterjee et al.*, *ibid.*, 1965, 67).

α-Kurchessine, isomeric with kurchessine, is also found in *H. antidysenterica*. Its close resemblance to kurchessine in chemical and physical properties has led to its formulations as the $C_{(3)}$ epimeride of kurchessine, *viz* 3α,20α-bisdimethylamino-5-pregnene (*Labler* and *Šorm*, *loc. cit.*).

Dihydrokurchessine, $C_{25}H_{46}N_2$, occurs in *P. terminalis*. It is 3β,20α-bisdimethylamino-5α-pregnane (*Knaack* and *Geissman*, *loc. cit.*).

Chonemorphine, $C_{23}H_{42}N_2$, is found in *Dictyophleba lucida*, *Malonetia bequaertiana* and *Chonemorpha* spp. It is a primary-tertiary (NMe_2) base which reacts with nitrous acid to form the known 20α-dimethylaminoe-5α-pregnan-3β-ol. The latter on Hofmann degradation, then hydrogenation, furnished 5α-pregnan-3β-ol. Chonemorphine is consequently assigned structure XXVIII ($R^1 = H$, $R^2 = Me$). This formulation has been confirmed by synthesis of the base of reduction of the oxime of 20α-dimethylamino-5α-pregnan-3-one (*Chatterjee* and *B. Das*, Chem. and Ind., 1959, 1445; 1960, 1247; *Das* and *P. P. Pillay*, J. Sci. Ind. Res., India, 1954, **13B**, 602, 701; *F. Khuong-Huu-Lainé*, *N. G. Bisset* and *Goutarel*, Ann. pharm. franç., 1965 **23**, 395; *Khuong-Huu et al.*, Bull. Soc. chim. Fr., 1965, 3035; *M.-M. Janot et al.*, *ibid.*, 1962, 111; *P.-L. Chien et al.*, J. org. chem., 1964, **29**, 315). *N*-**Acetylchonemorphine** occurs in *M. bequaertiana* (*Khuong-Huu-Lainé*, *Bisset* and *Goutarel*, *loc. cit.*).

Funtudiamine and **dictyodiamine,** both $C_{23}H_{42}N_2$, present in *F. latifolia,* and *D. lucida* and *P. terminalis*, respectively, have been shown, by similar studies, to be 3β-dimethylamino-20α-amino-5α-pregnane and 3β,20α-bis-methylamino-5α-pregnane (XXVIII; $R^1 = R^2 = H$), respectively (*Khuong-Huu et al.*, Bull. Soc. chim. Fr., 1964, 2169; *T. Kikuchi, S. Uyeo* and *T. Nishinaga*, Tetrahedron Letters, 1965, 1993, 3169; *Janot et al.*, Compt. rend., 1965, **260**, 6119).

Dictyophlebine, $C_{24}H_{44}N_2$, is another base found in the roots of *D. lucida.* It is a secondary-tertiary diacidic base whose structure has been elucidated chiefly by mass and n.m.r. spectrometry. Ruschig degradation afforded 20α-dimethylamino-5α-pregnan-3-one. A partial synthesis of the alkaloid from chonemorphine (XXVIII; $R^1 = H$, $R^2 = Me$), by treatment of the latter with ethyl chloroformate to yield 3-*N*-ethoxycarbonylchonemorphine, followed by reduction with lithium tetrahydridoaluminate, generated 3β-methyl-amino-20α-dimethylamino-5α-pregnane (XXVIII; $R^1 = R^2 = Me$), identical with the natural base (*Khuong-Huu et al.*, Bull. Soc. chim. Fr., 1965, 3035; *M. Truong-Ho et al., ibid.*, 1963, 2332; *Janot et al.*, Compt. rend., 1965, **260**, 6118).

Saracocine, $C_{26}H_{44}N_2O$, an alkaloid of *Sarcococca pruniformis*, contains an amidic C=O group, generating acetic acid on acid hydrolysis. Hofmann degradation of the alkaloid yielded a product having u.v. absorption characteristic of steroidal 3,5-dienes. The n.m.r. spectrum of the natural base pointed to the presence of one secondary and two tertiary *C*-methyl groups. On this and other evidence saracocine is formulated as XXIX (*J. M. Kohli, A. Zaman* and *A. R. Kidwai*, Tetrahedron Letters, 1964, 3309; *Chatterjee et al., loc. cit.; W. Vetter et al.*, Bull. Soc. chim. Fr., 1963, 1324).

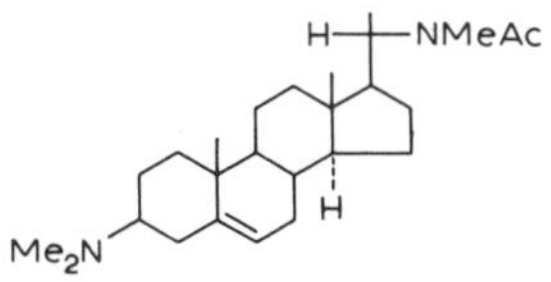

Saracocine

(XXIX)

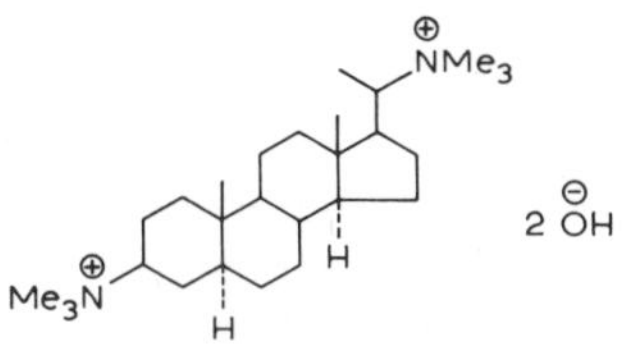

Malouetine

(XXX)

Malouetine, $C_{27}H_{54}N_2O_2$, occurs in the bark and roots of *M. bequaertiana* and has curare-type physiological activity. It is a bisquaternary ammonium hydroxide which on pyrolysis yields a methine base containing, on infrared evidence, a vinyl group. Hydrogenation of this compound led to the known 3β-dimethylamino-5α-pregnane. Treatment of the alkaloid with hydrogen chloride yielded the corresponding bisquaternary ammonium chloride, reduction with lithium tetrahydridoaluminate gave 3β,20α-bisdimethylamino-5α-pregnane, a known compound. These observations point to structure XXX for malouetine; it has been confirmed by a partial synthesis of the alkaloid from funtumafrine-C (p. 390), by successive conversion into the oxime, reduction, methylation, quaternisation and conversion to the dihydroxide by anion exchange.

Malouphylline, $C_{25}H_{42}N_2O_2$, is another constituent of *M. bequaertiana.* Although there are two nitrogen atoms present, the alkaloid is a monoacidic base, one being part of an acetamido group, this and an aldehyde group being detected by infrared absorption study. Reduction with lithium tetrahydridoaluminate yielded a diacidic secondary-tertiary base, which contains a primary alcoholic group. Ruschig deamination of the latter, followed by Hofmann degradation, yielded a known ketonic cyclic ether, XXXI (*B. Kamber et al.*, Helv., 1960, **43**, 347). The formation of this ether settles the stereochemistry of the alkaloid, XXXII, at position 20, it being known that an S_N2 displacement at this asymmetric centre is involved (*H. Favre et al.*, J. chem. Soc., 1953, 1115; *Cerny* and *A. Kasal*, Coll. Czech. chem. Comm., 1962, **27**, 2765). Malouphylline is consequently to be formulated as XXXII (*Janot et al.*, Bull. Soc. chim. Fr., 1962, 648):

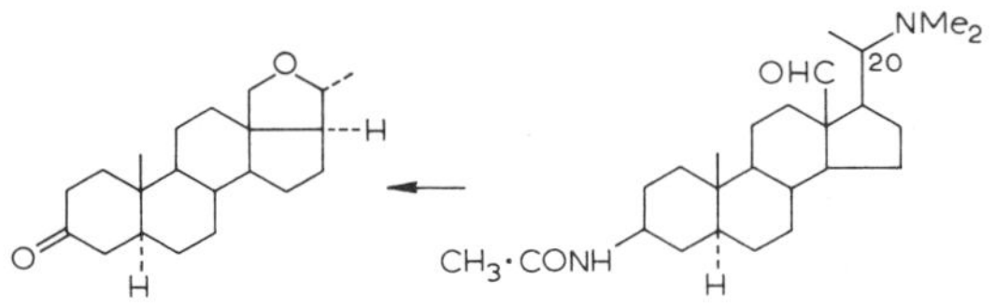

(XXXI)

Malouphylline

(XXXII)

The primary alcohol corresponding to malouphylline, (XXXII; CHO replaced by CH_2OH), named **malouphyllinine**, is also present in the same plant. It is readily obtainable from malouphylline by reduction with tetrahydridoborate (*Khuong-Huu-Lainé, Bisset* and *Goutarel, loc. cit.).*

Holarrhidine, $C_{21}H_{36}N_2O$, is found in the leaves of *Holarrhena antidysenterica*; it is a diacidic base, both amino groups being primary, since on Eschweiler–Clarke methylation a *N,N'*-tetramethyl derivative results. A primary alcohol group is present, because mild oxidation yields an aldehyde. A regioselective Hofmann degradation proved to be possible and led to a diene XXXIII, the driving force behind elimination of the 3-amino group presumably being formation of a conjugated diene. This diene was identical with one already obtained from **holarrhimine** (see below), by a similar procedure; its formation, together with a study of molecular rotation differences, allows holarrhidine to be formulated as XXXIV. It was the first steroidal alkaloid observed to have a 3α-amino group (*Labler* and *Cerny*, Coll. Czech. chem. Comm., 1959, **24**, 370; *Cerny*, *Labler* and *Šorm*, *ibid.*, p. 378).

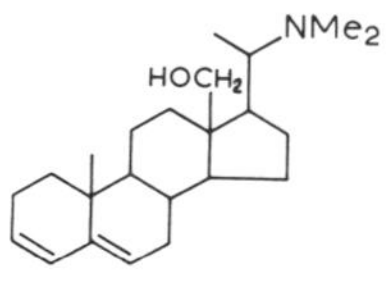

(XXXIII)

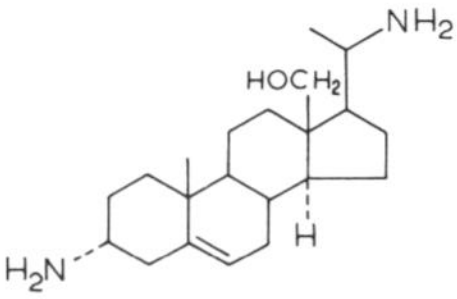

Holarrhidine
(XXXIV)

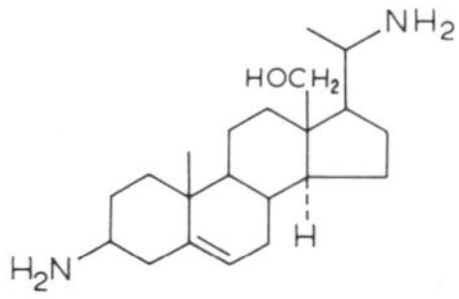

Holarrhimine
(XXXV)

Holarrhimine, $C_{21}H_{36}N_2O$, another base of *H. antidysenterica*, is a di-primary diamine, and is also a primary alcohol. Total methylation afforded an *N,N,N′,N′*-tetramethyl base, reducible to a dihydro derivative. Mild oxidation of the latter yielded an aldehyde, Wolff–Kishner reduction of which gave 3β,20α-bisdimethylamino-5α-pregnane, identified by synthesis from 3β-acetoxybisnorallocholanic acid (*Cerny* and *Šorm*, *ibid.*, 1955, **20**, 1473; *Labler*, *Cerny* and *Šorm*, *ibid.*, p. 1484). The same workers established unequivocally that the CH_2OH group is located at position 18 by relating the alkaloid to conessine (see below) by two different sequences (*idem*, *ibid.*, 1957, **22**, 76). Holarrhimine is therefore to be formulated as XXXV. **3-*N*-Methylholarrhimine, 20-*N*-methylholarrhimine** and ***N,N,N′,N′*-tetramethyl-holarrhimine** have also been isolated from *H. antidysenterica* (*Tschesche* and *K. Wiensz*, Chem. Ber., 1958, **91**, 1504); their structures have been elucidated largely by correlations with holarrhimine.

Several alkaloids which occur in *Pachysandra terminalis* belong to this subgroup. These include an unnamed base **"alkaloid D"** found to be identical with 3β,20α-bisdimethylamino5α-pregnane (*Knaack* and *Geissman*, *loc. cit.*). **Pachysamine-A,** $C_{24}H_{44}N_2$, contains a secondary *N*-methyl group and a tertiary *N*-dimethyl group; its n.m.r. spectrum reveals two angular methyl groups. Methylation affords 3α,20α-bisdimethylamino-5α-pregnane, and Ruschig degradation gives 20α-dimethylamino-5α-pregnan-3-one (funtumafrine-C, p. 390). Pachysamine-A is therefore to be formulated as XXXVI (R = H) (*M. Tomita*, *Uyeo* and *Kikuchi*, Tetrahedron Letters, 1964, 1641). **Pachysamine-B** is an amide, the acyl moiety being recognised as senecioyl by spectral study. It has been synthesised by acylation of pachysamine-A with senecioyl chloride, and therefore has structure XXXVI (R = Me_2C=CH · CO) (*idem*, *loc. cit.*):

Pachysamine-A (R = H)
Pachysamine-B (R = $Me_2C{=}CHCO$)
(XXXVI)

Epipachysamine-F ($R^1 = R^2 = Me, R^3 = R^4 = H$)
Epipachysamine-A ($R^1 = R^2 = R^3 = Me, R^4 = Ac$)
Epipachysamine-B ($R^1 = H, R^2 =$ nicotinyl, $R^3 = R^4 = Me$)
Epipachysamine-D ($R^1 = H, R^2 = COPh, R^3 = R^4 = Me$)
Epipachysamine-E ($R^1 = H$ $R^2 = CO{\cdot}CH{=}CMe_2, R^3 = R^4 = Me$)
(XXXVII)

Several alkaloids with the epimeric 3β configuration have also been encountered in the same plant. They are named **epipachysamine-A** to **-F,** and are all derived from 3β,20α-diamino-5α-pregnane. Alkaloid **A** (also called **saracodine**) is an *N*-acetyl amide, deacetylation of which yielded 3β-dimethylamino-20α-methylamino-5α-pregnane; methylation of the latter furnished 3β,20α-bisdimethylamino-5α-pregnane (*Kikuchi et al., ibid.,* 1964, 1817; *Chatterjee et al., ibid.,* 1965, 67). Epipachysamine-A is therefore XXXVII ($R^1 = R^2 = R^3 = Me$; $R^4 = Ac$). Alkaloid **B** furnished chonemorphine (see p. 392) and nicotinic acid on hydrolysis, and has been synthesised by acylation of the latter alkaloid using a mixed anhydride of nicotinic acid (*J. R. Vaughan* and *R. L. Osata,* J. Amer. chem. Soc., 1951, **73**, 3547). It is therefore XXXVII ($R^1 = H$, $R^2 =$ nicotinyl, $R^3 = R^4 = Me$) (*Kikuchi, Uyeo* and *Nishinaga,* Tetrahedron Letters, 1965, 1993). Alkaloid **D** also yielded chonemorphine on hydrolysis, accompanied by benzoic acid; it is therefore the benzoyl analogue of alkaloid B (XXXVII; $R^1 = H$, $R^2 = COPh$, $R^3 = R^4 = Me$) (*idem, ibid.,* p. 3169). Similarly alkaloid **E** is the senecioyl analogue (XXXVII; $R^1 = H$, $R^2 = CO{\cdot}CH{=}CMe_2$, $R^3 = R^4 = Me$) *(idem, ibid.)*. Alkaloid **F** was isolated as its *N*-acetyl derivative, and its structure rests in the first instance on mass spectral study of this derivative. Peaks at *m/e* 84 and 110 suggested the presence of a 3-dimethylaminopregnane skeleton, with A and B rings saturated, whilst those at 345 ($M - CH_3{\cdot}CO$) and 302 ($M - CH_3{\cdot}CHNHCO{\cdot}CH_3$) were in agreement with the presence of a 20-acetylamino group. Reduction of the derivative with lithium tetrahydridoaluminate, followed by Eschweiler–Clarke methylation, gave a product identical with that obtained by similar reduction of alkaloid A. Alkaloid **F** is thus XXXVII ($R^1 = R^2 = Me$, $R^3 = R^4 = H$) *(idem, ibid.)*.

Pachysandrine-A, $C_{33}H_{50}N_2O_3$, another *Pachysandra* base, was subjected to detailed i.r. and n.m.r. spectral study, which disclosed the presence of an acetoxyl group in the environment $-CH{\cdot}CHOAc{\cdot}CH-$, a conjugated amide C=O group, a CH_3NCO group, a phenyl group, a dimethylamino group and three *C*-methyl groups (one secondary, two tertiary). Hydrolysis of the alkaloid yielded acetic and benzoic acids and an *N,O*-deacyl derivative; oxidation of this converted it into a ketone the o.r.d. curve of which was similar to that of 5α-cholestan-4-one. Wolff–Kishner reduction of this ketone was accompanied by elimination of a methylamino group, leading to 20α-dimethylamino-5α-pregnane, the structure of which was confirmed by synthesis. The same ketone on oxidation with air afforded a diosphenol identical with that obtainable from 20α-dimethylamino-5β-pregnan-3-one by similar treatment. The acetoxy group in the alkaloid is therefore to be placed at position 4. The 3α,4β stereochemistry was established by n.m.r. spectral study and by a series of chemical transformations: the *O*-deacetyl alkaloid was treated with phosphoryl chloride–pyridine, then subjected to hydrolysis. A new hydroxy derivative resulted, a consequence of benzoyl *N*→*O* migration; its structure, XXXVIII, was settled by its oxidation to the same ketone obtained as described above. Consequently the deacyl alkaloid and the new hydroxy-base differ only in configuration at $C_{(4)}$. The acyl migration is known to proceed *via* an intermediate oxazoline, which is formed *via* a nucleophilic rear attack by the acyl group on $C_{(4)}$, and consequently migration is possible only when the groups involved are 1,2-diaxial (*J. Attenburrow, D. F. Elliott* and *G. F. Penny*, J. chem. Soc., 1948, 310).

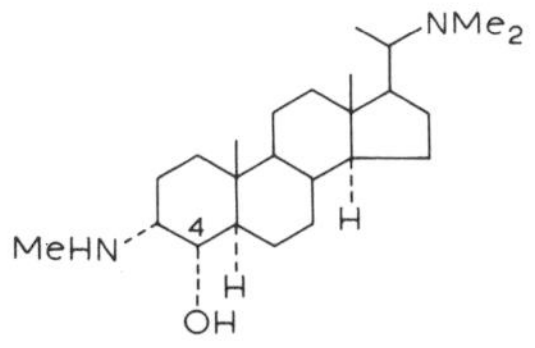

Pachysandrine-C
(XXXVIII)

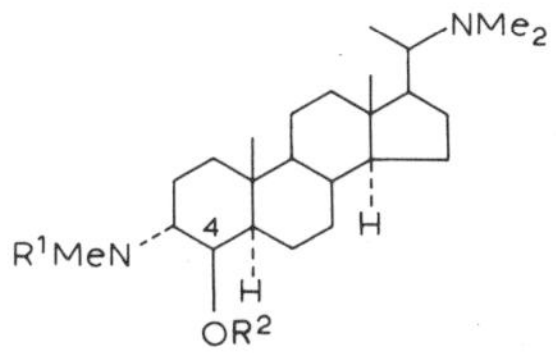

Pachysandrine-A (R^1 = COPh, R^2 = Ac)
Pachysandrine-B (R^1 = CO·CH=CMe$_2$, R^2 = Ac)
Pachysandrine-D (R^1 = H, R^2 = CO·CH=CMe$_2$)
(XXXIX)

Hydrolysis of the oxazoline yields a *cis*-1,2-aminocyclohexanol, XXXVIII, in which the 4-hydroxyl group has been inverted. Confirmation of this view was provided by the observation that this aminoalcohol on Eschweiler–Clarke methylation yields an oxazolidine, possible only if the MeNH and OH groups are *cis*. Pachysandrine-A must therefore be formulated as XXXIX (R^1=COPh, R^2=Ac) (*Tomita, Uyeo* and *Kikuchi*, Tetrahedron Letters, 1964, 1053).

Pachysandrine-B, $C_{31}H_{52}N_2O_3$, is closely similar to the foregoing base in its chemical behaviour, its spectral properties pointing to a similar structural type. Acetone and formaldehyde are formed on ozonolysis, total hydrolysis affords *N,O*-deacylpachysandrine-A, and partial hydrolysis *O*-deacylpachysandrine-B and acetic acid. The *O*-deacyl derivative has been partially synthesised by reaction of the *N,O*-deacyl product with senecioyl chloride, followed by mild basic hydrolysis. Pachysandrine-B is therefore XXXIX ($R^1 = CO \cdot CH{=}CMe_2$, $R^2 = Ac$).

Pachysandrine-C, $C_{24}H_{44}N_2O$, has been identified as the amino-alcohol XXXVIII, derivable from pachysandrine-A as detailed above (*Kikuchi et al.*, Tetrahedron Letters, 1964, 1817). The oxazolidine obtained from it by Eschweiler–Clarke methylation is also found in *Pachysandra terminalis.*

Pachysandrine-D, $C_{29}H_{50}N_2O_2$, on mild alkaline hydrolysis yields pachysandrine-C (XXXVIII), and its dihydro derivative proved to be identical with the acyl migration product of *O*-deacyldihydropachysandrine-B. These and spectral observations permit the alkaloid to be assigned structure XXXIX ($R^1 = H$, $R^2 = CO \cdot CH{=}CMe_2$) *(idem, cit.).*

Pachystermine-A, $C_{29}H_{48}N_2O_2$, and **pachystermine-B,** $C_{29}H_{50}N_2O_2$, major *Pachysandra* bases, have a ketone–secondary alcohol relationship: base A on reduction ($NaBH_4$) yields base B, and, conversely, B on oxidation (CrO_3) yields base A. Base A shows two C=O bands (1715 and 1735 cm^{-1}) in its i.r. spectrum, and its n.m.r. spectrum reveals the presence of NMe_2, $CHMe_2$, and three *C*-methyl groups (one secondary and two tertiary). The o.r.d. curve resembles that of 5α-cholestan-4-one (negative Cotton effect). Treatment of the alkaloid with alcoholic potash yielded a diosphenol already encountered in connection with pachysandrine-A (p. 396), and Wolff–Kishner reduction gave 20α-dimethylamino-5α-pregnane and 5α-pregnan-4-one. Reduction of the alkaloid with aluminium hydride furnished an amino-alcohol, XL, the structure being arrived at on mass spectrometric evidence.

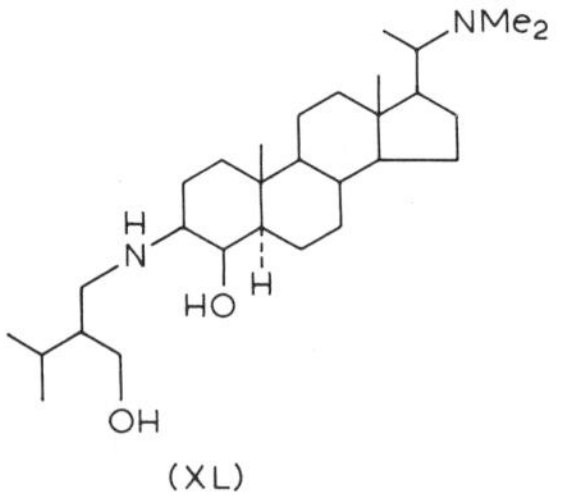

(XL)

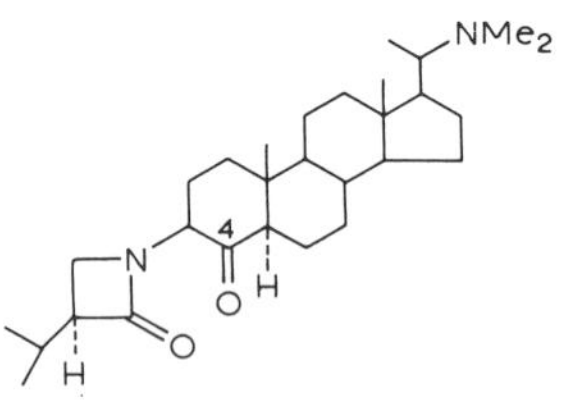

Pachystermine-A (XLI)

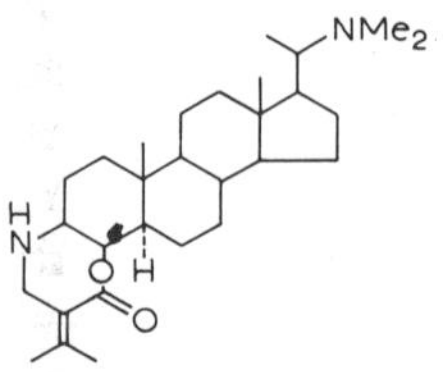

Pachysantermine-A (XLII)

These observations, coupled with the fact that the alkaloid is only a monoacidic base, are in harmony with the formulation of pachystermine-A as the β-lactam XLI, and pachystermine-B as the corresponding $C_{(4)}$ secondary alcohol. In the latter, the β-configuration of the hydroxyl group was deduced from n.m.r. spectral evidence relating to an amino-alcohol obtainable from the alkaloid by alkaline degradation: the $C_{(19)}$ methyl group signal appeared at low field, consonant with a 4β (axial) orientation for the hydroxyl group (*Kikuchi* and *Uyeo*, Tetrahedron Letters, 1965, 3473).

Pachysantermine-A, $C_{23}H_{48}N_2O_2$, is a minor *P. terminalis* alkaloid. It is a lactone, affording a diol on reduction with lithium tetrahydridoaluminate and having an i.r. carbonyl band at 1710 cm^{-1}. The plain, positive o.r.d. curve indicated that the carbonyl group was not in the steroidal skeleton, and the n.m.r. spectrum disclosed the presence of two allylic methyl groups, and two other allylic hydrogen atoms. Structure XLII for the alkaloid is in agreement with these observations; it was confirmed by *N*-methylation of the base followed by reduction ($LiAlH_4$), to yield the diol XL, already encountered as a transformation product of pachystermine-A (*idem, ibid.*, 1965, 3487).

(*d*) *Bases related to 3-aminoconanine*

A group of alkaloids occurring in *Holarrhena* species is derived from the pentacyclic base 3β-amino-5α-conanine (XLIII):

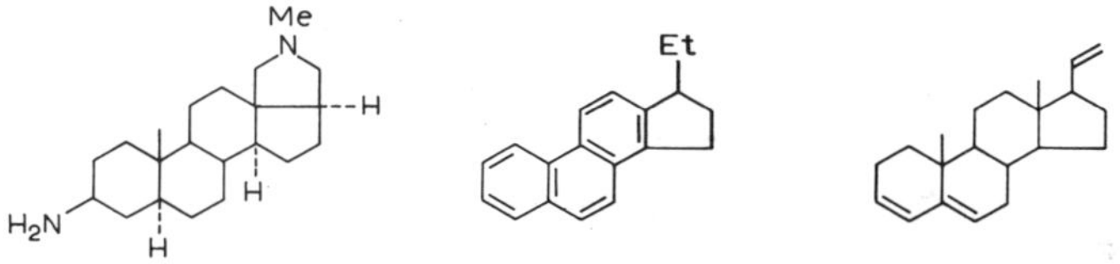

Me N H H_2N H H Et

(XLIII) (XLIV) (XLV)

The most familiar of these species is *H. antidysenterica*, a small Indian shrub known locally as "kurchi", and the most abundant alkaloid is conessine. For a review of the classification of kurchi alkaloids see *R. Tschesche* and *A. C. Roy*, Chem. Ber., 1956, **89**, 1288.

Conessine, $C_{24}H_{40}N_2$, occurs in the bark of all *Holarrhena* spp. examined, and was first correctly formulated as long ago as 1886 (*H. Warnecke*, Ber., 1886, **19**, 60). It is a ditertiary, diacidic, oxygen-free base, containing one olefinic linkage and three *N*-methyl groups; thus one nitrogen atom is present as an NMe_2 group and the other as a cyclic NMe group (*D. D. Kanga*, *P. R. Ayyar* and *J. L. Simonsen*, J. chem. Soc., 1926, 2123; *E. Späth* and *O. Hromatka*, Ber., 1930, **63**, 126). Conessine dihydriodide on pyrolysis in a reducing atmosphere affords a hydrocarbon mixture which on dehydrogenation with selenium yields 3′-ethyl-1,2-cyclopentenophenanthrene

(XLIV), accompanied by traces of the 3′-methyl analogue (*Diels* hydrocarbon). The former was identified by synthesis (*R. D. Haworth, J. McKenna* and *N. Singh*, Nature, 1948, **162**, 22; J. chem. Soc., 1949, 831; see also *B. Riegel, M. H. Gold* and *M. A. Kubico*, J. Amer. chem. Soc., 1943, **65**, 1772).

Hofmann degradation proved to be a valuable method of structural breakdown. Conessine dimethohydroxide on dry distillation afforded **apoconessine,** $C_{21}H_{29}NMe_2$, accompanied by trimethylamine and water. Apoconessine contains three ethylenic linkages, two of which are the result of the Hofmann degradation. One of these is a consequence of the elimination of the dimethylamino group and the other of rupture of the heterocyclic ring; two of the bonds are conjugated, on u.v. absorption evidence. Partial Hofmann degradation studies (*Haworth, McKenna* and *G. H. Whitfield*, J. chem. Soc., 1949, 3127) have revealed that the isolated double bond in the apo-base is the one formed by heterocyclic ring fission, since the intermediate desbase $C_{21}H_{30}{>}\overset{\oplus}{N}Me_2\overset{\ominus}{O}H$ and salts thereof are conjugated dienes and contain a heterocyclic ring. Emde degradation of apoconessine generates pregna-3,5,20-triene (XLV), identified by synthesis and by complete hydrogenation to pregnane and 5α-pregnane (*Haworth et al., ibid.*, 1951, 1736; *Haworth, McKenna,* and *Whitfield, ibid.*, 1953, 1102). Apoconessine must therefore be a dimethylaminopregna-3,5,20-triene, and partial structure XLVI, in which the $-C_2H_4NMe$ is part of a heterocycle, can be advanced.

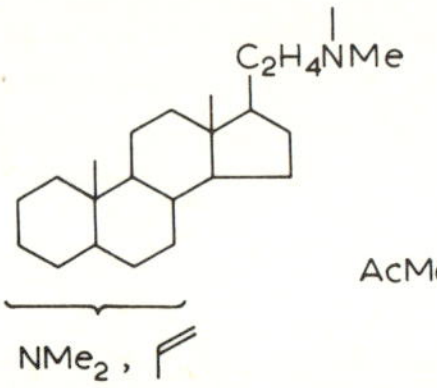

(XLVI) 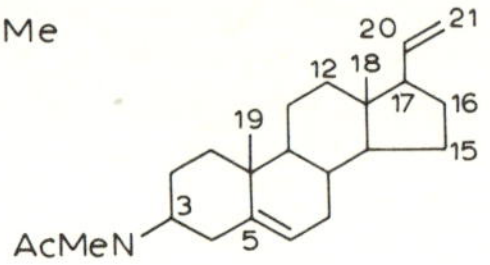 (XLVII) 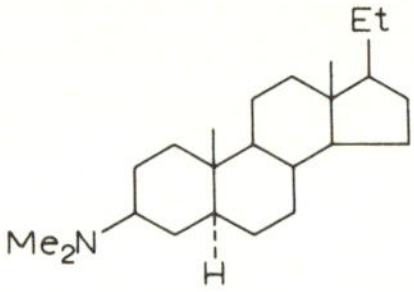 (XLVIII)

Partial demethylation of conessine with cyanogen bromide yielded **isoconessimine,** a minor kurchi base (see below), which is a secondary-tertiary base formed by elimination of a methyl group from the NMe_2 group of conessine. Acetylation of this base afforded an amide, which on successive Hofmann and Emde degradations yielded another amide, XLVII, in which the cyclic double bond and the nitrogen atom correspond to the same features in conessine *(idem, loc. cit.)*. This amide was synthesised, proving that conessine contains a 3β-dimethylamino-5,6-enoid system (*Haworth, McKenna* and *R. G. Powell, ibid.*, p. 1110). The amide was subjected to successive reduction, deacetylation and *N*-methylation, whereby it was transformed into 3β-dimethylamino-5α-pregnane (XLVIII), which was synthesised from 5α-pregnan-3-one *(idem, loc. cit.)*. Conessine is therefore derived from 3β-dimethylaminopregn-5-ene and must have a methylimino group at $C_{(20)}$ or $C_{(21)}$ (for numbering see formula XLVII), this group being part of a heterocycle terminating on some nearby carbon atom. Molecular models reveal that the only reasonable positions for the link-up are $C_{(12)}$, $C_{(15)}$ and $C_{(16)}$ (all β-orientated *N*-bonds), and $C_{(18)}$. Strong evidence has been adduced for believing that the point of union is $C_{(18)}$ (*Haworth et al.*, J. chem. Soc., 1953, 1115), and structures XLIX and L have consequently been proposed for conessine and apoconessine, respectively, the attachment of the methylimino group to $C_{(20)}$ being preferred mainly on the basis of Kuhn–Roth *C*-methyl analyses:

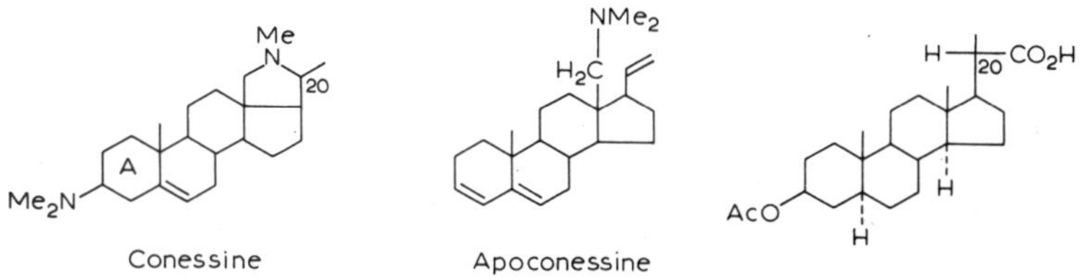

Conessine (XLIX)   Apoconessine (L)   (LI)

Treatment of conessine dimethiodide with alkali converts it into **heteroconessine,** a new base which yields the same partial Hofmann degradation product (conessimethine) as conessine *(Haworth et al., loc. cit.)*. The methine base reverts to heteroconessine in aqueous ethanol. Heteroconessine and conessine are best regarded as $C_{(20)}$-epimers of structure XLIX.

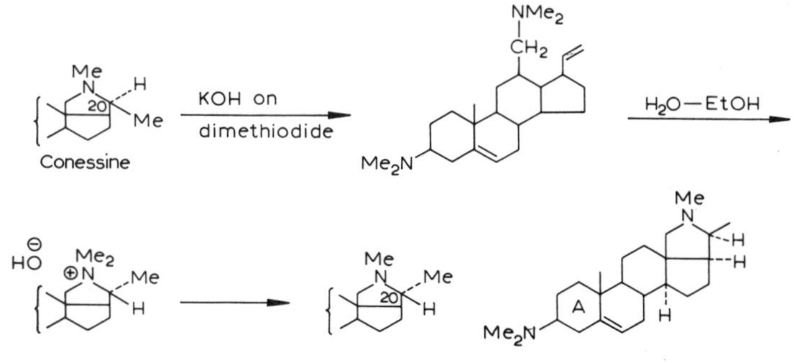

Conessine —KOH on dimethiodide→ … —H_2O—EtOH→

Heteroconessine   (+)-Conessine (LII)

Later investigations (*Haworth* and *McKenna*, Chem. and Ind., 1957, 1510; *Haworth* and *M. Michael*, J. chem. Soc., 1957, 4973) involving the oxidation of conessine and the pyrolysis of the products have confirmed the structural assignment, and substantiated the view that the double bond is 5,6; the latter feature was elegantly confirmed by the isolation of 2-methylcyclohexanone as an oxidation product, in which ring A has survived.

The absolute configuration of natural (+)-conessine has been settled by its transformation into steroidal derivatives such as 3β-dimethylamino-5α-pregnane (see above), which has been correlation with cholesterol, of known absolute configuration (*J. W. Cornforth, I. Youhotsky* and *G. Popják*, Nature, 1954, **173**, 536; *B. Riniker, D. Arigoni* and *O. Jeger*, Helv., 1954, **37**, 546). This link-up reveals the absolute configuration at all the asymmetric centres except $C_{(20)}$. The correlation of holarrhimine (p. 395) with 3β-acetoxybisnorallocholanic acid (LI), also of known absolute configuration, and the conversion of the alkaloid into conessine by steps not affecting the stereochemistry at $C_{(20)}$, prove that the $C_{(20)}$–N bond in conessine is α, and the methyl group at $C_{(20)}$ must therefore be β-orientated. Structure LII therefore represents natural (+)-conessine (*V. Cerny, L. Labler* and *F. Šorm*, Coll. Czech. chem. Comm., 1957, **22**, 76).

Structure LII for the alkaloid has been confirmed by several syntheses of conessine (*J. A. Marshall* and *W. S. Johnson*, J. Amer. chem. Soc., 1962, **84**, 1485; *G. Stork et al.*, *ibid.*, p. 2018; *D. H. R. Barton* and *A. N. Starratt*, J. chem. Soc., 1965, 2444). One such synthesis (*W. Nagata, T. Terasawa* and *T. Aoki*, Tetrahedron Letters, 1963, 869) takes the following course: the synthesis of the 13β-cyano-18-nor-5α-pregnane derivative (LIII) stereospecifically from 6-methoxy-1-tetralone has been described (*idem*, *ibid.*, p. 963). It was transformed in a single step into the pyrrolidine derivative LIV by simple reduction with lithium tetrahydridoaluminate at elevated temperature. Eschweiler–Clarke methylation of LIV gave its *N*-methyl derivative LV, identical with the dihydro derivative of **latifoline**, another *Holarrhena* base (see p. 408). Oxidation of the latter with chromic acid gave the corresponding 3-ketone, which on successive bromination, reaction with sodium iodide, and reduction with chromous chloride afforded (±)-conan-4-en-3-one (LVI). This was converted into its enamine LVII by reaction with dimethylamine (*Marshall* and *Johnson, loc. cit.*), and then reduced with sodium tetrahydridoborate to (±)-conessine (LII).

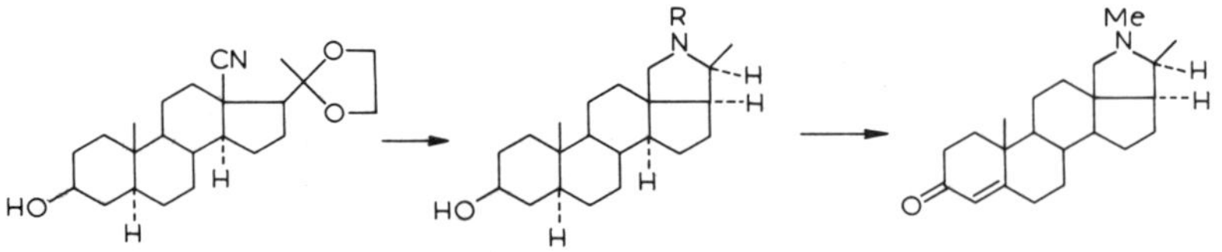

(LIII)

(LIV) (R = H)
(LV) (R = Me)

(LVI)

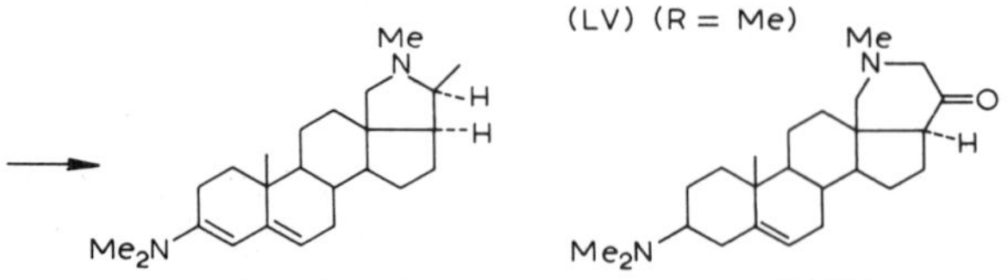

(LVII)

(LVIII)

Amongst interesting transformations of conessine is its behaviour towards iodine in tetrahydrofuran, in the presence of sodium bicarbonate. The product is a 3-piperidone, LVIII, a consequence of ring-expansion, its formation being explained by oxidation of conessine to an iminium iodide, followed by hydrolytic cleavage of the C=N bond to a methyl ketone, iodination of the latter, and recyclisation (*M.-M. Janot* and *R. Goutarel*, Bull. Soc. chim. Fr., 1962, 2234).

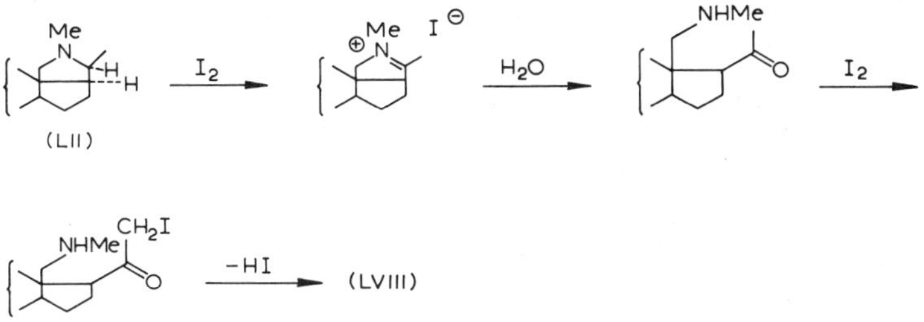

(LII) —I_2→ —H_2O→ —I_2→ —HI→ (LVIII)

Isoconessine and **neoconessine,** both artefacts, are formed by isomerisation of conessine by the action of sulphuric acid. The former is the result of a Westphalen rearrangement, accompanied by double bond migration, isoconessine being formulated as LIX (*P. Devissaguet et al.*, Tetrahedron Letters, 1966, 1073). Neoconessine has a spiro-structure in which a contraction of ring A has occurred (*Janot et al., ibid.*, 1966, 4375).

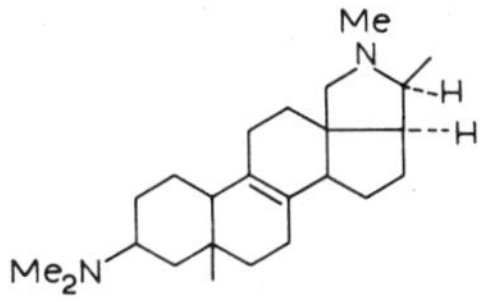

Isoconessine
(LIX)

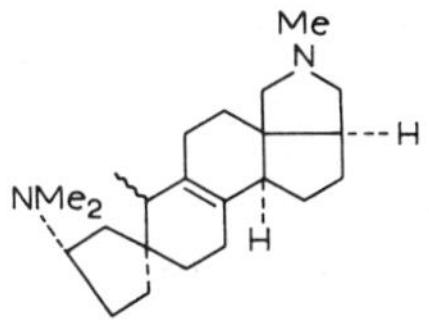

Neoconessine

3α-Aminoconan-5-ene (LX) is found in *H. antidysenterica*, its structure being established chiefly by mass spectrometry (*Cerny, L. Dolejs* and *Šorm*, Coll. Czech. chem. Comm., 1964, **29**, 1591). Also present is its *N,N*-dimethyl derivative conkuressine (LXI) (*Tschesche* and *P. Otto*, Chem. Ber., 1962, **95**, 1144; *Labler* and *Šorm*, Coll. Czech. chem. Comm., 1963, **28**, 2345).

Conessimine, $C_{23}H_{38}N_2$, and its isomer **isoconessimine** are both minor kurchi alkaloids, and are both secondary-tertiary bases converted into conessine by *N*-methylation. Isoconessimine is the product of partial demethylation of conessine with cyanogen bromide; here it was to be expected that the more exposed NMe_2 group would be attacked. Hence iso-

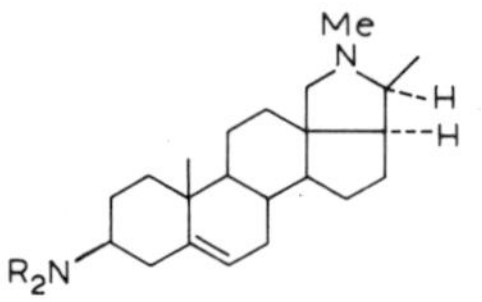

(LX) (R = H)
(LXI)(R = Me) Conkuressine

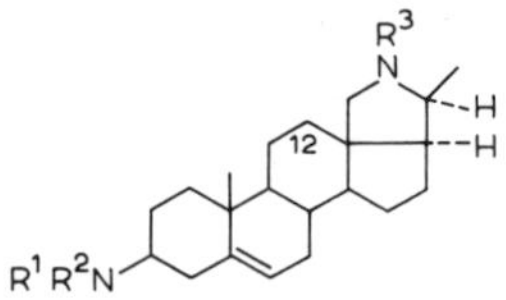

(LXII) (R^1 = H, R^2 = R^3 = Me) Isoconessimine
(LXIII) (R^1 = R^2 = Me, R^3 = H) Conessimine
(LXIV) (R^1 = R^3 = H, R^2 = Me) Conimine

conessimine is to be formulated as LXII (*S.* and *R. H. Siddiqui*, J. Indian chem. Soc., 1934, **11**, 787). Next, a selective reaction between conessine and hydrogen peroxide (one mol.) gave the 3-*N*-oxide; when this was demethylated in the same manner the cyclic NMe group was attacked, leading, after deoxygenation of the *N*-oxide group, to conessimine. The latter must therefore be assigned structure LXIII (*P. K. Bhattacharyya, B. D. Kulkarni* and *C. R. Narayana*, Chem. and Ind., 1962, 1377; see also *Cerny, Dolejs* and *Šorm, loc. cit.; Haworth, McKenna* and *Whitfield, loc. cit.*).

Conimine, $C_{22}H_{36}N_2$, is a disecondary base which on total *N*-methylation affords conessine. It is therefore structurally LXIV (*Siddiqui* and *Siddiqui, loc. cit.*).

Malouphyllamine, $C_{24}H_{40}N_2O$, a minor alkaloid of *Malouetia bequaertiana*, is a monoacidic base, one nitrogen atom being present as an *N*-acetyl group, on i.r. ($\nu_{C=O}$ 1642, 1653 cm $^{-1}$) and n.m.r. ($CO \cdot CH_3$ signal at 1.97δ) spectral evidence. That the alkaloid is derived from conanine was suggested by mass spectrometry (characteristic peak at M–15 and base peak at *m/e* 71), and supported by detailed n.m.r. spectral analysis. Structure LXV was advanced by *Janot, F. Lainé* and *Goutarel* (Bull. Soc. chim. Fr., 1962, 648) and confirmed by degradative studies on the base, and particularly by the establishment of an inter-relation with malouphylline, of known constitution (p. 394). Reduction of the latter yielded malouphyllinol ($-CH{=}O \rightarrow -CH_2$-OH), convertible to a toluene-4-sulphonate, which cyclised spontaneously to a quaternary ammonium salt LXVI:

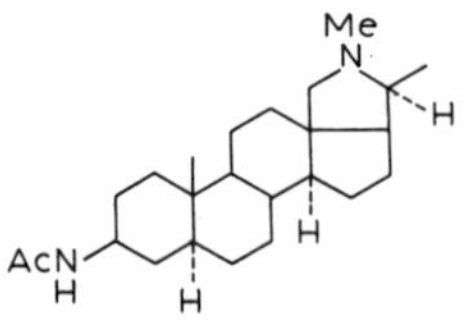

Malouphyllamine
(LXV)

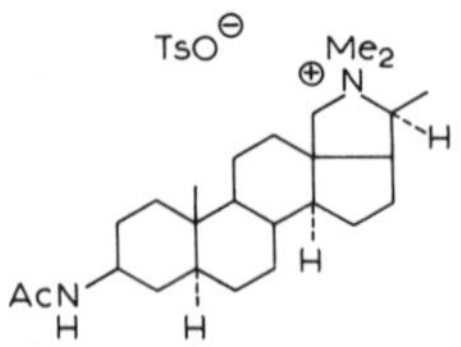

(LXVI)

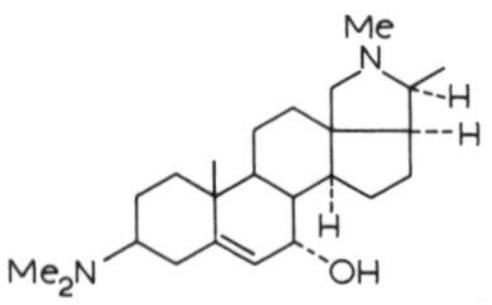

7α-Hydroxyconessine
(LXVII)

Anion exchange converted this to the corresponding iodide, which proved to be identical with malouphyllamine methiodide (*Janot, F. Khuong-Huu-Lainé* and *Goutarel, ibid.*, 1963, 641).

3α-Dimethylamino-5α-conanine has been identified in kurchi bark, and is identical with the minor product of hydrogenation of conkuressine (p. 404) (*Labler* and *Šorm, loc. cit.*).

7α-Hydroxyconessine, $C_{24}H_{40}N_2O$, is present in kurchi bark (*Tschesche* and *H. Ockenfels*, Ber., 1964, **97**, 2316) and is also formed, accompanied by the 7β epimer, by microbiological oxidation of conessine by *Aspergillus ochraceus* (*S. M. Kupchan et al.*, Tetrahedron Letters, 1963, 1767) or by *Cunninghamella echinulata* (*E. L. Patterson, W. W. Andres* and *R. E. Hartman*, Experientia, 1964, **20**, 256). Oxidation of the base yielded an α,β-unsaturated ketone ($\lambda_{max.}^{MeOH}$ 236.5 nm). The n.m.r. spectrum of the alkaloid, compared with that of conessine, showed that the $C_{(6)}$–H signal had moved to lower field (5.62δ) and appeared as a doublet, which observations are consistent with the placing of the hydroxyl group at $C_{(7)}$, and the formulation of the alkaloid as LXVII. Confirmation is provided by the formation of both epimers by allylic oxidation of conessine with *tert*-butyl perbenzoate (*Tschesche* and *Ockenfels, loc. cit.*). The 7β-epimer is probably present in kurchi bark.

Holaline, $C_{24}H_{42}N_2O$, a constituent alkaloid of *H. africana*, contains a hydroxyl group (tertiary), and is a ditertiary base containing >NMe and $-NMe_2$ groups. A conanine skeleton was revealed by the presence of the characteristic peak at *m/e* 71 in the mass spectrum. Dehydration with acetic anhydride–boron trifluoride yielded conessine: thus the hydroxyl group must be located in the angular position 5.

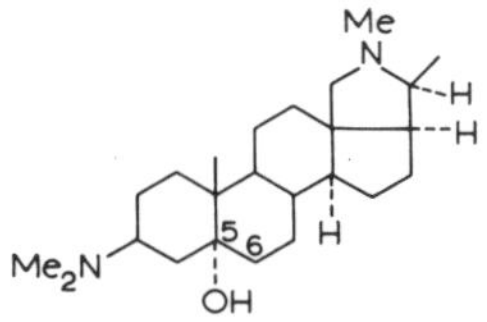

Holaline (LXVIII)

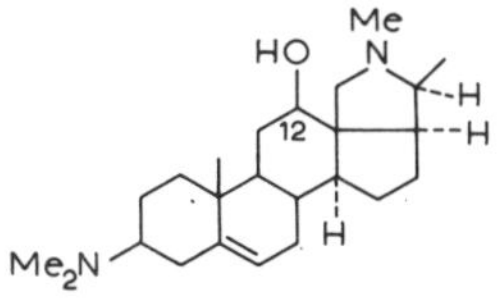

Holarrhenine (LXIX)

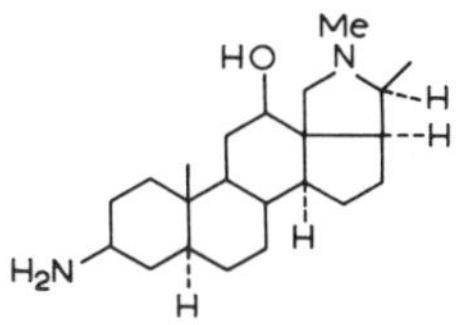

Funtessine (LXX)

Structure LXVIII was therefore proposed; it was confirmed by hydroxylation of conessine to 5,6-dihydroxydihydroconessine with potassium iodate. This diol yielded a monomesylate [at $C_{(6)}$], which with base afforded the 5α,6α-epoxide. Reduction of the latter ($LiAlH_4$) yielded the 5α-hydroxy compound LXVIII, identical with holaline (*Janot et al.*, Compt. rend., 1965, **260**, 6631).

Holarrhenine, $C_{24}H_{40}N_2O$, is a hydroxyconessine; it can be oxidised to a ketone (the hydroxyl group therefore being secondary), which is convertible to a dithioacetal. Treatment of the last with Raney nickel gives conessine (*F. L. Pyman*, J. chem. Soc., 1919, **115**, 163; *A. Uffer*, Helv., 1956, **39**, 1834). The ketone, on i.r. evidence, is a cyclohexanone, and is not α,β-unsaturated; this restricts to six the number of possible locations of the hydroxyl group in the alkaloid. The exact position of the group was revealed by Hofmann and Emde degradations, resulting in the known 12β-hydroxy-5α-pregnane (*L. van Hove*, Tetrahedron, 1959, **7**, 104); holarrhenine is therefore 12β-hydroxyconessine (LXIX). The corresponding –NHMe secondary base is **holarrheline,** another alkaloid of *H. africana* (*Janot et al., loc. cit.*).

Funtessine, $C_{22}H_{38}N_2O$, an alkaloid of the bark of *Funtumia latifolia*, is a primary-tertiary base which on total *N*-methylation affords dihydroholarrhenine; conversely the latter yielded funtessine on successive von Braun and Ruschig degradations. A conanine skeleton for the alkaloid was indicated by mass spectral study, and funtessine is therefore formulated as (LXX) (*Q. Khuong-Huu et al.*, Bull. Soc. chim. Fr., 1965, 1831).

Holarrhetine, from *H. africana*, is an ester-alkaloid which on alkaline hydrolysis affords holarrhenine (LXIX, p. 405) and pyroterebic acid, $Me_2C{=}CH\cdot CH_2\cdot CO_2H$. It is therefore the ester of holarrhenine and this acid. Similarly, **holafrine**, from the same source, yields, on identical treatment, 12β-hydroxyconessimine (LV, p. 402, β-OH at position 12) and the same acid, and thus has an analogous structure (*H. Rostock* and *E. Seebeck*, Helv., 1958, **41**, 11).

Kurcholessine, $C_{25}H_{44}N_2O_2$, is a further kurchi bark alkaloid which is a ditertiary base ($-NMe_2$ and $>$NMe groups) and also a diol, one hydroxyl group being tertiary and the other secondary.

Physicochemical studies are largely responsible for its formulation: its mass spectrum showed a peak at *m/e* 71 (*N*-methylpyrrolidine grouping in a conanine framework), and a base peak at 84 revealed that ring A has an NMe_2 group at $C_{(3)}$, and no substituents on $C_{(1)}$ and $C_{(2)}$. Further, a peak at 156, and the absence of one at 110, suggested that the hydroxyl groups and methyl group were located in the $C_{(4)}$–$C_{(7)}$ area of the molecule. The n.m.r. spectrum of the alkaloid contained signals corresponding to a $C_{(19)}$ methyl group and a $C_{(21)}$ methyl group (doublet); another methyl group gave rise to a doublet at markedly low field and it could be conjectured, with the help of biogenetic considerations, that this group was located at $C_{(4)}$, the low field position of the signal being explained by deshielding by nearby nitrogen and oxygen functions. The hydroxyl groups are not neighbouring, on i.r. evidence (no intramolecular hydrogen bonding). Kurcholessine forms a monoacetate, the secondary hydroxyl group having been acylated; dehydration of this ester afforded a product with, on n.m.r. evidence, a trisubstituted 5,6 double bond. This compound did not have the character of an enol acetate; it follows that one hydroxyl group in the alkaloid is located at the angular $C_{(5)}$ position, and the other at $C_{(7)}$. The absence of intramolecular hydrogen bonding of any type indicates that the hydroxyl groups must be *trans* and disposed relative to the dimethylamino group as shown in LXXI (*Tschesche, W. Meise* and *G. Snatzke*, Tetrahedron Letters, 1964, 1659).

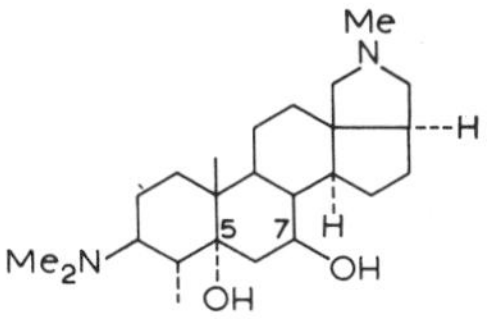

Kurcholessine (LXXI)

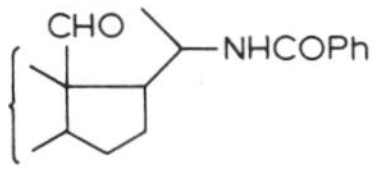

(LXXII)

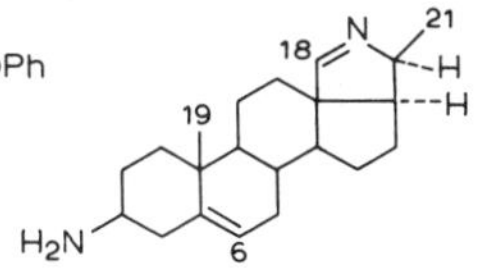

Conkurchine (LXXIII)

Conkurchine (irehline), $C_{21}H_{32}N_2$, an alkaloid of *Funtumia elastica*, is a primary amine and contains two double bonds. Its structure has been elucidated primarily through a link-up with holarrhimine (p. 395): *N*-acetylconkurchine, on treatment with benzoyl chloride and alkali, yielded an aldehyde of partial structure LXXII, as a consequence of ring-cleavage. Reduction of this aldehyde with tetrahydridoborate, followed by alkaline hydrolysis, afforded holarrhimine (XXXV). Conversely, oxidation of holarrhimine with chromic acid yielded conkurchine, *via* the intermediate aldehyde. The alkaloid must therefore be formulated as LXXIII, which structure is consonant with the n.m.r. spectrum of the base [methyl singlet (C_{19}), methyl doublet (C_{21}), olefinic H singlet (C_{18}-H) and poorly resolve doublet (C_6-H)] (*Khuong-Huu et al.*, Bull. Soc. chim. Fr., 1964, 1564, and earlier references there cited). **Conessidine,** $C_{22}H_{34}N_2$ is *N*-methylconkurchine (LXXIII; NH_2 replaced by NHMe), which fact was conveniently revealed by the observation that the alkaloid was formed by oxidation of 3-*N*-methylholarrhimine with chromic acid, (*idem, loc. cit.*; *Janot et al.*, Bull. Soc. chim. Fr., 1964, 1555).

A small group of alkaloids is related structurally to 5-conanene (LXXIV), and is conveniently reviewed here.

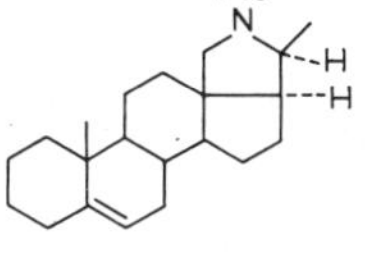

5-Conanene

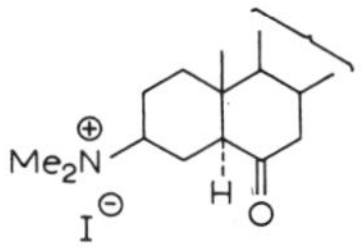

(LXXV)

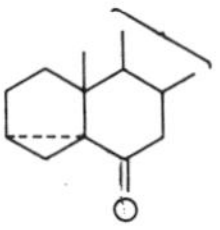

(LXXVI)

Latifoline, $C_{22}H_{35}NO$, found in the bark of *F. latifolia*, is an NMe tertiary base and a secondary alcohol. Its structure rests chiefly on its synthesis from conessine (LII; p. 401), which was reduced with tetrahydridoborate and then oxidised to 6-oxodihydroconessine. The 3-dimethylamino group was next selectively quaternised with methyl iodide, to give LXXV, base treatment of which afforded the 3,5-cyclo-6-oxo base LXXVI. Reduction of the latter to the 6α-alcohol, followed by acid-catalysed rearrangement, gave latifoline (LXXVII) (*Janot, Khuong-Huu* and *Goutarel*, Compt. rend., 1962, **254**, 1326; *R. Pappo*, J. Amer. chem. Soc., 1959, **81**, 1010; *J. Hara* and *Cerny*, Coll. Czech. chem. Comm., 1961, **26**, 2217).

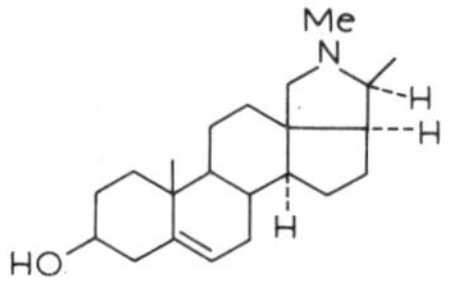

Latifoline (LXXVII) Latifolinine (LXXVIII) Funtuline (LXXIX)

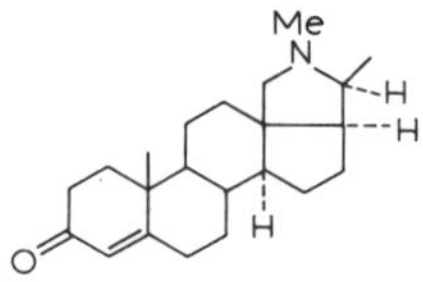

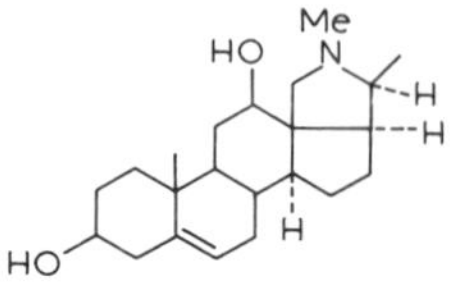

Latifolinine, $C_{22}H_{33}NO$, was perceived at an early stage to be closely akin to latifoline. It is an α,β-unsaturated ketone ($\lambda_{max.}$ 243 nm, $\nu_{max.}$ 1675 cm^{-1}) and its mass spectrum shows that it contains a conanine framework. It has been synthesised from conessine (p. 399) by von Braun demethylation to isoconessimine (LXII; p. 404), followed by Ruschig deamination to latifolinine (LXXVIII) (*Khuong-Huu, J. Yassi* and *Goutarel*, Bull. Soc. chim. Fr., 1963, 2486; *W. S. Johnson, V. J. Bauer* and *R. W. Franck*, Tetrahedron Letters, 1961, 72; *J. de Flineo et al., ibid.*, 1962, 1257; *A. Berths* and *J. Götz*, Ann., 1958, **619**, 96.).

Norlatifoline, also present in *F. latifolia*, is the secondary base (LXXVII; NMe replaced by NH) corresponding to latifoline. Its structure rests on the establishment of this relationship (*Khuong-Huu, Yassi* and *Goutarel, loc. cit.*; *Hora* and *Cerny, loc. cit.*).

Funtuline, $C_{22}H_{35}NO_2$, from *F. latifolia*, is a diol, and forms a diacetate. An n.m.r. spectral analysis revealed that both hydroxyl groups are secondary, that two *C*-methyl groups are present, one tertiary and one secondary, and that an olefinic proton exists within the molecule. A conanine skeleton was indicated by the mass spectrum of the alkaloid. The assigned structure LXXIX has been confirmed by synthesis from holarrhenine (LXIX; p. 406), which was *O*-acetylated, then subjected to von Braun degradation to yield a secondary base, LXXX. This on Ruschig degradation gave an α,β-unsaturated ketone LXXXI. This was converted to its acetate-enol acetate, which on reduction with tetrahydridoborate and hydrolysis afforded funtuline

(LXXIX) (*Goutarel et al.*, Bull. Soc. chim. Fr., 1964, 787; 1963, 2486).

Holadienine, $C_{22}H_{31}NO$, is a cross-conjugated dienone occurring in *H. africana* and *H. mitis*, formulated as LXXXII (*idem*, Compt. rend., 1965, **260**, 6631).

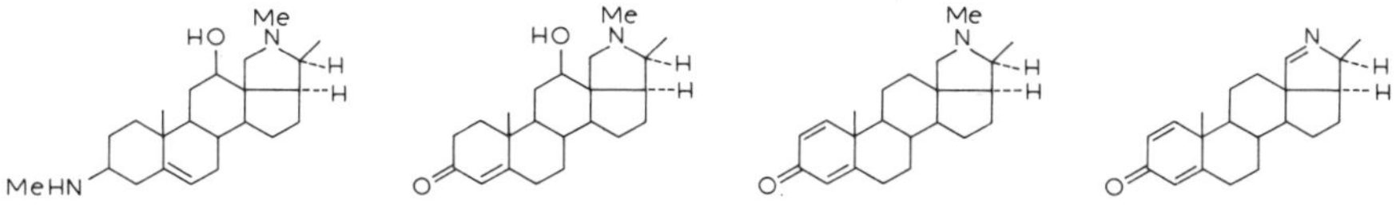

(LXXX) (LXXXI) Holadienine (LXXXII) Maingayine (LXXXIII)

Maingayine, from *Paravallaris maingayi*, is closely related structurally to holadienine, being 18-dehydro-*N*-demethylholadienine (LXXXIII) (*J. B. Davis et al.*, Chem. and Ind., 1970, 627).

Funtudienine, $C_{22}H_{31}NO$, is an amorphous alkaloid of *F. latifolia.* Its u.v. maximum at 280 nm (ε 22,000) revealed that it is a heteroannular dienone, and its mass spectrum (peaks at *m/e* 71 and M–15) suggested a conanine skeleton is present. N.m.r. spectroscopy pointed to tertiary and secondary methyl groups, and the presence of three olefinic protons, all at low field in agreement with their location along the dienone chromophore. Catalytic hydrogenation of the alkaloid (uptake $2H_2$) yielded a tetrahydrobase which proved, on i.r. absorption evidence, to be a cyclohexanone. Finally, Wolff–Kishner reduction of funtudienine gave cona-3,5-diene. These observations allow the alkaloid to be formulated as LXXXIV; this assignment has been confirmed by the following synthetic sequence.

latifoline (LXXVII: p. 408) → *O*-acetyllatifoline $\xrightarrow{CrO_3}$ *O*-acetyl-7-oxolatifoline $\xrightarrow[AcOH]{heat, H^{\oplus}}$ funtudienine (LXXXIV)

(*Khuong-Huu et al.*, Bull. Soc. chim. Fr., 1964, 2169).

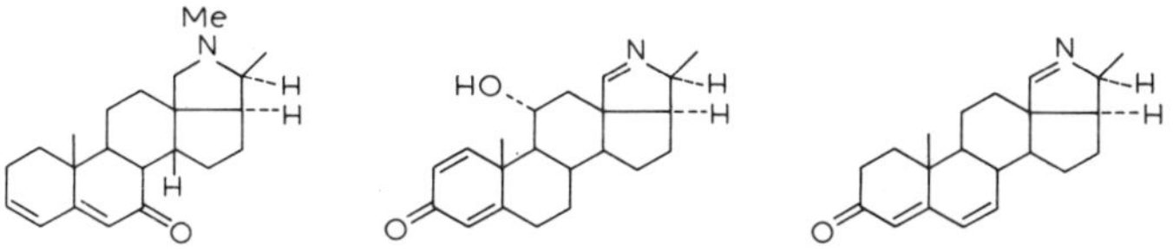

Funtudienine (LXXXIV) Holonamine (LXXXV) Malarboreine (LXXXVI)

Holonamine, $C_{21}H_{27}NO_2$, from kurchi bark, has been studied chiefly from the spectroscopic angle. It is a conjugated ketone ($\lambda_{max.}^{MeOH}$ 245.8 nm, ε 18,000) and a secondary alcohol. On acid treatment at elevated temperature it is rearranged to a mixture of aromatic species. Coupled with the fact that its i.r. spectrum shows no band characteristic of a CH_2 group adjacent to a C=O group (near 1410 cm^{-1}), this was taken to mean that the alkaloid is a cross-conjugated dienone. The position of the hydroxyl group was inferred from spectral comparisons with known 11α-hydroxypregna-1,4-dien-3-ones; particularly significant is the positive Cotton effect in the o.r.d. curve of holonamine, to be contrasted with the negative effect in unsubstituted pregnadienones. An additional double bond was located by n.m.r. spectral measurement, the splitting pattern characteristic of conkurchine-type bases (p. 407) being evident. Since the alkaloid contains no NMe group, this bond can be placed confidently in the "conkurchine position," and holonamine formulated as LXXXV (*Tschesche* and *Ockenfels*, Ber., 1964, **97**, 2316, 2326).

Malarboreine, $C_{21}H_{27}NO$, is a very minor alkaloid of *Malouetia arborea*. Its u.v. absorption ($\lambda_{max.}^{EtOH}$ 283 nm, ε 24,700) is indicative of a conjugated dienone system. That this chromophore is a 4,6-dien-3-one is strongly suggested by the close resemblance between the o.r.d. curves of the alkaloid and of ergosta-4,6,22-trien-3-one, and supported by n.m.r. spectral analysis. The latter revealed a methyl doublet, a methyl singlet, and an olefinic proton in a CH=N grouping, and mass spectrometry indicated a molecular weight of 309. Malarboreine has consequently been formulated as LXXXVI (*F. Sóti, Cerny* and *Šorm*, Tetrahedron Letters, 1967, 1437).

Malarborine, $C_{21}H_{31}NO$, is another alkaloid of the same source *(idem, loc. cit.)*. It contains a C=N linkage, and is a saturated ketone, formulated as LXXXVII. The structure has been confirmed by partial synthesis from holarrhimine (p. 395).

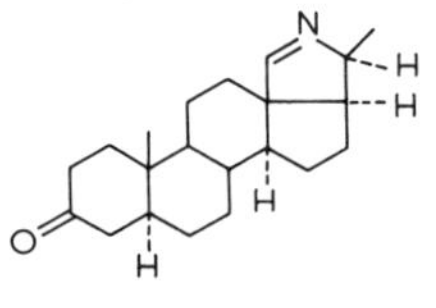

Malarborine (LXXXVII)

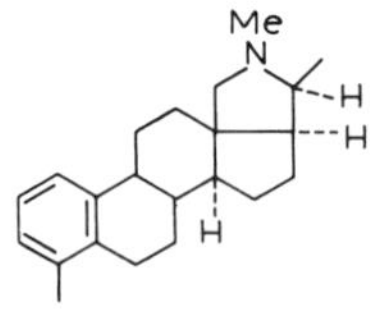

Holaromine (LXXXVIII)

Holaromine, $C_{22}H_{31}N$, possibly an artefact, has been isolated from *H. floribunda* (*Janot et al.*, Bull. Soc. chim. Fr., 1967, 4315; Ann. pharm. franç., 1967, **25**, 733).

(e) Bases related to 20-piperidyl-5α-pregnane

Plants of the genera *Solanaceae* and *Liliaceae* contain several alkaloids frequently referred to as solanum, veratrum and fritillaria bases, according to source. Most of them occur in the plant as glycosides, though a few have been isolated as free bases. They are characterised by a pregname (5α or 5β) skeleton containing a 5-methylpiperidine ring attached to $C_{(20)}$ in the steroid framework, *via* the 2-position of the heterocycle. Some have somewhat more complicated structures in that in this area of the molecule there appears an indolizidine unit (solanidine type) or an oxa-azaspirodecane unit (solasodine type). Two other sub-groups with abnormal steroid skeletons (the veratramine and cevine types) are discussed in the section on alkaloids containing a modified steroid skeleton (p. 427).

The carbohydrate fragment of the alkaloids is usually a tri- or tetrasaccharide, the monosaccharide units of which are identifiable by partial stepwise hydrolysis of the glycoalkaloid (*R. Kuhn et al.*, Ber., 1955, **88**, 289; *K. Schreiber*, Angew. Chem., 1955, **67**, 127).

(i) Simple bases

Veralkamine, $C_{27}H_{45}NO_2$, an alkaloid of *Veratrum album* (*J. Tomko et al.*, Pharm. Zentralhalle, 1960, **99**, 373), is a secondary base and contains two alcoholic hydroxyl groups, both secondary. Since the alkaloid gave a precipitate with digitonin it was presumed that one of the hydroxyl groups was in the 3β position in a steroid skeleton. Dehydrogenation of the base with selenium afforded 1,2-cyclopentenophenanthrene and 2-ethyl-5-methylpyridine. The absence of the usual angular methyl group at position 13 was suggested by the formation of the former, rather than its 3′-methyl derivative (Diels hydrocarbon). Partial hydrogenation gave a dihydro-base which on oxidation gave a diketone the i.r. spectrum of which showed two carbonyl bands, characteristic of a cyclohexanone and a cyclopentanone. Detailed spectral analysis placed the second hydroxyl group at $C_{(16)}$ and a second double bond at $C_{(12)}$–$C_{(13)}$, allowing formulation of veralkamine as LXXXIX (*Tomko* and *I. Bendik*, Coll. Czech. chem. Comm., 1962, **27**, 1404; *Tomko et al.*, Tetrahedron Letters, 1967, 3907). This structure and con-

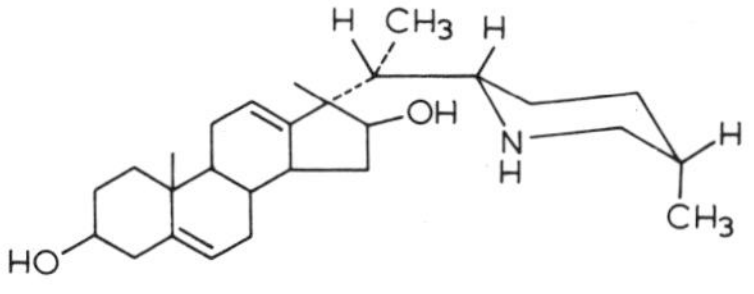

H CH3 H OH N H H CH3 HO

Veralkamine
(LXXXIX)

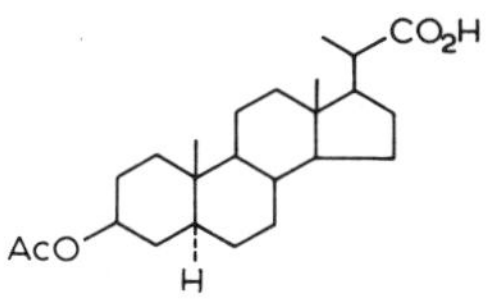

CO2H AcO H

(XC)

figuration have been confirmed by X-ray diffraction analysis of veralkamine hydriodide *(idem, ibid.)*.

Solacongestidine, $C_{27}H_{45}NO$, is the principal alkaloid of *Solanum congestiflorum*, where it occurs as a glycoside. It has been degraded chemically to 3β-acetoxybisnorallocholanic acid (XC) (*E. Bianchi et al.*, J. org. Chem., 1965, **30**, 754), by successive acetylation to an *O,N*,-diacetyl derivative, acid hydrolysis to a 26-acetylamino-22-oxo-derivative, and oxidation to the acid XC with chromic acid. It has also been synthesised from 3β-acetoxy-5-pregnene-20-one (XCI) by the route outlined below (*Schreiber* and *G. Adam*, Tetrahedron, 1964, **20**, 1707), leading to structure XCII for the alkaloid (for an alternative synthesis from solasodine see *Adam, D. Voigt* and *Schreiber*, J. prakt., Chem., 1971, **313**, 45).

CO·CH3
AcO
(XCI)
Li
N
CH3
HO–C
N
H
POCl3
py.
4 H2
hydrolysis
H
N
HO
(a) *N*-chlorosuccinimide
(b) base
N
25
23 24
O
H
Solacongestidine
(XCII)

Solafloridine, an accompanying base, is 16α-hydroxysolacongestidine (*Y. Sato et al.*, J. org. Chem. 1969, **34**, 1577). The base has been synthesised from 3β-acetoxypregna-5,16-dien-20-one in unexceptional steps (*H. Ripperger, F.-J. Sych* and *Schreiber*, Tetrahedron, 1972, **28**, 1619).

23-, and **24-oxosolacongestidine** are minor constituents of *S. congestiflorum*; their structures rest chiefly on interpretation of spectral data *(Sato et al., loc. cit.)*.

Verazine, $C_{27}H_{43}NO$, occurring in *Veratrum album*, is the Δ^5, $C_{(25)}$ epimeride of the foregoing alkaloid (*Tomko* and *A. Vassova*, Chem. Zvesti, 1964, **18**, 266). Spectral measurements have been used extensively in elucidation of its structure. (*Adam et al.*, Tetrahedron, 1967, **23**, 167; *Adam* and *Schreiber*, Tetrahedron Letters, 1963, 943). The alkaloid, XCIV, has been synthesised from Δ^5-tomatiden-3β-ol (XCIII; see p. 413) as follows (*Adam, Schreiber* and *Tomko*, Ann., 1967, **707**, 203).

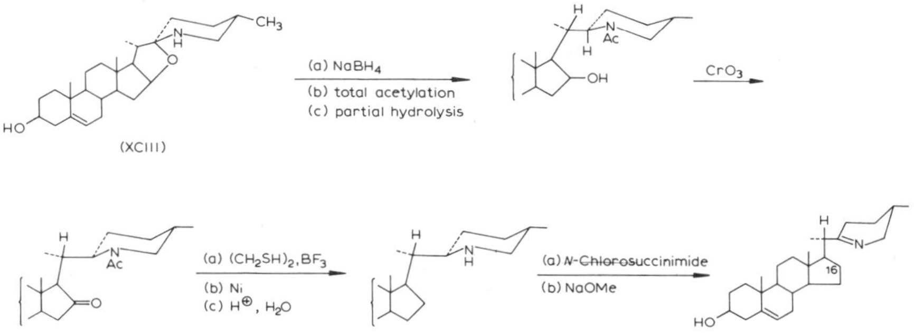

(XCIII)

(a) NaBH$_4$
(b) total acetylation
(c) partial hydrolysis

CrO$_3$

(a) (CH$_2$SH)$_2$,BF$_3$
(b) Ni
(c) H$^{\oplus}$, H$_2$O

(a) N-Chlorosuccinimide
(b) NaOMe

Verazine
(XCIV)

Tomatillidine, $C_{27}H_{41}NO_2$, is a minor alkaloid of *Solanum tomatillo* (*Bianchi, F. Diaz* and *J. A. Garbarino*, Gazz., 1960, **90**, 894). It is a secondary alcohol and a ketone, and also a tertiary base, but contains no *N*-methyl group. Two double bonds are present, C=C and C=N in type. A pointer to its structure was the observation that its dihydro derivative, on Wolff–Kishner reduction, afforded solacongestidine *(vide supra)*; tomatillidine is therefore a ketodehydrosolacongestinine. Oppenauer oxidation of the base gave an α,β-unsaturated ketone ($\lambda_{max.}^{EtOH}$ 240 nm, ε 17,000), behaviour suggesting strongly that the carbon–carbon double bond is 5,6. The position for the keto group was settled by total acetylation of the alkaloid to an *N,O*-diacetyl derivative; this reaction involves a shift of the C=N double bond into the neighbouring C–C bond, and since the product shows spectral characteristics of a conjugated enone this shift must have been into conjugation with the keto-group, which must therefore be placed at $C_{(24)}$. Tomatillidine is consequently to be formulated as XCV *(Bianchi et al., loc. cit.)*.

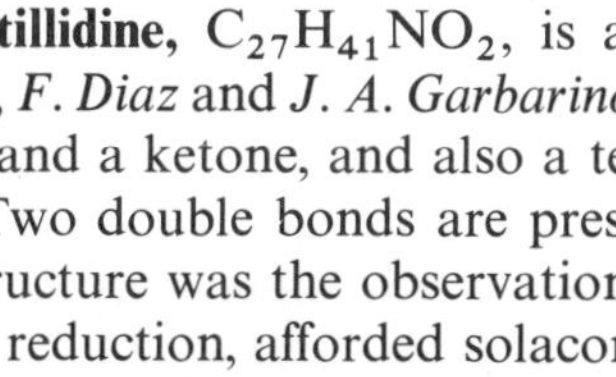

Tomatillidine
(XCV)

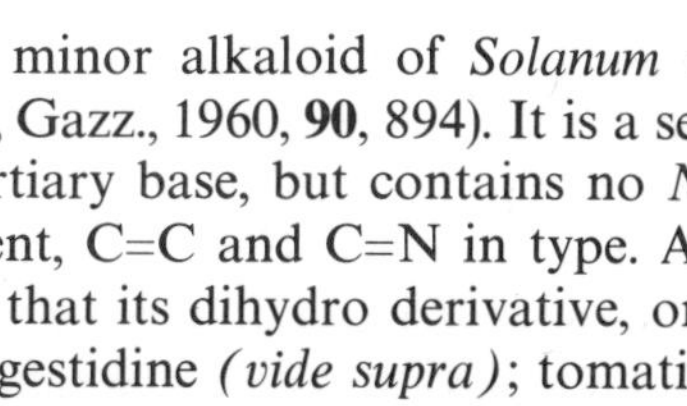

Solaphyllidine
(XCVI)

The configuration at $C_{(24)}$ has been settled by a synthetic sequence starting from 3β-acetoxypregn-5-en-20-ene (*Schreiber* and *Adam*, *loc. cit.*). **5α,6-Dihydrotomatillidine** also occurs in the same plant *(Bianchi et al., loc. cit.)*.

Solaphyllidine, $C_{29}H_{47}NO_5$, is the most abundant alkaloid of the leaves and berries of *S. hypomalacophyllum.* A careful analysis of its mass spectrum revealed the presence of a hydroxytetrahydropyridinium ion (*m/e* 114), whilst fragments at *m/e* 447 (M–42) and 429 (M–60) suggested the presence of an *O*-acetyl group in the molecule, the latter being confirmed by i.r. spectroscopy, which also revealed the presence of a hydroxyl group. The n.m.r. spectrum pointed to the presence of two angular *C*-methyl groups and two secondary methyl groups. Acetylation of the alkaloid yielded a triacetyl derivative, whilst reduction with lithium tetrahydridoaluminate afforded a tetraol, with generation of two new hydroxyl groups. An X-ray diffraction analysis of a single crystal of the alkaloid points to structure XCVI, though this is not the proven absolute configuration, it being assumed that the alkaloid is related to cholesterol (*A. Usubillaga et al.*, J. Amer. chem. Soc., 1970, **92**, 700).

Veralinine, $C_{27}H_{43}NO$, a minor alkaloid of *V. album* subsp. *lobelianum*, is a secondary amine and a secondary alcohol, and contains two unconjugated double bonds. Spectroscopic evidence pointed to the structure XCVII; this formulation was confirmed by a correlation of the base with veralkamine (p. 411), and the positions of the double bonds derived by molecular rotation difference studies (*Tomko et al.*, Tetrahedron, 1968, **24**, 6839).

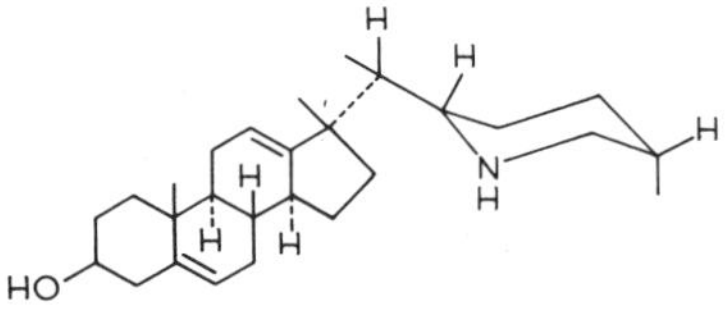

HO H H H H H N H H

Veralinine
(XCVII)

Veracintine, $C_{26}H_{41}NO$, from the above-ground parts of the same plant, contains two isolated double bonds, a secondary alcohol group and is a tertiary amine (not *N*-methyl). The alkaloid has a pyrroline ring attached to $C_{(20)}$, this moiety being detected by n.m.r. and mass spectroscopy; it does not therefore strictly belong to this subgroup, but is conveniently considered here. A base peak of *m/e* 82 in the mass spectrum is characteristic of a fragment generated by 20–22 bond fission to yield a pyrroline. Veracintine is formulated as XCVIII (*Tomko, V. Brazdova* and *Z. Voticky*, Tetrahedron Letters, 1971, 3041). A second alkaloid, $C_{26}H_{39}NO$, from the same source, has been shown by mass and n.m.r. spectral studies and optical rotatory dispersion to be 20-(2-methyl-1-pyrrolin-3-yl)-4-pregnen-3-one (XCVII, C=O at $C_{(3)}$ and $\Delta^{4,5}$). Confirmation of this formulation has been provided by a correlation with veracintine (*Vassova* and *Tomko*, Coll. Czech. chem. Comm., 1975, **40**, 695).

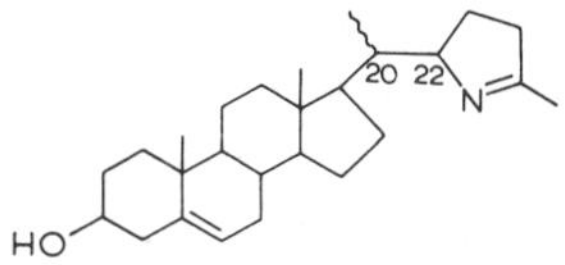

Veracintine
(XCVIII)

Etioline, $C_{27}H_{43}NO_2$, is a glycosidic base of the leaves of budding *V. grandiflorum.* It and an accompanying minor base are possible precursors of solanidine (p. 416), because on etiolation (deprivation of sunlight) the concentration of these bases decreases, whilst that of solanidine increases. The structure proposed for etioline is the result of a consideration of this possible biological relationship, of its i.r. spectrum (C=N and OH groups detected), of its n.m.r. spectrum (single olefinic proton), and of its formation of an *N,O,O*-triacetate (an enamine acetate) containing an additional olefinic proton signal in its n.m.r. spectrum. On Oppenauer oxidation etioline forms an α,β-unsaturated ketone, and on chromic acid oxidation a diketone results. The latter shows i.r. absorption characteristic of 5- and 6-membered ring carbonyl groups. The mass spectrum of the alkaloid, with a base peak at *m/e* 125, and other peaks at 124 and 98, appeared in harmony with the proposed structure when compared with the spectra of other *Veratrum* bases (*K. Kaneko et al., ibid.*, 1971, 4251), and etioline is formulated as 16α-hydroxyverazine [XCIV, α-OH at $C_{(16)}$].

Two minor alkaloids, **veralosidine** and **veralosinine**, have been isolated from the aerial part of *V. album* subsp. *lobelianum.* The former is a disecondary diol, and the latter its monoacetate. Veralosidine is formulated as the $C_{(25)}$ epimer of etioline (see above) and veralosinine as the 16-acetate (*A. M. Khashimov, R. Shakirov* and *S. Y. Yusonov*, C.A., 1972, **76**, 141109y).

(ii) Solanidanes

These bases, found mainly in species of *Solanum*, which include the common potato and tomato, are characterised structurally by the presence of a bicyclic octahydropyrrocoline unit fused on to ring D of a steroid framework.

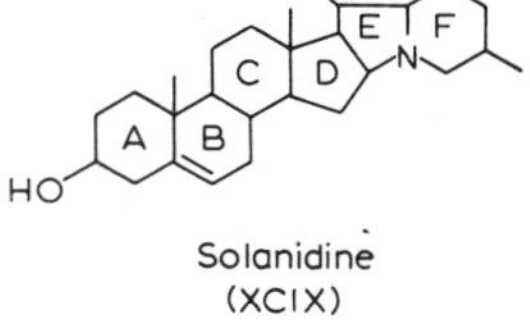

Solanidine
(XCIX)

Solanine, $C_{45}H_{73}NO_{15}$, and **solanidine,** $C_{27}H_{43}NO$, are respectively a glycoalkaloid, found in potato shoots, and the corresponding aglycone. Other glycobases related to solanidine are the **chaconines**: α-chaconine and α-solanine are trisaccharides, whilst β- and δ-chaconine and β-solanine are their partial hydrolysis products, also of natural occurrence (*Kuhn* and *I. Löw*, Angew. Chem., 1954, **66**, 639; *Schreiber*, Chem. Ber., 1954, **87**, 1007; Kulturplflanze, 1963, **11**, 422). Solanidine contains a tertiary amino group, an ethylenic linkage, and a secondary alcoholic group; the first is not an *N*-methyl group and is resistant to Hofmann degradation. It was therefore postulated at an early stage in structural investigations that the nitrogen atom was common to two rings (*A. Soltys* and *K. Wallenfels*, Ber., 1936, **69**, 811; *Soltys*, *ibid.*, 1933, **66**, 762; *C. Schöpf* and *R. Hermann*, *ibid.*, p. 298). Dehydration *via* the secondary hydroxyl group is facile, leading to solanidiene (solanthrene), $C_{27}H_{41}N$, a conjugated, heteroannular diene. On dehydrogenation with selenium solanidine affords the Diels hydrocarbon (*H. Rochelmeyer*, Arch. Pharm., 1936, **274**, 543; *L. C. Craig* and *W. A. Jacobs*, J. biol. Chem., 1943, **149**, 451), and the alkaloid forms a precipitate with digitonin; a steroidal structure was therefore anticipated. The chemical behaviour of the alcohol group coupled with that of the double bond and with molecular rotation difference studies left no doubt that the A/B bicyclic system of cholesterol was present in the molecule (*E. Geyer*, *Rochelmeyer* and *C. S. Shah*, Ber., 1938, **71**, 226; *Rochelmeyer*, Arch. Pharm., 1939, **277**, 340; *V. Prelog* and *S. Szpilfogel*, Helv., 1944, **27**, 390).

Accompanying the Diels hydrocarbon on dehydrogenation of the base was 2-ethyl-5-methylpyridine. Solanidine does not yield a $C_{(17)}$-ketone when successively *O*-acetylated and oxidised with chromic acid; this suggested that the basic part of the molecule is linked to the steroidal ring D in more than one position. *Prelog* and *Szpilfogel (loc. cit.)*, proposed, on these and other grounds, that solanidine was to be formulated as XCIX; this formulation was corroborated by *Jacobs* and *F. C. Uhle* (J. biol. Chem., 1945, **160**, 243), who transformed sarsasapogenic acid, an oxidation product of sarsasapogenin (*C.C.C.*, Vol. IIE, p. 29), into 5β-solanidan-3β-ol (C) by the four steps outlined below.

Solanidine has been converted into the last in two stages. These transformations in addition establish the stereochemistry of solanidine for rings A–D; further, since the side-chain in steroids at $C_{(17)}$ is β, the nitrogen atom in the alkaloid must be β. Moreover, the configuration at $C_{(25)}$ must be the same as in sarsasapogenin, and that at $C_{(20)}$ is, reasonably, the unhindered natural configuration found in steroids generally (for details see *W. E. Rosen* and *D. B. Rosen*, Chem. and Ind., 1954, 1581, and references there cited). The hydrogen atom at $C_{(22)}$ is very probably β, on evidence

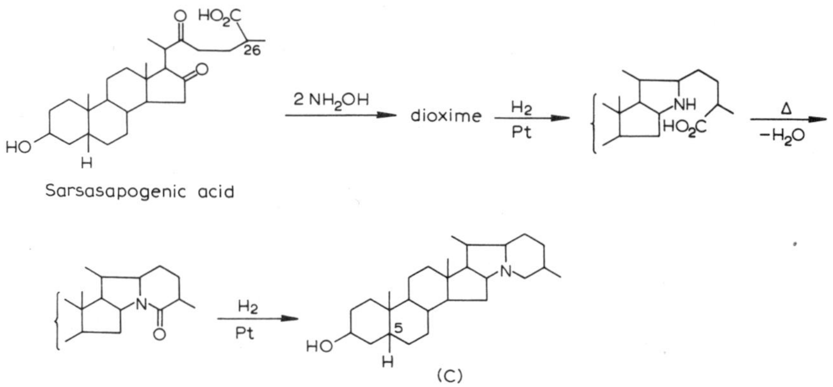

Sarsasapogenic acid

2 NH_2OH → dioxime $\xrightarrow[Pt]{H_2}$ → $\xrightarrow[-H_2O]{\Delta}$

$\xrightarrow[Pt]{H_2}$ (C)

emanating from a correlation between solanidane, solanidan-3β-ol and rubijervine (see p. 418; *idem, ibid.*; *F. L. Weisenborn* and *D. Burn*, J. Amer. chem. Soc., 1953, **75**, 259), and from a molecular model study. Solanidine is consequently represented by stereo-structure CI (*Sato* and *H. G. Latham*, Chem. and Ind., 1955, 444, J. Amer. chem. Soc., 1956, **78**, 3146).

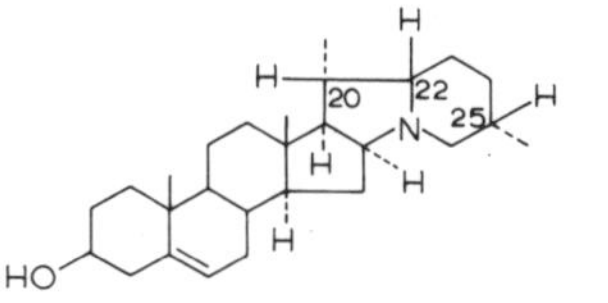

Solanidine (CI)

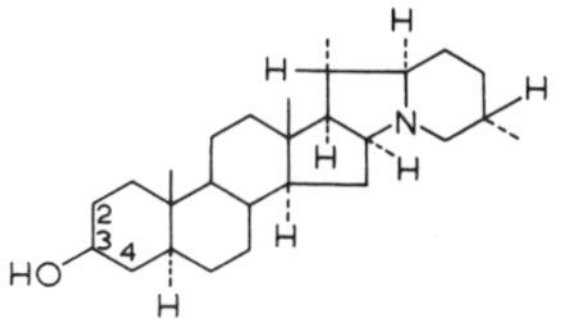

Demissidine (CII)

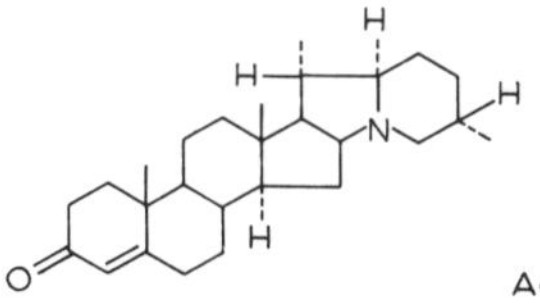

(CIII)

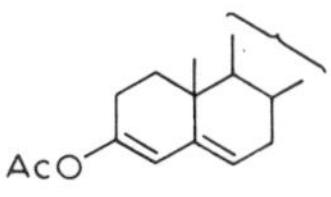

(CIV)

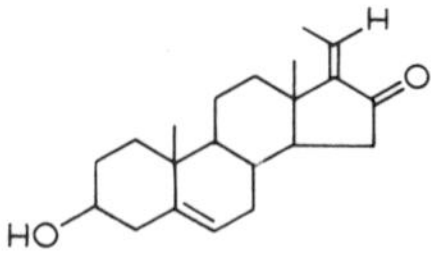

(CV)

This formulation has been confirmed by several partial syntheses of solanidine and compounds derived therefore (*Sato* and *N. Ikegawa*, J. org. Chem., 1961, **26**, 1945), and by a total synthesis of the alkaloid from totally synthetic demissidine (CII) *(vide infra)*. Oxidation of the latter (CrO_3) afforded demissidone, the 3-ketone, which on bromination gave the 2α, 4α-dibromoketone. Successive treatment with sodium iodide–acetone (dehydrobromination) and chromous chloride (halide hydrogenolysis) yielded solanid-4-en-3-one (CIII). The latter was converted into its enol acetate CIV, which on reduction with sodium tetrahydridoborate gave solanidine (CI) (*Schreiber* and *H. Rönsch*, Experientia, 1961, **17**, 491; Chem. Ber., 1964, **97**, 2362). Another total synthesis starts from 3β-hydroxy-*cis*-pregna-5,17(20)-dien-16-one (CV), which with ethyl (*S*)-5-nitro-2-methylpentanoate underwent Michael addition to give two epimeric ethyl (25*S*)-22-nitro-3β-hydroxy-16-oxocholest-5-en-en-26-oates (CVI).

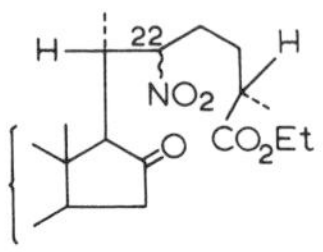

(CVI)

This on reduction and lactamisation afforded two epimeric lactams, which on reduction ($LiAlClH_4$) generated solanidine (CI) and its 22-epimeride (*S. V. Kessar, Y. P. Gupta* and *A. L. Rampal*, Tetrahedron Letters, 1966, 4319; *Kessar et al.*, Tetrahedron, 1971, **27**, 2153).

Demissine, $C_{50}H_{83}NO_{20}$, is a glycoalkaloid found in *S. demissum* leaves (*Kuhn* and *Löw*, Chem., Ber., 1947, **80**, 406). Acid hydrolysis affords glucose (2 mols), galactose (1 mol.), xylose (1 mol.), and **demissidine**, the aglycone, $C_{27}H_{45}NO$. It was observed that catalytic hydrogenation of solanidine (CI) yielded demissidine, which must, therefore, be solanidan-3β-ol(CII), since it is different from *Jacobs* and *Uhle*'s partially synthetic product (C) (see p. 417), the two being $C_{(5)}$-epimers. This formulation has been confirmed by conversion of tomatidine (p. 423) into demissidine (*Kuhn, Löw* and *H. Trischmann*, Angew. Chem., 1952, **64**, 397), and by a total synthesis of the base (*S. V. Kessar et al.*, Tetrahedron, 1971, **27**, 2153).

Hydroxysolanidines. **Rubijervine,** $C_{27}H_{43}NO_2$, occurs in the rhizomes of *Veratrum album* (*C. R. A. Wright* and *A. P. Luff*, J. chem. Soc., 1879, **35**, 405, 421; *Jacobs* and *Craig*, J. biol. Chem., 1943, **148**, 41). It is a disecondary diol, a tertiary base (not *N*-methyl), contains one olefinic bond, and is precipitated by digitonin (*idem, ibid.*, 1945, **159**, 617). Dehydrogenation products include 2-ethyl-5-methylpyridine and 1′-methyl-1,2-cyclopentenophenanthrene (*Sato* and *Jacobs, ibid.*, 1949, **179**, 623). That rubijervine is a hydroxysolanidine was indicated by the fact that its 3-benzoate was oxidisable to a ketone, which on Wolff–Kishner reduction afforded solanidine (*idem, loc. cit.*). Spectral studies revealed that the second hydroxyl group is at $C_{(12)}$. Further, the group cannot be located in rings A or B, because rubijervine is not a 1,2-glycol, the diketone oxidation product of rubijervine is not a 1,3-diketone, and the ketone mentioned above is not a conjugated enone. Next, one of the dehydrogenation products of the alkaloid is a phenol, $C_{18}H_{15}OH$, the formation of which rules out positions 23, 24 and 26 for the second hydroxyl group. The diketone described above almost certainly does not have one of its carbonyl groups at position 11, because $C_{(11)}$-ketones are well known to be chemically inert, which is contrary to the chemical behaviour of the diketone. Finally, on biogenetic grounds, the group is unlikely to be at $C_{(15)}$, and optical rotation studies suggest it is at $C_{(12)}$ and has the α-orientation *(idem, loc. cit.)*. Rubijervine is consequently 12α-hydroxysolanidine (CVII) (*S. W. Pelletier* and *D. M. Locke*, J. Amer. chem. Soc., 1957, **79**, 4531).

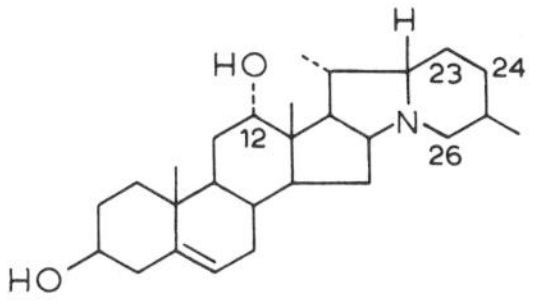

Rubijervine
(CVII)

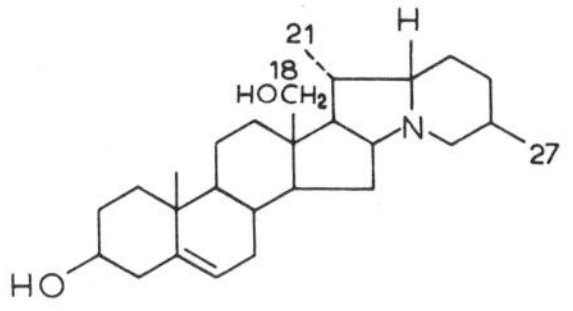

Isorubijervine
(CVIII)

Isorubijervine, also $C_{27}H_{43}NO_2$, is found in the same source and in *Veratrum viride* and *V. eschscholtzii*; in the rhizomes of the latter it occurs as its glucoside **isorubijervosine**, $C_{33}H_{53}NO_7$ (*M. W. Klohs* and co-workers, *ibid.*, 1953, **75**, 2133). Isorubijervine bears a resemblance to rubijervine, but one of its two hydroxyl functions is primary. Dehydrogenation yields 2-ethyl-5-methylpyridine and 1,2-cyclopentenophenanthrene, and the base is precipitated by digitonin. The nitrogen atom is tertiary, and not methylated. Vigorous oxidation (CrO_3) of dihydroisorubijervine affords a C_{27}-oxo-acid, the methyl ester of which (prepared using diazomethane) is very resistant to hydrolysis; the carboxyl group of the acid is evidently hindered and probably tertiary and angular, though it is evidently not located at position 10, since certain strophanthidin (*C.C.C.*, Vol. II D, p. 373) derivatives known to have such an angular methoxycarbonyl group are more easily hydrolysed than this ester. Isorubijervine forms a toluene-4-sulphonate (CIX), convertible by iodide ions to an iodide, both derivatives having the character of quaternary ammonium salts and being reducible to solanidine (CI). The primary alcohol group must therefore be within bonding distance of the nitrogen atom, the three possible positions being 18, 21 and 27 in structure CVIII. Of these, the first is preferred on steric grounds, and because of the tertiary nature of the $C_{(27)}$-oxo-acid (see above). Structure CVIII for the alkaloid explains these transformations, outlined below (*Jacobs* and *Sato*, J. biol. Chem., 1951, **191**, 63; *Pelletier* and *Jacobs*, J. Amer. chem. Soc., 1952, **74**, 4218; 1953, **75**, 4442; *D. Burn* and *W. Rigby*, Chem. and Ind., 1952, 668; J. chem. Soc., 1953, 963; *Burn* and *F. L. Weisenborn*, J. Amer. chem. Soc., 1953, **75**, 259; *Jacobs* and *Craig*, J. biol. Chem., 1943, **148**, 41; 1945, **159**, 617):

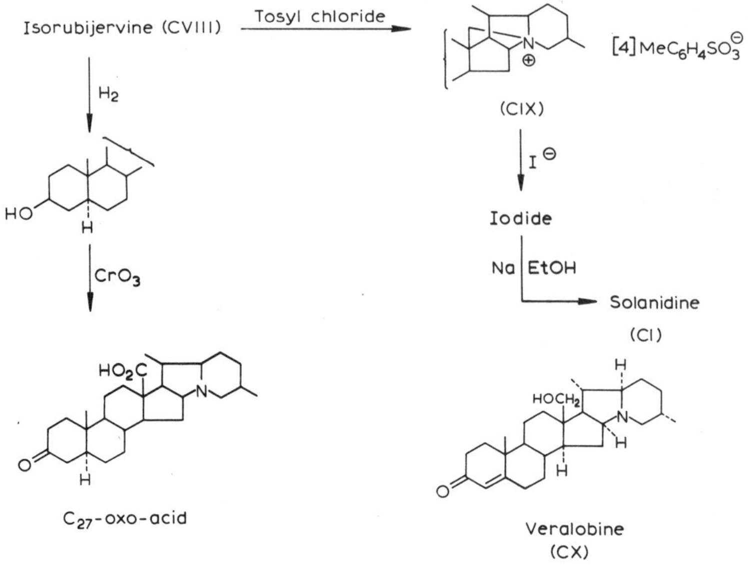

Confirmation of structure CVIII is provided by oxidation of isorubijervine to the corresponding aldehyde, followed by Wolf–Kishner reduction, which yields 5β-solanidan-3β-ol, a compound of known constitution (*Burn* and *Rigby*, *loc. cit.*; *Burn* and *Weisenborn*, *loc. cit.*). Finally, X-ray crystallographic studies and inter-relationships have revealed that both rubijervine and isorubijervine are 22*R*:*NS* in configuration (*E. Höhne et al.*, Tetrahedron, 1966, **22**, 673).

Veralobine, $C_{27}H_{41}NO_2$, present in *V. album* subsp. *lobelianum*, is an α,β-unsaturated ketone ($\lambda_{max.}^{EtOH}$ 242 nm), confirmed by its i.r. spectrum, and is also a primary alcohol. Its mass spectrum, with an intense peak at *m/e* 150, is consonant with an isorubijervine-type structure (*H. Budzikiewicz*, *ibid.*, 1964, **20**, 2267). The n.m.r. spectrum of the alkaloid has been analysed in detail, and, with the above observations is consistent with the formulation of veralobine as Δ^4-isorubijervone (CX) (*J. Tomko* and *A. Vassova*, Pharmazie, 1965, **20**, 385; *Tomko et al.*, Arch. Pharm., 1966, **299**, 347). This structure has been confirmed by the formation of verolobine by Oppenauer oxidation of isorubijervine.

Leptinidine, $C_{27}H_{43}NO_2$, occurs as the glycosides of its acetoxy-derivative, the **leptines**, in the leaves of *Solanum chacoense.* Hydrolysis of the acetyl group yields the **leptinines**, the glycosides of leptinidine. Leptinidine has a double bond and is a disecondary glycol. The presence of a cholesterol-type system was suggested by the precipitation of the base with digitonin, by molecular rotation difference measurements, by dehydrogenation (with copper-bronze) to a conjugated enone, and by acid-catalysed dehydration to a heteroannular diene. Oxidation followed by Wolff–Kishner reduction afforded solanidane. The above diene on dehydrogenation yielded 2-ethyl-3-hydroxy-5-methylpyridine; this places the second hydroxyl at $C_{(23)}$, and i.r. and n.m.r. investigations show it has the β-orientation. Leptinidine is consequently 23β-hydroxysolanidine (CXI) (*Kuhn* and *Löw*, Ber., 1961, **94**, 1088, 1096; 1962, **95**, 1748; *Schreiber* and *Ripperger*, *ibid.*, 1967, **100**, 1381). The alkaloid has been synthesised from tomatidenol (p. 425) (*idem*, Ber., 1969, **102**, 4080).

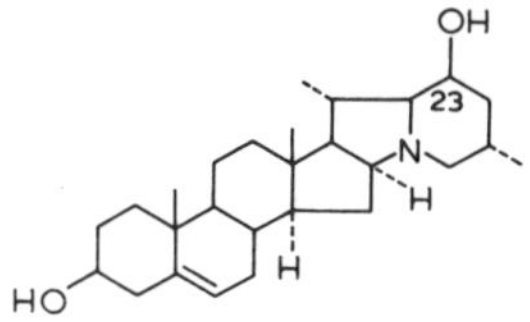

Leptinidine
(CXI)

Solanocapsine does not belong structurally to the solanidine group, but is conveniently discussed here. The alkaloid, $C_{27}H_{46}N_2O_2 \cdot H_2O$, occurs in *S. pseudocapsicum*, *S. capsicastrum* and *S. hendersonii* (*G. Barger* and *H. L. Fraenkel-Conrat*, J. chem. Soc., 1936, 1537). It yields Diels hydrocarbon and 2-ethyl-5-methylpyridine on dehydrogenation (*E. Schlittler* and *H. Uehlinger*, Helv., 1952, **35**, 2034). The nitrogen atoms are accounted for in terms of primary and secondary basic functions, and one of the oxygens is present as an ether function which, in conjunction with a hydroxyl group, is part of a hemiketal grouping. The alkaloid has been degraded to (−)-4-amino-3-methylbutyric acid and tigogenin lactone (*C.C.C.*, Vol. II E, p. 22) (*Schreiber* and *Ripperger*, Ann., 1962, **655**, 114) as follows. The primary amine group was blocked as the salicylidene anil, then the secondary amine group was acetylated. Acid hydrolysis eliminated the protecting group, and treatment of the product with nitrous acid afforded the 3β-alcohol. With hot acetic acid the acetate of this alcohol was dehydrated to an enamine, oxidation of which (CrO_3) yielded a macrocyclic lactam-lactone. Acid hydrolysis of the latter furnished tigogenin lactone (p. 422).

Similarly, the *N,N'*-diacetyl derivative of solanocapsine yielded (−)-4-amino-3-methylbutyric acid, the absolute configuration of which was settled by correlation with (+)-methylsuccinic acid, of known absolute stereochemistry. This establishes the configuration of the 25-methyl group as α, and solanocapsine is therefore formulated as CXII. Molecular rotation difference studies on the alkaloid, together with a comparison with a synthetic

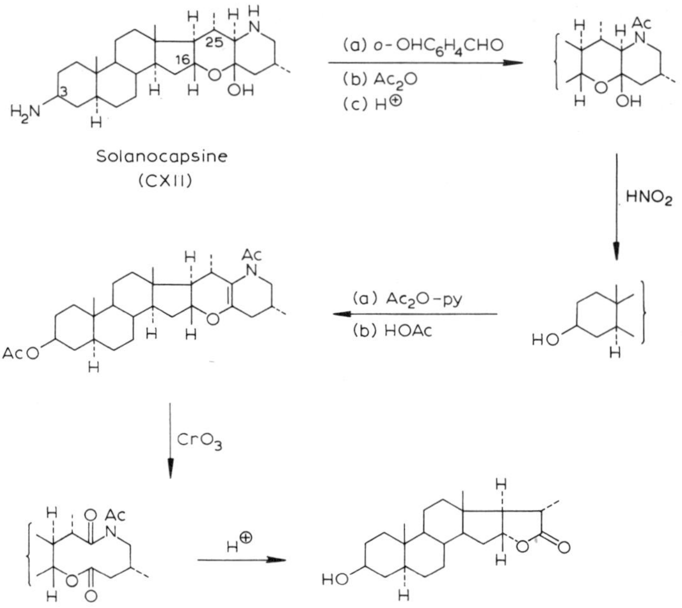

Solanocapsine
(CXII)

(a) o-$OHCC_6H_4CHO$
(b) Ac_2O
(c) $H^{\oplus}$

HNO_2

(a) Ac_2O-py
(b) HOAc

CrO_3

$H^{\oplus}$

Tigogenin lactone

sample of the 16β isomer, show that the configuration of the ether oxygen at $C_{(16)}$ is α (*idem, ibid.*, 1969, **723**, 159; *N. Nagai* and *Y. Sato*, Tetrahedron Letters, 1970, 2911). An X-ray analysis of *N*-2-bromobenzylidenesolanocapsine has been confirmed the structural assignment (*Höhne*, *Ripperger* and *Schreiber*, Tetrahedron, 1970, **26**, 3569), as has a synthesis from solafloridine (p. 412) (*Ripperger*, *F.-J. Sych* and *Schreiber*, Tetrahedron Letters, 1970, 5251; Tetrahedron, 1972, **28**, 1629).

(iii) Spirosolane alkaloids

This small group of alkaloids is characterised by the presence of an oxa-azaspirane bicyclic unit fused *via* a furan ring to ring D of the steroid framework, as in partial structure CXIII:

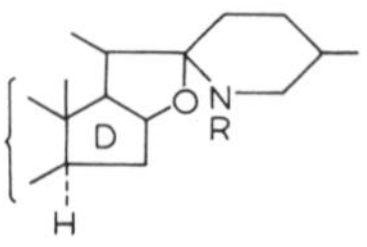

(CXIII)

Tomatine, $C_{50}H_{83}NO_{21}$, widely distributed in leaves of wild tomato plants (*Lycopersicum* spp.), yields on hydrolysis its aglycone **tomatidine,** $C_{27}H_{45}NO_2$, and the same monosaccharides as demissine (*T. D. Fontaine et al.*, Arch. Biochem., 1948, **18**, 467; 1950, **27**, 461; Ind. Eng. Chem., 1948, **26**, 2440; *Kuhn* and *Löw*, Chem. Ber., 1948, **81**, 552; *A. Gauhe, Kuhn* and *Löw, ibid.*, 1950, **83**, 448).

The aglycone is a secondary amine, is saturated, and is also a secondary alcohol; the inert character of the second oxygen atom suggested it was ethereal in character. Catalytic or chemical reduction of the base affords a diol, the second hydroxyl group being a consequence of the cleavage of a cyclic ether unit, rendered sensitive to such reducing conditions by a spiro union with a nitrogen ring. The reactions, molecular formula, and the precipitation of tomatidine by digitonin, suggested a 3β-hydroxysteroidal structure for the base (*J. S. Ard, Fontaine* and *R. M. Ma*, J. Amer. chem. Soc., 1951, **73**, 879). A close structural analogy between tomatidine and steroidal sapogenins (*C.C.C.*, Vol. II E, p. 6 *et seq.*) was apparent from the following transformations (*Sato, A. Katz* and *E. Mosettig, ibid.*, p. 880; *R. Kuhn* and *I. Löw*, Chem. Ber., 1952, **85**, 416).

$\{C_{27}H_{43}O$; HO–, >NH$\}$ Tomatidine —Ac_2O→ $\{C_{27}H_{43}O$; AcO–, >NAc$\}$ (CXIV) —$H^{\oplus}$→ $\{C_{27}H_{42}O$; AcO–, –NHAc$\}$ (CXV) $\{C_{27}H_{42}O$; AcO–, –$NAc_2\}$ (CXVI)

CXIV —CrO_3→ (CXVII); CXV —CrO_3→ (CXVII) and (CXVIII); CXVI —CrO_3→ (CXVIII)

AcO, H, O, $CO \cdot CH_3$

(CXVII) (CXVIII)

Mild acetylation of tomatidine affords an *N,O*-diacetate, CXIV, rearranged by acid to an isomeric diacetate, CXV, formation of which involves rupture of the nitrogen heterocycle in CXIV. More vigorous acetylation of the base generates a triacetyl derivative CXVI. Chromic acid oxidation of all three acetyl derivatives yields the acetoxylactone (CXVII) and 3β-acetoxy-5α-pregn-16-en-20-one (CXVIII) as outlined, these two products being known as oxidation products of acetyltigogenin (*C.C.C.*, Vol. II E, p. 7) and diacetyl-ψ-tigogenin (*ibid.*, p. 45), respectively. Because of these transformations structure CXIX was advanced for tomatidine, and substantiated by *Kuhn, Löw* and *Trischmann* (Angew. Chem., 1952, **64**, 397; *Kuhn* and *Löw*, *loc. cit.*), who observed that dihydrotomatidine (CXX) on dehydrogenation gave 2-ethyl-5-methylpyridine, and transformed tomatidine into demissidine (CII, see p. 417) as summarised below.

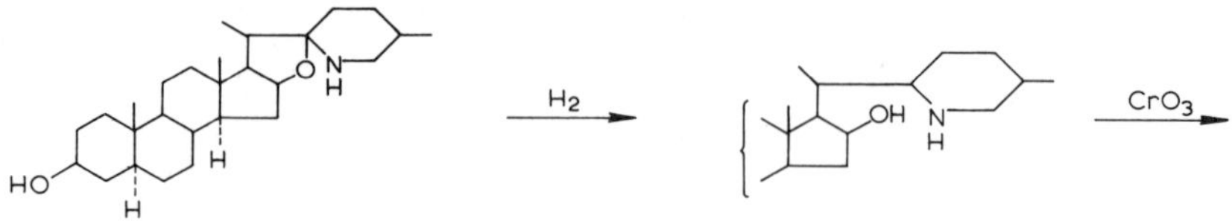

Tomatidine (CXIX) $\xrightarrow{H_2}$ (CXX) $\xrightarrow{CrO_3}$

$\xrightarrow{(1)\ 2H_2,\ (2)\ -H_2O}$ Demissidine (CII)

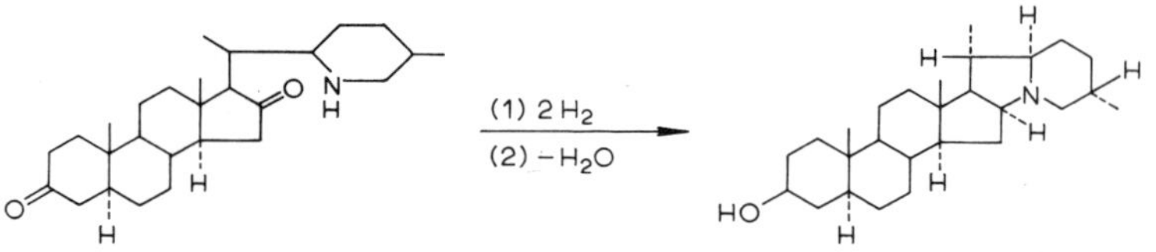

Configurational studies on tomatidine reveal that it corresponds in configuration to the sapogenin neotigogenin (*C.C.C.*, Vol. II E, pp. 22, 30), and is therefore more accurately represented by partial stereoformula CXXI (*Sato* and *Latham*, J. Amer. chem. Soc., 1965, **78**, 3146, 3150; *Uhle* and *J. A. Moore*, *ibid*., 1954, **74**, 6412; *Rosen* and *Rosen*, Chem. and Ind., 1954, 1581; *Sato et al*., J. Amer. chem. Soc., 1957, **79**, 6089). The base has been synthesised by two procedures: in the first (*Uhle* and *Moore*, *loc. cit*.), the 27-amine (CXXII) obtained by several simple steps from ψ-neotigogenin

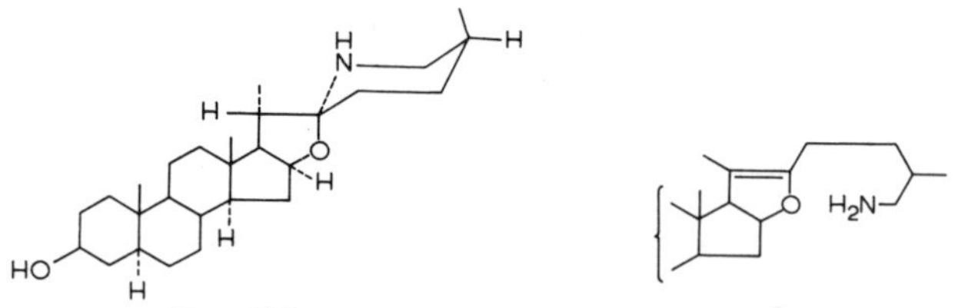

Tomatidine (CXXI)

(CXXII)

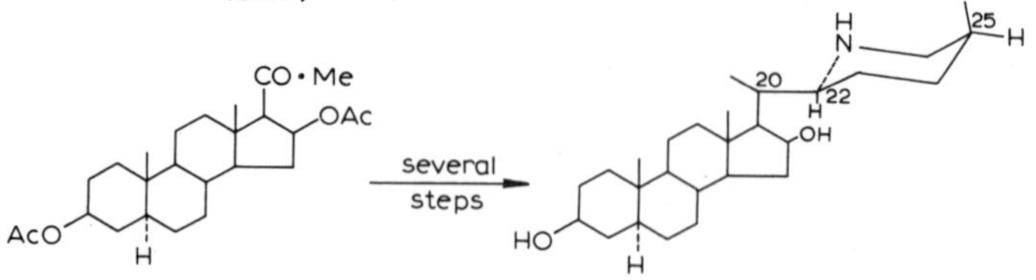

(CXXIII) $\xrightarrow{\text{several steps}}$ (CXXIV)

(*C.C.C.*, Vol. II E, p. 9) was cyclised with phosphoric acid. Also, 3β,16β-diacetoxy-5α-pregnan-20-one (CXXIII) was converted in several stages into the diol CXXIV, with the correct 20(*R*), 22(*S*), 26(*S*) stereochemistry.

This was transformed into its *N*-chloro derivative by *N*-chlorosuccinimide, which was cyclised with sodium methoxide to tomatidine (CXXI) (*Schreiber* and *Adam*, Ann., 1963, **666**, 155).

Tomatidenol, $C_{27}H_{43}NO_2$, is an alkaloidal constituent of *Solanum dulcamara*. It contains one double bond, catalytic hydrogenation of which affords tomatidine. The placing of the double bond between positions 5 and 6 has been confirmed by synthesis from diol CXXIV (*Schreiber* and *H. Rönsch*, *ibid.*, 1965, **681**, 196). The base occurs in the plant as a series of glycosides (the **solamarines**) (*P. M. Boll*, Acta chem. Scand., 1963, **17**, 1852, 2126; *Rönsch* and *Schreiber*, Phytochem., 1966, **5**, 1227), and is the $C_{(5,6)}$ unsaturated analogue of tomatidine (CXXI).

Solasonine and **solamargine** are two glycoalkaloids which afford on hydrolysis the aglycone **solasodine**, together with glucose, galactose and rhamnose, and glucose and rhamnose (two molecules), respectively (*Schreiber*, Kulturpflanze, 1963, **11**, 451; Planta Med., 1958, **6**, 435). They occur in *S. sodomeum*, *S. aviculare* and related species (*L. H. Briggs*, J. chem. Soc., 1942, 3).

Solasodine, $C_{27}H_{43}NO_2$, affords the Diels hydrocarbon on dehydrogenation, accompanied by 2-ethyl-5-methylpyridine, and yields a precipitate with digitonin (*Rochelmeyer*, Arch. Pharm., 1937, **275**, 336; *Briggs et al.*, J. chem. Soc., 1952, 3587). The base contains secondary alcoholic and amine groups and an ethylenic linkage, and its chemical reactions, rotational characteristics and spectral properties suggest strongly the presence of a cholesterotype A/B ring system (*Briggs* and *T. O'Shea*, *ibid.*, p. 1654; *Briggs et al.*, *ibid.*, 1950, 3013; *Briggs* and *Locker*, *ibid.*, p. 3020). *Briggs* and co-workers have proposed (*loc. cit.*) that solasodine is to be formulated as CXXV, which is also the gross structure of tomatidenol (see above), so that solasodine and tomatidenol must be stereoisomers.

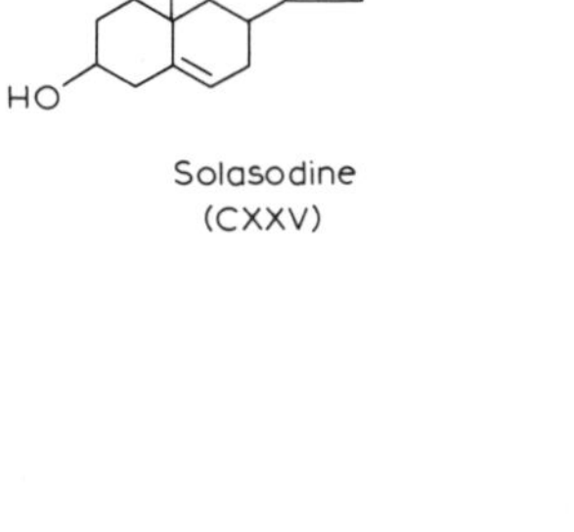

Solasodine
(CXXV)

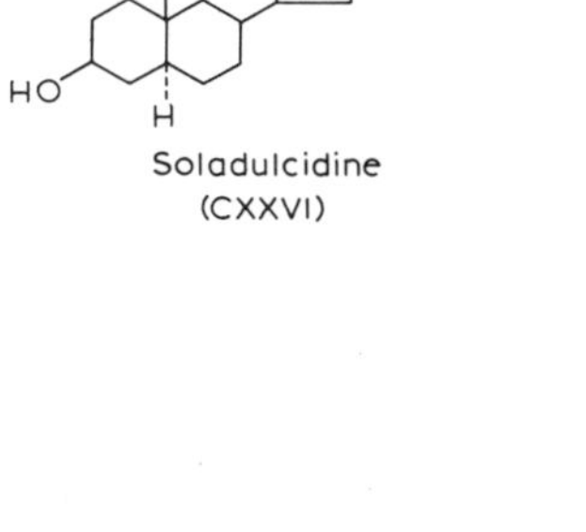

Soladulcidine
(CXXVI)

Likewise dihydrosolasodine (**soladulcidine**) has a gross structure CXXVI identical with that of tomatidine (CXXI), and these two bases are also stereoisomeric. The stereochemistry at positions 22 and 25 of solasodine and tomatidine has been revealed by several transformations of the bases. Firstly, soladulcidine has been degraded into (*R*)-(+)-5-methyl-2-piperidone, and tomatidine into (*S*)-(−)-5-methyl-2-piperidone, indicating that tomatidine and soladulcidine (and therefore solasodine) are epimeric at $C_{(25)}$. Next, differences in chemical behaviour, in molecular rotations, and in optical rotatory dispersion point to the existence of additional stereoisomerism at $C_{(22)}$ between tomatidine and soladulcidine (*L. Toldy*, Acta chim. Acad. Sci. Hung., 1958, **16**, 403; *Boll*, Acta chem. Scand., 1960, **14**, 783; 1963, **17**, 1176; *Schreiber*, Ann., 1965, **682**, 219). An X-ray analysis of tomatidine hydroiodide shows that tomatidine has the 25(*S*), 22β*N* configuration, and solasodine therefore is 25(*R*), 22α*N*, (*E. Höhne*, *Ripperger* and *Schreiber*, Tetrahedron, 1967, **23**, 3705). Several syntheses of solasodine have been described, which confirm the gross structure (*Uhle*, J. Amer. chem. Soc., 1953, **75**, 2280; *Schreiber*, *A. Walther* and *Rönsch*, Tetrahedron, 1964, **20**, 1939), and a stereospecific total synthesis has been described (*S. V. Kessar et al.*, *ibid.*, 1971, **27**, 2869).

Veramine, $C_{27}H_{41}NO_2$, a *Veratrum* base, yields on selenium dehydrogenation 2-ethyl-5-methylpyridine and Diels hydrocarbon, products typical of a steroidal alkaloid. It contains a secondary amine group, is a secondary alcohol and also contains an ether linkage. Two isolated double bonds are present, one being in the "cholesterol" position on molecular rotation difference evidence. A correlation with veralkamine (p. 411) enabled the other ethylenic bond to be placed at position 12–13. Veramine is formulated as CXXVII, the stereochemistry being deduced from o.r.d. measurements (*Adam et al.*, Tetrahedron Letters, 1968, 2815; Coll. Czech. chem. Comm., 1968, **33**, 4054).

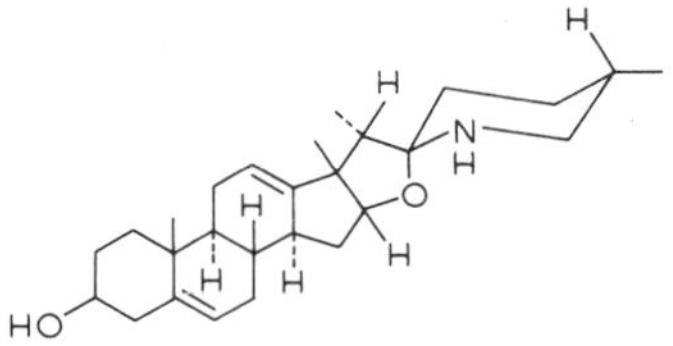

Veramine
(CXXVII)

(f) Biosynthesis of normal steroidal alkaloids

In comparison with other groups of alkaloids, relatively little effort has been directed towards investigating the biosynthesis of steroidal bases. But it is reasonably certain that the formation of these compounds *in vivo* follows the expected course, starting from acetylcoenzyme A, *via* mevalonic acid, isopentenyl pyrophosphate, farnesyl pyrophosphate and squalene as principal intermediates. Some of the less familiar intermediates have been found to accompany the bases in plants, the nitrogen atom or atoms being introduced into the steroid molecule at a late stage of the biosynthesis. *H. Sander* and *H. Grisebach* (Z. Naturforsch., 1958, **13b**, 755) have isolated labelled tomatine from seedlings of *Lycopersicon pimpinellifolium* fed with radioactive acetate. Similar results using the potato plant and labelled mevalonic acid have been reported by *A. R. Guseva* and collaborators (Biokhimiya, 1958, **23**, 412; 1960, **25**, 282; 1961, **26**, 723; Dokl. Akad. Nauk S.S.S.R., 1960, **133**, 228). Degradation of the radioactive products revealed that the distribution of labelled carbon atoms was as anticipated from earlier work on the biosynthesis of cholesterol and related steroids. There seems little doubt that many steroidal alkaloids and sapogenins, co-occurring in plants, are closely related biosynthetically (*R. Tschesche* and *F. Korte*, Angew. Chem., 1952, **64**, 633; *Tschesche*, Fortschr. Chem. Org. Naturstoffe, 1955, **12**, 131). The source of the nitrogen atom or atoms is probably transamination. The melting points and optical properties of members of this class of alkaloids are listed in Table 1 (pp. 428–430).

(g) Biological properties of normal steroidal alkaloids

A wide range of biological properties is shown by this group of bases. Conessine and kurchi extracts have been used in oriental countries for treatment of amoebic dysentery. *Veratrum* bases have been used as insecticides and hypotensive agents. Many *Solanum* bases show antibiotic activity against fungi and bacteria. Tomatine has been produced commercially for use in treatment of dermatitis. Comprehensive reviews of this aspect of the alkaloids have been published (*M. Alauddin* and *M. Martin-Smith*, J. Pharm. Pharmacol., 1962, **14**, 325, 469; *Martin-Smith* and *F. Sugrue*, *ibid.*, 1964, **16**, 569).

2. Alkaloids with a modified steroidal structure

Several families of alkaloids have a modified steroid structural framework. For discussion they are divided into three groups: *(a)* Salamander alkaloids, in which steroidal ring A has been modified, and there is no side-chain at position 17, or there is a simple small carbon unit there; *(b)* C-nor-D-homosteroidal alkaloids, which, as their description indicates, have a modified steroid system in which ring C has been contracted to a 5-membered ring and ring D has been expanded to a 6-membered ring; and

TABLE 1

NORMAL STEROIDAL ALKALOIDS

Alkaloid	*M.p. (°C)*	*$[\alpha]_D$(°)(solvent*
Funtumine	123	+95($CHCl_3$)
Funtumidine	182	+10($CHCl_3$)
Bokitamine	191	
Kurchiline	219	+46($CHCl_3$)
Kurchiphylline	184	−173($CHCl_3$)
Kurchiphyllamine	161	−211($CHCl_3$)
Kurchaline	185	−37($CHCl_3$)
Holadysine	120	−199($CHCl_3$)
Holadysamine	173	−78($CHCl_3$)
Holamine	135–136	+23($CHCl_3$)
Holaphylline	128	+24($CHCl_3$)
Holaphyllamine	amorphous; 260(decomp.)(·HCl)	+33(MeOH)(·HCl)
Holaphyllinol	226	−73($CHCl_3$)
Paravallarine	181	−52($CHCl_3$); −36° (EtO
Lanitine	184–186	
2*β*-Hydroxy-*N*-methylparavallarine	223–225	
Paravallaridine	231	−48($CHCl_3$)
20-Epi-*N*-methylparavallarine	191	−33($CHCl_3$)
Kibataline	171	−42($CHCl_3$)
Dictyolucidine	198	+30($CHCl_3$)
Dictyolucidamine	198	+7($CHCl_3$)
Jurubidine	182–186	−79($CHCl_3$); −49° (pyri
Paniculidine	228–244	−82($CHCl_3$)
Funtuphyllamine-A	173	+13($CHCl_3$)
Funtuphyllamine-B	214–215	+24($CHCl_3$)
Funtuphyllamine-C	172	+24($CHCl_3$)
Funtumaphrine-B	160	+43($CHCl_3$)
Funtumaphrine-C	176	+45($CHCl_3$)
Holafebrine	177	−61.5($CHCl_3$)
Irehamine	230	−33($CHCl_3$)
Irehine	174	−46(?)
Terminaline	243–244	+29($CHCl_3$:MeOH, 1:1
3*α*-Methoxy-20*α*-dimethylaminopregn-5-ene	151.5	−32($CHCl_3$)
Irehdiamine-A	148	−47($CHCl_3$–MeOH)
Irehdiamine-B	115–117	−56($CHCl_3$)
Irehdiamine-C	105–106	−47($CHCl_3$)
Kurchimine	104–106	−21($CHCl_3$)
Saracocine	235–236	
Kurchessine	132–133	−36($CHCl_3$)
α-Kurchessine	146.5–147.5	−31($CHCl_3$)
Dihydrokurchessine	109–110	+13($CHCl_3$)
Chonemorphine	148	+25($CHCl_3$)
N-Acetylchonemorphine	276–279	+15($CHCl_3$)
Funtudiamine	80	+6($CHCl_3$)
Dictyodiamine	115–116	+22($CHCl_3$)
Dictyophlebine	151	+20($CHCl_3$)

Table 1 *(continued)*

Alkaloid	*M.p. (°C)*	*[α]$_D$(°)(solvent)*
Malouetine	264 (dipicrate)	
Malouphylline	259–261	+32 ($CHCl_3$)
Malouphyllinine	257	−11 ($CHCl_3$)
Holarrhidine	181–182	−23 ($CHCl_3$)
Holarrhimine	183	−14 ($CHCl_3$)
3-*N*-Methylholarrhimine	163–164	−19 ($CHCl_3$)
20-*N*-Methylholarrhimine	163.5–166	−19 ($CHCl_3$)
N,*N*′-Tetramethylholarrhimine	227–229	−35 ($CHCl_3$)
3β,20α-Bisdimethylamino-5α-pregnane	109–110	+13 ($CHCl_3$)
Pachysamine-A	167–168	+20 ($CHCl_3$)
Pachysamine-B	171–173	+67 ($CHCl_3$)
Epipachysamine-A	203–205	−17 ($CHCl_3$)
Epipachysamine-B	260–262	+16 ($CHCl_3$)
Epipachysamine-D	245–248	+13 ($CHCl_3$)
Epipachysamine-E	210–212	+20 ($CHCl_3$)
Epipachysamine-F	250–253 (*N*-acetyl)	+6 ($CHCl_3$) (*N*-acetyl)
Pachysandrine-A	235–236	+80 ($CHCl_3$)
Pachysandrine-B	187–189	+93 ($CHCl_3$)
Pachysandrine-C	212–214	
Pachysandrine-D	184–185	+2 ($CHCl_3$)
Pachystermine-A	220–224	+24 ($CHCl_3$)
Pachystermine-B	256–258	−18 ($CHCl_3$)
Pachysantermine-A	260–263	+43 ($CHCl_3$)
Conessine	125–126	−2 ($CHCl_3$); +22° (EtOH)
Apoconessine	68–69	
Heteroconessine	130–131	
Isoconessine	oil; 325–326 (·2HI)	+97 (EtOH)
Neoconessine	128–129	+97 (EtOH)
3α-Aminoconan-5-ene	95–100	+27 (EtOH)
Conkuressine	86.5–87.5	+7 ($CHCl_3$); +18° (EtOH)
Conessimine	100	−22 ($CHCl_3$)
Isoconessimine	92	+30 (?)
Conimine	130	−30 (EtOH)
Malouphyllamine	219–220	+42 ($CHCl_3$)
3α-Dimethylamino-5α-conanine	(not obtained pure)	
7α-Hydroxyconessine	176–178	−61 ($CHCl_3$)
Holaline	267	+38 ($CHCl_3$:MeOH, 1:1)
Holarrhenine	197–198	−7 ($CHCl_3$)
Holarrheline	190	+18 (EtOH)
Funtessine	194–195	+49 ($CHCl_3$)
Holarrhetine	74–75	−4.6 (EtOH); −14.9° ($CHCl_3$)
Halofrine	116–117	−5.1 (EtOH); −19.1° ($CHCl_3$)
Kurcholessine	218.5–221.5	+5 (EtOH)
Conkurchine (irehline)	151	−52 ($CHCl_3$); −47° (EtOH)
Conessidine	124–125	−64 ($CHCl_3$)
Latifoline	129	−4 ($CHCl_3$)
Latifolinine	Oil	
Norlatifolinine	189–190	−32 ($CHCl_3$)
Funtuline	235	−6 ($CHCl_3$:MeOH, 4:1)
Holadienine	110	+80 ($CHCl_3$)

Table 1 *(continued)*

Alkaloid	*M.p. (°C)*	*$[\alpha]_D$(°) (solvent)*
Maingayine	150–152	+2.1 ($CHCl_3$)
Funtudienine	Amorphous	−247 ($CHCl_3$)
Holonamine	257–259	−14.8 (EtOH)
Malarboreine	169–177	+57 (EtOH)
Malorborine	164–165.5	
Veralkamine	119–123 and 165–169	−84 ($CHCl_3$)
Solacongestidine	169–174	+35.6 (?)
Solafloridine	162–165; 172–175	
23-Oxosolacongestidine	213–223	+33 ($CHCl_3$)
24-Oxosolacongestidine	158–162	+40.9 ($CHCl_3$)
Verazine	176–178	−91.7 ($CHCl_3$); −89.7° (EtOH)
Tomatillidine	219–222	−18.1 ($CHCl_3$)
5α,6-Dihydrotomatillidine	179–181	+21.4 ($CHCl_3$)
Solaphyllidine	165–170	−25.4 (MeOH)
Veralinine	124–126	−80 ($CHCl_3$)
Veracintine	196–201	+7.5 ($CHCl_3$)
Solanidine	222–224	−9.8 (MeOH)
Demissidine	221–222	+30 ($CHCl_3$)
Rubijervine	240–242	+19 (EtOH)
Isorubijervine	240–242	+7.7 (EtOH)
Leptinidine	247–248	−24 ($CHCl_3$)
Solanocapsine	216–217	+25 (MeOH); +26.3° (pyridine)
Tomatidine	210–211	−9 (MeOH; hydrochloride)
Tomatidenol	206	−45 ($CHCl_3$)
Solasodine	199	−92.4 (benzene); −97° (MeOH)
Soladulcidine	209–211	−50 ($CHCl_3$)
Veramine	Amorphous	−85.6 (EtOH); −93.9° ($CHCl_3$)

(c) Buxus alkaloids, most of which contain a cyclopropane ring fused on to a normal steroid system at positions 9 and 10.

(a) Salamander alkaloids

It has been known for many years that salamanders are venomous. The source of toxicity is the skin gland secretion, now known to contain several alkaloids. This secretion contains about 10% by weight of a mixture of alkaloids, amounting to about 40–50 mg per animal (*C. Schöpf* and *W. Braun*, Ann., 1934, **514**, 69). The composition of the alkaloid mixture varies slightly from one *Salamandra* subspecies to another. In all, about ten alkaloids have been isolated and identified; for convenience they are divided into two groups: (i) alkaloids containing an oxazolidine ring, and (ii) alkaloids not containing an oxazolidine ring. The alkaloids have been the subject of two reviews (*G. Habermehl* in Progress in Organic Chemistry, Vol. 7, Chapter 2, ed. *Sir James Cook* and *W. Carruthers*, Butterworths, London,

1968; *idem*, in "The Alkaloids", ed. *R. H. F. Manske*, Vol. IX, Chapter 9, Academic Press, London and New York, 1967). Mass spectrometry has played an important part in structure elucidation in this area (*Habermehl* and *G. Spiteller*, Ann., 1967, **706**, 213).

(i) Alkaloids containing an oxazolidine ring

Samandarine, $C_{19}H_{31}NO_2$, is saturated and is a secondary alcohol found in the venom of *Salamandra maculosa*. On oxidation it affords **samandarone**, a ketone also found in the same source. The second oxygen is ethereal in character. A Kuhn–Roth estimation reveals the presence of two *C*-methyl groups. The nitrogen atom is present as a part of a secondary amine; on reaction with methyl iodide *N*-methylsamandarine is first obtained, followed by its methiodide. Hofmann degradation of the latter yields a Hofmann base containing the same number of carbon atoms as the methiodide, indicating that the nitrogen atom is heterocyclic. Catalytic reduction of the Hofmann base (uptake H_2) yields a dihydro-base. The Hofmann base is an enol ether of an aldehyde, since it is cleaved by dilute acid into a hydroxyaldehyde, capable of aldehydo–lactol tautomerism; on oxidation it yields a lactone. Thus the nitrogen atom in the alkaloid is attached to a carbon atom carrying an ether linkage, or to a carbon atom adjacent to one carrying such a linkage, *i.e.* N–C–O– or N–C–C–O–. On warming the acid-treated Hofmann base with acetic anhydride a quaternary acetate results. This pictured as the result of recyclisation. Consequently

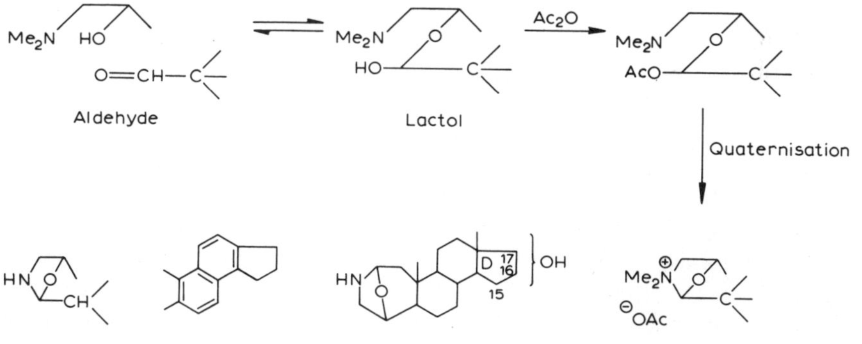

Aldehyde ⇌ Lactol —Ac_2O→ ↓ Quaternisation

(I) (II) (III)

the nitrogen and ether oxygen atoms are attached to the same carbon atom (*Schöpf* and *Braun*, *loc. cit.*; *Schöpf* and *K. Koch*, Ann., 1942, **552**, 37).

Reduction of samandarine with lithium tetrahydrido-aluminate yields a diol, samandiol (*Schöpf et al.*, Chem. Ber., 1950, **83**, 372). This diol is attacked by lead tetra-acetate (1 mol.) with formation of one mol. of formaldehyde, so that samandiol (which is not a primary alcohol) must contain the grouping $-NHCH_2 \cdot \underset{\displaystyle OH}{\underset{|}{CH}}-$. A partial formula, I, may therefore be written for samandarine. Samandiol on dehydrogenation yields 1,2-dimethylcyclopentenonaphthalene (II), identified by spectroscopy and synthesis (*Schöpf* and *D. Klein, ibid.*, 1954, **87**, 1638). Samandarine is therefore to be formulated with a BCD steroidal structure, ring A having been modified; three possibilities need to be considered, III, IV and V, the hydroxyl group being placed in ring D (positions 15, 16 or 17) because samandarone *(vide supra)* is a cyclopentanone ($\nu_{max.}$ 1740 cm^{-1}). The problem was resolved by an X-ray crystallographic analysis of samandarine hydrobromide (*Schöpf*, Experientia, 1961, **17**, 285), which revealed that the base is represented by VI, that is, IV with the hydroxyl group at $C_{(16)}$.

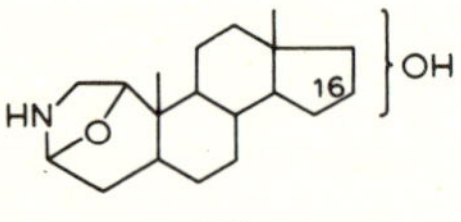

(IV) 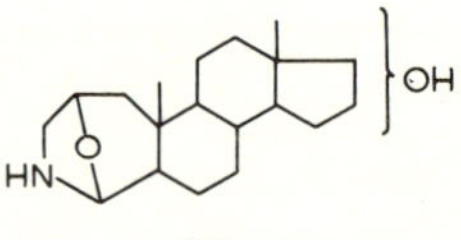 (V) 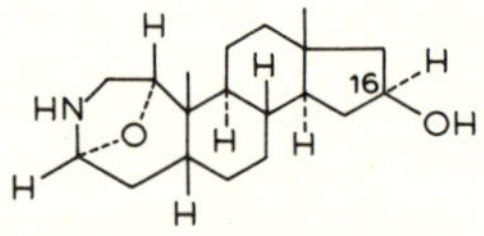 Samandarine (VI)

Samandarone, $C_{19}H_{29}NO_2$, is consequently VI [C=O at $C_{(16)}$]. It is reducible stereospecifically to samandarine by sodium and ethanol (*Schöpf* and *Koch, loc. cit.).* Its o.r.d. curve exhibits a large Cotton effect at 315 nm, in accord with the absolute configuration VI for samandarine (*E. Wölfel et al.*, Chem. Ber., 1961, **94**, 2361). *O*-**Acetylsamandarine** is also a salamander base. On alkaline hydrolysis it yields samandarine and acetic acid, the acetyl group being part of an ester rather than an amide group, because of the basic character of the compound and on i.r. evidence (*Habermehl*, Ann., 1964, **679**, 164).

Samandaridine, $C_{21}H_{31}NO_3$, is a lactone, identified as a δ-lactone on i.r. spectral measurement. Its structure, VIII, rests chiefly on an X-ray diffraction analysis of its hydrobromide (*idem*, Chem. Ber., 1963, **96**, 143), and its absolute configuration has been settled by partial synthesis. *N*-Acetylsamandarone reacted with glyoxylic acid to give the acid VII, together with the 15-isomer. The former was reduced with sodium tetrahydridoborate to the corresponding alcohol, then catalytically to the saturated hydroxy acid. Lactonisation and hydrolysis by mineral acid yielded samandaridine (VIII) (*idem, ibid.*, p. 840).

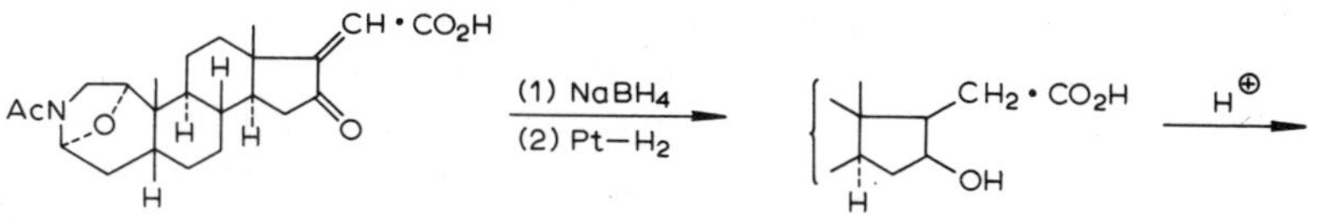

CH · CO_2H AcN O H H H H (VII) (1) $NaBH_4$ (2) Pt–H_2 CH_2 · CO_2H OH H $H^{\oplus}$

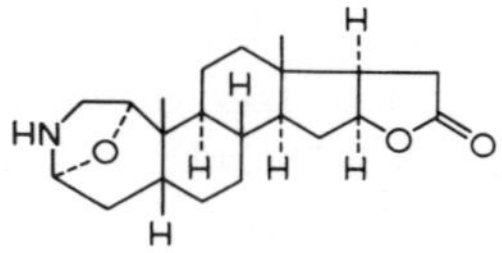

HN O H H H H H O O Samandaridine (VIII) HN O H H H O Samandenone (IX)

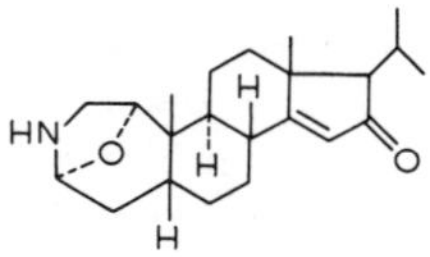

Samandenone, $C_{22}H_{33}NO_2$, is an α,β-unsaturated cyclopentenone, formulated chiefly on spectral evidence as IX (*idem, ibid.*, 1966, **99**, 1439).

Samandinine, $C_{24}H_{39}NO_3$, is formulated as X on i.r. and mass spectral evidence (*Habermehl* and *G. Vogel*, Toxicon, 1969, **7**, 163). Bands at 840–860 cm^{-1} were indicative of the presence of an oxazolidine ring:

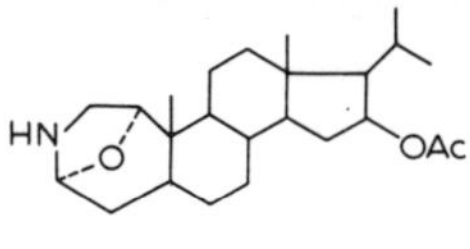

HN O OAc Samandinine (X) HN O OH H H (Xa)

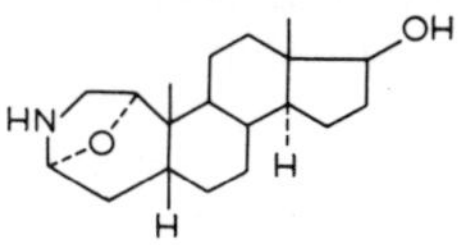

An alkaloid of structure Xa has recently been isolated from *Cyptobranchus maximus* (*S. Hara* and *M. Oka*, C. A., 1970, **73**, 56316g). It has been synthesised from 17β-acetoxy-1α-hydroxy-5β-androstan-3-one, itself prepared from 17β-acetoxyandrostan-1,4-dien-3-one, by a Schmidt reaction followed by reduction with lithium–ethylamine (*M. Benn* and *R. Shaw*, Canad. J. Chem., 1974, **52**, 2936).

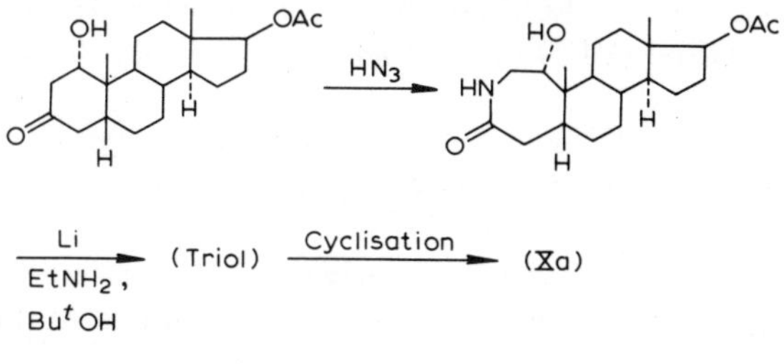

OH OAc O H H HN_3 HO OAc HN O H H

Li / $EtNH_2$, Bu^tOH → (Triol) Cyclisation → (Xa)

(ii) Alkaloids not containing an oxazolidine ring

Cycloneosamandione, $C_{19}H_{29}NO_2$, is a secondary alcohol and also a ketone; i.r. absorption shows the latter to be present in a five-membered ring. The alcohol group participates in a *tert*-methanolamine–iminocarbonyl tautomeric system $-\overset{|}{N}-\underset{|}{\overset{|}{C}}-OH \rightleftharpoons\ >NH\ O{=}C<$. N.m.r. spectroscopy reveals that only one angular methyl group is present, which observation led to the suggestion that the N−C−OH group may be involved in an angular position (*Schöpf* and *O. W. Müller*, Ann., 1960, **633**, 127); further, an aldehyde proton signal is found in the spectrum, suggesting that the usual $C_{(19)}$ methyl group has been replaced by a formyl group (*Habermehl* and *S. Göttlicher*, Chem. Ber., 1965, **98**, 1). This led to a formulation for the alkaloid, but this proved to be unsatisfactory stereochemically because the Wolff–Kishner reduction product of cycloneosamandione was not identical with synthetic 3-aza-A-homo-5α- or -5β-androstane (*C. W. Shoppee* and *G. Krueger*, J. chem. Soc., 1961, 3641). X-ray analysis revealed the reason for this unexpected result: the potential $C_{(19)}$ formyl group is α-oriented, and cycloneosamandione is to be formulated as XI (*Habermehl* and *Göttlicher*, *loc. cit.*). Its absolute configuration was settled by the identity of its o.r.d. curve with that of samandarone [VI, C=O at $C_{(16)}$].

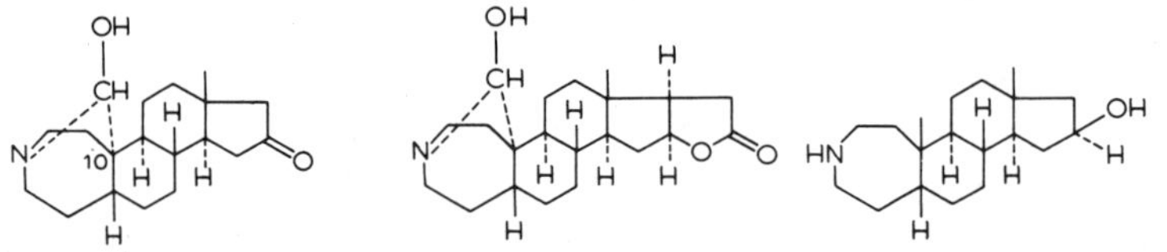

Cycloneosamandione (XI) Cycloneosamandaridine (XII) Samanine (XIII)

Cycloneosamandaridine, $C_{21}H_{31}NO_3$, is a secondary base and a *tert*-carbinolamine, the tautomeric system being identical with that in the foregoing base. A δ-lactone group is also present. Detailed infrared and mass spectral studies point to the alkaloid having structure XII, analogous to that of samandaridine (VIII) (*Habermehl* and *A. Haaf*, Chem. Ber., 1965, **98**, 3001).

Samanine, $C_{19}H_{33}NO$, is a minor salamander base, the structure of which has been elucidated almost entirely from spectral information. Hydroxyl and >NH functions were revealed by i.r. spectroscopy, confirmed by the observed formation of an *N,O*-diacetyl derivative. No i.r. band characteristic of an oxazolidine ring was present. The mass spectrum confirmed the molecular formula, and suggested that the hydroxyl group was not to be found in ring A. Structure XIII was advanced for the alkaloid (*idem*, Ann., 1969, **722**, 155), and has been confirmed by several syntheses (*idem, ibid.*; *Oka* and *Hara*, Tetrahedron Letters, 1969, 1193). A recent synthesis proceeds in outline as follows, the 2,3-seco-5β-steroid XIV being obtainable from 5β-cholestanone (*J. K. Paisley* and *L. Weiler, ibid.*, 1972, 261; *R. B. Rao* and *Weiler, ibid.*, 1973, 4971):

OHC, NC, OAc, H, H (XIV) —(1) $NaBH_4$ (2) TsCl-py.→ TsO, NC, H —B_2H_6→ TsO, H_2N, H

—$(PhCO)_2O$→ PhCO, N, OH, H, H —CrO_3→ O, H —PhCHO, $^{\ominus}OH$→ O, CHPh, H

—(1) $NaBH_4$ (2) Ac_2O-py→ OAc, CHPh, H —O_3, Zn→ OAc, O, H —Zn, HBr→ O, H —(1) Hydrolysis (2) $LiAlH_4$→ (XIII)

(iii) Biosynthesis of salamander bases

Biosynthetic studies leave no doubt that these bases are formed *in vivo* from acetate units *via* cholesterol (*Habermehl, Volkwein* and *Haaf*, Chem. Ber., 1968, **101**, 198). The unusual α-configuration at $C_{(10)}$ in the cyclo-bases XI and XII is explicable in terms of the enlargement of ring A by cleavage between $C_{(2)}$ and $C_{(3)}$ and insertion of a nitrogen atom. If a carbonyl group is present at $C_{(3)}$ when cleavage occurs, the cleaved fragment could experience a retro-Michael reaction. The intermediate aldehyde could then suffer

CHO, B, O, H → CHO, 10, H, O ⇌ CHO, H, O

an inversion at $C_{(10)}$. However, steroids containing a β-aldehyde group at $C_{(10)}$ do not suffer this type of ring cleavage.

(iv) Physiological action

All the salamander bases are highly toxic, causing convulsions, palpitation and paralysis. The central nervous system is affected, and the blood pressure raised. Lethal doses cause death as a consequence of respiratory paralysis (*E. S. Faust*, Arch. exptl. Pathol. Pharmakol., 1898, **41**, 229; *O. Gessner* and co-workers, *ibid.*, 1932, **167**, 638; 1935, **179**, 639; 1937, **187**, 378). The alkaloids also show novel cardioactive properties (*Schöpf*, Experientia, 1961, **17**, 285; *Habermehl*, Naturwiss., 1966, **53**, 123). The salamander alkaloids are listed in Table 2.

TABLE 2

SALAMANDER ALKALOIDS

Base	*M.p. (°C)*
Samandarine	188
Samandarone	190
Samandaridine	290
O-Acetylsamandarine	159
Samandenone	191
Samandinine	170
Cycloneosamandione	119
Cycloneosamandaridine	282
Samanine	197

(b) C-Nor-D-homosteroidal alkaloids

These bases are characterised by a C-nor-D-homosteroid framework XV, nearly all being found in the *Veratreae* tribe of the *Liliaceae*. They have been the subject of several reviews (*K. J. Morgan* and *J. A. Barltrop*, Quart. Reviews chem. Soc., 1958, **12**, 34; *S. M. Kupchan* and *A. W. By*, in "The Alkaloids", ed. *R. H. F. Manske*, Vol. X, Chapter 2, Academic Press, New York and London, 1968; *Kupchan, J. H. Zimmerman* and *A. Afonso*, Lloydia, 1961, **24**, 1; *L. F.* and *M. Fieser*, "Steroids", pp. 867–895, Reinhold, New York, 1959; *H.-G. Boit*, "Ergebnisse der Alkaloid-Chemie bis 1960", pp. 798–832, Akademie Verlag, Berlin, 1961; *Kupchan* and *W. E. Flacke*, in "Antihypertensive Agents", ed. *E. Schlittler*, pp. 229–258, Academic Press, New York, 1967; *K. S. Brown*, in "Chemistry of the Alkaloids", ed. *S. W. Pelletier*, van Nostrand–Reinhold, New York, 1970; *J. Tomko* and *Z. Voticky*, in "The Alkaloids", ed. *Manske*, Vol. XIV, Chapter 1, Academic Press, New York and London, 1973).

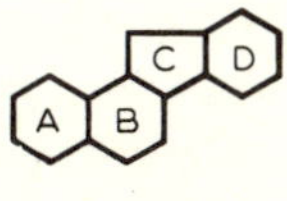

(XV) 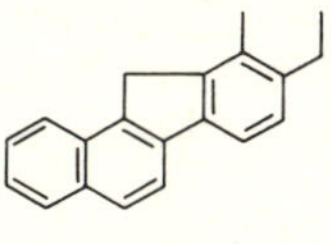 (XVI) 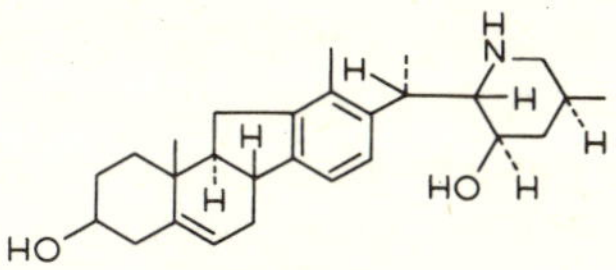 Veratramine (XVII)

For the purpose of discussion these alkaloids are subdivided into (i) alkamines (ii) ester-alkaloids and (iii) *Fritillaria* alkaloids.

(i) Alkamines

Veratramine, $C_{27}H_{39}NO_2$, occurs in the roots and rhizomes of *Veratrum viride* as its glucoside **veratrosine**, and as the aglycone in *V. grandiflorum* (*W. A. Jacobs* and *L. C. Craig*, J. biol. Chem., 1944, **155**, 565; *K. Saito*, Bull. chem. Soc. Japan, 1940, **15**, 22). It is a secondary amine containing two secondary alcohol groups, an isolated olefinic bond and a benzene ring ($\lambda_{max.}$ 268 nm); its triacetyl derivative can be nitrated. The alkaloid is precipitated by digitonin, suggesting, along with other behaviour, that a "cholesterol-type grouping" is present. Dehydrogenation yields 3-hydroxy-5-methylpyridine and a hydrocarbon $C_{20}H_{18}$, identified as 7-ethyl-8-methyl-1,2-benzofluorene (XVI) (*Jacobs*, *Craig* and *G. I. Lavin*, J. biol. Chem., 1941, **141**, 51; *L. Keller*, *C. Tamm* and *T. Reichstein*, Helv., 1958, **41**, 1633). Vigorous oxidation of veratramine affords benzene-1,2,3,4-tetracarboxylic acid. These and many other observations led to the proposals of structure XVII for veratramine (*Tamm* and *O. Wintersteiner*, J. Amer. chem. Soc., 1952, **74**, 3842; *J. Fried et al., ibid.*, 1951, **73**, 2970; *Fried* and *A. Klingsburg*, *ibid.*, 1953, **75**, 4929). This formulation is supported by the formation of ketone XVIII, in which the carbonyl group is inert chemically, by successive reduction, acetylation and oxidation. This same compound is formed from **jervine**, another *Veratrum* base (see below), by a related sequence, which finding exemplifies the close structural similarity between the two alkaloids (*Wintersteiner* and *H. Hosansky*, *ibid.*, 1952, **74**, 4474).

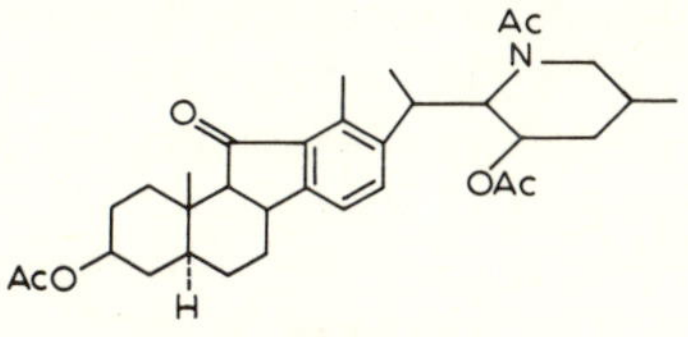

(XVIII) 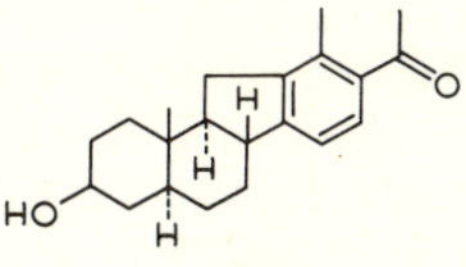 (XIX)

Structure XVII for the alkaloid has been confirmed by total synthesis from 16-acetyl-$\Delta^{12,14,16}$-5α-etiojervatriene-3β-ol (XIX), itself available from synthetic hecogenin (*C.C.C.*, Vol. II E, p. 33; *H. Mitsuhashi* and *K. Shibata*, Tetrahedron Letters, 1964, 2281; Chem. Pharm. Bull., Tokyo, 1968, **16**, 814), and by another synthesis (*W. S. Johnson et al.*, J. Amer. chem. Soc., 1967, **89**, 4524; Tetrahedron Letters, 1968, 2829; see also *T. Masamune* and co-workers, J. Amer. chem. Soc., 1967, **89**, 4521. This compound was converted to the epoxide XX by reaction with trimethylsulphonium ylid, then to aldehyde XXI with boron trifluoride. Reaction with potassium cyanide and

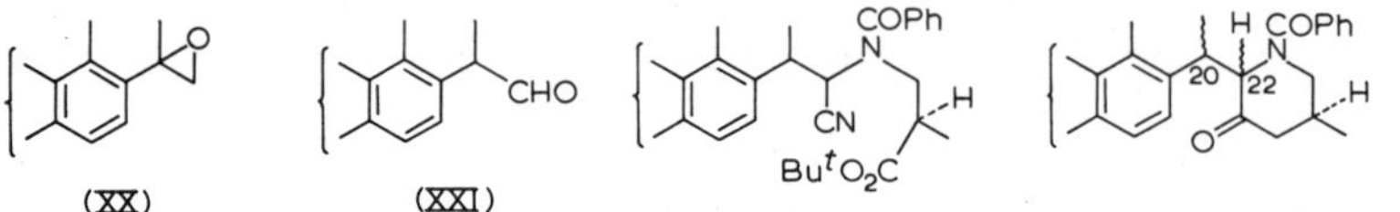

(XX) (XXI) (XXII) (XXIII)

(−)-*tert*-butyl 4-amino-3-methylbutyrate, followed by benzoylation, then gave XXII. Base-catalysed ($CH_3SO\overset{\ominus}{C}H_2$) cyclisation, with subsequent hydrolysis and decarboxylation, afforded XXIII as a mixture of stereoisomers. These were separated and one of them (the 20α-CH_3, 22β-H isomer) was converted into veratramine in eleven straightforward stages (*Johnson et al., ibid.*, 1967, **89**, 4523).

Jervine, $C_{27}H_{39}NO_3$, occurs as its glucoside in *V. album* and *V. viride* and also as the free base (*Jacobs* and *Craig*, J. biol. Chem., 1943, **148**, 51, and earlier references there cited). The chemistry of jervine is closely bound up with that of veratramine (see above). Jervine is a secondary amine, a secondary alcohol, an α,β-unsaturated ketone ($\lambda_{max.}$ 250 nm, log ε 4.2), and a cyclic ether. There is also present an isolated olefinic bond. Mild hydrogenation generates dihydrojervine (reduction of conjugated double bond; no u.v. absorption), and further hydrogenation affords tetrahydrojervine, which is saturated. The carbonyl group in all these compounds is inert chemically. Although jervine is not precipitated by digitonin, physicochemical evidence (*e.g.* molecular rotation differences) suggests strongly that the well-known cholesterol A/B bicyclic system is present. Further, jervine and dihydrojervine are both oxidisable to α,β-unsaturated ketones in a manner reminiscent of the oxidation of cholesterol to cholest-4-en-3-one (*Jacobs et al., ibid.*, 1947, **170**, 635; 1948, **175**, 57). Dehydrogenation of jervine (selenium) affords 2-ethyl-5-methylpyridine, 2-ethyl-3-hydroxy-5-methylpyridine (*Jacobs, Craig* and *Lavin, loc. cit.*; *Jacobs* and *Y. Sato*, J. biol. Chem. 1949, **181**, 55) and a hydrocarbon mixture with u.v. absorption characteristic of fluorenes. This suggested, assuming no ring-contraction, that ring C in jervine was 5-membered, and that a modified steroid framework was involved. Confirmation of this view was provided by energetic acetolysis ($Ac_2O + BF_3$) of jervine to give the indanone derivative XXIV (*Wintersteiner* and *M. Moore*, J. Amer. chem. Soc., 1953, **75**, 4938); the same degradation product was obtained from triacetyldihydroveratramine by oxidation with chromic acid and provides a link between the two alkaloids (*Tamm* and *Wintersteiner*, *loc. cit.*; *Wintersteiner* and *Hosansky*, *loc. cit.*). *N,O*-Diacetyl-

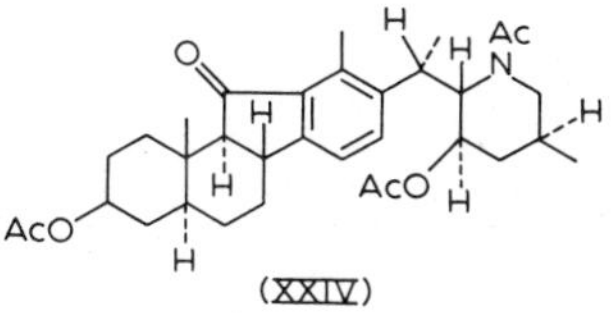

(XXIV)

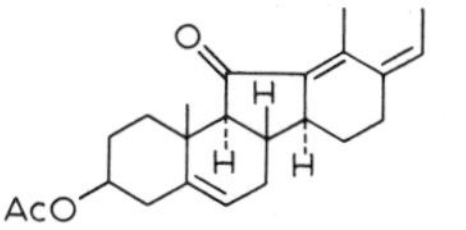

(XXV)

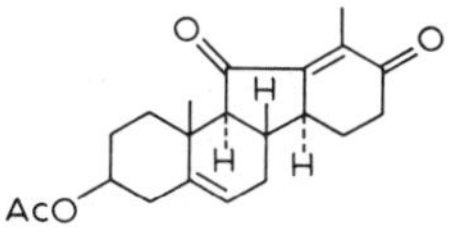

(XXVI)

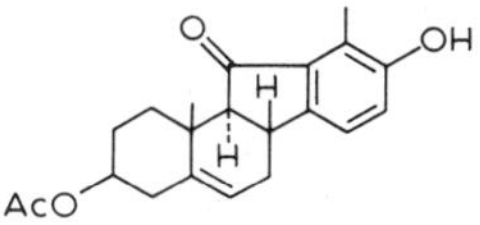

(XXVII)

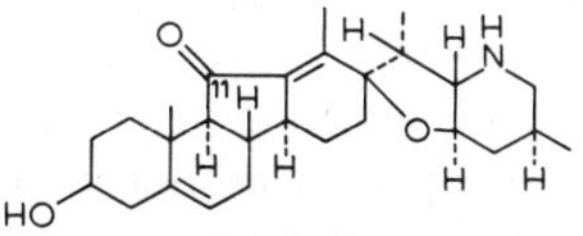

Jervine
(XXVIII)

jervine, when subjected to acetolysis under acid conditions, gave a nitrogen-free dienone, which itself on oxidation yielded acetaldehyde and an enedione; the latter underwent alkaline oxidation to a phenol. The dienone is formulated as XXV, the enedione as XXVI and the phenol as XXVII (*Fried et al.*, *loc. cit.*; *J. E. Herz* and *Fried*, J. Amer. chem. Soc., 1954, **76**, 5621; *Fried* and *Klingsburg*, *loc. cit.*). These and other results pointed to structure XXVIII for jervine; this formulation has been confirmed by a synthesis related to that of veratramine (see above) *(Masamune et al., loc. cit.)*. Reduction of ketone XIX with tetrahydridoborate provided the corresponding alcohol (mixture of isomers), which was converted to the corresponding bromides XXIX, which were separated. The higher-melting

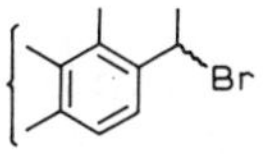

(XXIX)

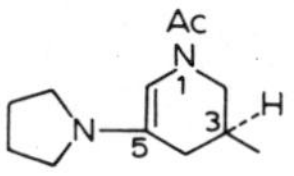

(XXX)

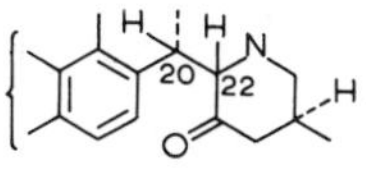

(XXXI)

bromide was treated with 1-acetyl-3(*S*)-methyl-5-piperidone enamine XXX, to yield compound XXXI with the desired 20α-methyl, 22β-H configuration [comparison with authentic material obtained by degradation of veratramine (XVII)]. This was subjected to successive reduction with lithium and ethylamine, catalytic hydrogenation and epoxidation, leading to epoxide XXXII.

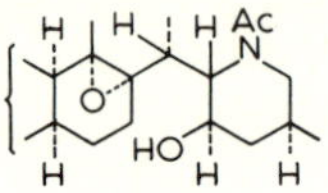

(XXXII) 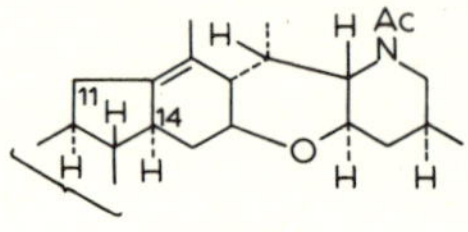 (XXXIII) 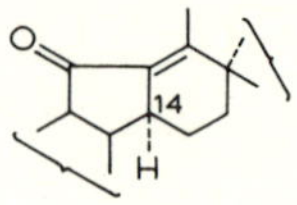 (XXXIV)

Exposure of this to alkali formed the tetrahydrofuran ring, yielding, after dehydration, the deoxojervine derivative XXXIII. Oxidation with chromic anhydride–pyridine afforded, in very poor yield, the desired 11-ketone XXXIV, the major product being the 14-alcohol. The former was converted into jervine (XXVIII) in seven conventional steps.

Veratramine and jervine show some unusual chemical reactions. Amongst these is the behaviour of jervine on catalytic hydrogenation (Pt–HOAc): accompanying the expected products was one isomeric with jervine and markedly more stable to acid than the alkaloid. It was conjectured that the oxygen of the tetrahydrofuran ring was no longer allylic. Base treatment of the product yielded a conjugated dienone, which itself was oxidisable to a 3-ketone, the u.v. spectrum of which was identical with that of the dienone. It was evident that the dienone had resulted from a shift of the "cholesterol" 5,6 double bond into the 7,8 position, allowing the jervine isomer to be formulated as XXXV and the dienone as XXXVI (*B. N. Iselin* and *Wintersteiner*, J. Amer. chem. Soc., 1955, **77**, 5318). For other interesting transformations of the alkaloids see *Kupchan* and *By*, *op. cit.*, and *Brown*, *op. cit.*

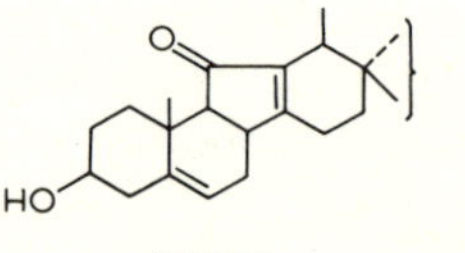

(XXXV) 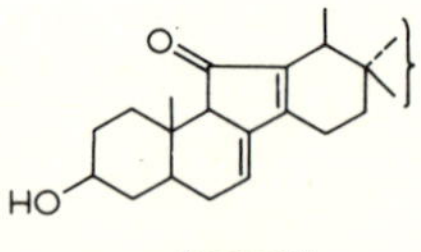 (XXXVI)

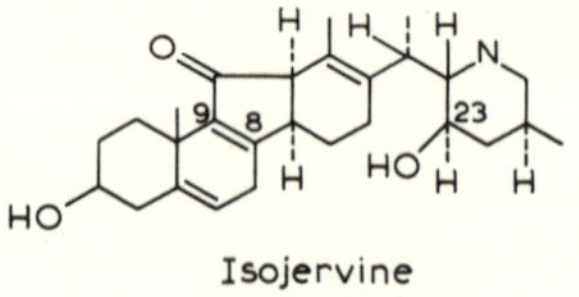

Isojervine
(XXXVII)

Isojervine is not an alkaloid, being formed by isomerisation of jervine with acid. Its formulation as XXXVII is based on its u.v. and i.r. absorption, and on the finding that it possessed a new secondary alcohol group, placed at $C_{(23)}$ because of the formation of 2-ethyl-3-hydroxy-5-methylpyridine as a selenium dehydrogenation product. The u.v. absorption is abnormal ($\lambda_{max.}$ 252 nm, ε 2900) and has been explained by steric inhibition of electronic interaction between the carbonyl group and the 8,9 double bond; selective reduction of the 5,6 double bond (the least substituted) yielded 5,6-dihydroisojervine with normal u.v. absorption ($\lambda_{max.}$ 238 nm, ε 9500). A plausible mechanism for this isomerisation of jervine is depicted below (*Jacobs* and *C. F. Huebner*, J. biol. Chem., 1947, 170, 635; *Jacobs* and *Sato*, *ibid.*, 1948, **175**, 57; *Iselin* and *Wintersteiner*, *loc. cit.*; *Wintersteiner* and *Moore*, Tetrahedron Letters, 1962, 795; J. org. Chem., 1964, **29**, 262, 270; 1965, **30**, 528; *Masamune et al.*, Bull. chem. Soc. Japan, 1962, **35**, 1749; J. org. Chem., 1964, **29**, 2282; *W. G. Dauben et al.*, *ibid.*, 1963, **28**, 293).

Jervine → Enolic form → (XXXVII)

11-Deoxojervine (cyclopamine), $C_{27}H_{41}NO_2$, occurs in the roots of *V. album* var. *grandiflorum maxim* and *V. californicum.* Its structure follows from its formation by Wolff–Kishner reduction of jervine, and from n.m.r. spectral measurements; it is formulated as XXVIII; CO at $C_{(11)}$ replaced by CH_2 (*Masamune et al.*, Tetrahedron Letters, 1964, 913; *Masamune*, *Takasugi* and *Mori*, *ibid.*, 1965, 489; *R. F. Keeler*, Phytochem., 1968, **7**, 303; 1969, **8**, 223; *Keeler* and *W. Binns*, *ibid.*, 1971, **10**, 1765).

Veratrobasine, $C_{27}H_{41}NO_3$, also found in *V. album*, is 11β-hydroxyjervine. Its structure is based on its formation by reduction of jervine (*Iselin*, *Moore* and *Wintersteiner*, J. Amer. chem. Soc., 1956, **78**, 403), and on an X-ray diffraction analysis (*G. N. Reeke*, *R. L. Vincent* and *W. N. Lipscomb*, *ibid.*, 1968, **90**, 1663). Veratrobasine is consequently formulated as XXVIII [β-OH at $C_{(11)}$) (*Kupchan* and *M. I. Suffness*, *ibid.*, p. 2730).

Verarine, $C_{27}H_{39}NO$, is an alkaloid of *V. album* subsp. *lobelianum.* A benzene ring in its structure was detected by u.v. spectroscopy. The base contains a hydroxyl group and a basic nitrogen atom, both secondary in character, and one double bond is present, not conjugated with the benzene ring. This unsaturation was shown to be of the "cholesterol type" by molecular rotation difference studies. The mass spectrum of the base is consistent with the presence of a C-nor-D-homo framework, and is similar to that of veratramine. Verarine is formulated as XXXVIII (*H. Budzikiewicz*, Tetrahedron, 1964, **20**, 2267; *J. Tomko* and *A. Vassova*, Chem. Zvesti, 1964, **18**, 266; *Tomko* and *S. Bauer*, Coll. Czech. chem. Comm., 1964, **29**, 2570).

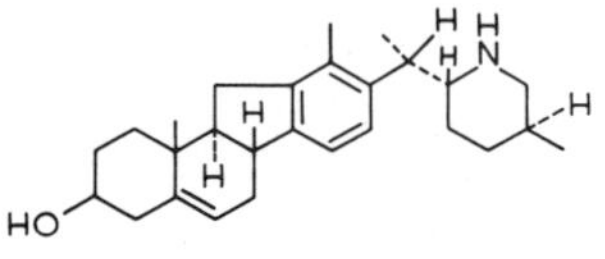

Verarine
(XXXVIII)

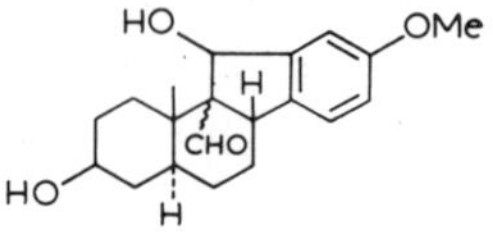

(XXXIX)

This structure has been confirmed by total synthesis (*J. P. Kutney* and co-workers, J. Amer. chem. Soc., 1968, **90**, 5332). The dihydroxyaldehyde XXXIX was synthesised by a multi-step route from β-naphthol; it was converted into its diacetate, then elimination of formic acid was effected, to give the alkene XL, hydroboration of which yielded the alcohol XLI. Birch reduction and hydrolysis then gave enone XLII, which was dehydrated to dienone XLIII, then methylated with methyl iodide and lithamide to XLIV ($R=CH_3$).

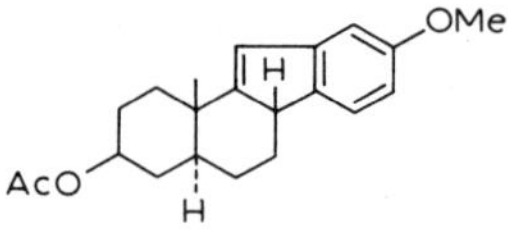

(XL) 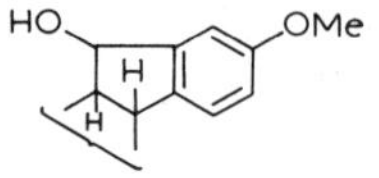 (XLI) 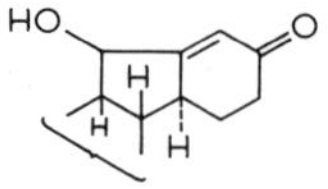 (XLII)

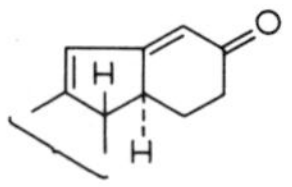

(XLIII) 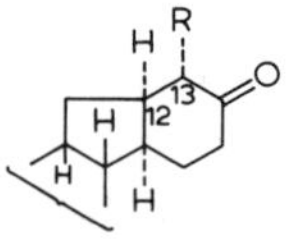 (XLIV) 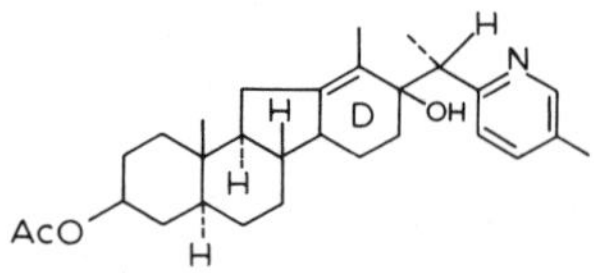 (XLV)

The 12,13 double bond was re-established by bromination–dehydrobromination and the product then acetylated and caused to react with 2-ethyl-5-methylpyridine, to yield the aldol XLV, as a mixture of epimers from which the desired isomer was extracted. Ring D in this was next aromatised and the pyridine ring selectively hydrogenated, to give a mixture of isomers. From this 3-*O*-acetyl-5α-dihydroverarine (XLVI) was separated. Then fol-

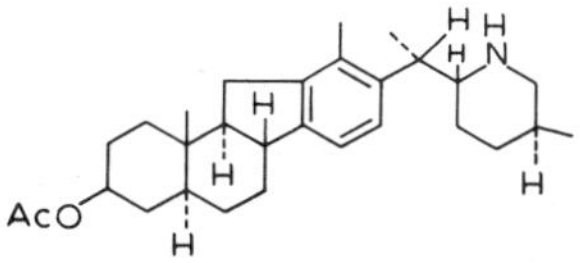

(XLVI)

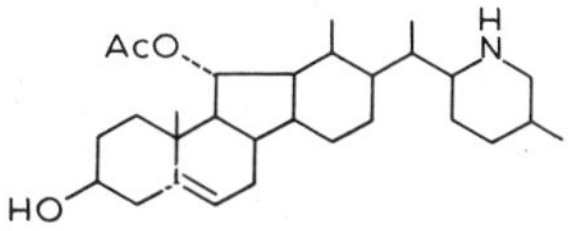

Muldamine
(XLVII)

lowed, in order, *N*-acetylation, selective hydrolysis of the *O*-acetyl group, oxidation to the 3-ketone, introduction of a 4,5 double bond, reduction to the homoallylic alcohol, and finally hydrolysis of the *N*-acetyl group, culminating in verarine (XXXVIII).

Muldamine, $C_{29}H_{47}NO_3$, found in *V. californicum*, shows saturated ester, hydroxyl and NH bands in its i.r. spectrum. Its mass spectrum shows a base peak at *m/e* 98, indicative of the presence of a methylpiperidine substituent at $C_{(20)}$. Its n.m.r. spectrum was compared in detail with those of jervine and jervane derivatives; these studies allowed structure XLVII to be proposed for muldamine (*Keeler*, Steroids, 1971, **18**, 741).

Veramarine, $C_{27}H_{43}NO_3$, is an amorphous alkaloid from the same source. It is a tertiary base (not *N*-methyl) containing three alcoholic hydroxyl groups; on acetylation it yields a diacetate, suggesting that one of the hydroxyl groups is hindered (tertiary). Catalytic hydrogenation furnishes a dihydro derivative (saturated), the double bond in the substrate being placed in the 5,6 position on molecular rotation difference evidence. Chromic acid oxidation of the dihydro-alkaloid led to a diketone, indicating that the other two hydroxyl groups are secondary. Molecular rotation difference measurements allowed the keto groups, and therefore the secondary hydroxyl groups in the alkaloid, to be positioned at $C_{(3)}$ and $C_{(16)}$. Detailed i.r., n.m.r., and mass spectral studies, and biogenetic considerations, which place veramarine structurally intermediate between the veratramine and cevine groups (see below), pointed to structure XLVIII for veramarine (*Tomko* and *Vassova*, Pharmazie, 1965, **20**, 385; *Tomko et al.*, Coll. Czech. chem. Comm., 1965, **30**, 3320). The stereochemistry of the base is based

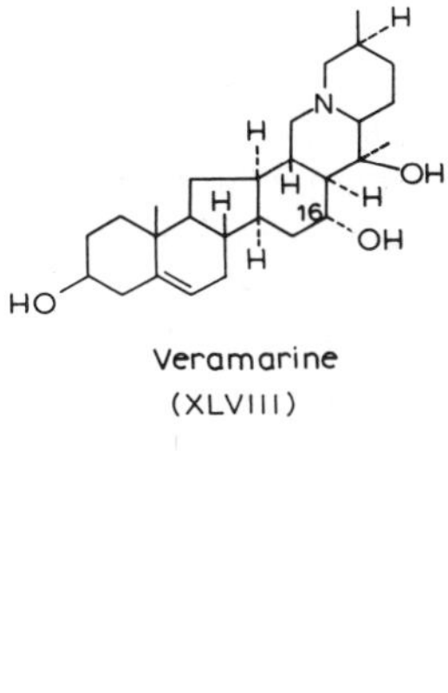

Veramarine
(XLVIII)

primarily on i.r. and n.m.r. spectral interpretations, the α-configuration (equatorial) for the 16-hydroxyl group being assigned from methanolysis kinetic studies (*S. Ito*, *T. Ogino* and *Tomko*, *ibid.*, 1968, **33**, 4429; *R. Hirschmann et al.*, J. Amer. chem. Soc., 1954, **76**, 4013).

The C-nor-D-homosteroidal alkaloids (alkamines) are listed in Table 3.

TABLE 3

C-NOR-D-HOMOSTEROIDAL ALKALOIDS (ALKAMINES)

Alkaloid	*M.p. (°C)*	*$[\alpha]_D$(°) (solvent)*
Veratramine	209.5–210.5	–70 (MeOH); –71 ($CHCl_3$)
Jervine	244–246	–150 (EtOH)
Isojervine	110–114 (decomp.)	–32 (EtOH)
11-Deoxojervine	237–238	–44.2 (EtOH)
Dihydrojervine	248–251	–82 (EtOH)
Veratrobasine	285–288 (decomp.)	–126 (pyridine)
Verarine	174–175	–68.5 (EtOH); –53.3 ($CHCl_3$)
Muldamine	210–211	–95 (EtOH:$CHCl_3$, 3:1)
Veramarine	Amorphous	–77 (EtOH); –85 ($CHCl_3$)

(ii) Ester-alkaloids

Veratrum spp. contain a group of alkaloids which occur as esters containing simple and more complex organic acid residues. On hydrolysis they yield an alkamine (one of six) and an acid or acids (often several of about nine). The acids commonly encountered are acetic, angelic, veratric, vanillic, *erythro*-3-acetoxy-2-hydroxy-2-methylbutyric, (–)-2-methylbutyric, (+)-2-hydroxy-2-methylbutyric, (+)-*threo*-2,3-dihydroxy-2-methylbutyric and (–)-*erythro*-2,3-dihydroxy-2-methylbutyric acids. The alkamines are highly hydroxylated C-nor-D-homosteroidal amines, the acid residues being attached to one or more nuclear hydroxyl groups. Details of early work in this area may be found in *T. A. Henry*, "The Plant Alkaloids", Churchill, London, 4th Edition, 1949, pp. 702–705, 709–711. The ester-alkaloids are summarised in Table 4; here we shall be concerned with structures of the alkamines.

Cevine, $C_{27}H_{43}NO_8$, is not a primary hydrolysis product of the ester-alkaloids, being formed from the "natural" alkamine **veracevine** (see Table 5) by isomerisation. However, because of its accessibility and stability, much structural investigation has been carried out on it, the results being of value in the elucidation of the structures of the other alkamines. It is a tertiary base containing, on i.r. evidence, no C=O group, and several hydroxyl groups. The exact number (seven) of the latter proved to be difficult to

TABLE 4

VERATRUM ESTER-ALKALOIDS

Parent alkamine	*Ester-alkaloid*	*Molecular formula*	*Positions of acyl groups*				*M.p. (°C)*	*References*
			3	*6*	*7*	*15*		
Zygadenine	Zygacine	$C_{29}H_{45}NO_8$	Ac					1
	Angeloylzygadenine	$C_{32}H_{49}NO_8$	Ang					2
	3-[(−)-2-methyl-butyryl]zygadenine	$C_{32}H_{51}NO_8$	Mb					3
	Veratroylzygadenine	$C_{36}H_{51}NO_{10}$	Ve				270–271	4,5
	Vanilloylzygadenine	$C_{35}H_{49}NO_{10}$	Va				258–259	4
Zygadenilic acid δ-lactone	Angeloylzygadenilic acid δ-lactone	$C_{32}H_{47}NO_8$				Ang*		6
Sabine	Sabadine	$C_{29}H_{47}NO_8$	Ac					3,7
Veracevine	Cevacine	$C_{29}H_{45}NO_9$	Ac				205–207	8
	Cevadine	$C_{32}H_{49}NO_9$	Ang				206	9–11
	Veratridine	$C_{36}H_{51}NO_{11}$	Ve				160–180	8–10
	Vanilloylveracevine	$C_{35}H_{49}NO_{11}$	Va				265–257	12
Germine	Germitetrine	$C_{41}H_{63}NO_{14}$	Ahmb		Ac	Mb	229–230	13–15
	Germitrine	$C_{39}H_{61}NO_{12}$	Mb		Ac	Hmb	216–219	16,17
	Neogermitrine	$C_{36}H_{55}NO_{11}$	Ac		Ac	Mb	234–235	17,18
	Germanitrine	$C_{30}H_{59}NO_{11}$	Ang		Ac	Mb	228–229	19,20
	Germinitrine**	$C_{39}H_{57}NO_{11}$					175–176	19
	Germerine	$C_{37}H_{59}NO_{11}$	Mb			Hmb	200–203	17, 21
	Germidine	$C_{35}H_{53}NO_{10}$	Ac			Mb	230–231	16, 17
	Neogermidine	$C_{35}H_{53}NO_{10}$			Ac	Mb	221–223	17, 22, 23
	Germbudine	$C_{37}H_{59}NO_{12}$	tDmb			Mb	160–164	23, 24
	Neogermbudine	$C_{37}H_{59}NO_{12}$	eDmb			Mb	149–152	14, 15
	Protoveratridine	$C_{32}H_{51}NO_9$	Mb				272–273	17, 21

Table 4 *(continued)*

Parent alkamine	*Ester-alkaloid*	*Molecular formula*	*Positions of acyl groups*				*M.p. (°C)*	*References*
			3	*6*	*7*	*15*		
Protoverine	Protoveratrine-A	$C_{41}H_{63}NO_{14}$	Hmb	Ac	Ac	Mb	267–269	13, 14, 25–27
	Protoveratrine-B	$C_{41}H_{63}NO_{15}$	tDmb	Ac	Ac	Mb	268–270	13, 14, 25–27
	Escholerine	$C_{41}H_{61}NO_{13}$	Ang	Ac	Ac	Mb	235–236	28, 29
	Desacetylproto-veratrine-A	$C_{39}H_{61}NO_{13}$	Hmb	Ac		Mb	191–192	14, 30
	Desacetylproto-veratrine-B	$C_{39}H_{61}NO_{14}$	tDmb	Ac		Mb	182–183	23, 30, 31

Ac = acetyl; Ang = angeloyl; Mb = (−)-3-methylbutyryl; Ve = veratroyl; Va = vanilloyl; Ahmb = *erythro*-3-acetoxy-2-hydroxy-2-methylbutyryl; Hmb = (−)-2-hydroxy-2-methylbutyryl; tDmb = (+)-*threo*-2,3-dihydroxy-2-methylbutyryl; eDmb = (−)-*erythro*-2,3-dihydroxy-2-methylbutyryl.

* At position 16.

**Acids are acetic, tiglic and angelic, positions uncertain.

References to Table 4.

1 *S. M. Kupchan, D. Lavie* and *R. D. Zonis*, J. Amer. chem. Soc., 1955, **77**, 689; *Kupchan, ibid.*, 1959, **81**, 1925.
2 *M. Suzuki et al.*, J. pharm. Soc., Japan, 1959, **79**, 619.
3 *H. Mitchner* and *L. M. Parks*, J. Amer. pharm. Assocn., 1959, **48**, 303.
4 *Kupchan* and *C. V. Deliwala*, J. Amer. chem. Soc., 1953, **75**, 1025.
5 *A. Stoll* and *F. Seebeck*, Helv., 1953, **36**, 1570.
6 *T. Tsukamoto* and *A. Yagi*, J. pharm. Soc. Japan, 1959, **79**, 1102; *Yagi* and *T. Kawasaki*, Chem. Pharm. Bull. (Tokyo), 1962, **10**, 519.
7 *Kupchan, N. Gruenfeld* and *N. Katsui*, J. med. pharm. Chem., 1962, **5**, 690; *H. Mohrle* and *H. Auterhoff*, Arch. Pharm., 1959, **292**, 337.
8 *Kupchan et al.*, J. Amer. chem. Soc., 1953, **75**, 5519; *Z. J. Vejdelek, K. Macek* and *B. Budesinsky*, Coll. Czech. chem. Comm., 1957, **22**, 98.
9 *Stoll* and *Seebeck*, Helv., 1952, **35**, 1270.
10 *S. W. Pelletier* and *W. A. Jacobs*, J. Amer. chem. Soc., 1953, **75**, 3248.
11 *Kupchan* and *A. Afonso*, J. Amer. pharm. Assocn., 1960, **49**, 242.
12 *D. M. Stuart* and *Parks*, *ibid.*, 1956, **45**, 252.
13 *H. A. Nash* and *R. M. Brooker*, J. Amer. chem. Soc., 1953, **75**, 1942.
14 *G. S. Myers et al.*, *ibid.*, 1956, **78**, 1621.
15 *Kupchan* and *C. I. Ayres*, J. Amer. pharm. Assocn., 1959, **48**, 440.
16 *J. Fried, H. L. White* and *O. Wintersteiner*, J. Amer. chem. Soc., 1950, **72**, 4621.
17 *Kupchan, ibid.*, 1959, **81**, 1921.
18 *Fried, P. Numerof* and *N. H. Coy*, *ibid.*, 1952, **74**, 3041.
19 *M. W. Klohs et al.*, *ibid.*, 1953, **75**, 4925.
20 *Kupchan* and *Afonso*, J. Amer. pharm. Assocn., 1959, **48**, 731.
21 *W. Poethke*, Arch. Pharm., 1937, **275**, 357, 571.
22 *Kupchan* and *Deliwala*, J. Amer. chem. Soc., 1954, **76**, 5545.
23 *Myers et al.*, *ibid.*, 1955, **77**, 3348.
24 *Kupchan* and *Gruenfeld*, J. Amer. pharm. Assocn., 1959, **48**, 737.
25 *Klohs et al.*, J. Amer. chem. Soc., 1952, **74**, 4473, 5107.
26 *Stoll* and *Seebeck*, Helv., 1953, **36**, 718.
27 *Kupchan* and *Ayres*, J. Amer. chem. Soc., 1960, **82**, 2252.
28 *Klohs et al.*, *ibid.*, 1954, **76**, 1152.
29 *Kupchan* and *Ayres*, J. Amer. pharm. Assocn., 1959, **48**, 735.
30 *Kupchan, Ayres* and *R. H. Hensler*, J. Amer. chem. Soc., 1960, **82**, 2616.
31 *Klohs et al.*, *ibid.*, 1953, **75**, 1953, **75**, 3595.

determine because of the inert (hindered) character of some of them. Since it is a saturated base, containing neither *N*- nor *O*-methyl groups, it must be heptacyclic. The eighth oxygen atom is ethereal in nature. Cevine is a powerful reducing agent, and can itself be reduced to dehydrocevine, by hydrogenolysis of this ether linkage. The reducing properties are explicable by the presence of a masked α-ketol grouping in cevine. This grouping arises by isomerisation of **cevagenine**, an intermediate between veracevine and cevine. Cevagenine contains a keto-group (*D. H. R. Barton* and *J. F. Eastham*, J. chem. Soc., 1953, 424; *Barton* and *C. J. W. Brooks*, Chem. and Ind., 1953, 1366).

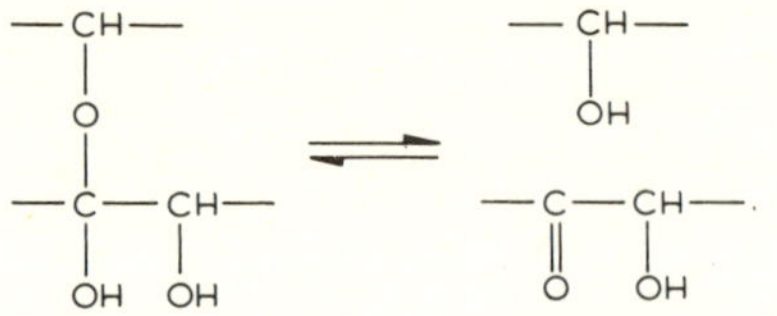

Cevine Cevagenine 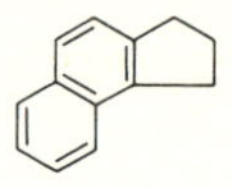 (XLIX)

Dehydrogenation of cevine with selenium affords a variety of products including β-picoline, 2-ethyl-5-methylpyridine, cevanthrol, $C_{23}H_{24}O$ (a phenol), cevanthridine, $C_{25}H_{27}N$, veranthridine, $C_{26}H_{25}N$, and a base $C_{20}H_{19}N$. Other products were a group of cyclopentenofluorenes, and 4,5-benzindane (XLIX). Closely related basic products are formed when cevine is heated with soda-lime or zinc dust. A careful analysis of these breakdown products led *Jacobs* and *Pelletier* (J. org. Chem., 1953, **18**, 765; J. Amer. chem. Soc., 1954, **76**, 2028; 1956, **78**, 1914) to suggest that cevine contained the skeleton L: in particular veranthridine is formulated as LI.

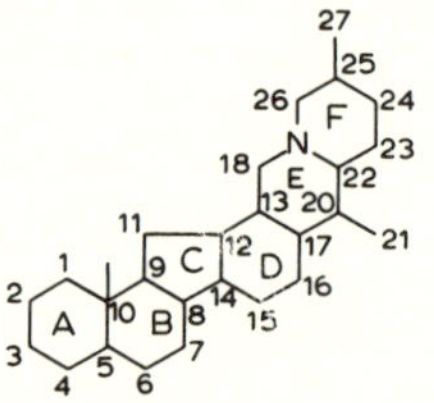

(L) 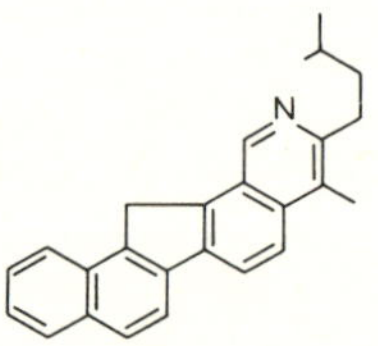 Veranthridine (LI) 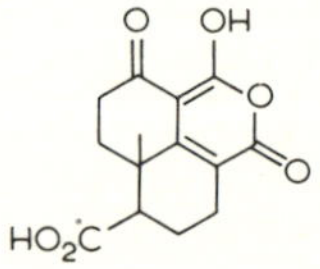 Decevinic acid (LII)

Oxidation of cevine with chromic acid led to several identifiable products (*Jacobs* and *Craig*, *ibid*., 1939, **61**, 2252; J. biol. Chem., 1940, **134**, 123; 1941, **141**, 253), including acetic, succinic, methylsuccinic and α,α-dimethylsuccinic acids. These were accompanied by several polycarboxylic acids, in particular a lactonic tricarboxylic acid $C_{14}H_{18}O_8$; this lost $2H_2O$ on treatment with alkali to yield decevinic acid, $C_{14}H_{14}O_6$. The latter was subsequently formulated as an acylglutaconic anhydride (LII) (*F. Gautschi et al*., Helv., 1954, **37**, 2280; 1955, **38**, 296; *Barton et al*., Experientia, 1954, **10**, 81) as a consequence of its dehydrogenation to 2-hydroxynaphthalic anhydride, and its behaviour on base treatment, which gave a γ-lactone, with loss of two mols. of carbon dioxide. This product, as its methyl ester, was allowed to react with methylmagnesium bromide, then dehydrogenated, to give 5-isopropyl-2-methylnaphthalene; it was also degraded to (+)-9-methyl-1-decalone. Consideration of the mode of derivation of decevinic acid and its congeners (rings A, B and C of cevine) indicated that cevine must be oxygenated at positions 3,4,9,12 and 14 in L, and the configuration at $C_{(10)}$ must be that of normal steroids (*O. Jeger et al.*, Helv., 1954, **37**, 2295). Another product of the oxidative degradation was the simple lactam LIII with the absolute configuration shown; its formation means that cevine is not oxygenated in ring F and that the configuration at $C_{(25)}$ is the same as in the other *Veratrum* bases (*idem, ibid.*, p. 2302). Further information about the position and orientation of the hydroxyl group in cevine has emanated from a study of specific oxidations of the various 1,2-glycol systems present with lead tetra-acetate and periodic acid. The results of these and extensive additional studies (see review references at beginning of section for details) point to the stereochemical formula LVI for cevine. This assign-

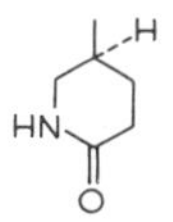

(LIII) 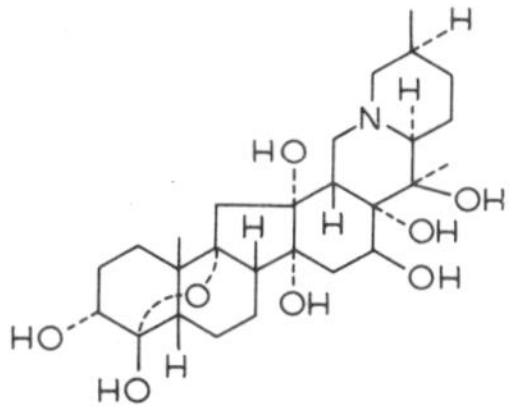 Cevine (LIV) Veracevine (LV) Cevagenine (LVI)

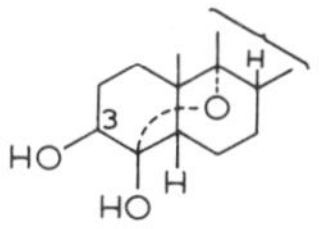

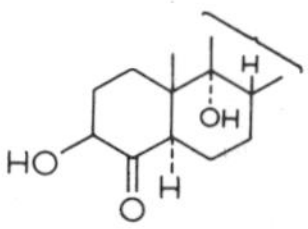

ment has been confirmed by an X-ray analysis of cevine hydroiodide (*W. T. Eeles*, Tetrahedron Letters, 1960, 24). Veracevine is consequently formulated as LV, the 3-epimer of cevine, and cevagenine as LVI, the α-ketol resulting from isomerisation of the hemiacetal group in cevine. It is noteworthy that in cevagenine the A/B ring junction is the more stable *trans*, an inversion, facilitated by the fact that the compound is an α-decalone, having occurred during the transformation LIV→LVI.

Germine, $C_{27}H_{43}NO_8$, is the alkamine of the largest group of *Veratrum* ester-alkaloids (see Table 4). Similarities in the behaviour of cevine and germine suggested the two were similarly constituted. In particular the presence in the latter of the same type of hemiacetal grouping was apparent, since germine is interconvertible with a ketonic isomer **isogermine** and a non-ketonic isomer **pseudogermine**, analogous, respectively, to cevagenine and veracevine (*Craig* and *Jacobs*, J. biol. Chem., 1943, **148**, 57; 1943, **149**, 451; *H. Jaffe* and *Jacobs*, *ibid.*, 1951, **193**, 325). Germine under forcing conditions yields a penta-acetate and under milder conditions a tetra-acetate. A notable difference from cevine is that germine is the true alkamine of a group of esters; this fact is bound up with a difference in the order of stability (pseudogermine > germine > isogermine) of the germine triad compared with that of the cevine triad (*Kupchan* and *C. R. Narayanan*, Chem. and Ind., 1956, 1092;

F. L. Weisenborn and *J. W. Bolger*, J. Amer. chem. Soc., 1954, **76**, 5543). Detailed chromic acid, lead tetra-acetate and periodic acid oxidations have revealed the positions and nature of the hydroxyl groups, and, along with other investigations, have culminated in structure LVII for germine (for principal references see *Morgan* and *Barltrop*, review reference on p. 436).

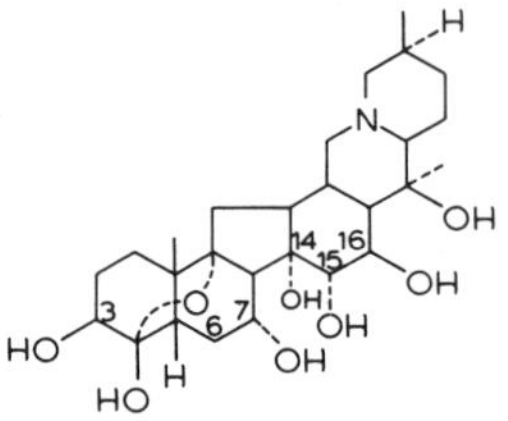

Germine (LVII)

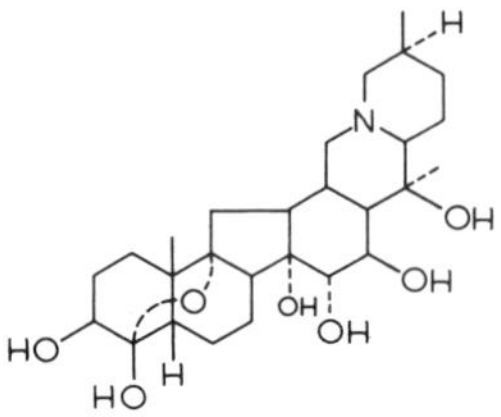

Zygadenine (LVIII)

Zygadenine, $C_{27}H_{43}NO_7$, the parent alkamine of several ester-bases, is a tertiary base containing neither *N*- nor *O*-methyl groups. It has the masked ketol (hemiacetal) group characteristic of this series (*F. V. Heyl* and *M. E. Herr*, J. Amer. chem. Soc., 1949, **71**, 1751; *Kupchan* and *Deliwala*, *ibid.*, 1953, **75**, 1025). Alkaline isomerisation yields first the free α-ketol (**isozygadenine**), then the hemiacetal **pseudoxygadenine**, the order of stability being pseudo-> iso->zygadenine, analogous to that with veracevine. Investigations conducted along lines similar to those described for cevine and germine point to structure LVIII for zygadenine (*Kupchan, Lavie* and *R. D. Zonis, ibid.*, 1955, **77**, 689; *Kupchan, ibid.*, 1956, **78**, 3546). The correctness of this assignment is substantiated by the fact that 7-dehydrogermine-3,16-diacetate (LVII; C=O at position 7 and OAc at positions 3 and 16) is reducible to zygadenine 3,16-diacetate by desulphurisation of its propylenedithioacetal with Raney nickel (*Kupchan, loc. cit.*).

Protoverine, $C_{27}H_{43}NO_9$, is the most highly hydroxylated *Veratrum* alkamine. It is formed by the hydrolysis of several ester-alkaloids (see Table 4) and was early recognised as a member of the cevine group, its two isomers being **isoprotoverine** (α-ketol) and **pseudoprotoverine** (hemiacetal) (*H. Auterhoff* and *F. Günther*, Arch. Pharm., 1955, **288**, 455; *Jacobs* and *Craig*, J. biol. Chem., 1942, **143**, 427; 1943, **149**, 271). Structure LIX for protoverine emerged from a detailed series of studies by *Kupchan* and co-workers (J. Amer. chem. Soc., 1960, **82**, 2242), and was confirmed by a link-up with germine (LVII): 7-dehydroprotoverine 14,15-acetonide 3,6,16-triacetate was reduced with calcium in liquid ammonia to a product identified as 7-dehydrogermine 14,15-acetonide 3,16-diacetate.

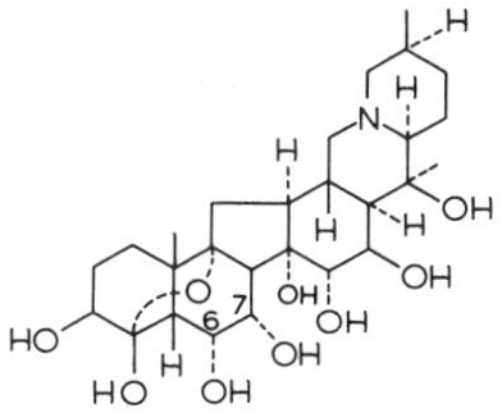

Protoverine
(LIX)

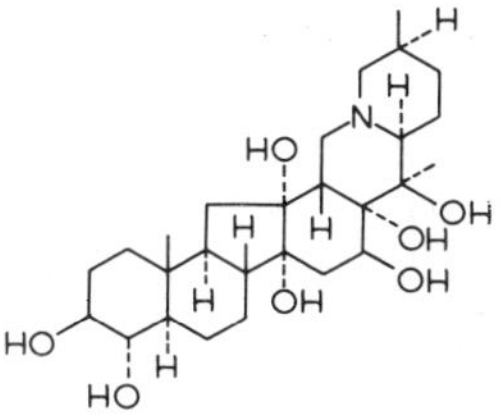

Sabine
(LX)

Sabine, $C_{22}H_{45}NO_7$, occurs as an alkamine in *Schoencaulon officinale* Gray, and is formed on hydrolysis of the ester-alkaloid **sabadine** (Table 4) (*Auterhoff* and *G. Möhrle*, Arch. Pharm., 1958, **291**, 288; 1959, **292**, 337; *Kupchan, A. D. J. Balon* and *E. Fujita*, J. org. Chem., 1962, **27**, 3103). Cevanthridine is formed as a dehydrogenation product; this pointed to a cevane-type framework. However, no masked α-ketol (hemiacetal) was present, sabine failing to isomerise under alkaline conditions and to show reducing properties (*Mitchner* and *L. M. Parks*, J. Amer. pharm. Assocn., 1959, **48**, 303). The structure LX advanced for sabine rests on lead tetra-acetate and periodate oxidation studies, on o.r.d. measurements, on comparisons with cevine, and on biogenetic considerations (*Kupchan, N. Gruenfeld* and *N. Katsui*, J. med. pharm. Chem., 1962, **5**, 690; *R. Criegee et al.*, Ann., 1956, **599**, 81).

Zygadenilic acid δ-lactone, $C_{27}H_{41}NO_7$, the alkamine found in nature as its 16-angeloyl ester, proved to be identical in all respects with the bismuth oxide oxidation product of pseudozygadenine (see above). This product had been encountered prior to the isolated of the alkamine, and had already been formulated as the ring A-contracted structure LXI (*B. Shimizu*, J. pharm. Soc. Japan, 1958, **78**, 444; 1959, **79**, 993; *T. Tsukamoto* and *A. Yagi*, *ibid.*, p. 1102; *Yagi* and *Kawasaki*, Chem. Pharm. Bull., Tokyo, 1962, **10**, 519).

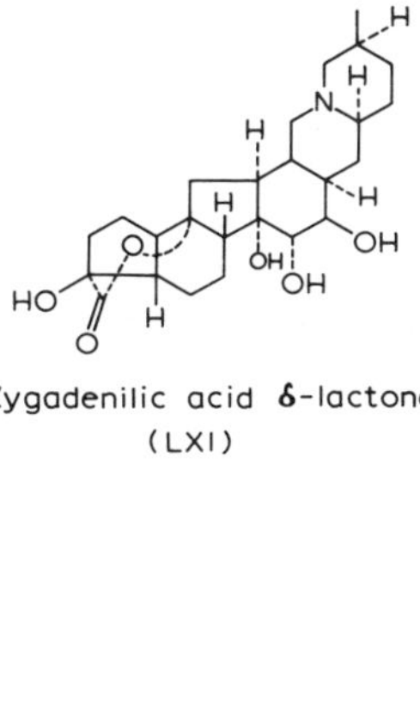

Zygadenilic acid δ-lactone
(LXI)

The veratrum alkaloids and related substances are listed in Table 5.

TABLE 5

VERATRUM ALKAMINES AND RELATED COMPOUNDS

Alkamine	*M.p. (°C)*	*$[\alpha]_D$(°) (solvent)*
Zygadenine	201–204	−48 (pyridine)
Zygadenylic acid δ-lactone		
Sabine	173–176	−33 (EtOH)
Veracevine	220–225	−33 ($CHCl_3$)
Germine	220	+5 (EtOH)
Protoverine	195–200	−12 (pyridine)
Cevine	165–170	−17.5 (EtOH)
Cevagenine	241–242 (decomp.)	−47.8 (EtOH)
Isogermine	250–252	−46.5 (pyridine)
Pseudogermine	—	—
Isozygadenine	—	—
Pseudozygadenine	169–171 (decomp.)	−33 ($CHCl_3$)
Isoprotoverine	254 (decomp.)	−42 (pyridine)

Determination of ester positions. Chemical methods involving selective hydrolysis of ester-alkaloids, and selective esterification of alkamines, have been used extensively in settling ester positions in the ester-alkaloids. For example, 7- and 16- esters are easily cleaved in neutral methanol, and many esters were acetates at position 7. An ester group (nearly always methylbutyryl) at $C_{(15)}$ prevents glycol oxidative cleavage and acetonide formation in ring D; likewise an ester at position 6 precludes glycol cleavage in ring B. Observations of this and related kinds usually enabled the ester positions to be diagnosed. The ester-alkaloid germitetrine is exemplary: total hydrolysis affords germine, two acetic acid molecules, one of (−)-3-methylbutyric acid, and one of (−)-*erythro*-2,3-dihydroxy-2-methylbutyric acid. Methanolysis gave initially a mono-deacetyl triester, then a completely deacetylated product identical with neogermbudine (see Table 4); molecular rotation differences showed that the first acetyl group was at position 7. Germitetrine is stable to periodate, whilst neogermbudine consumes one mole; consequently its second acetyl group is placed on the glycol of the dihydroxyl-methylbutyryl group. Germitetrine also reacts with one mole of chromic acid, to give a ketone transformed by base into a diosphenol. Neogermbudine must be acylated at positions 3 and 15, because periodate is not consumed; on $NaBH_4$ reduction of it germine 15-(−)-2′-methylbutyrate results, obtainable from several other ester-alkaloids. Germitetrine is therefore formulated as germine (LVII) in which the 3-OH is esterified with *erythro*-3-acetoxy-2-hydroxy-2-methylbutyric acid, the 7-OH with acetic acid and the 15-OH with (−)-2-methylbutyric acid (*Kupchan* and *Ayres*, J. Amer. pharm. Assocn., 1959, **58**, 440).

Amongst other ester-alkaloids, of recent vintage, are **germinaline,** from *V. album* subsp. *lobelianum*, which is a positional isomer of germitrine (*K. Samikov*, *R. Shakirov* and *S. Y. Yunusov*, C. A., 1972, **76**, 141099), and two veratroyl esters of germine, from the same source (*Tomko* and *A. Vassova*, Chem. Zvesti, 1971, **25**, 69).

(iii) Fritillaria alkaloids

The Chinese drug pei-mu consists of extracts of dried bulbs of *Fritillaria roylei (Liliaceae)*, and is used in the orient as a general medical panacea. Several alkaloids have in recent years been isolated from this and other *Fritillaria* extracts. Structurally they are intermediate between the alkamines and the highly hydroxylated ester-alkaloids. Some occur as free bases, others as glycosides.

Verticine, $C_{27}H_{45}NO_3$, occurs in *F. thunbergii*, *F. roylei* and related species. It is the main active principle of the drug pei-mu, and is a saturated, monoacidic, tertiary base containing three hydroxyl groups. Two of these are secondary, being oxidisable to a diketone and also easily acylated, whilst the third is tertiary. Dehydrogenation of verticine yielded 2,5-lutidine, 8-methyl-1,2-benzofluorene and veranthridine (LI), and suggested that the alkaloid was related structurally to the "ceveratrum" bases.

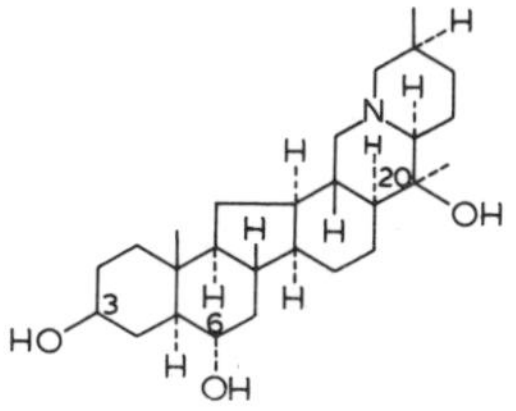

Verticine
(LXII)

Equilibration studies on the reduction products of the above diketone using sodium–ammonia and lithium tetrahydridoaluminate, reveal that in verticine the two secondary alcohol groups are equatorial. One of these was somewhat more resistant to oxidation with chromic oxide than the other; this was placed at $C_{(3)}$ and the other at $C_{(6)}$ chiefly on o.r.d. studies, largely on the two monoketones and the diketone. The tertiary hydroxyl group was located, along with an attendant methyl group, at position 20, partly because of intramolecular hydrogen bonding observed in the i.r. spectrum of 3,6-bisdeoxyverticine. Structure LXII has been advanced for verticine and supported by i.r. and n.m.r. spectral studies, by pK_a measurements on the alkaloid and its closely related derivatives, by biogenetic considerations and by X-ray diffraction analysis of verticinone (**fritillarine**, see below) methobromide (*S. Ito et al.*, Tetrahedron Letters, 1968, 5373). The base also occurs as its β-D-glucoside, the carbohydrate residue being attached at $C_{(3)}$ (*idem*, Chem. Pharm. Bull., Tokyo, 1961, **9**, 253; 1963, **11**, 1337, and earlier references there cited).

Fritillarine, $C_{27}H_{43}NO_3$, from the same sources, has been identified as verticinone, obtained from verticine by selective oxidation at $C_{(6)}$ and formulated as LXII (C=O at position 6), the formulation being supported by X-ray diffraction analysis of the methobromide *(idem, loc. cit.)*.

Imperialine, $C_{27}H_{43}NO_3$, a constituent of *F. imperialis*, and *F. roylei*, shows i.r. absorption characteristic of OH and ketonic C=O groups, the latter as part of a 6-ring. 8-Methyl-1,2-benzofluorene and veranthridine are dehydrogenation products. Molecular rotation data, u.v. spectral study, and biogenetic considerations led to the advancement of structure LXIII for imperialine, the position of the tertiary hydroxyl group being supported by mass spectral study (*R. N. Nuriddinov et al.*, C. A., 1964, **61**, 960; 1968, **68**, 69168g).

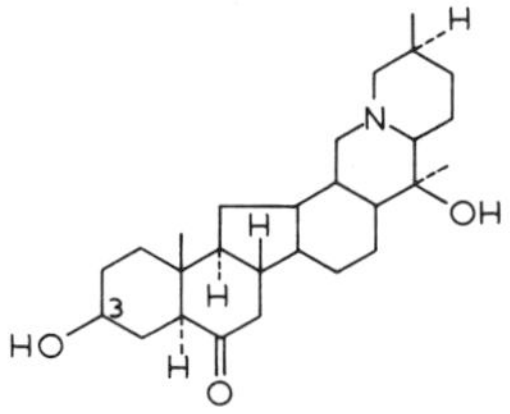

Imperialine
(LXIII)

Imperialone, $C_{27}H_{41}NO_3$, which is found in *Petilium eduardi*, is a diketone containing a single, tertiary hydroxyl group. It was identified as the oxidation product of imperialine, (LXIII; C=O at positions 3) (*Boit* and *L. Paul*, Ber., 1957, **90**, 723; *R. Shakirov, Nuriddinov* and *Yusunov*, C.A., 1964, **60**, 13280; 1966, **64**, 12746).

The Fritillaria alkaloids are listed in Table 6.

TABLE 6

FRITILLARIA ALKALOIDS

Alkaloid	*M.p. (°C)*	*$[\alpha]_D(°)$ (solvent)*
Verticine	223–224	−16.4 (EtOH); −20° ($CHCl_3$)
Fritillarine	212–213	−58.3 (EtOH); −54.5° ($CHCl_3$)
Imperialine	265–267	−34.7 (?)
Imperialone	228–231	

(c) Buxus alkaloids

The boxwood tree is a common garden shrub, a member of a small family *(Buxaceae)* of world-wide distribution. The tree has been used medicinally since very early times for a variety of disorders, including skin complaints, venereal disease, tuberculosis, cancer and malaria. Extracts of the plant contain a large number of alkaloids of closely related structure. Serious study of these bases began during the 1949–1950 period, when *E. Schlittler, K. Heusler* and *W. Friedrich* (Helv., 1949, **32**, 2209; 2226; 1950, **33**, 873, 878) succeeded in isolating seven alkaloids from extracts of *Buxus* leaves. Since that time investigations by several different groups have led to the isolation of a total of about eighty bases. All the alkaloids possess one of two modified steroid frameworks in which there are (with one exception) methyl or modified methyl substituents (sometimes two) at position 4 and an α-methyl group at position 14, the two skeletons being LXIV and LXV. [Irehine (p. 390) is an exception, but is not usually classed as a *Buxus* alkaloid].

(LXIV) (LXV)

From the structural viewpoint the alkaloids are seen to be between steroids and triterpenes. *Buxus* alkaloids may be monacidic (basic function at positions 3 or 20) or diacidic (basic functions at both positions 3 and 20); for discussion the bases are conveniently subdivided so. The principal *Buxus* alkaloids are listed in Tables 7 and 8. They have been reviewed in detail in several places (*V. Cerny* and *F. Šorm*, in "The Alkaloids", ed. *Manske*, Vol. IX, p. 375, Academic Press, London and New York, 1967; *Tomko* and *Voticky*, *op. cit.*, *Brown*, *op. cit.*, see p. 436).

(i) Monoacidic Buxus alkaloids

Monoacidic bases may have the nitrogen atom attached to positions 3 or 20. They are primary, secondary (NHMe) or tertiary (NMe_2) bases, some being amidic, the degree of substitution being denoted by a suffix letter according to the following convention; *N*-acyl groups are counted as NH:

Suffix letter	*Substitution on nitrogen*			
	at $C_{(3)}$		*at* $C_{(20)}$	
	R^1	R^2	R^3	R^4
K	Me	Me	–	–
L	–	–	Me	Me
M	Me	H	–	–
N	–	–	Me	H
O	H	H	–	–
P	—	–	H	H

This system of nomenclature has not, however, been adopted by all workers in the area. For molecular formulae and sources see Table 7. In the isolation of these bases advantage has been taken of their relatively weak basicity compared with the congeneric diacidic bases. Structure determinations have depended heavily on correlations with the latter [see section (*b*)].

TABLE 7

MONOACIDIC BUXUS ALKALOIDS

Alkaloid	*Molecular formula*	*M.p. (°C)*	$[\alpha]_D$*	*Source***
Buxandonine-L	$C_{24}H_{39}NO$	157–159	+24	a
Buxanine-M	$C_{32}H_{43}NO_2$	196–199	−38	a
Buxene-O	$C_{27}H_{41}NO_3$	202–204		a
Buxpiine-K (=cyclomicrobuxine)	$C_{25}H_{39}NO_2$	173	+159	b,c
Buxpsiine-K (=buxamideine-K, alkaloid C)	$C_{26}H_{39}NO$	180–183	+118	d
Buxtauine-M (=cyclomicrobuxinine)	$C_{24}H_{37}NO_2$	170	+153	b
Cyclobuxomicreine-K	$C_{25}H_{39}NO$	195–197	+37	c
Cyclobuxophylline-K	$C_{26}H_{41}NO$	194–196	−72	c
Cyclobuxophyllinine-M (=buxenone-M)	$C_{25}H_{39}NO$	181–182	−51	a,c
Cyclobuxosuffrine-K	$C_{25}H_{39}NO$	201–204	−62	c
Cyclobuxoviridine-L	$C_{26}H_{41}NO$	182–183	+16	c
Cyclomicrobuxeine-K	$C_{25}H_{37}NO$	141–142	+126	c
Cyclomicuranine-L	$C_{26}H_{43}NO_2$	209–211	−3	c
Cyclosuffrobuxine-K	$C_{25}H_{37}NO$	167–172	−92	c
Cyclosuffrobuxinine-M	$C_{24}H_{35}NO$	181–182	−51	c
trans-Cyclosuffrobuxinine-M	$C_{24}H_{35}NO$		−47	a
N-3-Methylbuxene-M	$C_{28}H_{43}NO_3$	180–182	−104	a
Cyclorolfeine	$C_{25}H_{41}NO_2$	253	+119	e
Cyclorolfoxazine	$C_{26}H_{41}NO_3$	239	+106	e

* Optical rotations (°) in chloroform
**Key to sources: a, *B. sempervirens*; b, *B. wallichiana*; c, *B. microphylla*; d, *B. balearica*; e, *B. rolfei*.

Several alkaloids of closely related structure have been isolated from *Buxus microphylla* varieties by *T. Nakano* and collaborators (Tetrahedron Letters, 1964, 1035, 1045; J. chem. Soc., 1965, 4512, 4537, 6688; *ibid.*, C, 1966, 1412; 1970, 590). They are represented by structures LXVI–LXXI, and are named as shown.

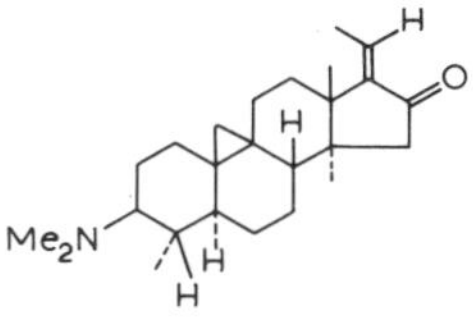

Cyclobuxosuffrine-K (LXVI)

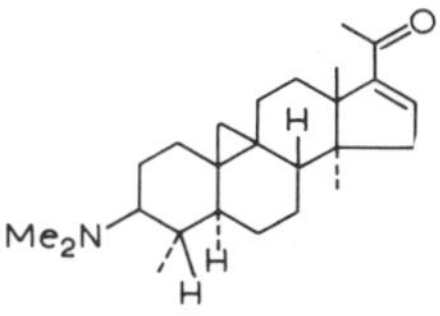

Cyclobuxomicreine-K (LXVII)

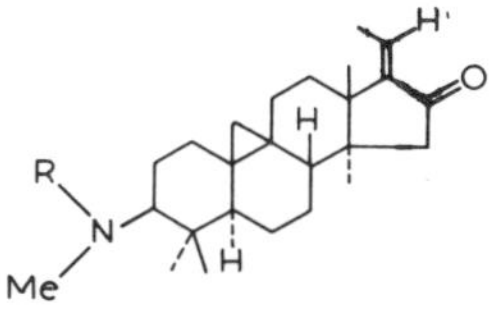

Cyclobuxophyllinine-K (R = Me)
Cyclobuxophyllinine-M (R = H)
Buxanine -M (R = PhCO)
(LXVIII)

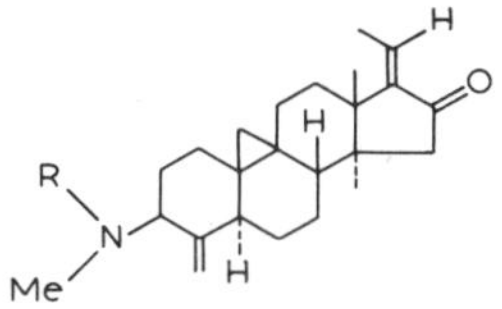

Cyclosuffrobuxine-K (R = Me)
Cyclosuffrobuxinine-M (R = H)
(LXIX)

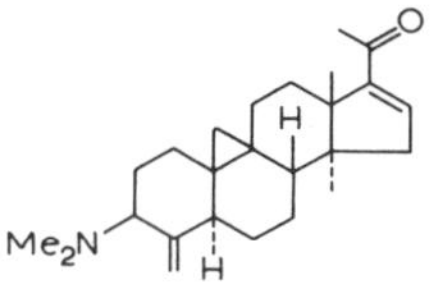

Cyclomicrobuxeine-K (LXX)

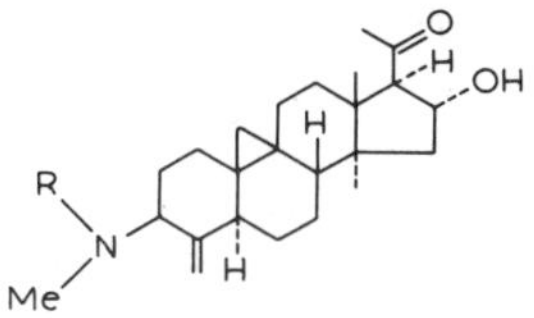

Cyclobuxophylline-K (R = Me)
Cyclomicrobuxinine-L (R = H)
(LXXI)

Their structures and configurations have been settled entirely by spectroscopic study, and by spectral and chemical correlations with diacidic bases of known structure and configuration [see Section *(ii)*, p. 459]. Mass spectrometry has been of value in establishing the presence of a 3β-dimethylamino group attached to an AB steroid system, and of a *C*-acetyl group ($Ac^{\oplus}$ ion, *m/e* 43) when present. I.r. spectroscopy revealed the existence of an α,β-unsaturated ketone group, in some cases in a five-membered ring; u.v. spectroscopy was helpful in a similar manner. N.m.r. spectroscopy gave information about the extent and positions of substitution on the conjugated and isolated C=C bonds, revealed the presence of a cyclopropyl CH_2 group (high field AB quartet), and enabled the stereochemistry of the $C_{(4)}$-methyl group to be deduced (where applicable) from a study of its deshielding effect on the cyclopropyl CH_2 signals. Finally, correlations with diacidic bases such as cyclobuxine-D [see Section *(ii)*] and others, of established structures, served to confirm proposed structures and configurations.

Buxene-O and **methylbuxene-M** are two alkaloids of *B. sempervirens* which are unusual in that they possess a urethane function at $C_{(3)}$. They are formulated as LXXII (=H) and LXXII (R=Me), respectively, and have been correlated with cyclobuxophyllinine (LIV; R=H) (*W. Döpke, R. Haertel* and *H. W. Feilhaber*, Tetrahedron Letters, 1969, 4423).

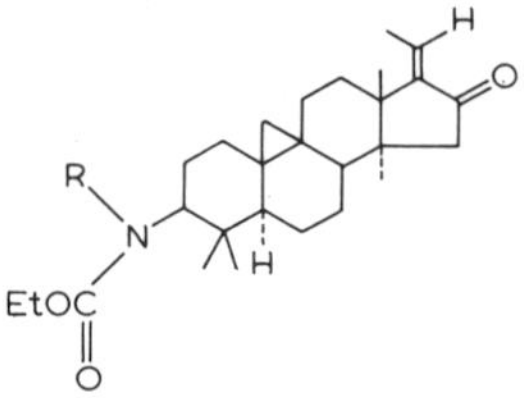

Buxene-O (R = H)
Methylbuxene-M (R = Me)
(LXXII)

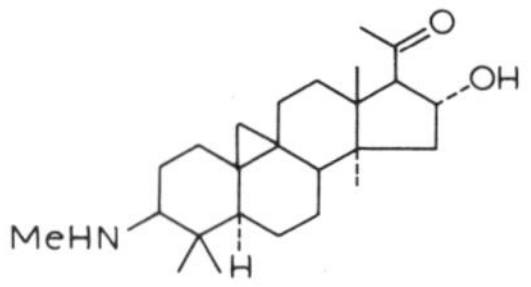

Cyclorolfeine
(LXXIII)

Cyclorolfeine, (LXXIII) has spectroscopic properties which reveal that it is an *N*-methyl secondary base, a methyl ketone and a secondary alcohol, the first two groups being located respectively at positions 3 and 17 in a steroid framework. The hydroxyl group is β to the carbonyl group and therefore at $C_{(16)}$, since the base is readily dehydrated to an α,β-unsaturated ketone. These and other data point to structure LXXIII for cyclorolfeine; it has been confirmed by correlation with the diacidic base cyclovirobuxine-D (see below) (*Goutarel* and co-workers, Bull. Soc. chim. Fr., 1966, 1216).

Buxamine-M is an amidic alkaloid, containing an *N*-benzoyl group. Its i.r. spectrum proved to be closely like that of cyclobuxophylline-K (LXVIII; R = Me), and it was observed to result when cyclobuxophyllinine-M (LXVIII; R = H) was *N*-benzoylated. It is therefore (LXVIII; R = PhCO) (*Döpke* and *Müller*, Pharmazie, 1966, **21**, 769).

Cyclorolfoxazine (LXXIV) has been assigned a structure solely on the basis of n.m.r., i.r. and mass spectral evidence. It is a minor base of *B. rolfei* (*Goutarel* and co-workers, *loc. cit.*).

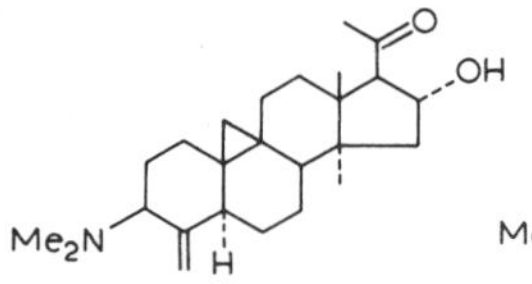

Cyclorolfoxazine
(LXXIV)

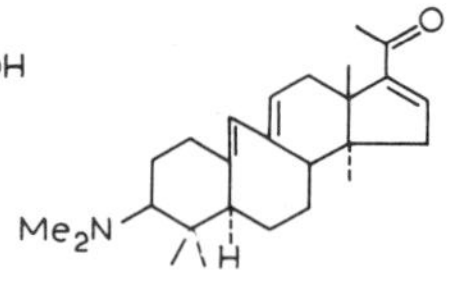

Buxpsiine
(LXXV)

Buxpsiine, isolated from *B. sempervirens*, is unusual in that it shows no cyclopropyl proton signals in its n.m.r. spectrum. Yet on taxonomic grounds it was felt to be closely related structurally to the monoacidic *Buxus* bases. It is an α,β-unsaturated methyl ketone containing three double bonds, one being conjugated with the carbonyl group, and two being conjugated with each other. Three olefinic protons are revealed by the n.m.r. spectrum. The positive multiple Cotton effect in the o.r.d. curve is in agreement with the presence of a Δ^{16}-20-keto unit. Structure LXXV has been proposed for the base; it is supported by a detailed analysis of its mass spectrum (*Tomko* and co-workers, Tetrahedron Letters, 1966, 915).

Another family of monoacidic bases, from the same or closely related sources, has the basic function at $C_{(20)}$. They are formulated and named below.

Cyclobuxoviridine-L (LXXVI)

Cyclomicuranine-L (LXXVII)

Buxandonine-L (LXXVIII)

Features of these molecules have been recognised by spectral investigation. Of particular interest is **buxandonine-L** (LXXVIII), which is exceptional in that it carries no alkyl group at position 4. Also, cyclomicuranine-L (LXXVII) shows a negative Cotton effect o.r.d. curve, typical of a 4,4-dimethyl-3-oxo-5α-steroid (*Nakano et al.*, J. chem. Soc. C, 1966, 1412; *Döpke* and *B. Müller*, Pharmazie, 1967, **21**, 666).

(ii) Diacidic Buxus alkaloids

Diacidic *Buxus* bases, much the commoner type of these alkaloids, have basic functions at both positions 3 and 20. They are generally more strongly basic than the monoacidic bases, and are consequently to be found in the more basic fractions of *Buxus* extracts. For review articles see introduction to section *(c)*. The basic groups may be primary, NHMe secondary, or NMe_2 tertiary, and a few are amidic in nature. The degree of substitution of the nitrogen atoms, as with the monoacidic bases, is denoted by suffix letters, *N*-acyl groups being counted as NH (see appended Table):

Suffix letter	*Substitution on nitrogen*			
	at $C_{(3)}$		*at* $C_{(20)}$	
	R^1	R^2	R^3	R^4
A	Me	Me	Me	Me
B	Me	Me	H	Me
C	H	Me	Me	Me
D	H	Me	H	Me
E	Me	Me	H	H
F	H	H	Me	Me
G	H	Me	H	H
H	H	H	H	Me
I	H	H	H	H

To date no naturally occurring members of subgroup I have been encountered. The diacidic bases, with their molecular formulae, melting points, optical activity and sources are summarised in Table 8.

TABLE 8

DIACIDIC *BUXUS* ALKALOIDS

Alkaloid	*Molecular formula*	*M.p. (°C)*	*[α]$_D$(°)**	*Source***
Buxaltine-H	$C_{35}H_{48}N_2O_3$	188–191		a
Buxamine-A	$C_{28}H_{48}N_2$	134	+40	g
Buxamine-E	$C_{26}H_{44}N_2$			i
Buxaminol-E	$C_{26}H_{44}N_2O$			i
Buxandrine-F	$C_{35}H_{52}N_2O_4$	189–290(decomp.)		a
Buxarine-F	$C_{33}H_{48}N_2O_3$	210–212	+98	a
Buxazidine-B	$C_{27}H_{46}N_2O_2$	234–236	−31	a
Buxazine	$C_{28}H_{48}N_2O_2$	238–239(decomp.)	+93	a
Buxeridine-C	$C_{34}H_{50}N_2O$	208–211	+14	a
Buxidienine-F	$C_{26}H_{44}N_2O_2$	237	+8	
Buxidine-F	$C_{33}H_{48}N_2O_3$	154–157	+67.5	
Buxiramine-D	$C_{27}H_{44}N_2O_2$	213–215		a
Buxitrienine-C	$C_{27}H_{44}N_2O$	192	+57	g
Buxocyclamine-A	$C_{27}H_{48}N_2$	187–188	+87	a
Cyclobuxine-B	$C_{26}H_{44}N_2O$	230–233	+119	h
Cyclobuxine-D	$C_{25}H_{42}N_2O$	235–237	+96.1	a
Cyclobuxoxazine-C	$C_{27}H_{46}N_2O_2$	234–236	+48 (EtOH)	h
Cyclokoreanine-B	$C_{27}H_{46}N_2O$	235–236	+109	d
Cyclomicrophylline-B (=cyclobaleabuxine, alkaloid-F)	$C_{27}H_{46}N_2O_2$	228–230	−90	f
Cyclomicrosine-C	$C_{34}H_{50}N_2O_3$	282–284	−33	b
Cycloprotobuxine-A	$C_{28}H_{50}N_2$	206–207	+76	b,c
Cycloprotobuxine-C	$C_{27}H_{48}N_2$	195	+40 (EtOH)	h
Cycloprotobuxine-F	$C_{26}H_{46}N_2$	163	+42	g
Cyclovirobuxeine-C	$C_{27}H_{46}N_2O$			e
Cyclovirobuxine-D	$C_{26}H_{46}N_2O$	205–210	+25 (EtOH)	h,i
Cyclovirobuxine-C	$C_{27}H_{48}N_2O$	201	+65	e
Cycloxobuxidine-F (=buxidine-F)	$C_{26}H_{44}N_2O_3$	227–230	+114	f
Cycloxobuxoxazine-C (=baleabuxoxazine-C)	$C_{27}H_{44}N_2O_3$	292	+116	c
16-Deoxybuxidienine-C	$C_{27}H_{46}N_2O$	200–201	+55	g
N-3-Acetylcycloprotobuxine-D	$C_{28}H_{48}N_2O$	221–224	+53	a
N-3-Benzoylbuxidienine-F (=*N*-3-benzoylbaleabuxidienine-F)	$C_{33}H_{48}N_2O_3$	286–288(decomp.)	−36	a
N-3-Benzoylcycloprotobuxoline-C	$C_{34}H_{52}N_2O_2$	252–255	+43	a
N-3-Benzoylcycloprotobuxoline-D	$C_{33}H_{50}N_2O_2$	235–238(decomp.)	+42	a
N-3-Benzoylcycloxobuxidine-F (=*N*-benzoylbaleabuxidine-F)	$C_{33}H_{48}N_2O_4$	274–276(decomp.)	+56	a
N-3-Benzoylcycloxobuxine-F (=buxatine)	$C_{33}H_{48}N_2O_2$	214–216(decomp.)	+90	a

Table 8 *(continued)*

Alkaloid	*Molecular formula*	*M.p. (°C)*	*[α]D(°)**	*Source***
N-3-Benzoylcycloxobuxoline-F	$C_{33}H_{48}N_2O_3$	255–256(decomp.)	+76	a
N-3-Benzoyldihydrocyclomicrophylline-F (=buxepidine)	$C_{33}H_{50}N_2O_3$	292–294	+19	a
N-3-Benzoyl-*O*-acetylcycloxobuxoline-F	$C_{35}H_{50}N_2O_4$	216–218	+114	a
N-3-Formylcyclovirobuxeine-B	$C_{28}H_{46}N_2O_2$	290	−157	e
N-3-Isobutyrylbaleabuxaline-F	$C_{30}H_{52}N_2O_4$	275	−32	c
N-3-Isobutyrylbuxidienine-F	$C_{30}H_{50}N_2O_3$	253	−67	c
N-3-Isobutyrylcycloxobuxidine-F (=*N*-isobutyrylbaleabuxidine-F)	$C_{30}H_{50}N_2O_4$	257	+71	c
N-3-Isobutyrylcycloxobuxidine-H	$C_{29}H_{48}N_2O_4$	285	+75	c
O-Tigloylcyclovirobuxeine-B	$C_{32}H_{52}N_2O_2$	178–183	−150	a
O-Vanilloylcyclovirobuxine-D	$C_{34}H_{52}N_2O_4$	210	±2	e
Pseudobaleabuxine-F	$C_{30}H_{50}N_2O_2$	236–240	+120.7	f

* Optical rotations in chloroform unless stated otherwise.

**a. *Buxus sempervirens* L.; b. *B. microphylla* Sieb. et Zucc. var. *suffruticosa* Makino; c. *B. balearica* Willd.; d. *B. Koreana* Nakai; e. *B. malayana* Ridl.; f. *B. balearica* Lam.; g. *B. madagascarica* Baillon; h. *B. wallichiana* Baillon; i. *B. microphylla* Sieb, et Zucc. var. *sinica*. Rehd. et Wils.

The diacidic *Buxus* bases are characterised in the main by one of five structural frameworks illustrated below.

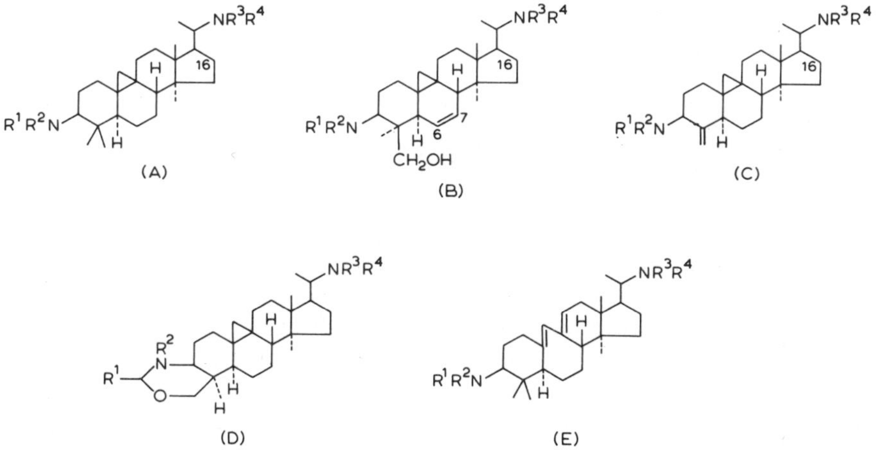

(A)

(B)

(C)

(D)

(E)

In structures (A), (B) and (C) an α-hydroxyl group is frequently found at position 16. Since the number of these alkaloids known is large, and structure determination has involved a great deal of spectroscopic study mainly of routine character, and correlations with certain of the better-known bases of known constitution, this discussion will concentrate on some of the latter. For details concerning other alkaloids see the detailed reviews by *Cerny* and *Šorm*, and *Tomko* and *Voticky (loc. cit.)*.

Cyclobuxine-D, the principal alkaloid of *B. sempervirens* leaves, forms an *N,N′,O*-triacetyl derivative, and is a disecondary (NHMe) amine and secondary alcohol. The n.m.r. spectrum revealed the presence of a $-CH_2 \cdot CH(OH) \cdot CH<$ grouping, one secondary and two tertiary *C*-methyl groups, a cyclopropyl methylene group and a $C{=}CH_2$ group, the last being confirmed by strong i.r. absorption near 890 cm^{-1}. That the $C{=}CH_2$ group and one basic function, and the secondary *C*-methyl and the other, are in close proximity was also apparent from comparative n.m.r. studies on the base and several simple derivatives. Cyclobuxine-D could be degraded to a diosphenol, in which one NHMe group and the $C{=}CH_2$ group were respectively transformed into carbonyl and enol groups. This product had $\lambda_{max.}^{EtOH}$ 296.5 nm, an unusually long wavelength for a diosphenol. Treatment of the product with hydrogen chloride generated a second diosphenol, $\lambda_{max.}^{EtOH}$ 277 nm, containing no cyclopropane CH_2 n.m.r. signals. Evidently the $O{=}C{-}C(OH){=}C<$ part of the first diosphenol is conjugated with the cyclopropane ring, and the latter could be reasonably placed, assuming a steroidal framework, as bridging positions 9 and 10. Dehydrogenation of cyclobuxine-D afforded a variety of products derived from anthracene and phenanthrene; on the other hand dehydrogenation of cyclobuxine-D lacking the cyclopropane ring yielded only phenanthrenes. A plausible explanation of the formation of anthracenes, involving 1,10 or 9,11 cleavage, followed by recyclisation at $C_{(6)}$ or $C_{(7)}$ with elimination of the cyclopropyl methylene group, can be offered. Oxidation of cyclobuxine-D, with and without prior protection of the basic functions, yielded cyclopentanone derivatives; in the latter case the cyclopentanone obtained readily suffered elimination of methylamine on treatment with a base to give a pair of geometrically isomeric *cisoid* enones. This observation is explicable in terms of the proximity of the $>$CHOH and one of the $-$NHMe groups in the alkaloid. These and related studies allowed cyclobuxine-D to be formulated

NHMe, OH —O→ NHMe, O —$^{\ominus}$OH→ H, O + $MeNH_2$

cis + trans

as LXXIX, exclusive of stereochemistry. Confirmation of this structure, and proof of the stereochemistry shown, was provided by a correlation of the alkaloid with the cyclopropanoid steroid cycloeucalenol (LXXX), of known structure and stereochemistry. For this purpose, cyclobuxine-D was reduced to its dihydroderivative (yielding an equatorial, 4α-methyl group), and

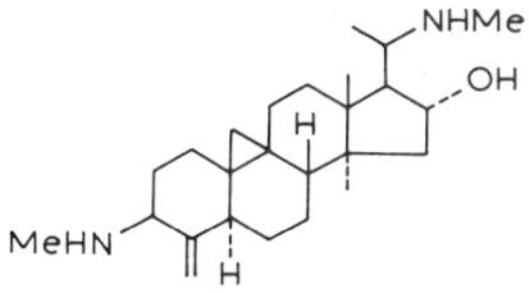

Cyclobuxine-D
(LXXIX)

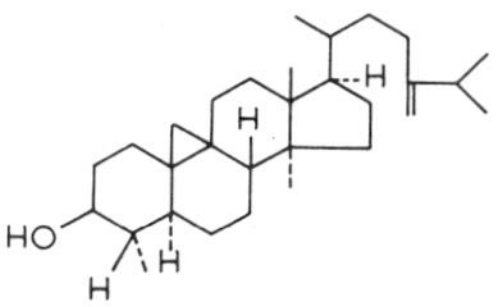

Cycloeucalenol
(LXXX)

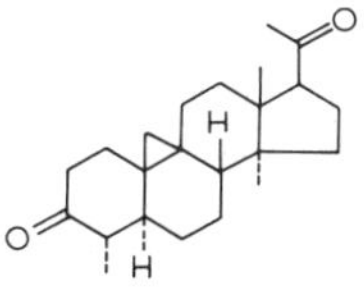

(LXXXI)

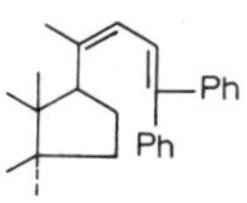

(LXXXII)

then subjected to Ruschig degradation and hydrogenation to yield diketone (LXXXI). Simultaneously cycloeucalenyl acetate was ozonised to the expected isopropyl ketone, reduced (CO→CHOH), dehydrated, oxonised, oxidised (CrO_3) and finally esterified. A modified Barbier–Wieland degradation of the ester yielded a tertiary alcohol which was dehydrated, subjected to allylic bromination and dehydrobrominated to a diene (LXXXII). This was hydrolysed (OAc→OH) and then oxidised (CrO_3) to a diketone identical in all respects with (LXXXI) (*Brown* and *Kupchan*, J. Amer. chem. Soc., 1962, **84**, 4590, 4592; 1964, **86**, 4414, 4424; *J. S. G. Cox*, *F. E. King* and *T. J. King*, J. chem. Soc., 1956, 1384). The configuration of the 3-methylamino group follows from the observation that the *N*,*N*′-dimethyl derivative of dihydrocyclobuxine-D is resistant to Hofmann degradation, behaviour well known with (equatorial) 3β-dimethylamino-steroids, and ascribed to the fact that these compounds are unable to adopt a *trans* anticoplanar conformation. The 3β-configuration was confirmed by the weak negative Cotton effect in the o.r.d. curve of the ozonolysis product of *N*,*N*′,*O*-triacetylcyclobuxine-D, which resembles that of 5α-cholestan-4-one, being therefore unaffected by the adjacent –NHAc group. This would be expected if this group were equatorial and therefore 3β. The α-orientation of the 16-hydroxyl group was deduced from molecular rotation differences on acetylation, and on n.m.r. evidence. Biogenetic grounds favour a 20α configuration for the other methylamino group. Finally, structure and configuration LXXIX is consonant with the results of an X-ray diffraction analysis of both *N,N'*-dimethylcyclobuxine-D and one of its monomethobromides (*C. G. Casinovi et al.*, C. A., 1964, **60**, 8729b). Cyclobuxine-D appears to be identical with "alkaloid-A" of *Heusler* and *Schlittler* (Helv., 1949, **32**, 2226).

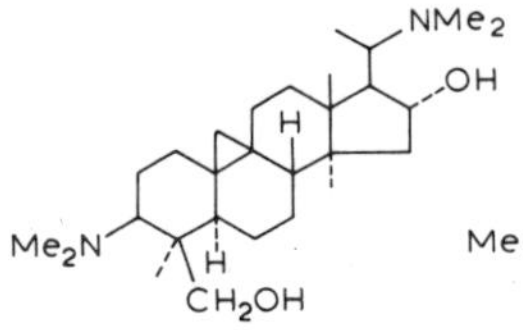

Cyclomicrophylline-A (LXXXIII) (LXXXIV) Cyclobuxoxazine-C (LXXXV)

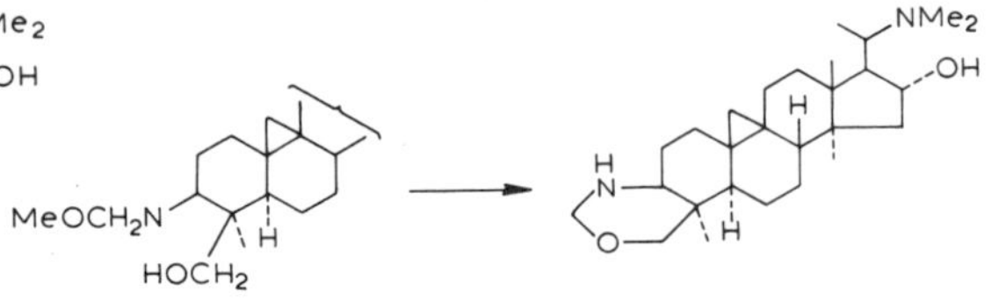

Cyclomicrophylline-A, from B. *microphylla* var. *suffruticosa* (*Nakano* and *S. Terao*, Tetrahedron Letters, 1964, 1035; J. chem. Soc., 1965, 4512), contains groupings $-\overset{|}{C}H-\underset{|}{\overset{|}{C}}-CH_2OH$ and $-CH_2\cdot\overset{OH}{\overset{|}{C}}H-\overset{|}{C}H$, and a cyclopropyl methylene group, on n.m.r. evidence. The alkaloid is a ditertiary (2 NMe_2) diacidic base containing a disubstituted double bond; this is placed in the 6,7-position on the following grounds. The cyclopropane ring in the alkaloid is opened by hydrogen chloride to yield an unconjugated diene; this on prolonged treatment with hydrogen chloride rearranges to a conjugated diene with u.v. absorption typical of steroidal 7,9(11)-dienes. The only possible location for the original double bond consistent with this behaviour is the 6,7-position. The proximity of the 4-CH_2OH and 3-NMe_2 groups was proved by Ruschig degradation of **cyclomicrophylline-C** (the 3-NHMe base corresponding to the alkaloid, found in the same source and yielding cyclomicrophylline-A on methylation) to a keto-alcohol, which on base treatment suffered retroaldol cleavage to formaldehyde and the 3-ketone. The latter was oxidisable to a 3,16-diketone which, as with cyclobuxine-D, lost methylamine readily on treatment with base, to give a *cis–trans* mixture of enones, reducible to an α-ethylketone which was identified by synthesis from cyclobuxine-D. The configuration of the CH_2OH and CH_3 groups at $C_{(4)}$ was settled by n.m.r. study of the AB quartet signals of the CH_2OH protons, the relatively low field position indicating an axial (β) hydroxymethyl group. In support, the 4-hydroxymethyl group and the 3β-NMe_2 group are involved in strong hydrogen bonding in several alkaloid derivatives; this is possible only if the former is β. Cyclomicrophylline-A is therefore formulated as LXXXIII (*Brown* and *Kupchan*, J. Amer. chem. Soc., 1964, **86**, 4430; *Nakano* and *Terao*, *loc. cit.*; Tetrahedron Letters, 1964, 1045; J. chem. Soc., 1965, 4537; *E. Wenkert* and *P. Beak*, Tetrahedron Letters, 1961, 358).

Cyclobuxoxazine-C, a minor alkaloid of B. *microphylla*, showed in its n.m.r. spectrum a pair of AB quartets centred at δ3.52 and 4.45 ppm, due to CH_2 groups both located close to an oxygen atom, one group being deshielded by adjacent nitrogen. The correctness of the latter statement was shown by the sensitivity of the deshielding to protonation of or acetylation of the nitrogen atom. This and other evidence pointed to the presence of an oxazine ring in the alkaloid; this was confirmed by the partial synthesis of the base from **dihydrocyclomicrophylline-F,** another minor alkaloid from the same source. This alkaloid, which has a 3β-NH_2 group, was treated with an excess of formaldehyde to yield the *N*-methoxymethyl base LXXXIV; this cyclised to cyclobuxoxazine-C (LXXXV) on exposure to alumina (*Nakano* and *Terao*, J. chem. Soc., 1965, 4537).

Buxamine-G, found in *B. sempervirens*, does not contain, on n.m.r. evidence, a cyclopropane ring, but contains two vinylic hydrogens, an *N*-methyl group, one secondary *C*-methyl group and four tertiary. Its u.v. absorption is characteristic of a heteroannular conjugated diene. A methyl ketone was obtained by Ruschig degradation of a primary amine group, using one equivalent of *N*-chlorosuccinimide, proving that an NH_2 group is located at position 20; the α-configuration of this group followed from circular dichroism studies. These and other observations, coupled with biogenetic considerations, led to the advancement of structure LXXXVI for the alkaloid (*D. Stauffacher*, Helv., 1964, **47**, 968; *Kupchan* and *W. L. Asbun*, Tetrahedron Letters, 1964, 3145; *D. Bertin* and *M. Legrand*, Compt. rend., 1963, **256**, 960). This has been confirmed by X-ray diffraction analysis of its di-

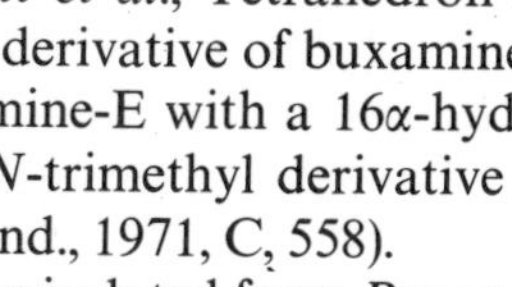

Buxamine-G
(LXXXVI)

hydro-iodide (*R. T. Puckett et al.*, Tetrahedron Letters, 1966, 3815). **Buxamine-E** is the 3-*N*-methyl derivative of buxamine-G (*Stauffacher, loc. cit.*), and buxaminol-E is buxamine-E with a 16α-hydroxyl group (*idem, ibid.*). **Buxamine-A** is the 3*N*,20*N*-trimethyl derivative of buxamine-G (*Goutarel* and co-workers, Compt. rend., 1971, C, 558).

For details of newer bases isolated from *Buxus* spp. see Specialist Periodical Reports, Chemical Society, Vols. 1–6 (1968–1976).

Guide to the Index

This index is constructed in a similar manner to the volume indexes of the first edition of the Chemistry of Carbon Compounds. However, to make the index easier to use, more descriptive entries have been made for the commonly occurring individual, and groups of chemicals.

The indexes cover primarily the chemical compounds mentioned in the text, and also include reactions and techniques, where named, and some sources of chemical compounds such as plant and animal species, oils, etc.

Chemical compounds have been indexed alphabetically under the names used by authors, editing being restricted to ensuring uniformity of entries under the same heading. In view of the alternative nomenclature that can often be used, a limited amount of cross-referencing has been done where it is considered to be helpful, but attention is particularly drawn to Convention 2 below.

For this and the succeeding volumes, the indexing conventions listed below have been adopted.

1. Alphabetisation

(a) The following prefixes have not been counted for alphabetising:

n-	*o-*	*as-*	*meso-*	D	*C-*
sec-	*m-*	*sym-*	*cis-*	DL	*O-*
tert-	*p-*	*gem-*	*trans-*	L	*N-*
	vic-				*S-*
		lin-			*Bz-*
					Py-

Some prefixes and numbering have been omitted in the index, where they do not usefully contribute to the reference.

(b) The following prefixes have been alphabetised:

Allo	Epi	Neo
Anti	Hetero	Nor
Cyclo	Homo	Pseudo
	Iso	

(c) A letter by letter alphabetical sequence is followed for entries, firstly for the main entry, followed by the descriptive entry. The only exception to this sequence is the placing of plural entries in front of the corresponding individual entries to prevent these being overlooked by a strict alphabetical sequence which could lead to a considerable separation of plural from individual entries. Thus "butanes" will come before *n*-butane, "butenes" before 1-butene, and 2-butene, etc.

2. *Cross references*

In view of the many alternative trivial and systematic names for chemical compounds, the indexes should be searched under any alternative names which may be indicated in the main body of the text. Only a limited amount of cross-referencing has been carried out, where it is considered that it would be helpful to the user.

3. *Esters*

In the case of lower alcohols esters are indexed only under the acid, *e.g.* propionic methyl ester, not methyl propionate. Ethyl is normally omitted *e.g.* acetic ester.

4. *Derivatives*

Simple derivatives are not normally indexed if they follow in the same short section of the text.

5. *Collective and plural entries*

In place of "— derivatives" or "— compounds" the plural entry has normally been used. Plural entries have occasionally been used where compounds of the same name but differing numbering appear in the same section of the text.

6. *Main entries*

The main entry of the more common individual compounds is indicated by heavy type. Where entries relate to sections of three pages or more, the page number is followed by "ff".

Index

α-Acetamidoacrylic acid, 119
2-Acetamidophenanthrene, 73
3-Acetamidophenanthrene, 73
5-Acetamino-anthracene, 72
2-Acetoacetnaphthalide, 61
α-Acetobutyrolactone, 205
Acetone dimethyl acetal, 317
Acetoxyacetylmethylmorphol, 273
17β-Acetoxyandrostan-1,4-dien-3-one, 433
3β-Acetoxybisnorallocholanic acid, 395, 402
14-Acetoxydihydrocodeinone, 291
8-Acetoxy-3,4-dimethoxyphenanthrene, 313
3β-Acetoxyethylindole-2-carbaldehyde, 232
17β-Acetoxy-1α-hydroxy-5β-androstan-3-one, 433
3-Acetoxy-2-hydroxy-2-methylbutyric acid, 444, 452
3β-Acetoxypregna-5,16-dien-20-one, 412
2α-Acetoxy-5α-pregnan-3-one, 383
3β-Acetoxypregn-5-en-20-ene, 413
3β-Acetoxy-5α-pregn-16-en-20-one, 423
8-Acetoxy-3,4,7-trimethoxyphenanthrene, 315
Acetylacetone, 61, 72, 73
9-Acetylacridine, 15
O-Acetylaknadinine, 312
Acetylaminoacridines, 21
N-Acetylanthranilic acid, 350
Acetylapoquinamine, 234
Acetylchlorodelpheline, 354
N-Acetylchonemorphine, **392**, 428
Acetylcoenzyme A, 427
N-Acetylconkurchine, 407
N-3-Acetylcycloprotobuxine-D, 460
N-Acetyldibenzo[*b,f*]azepine, 7
3-*O*-Acetyl-5α-dihydroverarine, 442
Acetylenedicarboxylic ester, 9, 45, 77, 78, 104. 121
16-Acetyl-5α-etiojervatriene-3β-ol, 437
Acetylhasubanol, 309
N-Acetylholadysine, 384
O-Acetyllatifoline, 409
N-Acetyl-L-leucine, 133
Acetylmethylmorphinol, 285
Acetylmethylmorphol, 271, 273
Acetylmethylmorphol-9,10-quinone, 273
1-Acetyl-3(*S*)-methyl-5-piperidone enamine, 440
N-Acetylnornicotine, 144
O-Acetyl-7-oxolatifoline, 409
O-Acetylphenyllobelol, 130
2-Acetylquinoline oxime, 218
O-Acetylsamandarine, **432**, 434, 436
N-Acetylsamandarone, 432
Acetylthebaol, 271
Acetyltigogenin, 423
N-Acetyltripiperideine, 145
Aconine, **358**
Aconitine, **358**, 359, 360, 362, 371
Aconitine alkaloids, 371
Aconitine-type alkaloids, 350, 358
Aconitum sp., 341, 350, 358, 362, 363
Aconitum chasmanthum, 366, 367, 368, 369
Aconitum excelsum, 363
Aconitum ferox, 366
Aconitum heterophyllum, 334, 341, 348, 370
Aconitum japonicum, 342, 345, 356
Aconitum kamtschaticum, 344
Acontium karakolicum, 333, 364
Aconitum leucostomum, 363
Aconitum lucidusculum, 333, 344, 345
Aconitum majimai, 345
Aconitum miyabei, 347
Aconitum napellus, 331, 369
Aconitum nasutum, 364
Acontium ranunculaefolium, 363
Aconitum sachalinense, 344

Aconitum salmatum, 348
Aconitum sanyoense, 342, 343
Aconitum spictatum, 367
Aconitum talassicum, 364, 365
Aconitum tasiromontanum, 342
Aconitum variegatum, 364, 365
Aconitum yesoensis, 344, 345
Aconosine, **364**, 371
Acridan, 7, 9, 28
Acridanone, 31
Acridanthione, 36
Acridarsines, 113
Acridarsinic acid, 111
Acridarsones, 111
Acridines, 29
–, chemical properties, 8
–, conversion to acridones, 36
–, hydrogenated, 28, 29
–, oxidation to acridones, 32
–, physical properties, 7
–, substituted, 15
–, 9-substituted, 29
Acridine, 1, 14, **15**
–, alkylation, 10
–, bond characteristics, 3
–, dimerisation, 16
–, electrophilic substitution, 13
–, nitration, 19
–, physical properties, 20
–, preparation, 4
–, protonation, 21
–, reaction with Grignard reagents, 14
–, reaction with sodamide, 19
–, reduction, 28, 29, 30, 31
–, resonance energy, 3
–, structure, 2, 3
–, ultraviolet spectra, 7
Acridine alkaloids, 257
Acridine-9-carboxylic ester, 27
Acridine dyes, 36
Acridine Orange L, 36
Acridine *N*-oxide, 8, 12
Acridine pigments, 38
β-Acridine-9-propionic ester, 15
Acridine salts, 15
Acridine-9-thiol, 35
Acridinium hydroxides, 11
Acridinium salts, 7, 10, 11, 29
9-Acridinylacetonitrile, 17
9-Acridinylhydrazine, 19
Acridones, 6, 29
–, reduction to acridines, 4
Acridone, 15, 16, 18, 19, 21, 24, **32**, 33, 190
–, reduction, 28
9-Acridones, 257
9-Acridone, 2, 13, 31, 210
Acridone alkaloids, 210, 214, 257
–, reactions, 259
–, with a C_5 side chain, 261
Acridone-2,7-disulphonic acid, 34
9-Acridonequinones, 35
9-Acridone-2-sulphonic acid, 34
Acriflavine, **22**, 37
Acrolein, 292
Acronychia sp., 178, 180, 187, 190, 197, 210, 211
Acronychia baueri, 34, 211, 263
Acronycidine, 180
Acronycine, 78, 191, **211**, 212, 257, 261, 263, 264, 265
Acrophyllidine, 187
Acrophylline, 187
Actinidia polygama, 158
Actinidine, **158**, 159, 160, 161, 162
Actinomycins, 35
Actinomycinol, **35**
Acutine, 178
Acutumidine, 311, **318**
Acutumine, 311, **318**
Acutuminine, 311, **318**
2-Acylaminobiphenyls, 41
Acylglutaconic anhydride, 448
Adenocarpine, **151**, 152, 153
Adenocarpus alkaloids, 152, 153
Adenocarpus commutatis, 151
Adenocarpus complicatus, 151, 152
Aegelenine, 180
Aegle sp., 180
Aegle marmelos, 208
Ailanthus sp., 175
Ajacine, 353
Ajaconine, 336, 338, **339**, 340
Aknadicine, 311, **312**, 317
Aknadilactam, 311, **312**
Aknadinine, 311, **312**, 317

Alatamine, 154
Alizarin Blue, 72
Alkaloids, aconitine, 358, 371
–, acridine, 257
–, acridone, 210
–, acronycine group, 263
–, adenocarpus, 152, 153
–, areca nut, 119, 120
–, atisine, 334, 340
–, bis-diterpene, 378
–, *buxus*, 430, 454, 455, 456, 459, 460
–, *celastraceae*, 154
–, cinchona, 225
–, containing oxazolidine rings, 430, 431
–, containing the quinoline ring system, 220
–, *daphniphyllum*, 374
–, diacidic *Buxus*, 459
–, diterpene, 323, 350
–, ester, 444, 445, 446, 452
–, *fritillaria*, 453, 454
–, from anthranilic acid, 216
–, from *Macrorungia longistrobus*, 217
–, *garrya*, 324
–, hasubunonine, 309, 311
–, hemlock, 120, 126
–, homomorphine, 319
–, lactonic, 370
–, *lobelia*, 128, 134, 135, 136, 137
–, monoterpenoid, 162
–, morphine group, 267
–, not containing oxazolidine rings, 430, 434
–, pomegranate root bark, 127
–, *prosopis*, 148, 150
–, quinoline, 171
–, 2-quinolone, 252
–, *rutacea*, 171, 172
–, salamander, 427, 430
–, *sedum*, 128, 134, 135, 136, 137
–, spirosolane, 422
–, steroidal, 381, 428, 429, 430
–, tobacco, 137, 144
–, tropane, 127
–, *veratrum*, 444, 445, 446, 452
–, with modified atisane skeleton, 341
–, with modified steroidal structure, 427
Alkaloid-A, **362**, 371, 378
Alkaloid-B, **362**, 371
Alkaloid CC-10, **320**
Alkaloid CC-20, **320**
Alkaloid D, **150**, 395
Alkamines, 437, 444, 445, 446, 452
–, esterification, 452
Alkenylcodeinones, 298
9-Alkoxy-10-alkyl-9,10-dihydroacridines, 9
1-Alkoxy-1-alkylphosphorins, 99
Alkoxyiodobenzenes, 262
1-Alkoxyphosphorins, 99, 100, 101, 102
Alkylacridines, 5, 17, 33
N-Alkylacridinium iodides, 10
N-Alkylacridinium salts, 8, 10
N-Alkylacridones, 12
1-Alkylarsenanes, 105, 106
1-Alkylarsenones, 106
9-Alkyl-10-aryl-9,10-dihydroacridines, 10
Alkylboranes, 283
N-Alkyl-9-cyano-9,10-dihydroacridines, 10
6-Alkyl-5,6-dihydrophenanthridines, 44, 45, 48
Alkyldihydrothebainones, 288
2-*n*-Alkyl-4-hydroxyquinolines, 221
Alkylphenanthrenes, 343
Alkylphenanthridines, 48, 49, 50
1-Alkyl-4-phosphorinanols, 88
2-Alkylpiperidines, 125
2-Alkyl-4-quinolones, 196
Allo-ψ-codeine, 273, 284, 304
trans-Allo-ψ-codeine, 284
Allopregnane, 381
Allosedamine, 132, 137
Aminacrine hydrochloride, 22
2-ω-Aminoacetylquinoline hydrochloride, 218
Aminoacridines, 19, 20, 21, 22
–, antibacterial properties, 22
–, basic dyestuffs, 36
–, diazonium salts, 21
–, physical properties, 20
–, protonation, 21
–, spectra, 20, 21
–, therapeutic properties, 22
3-Aminoacridine, 36
9-Aminoacridines, 35
9-Aminoacridine, 14, 19, 78
Aminoacridones, 34

8-Amino-6-*p*-aminophenylphenanthridinium chloride, 47
1-Aminoanthracene, 72
2-Aminoanthracene, 72
1-Aminoanthraquinone, 38, 72
2-Aminoanthraquinone, 72
Amino-9-arylphenanthridinium salts, 47
12-Aminobenz[*b*]acridine, 78
2-Aminobenzaldehyde, 24, 88, 107
2-Aminobenzanilides, 55
2-Aminobenzophenones, 214
1-Aminobenzotriazole, 6
1-(2-Amino-4-benzyloxy-5-methoxybenzyl)-6,7-dimethoxy-2-methyltetrahydroisoquinoline, 308
9-Amino-6-chloro-2-methoxyacridine, 23
14-Aminocodeinone, 291
3α-Aminoconan-5-ene, **403**, 429
3-Aminoconanine, 382
3β-Amino-5α-conanine, 399
3-Aminoconanine bases, 399
β-Aminocrotonic esters, 119
4-Amino-3-cyano-1-ethyltetrahydrophosphorin, 87
4-Amino-3-cyanophosphorinanes, 86
1,2-Aminocyclohexanol, 397
4-Amino-1-diethylaminopentane, 23
Amino-9,10-dihydroacridines, 19
2-Amino-9,10-dihydrophenanthrene, 73
6-Amino-1,10-dimethylphenanthridine, 51
3-Aminodiphenylamine, 19
2-(2′-Aminoethyl)-9,10-dihydrophenanthrene, 73
2-Amino-3′-hydroxybenzophenone, 223
3α-Amino-20α-hydroxy-5α-pregnane, 383
9-Amino-10-methylacridinium iodide, 22
2-Amino-*N*-methylbenzanilide, 55
4-Amino-3-methylbutyric acid, 421
4-Amino-3-methylbutyric ester, 438
1-(2-Amino-4,5-methylenedioxybenzyl)-6,7-dimethoxy-2-methyltetrahydroisoquinoline, 307
1-Aminophenanthrene, 74
2-Aminophenanthrene, 73
3-Aminophenanthrene, 73
4-Aminophenanthrene, 73
Aminophenanthridines, 52, 53
6-Aminophenanthridine, 42, 44, 48, 51, 52
Aminophenanthridones, 57
3-Amino-4-phenethylpyridine, 71
3-Amino-5α-pregnane, 382, 383
20-Amino-5α-pregnane, 382, 389
3α-Amino-5α-pregnan-20-one, 383
20α-Amino-5-pregnen-3β-ol, 390
γ-Aminopropyl 3-pyridyl ketone, 141
3β-Amino-5α,22α-*O*-spirostanes, 388
9-Amino-1,2,3,4-tetrahydroacridine, 22
4-Aminotetrahydro-1-phenylarsenin-3-carbonitrile, 108
δ-Aminovaleraldehyde, 127
Ammodendrine, **144**, 145
Ammondendron conollyi, 144
Amurine, 302, **307**
Amurinol, 308
tert-Amylhydroperoxide, 64
2-Amyl-4-methoxyquinoline, 172
2-Amylquinoline, 172, 193
Anabaseine, 142
Anabasine, **141**, 142, 144
Anabasis aphylla, 141
Anaferine, **155**
Anahygrine, **155**
Analgesic activity, relation to structure, 301
Anatabine, **142**, 143, 144
Anatalline, **143**, 144
Androcymbine, **319**, 320
Androcymbium melanthioides, 319
Angelic acid, 444
Angeloylzygadenilic acid, 445
Angeloylzygadenine, 445
Anhydroevoxine, 182
Anhydroignavinol, 342, 343
Anhydronupharamine, 164
Anhydro-oxobrowniine, 357
Anhydroperforine, 182
Aniba duckei, 164
Aniba rosaedora, 164
Anibine, **164**, 165
14-Anilinodihydrocodeinone, 292
2-Anilino-4-hydroxyfuran-3-carboxylic ester, 204
Anisic acid, 167

p-Anisidine, 23
Anopterine, 330, **333**, 334
Anopterus glandulosus, 333
Anopterus macleayanus, 333
Anopteryl alcohol, 334
Anthracenes, 462
Anthracene, 1, 2, 3
–, comparison with acridine, 9
Anthracene oil, 65
Anthranils, conversion to acridines, 5
Anthranil, 5, 6, 29
Anthranilic acids, 262
Anthranilic acid, 35, 37, 43, 171, 214, 215, 217, 222, 223, 224, 350, 352
–, derived alkaloids, 216
Anthraniloyllycoctonine, 352, 353
Antimalarials, 23
Antimonanes, 113
Antimonin, 114
Antimony, heterocyclic compounds, 113
Aplopappus hartwegi, 115
Apocinchenine, **229**, 240
Apocinchonine, **230**, 240
Apoconessine, **400**, 429
Apomorphine, 272, 273, 295
Apoquinamine, **234**, 240
Apoquinenine, 240
Apoquinine, 240
Arborinine, 190, 260, 263
Areca catechu, 119, 120
Arecaidine, **119**, 120
Arecaine, 119
Arecaitine aldehyde, 120
Areca nut alkaloids, 119, 120
Arecoline, **119**, 120
Arrow poisons, 350
Arsenanes, 105
Arsenane, **106**
Arsenic, heterocyclic compounds, 105
Arsenins, 112
4-Arsenones, 106, 107
Arylacridines, 33
N-Arylacridinium iodides, 10
α-Aryl-β-(9-acridyl)ethanols, 16
1-Arylarsenanes, 105, 106
Arylarsenins, 112
6-Aryl-5,6-dihydrophenanthridines, 48
2-Aryl-4-methoxyquinolines, 196
4-Aryl-1-methylphosphorinan-4-ols, 89
2-Aryl-*N*-methyl-4-quinolones, 196
Arylphenanthridines, 48, 50
4-Aryltetrahydro-1-methylphosphorins, 89
Aryltrimethylammonium hydroxides, 270
Aspergillus ochraceus, 405
Aspidospermidine alkaloids, 252
Astrocasia phyllantoides, 156, 157
Astrocasine, **156**, 157
Astrophylline, **157**
Atalanine, 192, 213, **259**
Atalaphylline, 191, **213**, 261, 264
Ataline, 192, **213**, 259
Atebrin, **23**
Atenine, 174
Atidine, **338**, 340
Atisine, 325, **334**, 335, 336, 339, 340
–, stereochemistry, 336, 337
Atisine alkaloids, 334, 340
Atlantia sp., 191, 192
Atlantia ceylanica, 213
Atlantia monophylla, 213
Avadharidine, 353
Aza-anthracenes, 253
1-Aza-anthracene, 64
Aza-anthraquinones, 62
1-Aza-anthraquinone, 64
2-Aza-anthraquinone, 68, 69
11-Azabenz[*a*]anthracene, 73
2-Azabicyclo[3.3.1]nonanes, 374
1-Azachrysene, 73
4-Azachrysene, 74
Aza compounds, 71
Azadicarboxylic ester, 77
3-Aza-A-homo-5-androstanes, 434
1-Azanaphthacene, 71, 72
9-Azaphenanthrene, 40
2-Azapyrene, 74
4-Azapyrene, 74
Azelaic acid, 146
9-Azido-1,2-benzofluorene, 81
2-Azido-2′-methoxydiphenylmethane, 6
Azima tetracantha, 147
Azimine, **147**
Azocarpine, **147**
Azomethine, 330
Azoxybenzoic acids, 6

Bacharaine, 175
Baeyer–Villiger oxidation, 149
Baleabuxoxazine-C, 460
Balfourodendron sp., 178, 180, 182, 187, 196
Balfourodendron riedelianum, 200, 209
Balfourodine, 210
Balfourodinium, 182
Balfourolone, **200**
Basic dyestuffs, 36
Baurella sp., 190, 191
Beckmann rearrangement, 80, 129, 227, 241, 274, 300
1,2-Benzacridine, 76
2,3-Benzacridine, 77
3,4-Benzacridine, 77
Benz[*a*]acridine, **76**
Benz[*b*]acridine, 76, 77
Benz[*c*]acridine, 76, 77
7*H*-Benz[*kl*]acridine, 78
Benz[*a*]acridine-12-carboxylic acid, 75
Benzacridinium salts, 77
Benz[*a*]acridone, 76
Benz[*b*]acridone, 77
Benzaldehyde, 29, 49, 223
1,2-Benzanthracene, 72, 73, 76
Benz[*a*]anthracene, 76
Benzazocines, 293
Benzene-1,2,3,4-tetracarboxylic acid, 437
Benzfluorenes, 299
Benzilic acid rearrangement, 39, 64, 313
4,5-Benzindane, 448
Benzisatin, 75
Benz[*f*]isoquinolines, 66, 67, 68
Benz[*g*]isoquinolines, 68, 69
Benz[*h*]isoquinolines, 69, 70, 71
Benzoacridines, 71, 75
Benzoacridones, 75
4,5-Benzocoumarandione, 76
Benzodiazepines, 223
1,2-Benzofluorene, 80
Benzo[*b*]fluorenone, 80
Benzo[*c*]fluorenone, 82
Benzoic acid, 19, 129, 132, 223, 342, 359, 396
–, reaction with diphenylamine, 4, 16
Benzoisoquinolines, 61
Benzo[*f*]isoquinoline, 61
Benzo[*h*]isoquinoline, 62
Benzo[*c*-2,7]naphthapyridines, 254
Benzophenanthridines, 71, 75
Benzo[*a*]phenanthridine, **80**
Benzo[*b*]phenanthridine, **80**
Benzo[*c*]phenanthridine, 79, 80, **81**
Benzo[*i*]phenanthridine, 80, **81**, 82
Benzo[*j*]phenanthridine, **82**
Benzo[*k*]phenanthridine, 82
Benzo[*c*]phenanthridine-11-carboxylic acid, 79
Benzo[*b*]phenanthrid-6-one, 80
Benzo[*c*]phenanthrid-6-one, 81
Benzo[*i*]phenanthrid-5-one, 81
Benzo[*j*]phenanthrid-6-one, 80, 82
Benzo[*k*]phenanthrid-6-one, **82**
Benzoquinolines, 60, 61, 62, 63
–, Diels–Alder adducts, 63
–, properties, 62
–, quaternary salts, 63
3,4-Benzo[*c*]quinoline, 40
5,6-Benzoquinoline, 63
6,7-Benzoquinoline, 64
Benzo[*f*]quinoline, 60, 61, 62, **63**, 64
Benzo[*g*]quinoline, 60, 62, **64**
Benzo[*h*]quinoline, 60, 62, **65**
Benzoylacetic acid, 130, 132
N-3-Benzoyl-*O*-acetylcycloxobuxoline-F, 461
9-Benzoylacridine, 15
Benzoylbaleabuxidienine-F, 460
N-Benzoylbaleabuxidine-F, 460
N-3-Benzoylbuxidienine-F, 460
10-Benzoyl-2-chloro-9-acridone, 31
N-Benzoylcinchotoxine oxime, 241
10-Benzoyl-9-cyano-9,10-dihydroacridine, 14
5-Benzoyl-6-cyano-5,6-dihydrophenanthridine, 49
N-3-Benzoylcycloxobuxidine-F, 460
N-3-Benzoylcycloxobuxine-F, 460
N-3-Benzoylcycloxobuxoline-F, 461
N-3-Benzoylcycloprotobuxoline-C,D, 460
N-Benzoyldihydrocinchotoxine, 239
N-3-Benzoyldihydrocyclomicrophylline-F, 461

N-Benzoylhomocincholoipon, 239
N-Benzoylhomomeroquinene ethyl ester, 241
N-Benzoylmeroquinene, 242, 243
6-Benzoylphenanthridine, 49
N-Benzoyl-2-piperidone, 142
1-Benzoyl-7-propionyl-*n*-heptane, 131
4-Benzoylpyridine-3-carboxylic acid, 68
N-Benzoyl-2-pyrrolidone, 141
9-Benzylacridine, 15
7-Benzylbenz[*c*]acridine, 76
Benzylideneaniline, 43
9-Benzylidene-10-methylacridinium salts, 12
9-Benzylidene-10-methyl-9,10-dihydroacridine, 12
Benzylisoquinolines, 306, 319
1-Benzyl-4-methyl-1-phenyl-phosphorinanium bromide, 85
Benzyloctahydroisoquinoline, 275
4-Benzyl-9-phenylacridine, 15
2-Benzylphenylstibonic acid, 114
4-Benzylpyridine, 68
Benzyne, 6, 43
Bernthsen reaction, 4
Biacridan, 16, 26
9,9′-Biacridene, 26
9,9′-Biacridine, **25**, 26
9,9′-Biacridinyl, **25**
10,10′-Biacridonyl, 8
Bicyclic octahydropyrrocoline units, 415
Bicycloatalaphylline, 264
Bicyclo-*N*-methylatalaphylline, 213
Bicyclo[2.2.2]octanes, 336
Bi(9,10-dihydroacridine), 10
Bi(9,10-dihydroacridinylidene), **26**
Bikhaconine, **367**, 371
Bikhaconitine, **367**, 371
Biosynthesis, cinchona alkaloids, 244
Biosynthesis, steroidal bases, 426
Biphenyls, 41
Biphenyl 2-isocyanate, 55
1,1′-Biphosphorinane-1,1′-dioxide, 86
1,1′-Biphosphorinane-1,1′-disulphide, 85
Biphosphorinylidene, 94, 95
2,3-Bipiperidyl, 157
2,3′-Bipiperidyl, 151
2,2′-Biquinolines, 40
Birch reduction, 327
9,9′-Bisacridanyl, 26
10,10′-Bisacridone ether, 8
Bis-2-butyl-3-methylborane, 327
2,5-Bis(2′-carboxyanilino)benzoquinone, 38
Bis(2-carboxyethyl)phenylarsine, 106
Bischler–Napieralski reaction, 41, 66, 69
Bis(2-cyanoethyl)phenylarsine, 108
3,6-Bisdeoxyverticine, 453
3β,20α-Bisdimethylamino-5α-pregnane, 392, 394, 395, 396
3α,20α-Bisdimethylamino-5-pregnene, 392
Bis-diterpene alkaloids, 378
3-(Bisethoxycarbonylamino) methylpyridine, 143
10-(1′,2′-Bismethoxycarbonyl) vinyl-9-acridone, 9
3β,20α-Bismethylamino-5α-pregnane, 393
Bismuth, heterocyclic compounds, 114
Bis(2-oxocyclohexyl)methane, 30
Boenninghausenia sp., 180
Bokitamine, **383**, 428
Boronia sp., 178
N-Bromoacetylhomosecodaphniphyllic ester, 376
1-Bromoacridine, 18
2-Bromoacridine, **18**
3-Bromoacridine, 18
4-Bromoacridine, 18
9-Bromoacridine, 12, 18
9-Bromoacridine *N*-oxide, 12
N-Bromoacridinium bromide, 18
N-2-Bromobenzylidenesolanocapsine, 422
3-Bromo-4-chloroquinoline, 43
1-Bromocodeinone, 276
14-Bromocodeinone, 274, 286, 290, 291, 305, 309
Bromocodide, 290
N-Bromoconiine, 125
1-Bromodihydrocodeinone, 281
7-Bromodihydrocodeinone dimethyl acetal, 304
1-Bromodihydromorphine, 301
1-Bromohydrosinomenilone, 305
3-Bromomesobenzanthrone, 38
1-Bromo-7-methoxycodeinone, 304

1-(2-Bromo-4,5-methylenedioxybenzyl)-7-hydroxy-6-methoxy-2-methyltetrahydroisoquinoline, 307
2-(Bromomethyl)phenethyl bromide, 109
Bromophenanthridines, 48, 51
2-Bromophenanthridone, 57
2-(4′-Bromophenyl)-2-phenyltetrahydroisophosphinolinium bromide, 91
3-Bromopyridine, 139
1-Bromosinomeneine, 304
1-Bromosinomeneine ketone, 305
1-Bromosinomeneinic acid, 305
1-Bromosinomenilic acid, 305
1-Bromosinomenilone, 305
1-Bromo-Δ^7-thebainone, 276
Browniine, **356**, 357, 368
Bucharaine, **199**
Bucharamine, 187
Bucharidine, 175, **199**
Bucharine, **199**
Bukharidine, **199**
1-Buten-3-ynyl-8-hydroxy-2-(*cis*-2-penten-4-ynyl)-azaspiro[5.5]undecane, 165
1-Butyl-1,2-dihydro-2,4,6-triphenylphosphorin, 94
Butylphosphorins, 98, 100, 101, 102
N-Butyryl-2-piperidone, 125
Buxaltine-H, 460
Buxamideine-K, 456
Buxamine-A, 460, **465**
Buxamine-E, 460, **465**
Buxamine-G, **465**
Buxamine-M, **458**
Buxaminol-E, 460, 465
Buxandonine, 456
Buxandonine-L, **459**
Buxandrine-F, 460
Buxanine-M, 456, 457
Buxarine-F, 460
Buxatine, 460
Buxazidine-B, 460
Buxazine, 460
Buxene-O, 456, **457**
Buxenone-M, 456
Buxepidine, 461
Buxeridine-C, 460
Buxidienine-F, 460
Buxidine-F, 460
Buxiramine-D, 460
Buxitrienine-C, 460
Buxocyclamine-A, 460
Buxpiine-K, 456
Buxpsiine-K, 456, **458**
Buxtauine-M, 456
Buxus alkaloids, 430, 454, 455, 456, 459, 460
Buxus microphylla, 456, 464, 465
Buxus sempervirens, 390, 457, 458, 462, 465

Cadaverine, 127
Caffeic acid, 120
Caledon Red BN, 37
Calycanine, 249
Calycanthaceous alkaloids, biosynthesis, 251
Calycanthidine, **250**
Calycanthine, **248**, 250, 251
Calycanthus sp., 250
Calycanthus alkaloids, 171
Calycanthus floridus, 250, 251
Calycanthus occidentalis, 250
Camellia sp., 172
Cammaconine, **365**, 371
Camptothecea accuminata, 246
Camptothecea alkaloids, 171
Camptothecin, **246**, 247, 248
Cancentrine, 267, **321**
Caproic acid, 149
β-Carboline, 249
2-Carboxy-6-carboxymethyl-1-methylpiperidine, 131
4-Carboxyphenoxyacetic acid, 167
5(4′-Carboxyphenyl)-5,10-dihydro-3-methylacridarsine, 111
Cardiac inhibitors, 350
Carica papaya, 146, 147
Carnavoline, **148**
Carpaine, **146**, 147
ψ-Carpaine, **147**
Carpamic acid, 146
Carpyrinic ester, 146
Casimiroa sp., 174, 178, 180
Casimiroa edulis, 198
Casimiroine, 174
Cassia carnavol, 148, 149
Cassia excelsa, 147
Cassine, **147**, 148

Celastraceae alkaloids, 154
Cepharamine, 311, **312**, 313
Cevacine, 445
Cevadine, 445
Cevagenine, **447**, 449, 452
Cevanthridine, 448, 451
Cevanthrol, 448
Ceveratrum bases, 453
Cevine, **444**, 447, 448, 449, 452
Chaconines, 416
Chasmaconitine, **367**, 371
Chasmanine, **368**, 369, 371
Chasmanthinine, **367**, 368, 371
Chavicic acid, 116
Chavicine, **116**
Chelidonine, 81
Chemiluminescence, 7, 27, 28
Chimonanthine, **250**
Chimonanthus fragrans, 250, 251
Chloritaena sp., 180
2-Chloroacridarsinic acid, 111
1-Chloroacridine, **18**, 19
2-Chloroacridine, 18
3-Chloroacridine, 18, 19
4-Chloroacridine, 18
9-Chloroacridines, 262
9-Chloroacridine, 4, 12, 13, 15, 18, 23, 25, 31, 32, 33, 35, 36
1-Chloro-2-aminoanthraquinone, 72
9-Chloroaminofluorenes, 42
1-Chloroanthraquinone, 37
1-Chloroanthraquinone-2-carboxylic acid, 37
1-Chloroantimonane, 113
1-Chloroarsenane, 106
5-Chloro-6-azachrysene, 74
12-Chloro-1-azanaphthacene-6,11-dione, 72
12-Chlorobenz[*b*]acridine, 77
2-Chlorobenzoic acid, 75
6-Chlorobenzo[*c*]phenanthridine, 81
9-Chlorobenzo[*g*]quinoline, 64
10-Chlorobenzo[*g*]quinoline, 60
o-Chlorobenzylideneaniline, 43
5-Chloro-3-*p*-chlorophenylanthranil, 6
α-Chlorocodide, **283**, 284, 290, 304
β-Chlorocodide, **283**, 284, 290
2-Chloro-5-cyano-5,10-dihydroacridarsine, 111
6-Chloro-3-cyano-2,4-dihydroxypyridine, 118
6-Chloro-9(4′-diethylamino-1′-methylbutyl)amino-2-methoxyacridine dihydrochloride, 23
5-Chloro-5,10-dihydroacridarsine, 111
1-Chloro-1,4-dihydroantimonin, 114
5-Chloro-5,10-dihydrodibenz[*b,e*]-antimonin, 114
5-Chloro-5,10-dihydrobenzo[*b,e*]-phosphorin, 96
5-Chloro-5,10-dihydromethylacridarsines, 111
9-Chloro-4,5-dimethylacridine, 5
5-Chloro-4′-methoxydiphenylamine-2-carboxylic acid, 23
9-Chloro-10-methylacridinium methosulphate, 33
9-Chloro-10-methylacridinium salts, 27
3-Chloro-3-methylbutyne, 212
3-Chloro-6-methylphenanthridine, 46
1-Chloro-2-naphthylamine, 60
9-Chloronitroacridines, 19
7-Chloro-2-nitrofluorenone, 55
8-Chloro-3-nitrophenanthridone, 55
2-Chloro-5-nitropyridine, 124
6-Chlorophenanthridine, 44, 48, 51, 52
9-Chlorophenanthridine, 47
10-Chlorophenanthridine, 44
2-Chlorophenanthridone, 57
3-Chlorophenanthridone, 46
1-Chlorophosphorinane-1-oxide, 86
N-Chloro-2-*n*-propylpiperidine, 125
4-Chloroquinaldine, 194
Chloroquine, 23
4-Chloroquinoline, 118
6-Chlororicinine, 118
N-Chlorosuccinimide, 57, 125, 465
1-Chlorotetrahydroarsinoline, 108
Choisya sp., 180, 182
Choisyine, 182
5α-Cholestan-4-one, 397, 398
5β-Cholestanone, 435
Cholest-4-en-3-one, 438
Cholesterol, 414, 416, 435, 438
Chonemorpha sp., 392
Chonemorphine, **392**, 396, 428

Chrysenequinone, 80
Cinchamidine, 231
Cinchene, 229
Cinchenine, 235, 240
Cincholoipon, **226**
Cincholoiponic acid, **226**, 227, 235, 240
Cinchomeronic acid, 68, 226
Cinchona sp., 225, 244
Cinchona alkaloids, 171, 225
–, biosynthesis, 244
–, configurations, 238
–, derived from indole, 232
–, synthetical studies, 239
Cinchona bark, 171
Cinchona bases, physical properties, 239
–, stereochemistry, 234
Cinchona legeriana, 246
Cinchonamine, **232**, 233, 240, 244
Cinchona succirubra, 230, 232
Cinchonidine, 225, **230**, 235, 236, 238, 240, 246
β-Cinchonidine, 230
Cinchonidone, 246
Cinchoninal, **226**
Cinchonine, 225, **226**, 227, 228, 229, 230, 233, 235, 236, 238, 240, 246
Cinchoninic acid, 226, 227, 229, 230
Cinchoninine, **226**, 228, 240
Cinchotenine, **226**, 231, 240
Cinchotine, 231
Cinchotoxine, **227**, 228, 235, 240
Citrus sp., 172, 174, 180
C. I. Vat Green, 3, 38
C. I. Vat Red 35, 37
C. I. Vat Violet 14, 37
Claisen rearrangement, 207
Clemmensen reduction, 303, 310
Coal-tar, 63, 74
–, acridine content, 2
Codeimethines, **284**, 285, 294
α-Codeimethine, 271, 299
β-Codeimethine, 271, 291
Codeine, 267, 268, 273, 289, 303, 321
–, commercial preparation, 269
–, exhaustive methylation, 284
–, structure, 270, 279
ψ-Codeine, 273, 284, 304
trans-Codeine, **282**
trans-ψ-Codeine, 284
Codeine methyl ether, 287, 290
β-Codeinethine, 272
Codeinone, 270, 272, 290, 293, 294, 303
ψ-Codienone, 273, 297
Codeinone dimethyl acetal, 278
Codeinone methiodide, 271
Codiran, 279, 285
Colchicum cornigerum, 319, 320
Collidine, 327
β-Collidine, 226, 241
Combes synthesis, 61, 72
Compositae sp., 323
Cona-3,5-diene, 409
5-Conanene, 407
Conan-4-en-3-one, 402
Conanine, 404
Condelphine, **365**, 366, 371
Conessidine, **407**, 429
Conessimethine, 401
Conessimine, **403**, 404, 429
Conessine, 395, **399**, 400, 401, 402, 403, 405, 406, 408
Confusameline, 182
Conhydrine, 120, **121**, 122, 123, 125, 126
Conhydrinone, **121**
α, β and δ-Coniceine, **125**, 126
β-Coniceine, 121, 122
γ-Coniceine, 120, **124**, 125, 126
ε-Coniceine, **126**
Coniine, **120**, 122, 123, 124, 126
Coniine[^{15}N], 121
Conimine, **404**, 429
Conium maculatum, 120, 124
Conkurchine, **407**, 429
Conkuressine, 403, 405
Conopharyngia pachysiphon, 390
Conopharyngine, **390**, 428
Conquinamine, **234**, 240
Conrad–Limpach synthesis, 61
Consolida regalis, 355
Conyrine, **120**, 124, 126
Corynantheal, 244, 246
Corynanthé alkaloids, 244
Cotton effect, 331
Croton linearis, 306
Cryptobranchus maximus, 433

Cryptolepine, **220**
Cryptolepis sanquinolenta, 220
Cryptolepis triangularis, 220
Cryptophenols, 259
Cuauchichicine, **328**, 329, 330
Cunninghamella echinulata, 405
Cupreine, **232**, 240
Cuscohydrine, 155, 156
Cuspareine, 172, 193, **194**, 195
Cusparia sp., 180
Cusparia trifoliata, 193
Cusparine, 172, **193**
2-Cyanoacetophenone, 275
Cyanoacetyl chloride, 118
9-Cyanoacridine, 14, 19
5-Cyano-5,10-dihydroacridarsine, 111
9-Cyano-9,10-dihydroacridine, 14
Cyanomethylaniline, 17
4-Cyanomethyl-5,6-dimethoxy-1,2-naphthaquinone, 275
1-Cyanomethyl-7,8-dimethoxy-2-oxo-1,2,3,4-tetrahydronaphthalene, 312
3-Cyano-1-methyl-2-pyridone, 117
13β-Cyano-18-nor-5α-pregnanes, 402
2-Cyanopyridine, 121
Cyclobaleabuxine, 460
Cyclobuxine-B, 460
Cyclobuxine-D, 457, 460, **462**, 463, 464
Cyclobuxomicreine-K, 456
Cyclobuxophylline-K, 456, 458
Cyclobuxophylline-M, 456, 458
Cyclobuxosuffrine-K, 456
Cyclobuxoviridine-L, 456
Cyclobuxoxazine-C, 460, **465**
Cycloeucalenol, 463
Cycloeucalenyl acetate, 463
Cycloheptanones, 351
Cyclohept[*b*]indole-10(5*H*)-one, 13
Cyclohexane-1,3-dione, 255
Cyclohexanone, 5, 29, 35, 406
Cyclohexene, 6
Cyclokoreanine-B, 460
Cyclomicrobuxeine-K, 456
Cyclomicrobuxine, 456
Cyclomicrobuxinine, 456
Cyclomicrophylline-A, **464**
Cyclomicrophylline-B, 460
Cyclomicrophylline-C, **464**
Cyclomicrosine-C, 460
Cyclomicuranine, 456
Cyclomicuranine-L, 459
Cycloneosamandaridine, **434**, 436
Cycloneosamandione, **434**, 436
Cyclo-octatetraene, 91
Cyclopamine, **441**, 444
Cyclopenase, 224
Cyclopenine, **222**, 223, 224
Cyclopenol, **222**, 223, 224
Cyclopentadienes, 95
Cyclopentanones, 328, 368, 372, 432, 462
4*H*-Cyclopenta[*def*]phenanthrene-4-one, 74
Cyclopentenofluorenes, 448
1,2-Cyclopentenophenanthrene, 411, 419
Cyclopropanecarbonitrile, 141
Cycloprotobuxine-A, C, F, 460
Cyclopropyl 3-pyridyl ketimine, 141
Cyclorolfeine, 456, **458**
Cyclorolfoxazine, 456, **458**
Cyclosuffrobuxine-K, 456
Cyclosuffrobuxinine-M, 456
Cyclovirobuxeine-C, 460
Cyclovirobuxine-C, 460
Cyclovirobuxine-D, 458, 460
Cycloxobuxidine-F, 460
Cycloxobuxoxazine-C, 460

Daphmacrine, **375**, 378
Daphmacrine methiodide, 374
Daphmacropodine, **375**, 378
Daphnilactone-A, **376**, 378
Daphnilactone-B, **376**, 378
Daphniphyllidine, **375**, 378
Daphniphylline, **374**, 375, 376, 378
Daphniphyllum alkaloids, 350, 374
Daphniphyllum macropodum, 374, 377
Decatropis sp., 180
Decatropis bicolor, 203
Decevinic acid, 448
De-ethyldesoxycamptothecin, 248
Dehydroanabasine, 142
14-Dehydrobrowniine, **357**
Dehydrocevine, 447
Dehydrocinchonamine, 233
Dehydroconiine, 124

Dehydrocusparine, 193
14-Dehydrodelcosine, **356**, 357
Dehydrodelpheline, 372
Dehydrodelphisine, 368
18-Dehydro-*N*-demethylholadienine, 409
Dehydrodeoxometaphanine, 313
3,4-Dehydro-8,10-diethyllobelidiol, 132, 134
3,4-Dehydro-8-ethyl-10-phenyllobelidiol, 132, 134
3,4-Dehydro-8-ethyl-10-phenyllobelionol, 131, 134
3,4-Dehydro-8-ethyl-10-phenyllobelidione, 134
7-Dehydrogermine-3,16-diacetate, 450
Dehydroheteratisine, 372
Dehydroisolongistrobine, **218**, 219
Dehydro-8-methyl-10-ethyllobelidiol, 134
4,5-Dehydro-8-methyl-10-phenyllobelidiol, 134
Dehydronormetathebainone, 306
Dehydronornicotine, 140
Dehydro-oxodelpheline, 372
7-Dehydroprotoverine-3,6,16-triacetate, 450
Dehydrosongorine, 333
Dehydroxyskytanthine, 161
Delavaine, 302, **309**, 311, 316
Delcosine, **355**, 356, 357
Delnudine, **348**, 349
Delphatine, **352**, 357
Delphelatine, 354
Delpheline, **352**, 354, 357
Delphinine, **360**, 362, 366, 371
Delphinium sp., 339, 350
Delphinium ajacis, 352, 355
Delphinium barbeyi, 352, 354
Delphinium brownii, 352, 356
Delphinium cardinale, 341, 357
Delphinium confusum, 366
Delphinium consolida, 352, 355, 356
Delphinium denudatum, 339, 341, 348
Delphinium elatum, 352, 354
Delphinium occidentale, 354
Delphinium orientale, 355
Delphinium staphisagria, 348, 360, 368, 378
Delphisine, **368**, 369, 371
Delphonine, 361, 371
Delsemine, 353
Delsoline, **356**, 357
Deltaline, **354**, 355, 357
Deltamine, 355
Demethoxydihydrosinomenine, 303
7-*O*-Demethyl evolitrine, 180
O-Demethylgalipine, 194
O-Demthylheteratisine, 373
N-Demethylkurchiphylline, 384
De-*N*-methylnoracronycine, 212
O-Demethylquinone, 232
Demissidine, **417**, 418, 424, 430
Demissidone, 417
Demissine, **418**, 430
Dendrobates pumilio, 255
Dentrobates histrionicus, 165
Denudatine, **339**, 340
11-Deoxojervine, **441**, 444
Deoxyaconitine, **363**, 371
Deoxyacutumine, 318
16-Deoxybuxidienine-C, 460
Deoxycarpyrinic acid, 146
Deoxycinchonidine, 235
Deoxycinchonine, 235
Deoxycodeines, 281, **285**, 290, 297
Deoxymethylenelycoctonine, 350
Deoxymorphines, 285, 286
Deoxynupharamine, 163
Deoxynupharidine, 164
Deoxypseudoaconitine, 367
Deoxyquinidine, 243
Deoxyquinine, 243
Deoxyuzurimine-*N*-oxide, 376
Depressor bases, 350
Desacetylprotoveratrine A and B, 446
N-Desethylneoline, 370
Des-*N*-methylacronycine, 261, 264
Des-*N*-methylnoracronycine, 261, 264
Desoxyheteratisine, 373
Desoxyprosopine, 148, 149
Desoxytecostanine, 161, 162
16*R*,16-Deuterokauran-15-one, 329
15-Deuterokaur-16-en-15β-ol, 329
7,8-Diacetoxy-3,4-dimethoxyphenanthrene, 313
4,6-Diacetoxy-3-methoxyphenanthrene, 271

4,8-Diacetoxy-3-methoxyphenanthrene, 273
3β,16β-Diacetoxy-5α-pregnan-20-one, 425
1,3-Diacetoxy-2,5,6-trimethoxyphenanthrene, 316
Diacetylallocinchonamine, 232
N,O-Diacetyljervine, 439
Diacetyl-ψ-tigogenin, 423
Diacidic *Buxus* alkaloids, 459
Diacyl peroxides, 15
1,1-Dialkoxyphosphorins, 99, 100, 101, 102
1,1-Dialkoxyphosphorinium salts, 100
1-Dialkylaminophosphorins, 100, 101, 102
10,10′-Dialkyl-9,9′-biacridinium dinitrates, 28
1,1-Dialkylphosphorins, 99, 100
3,4-Dialkylquinolines, 58
3,6-Diaminoacridines, 22, 36
3,8-Diamino-6-*p*-aminophenyl-5-phenanthridinium chloride, 47
2,2′-Diamino-4,4′-dimethylaminodiphenylmethane, 37
3,9-Diamino-7-ethoxyacridine lactate, 22
3,6-Diamino-10-methylacridinium chloride, 22
3,8-Diamino-5-methyl-6-phenylphenanthridinium bromide, 47
3,20-Diamino-5α-pregnane, 382, 391, 396
4,6-Dianilinoisophthalic acid, 39
2,6-Diaryl-4-methylpyrylium salts, 97
1,1-Diarylphosphorins, 97
1,5-Diazabicyclo[4.3.0]non-5-ene, 114
1,10-Diazachrysene, 64
1,8-Diazadibenzofluorenone, 39
Diazonium salts, reaction with 9-methylacridine, 17
Diazonium tetrafluoroborates, 99
5*H*-Dibenzo[*b,f*]azepine, 27
Dibenzocycloheptatriene, 27
Dibenzo[1,7]phenanthroline, 40
Dibenzo[1,10]phenanthroline, 40
Dibenzo[4,7]phenanthroline, 39
Dibenzo[*b,d*]pyridine, 40
4*H*-Dibenzoquinolizines, 45
Dibenzoquinolizinetetracarboxylic ester, 45
Dibenzothiazepines, 42
1,7-Dibenzoyl-*n*-heptane, 129
Dibenzoylmethane, 61
Dibenzoylsinomenol, 303
Dibenzoyltartaric acid, 132
9,10-Dibenzyl-9,10-dihydroacridine, 15
7,12-Dibenzyl-7,12-dihydrobenz[*a*]-acridine, 76
Dibenzyl-7,12-dihydrobenz[*c*]acridine, 76
4,5-Dibenzylideneoctahydroacridine, 30
9,10-Dibenzyl-9-phenyl-9,10-dihydroacridine, 15
Diborane, 281, 283
2,7-Dibromoacridine, 18
2,7-Dibromo-9-acridone, 34
Dibromocotinine, 138
3,9-Dibromomesobenzanthrone, 38
1,5-Dibromopentane, 84, 85, 86
2,4-Dibromophenanthridones, 57
1,1′-Dibromotetrahydro-ψ-morphine, 301
Dibromoticonine, 138
1,1-Dibutyl-1,4-dihydrostannabenzene, 114
1,1-Di-*n*-butyl-2,6-dihydroxyphosphorinanium chloride, 89
3,5-Di-*tert*-but-2-ylfuryl ketone, 98
N,N′-Di(2′-carboxyphenyl)-*p*-phenylenediamine, 39
Dicentra canadensis, 321
3,9-Dichloroacridine, 18, 19
3,7-Dichloroacridone, 6
Dichloroarsines, 105
2,4-Dichlorobenzoic acid, 23
2,4-Dichlorobenzo[*g*]quinoline, 62
Dichloro-*N,N*-diethylphosphoramidate, 86
2,5-Dichloro-5,10-dihydroacridarsine, 111
6,8-Dichloro-2-methoxyacridine, 23
2,4-Dichlorophenanthridones, 57
Dichlorophenylphosphine, 91
Dichlorophosphines, 84
2-Dichlorophosphinodiphenylmethane, 96
Dichlorostilbenes, 113
Dictamnine, 180, **203**, 204, 205, 214, 215, 216
Dictamnus sp., 174, 180, 182, 187
Dictamnus alba, 208, 214
Dictyodiamine, **393**, 428

Dictyolucidamine, **387**, 428
Dictyolucidine, **387**, 428
Dictyophleba lucida, 387, 392, 393
Dictyophlebine, **393**, 428
Di(cyanoethyl)phosphines, 86
Dicyclohexyl-2,6-dihydroxyphosphorinamium chloride, 89
Diels–Alder adducts, anthracene, 9
Diels–Alder reaction, 43, 293
Diels hydrocarbon, 416, 421, 425, 426
Dienamines, 293
4,6-Diethoxy-3,7-dimethoxyphenanthrene, 303
1-Diethylamino-4-aminopentane, 23
1,2-Diethylcyclohexane, 235
8,10-Diethyllobelidiol, 134
8,10-Diethyllobelidione, 132
8,10-Diethyllobelionol, 134
8,10-Diethylnorlobelidione, 134
8,10-Diethylnorlobelionol, 134
Digitonin, 416, 419, 425, 437, 438
2,4-Dihalogenophenanthridones, 57
1,1-Dihalogeno-2,4,6-triphenylphosphorins, 99
5,10-Dihydroacridarsines, 110, 111
9,10-Dihydroacridines, 11, 26, 27
–, 9-substituted, 14
9,10-Dihydroacridine, 4, 6, 7, 8, 13, 27, **28**, 29, 33, 78
9,10-Dihydroacridine-9-thione, 35, 36
9,10-Dihydroacridone, 15
Dihydroacronycine, 259, 260, 263
Dihydroajaconine, 339
Dihydroallo-ψ-codeine, 281
Dihydroammodendrine, 144
Dihydroarecaidine, 119
5,6-Dihydroarsanthridines, 110, 111
Dihydroastrocasine, 156
Dihydroatisine, 335, 338
5,6-Dihydrobenz[*c*]acridine, 77
5,12-Dihydrobenzacridine, 77
7,12-Dihydrobenz[*a*]acridine, 76
5,6-Dihydrobenz[*h*]isoquinoline, 71
5,6-Dihydrobenzo[*f*]quinoline, 63, 64
Dihydrocinchonamine, 233
Dihydrocinchonidine, **231**, 240, 241
Dihydrocinchonine, **226**, 231, 237, 240, 241
Dihydrocinchotoxine, 239, 241
Dihydrocodeine, 270, 278, 279, 285
Dihydro-ψ-codeine, 281
Dihydrocodeinemethine, 285
Dihydrocodeinone, 278, 288, 297, 303
trans-Dihydrocodeinone, 282
Dihydrocodeinone enol acetate, 291, 297
Dihydrocodeinone oxime, 274
Dihydrocuauchichicine, 329
Dihydrocupreine, **232**
9,10-Dihydro-9-cyanoacridine, 26
Dihydrocyclobuxine-D, 463
Dihydrocyclomicrophylline-F, **465**
Dihydrodaphmacrine, 375
Dihydrodeoxycodeines, **286**
Dihydrodeoxycodeine-C, 286
Dihydrodeoxycodeine methyl ether, 276
Dihydrodeoxymorphine, 286, 301
5,10-Dihydrodibenz[*b,e*]antimonins, 113
5,11-Dihydrodibenz[*b,e*]1,4-oxazepine, 13
1,4-Dihydro-1,1-dibutylstannabenzene, 97, 112
Dihydro-β-diethylthiocodide, 290
3,4-Dihydro-5,7-dimethoxy-2-quinolone, 211
9,10-Dihydro-9,10-dimethylacridine, 29
5,6-Dihydro-5,5-diphenyldibenzo[*b,d*]-phosphorinium bromide, 95
1,2-Dihydro-1,1-diphenylphosphorinium perchlorate, 93
Dihydroepistephamiersine, 316
Dihydrofuro[2,3-*b*]quinoline, 204
Dihydrogarryfoline, 329
Dihydrohetisine, 341
Dihydroholaphylline, 391
Dihydroholarrhenine, 406
Dihydrohydroxycodeone, 278
1,2-Dihydro-4-hydroxy-3-methoxy-1-methylquinoline, 217
Dihydroindolinocodeinone, 310
Dihydroirehdiamine, 391
Dihydroisocodeine, 279
Dihydroisohistrionicotoxin, 165
Dihydroisorubijervine, 419
Dihydrojervine, 438, 441, 444
β-Dihydrokaurene, 331
Dihydrokurchessine, 392, 428
Dihydrometaphanine, 313

9,10-Dihydro-10-methylacridine-9-thione, 36
5,6-Dihydro-6-methylarsanthridine, 111
5,10-Dihydro-5-methyldibenz[*b,e*]-antimonin, 114
5,6-Dihydro-5-methyldibenzo[*b,d*]phosphorin-5-oxide, 96
Dihydromorphinones, 288
Dihydronapelline, 333
Dihydronorsalutaridine, 302, **306**, 307
4,5-Dihydro-5-oxo-4-azapyrene, 74
1,4-Dihydro-4-oxo-1-phenylphosphorin-1-oxide, 94
1,4-Dihydro-4-oxo-1,2,6-triphenylphosphorin-1-oxide, 94
5,6-Dihydrophenanthridine, 44, 47, 57
5,10-Dihydro-5-phenylacridarsine, 111
2,3-Dihydro-2-phenyl-1*H*-naphtharsenin, 112
5,6-Dihydro-5-phenylphosphinolino-[4,3-*b*]indole, 93
5,6-Dihydro-5-phenylphosphinolino-[4,3-*b*]quinoline, 93
5,6-Dihydro-5-phosphaphenanthrene, 96
Dihydrophosphorins, 93, 95
Dihydrophosphorinium salts, 103
Dihydropinidine, 150
Dihydroprometaphanine, 315
Dihydropseudokobusindione, 345
Dihydroquinidine, **231**, 240, 241
Dihydroquinine, **241**
3,4-Dihydro-2-quinolone, 252
Dihydroquinone, **231**, 240
Dihydrosalutaridine, 302, 305, **306**, 307
Dihydrosalutaridinol, 307
Dihydrosinomenine, 303
Dihydrosolasodine, 426
Dihydrosongorine, 333
1,2-Dihydrotetraphenylphosphorin, 94
1,2-Dihydrotetraphenylphosphorin oxide, 94
Dihydrothebaine, 278, 287, 288, 291, 297, 303
β-Dihydrothebaine, **286**, 289
Dihydrothebaine-φ, 289, 305
Dihydrothebaine Δ^6-enol methyl ether, 287
Dihydrothebainol 6-methyl ether, 287
Dihydrothebainone, 275, 276, 277, 279, 281, 286, 289, 291, 297, 301, 303
β-Dihydrothebainone, 280, 281, 289
Dihydrothebainone Δ^5-enol methyl ether, 287
Dihydrotomatidine, 424
5α,6-Dihydrotomatillidine, **413**, 430
9,10-Dihydro-9,9,10-trimethylacridine, 29
Dihydroveatchine, 324, 325, 327
Dihydroxy-9-acridones, 35
1,3-Dihydroxyacridone, 212, 263
5,6-Dihydroxydihydroconessine, 405
Dihydroxydihydroindole, 233
2,3-Dihydroxy-4-(3-hydroxyphenyl)-quinoline, 223
1,3-Dihydroxy-*N*-methylacridone, 264
2,3-Dihydroxy-2-methylbutyric acid, 444, 452
4,4-Dihydroxy-1-methyl-1-phenyl-arsenanium iodide, 107
2,4-Dihydroxy-6-methylpyridine-3-carboxylic acid, 118
2,3-Dihydroxy-4-phenylquinoline, 223
2,6-Dihydroxyphosphorinanes, 88
3,16-Dihydroxypregnadienes, 384
3,20-Dihydroxypregna-5,6-dienes, 384
2,6-Dihydroxy-1-(1-propylbutyl)phosphorinane-1-oxide, 89
3,4-Dihydroxypyridine, 118
2,4-Dihydroxyquinolines, 199, 215
Dimethacrin, **29**
2,4-Dimethoxyacridone, 6
2,4-Dimethoxyaniline, 198
(1,2-Dimethoxycarbonylvinyl)acridinium methoxide, 9
1,3-Dimethoxy-4-methoxycarbonyl-*N*-methylacridone, 263
1,3-Dimethoxy-*N*-methylacridone, 260, 263
1,3-Dimethoxy-10-methylacridone, 34
4,4-Dimethoxy-1-methyl-1-phenyl-arsenanium iodide, 107
3,4-Dimethoxyphenanthrene, 272
3-(2,4-Dimethoxyphenyl)anthranil, 6
2-(2′,3′-Dimethoxyphenyl)cyclohex-2-ene-1-one, 277
2,4-Dimethoxyquinolines, 205
4,5-Dimethylacridine, 5
4,5-Dimethylacridine-9-carboxylic acid, 5

9,10-Dimethylacridinium iodide, 12, 29
3-Dimethylallyl-4,8-dimethoxyquinolone, 209
γ,γ-Dimethylallyl pyrophosphate, 215
3-γ,γ-Dimethylallylquinoline, 199
Dimethylallylquinolone, 215
3-(3′,3′-Dimethylallyl)-4,6,8-trimethoxy-1-methyl-2-quinolone, 175
4-Dimethylaminoacridine, 34
9-Dimethylaminoacridine, 21
3β-Dimethylamino-20α-amino-5α-pregnane, 393
4-Dimethylaminoaniline, 17
4-Dimethylaminobenzaldehyde, 51
3α-Dimethylamino-5α-conanine, **405**, 429
3β-Dimethylamino-20α-methylamino-5α-pregnane, 396
2-Dimethylamino-2′-methylbiphenyl, 41
2-Dimethylaminononane, 151
6-*N,N*-Dimethylaminophenanthridine, 52
9-(4′-Dimethylaminophenylacridines), 34
6-*p*-Dimethylaminophenyl-5-methylphenanthridinium chloride, 47
3α-Dimethylamino-5α-pregnane, 382
3β-Dimethylamino-5α-pregnane, 393, 400, 402
20α-Dimethylamino-5α-pregnane, 397, 398
3β-Dimethylamino-5α-pregnan-2α-ol, 383
20α-Dimethylamino-5α-pregnan-3β-ol, 392
20α-Dimethylamino-5α-pregnan-3-one, 392, 393, 395, 398
Dimethylaminopregna-3,5,20-triene, 400
3α-Dimethylaminopregn-5-ene, 385
3β-Dimethylaminopregn-5-ene, 401
Dimethylaminopropionic acid, 274
3-(3′-Dimethylamino-*n*-propyl)-phenanthrene, 73
4-(3-Dimethylamino-*n*-propyl)-phenanthrene, 73
1,2-Dimethylarsenane, 105
1,12-Dimethylbenzo[*k*]phenanthridine, 82
1,3-Dimethylbenzo[*f*]quinoline, 61
2,4-Dimethylbenzo[*g*]quinoline, 61, 65
6,8-Dimethylbenzo[*g*]quinoline, 65
2,6-Dimethyl-*N*-benzylpiperidine, 151
10,10′-Dimethyl-9,9′-bi(9,10-dihydroacridinylidene), 27
10,10′-Dimethyl-9,9′-bisacridinium dinitrate, 27
O,O-Dimethylchasmanine, 369
N,N′-Dimethylcyclobuxine-D, 464
1,2-Dimethylcyclopentenonaphthalene, 432
9,9′-Dimethyl-9,10-dihydroacridine, 29
10,10′-Dimethyl-9,9′-dihydrobiacridine, 28
N,N-Dimethylethylenediamine, 274
4,5-Dimethylfluorenone, 54
2,3-Dimethylindole, 233
Dimethylketene, 9, 46
2,4-Dimethylnaphtho[2,3-*h*]quinoline, 72
8,10-Dimethylnaphtho[2,1-*g*]quinoline, 73
4,4-Dimethyl-3-oxo-5α-steroids, 459
1,7-Dimethylphenanthrene, 326, 341
1,8-Dimethylphenanthrene, 343
1,10-Dimethylphenanthridine, 51
3,6-Dimethylphenanthridine, 50
1,10-Dimethylphenanthridone, 54
Di(2,2′-methylphenyl)amine-2-carboxylic acid, 5
9-(Dimethylphenyliminomethyl)acridine, 16
1,7-Dimethyl-6-*n*-propylphenanthrene, 343, 344
2,3-Dimethylquinoline, 2
2,4-Dimethylquinoline, 65
2,5-Dimethylquinoline, 221
1,2-Dimethyl-4-quinolone, 197
2,5-Dimethyl-5,6,7,8-tetrahydroquinoline, 221
Dimidium bromide, 47
Dinitroacridines, 19
Dinitro-9-acridones, 34
7,9-Dinitrobenzo[*f*]quinoline, 64
6,6′-Dinitro-2,2′-diphenic acid, 122, 124, 132, 142
3,8-Dinitro-5-methyl-6-phenylphenanthridinium chloride, 47
1,1-Di-*n*-octyl-2,6-dihydroxyphosphorinanium chloride, 89
Diosphenols, 462
1,2-Dioxetan, 27
Diphenamic acid, 54

Diphenylamine, 16, 24, 28, 31
–, reaction with carboxylic acids, 4
Diphenylaminecarboxylic acids, 262
Diphenylamine-2-carboxylic acid, 2, 31
Diphenylbenzoquinolines, 61
1,3-Diphenylbenzo[*f*]quinoline, 63
2,4-Diphenylbenzo[*g*]quinoline, 63
1,1-Diphenyl-4-bromotetrahydrophosphorinium chlorate, 90
Diphenylketene, 95
8,10-Diphenyllobelidiol, 129, 134
8,10-Diphenyllobelidione, 129, 134
8,10-Diphenyllobelionol, 129, 134
4-(Diphenylmethylene)-1,4-dihydro-1,2,6-triphenylphosphorin-1-oxide, 95
8,10-Diphenylnorlobelidiol, 134
N,N'-Diphenyl-*p*-phenylenediamine-2,5-dicarboxylic acid, 38
1,1-Diphenylphosphinoline, **103**
1,1-Diphenylphosphorin, **103**
1,1-Diphenylphosphorinanium bromide, 85
3',2-Dipiperidyl, 142
3',2-Dipyridyl, 142, 144
1,1'-Disinomenine, 305
Diterpenoid alkaloids, 323, 341, 350
Di-*o*-tolylamine, 5
Döbner–Miller synthesis, 61, 71
Dubamine, 172
Dubinidine, 182
Dubinine, 182
Duboisia hopwoodii, 140
Dyestuffs, acridine, 36

Echinine, **216**, 217
Echinops sp., 216
Echinopsidine, **217**
Echinopsine, **217**
Echinops ritro, 217
Echinorine, **216**, 217
Eduleine, 178
Eduline, 178
Edulinine, 174
Edulitine. 174. **198**
Elatidine, **352**, 357
Elatine, **352**, 353
Eldeline, 354
Emde degradation, 400
Emde ring-opening, 58
Ephedrine, 236, 237
ψ-Ephedrine, 236
Epicinchonidine, 236, 238
Epicinchonine, 236, 238
Epihydroquinidine, 237
Epimeloscine, **252**, 253
20-Epi-*N*-methylparavallarine, **386**, 428
3-Epinupharamine, 164
Epipachysamine-A to -F, **396**, 429
Epiquinamine, 234
Epiquinidine, **231**, 236, 238, 240, 243
Epiquinine, **231**, 236, 238, 240, 243
Epistephamiersine, 311, 316
15-Epiveatchine, 329
Equisetum arvense, 115
Ergosta-4,6,22-trien-3-one, 410
Eriostemnon sp., 174, 180
Escholerine, 446
Escolloniaceae sp., 333
Esenbeckia sp., 180
Ester-alkaloids, 437, 444, 445, 446, 452
–, ester positions, 452
Ethacridine, **22**
3-*N*-Ethoxycarbonylchonemorphine, 393
N-Ethoxycarbonylnornicotyrine, 140
3-Ethoxycarbonyl-1-phenyl-4-arsenone, 106
1-Ethoxycarbonylpyrrole, 140
9-Ethoxy-10-methyl-9-phenyl-9,10-dihydroacridine, 12
6-Ethoxy-2-nitrophenanthridine, 56
Ethoxyphosphorins, 100, 101, 102
γ-Ethoxypropylmagnesium bromide, 139
3-Ethoxy-4,6,8-triethoxyphenanthrene, 312
6-Ethoxy-3,4,8-trimethoxyphenanthrene, 309
Ethylacridines, 17
Ethylarsenanes, 106
1-Ethylbisminane, 114
3'-Ethyl-1,2-cyclopentenophenanthrene, 399
14-Ethyl-3,4-dimethoxy-6-oxohexahydrophenanthrene, 294
3-Ethyl-1,8-dimethylphenanthrene, 343
7-Ethyl-1,9-dimethylphenanthrene, 332

Ethyl 2-ethoxycarbonylaminobiphenyl, 55
O-Ethylhomostephanol, 312
2-Ethyl-3-hydroxy-5-methylpyridine, 421, 438, 441
3-Ethyl-4-hydroxy-2-quinolone, 204
β-Ethylindole, 249
6-Ethyl-1-methyl-3-azaphenanthrene, 335
7-Ethyl-1-methyl-3-azaphenanthrene, 324, 332
7-Ethyl-8-methyl-1,2-benzofluorene, 437
4-Ethyl-3-methylhexane, 238
10-Ethyl-8-methyllobelidione, 131
10-Ethyl-8-methyllobelidiol, 131, 134
6-Ethyl-1-methylphenanthrene, 335
7-Ethyl-1-methylphenanthrene, 324
2-Ethyl-5-methylpyridine, 411, 416, 418, 419, 421, 424, 425, 426, 438, 442, 448
3-Ethyl-4-methylpyridine, 226
8-Ethylnorlobelol, 137
Ethylphenanthridines, 50
6-Ethylphenanthridine, 50
8-Ethyl-10-phenyllobelidiol, 131, 134
Ethyl-4-phosphorinanols, 89
1-Ethyl-4-phosphorinanone, 87
Ethylpyridine, 226, 233
Ethyl(2-pyridyl)methanol, 123
3-Ethylquinuclidine, 232
1-Ethyltetrahydro-1-phenylbenzo[*h*]phosphinolinium bromide, 92
Ethyltetrahydrophosphinoline, 91
α-Ethylthiocodide, 290
β-Ethylthiocodide, 290
δ-Ethylthiocodide, 290, 297
8-Ethylthiodihydrothebainone, 290
5-Ethyl-3,4,8-trimethoxyphenanthrene, 295
Etioline, 415
Euonine, 154
Euonymine, 154
Euonymus alatus, 153
Eunonymus europaeus, 153
Eunonymus sieblodiana, 153
Evellerine, 182
Evocarpine, 178
Evodia sp., 178, 180, 182, 190, 191, 211
Evodia alata, 210
Evodine, 182
Evolatine, 182
Evolitrine, 180, 205
Evonimine, 154
Evonine, 154
Evoninic acid, **153**
Evoprenine, 191, 261, 265
Evoxanthidine, 191, 260, 263
Evoxanthine, 191, **211**, 263
Evoxine, 182, **208**
Evoxoidine, 182
Excelsine, **363**, 371
Exovanthine, 259, 260, 262

Fabiana imbricata, 220, 221
Fabianine, **220**, 221
Fagara sp., 174, 180, 190, 192, 210
Fagara capensis, 203
Fagara chalybea, 203
Fagara xanthoxyloides, 203
β-Fagarine, 180
γ-Fagarine, 180, **203**, 205
Farnesyl pyrophosphate, 427
Fischer indole synthesis, 93
Flavinantine, 302, **308**
Flavinine, 302, **309**
Flavothebaone, 298, **299**
Flindersia sp., 175, 180, 182, 187
Flindersia australis, 208
Flindersia ifflaiana, 206, 207
Flindersiamine, 180
Flindersine, 175, **208**
Fluorenes, 42, 95, 438
Fluorenols, 52
Fluorenones, 53, 54
Fluorescence spectra, acridones, 258
Foliasidine, 175
Folicanthine, **250**, 251
Folidine, **198**
Folifidine, 174
Folimidine, 178, **197**, 207
Folimine, 174, **198**, 207
Foliminine, 182, **207**
Foliosidine, **198**
Foliosine, **207**
1-Formamido-2-phenylnaphthalene, 79
2-Formamido-1-phenylnaphthalene, 80
9-Formylacridine, 17

Formylamino-2-phenylcyclohexane, 58, 59
N-3-Formylcyclovirobuxeine-B, 461
7-Formyl-9,10-dimethylbenz[*c*]-acridine, 76
N-Formylnornicotine, 144
6-Formylphenanthridine, 44, 50
Friedel–Crafts reaction, 41, 69, 108
Friedlander reaction, 93
Fritillaria alkaloids, 437, 453, 454
Fritillaria bases, 411
Fritillaria imperialis, 454
Fritillaria roylei, 453, 454
Fritillaria thunbergii, 453
Fritillarine, **453**, 454
Funtamine, 385
Funtessine, **406**, 429
Funtudiamine, **393**, 428
Funtudienine, **409**, 430
Funtuline, **408**, 429
Funtumafrines, **390**, 428
Funtumia africana, 389, 390
Funtumia elastica, 389, 390, 391, 407
Funtumia latifolia, 382, 383, 390, 391, 393, 406, 408, 409
Funtumidine, **383**, 428
Funtumine, **382**, 383, 428
Funtuphyllamines, **389**, 390, 428
Furoquinolines, 180, 201
–, biosynthesis, 216
Furoquinolones, 201

Galactose, 418, 425
Galipea sp., 172, 174
Galipea officinalis, 193
Galipine, 172, **193**, 196, 215
Galipoidine, **195**
Galipoline, 172, 193, **194**
Garrya alkaloids, 324, 325, 335, 341, 342, 343, 345
–, stereochemistry, 329
Garryaceae sp., 323
Garrya laurifolia, 328, 329
Garrya veatchii, 324
Garryfoline, **329**, 330, 331, 335
Garryine, **324**, 325, 326, 327, 328, 330
Geijera sp., 175, 180, 182
Genista hystrine, 145
Gentiana sp., 145
Gentianine, **145**
Gentia picrin, 146
Geraniol, 244
Geranyl pyrophosphate, 199
Germanitrine, 445
Germerine, 445
Germidine, 445
Germinaline, **452**
Germine, 445, **449**, 450, 452
–, veratroyl esters, 452
Germitetrene, 445
Germitetrine, 452
Germitrine, 445
Girgensolinia diptera, 115
Glutaraldehyde, 88, 131, 132
Glycomis sp., 180, 190, 191
Glycoperine, 182
Glycosmis pentaphylla, 263
Gomberg reaction, 140
Gravacridon chlorine, 192, 212
Gravacridondiol, 192, 212
Gravacridonol chlorine, 192, 212
Graveoline, 178, **196**
Graveolinine, 172, **196**
Gregory process, 269
Grignard reaction, 295
Grignard reagents, 29, 33, 88, 288, 291
Guvacine, **120**
Guvacoline, **120**

Halfordia sp., 174, 180
Halfordia kendack, 198
Halfordia scleroxyla, 167
Halfordamine, 174, **198**
Halfordine, **167**, 168
Halfordinine, 182, 203
Halfordinol, **167**, 168
Halfordinone, **167**, 168
Halogenoacridines, 18
Halogenocodides, 270, 273, 283
Halogenomorphides, 270, 283
Halogenophenanthridines, 51
2-Halogenophenanthridones, 57
Haplamine, 175, **199**
Haplofilidine, **208**
Haplophine, 180
Haplophydine, 182

Haplophyllidine, 182
Haplophyllum sp., 172, 174, 175, 178, 180, 182, 187, 189, 198, 203
Haplophyllum actifolium, 197
Haplophyllum bucharicum, 199
Haplophyllum foliosum, 197, 198, 207, 210
Haplophyllum hispanicum, 208
Haplophyllum perforatum, 199, 208
Haplophyllum tuberculatrum, 197
Haplopine, 180, 203, 205, **208**
Hasubanonine, 267, **309**, 310, 311, 312
Hecogenin, 438
Helietta sp., 180
Hemlock alkaloids, 120
Hernandia bivalvis, 316
Hernandifoline, 311, **316**, 317
Hernandine, 316
Hernandoline, 311, **312**
Hernandolinol, 312
Hertisinone, **341**, 349
Hesperethusa sp., 175
Hesperethusa crenulata, 199
Heteratisine, **370**, 371, 372, 373
Heteroconessine, **401**, 429
Heterocycles, containing bridgehead phosphorus, 104
Heterocyclic phosphorus compounds, 83
Heterophyllidine, **373**
Heterophylline, **373**
Heterophyllisine, **373**
Hetidine, **348**, 349
Hetisine, **341**, 342, 345, 349
Hetisine methohydroxide, 342
Hexafluro-2-butyne, 104
Hexahydrocinchomeronic acid, 227
Hexahydrocyclohept[*b*]indol-10[5*H*]one, 30
Hexahydrophenanthridines, 58, 59
Hexahydrophenanthridones, 58
Hexahydrotetraoxo-5,12-diazapentacene, 38
Histrionicotoxin, **165**
Hofmann degradation, 123, 146, 271, 273, 274, 285, 294, 295, 299, 310, 321, 330, 400, 416, 431
Holadienine, **409**, 429
Holadysamine, **384**, 385, 428
Holadysine, **384**, 428
Holafebrine, **390**, 428
Holafrine, **406**, 429
Holaline, **405**, 429
Holamine, **384**, 385, 428
Holaphyllamine, **385**, 428
Holaphyllidine, **385**, 428
Holaphylline, **385**, 428
Holaphyllinol, **385**, 428
Holaromine, **410**, 430
Holarrheline, **406**, 429
Holarrhena sp., 382, 399
Holarrhena africana, 385, 405, 409
Holarrhena antidysenterica, 383, 384, 391, 392, 394, 395, 399, 403
Holarrhena febrifuga, 390
Holarrhena floribunda, 410
Holarrhena mitis, 409
Holarrhenine, **406**, 429
Holarrhetine, **406**, 429
Holarrhidine, **394**, 429
Holarrhimine, 394, **395**, 407, 410, 429
Holonamine, **409**, 410, 430
Homochasmanine, **396**, 371
Homocincholoipon, 241
Homodaphniphyllic acid, 375, 376
Homodaphniphyllic ester, 378
Homomorphine, 267
Homomorphine alkaloids, 319
Homosecodaphniphyllic acid, 376
Homosecodaphniphyllic ester, 378
Homostephanoline, 311, **312**
Hortia sp., 180
Huang–Minlon reduction, 332, 338
9-Hydrazinoacridine, 4
Hydrinic acid, 138
Hydroacridines, 28, 29
Hydrocinchoninone, 241
14-Hydrocodeinone, 274
Hydroxyacridines, 2, 21, 23, 24, 25
1-Hydroxyacridine, 24
3-Hydroxyacridine, 24
4-Hydroxyacridine, 24, 25
9-Hydroxyacridine, 24
Hydroxyacridones, 34, 35
N-Hydroxyacridone, 8
1-Hydroxyarsenane-1-oxide, 106
3-Hydroxybenzaldehyde, 241

9-Hydroxycamptothecin, 246
Hydroxycodeine, 273
14-Hydroxycodeinone, 290, 291
12β-Hydroxyconessimine, 406
7α-Hydroxyconessine, **405**, 429
12β-Hydroxyconessine, 406
Hydroxyconiine, 122
14-Hydroxydihydrocodeinone, 278
4-Hydroxy-2,3-dimethoxy-10-methyl-acridone, 210
6-Hydroxy-3,8-dinitro-6-phenyl-5,6-dihydrophenanthridine, 47
4-Hydroxy-1,1-diphenylphosphorinanium bromide, 85
2-Hydroxy-2′-formaldiphenylamine, 13
12β-Hydroxyfuntumine, 383
4-Hydroxyiminophosphorinane oxides, 88
7-Hydroxyisoquinoline, 241
11β-Hydroxyjervine, 441
9α-Hydroxyjurubidine, 388
Hydroxylamine-*O*-sulphonic acid, 11
Hydroxylunacridine, 174, 200, 209
Hydroxylunidine, 174, 200
Hydroxylycoctonine, 351
3-Hydroxy-4-methoxycinnamic acid, 316
1-Hydroxy-3-methoxy-10-methylacridone, 192
1-Hydroxy-*N*-methylacridone, 260
1-Hydroxy-3-methylbutan-2-one, 167
2-Hydroxymethylbutyric acid, 444
2-Hydroxymethylcyclohexanone, 58
7-Hydroxy-4-*O*-methyldehydrometa-thebainone methoperchlorate, 307
9-Hydroxymethyl-9,10-dihydroacridine, 26
9-Hydroxymethyl-9,10-dihydro-anthracene, 27
7-Hydroxy-4-*O*-methylmorphothebaine methoperchlorate, 307
2α-Hydroxy-*N*-methylparavallarine, 386, 428
3-Hydroxy-5-methylpyridine, 437
3-Hydroxy-1-methyl-4-pyridone, 118
4-Hydroxy-1-methyl-2-pyridone, 117
2-Hydroxynaphthalic anhydride, 448
2-Hydroxy-1-naphthoylformanilide, 76
Hydroxyphenanthridines, 53
6-Hydroxyphenanthridine, 44, 48
N-β-(4-Hydroxyphenyl)ethylacetinidium chloride, 159, 162
N-β-(4-Hydroxyphenyl)ethyltecos-tidinium chloride, 159, 162
4-Hydroxy-1-phosphabicyclo[2.2.2]-octane ethobromide, 104
1-Hydroxyphosphorinane-1-oxide, 86
3β-Hydroxypregna-5,17(20)-dien-16-one, 418
12β-Hydroxy-5α-pregnane, 406
3β-Hydroxy-5α-pregnan-20-one, 389
20α-Hydroxy-5α-pregnan-3-one, 383
20β-Hydroxy-5α-pregnan-3-one, 383
17α-Hydroxyprogesterone, 387
5-Hydroxy-2-*n*-propylpiperidine, 124
2-(1-Hydroxypropyl)pyridine, 121, 122
3-Hydroxypseudane, 222
3-Hydroxypyridine, 146, 148
3-Hydroxy-4-pyridone, 118
4-Hydroxyquinoline-2-carboxylic acid, 171
Hydroxyskytanthines, **160**, 162
16α-Hydroxysolacongestidine, 412
Hydroxysolanidines, 418
12α-Hydroxysolanidine, 419
23β-Hydroxysolanidine, 421
Hydroxytetrahydropyridinium ions, 414
1-Hydroxytetraphenylphosphorin, 94
1-Hydroxytetraphenylphosphorin hydrate, 97
6-Hydroxythebaine, 268
Hydroxywilfordic acid, 153
Hypaconitine, **363**, 371
Hypognavine, **343**, 344, 349
Hypognavinol, **343**, 344
Hystrine, **145**

Ifflaiamine, 187, 206, 207
Ignavine, **342**, 343, 345, 346, 349
9-Imino-10-methylacridan, 22
9-Imino-10-methylacridine, 20
9-Imino-10-methyl-9,10-dihydroacridine, 78
Immonium salts, 347
Imperialine, **454**
Imperialone, **454**
Indaconitine, **366**, 371
Indanones, 318
Indanthrene Olive Green B, 38
Indanthrene Olive T, 38

Indanthrene Red RK, 37
Indanthrene Red Violet RRK, 37
Indeno[2,1-*f*]quinoline-11-one, 74
Indoles, autoxidation to quinolones, 247
Indole alkaloids, 244
Indolines, 30
Indolinocodeine, **291**, 309
Indolizidine units, 411
1-Iodoarsenane, 106
Iodoconiine, 122, 126
2-Iodophenanthridone, 57
Irehamine, **390**, 428
Irehdiamines, **391**, 392, 428
Irehine, **390**, 407, 428, 429
Isatin, 29, 77, 88, 107
Isoarsinolines, 109
Isoatisine, 325, 334, 336, 340
Isobalfourodine, 189, **209**
N-3-Isobutyrylbaleabuxaline-F, 461
N-Isobutyrylbaleabuxidine-F, 461
N-3-Isobutyrylbuxidienine-F, 461
N-3-Isobutyrylcycloxobuxidine, 461
Isocinchonines, **229**, 240
Isocodeine, 304
Isocodeine methyl ether, 287
Isoconessimine, **400**, 401, 403, 408, 429
Isoconessine, **403**, 429
Isocuauchichicine, **329**, 330
Isocusparine, **193**
Isodehydrostephine, 318
Isodeoxysongorine, 332
Isodictamnine, 187, **203**, 204
Isofuntumidine, 383
Isogalipine, 193, 194, 196
Isogermine, **449**, 450, 452
Isohypognavine, **345**, 349
Isohypognavinol, **345**, 349
Isojervine, **441**, 444
Isojuribidine, **388**, 428
Isojuripidine, **388**, 428
Isolobinanidine, 134
Isolobinanine, 134
Isolobinine, 134
Isolongistrobine, **218**, 219
Isolycoctonine, 351
Isomacrorine, **217**, 218, 219
Isomaculosidine, 182, 187, **208**
Iso-2-methylconidine, 126
Isomorphines, 301
Isonapelline, 330, **331**, 332
Isoneopine, 291
Isonicoteine, **144**
Iso-orensine, **152**, 153
Isopaniculidine, **388**, 428
Isopelletierine, **127**, 128
7-Isopentenyloxy-γ-fagarine, 180
3-Isopentenyloxy-7,8-methylenedioxy-4-methoxy-2-quinolone, 175
Isopentenyl pyrophosphate, 427
5-Isopropyl-2-methylnaphthalene, 448
6-Isopropyl-1-methylphenanthrene, 340
Isoprosopinine, **149**, 150
Isoprotoverine, **450**
Isoptelefolidine, 175
Isopyrokarakolidine, 364, 371
Isoquinoline, 227
α-(4-Isoquinolyl)-*o*-aminocinnamic acid, 79
Isorubijervine, **419**, 420, 430
Isorubijervone, 420
Isorubijervosine, **419**
Isosalutaridine, 308
Isosinomenine, 301, 302, **305**
Isosongorine, 330, **331**, 332
Isotalatizidine, **365**, 366, 371
Isothebaine, 269
Isovincoside lactam, 248
Isozygadenine, **450**, 452

Japonin, 178
Jervine, **437**, 438, 439, 440, 441, 444
Jesaconitine, **362**, 371
Julocrotine, **169**
Julocroton montevidensis, 169
Jurubidine, **388**, 428
Jurubine, **388**, 428

Karacoline, **364**, 371
Karakolidine, **364**, 371
Kaurane, 324
Kaurene, 330, 331
Kaur-16-en-19-oic acid, 328
Kaur-16-en-15-ols, 329
Ketodehydrosolacongestinine, 413
Khaplofoline, 189, **210**
Kibatalia arborea, 390

Kibatalia gitingensis, 386
Kibataline, **387**, 428
Knorr synthesis, 61
Kobusine, 344, 345, 349
Kokusagine, 180
Kokusaginine, 180, **203**, 205
Kreysigia multiflora, 319
Kreysiginine, 319, **320**
Kurchaline, **384**, 428
Kurchessine, **392**, 428
α-Kurchessine, **392**, 428
Kurchi, 399
Kurchiline, **383**, 384, 428
Kurchimine, **392**, 428
Kurchiphyllamine, **384**, 428
Kurchiphylline, **384**, 428
Kurcholessine, **406**, 429
Kynurenic acid, 171, 193

Lactonic alkaloids, 350, 370
Lantine, **386**, 428
Lappaconidine, **363**, 364, 371
Lappaconine, **363**, 371
Lappaconitine, **363**, 371
Latifoline, **402**, 408, 429
Latifolinine, **408**, 429
Laudanosine, 306
Lehmstedt–Tanasescu reaction, 6
Lelobanidine, **130**
Lelobanine, **131**
Lemobiline, 187
Lemonia sp., 190
Lepidine, 226, 229, 249
Leptines, **420**, 421, 430
Leptinidine, **420**, 421, 430
Leptinines, **420**
Leucaena glauca, 118, 119
Leucaenine, **118**, 119
Leucinol, 118
Liliaceae sp., 411, 436
9-Lithiomethyleneacridine, 17
Lobelanidine, **129**, 134
Lobelanine, **129**, 134
Lobelia alkaloids, 128, 134, 135, 136, 137
Lobelia cardinalis, 133, 254
Lobelia inflata, 128, 131, 133
Lobelia syphilitica, 132, 136, 254
Lobelidione, 132
Lobeline, **129**, 134
Lobelinic acid, 129
Lobinaline, **133**, 254
Lobinanidine, **132**, 134
Lobinanine, **131**, 134
Lobinine, **131**, 132, 134
Loganin, 244
Loiponic acid, **226**, 227, 235, 240
Lolium perenne, 253
Longistrobine, **218**, 219
Lucidusculine, 330, 333
Lucigenin, **27**, 28
Lunacridine, 174, **200**, 202
Lunacrine, 187, **201**, 202
Lunacrinium, 180
Lunacrinol, **209**
Lunamarine, 178, **196**
Lunamarinium, 172
Lunasia-1, **196**
Lunasia II, **209**
Lunasia sp., 172, 174, 175, 182, 187, 196, 200
Lunasia amara, 200, 201, 209
Lunasia Base-1, 178
Lunasia quercifolia, 202
Lunasine, 180, **202**
Lunidine, 174, **200**
Lunidonine, 175, **200**
Lunine, 187, 200
Luninium, 182
2,5-Lutidine, 453
Lycaconitine, 353
Lycoctonal, 351
Lycoctonam methyl ether, 357
Lycoctonine, **350**, 351, 352, 357, 369
Lycoctonine alkaloids, 350, 351, 353, 357
Lycopersicon pimpinellifolium, 427
Lycopersicum sp., 423
Lycorine, 40
Lysine, 128, 254
Lythramine, **166**
Lythranidine, **166**
Lythranine, **166**
Lythrum anceps, 166

Macrodaphine, **376**, 378
Macrodaphniphyllidine, 378

Macrorine, **217**, 218
Macrorungia longistrobus, 217, 218
Macrorungine, 217, 218
Maculine, 180, 203, 205
Maculosidine, 180
Maculosine, 182
Maingayine, **409**, 430
Malarboreine, **410**, 430
Malarborine, **410**, 430
Maleic anhydride, 9, 63, 77, 274, 292
Malouetia arborea, 410
Malouetia bequaertiana, 389, 392, 393, 394, 404
Malouetine, **393**, 394, 429
Malouphyllamine, **404**, 405, 429
Malouphylline, **394**, 395, 404, 429
Malouphyllinine, **394**, 395, 429
Malouphyllinol, 404
Mappia foetida, 246, 248
Mappicine, **248**
Maytenus ovatus, 155
Maytine, **155**
Maytoline, **155**
Medicosma sp., 180, 182
Medicosmine, 182
Melicope sp., 180, 182, 190
Melicope fareana, 211
Melicope perspicuinerva, 203
Melicopicine, 190, **211**, 259, 260, 262, 263
Melicopidine, 32, 190, **211**, 259, 260, 262, 263
Melicopine, 32, 191, **211**, 259, 260, 262
Melodinus alkaloids, 171
Melodinus scandens, 252
Meloscandonine, **252**, 253
Meloscine, **252**, 253
Meloscine alkaloids, 253
Menispermum dauricum, 311
Mepacrine, **23**
Meroquinene, **226**, 228, 229, 235, 240
Meroquinene nitrile, 227
Mesaconitine, **359**, 362, 363, 371
Mesobenzanthrone, 38
Mesochimonanthine, **250**
Mesodihydrobenzacridines, 76
Metaphanine, 311, **313**, 314, 316
Metaphaninedithiol acetal, 313
Metathebainone, 289, 294, 301, 304
Methoxyacridines, 24, 25
9-Methoxyacridine, 33
4-Methoxy-2-*n*-amylquinolone, 193
2-(2′-Methoxyanilino)benzaldehyde, 24
8-Methoxybenz[*f*]isoquinoline, 66
9-Methoxycamptothecin, 246
Methoxycarbonylphenyl esters, 31
β-Methoxycarbonylpropionyl chloride, 219
2-Methoxy-6-chloroacridone, 23
Methoxycinchonidine, 230
7-Methoxydeoxyneopine, 291
6-Methoxydictamnine, 180
9-Methoxy-9,10-dihydroacridine, 9
3α-Methoxy-20α-dimethylaminopregn-5-ene, **391**, 428
4-Methoxy-1,1-diphenylphosphorinanium bromide, 85
Methoxy-2-fluroenol, 53
Methoxy-9-fluorenols, 53
6-Methoxyhydroxylunidine, 176
6-Methoxylepidine, 242
6-Methoxylunidine, 176
6-Methoxylunidonine, 176
4-Methoxy-3-(3-methylbut-2-enyl)-2-quinolones, 205
4-Methoxy-2-[2′-(3″,4″-methylenedioxyphenyl)ethyl]quinoline, 193
8-Methoxy-1-methyl-4-quinoline, 197
4-Methoxy-1-methyl-2-quinolone, 175, 199, 203
2-(6′-Methoxy-1-naphthyl)ethylamine, 66
Methoxyphenanthridines, 53
5-Methoxyphenanthridine, 55
6-Methoxyphenanthridine, 51
Methoxyphenylanthranils, 6
4-Methoxy-2-phenylquinoline, 172
Methoxyphosphorins, 98, 99, 100, 101, 102
4-Methoxyphthalic acid, 297
5-Methoxy-α-picoline, 124
3α-Methoxy-5α-pregnane, 391
5-Methoxy-2-*n*-propylpyridine, 124
3-Methoxy-Δ^2-pseudane, 222
3-Methoxy-4-pyridone, 119
4-Methoxyquinaldine, 193, 194
4-Methoxyquinolines, 196

6-Methoxyquinoline-4-carboxylic acid, 230
6-Methoxy-4-quinolyllithium, 243
6-Methoxytecleanthine, 192, **211**
6-Methoxy-1-tetralone, 402
7-Methoxythebainone, 303
1-Methyl-2-acetoacetnaphthalide, 61
N-Methylaconitine, 362
2-Methylacridarsinic acid, 111
Methylacridines, 2, 17
9-Methylacridine, 7, 10, 14, **16**, 17, 29
9-Methylacridine-diphenylamine, 16
9-Methylacridine-9-methyl-9,10-dihydroacridine, 16
N-Methylacridinium chloride, 10
N-Methylacridinium iodide, 11, 12
N-Methylacridones, 25, 32, 33
N-Methylacridone, 11, 12, 24, 25, 27, 190, 260
10-Methylacridone, 12, 32, 33
10-Methyl-9-acridonephosphoryl chloride, 27
2-Methylaminobenzophenones, 34
3β-Methylamino-20α-dimethylamino-5α-pregnane, 393
2-Methylamino-2′-methylbiphenyl, 58
6-*N*-Methylaminophenanthridines, 52
20α-Methylamino-5α-pregnan-3β-ol, 389
3α-Methylaminopregn-5-ene, 384
3β-Methylaminopregn-5-ene, 385
3β-Methylaminopregn-5-en-20β-ol, 385
N-Methylammodendrine, 144
N-Methylanabasine, 142, 144
N-Methylanatabine, **143**
O-Methylandrocymbine, 320
N-Methylanthranilic acid, 220
1-Methylantimonane, 113
Methylarsenanes, 106
1-Methylaresenane, 106
N-Methylatalaphylline, 213, 261
O-Methylbalfourodinium, 200
8-Methyl-1,2-benzofluorene, 453, 454
1-Methylbenzo[*k*]phenanthridine, 80, 82
N-Methylbenzo[*a*]phenanthridone, 80
10-Methylbenzo[*g*]quinoline, 60
N-Methylbicycloatalaphylline, 261, 264
O-Methylbicycloatalaphylline, 264
5-Methyl-2,3′-bipyridine, 144
3-Methylbut-2-enylmalonic ester, 199
Methylbuxene-M, 456, **457**
N-Methylcarbostyril, 194
N-Methylcarpamic ester, 146
O-Methylchasmanine, 369
2-Methylconidine, 126
N-Methylconiine, 120, 121
N-Methylconkurchine, 407
2-Methylcyclohexanone, 402
1′-Methyl-1,2-cyclopentenophenanthrene, 418
9-Methyl-1-decalone, 448
5-Methyldibenzo[1,4]diazepine, 11
10-Methyl-9,10-dihydroacridine, 12
Methyldihydro-ψ-codeinone, 297
5-Methyldihydromorphinone, 288
N-Methyldihydroparavallarine, 386
N-Methyl-5,6-dihydrophenanthridine, 41, 57
N-Methyl-3′,2-dipyridyl, 144
3,4-Methylenedioxycinnamylidene malonic acid, 115
9-Methylene-10-methyl-9,10-dihydroacridine, 12
Methylevoxine, 182
N-Methylflavinine methiodide, 309
N-Methylflindersine, 175
Methylgranatic acid, 130
O-Methylhalfordinol, 167
3-*N*-Methylholarrhimine, 395, 407, 429
20-*N*-Methylholarrhimine, **395**, 429
O-Methylhydroxyluninium ion, 182
1-Methyl-2-imidazole-5-carboxylic ester, 219
1-Methyl-4-iminoquinoline, 217
3-Methylisoindol-1-one, 275
Methylisopelletierine, 122, **127**, 128
Methyllycaconitine, 353
18-*O*-Methyllycoctonine, 352
N-Methyl-2-methoxyaniline, 199
Methylmorphenol, 271, 285, 299
N-Methylmorphinan, 275
1-Methyl-2-naphthylamine, 60
N-Methyl-3-nicotinyl-2-pyrrolidone, 139
8-Methylnorlobelol, 133, 137
N-Methylnupharamine, 163, 164
1-Methyl-4-oxo-1-phenylarsenanium iodide, 107

N-Methylparavallarine, 387
Methylpelletierine, **127**, 128
10-Methyl-9-phenacyl-9,10-dihydro-
acridine, 11
1-Methylphenanthrene, 335
1-Methylphenanthrene-7-carboxylic acid,
326
Methylphenanthridines, 51
6-Methylphenanthridine, 44, 46, 49, 50
N-Methylphenanthridinium bromide, 57
5-Methylphenanthridium iodide, 45
N-Methylphenanthridone, 45, 55, 56
3-Methylphenanthridone, 50
5-Methylphenanthridones, 45
N-Methyl-9-phenylacridinium iodide, 11,
12
8-Methyl-10-phenyllobelidiol, 134
8-Methyl-10-phenyllobelionol, 134
1-Methyl-1-phenylphosphorinanium
iodide, 85
Methyl phenyl sulphone, 219
Methylphosphorins, 98, 100, 101, 102,
103
1-Methylphosphorinane-1-oxide, 85
Methyl-4-phosphorinanols, 89
Methyl-4-phosphorinanones, 87, 88
N-Methylpipecolinic acids, 132
N-Methylpiperidine, 115
1-Methylpiperidine-2,6-diacetic acid, 129
1-Methylpiperidine-2,6-dicarboxylic acid,
129
5-Methyl-2-piperidone, 426
2-Methyl-6-(2-propenyl) piperidine, 150,
151
1-Methyl-2-propionylpiperidine, 122
2-Methyl-6-*n*-propylpiperidine, 150
2-Methyl-6-*n*-propylpyridine, 150
O-Methylpteleofolium ion, 180
5-Methylpyridine-3,4-dicarboxylic acid,
158
1-Methylpyridinium-3-carboxylate, 117
N-Methylpyrrolidine, 138
1-Methyl-2-pyrrolidine, 139
N-Methyl-2-pyrrolidone, 139
20-(2-Methyl-1-pyrrolin-3-yl)-4-pregnen-
3-one, 414
3-*N*-Methyl-2,4-quinazolinedione, 223
8-Methyl-5*H*-quino[2,3,4-*k*]acridine, 40
4-Methylquinoline, 16, 226
N-Methyl-4-quinolone, 217
1-Methyl-2-quinolone, 174, 195
1-Methyl-4-quinolone, 196
1-Methyl-4-quinolone-3-carboxylic acid,
211, 259
4-*O*-Methylsalutaridine methoperchlorate,
307
N-Methylsamandarine, 431
N-Methylsuccinylanthranilic acid, 350
1-Methyltetrahydroarsinoline, 109
2-Methyltetrahydroisoarsinoline, 109
11-Methyltetrahydronaphtho[2,1-*g*]-
isoquinoline, 73
Methylthiophosphonic dibromide, 85
1-Methyl-4(3′-trifluoromethylphenyl)-
phosphorinan-4-ol, 90
N-Methyltryptamine, 249
Methyl vinyl ketone, 292, 293
Metopon, 288
Mevalonic acid, 215, 216, 427
Michael addition, 290
Miersine, 311, **316**
Mimosa pudica, 119
Mimosine, **119**
Miyaconitine, **347**, 348, 349
Miyaconitinone, **348**, 349
Monastral pigments, 38
Monnieria sp., 180, 190
Monoacetyldelcosine, **356**, 357
Morphinan, 275
Morphinandienone, 267
Morphine, 268, 273, 295, 303
–, isolation, 269
–, rearrangement, 272
–, structure, 270, 279
–, total synthesis, 275
ψ-Morphine, **301**
trans-Morphine, 281, **282**
Morphine alkaloids, 267
–, molecular rearrangement, 294
–, stereochemistry, 278
Morphine *N*-oxides, 270
Morphothebaine, 272, 294, 304
Morphothebaine dimethyl ether, 294
Muldamine, **443**, 444
Multifloramine, 320
Murraya sp., 180

Myosmine, **140**, 141, 144

Napelline, 330, **331**, 332, 333
Naphthaldehydes, 66
1*H*-Naphth[1,8-*c,d*]arsenine, 112
Naphthoisoquinolines, 71
Naphtho[2,1-*g*]isoquinoline, **73**
Naphtho[2,1,8-*def*]isoquinoline, **74**
β-Naphthol, 442
Naphthoquinolines, 71
Naphtho[1,2-*f*]quinoline, **73**
Naphtho[1,2-*g*]quinoline, 72, **73**
Naphtho[1,2-*h*]quinoline, **74**
Naphtho[2,1-*f*]quinoline, **73**
Naphtho[2,1-*g*]quinoline, **73**
Naphtho[2,1-*h*]quinoline, **73**
Naphtho[2,3-*f*]quinoline, **72**
Naphtho[2,3-*g*]quinoline, **72**
Naphtho[2,3-*h*]quinoline, **72**
1-Naphthylamine, 60, 65
2-Naphthylamine, 37, 60, 61, 63, 75
N-2-Naphthylanthranilic acid, 75
1-Naphthylethylamines, 69
Neber rearrangement, 218
Nemuaron vieillardii, 308
Neoconessine, **403**, 429
Neodaphniphylline, 378
Neo-euonymine, 154
Neo-evonine, 154
Neogermbudine, 445, 452
Neogermidine, 445
Neogermitrine, 445
Neohydroxylunine, 189
Neoline, **369**, 370, 371
Neopelline, **369**, 370, 371
Neopine, 268, 278, 282, 290, 291
Neopinone, 290, 294
Neotigogenin, 424
ψ-Neotigogenin, 424
Neoyuzurimine, 378
Nepetalinic acids, 160
Nicotelline, **143**, 144
Nicotiana sp., 137
Nicotiana rustica, 138
Nicotiana tabacum, 137, 138, 143, 144
Nicotinamide, 167
Nicotine, **138**, 139, 140, 144
Nicotinic acid, 62, 117, 138, 139, 141, 142, 143, 396
Nicotinic acid methyl betaine, 117
Nicotinonitrile, 139
Nicotinylacetic ester, 139
Nicotinyl chloride, 141
Nicotinylsuccinic ester, 139
Nicotyrine, **138**, 144
Nigrifactin, **165**, 166
Nitramin, **224**
Nitraria schoberi, 224
Nitroacridines, 8, 19, 32
9-Nitroacridine *N*-oxide, 12
Nitroacridones, 8, 19, 31, 32, 33, 34
7-Nitrobenzo[*f*]quinoline, 64
o-Nitrobenzoyl chloride, 205
2-Nitrobenzyl bromide, 219
14-Nitrocodeinone, 291
2-Nitrodibenzo[*b,f*]-1,4-thiazepine, 42
3′-Nitrodiphenylamine-2-carboxylic acid, 31
2-Nitrofluorenol, 52
2-Nitrofluorenone, 54
3-Nitrofluorenone, 55
22-Nitro-3β-hydroxy-16-oxocholest-5-en-26-oic ester, 418
9-(Nitromethyl)-9,10-dihydroacridine, 14
5-Nitro-2-methylpentanoic ester, 418
Nitrones, 16
Nitrophenanthridines, 46, 48, 52, 53, 57
8-Nitrophenanthridine, 42
Nitrophenanthridinium salts, 47
Nitrophenanthridones, 46, 54, 57
2-Nitrophenanthridone, 55, 56
4-Nitrophenanthridone, 56
N-Nitroso-*N*-(3-pyridyl) isobutyramide, 140
Nitrotetrahydrophenanthridines, 58
Noracridones, 257
Noracronycine, 259, 260, 261, 263, 264, 265
Noralkaloids, 259
Norallosedamine, 132, 137
Norcotinone, 139
Nordihydroacronycine, 211
Nordihydrothebainol, 277
C-Nor-*D*-homosteroidal alkaloids, 427, 436, 444
17-Nor-16-ketones, 331

Norlatifoline, **408**, 429
Norlobelanidine, 134
Norlobelanine, 134
Normacrorine, **217**, 218
Nornicotine, 139, **140**, 144
Nornicotyrine, 138, **140**
Nororixine, 176
Norsedamine, 132
Norsinoacutine, 302, **306**, 309
Norsinoacutinols, 306
Nudaurine, 302, 306, **308**
Nudifloric acid, 158
Nudifluorine, **157**, 158
Nupharamine, **162**, 163, 164
Nuphar japonicum, 164
Nuphar luteus, 164

Octahydroacridines, 29, **30**
Octahydro-1-azanaphthacene, 72
Octahydrobenzo[*f*]quinolines, 64
Octahydrobenzo[*h*]quinolines, 65
Octahydrophenthridines, 58, 59
Ocutine, **197**
Opium, 269
Orensine, **151**, 153
Organo-antimony compounds, 113
Organo-arsenic compounds, 105
Organo-bismuth compounds, 114
Organo-lithium compounds, 14
Organo-phosphorus compounds, 83
Oricia sp., 175
Oricia suaveolens, 199
Oricine, 175, **199**
Oripavine, 268
Orixa sp., 172, 174, 178
Orixa japonica, 201
Orixine, 172, **201**
Orixine sp., 176
Orixinone, 172, **201**
Oxa-azaspirane bicyclic units, 422
Oxa-azaspirodecane units, 411
Oxaziridines, 47
Oxazolidines, 397
Oxiranes, 237
Oxoatisinedicarboxylic acids, 336
Oxobrowniine, 356, 357
Oxodecahydroisoquinolines, 242
19-Oxodehydroheteratisine, 373

Oxodelavaine, 311, **316**
α-Oxodelphonine, 361
6-Oxodihydroconessine, 408
α-Oxodihydromorphilactonate, 279
Oxogarryine, 326
16-Oxohasubanonine, 316
Oxoheteratisine, 372
6-Oxonicotinic acid, 158
Oxonine, **359**
Oxonitine, 359
5-Oxo-octanal, 121
Oxosolacongestidines, **412**, 430
Oxostephamiersine, 311, **316**
12-Oxotetradecanoic acid, 146
2-Oxoundecylquinol-4-one, 215
Oxoveatchine, 326

Pachysamine-A, **395**, 429
Pachysamine-B, **395**, 429
Pachysandra terminalis, 390, 392, 395, 398, 399
Pachysandrine-A, **396**, 429
Pachysandrine-B, **397**, 429
Pachysandrine-C, **398**, 429
Pachysandrine-D, **398**, 429
Pachysantermine-A, **399**, 429
Pachystermine-A, **398**, 429
Pachystermine-B, **398**, 429
Pallidine, 302, **308**
Pamaquin, 23
Paniculidine, **388**, 428
Paniculine, **388**, 428
Papaver sp., 305, 307
Papaver bracteatum, 269
Papaver orientale, 269
Papaver somiferum, 269
Paravallaridine, **386**, 428
Paravallarine, **385**, 386, 428
Paravallario microphylla, 385, 386
Paravallaris maingayi, 409
Peganum sp., 172
Pei-mu, 453
Pelletierine, **127**, 128
Penicillium crustosum, 224
Penicillium cyclopium, 222
Penicillium puberulum, 223
Penicillium viridicatum, 222, 224
Pentacene, 39

Pentamethylenearsines, 105
Pentamethylenephosphanes, 84
Pentaphene, 39
Perforine, 182, **208**
Perhydroacridines, stereochemistry, 31
α-Perhydroacridine, 30, 31
β-Perhydroacridine, 30, 31
γ-Perhydroacridine, 31
Perhydrobenzo[*h*]quinolines, 65
Perhydrophenanthridines, 59
Perhydroquinolines, 254
Perlolidine, 253, **254**
Perloline, **253**, 254
Peroxylauric acid, 209
Petilium eduardi, 454
Pfitzinger reaction, 5, 29, 77, 107
Pheblalium sp., 180
α-Phenacylpiperidines, 254
Phenalene, 74
Phenanthraquinone, 54
Phenanthrene, 40, 462
Phenanthridines, 42, 43
–, chemical properties, 44
–, hydrogenated, 57, 58
–, oxidation, 53
–, physical properties, 43, 44
–, synthesis, 41
Phenanthridine, 40, 43, **49**
–, electrophilic substitution, 48
–, nitration, 48
–, nucleophilic substitution, 48
–, oxidation, 46
–, ozonolysis, 46
–, physical properties, 43, 44
–, reaction with perphthalic acid, 47
–, reduction, 57
Phenanthridine-6-carboxylic acid, 44, 49
Phenanthridine methiodide, 49
Phenanthridine-*N*-oxide, 47, 49
Phenanthridinethione, 56
Phenanthridinols, 53
3-Phenanthridinol, 53
6-Phenanthridinol, 53
Phenanthridinium 1553, **47**
Phenanthridinium perchlorate, 57
Phenanthridium salts, 44, 46, 47, 49, 57
Phenanthridones, 42, 53
Phenanthridone, 43, 44, 45, 46, 47, 50, 51, 53, 54, **55**, 56
Phenanthridone, (*continued*)
–, reduction, 57
Phenanthridonecarboxylic acids, 57
Phenanthridone sodium salt, 56
Phenethylisoquinoline alkaloids, 319
Phenidium chloride, **47**
9-Phenoxyphenanthridine, 43
6-Phenylacridarsine, 113
5-Phenyl-10-acridarsone-5-oxide, 111
Phenylacridines, 17
9-Phenylacridine, 4, 14, 15, 33
Phenylacridine picrates, 16
Phenylalanine, 224, 255
1-Phenylamino-2-methylacridone, 40
3-Phenylanthranils, 6
N-Phenylanthranilic acid, 18
1-Phenylantimonane, 113
Phenylarsenanes, 106
1-Phenyl-4-arsenone, 106, 107
Phenylarsinebismagnesium bromide, 109
2-Phenyl-1-azafluorenone-4-carboxylic acid, 63, 64
3-Phenylbenzo[*f*]quinoline-1-carboxylic acid, 63
3-Phenylbenzotriazin-4-one, 6
N-Phenyl-1,8-diaminonaphthalene, 78
9-Phenyl-9,10-dihydroacridine, 11, 14
10-Phenyl-9,10-dihydroacridine-9-thione, 36
Phenyldihydrothebaine, 274, 295, 297
m-Phenylenediamine, 22
2-Phenylethylpyridine, 62
9-Phenyl-9-fluorenylamine, 42
14-Phenylhydroxylaminocodeinone, 292
Phenyliminobenzophenone, 42
8-Phenyllobelol, 132, 137
6-Phenyl-*N*-methylphenthridium methosulphate, 45
Phenylnaphthalenes, 299, 300
2-Phenyl-2*H*-naphth[1,8-*cd*]isothiazole 1,1-dioxide, 78
N-Phenyl-1-naphthylamine, 76
8-Phenylnorlobelol, 132, 137
Phenylphenanthridines, 51
6-Phenylphenanthridine, 42, 48
6-Phenylphenanthridium compounds, 40
Phenyl-(3-phenylpropyl)arsinic acid, 108
Phenylphosphorins, 98, 100, 101, 102, 103

Phenyl-4-phosphorinanols, 89
Phenyl-4-phosphorinanones, 87, 88
Phenylpropanecarboxylic acids, 215
3-Phenylpyridine, 62
2-Phenylquinolines, 214, 215
2-Phenyltetrahydroisophosphinoline, 91
2-Phenyltetrahydro-1*H*-phosphorino-[4,3-*b*]indole, 88
2-Phenyltetrahydrophosphorino-[4,3-*b*]quinoline, 88
Phloroglucinol, 352
Phoma terrestris, 65
Phomazarin, **65**
Phosphabarrelenes, 104
Phosphabenzenes, 97
Phosphabicyclo[2.2.2]octa-2,5,7-trienes, 104
Phosphacyclohexanes, 84
9-Phosphafluorene, 96
Phosphaphenanthrenes, 97
4-Phosphinanones, 94
Phosphine oxides, 84, 85
Phosphine sulphides, 84
5-Phosphoniaspiro[4.5]decanes, 105
6-Phosphoniaspiro[5.5]undecanetetraol chloride, 89
Phosphonium salts, 84
Phosphorins, 97, 98, 99, 100, 101, 102, 103, 104
Phosphorin, **97**
Phosphorinanes, 83, 84
Phosphorinane, **86**
Phosphorinane oxides, 85
Phosphorinane-1-oxide, 86
Phosphorinanium salts, 85
Phosphorinanols, 88, 89
Phosphorinanones, 86, 88
–, 2,6-disubstituted, 87
Phosphorinium tetrafluoroborates, 102
Phosphorus heterocyclic compounds, 83, 104
Photochemical cyclisation, azastilbenes, 71
δ-Phthalimidobutyl *n*-propyl ketone, 125
Phthalocyanines, 38
Phthalonic acid, 156
3,4-Phthaloylacridone dyestuffs, 37
Phyllocladene, 325, 331
α-Picoline, 121, 124, 125, 132
β-Picoline, 118, 138, 448
γ-Picoline, 226
Picolinic acid, 120
Picolinic ester, 122
α-Picolyllithium, 127, 155
Pictet–Gams reaction, 67, 69
Pictet–Spengler reaction, 70
Pilo-kea, 197
Pilokeanine, 178, **197**
Pimanthrene, 326, 341
Pinidine, **150**, 151
Pinus sabiniana, 150
Pipecolic acid, 133
Pipecolinic acid, 121, 123, 141
Piperic acid, **115**, 116
Piperideine, 132
Piperidine, 146
–, natural occurrence, 115
Piperidine alkaloids, 115
Piperidine-2-carboxylic acid, 121
Piperidine-3,4-dicarboxylic acid, 227
α-Piperidone, 116
3-Piperidones, 403
20-Piperidyl-5α-pregnane, 382
20-Piperidyl-5α-pregnane bases, 411
(2-Piperidyl)-2-propanone, 133
β-(2-Piperidyl)propionic acid, 125
Piperine, **115**, 116
Piper longum, 116, 117
Piperlongumic acid, 116
Piperlongumine, **116**, 117
Piper nigrum, 115
Piperyl chloride, 115
N-Piperylpiperidine, 115
Piplartine, **117**
Pitavia sp., 180
Pivalic acid, 76
Platydesma sp., 178, 182
Platydesma campanulata, 197
Platydesmine, 182, 203, 215, 216
Podocarpic acid, 337
Pomegranate bark alkaloids, 127, 128
Pomeranz–Fritsch synthesis, 66, 241
Poncirus sp., 180
5α-Pregnane, 381, 400
5α-Pregnane-3,20-dione, 383
5α-Pregnan-3β-ol, 392
5α-Pregnan-3-one, 391, 401

Pregna-3,5,20-triene, 400
Pregnenolone, 385
O-Prenylacridone alkaloids, 213
3-Prenyl-2-quinolones, 216
Preskimmianine, 174, **208**
Proflavine, **22**, 37
Progesterone, 385, 391
Prometaphanine, 311, **315**, 316
2-Propenylpiperidine, 125
2-Propenylpyridine, 121
Propiolic ester, 45, 96, 293
2-Propionylpyridine, 121, 122
6-Propylphenanthridine, 48
2-*n*-Propylpiperidine, 120
2-*n*-Propylpyridine, 120
Prosafrine, **149**, 150
Prosafrinine, **149**, 150
Prosophylline, **149**, 150
Prosopine, **148**, 149, 150
Prosopinine, **148**, 149, 150
Prosopinone, **148**, 149
Prosopis alkaloids, 148, 150
Prosopis africana, 148
Prostephabyssine, 311, **316**
Protocatechuic acid, 193
Protostephanine, **297**
Protoveratridene, 445
Protoveratrine A and B, 446
Protoverine, 446, **450**
Pseudaconine, **366**, 371
Pseudaconitine, **360**, 366, 367, 371
Pseudanes, 221, 222
Pseudane-*N*-oxides, 222
Pseudobaleabuxine-F, 461
Pseudo-bases, 10, 11, 12, 45, 47
Pseudoconhydrine, 120, **123**, 124, 126
Pseudoconiceine, **126**
Pseudogermine, **449**, 450, 452
Pseudokobusine, **345**, 349
Pseudomonas aeruginosa, 221
Pseudopelletierine, 127
Pseudoprotoverine, **450**
Pseudozygadenine, **450**, 451, 452
Pschorr reaction, 55, 68, 71
Ptelea sp., 174, 175, 176, 182, 187, 189, 209
Pteleatinium ion, 180
Ptelea trifoliata, 208
Ptelecortine, 174
Ptelefolidine, 175
Ptelefolidone, 187
Ptelefoline, 174, 175
Ptelefolone, 187
Pteleforin, 189
Ptelefructine, 175
Pteleine, 180, 205
Pteleoline, 175
Ptelertinium chloride, 208
Pumiliotoxin-C, **255**
Punica granatum, 127
Pyrans, 215
Pyranoquinolines, 209
Pyrene, 74
Pyridine, 68, 117, 118, 121, 138, 193
–, natural occurrence, 115
Pyridine alkaloids, 115
Pyridine-2-carbaldehyde, 121
Pyridine-3-carbaldehyde, 141
Pyridine-2,4-dicarboxylic acid, 143
Pyridine-2,6-dicarboxylic acid, 146, 150
Pyridine-3,4-dicarboxylic acid, 226
Pyridine-2,3,4-tricarboxylic acid, 226
Pyridine-2,3,6-tricarboxylic acid, 162
2-Pyridylacetone, 127
3-Pyridyllithium, 141
2-(3-Pyridyl)piperidine, 141
1-(2′-Pyridyl)propan-2-ol, 126, 127
N-(3-Pyridyl)pyrrole, 138
2-(3-Pyridyl)pyrrole, 138, 140
2-(3-Pyridyl)pyrrolidine, 140
2-(3-Pyridyl)tetrahydropyridine, 142
Pyroaconitine, **360**, 367
Pyrobikhaconitine, 367
Pyrocusparine, 193
Pyrodelphinine, 360, 361
Pyrodiacetylneoline, 370
Pyrojesaconitine, 362
Pyrokarakolidine, 364
Pyro-oxonitine, 359, 361
Pyroterebic acid, 406
2-Pyrrolidone, 141
Pyrrolines, 414
Pyrrolo[1,2-*f*]phenanthridine, 45
Pyrrolo[1,2-*f*]phenanthridine-1-carboxylic ester, 45
Pyrylium salts, 67, 97

Pyrylium tetrafluoroborates, 97

Quinacridines, 38
–, angular, 39
Quinacridine pigments, 36
Quinacridones, 38
Quinacrine, **23**
Quinaldine, 172, 193
Quinamicine, **234**, 240
Quinamine, **233**, 234, 240
Quinene, 235
Quinicine, 231
Quinidine, 225, **230**, 231, 232, 235, 236, 238, 240, 243
Quinine, 225, **230**, 231, 232, 235, 236, 237, 238, 240
–, biosynthesis, 244, 246
–, total synthesis, **241**, 242, 243
Quininic acid, 230
Quininone, 231, 240
Quino[2,3-*a*]acridine, 40
Quino[2,3-*b*]acridine-7,14-dione, 38
Quino[3,2-*b*]acridine-12,14-dione, 39
Quino[3,2-*b*]acridine-13,14-dione, 39
Quinolines, 61
–, alkaloids, 193
–, from fungi and micro-organisms, 171, 221
Quinoline, 41, 172, 193, 194, 226, 227, 249
Quinoline alkaloids, 171, 233, 244
–, biosynthesis, 214
–, from tryptophan, 225
Quinoline-4-carboxylic acid, 226, 227
Quinoline-2,3-dicarboxylic acid, 2, 8
Quinoline-3,4-dicarboxylic acid, 41, 46
Quinolinic acid, 62
Quinolones, alkaloids, 195, 252
2-Quinolones, 174, 195
–, pentenyl substituted, 197
4-Quinolones, 178, 195
Quinol-4-one-3-carboxylic acid, 214
Quinotoxine, **231**, 240, 241, 242
Quinuclidine, 233
Quitenine, **231**, 240

Ranunculaceae sp., 323
Rauwolfia verticillata, 168
Ravenia sp., 174, 175, 180, 187, 190
Ravenia spectabilis, 207
Ravenine, 175, 206, 207
Ravenoline, 175, **207**
Reformatsky reaction, 88
Reissert compounds, 45
Reissert reaction, 14, 49
Remijia sp., 225
Remijia pedunculata, 232
Reticuline, 302, **308**
Retroaldol reaction, 221
Retro-Michael reaction, 435
Rhamnose, 425
Rhizama nupharis, 162
Rhizoctonia leguminicola, 168
Ribaline, 187
Ribalinidine, 189
Ribalinine, 189
Ribalinium, 182
Ricinidine, 117
Ricinine, **117**, 118
Ricininic acid, 117
Ricinus communis, 117
Rivanol, **22**
Robustine, 180, 203
Robustinine, 174
Rosaceae sp., 323
Rubijervine, 417, **418**, 419, 420, 430
Ruschig degradation, 382, 383, 391, 393, 406, 464, 465
Ruta sp., 172, 174, 178, 182, 189, 190, 191, 192, 209
Ruta angustifolia, 196
Rutacridone, 191, **212**
Rutaceae sp., 257
Rutacea alkaloids, 171, 172
Rutaceous quinolone alkaloids, biosynthesis, 214
Ruta graveolens, 196, 212
Rutamine, 178
RW47, 168

S. 897, 47
Sabadine, 445, **451**
Sabine, 445, **451**, 452
Salamander alkaloids, 427, 430
Salamander bases, biosynthesis, 435
–, physiological action, 435

Salamandra maculosa, 431
Salicylidene anil, 421
Salutaridine, 267, 283, 291, 302, **305**, 306
Salutaridinols, 306
Samandaridine, **432**, 434, 436
Samandarine, **431**, 432, 434, 436
Samandarone, **431**, 432, 436
Samandenone, **433**, 436
Samandinine, **433**, 436
Samandiol, 431, 432
Samanine, **434**, 436
Sandmeyer reaction, 18, 44
Sanguinarine, 81
Santiaguine, **152**, 153
Sapogenins, 423, 427
Saracocine, **393**, 428
Saracodine, **396**, 429
Sarcococca pruniformis, 391, 392, 393
Sarsasapogenic acid, 416
Sarsasapogenin, 416
Scandine, **252**, 253
Schmidt reaction, 53, 57, 80, 433
Schoencaulon officinale, 451
Scopolinic acid, 129
Secodaphniphylline, 374, 376, 378
Secologanin, 244
Sedamine, **132**, 137
Sedinine, 134
Sedinone, 134
Sedridine, 133, 137
Sedum sp., 132
Sedum acre, 133
Sedum alkaloids, 129, 134, 135, 136, 137
Senecioyl chloride, 395
Shikimic acid, 214
Sinoacutine, 302, **306**, 308
Sinomenine, 267, 275, 286, **302**, 303, 304
Sinomenine methyl ether, 304
Sinomeninic acid, 305
Sinomeninone, 304
Sinomenium acutum, 311
Sinomenol, 303
Skatole, 249
Skimmia sp., 178, 180, 182
Skimmia japonica, 214, 215
Skimmianine, 180, **203**, 205, 214, 216
Skraup reaction, 60, 63, 65, 71, 72, 73, 74
Skytanthines, **160**, 162
Skytanthus acutus, 160
Solacongestidine, **412**, 413, 430
Soladulcidine, **426**, 430
Solafloridine, **412**, 422, 430
Solamargine, **425**
Solamarines, 425
Solanaceae sp., 411
Solanidanes, 415
Solanidane, 417, 421
Solanidan-3β-ol, 416, 417, 418, 420
Solanid-4-en-3-one, 417
Solanidiene, 416
Solanidine, **416**, 430
Solanine, **416**, 430
Solanocapsine, **421**, 430
Solanthrene, 416
Solanum sp., 415
Solanum aviculare, 425
Solanum bases, 411
–, biological properties, 427
Solanum capsicastrum, 421
Solanum chacoense, 420
Solanum congestiflorum, 412
Solanum demissum, 418
Solanum dulcamara, 425
Solanum hendersonii, 421
Solanum hypomalacophyllum, 413
Solanum paniculatum, 388
Solanum pseudocapsicum, 421
Solanum sodomeum, 425
Solanum tomatillo, 413
Solanum tonrum, 388
Solaphyllidine, **413**, 430
Solasodine, 412, **425**, 426, 430
Solasonine, **425**
Songoramine, 330, 333
Songorine, **331**, 332
Spathelia sp., 175
Spathelia sorbifolia, 199
Spectabiline, 187, **207**
Spiradine-A, **346**, 349
Spiradine-B, **346**, 349
Spiradine-C, **346**, 349
Spiradine-D, **347**, 349
Spiradine-F, **340**
Spiradine-G, **340**
Spiraea japonica, 340, 346
Spirobisarsinolinium compounds, 109

Spirobistetrahydrophospholinium bromide, 92
Spirosolane alkaloids, 422
Squalene, 427
Staphidine, **378**, 379
Staphimine, **378**, 379
Staphinine, **378**, 379
Staphisine, **348**, 349
Stephaboline, 311, **316**
Stephabyssine, 311, **316**
Stephamiersine, 311, **316**
Stephania abyssinica, 311, 316
Stephania delavayi, 311
Stephania hernandifolia, 311, 312
Stephania japonica, 311
Stephasunoline, 311, **316**
Stephavanine, 311, **318**
Stephine, 318
Stephisoferuline, 311, **316**
Stephuline, 316
Stereochemistry, cinchona bases, 234
–, morphine alkaloids, 278
Steroidal alkaloids, 381, 428, 429, 430
–, biological properties, 427
–, biosynthesis, 426
Steroidal bases, 381
Steroidal sapogenins, 423
Stevane-B, 331
Steviol, 330, 331
Stieglitz rearrangement, 42
Streptomyces sp., 165
Strictosamide, 248
Strophanthidin, 419
Structure and analgesic activity, 301
Styrene derivatives, 16
3-Styrylisoquinoline, 80
4-Styrylisoquinoline, 79
4-Styryl-2-methylbenzo[*g*]quinoline, 65
Styrylpyridines, 61
3-Styrylpyridines, 68
4-Styrylquinoline, 81
Succindialdehyde, 116
Succinic acid, 156, 448
N-Succinylanthranilic acid, 350
Sweroside, 244
Swertiamarin, 146
Syphilobine-A, **136**, 137, 254
Syphilobine-F, **136**, 137, 254
Tacrine, **22**
Talatizamine, **364**, 365, 371
Talatizidine, **365**, 366, 371
Tautomerism, hydroxyacridines, 24
Teclea sp., 180, 190, 191, 192
Teclea boviviniana, 211
Tecleanone, 34
Tecleanthine, 191, 260, 262
Tecomanine, **161**, 162
Tecoma stans, 159, 161
Tecostanine, **161**, 162
Tecostidine, **159**, 162
Terminaline, **390**, 428
Terpenoid pyridine and piperidine alkaloids, 162
Tetra-acetoxy-2,2′-biphenanthryl, 301
Tetrabromoacridine, 18
Tetrabromo-9-acridone, 34
Tetracene, 77
Tetrachloro-9-acridone, 34
Tetrachlorodihydrophosphorin, 95
1,2,3,4-Tetrahydroacridine, 5, **29**, 30, 31
1,4,5,8-Tetrahydroacridine, **29**
1,2,3,4-Tetrahydroacridine-9-carboxylic acid, 29
1,2,3,4-Tetrahydroacridine *N*-oxide, 30
Tetrahydro-9-acridone, 35
Tetrahydroarsenins, 108
Tetrahydroarsinolines, 108
Tetrahydroatidine, 338
Tetrahydroatisine, 335
Tetrahydrobenz[*h*]isoquinoline, 70
Tetrahydrobenzoquinolines, 62, 63, 64, 65
Tetrahydrobenzo[*g*]quinoline, 65
Tetrahydro-9,9′-biacridine, 16, **26**, 27
Tetrahydrocodeimethine, 279, 285
Tetrahydrodeoxycodeine, 275, 286, 290, 303, 310
Tetrahydrodeoxyindolinocodeine, 312, 313
Tetrahydrodeoxymorphine, 286
Tetrahydro-2,5-diphenylarsenino[4,3-*c*]-pyrazolone, 107
Tetrahydro-1,2-diphenylphosphinolinium bromide, 92
Tetrahydro-1,1-diphenylphosphorinanium perchlorate, 90

Tetrahydroepiveatchine, 328, 329
1-(3-Tetrahydrofuryl)-4,8-dimethyl-8-hydroxynonane, 162
1-(3-Tetrahydrofuryl)-pentan-4-one, 162
Tetrahydrohexaphenyl-4,4'-biphosphorinylidene-1,1'-dioxide, 94
Tetrahydro-2-hydroxytetraphenylphosphorin 1-oxide, 97
Tetrahydroisoarsinolines, 108
Tetrahydroisocodeimethine, 279
Tetrahydro-α-isomorphitetrol, 279
Tetrahydroisophosphinolines, 91
Tetrahydromeloscine, 252
Tetrahydro-8-methoxy-1,7-dioxobenz-[*i,j*]arsinolizine, 110
Tetrahydro-7-methoxy-1-methyl-4-arsinolone, 110
Tetrahydro-1-methylarsinoline, 108
Tetrahydro-1-methylnicotinic ester, 119
Tetrahydromorphitetrol, 279
1,2,5,6-Tetrahydronicotinic acid, 120
Tetrahydrophenanthridines, 58, 59
Tetrahydro-2-phenylarsenino[4,5-*b*]-quinoline, 107
Tetrahydro-1-phenylarsinoline, 108
Tetrahydro-1-phenyl-4-phosphinolone, 93
Tetrahydrophosphinolines, 91
Tetrahydrophosphorins, 89, 90, 91
Tetrahydropiperlongumine, 116
Tetrahydropyridines, 119
Tetrahydrosalutaridinol, 307
Tetrahydrospiro(arsinoline-1,2'-isoarsindolium)bromide, 109
Tetrahydrospirobi(arsinolinium)-bromide, 109
Tetrahydrospirobi(isoarsinolinium)-bromide, 109
Tetrahydrothebaine, 287
Tetrahydro-1,1,2-triphenylphosphinolinium tetrafluoroborate, 92
Tetrahydroveatchine, 324
3,4,6,8-Tetramethoxyphenanthrene, 294
3,6-Tetramethyldiaminoacridine, 37
Tetramethylholarrhimine, **395**, 429
Tetramethyllappaconine, 363
Tetranitromethane, 291
Tetraphenylbiphosphine, 85
Tetraphenylindone, 95
Tetraphenyl-5-phosphoniaspiro[4,5]deca-1,3-diene bromide, 105
Thamnosma sp., 180, 190
Thebaine, 267, 268, 272, 274, 275, 283, 291, 294, 295, 297, 306
–, Diels–Alder adducts, 298, 302
–, Diels–Alder reaction, 292
–, iodination, 292
–, reaction with metal carbonyls, 300
–, reduction, 286, 287, 288, 289
–, structure, 270
–, synthesis, 278
Thebaine methiodide, 271
Thebainones, 288, **289**, 294
Thebainone, 301, 303
α-Thebainone, 289
β-Thebainone-A, 281
β-Thebainone-*A*-methine, 288
Thebenidine, **74**
Thebenine, 74, 272, 274, 295, 304, 307
Thebenone, 279, 280
β-Thebenone, 280
Thermopsis rhombifolia, 115
Thioacridone, 26, 35
9-Thioacridones, 32
Thiocodides, 290
1-Thiocyanatoarsenane, 106
9-Thioxo-9,10-dihydroacridine, 26
Tiglic acid, 334
O-Tigloylcyclovirobuxeine-B, 461
Tigogenin lactone, 421
Timonius kaniensis, 232
Tobacco alkaloids, 137, 144
Toddalia sp., 180
5-*o*-Toluoylquinoline, 72
1-*o*-Tolyl-7-methylisatin, 5
Tomatidenol, 421, **425**, 430
Tomatiden-3β-ol, 412
Tomatidine, 418, **423**, 424, 426, 430
Tomatillidine, **413**, 430
Tomatine, 423, 427
Trewia nudifloria, 157
1,2,3-Triacetoxy-5,6-dimethoxyphenanthrene, 316
N,N',O-Triacetylbuxine-D, 463
Triacetyldihydroveratramine, 439
Triacetylneoline, 370
Trialkylacetic acids, 76

2,4,7-Tribromo-9-acridone, 34
2,4,6-Tri-*tert*-butylphorphorin, 98
Tricarbonyliron, 300
Triethyl phosphite, 94
1,1,1-Trifluorophosphorinane, 85
Trigonelline, **117**, 138
3,4,5-Trihydroxyphenanthrene, 271
3,4,5-Trimethoxybenzaldehyde, 116
3,4,5-Trimethoxycinnamic acid, 116, 117
6,7,8-Trimethoxydictamnine, 203
5,6,5′-Trimethoxydiphenic acid, 297
1,2,3-Trimethoxy-*N*-methylacridone, 260, 263
1,3,5-Trimethoxy-10-methylacridone, 192, 211
2,3,4-Trimethoxy-10-methylacridone, 210
1,5,6-Trimethoxy-2,3-methylenedioxy-10-methylacridone, 192
4,7,8-Trimethoxy-1-methyl-2-quinolone, 175, 199
Trimethoxyphenanthrenes, 272
3,4,6-Trimethoxyphenanthrene, 288
3,4,5-Trimethoxyphenylpropionic acid, 116
3,4,5-Trimethoxyphthalic acid, 319
4,6,8-Trimethoxy-2-quinolone, 198
3,4,6-Trimethoxy-8-vinylphenanthrene, 294
7,9,10-Trimethylbenz[*c*]acridine, 76
O,O,O-Trimethylneoline, 369
Trimethylsulphonium ylid, 438
Triphenylmethyl perchlorate, 57
1,6,9-Triphenyl-1-phosphatricyclo[4.3.0]-nona-1,3,8-triene[2,3,4,5-tetracarboxylate, 104, 105
1,2,5-Triphenylphosphole, 104
2,4,6-Triphenylphosphorin, 94, 103
2,4,6-Triphenylpyrylium tetrafluoroborate, 97
Tripiperideine, 151
Tris(hydroxymethylene)phosphine, 97
Tris(trimethylsilyl)phosphine, 97
Tris(triphenylphosphine)rhodium chlorate, 29
Trityl tetrafluoroborate, 100
α-Truxilic acid, 152
Trypaflavine, **22**
Trypanocides, 47
Trypanosoma congolense, 47
Tryptamine, 247
Trypterygium wilfordii, 153
Tryptophan, 171, 214, 244, 251
Tryptophan alkaloids, 225

Vakognavine, 346, **348**, 349
Valeriana officinalis, 159
Valerianine, **159**, 162
Vanillic acid, 318, **444**
Vanillin, 194
O-Vanilloylcyclovirobuxine-D, 461
Vanilloylveracevine, 445
Vanilloylzygadenine, 445
Vat Black 25, 38
Vat dyestuffs, 36, 37, 38
Veatchine, **324**, 325, 326, 327, 328, 330
Veracevine, **444**, 445, 449, 452
Veracintine, **414**, 430
Veralinine, **414**, 430
Veralkamine, **411**, 426, 430
Veralobine, **420**
Veralosidine, 415
Veralosinine, 415
Veramarine, **443**, 444
Veramine, **426**, 430
Veranthridine, 448, 453, 454
Verarine, **441**, 442, 443, 444
Veratraldehyde, 193, 194
Veratramine, **437**, 438, 439, 440, 442, 443, 444
Veratric acid, 194, 366, 367, 444
Veratridine, 445
Veratrobasine, **441**, 444
Veratrosine, **437**
Veratroylzygadenine, 445
Veratrum album, 411, 412, 414, 415, 418, 420, 438, 441, 452
Veratrum bases, 411
–, biological properties, 427
Veratrum californicum, 441, 443
Veratrum eschscholtzii, 419
Veratrum grandiflorum, 415, 437
Veratrum viride, 419, 437, 438
Verazine, **412**, 430
Verticine, **453**, 454
Verticinone, 453

Vespris sp., 178, 180, 190
Vespris ampody, 34, 196
Vinicoside, 244
3-Vinylpiperidine, 227
3-Vinylpiperidyl-4-acetonitrile, 227
3-Vinylquinuclidine, 232
β-Vinyl-α-quinuclidone oxime, 228
Viridicatine, **222**, 223, 224
Viridicatol, 222, 223, 224

Westphalen rearrangement, 403
Wilfordic acid, **153**
Wilfordine, 154
Withania somnifera, 155
Wittig reaction, 83, 90, 327
Wolff–Kishner reduction, 121, 149, 330, 331, 347, 360, 418, 434
Wurtz–Fittig reaction, 91

Xanthevodine, 191, **211**, 260, 263
Xanthoxoline, 191, **211**, 260, 263
Xylose, 418

Ylides, 103
Yuzurimine, 374, 376, **377**, 378
Yuzurimine-A, **377**
Yuzurimine-B, **377**
Yusurimine C and D, 378
Yuzurine, **377**, 378

Zanthoxylum sp., 175, 180, 182
Zanthoxylum arnottianum, 203
Zanthoxylum decaryi, 203
Zanthoxylum tsihanimposa, 203
Zeanic acid, **225**
Zygacine, 445
Zygadenilic acid, 445
Zygadenilic δ-lactone, **451**, 452
Zygadenine, 445, **450**, 452
Zygadenine 3,16-diacetate, 450